Edited by
Christopher Barner-Kowollik,
Till Gruendling,
Jana Falkenhagen, and
Steffen Weidner

**Mass Spectrometry
in Polymer Chemistry**

Related Titles

Schlüter, D. A., Hawker, C., Sakamoto, J. (eds.)

Synthesis of Polymers

New Structures and Methods

Hardcover
ISBN: 978-3-527-32757-7

Harada, A. (ed.)

Supramolecular Polymer Chemistry

Hardcover
ISBN: 978-3-527-32321-0

Lendlein, A., Sisson, A. (eds.)

Handbook of Biodegradable Polymers

Isolation, Synthesis, Characterization and Applications

Hardcover
ISBN: 978-3-527-32441-5

Knoll, W., Advincula, R. C. (eds.)

Functional Polymer Films

2 Volume Set

2011
Hardcover
ISBN: 978-3-527-32190-2

Chujo, Y. (ed.)

Conjugated Polymer Synthesis

Methods and Reactions

2011
Hardcover
ISBN: 978-3-527-32267-1

Mathers, Robert T., Meier, Michael A. R. (eds.)

Green Polymerization Methods

Renewable Starting Materials, Catalysis and Waste Reduction

2011
Hardcover
ISBN: 978-3-527-32625-9

Loos, K. (ed.)

Biocatalysis in Polymer Chemistry

2011
Hardcover
ISBN: 978-3-527-32618-1

Xanthos, Marino (ed.)

Functional Fillers for Plastics

Second, updated and enlarged edition

2010
Hardcover
ISBN: 978-3-527-32361-6

Leclerc, Mario, Morin, Jean-Francois (eds.)

Design and Synthesis of Conjugated Polymers

2010
Hardcover
ISBN: 978-3-527-32474-3

Cosnier, S., Karyakin, A. (eds.)

Electropolymerization

Concepts, Materials and Applications

2010
Hardcover
ISBN: 978-3-527-32414-9

Edited by
Christopher Barner-Kowollik, Till Gruendling,
Jana Falkenhagen, and Steffen Weidner

Mass Spectrometry in Polymer Chemistry

WILEY-VCH Verlag GmbH & Co. KGaA

The Editors

Prof. Dr. C. Barner-Kowollik
Karlsruhe Institute of
Technology (KIT)
Engesserstr. 18
76128 Karlsruhe
Germany

Dr. Till Gruendling
Karlsruhe Institute of Technology (KIT)
Engesserstr. 18
76128 Karlsruhe
Germany

Dr. Jana Falkenhagen
Federal Institute for
Mat. Research & Testing (BAM)
Richard-Willstätter-Str. 11
12489 Berlin
Germany

Dr. Steffen Weidner
Federal Institute for
Mat. Research & Testing (BAM)
Richard-Willstätter-Str. 11
12489 Berlin
Germany

Cover:
Wiley-VCH thanks Gene Hart-Smith for the permission to use the cover illustration.

All books published by **Wiley-VCH** are carefully produced. Nevertheless, authors, editors, and publisher do not warrant the information contained in these books, including this book, to be free of errors. Readers are advised to keep in mind that statements, data, illustrations, procedural details or other items may inadvertently be inaccurate.

Library of Congress Card No.: applied for

British Library Cataloguing-in-Publication Data
A catalogue record for this book is available from the British Library.

Bibliographic information published by the Deutsche Nationalbibliothek
The Deutsche Nationalbibliothek lists this publication in the Deutsche Nationalbibliografie; detailed bibliographic data are available on the Internet at http://dnb.d-nb.de.

© 2012 Wiley-VCH Verlag & Co. KGaA, Boschstr. 12, 69469 Weinheim, Germany

All rights reserved (including those of translation into other languages). No part of this book may be reproduced in any form – by photoprinting, microfilm, or any other means – nor transmitted or translated into a machine language without written permission from the publishers. Registered names, trademarks, etc. used in this book, even when not specifically marked as such, are not to be considered unprotected by law.

Cover Design   Formgeber, Eppelheim

Typesetting   Thomson Digital, Noida, India
Printing and Binding   Fabulous Printers Pte Ltd, Singapore

Printed in Singapore
Printed on acid-free paper

Print ISBN: 978-3-527-32924-3
ePDF ISBN: 978-3-527-64184-0
oBook ISBN: 978-3-527-64182-6
ePub ISBN: 978-3-527-64183-3
Mobi ISBN: 978-3-527-64185-7

Contents

List of Contributors *XIII*

Introduction *1*
*Christopher Barner-Kowollik, Jana Falkenhagen,
Till Gruendling, and Steffen Weidner*
References *4*

1 Mass Analysis *5*
Gene Hart-Smith and Stephen J. Blanksby
1.1 Introduction *5*
1.2 Measures of Performance *5*
1.2.1 Mass Resolving Power *6*
1.2.2 Mass Accuracy *8*
1.2.3 Mass Range *9*
1.2.4 Linear Dynamic Range *9*
1.2.5 Abundance Sensitivity *10*
1.3 Instrumentation *12*
1.3.1 Sector Mass Analyzers *12*
1.3.2 Quadrupole Mass Filters *15*
1.3.3 3D Ion Traps *17*
1.3.4 Linear Ion Traps *19*
1.3.5 Time-of-Flight Mass Analyzers *20*
1.3.6 Fourier Transform Ion Cyclotron Resonance Mass Analyzers *22*
1.3.7 Orbitraps *24*
1.4 Instrumentation in Tandem and Multiple-Stage Mass Spectrometry *25*
1.5 Conclusions and Outlook *29*
References *30*

2 Ionization Techniques for Polymer Mass Spectrometry *33*
Anthony P. Gies
2.1 Introduction *33*
2.2 Small Molecule Ionization Era *34*

2.2.1	Electron Ionization (EI)	34
2.2.2	Chemical Ionization (CI)	36
2.2.3	Pyrolysis Mass Spectrometry (Py-MS)	37
2.3	Macromass Era of Ionization	38
2.3.1	Field Desorption (FD) and Field Ionization (FI)	38
2.3.2	Secondary Ion Mass Spectrometry (SIMS)	40
2.3.3	Fast Atom Bombardment (FAB) and Liquid Secondary Ion Mass Spectrometry (LSIMS)	42
2.3.4	Laser Desorption (LD)	43
2.3.5	Plasma Desorption (PD)	44
2.3.6	Other Ionization Methods	45
2.4	Modern Era of Ionization Techniques	45
2.4.1	Electrospray Ionization (ESI)	46
2.4.2	New Trends	48
2.4.3	Atmospheric Pressure Chemical Ionization (APCI)	49
2.4.4	New Trends	49
2.4.5	Matrix-Assisted Laser Desorption/Ionization (MALDI)	49
2.4.6	New Trends	52
2.5	Conclusions	53
	References	53
3	**Tandem Mass Spectrometry Analysis of Polymer Structures and Architectures**	**57**
	Vincenzo Scionti and Chrys Wesdemiotis	
3.1	Introduction	57
3.2	Activation Methods	59
3.2.1	Collisionally Activated Dissociation (CAD)	59
3.2.2	Surface-Induced Dissociation (SID)	60
3.2.3	Photodissociation Methods	60
3.2.4	Electron Capture Dissociation and Electron Transfer Dissociation (ECD/ETD)	61
3.2.5	Post-Source Decay (PSD)	62
3.3	Instrumentation	62
3.3.1	Quadrupole Ion Trap (QIT) Mass Spectrometers	63
3.3.2	Quadrupole/time-of-flight (Q/ToF) Mass Spectrometers	69
3.3.3	ToF/ToF Instruments	72
3.4	Structural Information from MS^2 Studies	75
3.4.1	End-Group Analysis and Isomer/Isobar Differentiation	75
3.4.2	Polymer Architectures	75
3.4.3	Copolymer Sequences	76
3.4.4	Assessment of Intrinsic Stabilities and Binding Energies	77
3.5	Summary and Outlook	78
	References	79

4	**Matrix-Assisted Inlet Ionization and Solvent-Free Gas-Phase Separation Using Ion Mobility Spectrometry for Imaging and Electron Transfer Dissociation Mass Spectrometry of Polymers** *85*
	Christopher B. Lietz, Alicia L. Richards, Darrell D. Marshall, Yue Ren, and Sarah Trimpin
4.1	Overview *85*
4.2	Introduction *87*
4.3	New Sample Introduction Technologies *92*
4.3.1	Laserspray Ionization – Ion Mobility Spectrometry-Mass Spectrometry *95*
4.3.2	Matrix Assisted Inlet Ionization (MAII) *99*
4.3.3	LSIV in Reflection Geometry at Intermediate Pressure (IP) *100*
4.4	Fragmentation by ETD and CID *102*
4.5	Surface Analyses by Imaging MS *103*
4.5.1	Ultraf Fast LSII-MS Imaging in Transmission Geometry (TG) *105*
4.5.2	LSIV-IMS-MS Imaging in Reflection Geometry (RG) *106*
4.6	Future Outlook *109*
	References *110*
5	**Polymer MALDI Sample Preparation** *119*
	Scott D. Hanton and Kevin G. Owens
5.1	Introduction *119*
5.2	Roles of the Matrix *120*
5.2.1	Intimate Contact *121*
5.2.2	Absorption of Laser Light *121*
5.2.3	Efficient Desorption *122*
5.2.4	Effective Ionization *123*
5.3	Choice of Matrix *125*
5.4	Choice of the Solvent *125*
5.5	Basic Solvent-Based Sample Preparation Recipe *127*
5.6	Deposition Methods *127*
5.7	Solvent-Free Sample Preparation *130*
5.8	The Vortex Method *132*
5.9	Matrix-to-Analyte Ratio *134*
5.10	Salt-to-Analyte Ratio *136*
5.11	Chromatography as Sample Preparation *138*
5.12	Problems in MALDI Sample Preparation *140*
5.13	Predicting MALDI Sample Preparation *142*
5.14	Conclusions *143*
	References *144*
6	**Surface Analysis and Imaging Techniques** *149*
	Christine M. Mahoney and Steffen M. Weidner
6.1	Imaging Mass Spectrometry *149*
6.2	Secondary Ion Mass Spectrometry *150*

6.2.1	Static SIMS of Polymers	150
6.2.1.1	The Fingerprint Region	151
6.2.1.2	High-Mass Region	162
6.2.2	Imaging in Polymer Blends and Multicomponent Systems	168
6.2.3	Data Analysis Methods	171
6.2.4	Polymer Depth Profiling with Cluster Ion Beams	174
6.2.4.1	A Brief Discussion on the Physics and Chemistry of Sputtering and its Role in Optimized Beam Conditions	180
6.2.5	3-D Analysis in Polymer Systems	182
6.3	Matrix-Assisted Laser Desorption Ionization (MALDI)	184
6.3.1	History of MALDI Imaging Mass Spectrometry	184
6.3.2	Sample Preparation in MALDI Imaging	185
6.3.3	MALDI Imaging of Polymers	188
6.3.4	Outlook	192
6.4	Other Surface Mass Spectrometry Methods	192
6.4.1	Desorption Electrospray Ionization	192
6.4.2	Plasma Desorption Ionization Methods	194
6.4.3	Electrospray Droplet Impact for SIMS	194
6.5	Outlook	196
	References	196
7	**Hyphenated Techniques**	**209**
	Jana Falkenhagen and Steffen Weidner	
7.1	Introduction	209
7.2	Polymer Separation Techniques	210
7.3	Principles of Coupling: Transfer Devices	214
7.3.1	Online Coupling Devices	214
7.3.2	Off-Line Coupling Devices	218
7.4	Examples	220
7.4.1	Coupling of SEC with MALDI-/ESI-MS	220
7.4.2	Coupling of LAC/LC-CC with MALDI-/ESI-MS	224
7.5	Conclusions	228
	References	228
8	**Automated Data Processing and Quantification in Polymer Mass Spectrometry**	**237**
	Till Gruendling, William E. Wallace, Christopher Barner-Kowollik, Charles M. Guttman, and Anthony J. Kearsly	
8.1	Introduction	237
8.2	File and Data Formats	237
8.3	Optimization of Ionization Conditions	239
8.4	Automated Spectral Analysis and Data Reduction in MS	241
8.4.1	Long-Standing Approaches	242
8.4.2	Some New Concepts	243
8.4.3	Mass Autocorrelation	243

8.4.4	Time-Series Segmentation *245*
8.5	Copolymer Analysis *248*
8.6	Data Interpretation in MS/MS *251*
8.7	Quantitative MS and the Determination of MMDs by MS *252*
8.7.1	Quantitative MMD Measurement by MALDI-MS *253*
8.7.1.1	Example for Mixtures of Monodisperse Components *256*
8.7.1.2	Example for Mixtures of Polydisperse Components *257*
8.7.1.3	Calculating the Correction Factor for Each Oligomer *260*
8.7.1.4	Step by Step Procedure for Quantitation *261*
8.7.1.5	Determination of the Absolute MMD *262*
8.7.2	Quantitative MMD Measurement by SEC/ESI-MS *266*
8.7.2.1	Exact Measurement of the MMD of Homopolymers *266*
8.7.2.2	MMD of the Individual Components in Mixtures of Functional Homopolymers *270*
8.7.3	Comparison of the Two Methods for MMD Calculation *273*
8.7.4	Simple Methods for the Determination of the Molar Abundance of Functional Polymers in Mixtures *274*
8.8	Conclusions and Outlook *276*
	References *276*

9 Comprehensive Copolymer Characterization *281*
Anna C. Crecelius and Ulrich S. Schubert

9.1	Introduction *281*
9.2	Scope *282*
9.3	Reviews *282*
9.4	Soft Ionization Techniques *283*
9.4.1	MALDI *283*
9.4.2	ESI *292*
9.4.3	APCI *294*
9.5	Separation Prior MS *297*
9.5.1	LC-MS *297*
9.5.2	Ion Mobility Spectrometry-Mass Spectrometry (IMS-MS) *299*
9.6	Tandem MS (MS/MS) *301*
9.7	Quantitative MS *303*
9.8	Copolymers for Biological or (Bio)medical Application *304*
9.9	Software Development *307*
9.10	Summary and Outlook *309*
	References *309*

10 Elucidation of Reaction Mechanisms: Conventional Radical Polymerization *319*
Michael Buback, Gregory T. Russell, and Philipp Vana

10.1	Introduction *319*
10.2	Basic Principles and General Considerations *320*
10.3	Initiation *321*

10.3.1	Radical Generation *321*
10.3.1.1	Thermally Induced Initiator Decomposition *321*
10.3.1.2	Photoinduced Initiator Decomposition *331*
10.3.1.3	Other Means *334*
10.3.2	Initiator Efficiency *335*
10.4	Propagation *335*
10.4.1	Propagation Rate Coefficients *336*
10.4.2	Chain-Length Dependence of Propagation *340*
10.4.3	Copolymerization *342*
10.5	Termination *347*
10.6	Chain Transfer *351*
10.6.1	Transfer to Small Molecules *351*
10.6.2	Acrylate Systems *356*
10.7	Emulsion Polymerization *364*
10.8	Conclusion *365*
	References *365*

11 Elucidation of Reaction Mechanisms and Polymer Structure: Living/Controlled Radical Polymerization *373*
Christopher Barner-Kowollik, Guillaume Delaittre, Till Gruendling, and Thomas Paulöhrl

11.1	Protocols Based on a Persistent Radical Effect (NMP, ATRP, and Related) *374*
11.2	Protocols Based on Degenerative Chain Transfer (RAFT, MADIX) *386*
11.3	Protocols based on CCT *393*
11.4	Novel Protocols and Minor Protocols *397*
11.5	Conclusions *398*
	References *399*

12 Elucidation of Reaction Mechanisms: Other Polymerization Mechanisms *405*
Grażyna Adamus and Marek Kowalczuk

12.1	Introduction *405*
12.2	Ring-Opening Polymerization Mechanisms of Cyclic Ethers *406*
12.3	Ring-Opening Polymerization Mechanisms of Cyclic Esters and Carbonates *408*
12.4	Ring-Opening Metathesis Polymerization *423*
12.5	Mechanisms of Step-Growth Polymerization *425*
12.6	Concluding Remarks *430*
	References *431*

13 Polymer Degradation *437*
Paola Rizzarelli, Sabrina Carroccio, and Concetto Puglisi

13.1	Introduction *437*
13.2	Thermal and Thermo-Oxidative Degradation *438*

13.3	Photolysis and Photooxidation	*449*
13.4	Biodegradation	*454*
13.5	Other Degradation Processes	*455*
13.6	Conclusions	*457*
	References	*461*

14   Outlook *467*
Christopher Barner-Kowollik, Jana Falkenhagen, Till Gruendling, and Steffen Weidner

Index *469*

List of Contributors

Grażyna Adamus
Polish Academy of Sciences
Center of Polymer and Carbon Materials
34 M. Curie-Sklodowska Street
41-800 Zabrze
Poland

Christopher Barner-Kowollik
Karlsruhe Institute of Technology (KIT)
Institut für Technische Chemie und Polymerchemie
Macromolecular Chemistry
Engesserstr. 18
76128 Karlsruhe
Germany

Stephen J. Blanksby
School of Chemistry
University of Wollongong
Wollongong, NSW 2522
Australia

Michael Buback
Georg-August-Universität Göttingen
Institut für Physikalische Chemie
Tammannstr. 6
37077 Göttingen
Germany

Sabrina Carroccio
National Research Council (CNR)
Institute of Chemistry and Technology of Polymers (ICTP)
Via Paolo Gaifami 18
95126 Catania
Italy

Anna C. Crecelius
Friedrich-Schiller-University Jena
Laboratory of Organic and Macromolecular Chemistry (IOMC)
Humboldtstr. 10
07743 Jena
Germany

Guillaume Delaittre
Karlsruhe Institute of Technology (KIT)
Institut für Technische Chemie und Polymerchemie
Macromolecular Chemistry
Engesserstr. 18
76128 Karlsruhe
Germany

Jana Falkenhagen
Bundesanstalt für Materialforschung und -prüfung (BAM)
Federal Institute for Materials Research and Testing
Richard-Willstätter-Strasse 11
12489 Berlin
Germany

List of Contributors

Anthony P. Gies
Vanderbilt University
Department of Chemistry
7330 Stevenson Center
Station B 351822
Nashville, TN 37235
USA

Till Gruendling
Karlsruhe Institute of Technology (KIT)
Institut für Technische Chemie und Polymerchemie
Macromolecular Chemistry
Engesserstr. 18
76128 Karlsruhe
Germany

Charles M. Guttman
National Institute of Standards and Technology
Polymers Division
Gaithersburg, MD 20899
USA

Scott D. Hanton
Intertek ASA
7201 Hamilton Blvd. RD1, Dock #5
Allentown, PA 18195
USA

Gene Hart-Smith
School of Biotechnology and Biomolecular Sciences
University of New South Wales
Sydney, NSW 2052
Australia

Anthony J. Kearsley
National Institute of Standards and Technology
Applied and Computational Mathematics Division
Gaithersburg, MD 20899
USA

Marek Kowalczuk
Polish Academy of Sciences
Center of Polymer and Carbon Materials
34 M. Curie-Sklodowska Street
41-800 Zabrze
Poland

Christopher B. Lietz
Wayne State University
Department of Chemistry
5101 Cass Ave
Detroit, MI 48202
USA

Christine M. Mahoney
National Institute of Standards and Technology
Material Measurement Laboratory
Surface and Microanalysis Science Division
100 Bureau Drive, Mail Stop 6371
Gaithersburg, MD 20899-6371
USA

Darrell D. Marshall
Wayne State University
Department of Chemistry
5101 Cass Ave
Detroit, MI 48202
USA

Kevin G. Owens
Drexel University
Chemistry Department
3141 Chestnut Street
Philadelphia, PA 19104
USA

Thomas Paulöhrl
Karlsruhe Institute of Technology (KIT)
Institut für Technische Chemie und Polymerchemie
Macromolecular Chemistry
Engesserstr. 18
76128 Karlsruhe
Germany

Concetto Puglisi
National Research Council (CNR)
Institute of Chemistry and Technology
of Polymers (ICTP)
Via Paolo Gaifami 18
95126 Catania
Italy

Yue Ren
Wayne State University
Department of Chemistry
5101 Cass Ave
Detroit, MI 48202
USA

Alicia L. Richards
Wayne State University
Department of Chemistry
5101 Cass Ave
Detroit, MI 48202
USA

Paola Rizzarelli
National Research Council (CNR)
Institute of Chemistry and Technology
of Polymers (ICTP)
Via Paolo Gaifami 18
95126 Catania
Italy

Gregory T. Russell
Department of Chemistry
University of Canterbury
20 Kirkwood Ave.
Upper Riccarton, Christchurch 8041
New Zealand

Ulrich S. Schubert
Friedrich-Schiller-University Jena
Laboratory of Organic and
Macromolecular Chemistry (IOMC)
Humboldtstr. 10
07743 Jena
Germany

Vincenzo Scionti
University of Akron
Department of Chemistry
302 Buchtel Common
Akron, OH 44325
USA

Sarah Trimpin
Wayne State University
Department of Chemistry
5101 Cass Avenue
Detroit, MI 48202
USA

Philipp Vana
Georg-August-Universität Göttingen
Institut für Physikalische Chemie
Tammannstr. 6
37077 Göttingen
Germany

William E. Wallace
National Institute of Standards and
Technology
Chemical and Biochemical Reference
Data Division
Gaithersburg, MD 20899
USA

Steffen M. Weidner
Bundesanstalt für Materialforschung
und -prüfung (BAM)
Federal Institute for Materials Research
and Testing
Richard-Willstätter-Strasse 11
12489 Berlin
Germany

Chrys Wesdemiotis
University of Akron
Department of Chemistry
302 Buchtel Common
Akron, OH 44325
USA

Introduction

Christopher Barner-Kowollik, Jana Falkenhagen, Till Gruendling, and Steffen Weidner

The first mass spectrometric experiment was arguably conducted by J. J. Thomson in the late 19th century, when he measured mass-to-charge ratios (m/z) in experiments that would eventually lead to the discovery of the electron [1]. By 1912, Thomson's investigations into the mass of charged atoms resulted in the publication of details of what could be called the first mass spectrometer [2, 3]. Interestingly, Thomson also employed one of the first man-made polymeric materials in his design of the parabola spectrograph: a material with trade name Ebonite or Vulcanite, a highly crosslinked natural rubber, which, although it was fairly brittle, provided an excellent electrical insulator and could easily be milled into shape. At the time, Thomson was most likely unaware of its chemical identity, as Staudinger's ground breaking macromolecular hypothesis was not to be established until a few years later [4]. By 1933 – the same year in which German chemist Otto Röhm patented and registered Plexiglas as a brand name – F. W. Aston had firmly established mass spectrometry as a field of analytical chemistry. Using the technique, he ascertained the isotopic abundances of essentially all of the chemical elements [5]. Thomson, Aston, and Staudinger were later to receive the Nobel Prize for their individual achievements.

Today, mass spectrometry provides the synthetic polymer chemist with one of the most powerful analytical tools to investigate the molecular structure of intact macromolecules. The development of technology that would be able to achieve this task was not realized until the late 1980s. Indeed, the mass analyzers themselves were not the key problem, as they were already fairly advanced at the time. An ionization technique that allowed the entire synthetic macromolecule to be transferred into the gas phase as ions without fragmentation could, however, only be realized in the late 1980s. The application of traditional MS ionization techniques requiring thermal evaporation of the sample to the large and entwined macromolecules was considered quite impossible, although notable attempts existed at employing the more traditional ionization techniques to polymeric material [6–8]. This perception had to undergo a drastic revision in 1988 and thereafter, the years in which electrospray ionization mass spectrometry (ESI) [9] and matrix assisted laser desorption and ionization (MALDI) [10, 11] were first reported of being capable to ionize proteins and synthetic polymers. Largely on the back of the work of four researchers, Karas [12–14], Hillenkamp [10, 12, 13], Tanaka [11], and Fenn [9, 15, 16], these new soft ionization mass spectrometry techniques commenced their success story initially in the field of biochemistry and later in synthetic polymer chemistry. Since the early 1990s, soft ionization mass spectrometry techniques have become an important part of polymer chemistry, ranging from unraveling polymerization mechanisms, assessing copolymer structures to studying the degradation of polymeric materials on a molecular level. However, a case can nevertheless be made that mass spectrometry is an underutilized tool in polymer chemistry compared to its high potential [17]. Such a notion is underpinned by an analysis of the current literature: of the approximately 10 000 studies conducted upon – for example – polyacrylates (which are readily ionizable) from 2000 until 2010, NMR spectroscopy played a significant role in ∼15% of these studies, whereas soft ionization mass spectrometry played a significant role in only about 3% of these studies. This is despite of the fact that soft ionization mass spectrometry technology has – due to its dominance as a highly applicable analytical tool in the biological sciences – become almost as readily available as NMR. To date, mass spectrometry remains the only technique with the power to isolate (provided the correct mass analyzer is employed) and image individual polymer chains on a routine basis.

Although there have been some notable books addressing the field of mass spectrometry applied to synthetic polymers [18–20], no publication especially dedicated to the needs of synthetic polymer chemists exists, which could aid in the selection of appropriate mass spectrometric tools. Specifically, most books on polymer mass spectrometry do not engage with the topics of living/controlled radical polymerization methods and their mechanistic underpinnings or the mass spectrometric investigation of polymerization processes in general. In addition, an update on the current situation of polymer mass spectrometry is required. With the present compilation, we wish to close this critical gap in the literature and provide a state-of-the-art overview on the applications of mass spectrometry in molecular polymer chemistry to the reader. In this edited publication, a series of leading researchers in the field will present their expert perspectives on several – in our view – important topics in contemporary mass spectrometry. It is thus no surprise that the large majorities of authors contributing to the present book are chemists by training, as we have attempted to provide a book that addresses the analytical requirements posed in contemporary polymer chemistry.

The book opens with an overview of the available mass analyzers. Special consideration is given by the authors Steven Blanksby and Gene Hart-Smith to their uses in polymer chemistry. Various ionization techniques applicable in polymer mass spectrometry are then explored in-depth by Anthony Gies. Chrys Wesdemiotis subsequently takes a close look at tandem mass spectrometry, a highly important tool for the elucidation of polymer structure and one of the major contemporary fields of development in the mass spectrometry sector. Sarah Trimpin and colleagues follow with their contribution, describing gas-phase ion-separation procedures as applied to synthetic polymers. The ionization process of polymers via the MALDI approach requires a careful design of the sample preparation procedures. Scott Hanton and Kevin Owens therefore provide a close look at how polymer samples are best prepared. Synthetic polymers are not only important materials in their own right, but are also frequently employed to (covalently) modify variable surfaces. Surface analysis is notoriously challenging and a range of techniques have to be employed to map the chemical characteristics of surface-bound macromolecules. Christine Mahoney and Steffen Weidner provide a detailed description of the part which surface-sensitive mass spectrometric techniques play in elucidating a polymer surface's structure.

Soft ionization mass spectrometry techniques can be especially powerful when combined with chromatographic techniques such as size exclusion chromatography (SEC), liquid adsorption chromatography at critical conditions (LACCCs) or both. Jana Falkenhagen and Steffen Weidner explore the wide variety of so-called hyphenated techniques and impressively demonstrate the information depth that can be attained by employing such technologies. While arguably the majority of molecular weight determination is carried out via SEC often equipped with refractive index as well as light scattering detectors, Till Gruendling, William Wallace, and colleagues demonstrate how MALDI-MS as well as SEC coupled to ESI-MS can be employed to deduce absolute molecular weight distributions. The chapter also provides an overview of contemporary automated data processing techniques for mass spectrometric data. Most polymers generated are arguably copolymers and it thus is mandatory to dedicate an entire chapter to the analysis of copolymers via mass spectrometry – Ulrich Schubert and Anna Crecelius provide an in-depth analysis. The field of living/controlled radical polymerization provides fascinating high precision avenues for the construction of complex macromolecular architectures and enables the generation of polymers with a high degree of end-group fidelity, which are often employed in cross-discipline applications (e.g., in biosynthetic conjugates). Soft ionization mass spectrometry plays an integral part in unraveling the mechanism of living/controlled radical polymerization processes as well as in the characterization of macromolecular building blocks: Christopher Barner-Kowollik and colleagues take a close look at the current state-of-the-art. Similarly, polymers generated via conventional radical polymerization can be readily investigated via mass spectrometry. Here, especially the investigation of the initiation process and of the generated end-group type is of high importance – Michael Buback, Greg Russell, and colleagues report. Finally, Grazyna Adamus and Marek Kowalczuk survey the field of mass spectrometry applied to polymers prepared via nonradical methods such as coordination polymerization, polycondensation, and polyaddition. The question of polymer stability and a detailed understanding of polymer degradation processes on a molecular level are of paramount importance for an evaluation of the performance of a polymer in chemical applications or as a material. Sabrina Carroccio and colleagues survey the field of soft ionization mass spectrometry applied to the molecular study of degradation processes at the book's conclusion.

With the above spectrum, we hope to have covered most of what constitutes modern mass spectrometry applied to questions of organic polymer chemistry. The final chapter provides an outlook and evaluation – from our perspective – of what the important advances in mass spectrometry technology related to polymer chemistry could be and which important chemical questions are yet to be addressed by soft ionization techniques.

Karlsruhe and Berlin, February 2011
Christopher Barner-Kowollik
Jana Falkenhagen
Till Gruendling
Steffen Weidner

References

1. Thomson, J.J. (1897) *Philos. Mag.*, **5**, 293.
2. Thomson, J.J. (1912) *Philos. Mag.*, **24**, 209.
3. Thomson, J.J. (1913) *Proc. R. Soc. Lond. A*, **89**, 1.
4. Staudinger, H. (1920) *Ber. Dtsch. Chem. Ges.*, **53**, 1073.
5. Aston, F.W. (1933) *Mass Spectra and Isotopes*, Edward Arnold, London.
6. Achhammer, B.G., Reiney, M.J., Wall, L.A., and Reinhart, F.W. (1952) *J. Polym. Sci.*, **8**, 555.
7. Hummel, D.O., Düssel, H.J., and Rübenacker, K. (1971) *Makromol. Chem.*, **145**, 267.
8. Lattimer, R.P., Harmon, D.J., and Hansen, G.E. (1980) *Anal. Chem.*, **52**, 1808.
9. Meng, C.K., Mann, M., and Fenn, J.B. (1988) *Z. Phys. D: At. Mol. Clusters*, **10**, 361.
10. Karas, M. and Hillenkamp, F. (1988) *Anal. Chem.*, **60**, 2299.
11. Tanaka, K., Waki, H., Idao, Y., Akita, S., Yoshida, Y., and Yoshida, T. (1988) *Rapid Commun. Mass Spectrom.*, **2**, 151.
12. Karas, M., Bachmann, D., Bahr, U., and Hillenkamp, F. (1987) *Int. J. Mass Spectr. Ion Proc.*, **78**, 53.
13. Karas, M., Bachmann, D., and Hillenkamp, F. (1985) *Anal. Chem.*, **57**, 2935.
14. Karas, M. and Bahr, U. (1985) *Trends Anal. Chem.*, **5**, 90.
15. Yamashita, M. and Fenn, J.B. (1984) *J. Phys. Chem.*, **88**, 4451.
16. Yamashita, M. and Fenn, J.B. (1984) *J. Phys. Chem.*, **88**, 4671.
17. Hart-Smith, G. and Barner-Kowollik, C. (2010) *Macromol. Chem. Phys.*, **211**, 1507.
18. Li, L. (2010) *MALDI Mass Spectrometry for Synthetic Polymer Analysis (Chemical Analysis: A Series of Monographs on Analytical Chemistry and Its Applications)*, John Wiley & Sons, Hoboken, NJ.
19. Pasch, H. and Schrepp, W. (2003) *MALDI-TOF Mass Spectrometry of Synthetic Polymers*, Springer, Berlin.
20. Montaudo, G. and Lattimer, R.P. (2001) *Mass Spectrometry of Polymers*, CRC Press, Boca Raton, FL.

1
Mass Analysis

Gene Hart-Smith and Stephen J. Blanksby

1.1
Introduction

Modern day mass analyzer technologies have, together with soft ionization techniques, opened powerful new avenues by which insights can be gained into polymer systems using mass spectrometry (MS). Recent years have seen important advances in mass analyzer design, and a suite of effective mass analysis options are currently available to the polymer chemist. In assessing the suitability of different mass analyzers toward the examination of a given polymer sample, a range of factors, ultimately driven by the scientific questions being pursued, must be taken into account. It is the aim of the current chapter to provide a reference point for making such assessments.

The chapter will open with a summary of the measures of mass analyzer performance most pertinent to polymer chemists (Section 1.2). How these measures of performance are defined and how they commonly relate to the outcomes of polymer analyses will be presented. Following this, the various mass analyzer technologies of most relevance to contemporary MS will be discussed (Section 1.3); basic operating principles will be introduced, and the measures of performance described in Section 1.2 will be summarized for each of these technologies. Finally, an instrument's tandem and multiple-stage MS (MS/MS and MS^n, respectively) capabilities can play a significant role in its applicability to a given polymer system. The capabilities of different mass analyzers and hybrid mass spectrometers in relation to these different modes of analysis will be summarized in Section 1.4.

1.2
Measures of Performance

When judging the suitability of a given mass analyzer toward the investigation of a polymer system, the relevant performance characteristics will depend on the scientific motivations driving the study. In most instances, knowledge of the following measures of mass analyzer performance will allow a reliable assessment to be made: mass resolving power, mass accuracy, mass range, linear dynamic range, and abundance

1.2.1
Mass Resolving Power

Mass analyzers separate gas-phase ions based on their mass-to-charge ratios (m/z); how well these separations can be performed and measured is defined by the instrument's mass resolving power. IUPAC recommendations allow for two definitions of mass resolving power [1]. The "10% valley definition" states that, for two singly charged ion signals of equal height in a mass spectrum at masses M and $(M - \Delta M)$ separated by a valley which, at its lowest point, is 10% of the height of either peak, mass resolving power is defined as $M/\Delta M$. This definition of mass resolving power is illustrated in portion A of Figure 1.1. The "peak width definition" also defines mass resolving power as $M/\Delta M$; in this definition, M

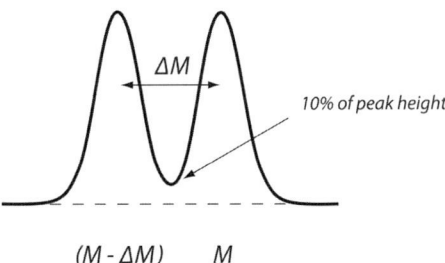

(A) 10 percent valley definition

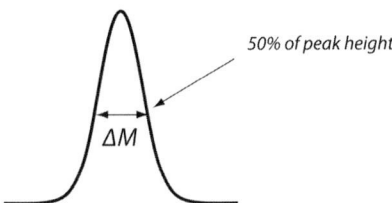

(B) FWHM definition

Figure 1.1 Methods of calculating mass resolving power. Portion (A) illustrates calculation via the 10% valley definition. Portion (B) illustrates calculation via the FWHM definition.

refers to the mass of singly charged ions that make up a single peak, and ΔM refers to the width of this peak at a height which is a specified fraction of the maximum peak height. It is recommended that one of three specified fractions should always be used: 50%, 5%, or 0.5%. In practice, the value of 50% is frequently utilized; this common standard, illustrated in portion B of Figure 1.1, is termed the "full width at half maximum height" (FWHM) definition. The mass resolving power values quoted for the mass analyzers described in this chapter use the FWHM criterion.

In the context of polymer analysis, the mass resolving power is important when characterizing different analyte ions of similar but nonidentical masses. These different ions may contain separate vital pieces of information. An example of this would be if the analytes of interest contain different chain end group functionalities; characterization of these distinct end groups would allow separate insights to be gained into polymer formation processes. Whether or not this information can be extracted from the mass spectrum depends on the resolving power of the mass analyzer. The importance of mass resolving power in this context has been illustrated in Figure 1.2 using data taken from a study conducted by Szablan et al., who were interested in the reactivities of primary and secondary radicals derived from various photoinitiators [2]. Through the use of a 3D ion trap mass analyzer, these authors were able to identify at least 14 different polymer end group combinations within a m/z window of 65. This allowed various different initiating radical fragments to be identified, and insights to be gained into the modes of termination that were taking place in these polymerization systems. It can be seen that the mass resolving power of the 3D ion trap allowed polymer structures differing in mass by 2 Da to be comfortably distinguished from one another.

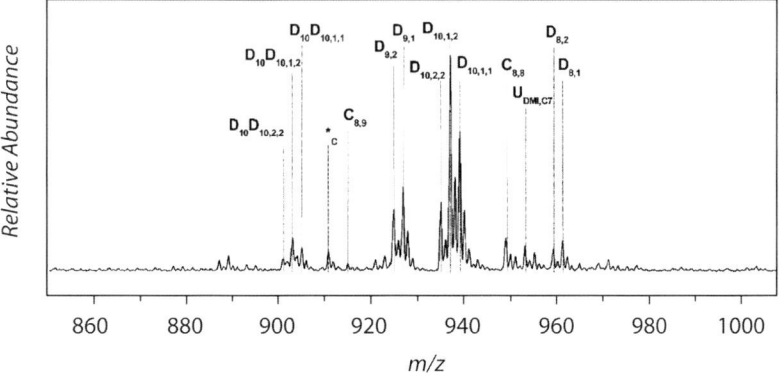

Figure 1.2 A 3D ion trap-derived mass spectrum of the polymer obtained from an Irgacure 819-initiated pulsed laser polymerization of dimethyl itaconate, adapted from Figure 12 of Szablan et al. [2].

1.2.2
Mass Accuracy

Mass accuracy refers to the m/z measurement error – that is, the difference between the true m/z and the measured m/z of a given ion – divided by the true m/z of the ion, and is usually quoted in terms of parts per million (ppm). For a single reading, the term "mass measurement error" may be used [3]. It is usual for mass accuracy to increase with mass resolving power, and a higher mass accuracy increases the degree of confidence in which peak assignments can be made based upon the m/z. This lies in the fact that increases in mass accuracy will result in an increased likelihood of uniquely identifying the elemental compositions of observed ions.

When attempting to identify peaks in mass spectra obtained from a polymer sample, it is common for different feasible analyte ions to have similar but non-isobaric masses. If the theoretical m/z's of these potential ion assignments differ by an amount lower than the expected mass accuracy of the mass analyzer, an ion assignment cannot be made based on m/z alone. Ideally such a scenario would be resolved through complementary experiments using, for example, MS/MS or alternate analytical techniques, in which one potential ion assignment is confirmed and the others are rejected. However if such methods are not practical, the use of a mass analyzer capable of greater mass accuracy may be necessary. An example of the use of ultrahigh mass accuracy data for this purpose can be found in research conducted by Gruendling et al., who were investigating the degradation of reversible addition-fragmentation chain transfer (RAFT) agent-derived polymer end groups [4]. These authors initially used a 3D ion trap instrument to identify a peak at m/z 1275.6 for which three possible degradation products could be assigned. To resolve this issue, the same sample was analyzed using a Fourier transform ion cyclotron resonance (FT-ICR) mass analyzer. As illustrated in Figure 1.3, the ultrahigh mass accuracy obtained using FT-ICR allowed two of the potential ion assignments to be

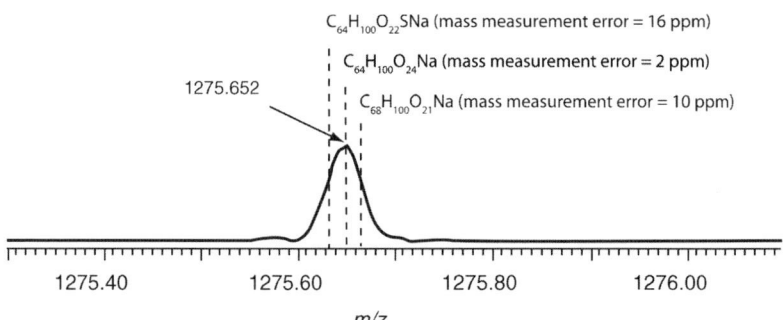

Figure 1.3 An FT-ICR-derived signal from the degradation product of a RAFT end group containing polymer chain. The gray chemical formulas describe potential ion assignments ruled out based on higher than expected mass measurement errors. The black chemical formula describes the ion assignment confirmed via an acceptable mass measurement error. Image adapted from Figure 2 of Gruendling et al. [4].

ruled out based on higher than expected mass measurement errors; the mass measurement error of the third ion was reasonable, allowing a specific degradation product to be confirmed.

1.2.3
Mass Range

The mass range is the range of m/z's over which a mass analyzer can operate to record a mass spectrum. When quoting mass ranges, it is conventional to only state an upper limit; it is, however, important to note that for many mass analyzers, increasing the m/z's amenable to analysis will often compromise lower m/z measurements. As such, the mass ranges quoted for the mass analyzers described in this chapter do not necessarily reflect an absolute maximum; they instead provide an indication of the upper limits that may be achieved in standard instrumentation before performance is severely compromised.

The mass range is frequently of central importance when assessing the suitability of a given mass analyzer toward a polymer sample. For many mass analyzers, there is often a high likelihood that the polymer chains of interest are of a mass beyond the mass range; this places a severe limitation on the ability of the mass spectrometer to generate useful data. Because mass analyzers separate ions based on their m/z's, the generation of multiply charged ions may alleviate this issue. Relatively high mass resolving powers are, however, required to separate multiply charged analyte ions, and efficient and controlled multiple charging of polymer samples is generally difficult to achieve. As such, the generation of multiply charged ions is not a reliable method for overcoming mass range limitations, and for many studies, mass range capabilities will ultimately dictate a mass analyzer's suitability.

1.2.4
Linear Dynamic Range

The linear dynamic range is the range over which the ion signal is directly proportional to the analyte concentration. This measure of performance is of importance to the interpretation of mass spectral relative abundance readings; it can provide an indication of whether or not the relative abundances observed in a mass spectrum are representative of analyte concentrations within the sample. The linear dynamic range values quoted within this chapter represent the limits of mass analysis systems as integrated wholes; that is, in addition to the specific influence of the mass analyzer on linear dynamic range, the influences of ion sampling and detection have been taken into consideration. In many measurement situations, however, these linear dynamic range limits cannot be reached. Chemical- or mass-based bias effects during the ionization component of an MS experiment will frequently occur, resulting in gas-phase ion abundances that are not representative of the original analyte concentrations. When present, such ionization bias effects will generally be the dominant factor in reducing linear dynamic range. Only in the instances in which ionization bias effects can be ruled out can the linear dynamic range values quoted in this chapter provide an indication of the trustworthiness of mass spectral abundance data.

In most polymer analyses, ionization bias effects will be prevalent. There are, however, specific scenarios in which ionization bias effects can rightfully be assumed to be minimal. One example can be found in free radical polymerizations in which propagating chains are terminated via disproportionation reactions. When considering such a system, it can be noted that disproportionation products are produced in equal abundances, but identical reaction products may also be generated from other polymerization mechanisms; accurate relative abundance data are therefore needed to infer the extent to which these other mechanisms are occurring. Because the products in question are chemically similar and have similar masses, depending on the chosen ionization method, it may be possible to conclude that these chains will not experience chemical- or mass-based ionization bias relative to each other. Under these circumstances, the linear dynamic range of the mass analysis system is crucial to the determination of accurate relative abundances for these products. This scenario can be seen in research conducted by Hart-Smith *et al.* [5], who used a 3D ion trap instrument to analyze acrylate-derived star polymers. The mass spectrum illustrated in Figure 1.4, taken from this research, shows two peaks, A and B, which correspond to disproportionation products. Based on the comparatively high relative abundance of peak B and the linear dynamic range of the 3D ion trap, these authors were able to infer that another mechanism capable of producing peak B, intermolecular chain transfer, was up to two times more prevalent than disproportionation in the polymerization under study.

1.2.5
Abundance Sensitivity

Abundance sensitivity refers to the ratio of the maximum ion current recorded at an m/z of M to the signal level arising from the background at an adjacent m/z of

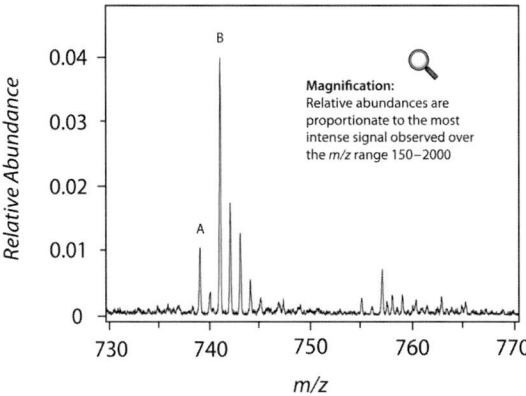

Figure 1.4 A 3D ion trap-derived mass spectrum of star polymers obtained from a RAFT-mediated polymerization of methyl acrylate, adapted from Figure 6.3.8 of Hart-Smith *et al.* [5].

($M + 1$). This is closely related to dynamic range: the ratio of the maximum useable signal to the minimum useable signal (the detection limit) [1]. Abundance sensitivity, however, goes beyond dynamic range in that it takes into account the effects of peak tailing. By considering the abundance sensitivity of a mass analyzer, one can obtain an indication of the maximum range of analyte concentrations capable of being detected in a given sample.

In the analysis of polymer samples, it is often the case that the characterization of low abundance species is of more importance than the characterization of high abundance species. For example, it is well established that polymer samples generated via RAFT polymerizations will often be dominated by chains which contain end groups derived from a RAFT mediating agent; if novel insights are to be gained into these systems, it is often required that lower abundance polymer chains are characterized. This can be seen in work conducted by Ladavière et al. using a time-of-flight (TOF) mass analyzer [6]. The spectrum shown in Figure 1.5, taken from this research, indicates the presence of chains with thermal initiator derived end groups (IU_x^{Na}, IY_x^K, and IY_x^{Na}) and chains terminated via combination reactions (C_x^{Na}), in addition to the dominant RAFT agent-derived end group containing chains. The peaks associated with termination via combination are one order of magnitude lower

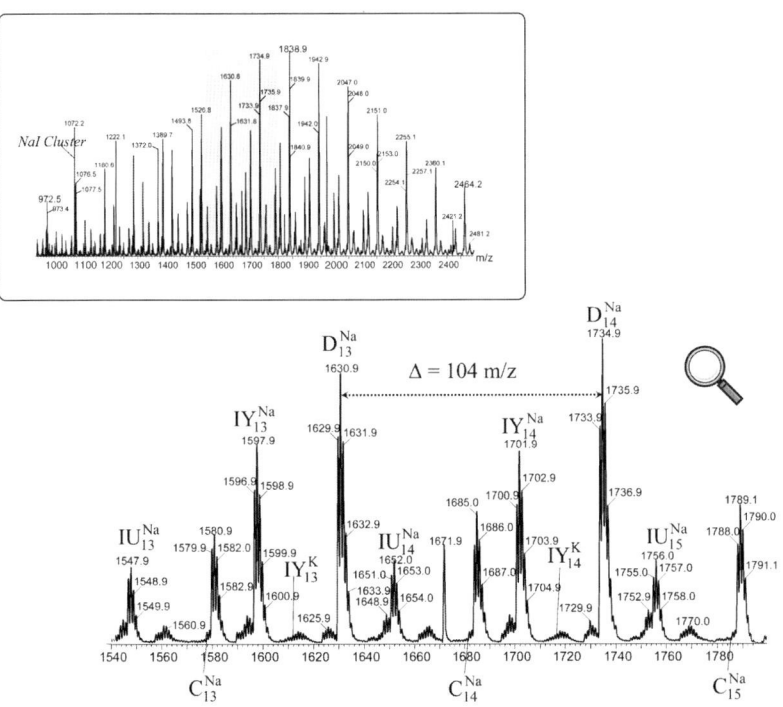

Figure 1.5 An electrospray ionization-TOF-derived mass spectrum of the polymer obtained from a RAFT-mediated polymerization of styrene, adapted from Figure 1 of Ladavière et al. [6].

than the most abundant peak within the spectrum and are clearly discernable from baseline noise. When attempting to characterize low abundance chains in such a manner, the abundance sensitivities listed in this chapter can provide some indication of the extent to which this can be achieved when using a given mass analyzer.

It is, however, important to note that the ability to observe relatively low abundance chains will also be influenced by components of the MS experiment other than the mass analyzer. The ionization method being used may, for example, be inefficient at ionizing the chains of interest, reducing the likelihood of their detection. The method used to prepare the polymer sample for ionization may also have an impact; for instance evidence suggests that issues associated with standard methods of polymer sample preparation for matrix-assisted laser desorption/ionization (MALDI) experiments reduce the capacity to detect relatively low abundance species [6, 7], and that these issues significantly outweigh the influence of mass analyzer abundance sensitivities [7]. The mass analyzer abundance sensitivities quoted in this chapter should therefore be contemplated alongside other aspects of MS analysis, such as those mentioned above, when designing experimental protocols for the detection of low abundance polymer chains.

1.3
Instrumentation

Since the early twentieth century, when the analytical discipline of MS was being established, many methods have been applied to the sorting of gas-phase ions according to their m/z's. The following technologies have since come to dominate mass analysis in contemporary MS and are all available from one or more commercial vendors: sector mass analyzers, quadrupole mass filters, 3D ion traps, linear ion traps, TOF mass analyzers, FT-ICR mass analyzers, and orbitraps. This section presents the basic operating principles of these instruments and summarizes their performance characteristics using the measures of performance discussed in Section 1.2. As cost and laboratory space requirements are often a determining factor in the choice of instrumentation, these characteristics are also listed.

For each mass analyzer presented in this section, the summarized performance characteristics do not necessarily represent absolute limits of performance. The use of tailored mass analysis protocols in altered commercial instrumentation, or instrumentation constructed in-house, can often allow for performance beyond what would typically be expected. The listed figures of merit, therefore, represent a summary of optimal levels of performance that should be capable of being readily accessed using standard commercially available instrumentation.

1.3.1
Sector Mass Analyzers

Sector mass analyzers are the most mature of the MS mass analysis technologies, having enjoyed widespread use from the 1950s through to the 1980s. The

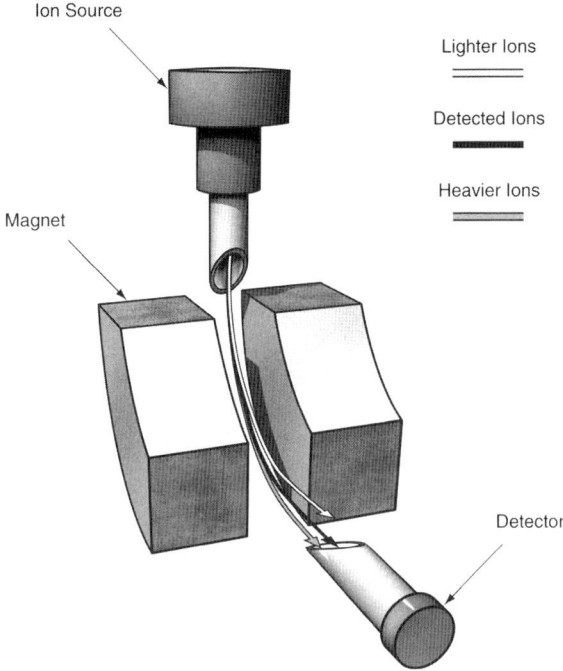

Figure 1.6 An illustration of the basic components of a magnetic sector mass analyzer system, and the means by which it achieves m/z-based ion separation.

illustration in Figure 1.6 demonstrates the basic operating principle of magnetic sectors, which are employed in all sector mass analyzers. Magnetic sectors bend the trajectories of ions accelerated from an ion source into circular paths; for a fixed accelerating potential, typically set between 2 and 10 kV, the radii of these paths are determined by the momentum-to-charge ratios of the ions. In such a manner, the ions of differing m/z's are dispersed in space. While dispersing ions of different momentum-to-charge ratios, the ions of identical momentum-to-charge ratios but initially divergent ion paths are focused in a process called direction focusing. These processes ensure that, for a fixed magnetic field strength, the ions of a specific momentum-to-charge ratio will follow a path through to the ion detector. By scanning the magnetic field strength, the ions of different m/z can therefore be separated for detection.

When utilizing a magnetic sector alone, resolutions of only a few hundred can be obtained. This is primarily due to limitations associated with differences in ion velocities. To correct for this, electric sectors can be placed before or after the magnetic sector in "double focusing" instruments, as illustrated in Figure 1.7. Electric sectors disperse ions according to their kinetic energy-to-charge ratios, while also providing the same type of direction focusing as magnetic sectors. Through the careful design of two sector instruments, these kinetic energy dispersions can be corrected for by the momentum dispersions of the magnetic

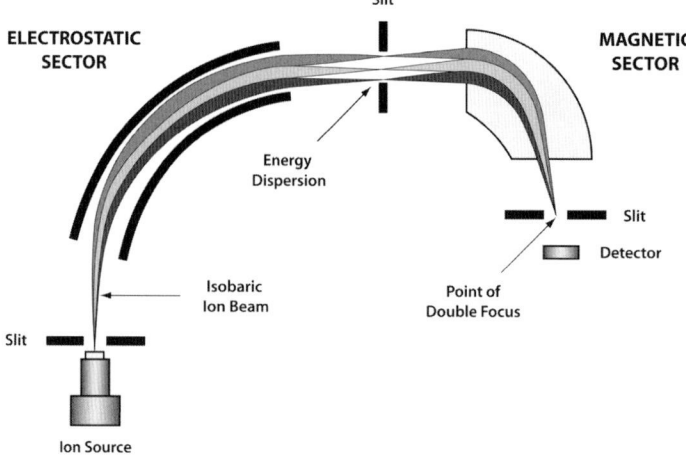

Figure 1.7 The operating principles of a double focusing sector mass analyzer.

sector. This results in velocity focusing, where ions of initially differing velocities are focused onto the same point. As both sectors also provide direction focusing, differences in both ion velocities and direction are accounted for in this process of double focusing.

The performance characteristics of double focusing sector instruments, as listed in Table 1.1, are unrivaled in terms of linear dynamic range and abundance sensitivity, while excellent mass accuracy and resolution are also capable of being obtained [8]. Despite these high-level performance capabilities, which have largely been established in elemental and inorganic MS, the use of sector mass analyzers in relation to other instruments has declined. This is because the applications of MS to biological problems, which have driven many of the contemporary advances in mass analyzer design, do not place an emphasis on obtaining ultrahigh linear dynamic ranges or abundance sensitivities. When coupled with the prohibitive size and cost of sector mass analyzers, this has seen other mass analyzer technologies favored by commercial producers of MS instrumentation. As such, sector mass analyzers have not been widely implemented in the analysis of macromolecules, such as synthetic polymers.

Table 1.1 Typical figures of merit for double focusing sector mass analyzers.

Mass resolving power	100 000
Mass accuracy	<1 ppm
Mass range	10 000
Linear dynamic range	1×10^9
Abundance sensitivity	$1 \times 10^6 - 1 \times 10^9$
Other	High cost and large space requirements

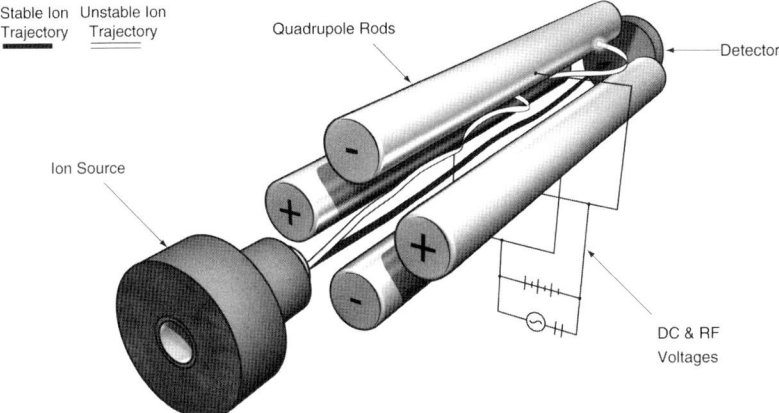

Figure 1.8 An illustration of the basic components of a quadrupole mass filter system, and the means by which it achieves m/z-based ion separation.

1.3.2
Quadrupole Mass Filters

Since the 1970s quadrupole mass filters have been perhaps the most widely utilized mass analyzer. The basic features of this method of mass analysis are illustrated in Figure 1.8. Quadrupole mass filters operate via the application of radio frequency (RF) and direct current (DC) voltages to four rods: the combination of RF and DC voltages determine the trajectories of ions of a given m/z within the mass filter; stable ion trajectories pass through to the detector while ions of unstable trajectories are neutralized by striking the quadrupole electrodes. By increasing the magnitude of the RF and DC voltages, typically while keeping the ratio of these two different voltages constant, the ions of differing m/z can sequentially pass through the mass filter for detection.

In discussing the operation of quadrupole mass filters, Mathieu stability diagrams are often of great utility. These diagrams, an example of which has been shown in Figure 1.9, allow one to obtain a ready visualization of the ions which will pass through to the detector and the ions which will not. The equations of ion motion in a quadrupole mass filter are second-order differential equations – this is because the RF voltages applied during mass analysis are time varying – and Mathieu stability diagrams are graphical representations of general solutions to these second-order differential equations. They are produced by plotting a parameter related to the RF voltage, q, against a parameter related to the DC voltage, a. These parameters are also determined by the frequency of the RF voltage, the size of the quadrupole rods and the m/z's of the ions under scrutiny. As the size of the quadrupole rods remain unchanged and the frequency of the RF voltage is usually held constant, one can therefore readily observe voltage combinations that will lead to stable trajectories for ions of a specified m/z. These areas in the Mathieu stability diagram are termed stability regions, and are labeled A, B, C, and D in Figure 1.9.

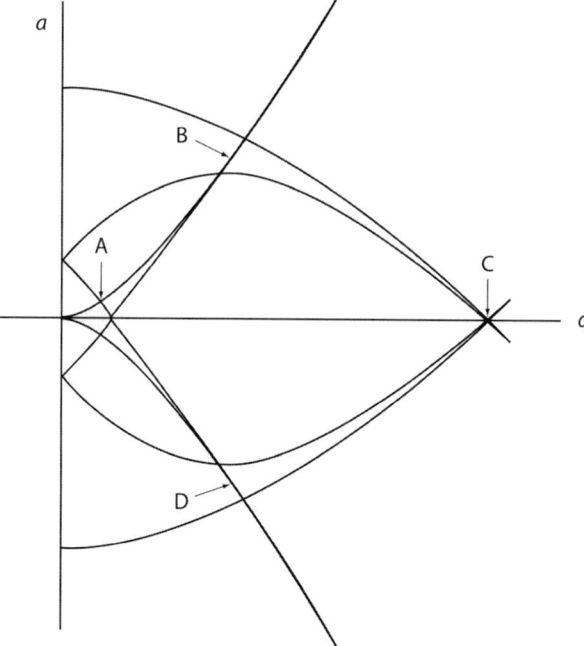

Figure 1.9 The Mathieu stability diagram. Stability regions are labeled A, B, C, and D.

Typical figures of merit for quadrupole mass filters have been listed in Table 1.2. Though quadrupole mass filters are a mainstay of contemporary mass analysis, they are low-performance instruments when judged in terms of mass resolving power, mass accuracy, and mass range. The vast majority of quadrupole mass filters operate within the first stability region, labeled A in Figure 1.9, and improvements in mass resolving power have been demonstrated when operating in higher stability regions [9, 10]. These improvements, however, come at the expense of mass range. Likewise, mass range extensions, which have been achieved through reductions in the operating frequency of the RF voltage [11–14], come at the expense of mass resolving power. The inability to maximize mass range, mass resolving power, or mass accuracy without compromise has ensured that, when operating quadrupole mass filters under standard conditions, these performance characteristics remain

Table 1.2 Typical figures of merit for quadrupole mass filters.

Mass resolving power	100–1000
Mass accuracy	100 ppm
Mass range	4000
Linear dynamic range	1×10^7
Abundance sensitivity	$1 \times 10^4 - 1 \times 10^6$
Other	Low cost and low space requirements

modest. Despite these limitations, quadrupole mass filters are capable of producing excellent linear dynamic ranges and abundance sensitivities. Along with their low cost, ease of automation, low ion acceleration voltages, and small physical size, these performance capabilities have contributed to the continued popularity of these instruments.

1.3.3
3D Ion Traps

Quadrupole ion traps are close relatives of the quadrupole mass filter and may be employed as 2D or 3D devices. The present section focuses upon the 3D ion trap, an example of which has been illustrated in Figure 1.10. The operating principles of 3D ion traps are similar to those of quadrupole mass filters. 3D ion traps, however, apply their electric fields in three dimensions as opposed to the two dimensions of mass filters; this is achieved through the arrangement of electrodes in a sandwich geometry: two end-cap electrodes enclose a ring electrode. This arrangement allows ions to be trapped within the electric field. When considering the operating principles of 3D ion traps, the Mathieu stability diagram may once again be used to visualize the

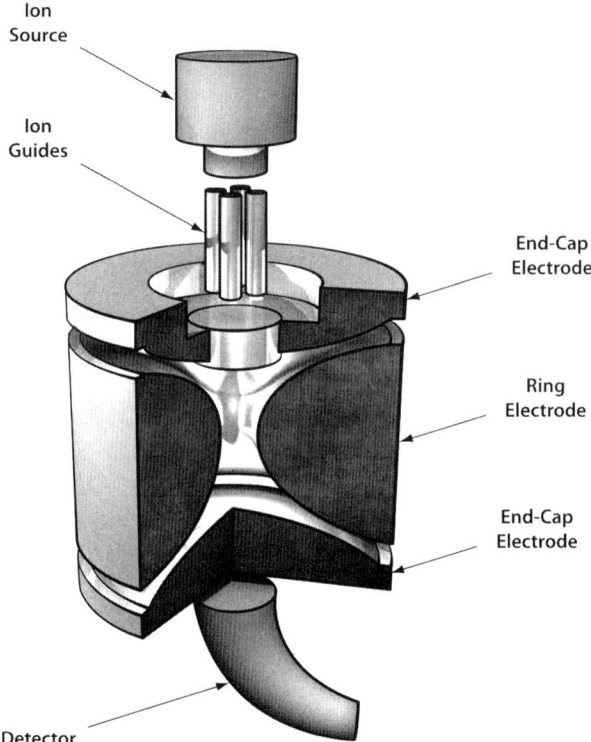

Figure 1.10 An illustration of the basic components of a 3D ion trap system.

Table 1.3 Typical figures of merit for 3D ion traps.

Mass resolving power	1000–10 000
Mass accuracy	50–100 ppm
Mass range	4000
Linear dynamic range	$1 \times 10^2 - 1 \times 10^3$
Abundance sensitivity	$1 \times 10^2 - 1 \times 10^3$
Other	Low cost and low space requirements

ions selected for detection. Unlike quadrupole mass filters, however, it is the unstable ions that are detected in standard 3D ion trap mass analysis. Mass selective instability is introduced by scanning the RF voltage applied to the device; as the voltage increases, the ions of sequentially higher m/z's are selected for detection by being ejected through an end-cap opening.

3D ion traps are generally capable of achieving moderate levels of performance in terms of mass resolving power, mass accuracy, and mass range, as can be seen in the figures of merit listed in Table 1.3. Innovative modes of operation can, however, allow these performance characteristics to be improved. For example, mass range extensions can be achieved by using resonance ejection, in which resonance conditions are induced by matching the frequency of ion oscillations in the trap with the frequency of a supplementary potential applied to the end-cap electrodes. Large enough amplitude of the resonance signal will allow ions to be ejected from the trap. Mass ranges of approximately 70 000 have been observed in conventional 3D ion traps using resonance ejection [15, 16], though this mode of operation is not readily supported by commercial instrumentation. The substantial lowering of RF voltage scan rates is another method by which 3D ion trap performance can be improved. Using this technique, resolutions of up to 10^7 have been achieved [17]. Such high levels of performance, however, come at the expense of analysis time and are generally performed over narrow mass ranges. As such, practical operating conditions result in significantly lower mass resolving powers.

The linear dynamic ranges of ion trapping devices, such as 3D ion traps, are limited by mass discrimination effects associated with ion/ion interactions or charge transfer to background gases. The extents to which these effects occur are influenced by ion trap storage capacities. In 3D ion traps, mass discrimination effects can ultimately lead to quite low performance. Though methods based around the selective accumulation of specific ions have been shown to increase linear dynamic range up to at least 10^5 [18], these methods rely upon preselection of ions for analysis and are therefore unlikely to be practical for most polymer studies.

The abundance sensitivities of 3D ion traps are also relatively low. As with the linear dynamic range, the abundance sensitivity is related to the ion storage capacity when an ion trapping device is employed. The ion storage capacity of a specific device will depend on its dimensions and operating parameters, but in general, commercial 3D ion traps can be estimated to be capable of trapping 10^6-10^7 ions [19]. Though attempts at increasing abundance sensitivity have been made [20, 21], the inherent

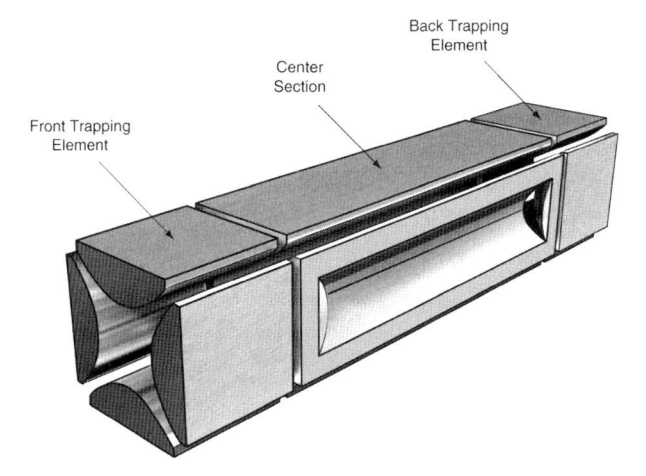

Figure 1.11 An illustration of the basic components of a radial ejection linear ion trap.

limitations of 3D ion trap storage capacity ensure that the abundance sensitivities of these instruments remain a weakness.

Despite their relatively modest performance capabilities, commercial ion traps are highly robust, have impressive MS/MS and MS^n capabilities (as expanded upon in Section 1.4) and are also remarkable for their attractively low size and cost. As such, 3D ion traps continue to see widespread use as workhorse-type instruments.

1.3.4
Linear Ion Traps

The operation of a quadrupole ion trap as 2D device – a linear ion trap – was first described in the late 1960s, but it is only in recent years that linear ion traps have emerged as a prominent form of mass analyzer. In contemporary stand-alone linear ion traps [22], an example of which has been illustrated in Figure 1.11, ions are trapped radially in a central section by an RF voltage, and axially by static DC potentials applied to end trapping elements. As with 3D ion traps, ions associated with unstable regions in the Mathieu stability diagram are selected for detection. The mass selective ejection of ions occurs radially through slots in central section rods and is achieved via the application of alternating current (AC) voltages. In addition to these stand-alone radial ejection devices, axial ejection linear ion traps have also found utility in contemporary MS by enhancing the performance of triple quadrupole (QqQ) mass spectrometers [23]; the basic capabilities of QqQ instruments will be discussed in Section 1.4.

Typical figures of merit for linear ion traps have been listed in Table 1.4. The mass resolving powers, mass accuracies, and mass ranges of linear ion traps are controlled by many of the processes associated with 3D ion trap mass analysis; as such, the capabilities of these two forms of mass analyzer are comparable when using these measures of performance. Linear ion traps do, however, feature ion storage capacities

Table 1.4 Typical figures of merit for linear ion traps.

Mass resolving power	1000–10 000
Mass accuracy	50–100 ppm
Mass range	4000
Linear dynamic range	1×10^3–1×10^4
Abundance sensitivity	1×10^3–1×10^4
Other	Low cost and low space requirements

that are over an order of magnitude higher than those of 3D ion traps [24]; the associated decreases in mass discrimination [22] suggest that greater linear dynamic range capabilities should be expected from these instruments. The trapping efficiencies of linear ion traps have also been demonstrated to be superior to those of 3D ion traps [22]. This advantage, in concert with their superior ion storage capacities, leads to relatively high abundance sensitivities. Like 3D ion traps, linear ion traps also feature high levels of robustness, excellent MS/MS and MS^n capabilities, favorably small size, and relatively low costs. When coupled to their superior performance capabilities, these features suggest that linear ion traps will likely supplant 3D ion traps as the dominant technology in quadrupole ion trap mass analysis.

1.3.5
Time-of-Flight Mass Analyzers

The 1980s witnessed the development of the revolutionary pulsed ionization method of MALDI. The mass analysis technique that saw the greatest increase in prominence as a result of this development was the TOF process, which requires a well-defined start time and is therefore ideally suited to being interfaced with pulsed ion sources. Though various important advances have been made to the TOF process since the development of MALDI, the basic operating principles underlying this method of mass analysis remain conceptually simple. These basic principles of operation can be seen in the illustration shown in Figure 1.12. All TOF mass analyzers rely upon the acceleration of ions obtained from an ion source through a fixed potential into a drift region of a set length. This process of ion acceleration results in all ions of the same charge obtaining the same kinetic energy, and as kinetic energy is equal to 0.5 mv^2, with m representing the mass of the ion and v the velocity of the ion, lower mass ions will obtain a greater velocity than higher mass ions. Lower mass ions will therefore traverse the distance of the drift region in a shorter amount of time than heavier ions, resulting in the separation of ions according to their m/z. As the length of the drift region is known, ion velocities can be determined by measuring the time they take to reach the detector, allowing the m/z of the ions to be determined.

An important source of error in a TOF experiment stems from small differences in the kinetic energies of ions of the same m/z; when MALDI ion sources are used, these kinetic energy distributions can be traced to aspects inherent to the complex processes involved in gas-phase ion generation [25]. To correct for these differences, almost all

1.3 Instrumentation | 21

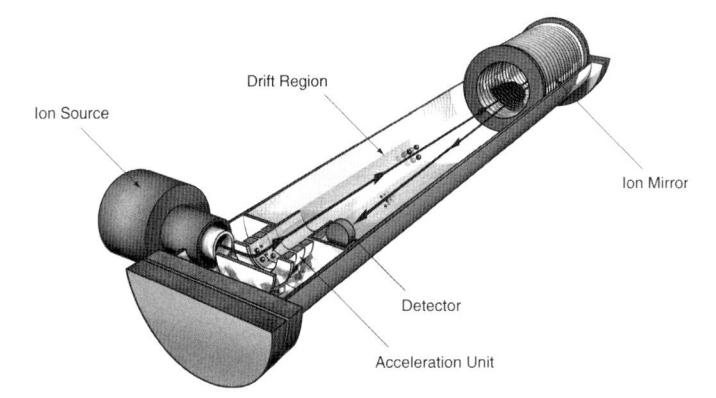

Figure 1.12 An illustration of the basic components of an orthogonal acceleration TOF mass analysis system featuring an ion mirror, and the means by which it achieves m/z-based ion separation.

TOF mass analyzers employ a single ion mirror, as illustrated in Figure 1.12. These reflectron TOF instruments operate by sending ions down one flight distance toward an electrostatic mirror, which then reflects the ions down a second flight distance toward a detector. In addition to compensating for differences in ion kinetic energies, the use of an ion mirror has the additional advantage of increasing the total flight distance without having to significantly increase the size of the mass spectrometer. These improvements lead to significantly increased mass resolving power and mass accuracy [26]. Kinetic energy distributions can also be corrected for through the process of delayed extraction, in which MALDI is performed in the absence of an electric field; ions are subsequently extracted using a high voltage pulse after a predetermined time delay. This process of delayed extraction has also been demonstrated to produce significant improvements in mass resolving power and mass accuracy [27–29].

Another major advance in contemporary TOF mass analysis has been the invention of orthogonal acceleration for coupling to continuous ionization sources [30], the basic principles of which are also illustrated in Figure 1.12. This technique makes use of independent axes for ion generation and mass analysis; a continuous ion source fills an acceleration region, and when full, an orthogonal acceleration process sends the ions into the TOF drift region. While the ions are being separated in the drift region, a new set of ions is collected in the acceleration region; this produces great experimental sensitivity. Importantly, orthogonal acceleration TOF has allowed ionization methods other than MALDI, most notably electrospray ionization (ESI), to benefit from the strong performance characteristics of TOF mass analysis.

Typical figures of merit for TOF mass analyzers are given in Table 1.5. Most contemporary TOF mass analyzers are reflectron instruments; the quoted mass resolving power and mass accuracy values are based upon the capabilities of these mass spectrometers. The mass accuracy values obtained from contemporary instrumentation have also benefited from the development of increasingly fast electronics; the nanosecond time resolution that is now routinely available contributes to the potential for achieving excellent mass accuracy. Increasingly fast electronics have also

Table 1.5 Typical figures of merit for TOF mass analyzers.

Mass resolving power	1000–40 000
Mass accuracy	5–50 ppm
Mass range	>100 000
Linear dynamic range	1×10^6
Abundance sensitivity	1×10^6
Other	Moderate cost and moderate space requirements

contributed toward increases in linear dynamic range. As there is a trade-off between the speed of the electronics and dynamic range, the capabilities of digital electronics improve the point at which this trade-off occurs. In terms of abundance sensitivity, the phenomenon of detector ringing as a result of higher abundance ion detection can have an adverse impact [31]. This problem is, however, minor when compared to the issues related to ion storage capacities in ion trapping instruments; TOF mass analyzers therefore typically feature higher levels of abundance sensitivity when compared to ion trapping devices. Of particular importance to polymer chemists is the mass range of TOF mass analyzers, which is theoretically unlimited. Mass ranges of 2000 kDa have been demonstrated in a cryodetection MALDI-TOF instrument [32], and in practice, mass ranges of >70 000 can be readily achieved using commercial instrumentation. In combination, these performance characteristics make TOF mass analysis an incredibly attractive option for many polymer studies.

1.3.6
Fourier Transform Ion Cyclotron Resonance Mass Analyzers

Mass analysis by FT-ICR was first described in 1974 [33], and the method has since grown to become the state of the art in terms of mass resolving power and mass accuracy capabilities. The basic operating principles of FT-ICR mass analysis are illustrated in Figure 1.13. In a similar manner to magnetic sector-based mass analysis, FT-ICR utilizes a magnetic field in its determinations of m/z. The kinetic energies of the ions measured by FT-ICR are, however, significantly lower than those analyzed by magnetic sector mass analyzers; this has the important consequence that, rather than being deflected by the magnetic field, the ions are trapped within the magnetic field. These trapped ions orbit with "cyclotron" frequencies that are inversely proportional to their m/z. Following the trapping of ions, RF voltages on excitation plates held perpendicular to the magnetic field are swept through a range of frequencies; this causes the sequential resonance excitation of ions into higher radii orbits. The oscillating field generated by these ion ensembles induces image currents in circuits connected to detection plates; the resultant time-domain signals of ion motion are converted into frequency-domain signals via a Fourier transform, which leads to the generation of a mass spectrum. If low pressures are maintained within the FT-ICR cell, the cyclotron motion can be held for many cycles, reducing the uncertainty of the frequency measurements and thereby allowing m/z to be determined with great accuracy.

BEFORE EXCITATION

DURING EXCITATION

Excitation Plates

RF On

AFTER EXCITATION Detection Plates

Time-Domain Signal

FT

Mass Spectrum

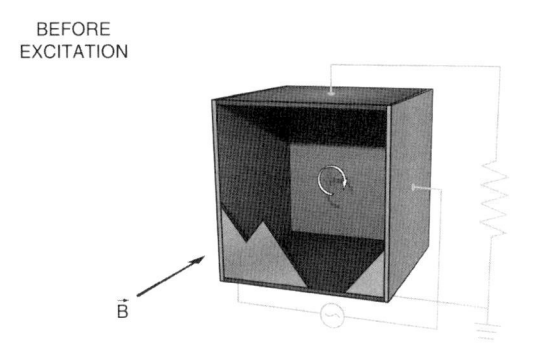

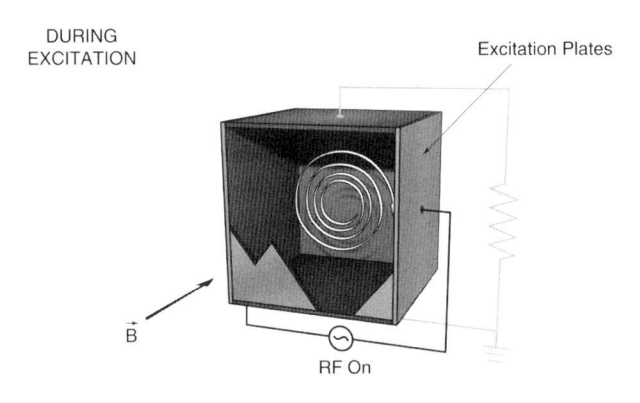

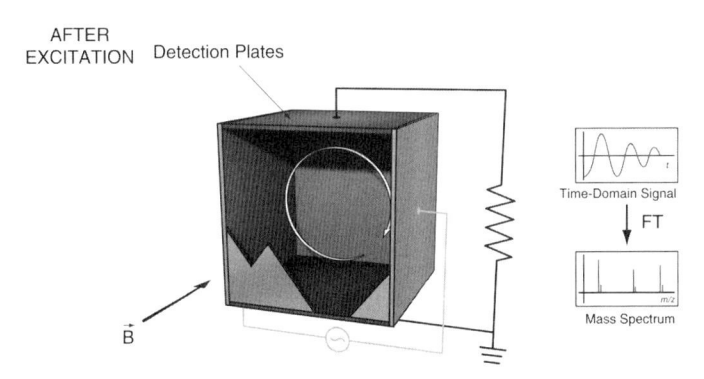

Figure 1.13 The operating principles of FT-ICR mass analysis. White arrows represent illustrative ion paths.

A summary of typical figures of merit for FT-ICR mass analyzers is given in Table 1.6. The mass resolving power and mass accuracy capabilities of these instruments are unparalleled in contemporary MS, and significant opportunities exist for these capabilities to be improved even further [34]. FT-ICR instruments also

1 Mass Analysis

Table 1.6 Typical figures of merit for FT-ICR mass analyzers.

Mass resolving power	10 000–1 000,000
Mass accuracy	1–5 ppm
Mass range	>10 000
Linear dynamic range	$1 \times 10^3 – 1 \times 10^4$
Abundance sensitivity	$1 \times 10^3 – 1 \times 10^4$
Other	High cost and large space requirements

offer increased mass ranges relative to other ion trapping devices. Ion storage capacities, however, remain to be a determining factor in limiting linear dynamic ranges and abundance sensitivities. Despite this, FT-ICR mass analyzers are the highest quality option for the analysis of polymer samples when ultrahigh mass accuracy and resolving power are required. These advantages do, however, come at a premium in terms of instrument cost and laboratory space requirements.

1.3.7
Orbitraps

Orbitraps represent the most recently developed form of mass analyzer in widespread contemporary usage, having been first described in 2000 [35]. The general principles of operation associated with orbitrap mass analysis are illustrated in Figure 1.14. These mass analyzers, like FT-ICR instruments and quadrupole ion traps, function as ion trapping devices. Unlike these other mass analyzers, however, orbitraps perform their trapping functions in the absence of magnetic or RF fields and instead utilize a purely electrostatic field generated by an outer barrel-like electrode and an inner axial spindle. Ions are injected tangentially into this field. The electrodes are carefully shaped such that the electrostatic attractions of the ions to

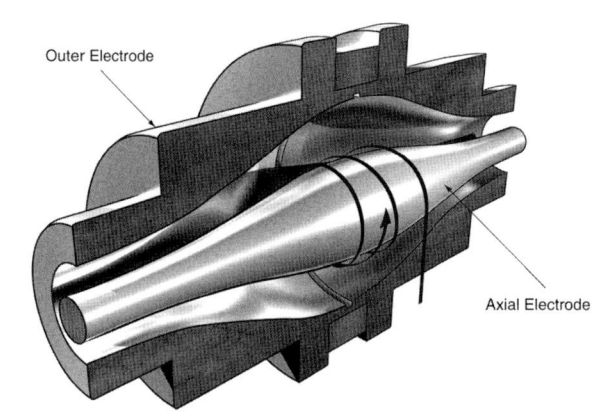

Figure 1.14 An illustration of the basic components of an Orbitrap mass analyzer. The black arrow represents an illustrative ion path.

Table 1.7 Typical figures of merit for orbitrap mass analyzers.

Mass resolving power	10 000–150 000
Mass accuracy	2–5 ppm
Mass range	6000
Linear dynamic range	$1 \times 10^3 – 1 \times 10^4$
Abundance sensitivity	1×10^4
Other	Moderate cost and low space requirements

the inner electrode are balanced by their centrifugal forces, causing them to orbit around the spindle, while the axial field causes the ions to simultaneously perform harmonic oscillations along the spindle at a frequency proportional to $(m/z)^{0.5}$. Image currents induced by the oscillating ions are detected by the outer wall of the barrel-like chamber, and in a similar manner to FT-ICR, the resultant time-domain signals of ion motion are converted into frequency-domain signals via a Fourier transform, which allows a mass spectrum to be produced.

A summary of typical figures of merit associated with orbitrap mass analyzers is presented in Table 1.7. As with FT-ICR, the mass resolving powers obtained using orbitraps are proportional to the number of harmonic oscillations that are detected. As the maximum acquisition times in orbitraps are more limited than those of FT-ICR instruments, their mass resolving power ceilings are not as high. Nevertheless, orbitraps are still capable of achieving mass resolving powers of up to 150 000 [36], which places them among the most powerful instruments available today. The mass accuracy values capable of being obtained using orbitraps approach those of FT-ICR instruments; mass accuracies within 2 ppm can be expected when internal calibration is performed [37]. Though issues relating to ion storage capacity still place significant limitations on the capabilities of orbitraps in relation to linear dynamic range and abundance sensitivity, they nonetheless feature larger ion storage capacities [35–38] and greater space charge capacities at higher mass [39] when compared to FT-ICR mass analyzers and 3D ion traps. As such, orbitraps have been shown to compare favorably with these instruments when judged by these particular measures of performance [36]. A major advantage of orbitraps is that their functioning does not require the use of superconducting magnets; they are therefore significantly less costly than FT-ICR instruments and have far more modest laboratory space requirements. These factors may ultimately ensure that orbitraps become favored over FT-ICR instruments for ultrahigh mass resolving power and mass accuracy polymer in analyses.

1.4
Instrumentation in Tandem and Multiple-Stage Mass Spectrometry

Developments in ion source, mass analyzer, and detector technologies have significantly improved the performance characteristics of modern mass spectrometers (*vide supra*). While these improvements have contributed greatly to the utility of MS in polymer science by more sensitive and accurate measurement of m/z, understanding the structural connectivity of molecular species cannot be established by molecular mass alone no matter how accurately measured. In other disciplines, MS/MS as well as MS^n have greatly expanded the scope and utility of MS by providing structural elucidation. Nowhere is this more apparent than in proteomics where the sequences (i.e., molecular structure) of peptide biopolymers are now established, almost exclusively, by this approach. Comparatively, the implementation of MS/MS and MS^n in polymer characterization has been modest and this can be attributed to two key reasons: (i) the yield of product ions observed is often low and (ii) the greater heterogeneity in synthetic polymer samples makes interpretation of the data challenging [40, 41]. While MS/MS and MS^n of polymers is the topic of a further chapter of this book, here we discuss the key performance criteria of different combinations of mass analyzers and highlight some contemporary developments that may play a role in overcoming some of the current challenges in generating information-rich tandem mass spectra of polymers.

MS/MS involves two stages of MS: precursor ions are mass-selected in the first stage (MS-I) of the experiment and are induced to undergo a chemical reaction that changes their mass or charge, leaving behind a product ion and a neutral fragment (or possibly another product ion if the precursor ion was multiply charged); in the second stage (MS-II), the product ions generated from these chemical reactions are mass analyzed. Some instruments allow this process to be repeated multiple times in MS^n experiments, where n refers to the number of stages of MS performed. The chemical reactions that proceed between the different MS stages are most frequently unimolecular dissociation reactions initiated by an increase in internal energy. Such ion activation is most commonly affected by collision-induced dissociation (CID), whereby the mass-selected precursor ion undergoes an energetic collision with an inert, stationary gas (e.g., N_2, Ar, or He) [42]. The amount of energy imparted in the collision is related to the translational kinetic energy of the precursor ion (most easily considered as the product of the number of charges on the ionized molecule and the accelerating potential applied within the instrument) and its mass. Thus, for ionized oligomers if the mass increases without a concomitant increase in charge, the internal energy will be less resulting in a lower product ion abundance.

Various different scan types can be executed in MS/MS and MS^n experiments, and each type can be used to extract different pieces of information from the sample under investigation. These scan types, as summarized in Figure 1.15 for MS/MS experiments, depend upon the static (i.e., transmission of a single m/z) or scan (i.e., analysis of all m/z) status of each stage of the experiment. How these different scan types apply to investigations concerning polymer systems are expanded upon below.

Selected reaction monitoring is generally implemented for the purposes of selective ion quantification and involves passing a known precursor ion through MS-I, and a known product ion (or ions) through MS-II. In this manner, precursor ions of indistinguishable mass to other ions generated from the sample can be selectively monitored if they produce unique product ions. Added specificity can be obtained by undertaking multiple-reaction monitoring experiments where, for example, several product ions are monitored in MS-II. Such analyses are usually

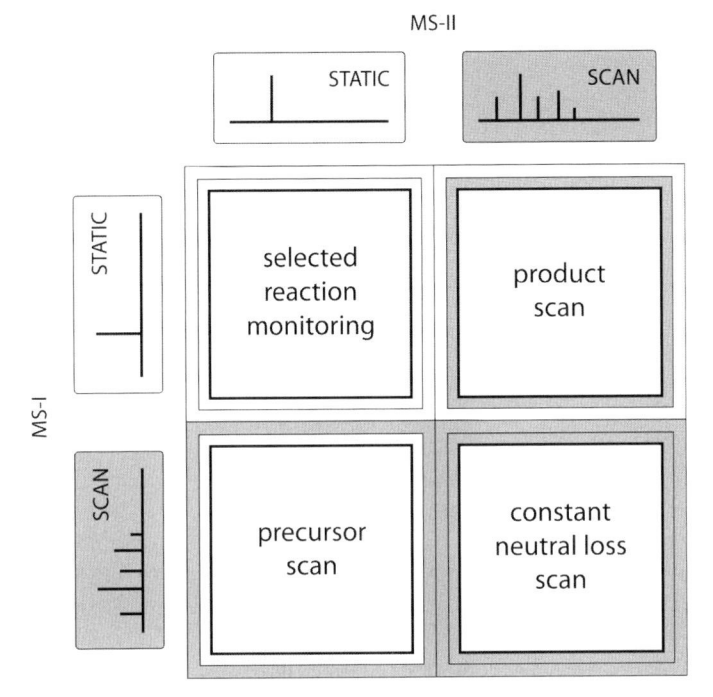

Figure 1.15 A summary of the scan types available to MS/MS experiments.

performed as a mechanism for highly selective and sensitive detection in online liquid or gas chromatography (i.e., LC-MS/MS and GC-MS/MS). Though these experiments can be of enormous utility for various purposes, for example, in the quantification of pharmaceutical compounds in human plasma, selected reaction monitoring has limited value in the majority of conventional MS investigations into polymerization systems. This lies in the fact that polymerization experiments typically couple poorly to conventional chromatographic methods and often fragmentation of the systems in question has not previously been established.

As with selected reaction monitoring, the precursor ion scan type requires knowledge of the preferred modes of fragmentation for the precursor ions. This particular scan type allows for the selective detection of particular classes of precursor ions, namely, precursor ions which dissociate to form common product ions. The precursor scan type can be of use in the analysis of polymer samples if the fragmentation behavior for a given ion obtained from the sample is known. Using this information, other structurally similar ions can be identified, potentially simplifying the identification of products in the sample. In a similar way, the constant neutral loss scan monitors a neutral mass loss between precursor and product ion and thus can also be used for monitoring for molecules with a similar structural motif in the presence of other, more abundant but structurally unrelated compounds. In a neutral loss scan, MS-I and MS-II are scanned to maintain a constant m/z difference associated with the neutral fragment of interest. In a similar manner to the precursor scan, the constant neutral loss scan can be used to identify ions of a common structural background, which can potentially aid in product identification. While this combination of scan types has been used to great effect in other fields, notably lipid MS [43], it has yet to be widely applied in polymer characterization because of the wide range of polymer structures and only a limited knowledge of fragmentation behavior.

Of the four scan types in MS/MS, the product ion scan is the most direct method of obtaining structural information from a previously unidentified ion and is thus perhaps the most widely applicable to samples associated with investigations into polymer systems. The product scan type isolates a specific precursor ion and analyzes all of the resulting product ions; careful interpretation of the fragmentation pathways indicated by the product ions can allow insights into the structural make-up of the precursor ion.

The instruments capable of undertaking MS/MS experiments can be classified into two groups: tandem-in-space instruments and tandem-in-time instruments. Tandem-in-space instruments require separate mass analyzers to be utilized for each MS stage and are associated with beam-type technology such as sector mass analyzers, quadrupole mass filters, and TOF mass analyzers. The earliest MS/MS instruments used sectors for the two MS stages [44], and the promise shown by these instruments led to the development of the cheaper and more user-friendly QqQ [45]. These QqQ instruments, which are still heavily in use today, operate the first and last quadrupoles as the actual mass filters, with the middle quadrupole acting as a CID cell. Since the introduction of QqQ instruments, various hybrid instruments, in which distinct mass analysis methodologies are applied at each stage of the MS/MS experiment, have been developed. Of these instruments, the quadrupole-TOF (Q-TOF) mass spectrometer [46] has become a mainstay of MS/MS technology. Q-TOF instruments have the disadvantage of not being able to perform the precursor and constant neutral loss scans that QqQ instruments are capable of, as the TOF method of separating ions in time is not conducive to these screening-type scans; however, these instruments remain advantageous for the speed in which MS/MS experiments can be conducted and their ability to provide accurate mass measurements on product ions. In contrast to the tandem-in-space instruments, tandem-in-time instruments separate the different MS stages by time, with the various stages of MS/MS being performed in one mass analyzer. These instruments are associated with ion trapping technology such as quadrupole ion traps, FT-ICR mass analyzers, and orbitraps. Though tandem-in-time instruments are incapable of performing precursor ion or constant neutral loss scans, they have the advantage of being able to readily perform MS^n experiments, as the separate MS stages do not require the implementation of multiple mass analyzers. This can be advantageous in providing detailed structural information by elucidation of secondary or even tertiary product ions. More recently, hybrid instruments have been developed that combine both in-space with in-time mass analysis leading to a fascinating array of combinations, as summarized in Figure 1.16.

While a detailed discussion of the relative capabilities of the instrument geometries listed in Figure 1.16 is beyond the scope of this chapter (see Ref. [47]), one key criterion that should be considered in the context of MS/MS is the collision energy

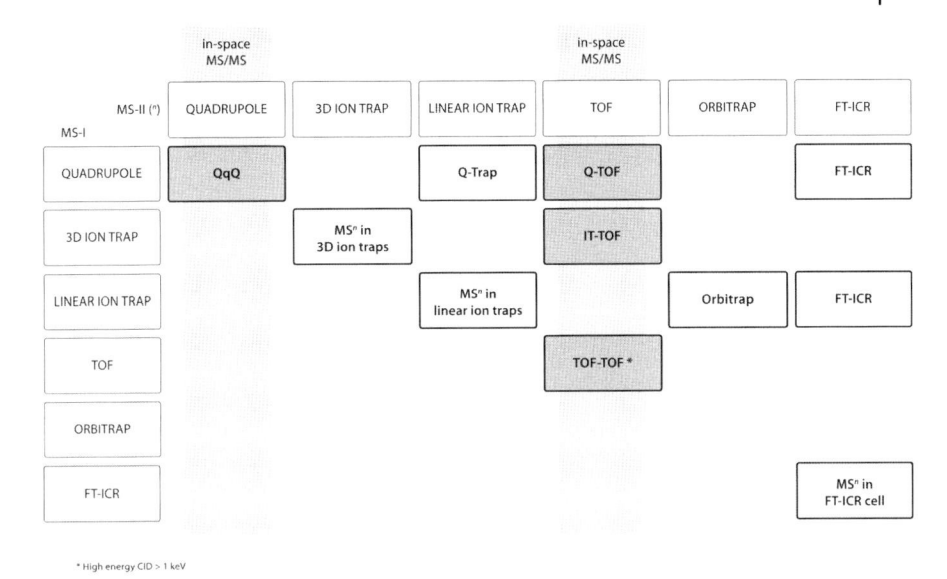

Figure 1.16 A summary of mass analyzers and mass analyzer combinations used for MS/MS and MS^n.

regime (an excellent tutorial on the full range of figures of merit in ion activation is found in Ref. [48]). Depending on the instrument configuration, CID is generally considered as either high or low energy depending on whether the translational energy of the precursor ion is $>$ or <1 keV, respectively. Early sector-based mass spectrometers operated in the keV energy range, and thus the mechanism of ion activation and the amount of internal energy imparted to a precursor ion was significantly different to that of a contemporary QqQ or Q-TOF geometry instrument (typically <100 eV). Interestingly, more recent developments in TOF-TOF geometries have, almost serendipitously, provided renewed interest in the high-energy CID regime. This holds out some promise for polymer analysis where oligomers can be of high mass and may be only singly charged [49]. Furthermore, the current research into alternative approaches to ion activation and their implementation into commercial instrumentation (e.g., electron capture dissociation, electron transfer dissociation, and photodissociation) augurs well for future development of tandem mass spectrometric methods for polymer science.

1.5
Conclusions and Outlook

Recent and continuing advances in mass analyzer design have ensured that the field of polymer chemistry remains well positioned to see increased benefits from MS. Workhorse-type instruments, such as quadrupole ion traps, are now well established, and their role in providing readily accessible, informative polymer characterizations is unlikely to be displaced in the near future. With the growing accessibility of high performance mass analysis options, a general increase in the depth of mass spectrometric-based polymer characterizations can, however, be envisaged. In this regard, the benefits of ultra high mass accuracy and mass resolving power, previously obtained solely from FT-ICRs for polymer samples, are likely to become more broadly appreciated with the advent of orbitrap mass analysis. The increasing power of TOF technologies is also of note; progressively higher mass accuracies and resolving powers are being achieved over broad mass ranges. Along with their potential to produce excellent abundance sensitivities, these improving performance characteristics suggest that TOF mass analyzers will continue to play an important role in polymer chemistry. Though continued progress in individual mass analyzer design can be expected, the dominant trend in contemporary MS instrument development has been in mass analyzer hybridization. The specific benefits of these hybrid instruments will become more apparent as the role of MS/MS and MS^n in polymer chemistry is further established. Given this suite of new and developing technologies, an intriguing array of mass analysis options is now primed to be more thoroughly explored by the polymer chemistry community.

References

1 McNaught, A.D. and Wilkinson, A. (1997) BT IUPAC Compendium of Chemical Terminology, Blackwell Scientific Publications, Oxford.

2 Szablan, Z., Junkers, T., Koo, S.P.S., Lovestead, T.M., Davis, T.P., Stenzel, M.H., and Barner-Kowollik, C. (2007) Mapping photolysis product radical reactivities via soft ionization mass spectrometry in acrylate, methacrylate, and itaconate systems. *Macromolecules*, **40** (19), 6820–6833.

3 Brenton, A.G. and Godfrey, A.R. (2010) Accurate mass measurement: Terminology and treatment of data. *J. Am. Soc. Mass Spectrom.*, **21** (11), 1821–1835.

4 Gruendling, T., Pickford, R., Guilhaus, M., and Barner-Kowollik, C. (2008) Degradation of RAFT polymers in a cyclic ether studied via high resolution ESI-MS: Implications for synthesis, storage, and end-group modification. *J. Polym. Sci. Pol. Chem.*, **46** (22), 7447–7461.

5 Hart-Smith, G., Chaffey-Millar, H., and Barner-Kowollik, C. (2008) Living star polymer formation: detailed assessment of poly(acrylate) radical reaction pathways via ESI-MS. *Macromolecules*, **41** (9), 3023–3041.

6 Ladavière, C., Lacroix-Desmazes, P., and Delolme, F. (2009) First systematic MALDI/ESI mass spectrometry comparison to characterize polystyrene synthesized by different controlled radical polymerizations. *Macromolecules*, **42** (1), 70–84.

7 Hart-Smith, G., Lammens, M., Du Prez, F.E., Guilhaus, M., and Barner-Kowollik, C. (2009) ATRP poly(acrylate) star formation: A comparative study between MALDI and ESI mass spectrometry. *Polymer*, **50** (9), 1986–2000.

8 Moens, L. and Jakubowski, N. (1998) Double-focusing mass spectrometers in ICPMS. *Anal. Chem.*, **70** (7), 251A–256A.

9 Du, Z.H., Olney, T.N., and Douglas, D.J. (1997) Inductively coupled plasma mass spectrometry with a quadrupole mass filter operated in the third stability region. *J. Am. Soc. Mass Spectrom.*, **8** (12), 1230–1236.

10 Ying, J.F. and Douglas, D.J. (1996) High resolution inductively coupled plasma

10 mass spectra with a quadrupole mass filter. *Rapid Commun. Mass Spectrom.*, **10** (6), 649–652.

11 Winger, B.E., Lightwahl, K.J., Loo, R.R.O., Udseth, H.R., and Smith, R.D. (1993) Observation and implications of high mass-to-charge ratio ions from electrospray-ionization mass-spectrometry. *J. Am. Soc. Mass Spectrom.*, **4** (7), 536–545.

12 Lightwahl, K.J., Winger, B.E., and Smith, R.D. (1993) Observation of the multimeric forms of concanavalin-A by electrospray-ionization mass-spectrometry. *J. Am. Chem. Soc.*, **115** (13), 5869–5870.

13 Lightwahl, K.J., Schwartz, B.L., and Smith, R.D. (1994) Observation of the noncovalent quaternary associations of proteins by electrospray-ionization mass-spectrometry. *J. Am. Chem. Soc.*, **116** (12), 5271–5278.

14 Collings, B.A. and Douglas, D.J. (1997) An extended mass range quadrupole for electrospray mass spectrometry. *Int. J. Mass Spectrom. Ion. Process*, **162** (1–3), 121–127.

15 Kaiser, R.E., Cooks, R.G., Stafford, G.C., Syka, J.E.P., and Hemberger, P.H. (1991) Operation of a quadrupole ion trap mass-spectrometer to achieve high mass charge ratios. *Int. J. Mass Spectrom. Ion. Process*, **106**, 79–115.

16 Stephenson, J.L. and McLuckey, S.A. (1996) Ion/ion proton transfer reactions for protein mixture analysis. *Anal. Chem.*, **68** (22), 4026–4032.

17 Schwartz, J.C., Syka, J.E.P., and Jardine, I. (1991) High-resolution on a quadrupole ion trap mass-spectrometer. *J. Am. Soc. Mass Spectrom.*, **2** (3), 198–204.

18 Duckworth, D.C., Barshick, C.M., Smith, D.H., and McLuckey, S.A. (1994) Dynamic-range extension in glow-discharge quadrupole ion-trap mass-spectrometry. *Anal. Chem.*, **66** (1), 92–98.

19 March, R.E. and Hughes, R.J. (1989) BT Quadrupole Storage Mass Spectrometry, John Wiley and Sons, New York.

20 Garrett, A.W., Hemberger, P.H., and Nogar, N.S. (1995) Selective ionization for elemental analysis with an ion trap mass spectrometer. *Spectrochimica Acta Part B*, **50** (14), 1889–1895.

21 Eiden, G.C., Barinaga, C.J., and Koppenaal, D.W. (1996) Plasma source ion trap mass spectrometry: Enhanced abundance sensitivity by resonant ejection of atomic ions. *J. Am. Soc. Mass Spectrom.*, **7** (11), 1161–1171.

22 Schwartz, J.C., Senko, M.W., and Syka, J.E.P. (2002) A two-dimensional quadrupole ion trap mass spectrometer. *J. Am. Soc. Mass Spectrom.*, **13** (6), 659–669.

23 Hager, J.W. (2002) A new linear ion trap mass spectrometer. *Rapid Commun. Mass Spectrom.*, **16** (6), 512–526.

24 Douglas, D.J., Frank, A.J., and Mao, D.M. (2005) Linear ion traps in mass spectrometry. *Mass Spectrom. Rev.*, **24** (1), 1–29.

25 Karas, M. and Kruger, R. (2003) Ion formation in MALDI: The cluster ionization mechanism. *Chem. Rev.*, **103** (2), 427–439.

26 Mamyrin, B.A., Karataev, V.I., Shmikk, D.V., and Zagulin, V.A. (1973) Mass-reflectron a new nonmagnetic time-of-flight high-resolution mass-spectrometer. *Zhurnal Eksperimentalnoi Teoreticheskoi Fiziki*, **64** (1), 82–89.

27 Vestal, M.L., Juhasz, P., and Martin, S.A. (1995) Delayed extraction matrix-assisted laser-desorption time-of-flight mass-spectrometry. *Rapid Commun. Mass Spectrom.*, **9** (11), 1044–1050.

28 Brown, R.S. and Lennon, J.J. (1995) Mass resolution improvement by incorporation of pulsed ion extraction in a matrix-assisted laser-desorption ionization linear time-of-flight mass-spectrometer. *Anal. Chem.*, **67** (13), 1998–2003.

29 Colby, S.M., King, T.B., and Reilly, J.P. (1994) Improving the resolution of matrix-assisted laser desorption/ionization time-of-flight mass-spectrometry by exploiting the correlation between ion position and velocity. *Rapid Commun. Mass Spectrom.*, **8** (11), 865–868.

30 Dawson, J.H.J. and Guilhaus, M. (1989) Orthogonal-acceleration time-of-flight mass spectrometer. *Rapid Commun. Mass Spectrom.*, **3** (5), 155–159.

31 Myers, D.P., Mahoney, P.P., Li, G., and Hieftje, G.M. (1995) Isotope ratios and abundance sensitivity obtained with an inductively-coupled plasma-time-of-flight

mass-spectrometer. *J. Am. Soc. Mass Spectrom.*, **6** (10), 920–927.

32 Aksenov, A.A. and Bier, M.E. (2008) The analysis of polystyrene and polystyrene aggregates into the mega Dalton mass range by cryodetection MALDI TOF MS. *J. Am. Soc. Mass Spectrom.*, **19** (2), 219–230.

33 Comisarow, M.B. and Marshall, A.G. (1974) Fourier-transform ion-cyclotron resonance spectroscopy. *Chem. Phys. Lett.*, **25** (2), 282–283.

34 Zhang, L.K., Rempel, D., Pramanik, B.N., and Gross, M.L. (2005) Accurate mass measurements by Fourier transform mass spectrometry. *Mass Spectrom. Rev.*, **24** (2), 286–309.

35 Makarov, A. (2000) Electrostatic axially harmonic orbital trapping: a high-performance technique of mass analysis. *Anal. Chem.*, **72** (6), 1156–1162.

36 Hu, Q., Noll, R.J., Li, H., Makarov, A., Hardman, M., and Cooks, R.G. (2005) The Orbitrap: A new mass spectrometer. *J. Mass Spectrom.*, **40** (4), 430–443.

37 Makarov, A., Denisov, E., Kholomeev, A., Baischun, W., Lange, O., Strupat, K., and Horning, S. (2006) Performance evaluation of a hybrid linear ion trap/Orbitrap mass spectrometer. *Anal. Chem.*, **78** (7), 2113–2120.

38 Hardman, M. and Makarov, A.A. (2003) Interfacing the orbitrap mass analyser to an electrospray ion source. *Anal. Chem.*, **75** (7), 1699–1705.

39 Perry, R.H., Cooks, R.G., and Noll, R.J. (2008) Orbitrap mass spectrometry: Instrumentation, ion motion and applications. *Mass Spectrom. Rev.*, **27** (6), 661–699.

40 Weidner, S.M. and Trimpin, S. (2008) Mass spectrometry of synthetic polymers. *Anal. Chem.*, **80** (12), 4349–4361.

41 Crecelius, A.C., Baumgaertel, A., and Schubert, U.S. (2009) Tandem mass spectrometry of synthetic polymers. *J. Mass Spectrom.*, **44** (9), 1277–1286.

42 McLuckey, S.A. (1992) Principles of collisional activation in analytical mass-spectrometry. *J. Am. Soc. Mass Spectrom.*, **3** (6), 599–614.

43 Blanksby, S.J. and Mitchell, T.W. (2010) Advances in mass spectrometry for lipidomics. *Ann. Rev. Anal. Chem.*, **3**, 433–465.

44 Kondrat, R.W. and Cooks, R.G. (1978) Direct analysis of mixtures by mass-spectrometry. *Anal. Chem.*, **50** (1), 81A–92A.

45 Yost, R.A. and Enke, C.G. (1978) Selected ion fragmentation with a tandem quadrupole mass-spectrometer. *J. Am. Chem. Soc.*, **100** (7), 2274–2275.

46 Glish, G.L. and Goeringer, D.E. (1984) Tandem quadrupole-time-of-flight instrument for mass-spectrometry mass-spectrometry. *Anal. Chem.*, **56** (13), 2291–2295.

47 Glish, G.L. and Burinsky, D.J. (2008) Hybrid mass spectrometers for tandem mass spectrometry. *J. Am. Soc. Mass Spectrom.*, **19** (2), 161–172.

48 McLuckey, S.A. and Goeringer, D.E. (1997) Slow heating methods in tandem mass spectrometry. *J. Mass Spectrom.*, **32** (5), 461–474.

49 Gies, A.P., Vergne, M.J., Orndorff, R.L., and Hercules, D.M. (2007) MALDI-TOF/TOF CID study of polystyrene fragmentation reactions. *Macromolecules*, **40** (21), 7493–7504.

2
Ionization Techniques for Polymer Mass Spectrometry
Anthony P. Gies

2.1
Introduction

Over the past 60 years, there has been a rapid expansion in the application of mass spectrometry (MS) to synthetic polymer analysis. This is mainly due to the development of macromolecular ionization techniques such as matrix-assisted laser desorption/ionization (MALDI) and electrospray ionization (ESI), as well as the instruments becoming more affordable. Historically, size exclusion chromatography (SEC), vapor pressure osmometry (VPO), nuclear magnetic resonance (NMR), light scattering, infrared spectroscopy, and ultraviolet/visible spectroscopy have been used for polymer characterization. However, these "classical" methods have the disadvantage of being *averaging* techniques, which provide general information about the "average" polymer mixture as a whole, instead of providing selective information about individual oligomers within the mixture. Further, these classical *averaging* techniques are rarely capable of providing information about the various oligomers and impurities that may be present within the polymer mixture. In contrast, MS can be used for molecular mass determination, architectural elucidation, end-group analysis, quantification at trace levels, analysis of complex mixtures, and determination of degradation mechanisms [1]. Additionally, most of the classical techniques are "relative" methods that rely on calibration standards, such as polystyrene, poly(ethylene glycol), and poly(methyl methacrylate), which in many cases do not possess properties similar to the specific polymer that is being studied, and inferences must be made to yield "estimated" characterization information. Alternatively, MS is an "absolute" method which does not rely upon polymers' standards for calibration; this is an important advantage when standards do not exist for the polymer of interest.

The purpose of the mass spectrometer is to act as a tool for converting the sample into measureable products that are indicative of the polymer of interest [2]. The products analyzed by the mass spectrometer are gaseous ions (positive and/or negative), whose masses and relative abundances are displayed in the mass spectrum. Figure 2.1 shows the three basic components of a mass spectrometer: (A) a *source* for converting the analyte into gaseous ions; (B) a *mass analyzer* for separating

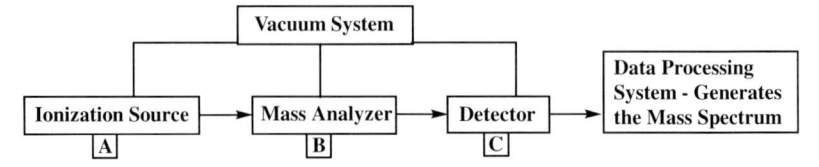

Figure 2.1 Block diagram showing the basic components of a mass spectrometer.

ions based upon their mass-to-charge ratio; and (C) a *detector* to convert analyte ions into an electrical signal for computer processing. In order for this process to work, it is essential for the analyte to yield charged species in the gas phase [2]; these can be formed in a wide variety of ion sources and will form the basis of our discussion.

2.2
Small Molecule Ionization Era

During the early development of polymer MS ionization techniques (1950s–1960s – the "small molecule ionization era"), the application of MS to synthetic polymers was limited due to their low volatility and thermal instability of these materials. Ion sources used during this era generally consisted of electron ionization (EI) and chemical ionization (CI).

2.2.1
Electron Ionization (EI)

Electron ionization, initially devised by Dempster [3] and improved by Bleakney [4] and Nier [5], was the earliest commercialized technique used for polymer analysis.

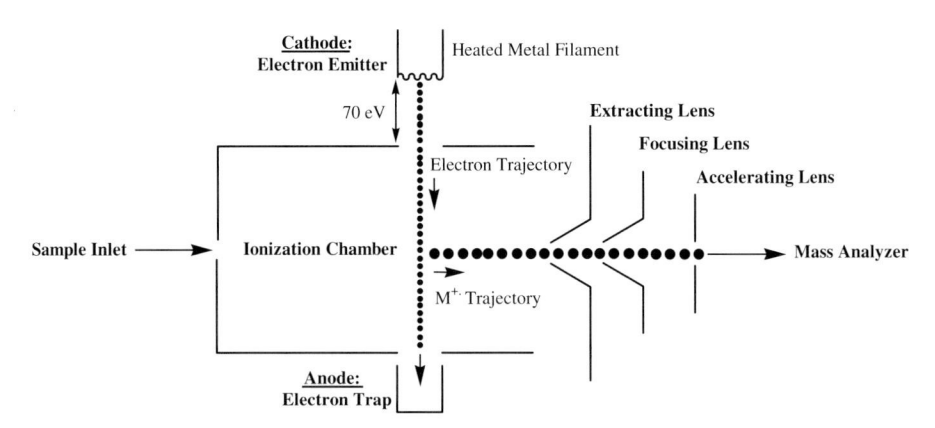

Figure 2.2 Schematic diagram of an electron ionization (EI) source.

In EI (Figure 2.2), the sample is initially vaporized and, upon entering the EI source, the gaseous sample molecules (under low-pressure conditions of $\sim 10^{-5}$ to 10^{-6} Torr) collide with an electron beam of $\sim$70 eV kinetic energy. (The electron beam is generated by "boiling off" high-energy electrons from an incandescent filament and forcing them to travel through the ion chamber to an anode ("electron trap") on the opposite side [2].) Energy transferred to the analyte molecule will primarily be stored in internal modes, such as vibration and electronic excitation; each of these modes is capable of several electron volts uptake. (Translational and rotational excitation cannot store significant amounts of energy.) If the electron, in terms of energy transfer, collides effectively with the neutral, the energy transferred can exceed the ionization energy (generally in the range of 7–15 eV) of the neutral. The result of collision of the analyte molecules (M) with the electron beam causes analyte molecule ejection of an electron, leading to the formation of intact molecular radical cations, $M^{+\cdot}$:

$$M + e^- \rightarrow M^{+\cdot} + 2e^-$$

This process has a low ion yield (0.001%) [2] and leaves the newly created molecular radical cations with a range of internal energies; therefore, the radical cations with high energies break apart and form a number of fragment ions through decomposition reactions. As a result of this ionization, the EI mass spectrum usually consists of a molecular ion peak and a number of fragment ion peaks. The fragmentation pattern thus obtained is an indication of the structure of the sample molecule – it is a "fingerprint" of the compound being analyzed. The degree of fragmentation can be lowered by reducing the electron energy.

Limitations

Although EI is a simple and sensitive technique for the analysis of small molecules (<1000 Da), it suffers from the drawback that many compounds are not stable under EI conditions. For example, when 2–8 eV of energy is transferred to a molecule during ionization, certain types of molecules fragment so extensively that the molecular ion (the intact analyte molecule) is either absent or of very little significance. Without the presence of a molecular ion, it is a difficult task to determine the molecular mass of an unknown compound. A further pitfall of EI is its incompatibility with thermally labile and nonvolatile compounds. For example, the thermal energy, required to disrupt the intermolecular bonding in polar compounds and volatize them, may also induce decomposition. In some cases, chemical derivatization of polar analytes may be used to "block" polar functional groups from participating in intermolecular hydrogen bonding and help to overcome the difficulties in volatilizing the intact analyte molecule. However, chemical derivatization requires additional sample-handling steps and has potential for introducing an uncertainty, when determining the molecular mass of the original compound, due to the possibility of incomplete derivatization or not knowing the exact number of derivatization sites [1].

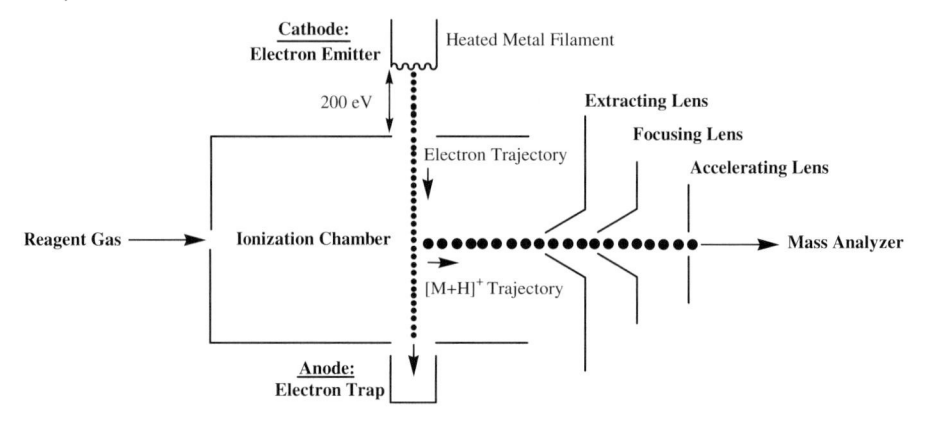

Figure 2.3 Diagram of a chemical ionization (CI) source.

2.2.2
Chemical Ionization (CI)

Chemical ionization (Figure 2.3), introduced as an analytical technique by Munson and Field in the mid-1960s [6–9], is similar to EI, except that a reagent gas is used to promote ionization of the analyte molecules instead of an electron beam. Additionally, the source is kept at a higher pressure (0.1–2 Torr) than an EI source (10^{-5} Torr) [2]. The ionization process in CI involves ion–molecule reactions of the sample molecules with reagent gas ions. The reagent gas (RH), for example, methane, is introduced into the source and is ionized by EI to form reactant ions (RH_2^+, for example, CH_5^+ for methane) that react with analyte molecules (M) to form stable protonated analyte ions, which represent undissociated sample molecules. (The 10^3–10^4-fold excess of reagent gas also shields the analyte molecules from competing for direct EI.)

$$RH^+ + RH \rightarrow RH_2^+ + R \text{(reagent ion formation)}$$

$$RH_2^+ + M \rightarrow RH + MH^+ \text{(proton transfer)}$$

The most common CI process is a proton transfer from the reagent gas to form MH^+ ions; however, the electrophilic addition of the reagent ion can also occur to form $[M + RH_2]^+$. The ion formation process occurs with controlled internal energy deposition, because the energy is attenuated via collisions with the neutral reagent gas molecules; therefore, molecules ionized with CI generally exhibit less fragmentation than EI. Also, the ions formed by CI are called "even-electron quasi-molecular ions" and, unlike EI-formed molecular ions, their masses are different from those of their neutral source molecules [1]. Two important features of CI are that (1) the extent of fragmentation can be controlled by a proper choice of reagent gas and (2) the feasibility of the selective ionization of a specific compound. Through proper selection of the reagent gas, only those compounds with a high proton affinity greater than that of the reagent gas will be ionized [1]. For example, amines and ethers usually give abundant MH^+ ions, while saturated hydrocarbons generally yield $(M-1)^+$ ions [2]. Both types of ions are useful for indicating the molecular weight, because they are often present in the CI spectra of compounds whose EI spectra show *no* molecular ions [2]. Chemical ionization produces a much simpler mass spectrum than EI since it contains primarily the molecular ion signal and only a few fragment ions. Thus, the primary use of CI is to confirm or determine the molecular mass of volatile compounds and act as a complimentary ionization technique to EI.

Limitations

Like EI, a major limitation of CI is its inability to handle nonvolatile compounds, and thus, it is restricted to compounds with molecular masses below 1000 Da. Another obvious limitation is the lack of structural information in the CI spectrum. An innovative approach to circumvent this difficulty is to acquire both EI and CI spectra on the same injection of the sample; this is accomplished by evacuating the reagent gas from the chamber on every alternate scan.

2.2.3
Pyrolysis Mass Spectrometry (Py-MS)

The main problem with using EI and CI is that the sample must first be vaporized by heating. It is important to note that polymers with masses greater than a few hundred Daltons are not readily volatile; therefore, it is difficult to introduce polymers into a mass spectrometer in the gas phase. This would appear to limit the utility of these techniques to that of monomer analysis. However, high-molecular-weight polymers can be made amenable to EI and CI analysis through the use of a pyrolyzer. (Figure 2.4 shows a modern furnace pyrolyzer interfaced with a GC/MS.) This works by thermally degrading a polymer (generally at temperatures between 400 and 600 °C) into its unimer, dimer, trimer, and so on and introducing the low molecular weight degradation products into an EI or CI source for MS analysis.

Limitations

Though pyrolysis was a valuable breakthrough for identifying the basic components of polymers, it still was not the ideal analysis technique for studying intact synthetic polymers. For example, in the early stages of pyrolysis-mass spectrometry (Py-MS), pyrolysis data lacked reproducibility and consistency of results in characteristic degradation of polymeric materials. This was primarily solved by the introduction of Curie-point pyrolyzers and other advanced pyrolyzers (such as the microfurnace) [10–12]. Another limitation of pyrolysis was that it only provided indirect analysis information since it required degradation of the polymer chains before analysis. Many times the really useful information that is generated during Py-MS analysis is only observed as trace level peaks which are difficult to interpret or yield

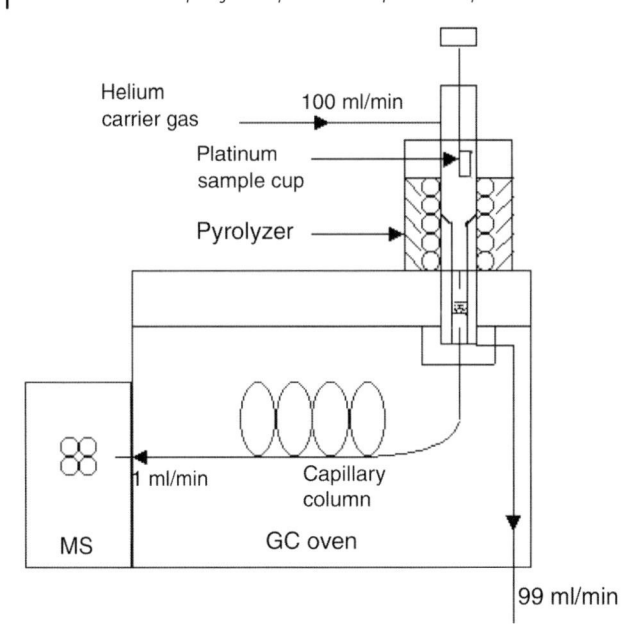

Figure 2.4 Conceptual representation of a pyrolysis-gas chromatography/mass spectrometry (Py-GC/MS) setup.

questionable results. Ideally, polymer chemists would like information about the intact oligomers and their distribution; this is not possible by pyrolysis, and therefore prompted the search for new direct ionization methods for analyzing higher molecular weight species.

2.3
Macromass Era of Ionization

The next era of polymer MS is the "macromass" ionization era (1970s–1980s). This era led to the ability to acquire mass spectra of intact polymer oligomers by direct analysis without the use of a pyrolyzer [13]. There were a number of ionization techniques introduced in this era for polymer analysis: field desorption (FD), fast atom bombardment (FAB), secondary ion mass spectrometry (SIMS), and laser desorption mass spectrometry (LD-MS). These methods involved bombardment of a prepared sample with fast atoms, ions, or laser photons, or in the case of FD, a high electric field combined with gentle heating.

2.3.1
Field Desorption (FD) and Field Ionization (FI)

Field desorption mass spectrometry (FD-MS) was introduced by Beckey in 1969 [14]. This technique is derived from field ionization (FI) MS which was originated by

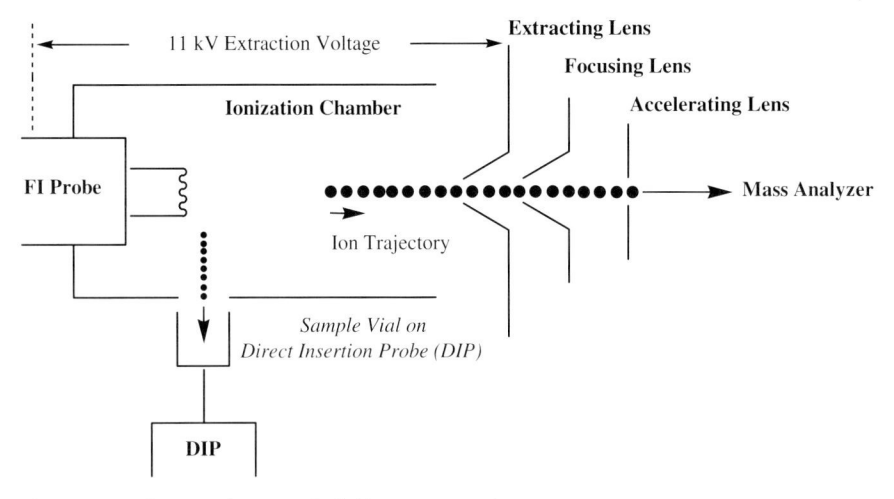

Figure 2.5 Schematic diagram of a field ionization (FI) source.

Inghram and Gomer in 1954 [15]. The main differences between FD and FI are that, in FI (Figure 2.5), analyte molecules must first be desorbed into the vapor state, from a sample cup, before they enter an electric field between two electrodes; in FD (Figure 2.6), the sample is desorbed directly from the surface of a probe or electrode when it enters a high electric field of 10^7–10^8 V cm^{-1}. Due to its simpler design, FD is a much more sensitive technique for polymer analysis than FI. Further, FD circumvents many of the sample volatility limitations associated with other techniques in the MS of synthetic polymers. It is especially useful for obtaining mass spectra of nonvolatile polymers such as low-molecular-weight polyethylene up to m/z 3500 [13]. In FD ionization, an intense electric field (on the order of 10^8 V cm^{-1}) is created by applying a voltage of 10–20 kV to an "emitter" (a thin tungsten wire supported by two metal posts with a ceramic base), on which carbon microneedles, or "whiskers" have been grown. These whiskers (approximately of 1 μm diameter) are grown on a thin wire filament by pyrolyzing the vapors of benzonitrile [1]. A polymer

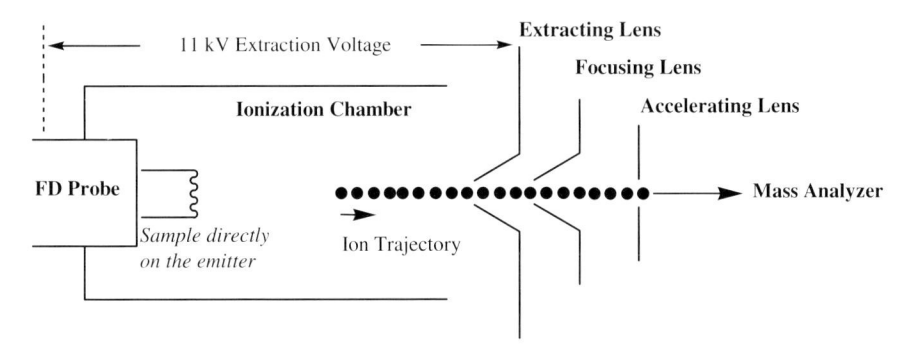

Figure 2.6 Diagram of a field desorption (FD) source.

solution is deposited via a syringe onto the emitter, and the emitter is placed in the mass spectrometer source (approximately 1 mm away from the cathode). The emitter is held at high positive voltage a few millimeters away from a counter electrode held at a potential about 10 kV more negative than the emitter. This setup produces high field strengths at the tips of the whiskers, causing ionization and desorption of primarily intact polymer oligomers ions. Field desorption of the analyte molecules produces ions mainly of the $[M + H]^+$ and $[M + Na]^+$ charge states. The ion–molecule reactions of M^+ with the sample molecules produce $[M + H]^+$ ions, and the presence of sodium impurities (or by intentionally introduction during sample preparation) produces the $[M + Na]^+$ ions [1]. The influence of the strong electric field desorbs these ions from the emitter surface and draws them toward the extraction lens for introduction into the mass analyzer. Since only a small excess internal energy is imparted to the analyte molecule, FD is considered a "soft" ionization method, with little or no fragmentation seen in FD mass spectra, an advantage when determining polymer molecular weight.

Applications of FD-MS to the direct analysis of synthetic polymers date back to the 1970s. In 1979, Matsuo *et al.* [16] reported mass spectra of polystyrene up to 11 000 Da with this technique. Additionally, series of poly(ethylene glycol)s, poly(propylene glycol)s, and poly(tetrahydrofuran)s were examined by Lattimer and Hansen [17]. The fact that hydrocarbon polymers like polybutadiene, polyisoprene, and even polyethylene can be ionized by FD [18] has maintained interest in this technique. It is also worth noting that FD-MS is versatile enough to handle polar and nonpolar types of polymers.

Limitations

Despite the success of FD in analyzing nonvolatile compounds, it has not enjoyed wide popularity, mainly owing to operational difficulties in making reproducible and effective emitters [1]. Two other factors have hindered the acceptance of FD: (1) it is a very demanding technique that requires an experienced operator to achieve useable polymer mass spectra and (2) sample ion currents are low and short lived [19]. This technique was practically abandoned soon after the discovery of FAB, in 1981. However, it is worth mentioning that FD-MS has a higher mass limit (10 000–15 000 Da for polystyrene [20, 21]) than either EI- or CI-MS and has the advantage of virtually no background chemical noise.

2.3.2
Secondary Ion Mass Spectrometry (SIMS)

Static SIMS was introduced in 1969 by Benninghoven as a method to produce quasi-molecular ions (e.g., $[M + Ag]^+$), as opposed to *dynamic* SIMS, which is used for elemental analysis and depth profiling of materials [22, 23]. The name *static SIMS* is used to distinguish the set of operating parameters that allows only the surface layers of a solid sample (~2 nm) to be analyzed from the parameters of *dynamic SIMS* used to "burrow" into internal layers of a solid sample ("depth profiling") [13]. *Static* SIMS,

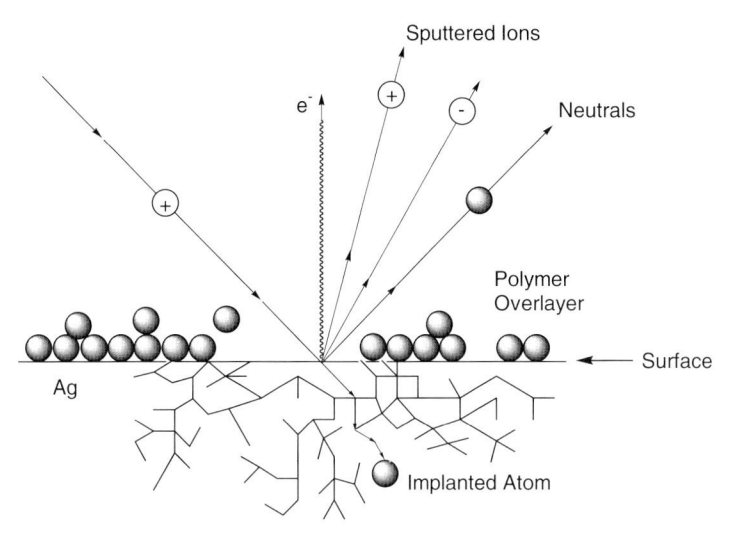

Figure 2.7 Illustration of the secondary ion mass spectrometry (SIMS) process.

as shown in Figure 2.7, is achieved by using low primary-ion dose densities at ultralow pressure (10^{-10} Torr) [13]. Higher ion fluxes are used in *dynamic* SIMS to effect depth profile measurements [13]. For one mode of *static* SIMS analysis, a polymer solution is placed on a metal substrate target [24]. The solution evaporates leaving a solid, thin polymer overlayer. The target is bombarded by a primary ion beam, typically argon or cesium ions, and the primary ion beam implants primary ions and disrupts the polymer surface leading to the sputtering of secondary particles. The sputtered particles include secondary positive ions, negative ions, electrons, and neutrals. More than 99% of the sputtered particles are neutrals. Either positive or negative ions are extracted and sent to the mass analyzer. A second mode of SIMS operation is to ion-bombard thick films directly. This produces small fragment ions (<500 Da) that can be useful for polymer identification, impurity detection, and to study surface modification and other surface reactions.

Limitations

SIMS ionization is "harsher" than FD-MS; more fragment parks appear in SIMS mass spectra. Typically, quasi-molecular ion peaks up to 10 000 Da can be seen in *static* SIMS mass spectra acquired with a time-of-flight (TOF) mass analyzer; these results are comparable to FD-MS [25]. Similar to FD-MS, *static* SIMS mass spectra of polymer oligomers with masses greater than about 12 000 Da contain only fragment ion peaks. The fragment ions from FD-MS are due to pyrolysis of the polymer from heating; however, in SIMS the fragment ions are formed from bombardment by the primary ion beam [25]. Information such as monomer units, end groups, and fragmentation patterns can be determined from fragment peaks [26]. SIMS has enjoyed a wide popularity as a surface characterization technique. However, its major disadvantage, with respect to analytical applications, is the short-lived sample current.

2.3.3
Fast Atom Bombardment (FAB) and Liquid Secondary Ion Mass Spectrometry (LSIMS)

Introduced in 1981, FAB is an ionization technique similar to SIMS [27, 28]. However, unlike SIMS, FAB is considered a "soft" ionization technique because the intact analyte ion undergoes relatively little fragmentation [13]. The main differences between FAB and SIMS are that, in FAB (Figure 2.8), the sample is dissolved in a polar liquid matrix (e.g., glycerol), which results in a sample current that lasts much longer than observed in SIMS. The second difference is that FAB uses a beam of high-energy (6–10 keV) *atoms* rather than *ions*, which are used in SIMS. When first introduced, the MS community immediately recognized FAB as being a major breakthrough in the analysis of condensed-phase samples. The FAB process begins by mixing a polymer solution with a liquid matrix, such as glycerol, and placing the resulting mixture onto a substrate. Next, a FAB gun, using argon or xenon gas atoms, produces a beam of fast *atoms* [1]. (Xe is the preferred gas because the impact of its massive atoms enhances the yields of secondary ions). The atom gun uses a three-step process for forming fast atoms: (1) neutral gas atoms are first ionized by collisions with electrons, (2) accelerated to the requisite potential (2–10 kV), and (3) neutralized in a dense cloud of the excess neutral gas atoms by resonance electron capture; thus, fast ions are converted to fast atoms. The sample is then bombarded with these neutral fast atoms. The bombarding atoms ionize the sample, and ions are extracted and mass analyzed. It should be noted that

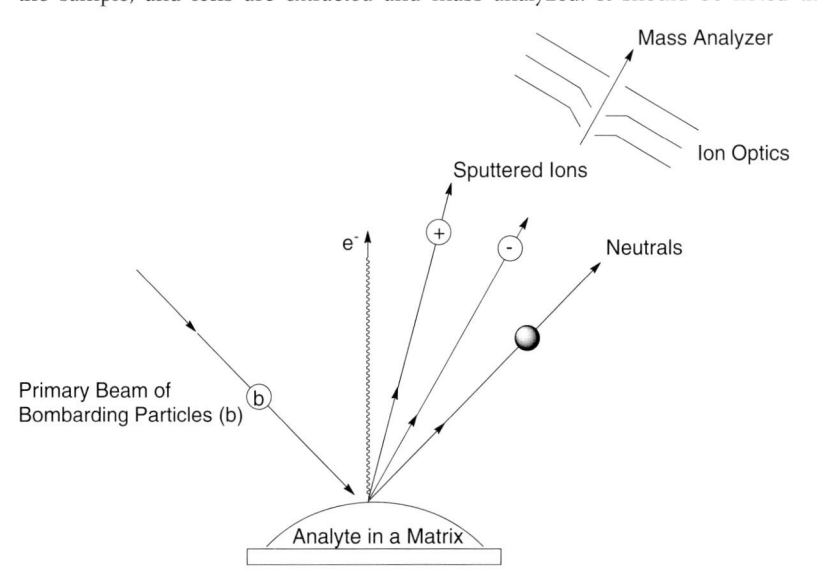

Figure 2.8 Sketch illustrating the fast atom bombardment (FAB) ionization process.

the collision rate within a few molecular dimensions of the surface (termed the "selvedge") is extraordinarily high causing estimated pressures of 1000 atm and temperatures near 1000 °C [13]. This leads to ion/molecular reactions and collisions which generate protonated molecules (and other adduct ions) that have been sputtered from solution and robbed of excess energy through multiple collisions [13]. The liquid matrix attenuates the damage to the sample by absorbing the impact of the bombarding fast atoms, reduces the binding energy of the sample molecules, and, most importantly, provides a medium in which the ionization of the sample can be facilitated. It is worth noting that the proper choice of a liquid matrix is crucial to obtain a good-quality FAB mass spectrum and to optimize ion current. Further, the use of a liquid matrix continuously replenishes the sample to avoid depletion, a potential problem in SIMS when used with scanning instruments, such as quadrupole and magnetic sectors.

If primary *ions* (such as Cs^+) instead of neutrals are used to provide the energy for secondary ion ejection from the liquid matrix, the technique is called "liquid secondary ion mass spectrometry" (LSIMS) [29]. The main difference between FAB and LSIMS is in the way that the primary beam is generated. In LSIMS, the current generation of ion guns contains a pellet of cesium aluminum silicate, which when heated to an incandescent temperature, emits Cs^+ ions. These ions are then accelerated through an electric field to produce a beam of high-energy (25–40 keV) Cs^+ ions. LSIMS offers certain advantages over FAB. Because of the high-translational-energy ion beams used in LSIMS, high-mass ions are easily sputtered. Furthermore, it is easier to obtain a well-defined electrically focused primary ion beam (which is not possible with a neutral atom beam), resulting in enhanced secondary ion yields. Also, there is no gas load in a LSIMS ion source, and thus, background noise is lower. These advantages lead to an improvement in the detection sensitivity and high-mass analysis capability. As a result of these advantages, more mass spectrometers are outfitted with LSIMS sources than FAB sources.

Limitations

The upper mass limit for a routine FAB operation is 4000 Da; above this mass, the molecular ion signal usually is very weak. Also, it should be noted that the conditions of the FAB processes have the potential to promote unwanted reactions between analyte and matrix. Nowadays, many of the tasks performed in the past by FAB are performed by MALDI and ESI. This is because MALDI and ESI can analyze the same classes of polymers at much higher mass ranges and faster throughput times.

2.3.4
Laser Desorption (LD)

Another ionization technique developed in the late 1970s and early 1980s was LD-MS. For LD ionization, a sample is deposited onto a substrate, sometimes with an added salt, and is then ablated with a submicrosecond laser pulse. Energy transfer to the analyte is achieved with lasers emitting in the ultraviolet range to cause electronic excitation (or in the infrared to cause vibrational excitation) and ionization of the analyte [13]. Ionization occurs mainly by cation attachment to the desorbed molecules. The ions are extracted into a mass analyzer. Laser desorption found commercial application in the LAMMA (laser micro mass analyzer) instrument by Leybold-Heraeus, Germany [30]. This instrument was mainly used for the elemental analysis of biological tissue (e.g., Ca-, K-, and Na-distribution).

Limitations

The extent of energy transfer is difficult to control and often leads to excessive thermal degradation. Another pitfall of LD is that not all compounds absorb radiation at the laser wavelength. As a consequence, LD is applicable to a limited number of compounds, usually those with molecular mass of <1000 Da, and has not gained acceptance for synthetic polymer characterization.

2.3.5
Plasma Desorption (PD)

The principle of desorption by bombardment of organic samples with the fission products of the ^{252}Cf nuclear decay, later called plasma desorption (PD), was published by Macfarlane in 1974 [31]. Subsequently, the groups of Sundqvist and Roepstoff greatly improved the analytical potential of this technique by the addition of nitrocellulose, which not only cleaned up the sample but was also suspected of functioning as a signal-enhancing matrix [32]. In this ionization technique (Figure 2.9), the sample deposited on a small aluminized nylon foil is exposed in the source to fission fragments of ^{252}Cf having an energy of several megaelectronvolts [19]. When the radionuclide ^{252}Cf decays, it produces two fission products (e.g., $^{142}Ba^{18+}$ and $^{106}Tc^{22+}$), which move apart in opposite directions. One portion of the nuclear fission fragments moves toward a timing trigger for the TOF clock and serves as a time marker for the fission event and hence the onset of ionization. The other portion of the nuclear fission fragments travels toward a sample, which is

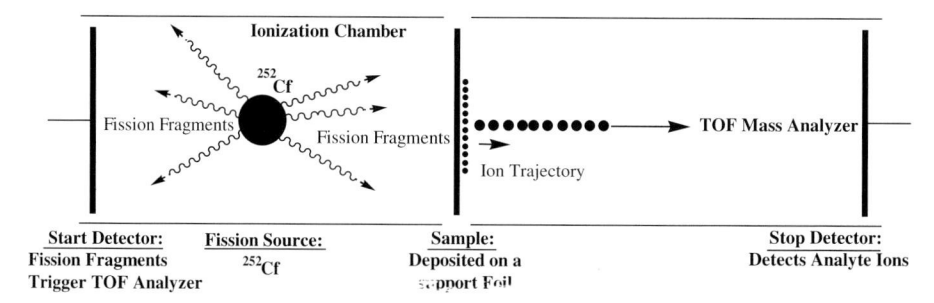

Figure 2.9 Conceptual diagram of a ^{252}Cf plasma desorption (PD) source interfaced with a time-of-flight mass spectrometer.

coated on a very thin support foil. Each highly energetic ion passes through the foil and, on impact with the sample, causes a localized "hot spot" (about 10 000 K) [33]. The sudden uptake of energy into the sample results in rapid desorption before decomposition can occur. Subsequently, in the positive ion mode, most substances give protonated molecules, $[M + H]^+$, with little, if any, observed fragmentation.

Limitations

The sample ion flux is usually low; therefore, to obtain sufficiently high ion statistics, a large number of scans (5×10^5–10^6) are summed and averaged. A problem with this method is the quenching of the spectrum by involatile compounds present on the foil as impurities. This can be overcome by first coating the foil with a thin layer of absorbant, which is selective to the given class of analyte, followed by a washing step to remove impurities from the surface. Plasma desorption has allowed the observation of ions in the mass range of 10 000 Da [34]. However, nowadays it is of limited use and has been replaced mainly by MALDI [19].

2.3.6
Other Ionization Methods

Several other ionization techniques were introduced in this era that had some success for polymer MS, including desorption chemical ionization (DCI) [35], electrohydrodynamic ionization (EHI) [36–38], thermospray ionization (TSP) (Figure 2.10) [39], and atmospheric pressure chemical ionization (APCI) using a ^{63}Ni-based source [40, 41]. However, none of these methods readily produces oligomer ions in the high-mass range. The next step in the advancement of polymer MS was to develop ionization techniques that would readily ionize molecules larger than 10 000 Da.

2.4
Modern Era of Ionization Techniques

The "modern era" of polymer MS (late 1980s to present) began with the development of ESI [42] and MALDI [43, 44] in the late 1980s. These two ionization techniques

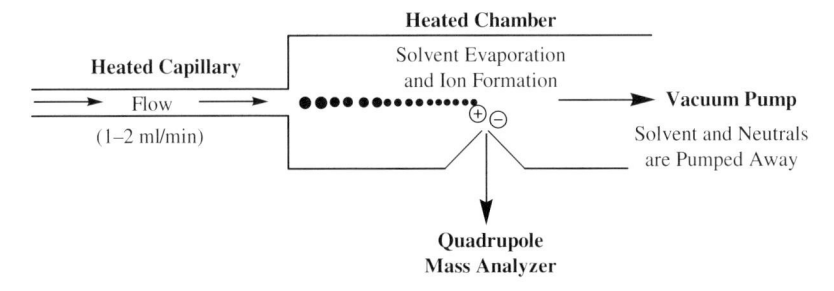

Figure 2.10 Sketch of a thermospray (TSP) source.

significantly enhanced synthetic polymer characterization by increasing the molecular weight range for MS well into the mega-Dalton range [45]. Whereas chain entanglement presented a huge problem for previous ionization techniques, MALDI and ESI kept the polymer dilute, thereby minimizing this effect. To emphasize the impact these techniques had upon MS of macromolecules, the 2002 Nobel Prize in Chemistry was awarded to John Fenn and Koichi Tanaka for their developments of electrospray [42, 46] and MALDI [43], respectively.

2.4.1
Electrospray Ionization (ESI)

ESI was first reported by Malcolm Dole in 1968, to examine polystyrene oligomers [47]. However, Dole was using a rather primitive mass analyzer, which did not compensate for droplet freezing during solvent evaporation. These problems were overcome by John Fenn in 1984 [42]. Fenn based his ESI source on the general design originated in the 1960s by Dole, but added a drying gas to help evaporate solvent from the ions [42]. The electrospray ionization process begins by forcing a dilute polymer solution through a needle held at high potential (approximately 4 kV) at atmospheric pressure. A schematic of an ESI source interfaced to a mass analyzer is shown in Figure 2.11. The high potential on the needle produces solvent droplets that are highly charged. The electric field is important for several reasons, one being that it keeps the droplets from freezing (during endothermic loss of solvent by evaporation) by causing the charged droplet to endure many collisions through which some translational energy is converted to internal energy, thereby warming the droplet [13, 48]. At the onset of electrospray, the electric field establishes an electrostatic force sufficient to pull liquid and droplets out of the nozzle toward the ground plate, thereby forming an elliptically shaped fluid cone into the so-called *Taylor cone*, which tapers to a relatively sharp point [13]. As charged aerosol droplets of solution are ejected from the *Taylor cone*, the droplets shrink as the solvent evaporates and retain ions (which are relatively involatile) within them. Droplet shrinkage continues, as shown in Figure 2.11b, until the surface charge on the droplets exceeds the Rayleigh limit of instability (the point at which repulsive forces between like charges in an electrolytic solution overcome the cohesive forces of the solvent), then the droplets burst into even smaller droplets through a series of coulombic "explosions." Continued explosions reduce the droplet size until fully desolvated charged polymer molecules are ejected. For positive ESI, protons presumably associated with sites of high Lewis basicity [13]. The electrosprayed ions are then introduced into a mass analyzer. Positively charged electrosprayed ions generally consist of $[M + zH]^{z+}$ or $[M + z\text{Cation}]^{z+}$ or various combinations of the two, where z is the number of charged adducts.

ESI has two major advantages over other ionization techniques: (1) it drastically increases the upper limit for m/z measurement, making it amenable to a large variety of mass spectrometers [2] and (2) the ability to directly couple liquid chromatographic methods such as SEC or HPLC to MS [20]. The eluent from the chromatographic

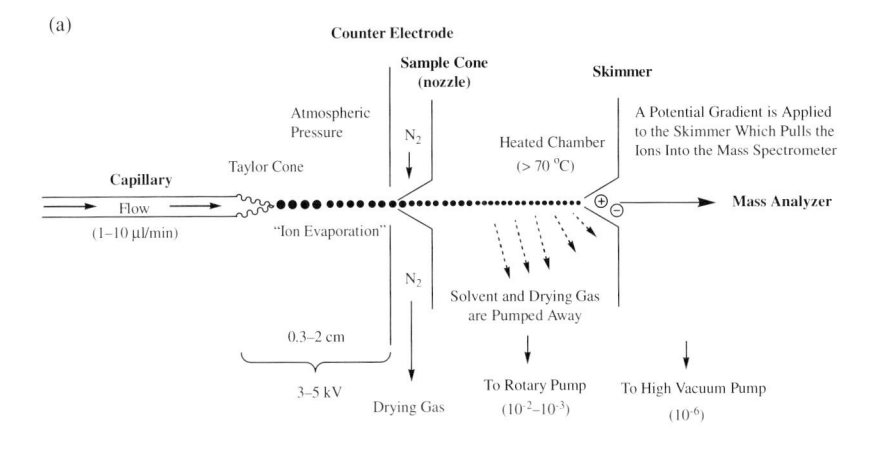

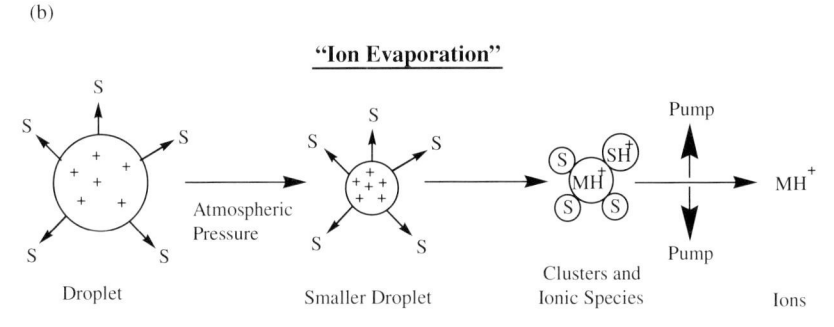

Figure 2.11 Schematic diagrams of (a) a typical electrospray ionization (ESI) source along with (b) a conceptualization of the ion evaporation model for ion formation.

column flows directly into the ESI source for online separation. Polymers are separated according to chemical structure or molecular size by chromatography, and molecular weight information can be obtained by MS.

Limitations

Since ESI generally produces highly charged species (sometimes up to 25 charge states), the polydispersity of polymeric samples can present at problem due to the variety of overlapping mass peaks produced. To increase reliability, interpretability of the spectra, and discrimination of higher mass species, it is generally necessary to use a chromatographic preseparation before polymer sample introduction into the ESI source. Optimum conditions for obtaining ESI mass spectra vary significantly depending on the solvent system used to deliver that analyte to the interface. The composition of the buffer solution can also greatly influence the appearance of the mass spectrum. Solutions of millimolars of Na^+ do not cause deterioration of chromatographic performance or fouling of the ion source although such additives may attenuate the ESI signal. Solvent adducts can also "dilute" the analytical signal by splitting the mass peaks. Increasing the electric field between electrodes in the interface often breaks up ion/solvent adducts. It is generally necessary to seek a balance between dissociation of solvent adducts and conditions under which covalent bonds in the analyte can undergo dissociation. In general, electrolyte solutions more concentrated than 100 mM will adversely affect an ESI analysis.

2.4.2
New Trends

Trends observable in ESI are the reduction in flow rate and droplet size. This was initially pioneered in 1994 by Wilm and Mann [49], and the latest advances in "nanospray" have reduced the flow rate to nl min^{-1} and the droplet size also to the nm range to allow for characterization in the subpmol range of detection. Use of the nanospray ESI source offers several analytical advantages over conventional electrospray: (1) the consumption of analyte solution is only nanoliters per minute, so very little sample solution is required to yield useful mass spectral information for lengthy experiments [50]; (2) since the bulk flow of analyte solution is so low, nanospray is more tolerant of electrolytes than conventional electrospray; and (3) nanospray has a much higher ion transfer efficiency than conventional ESI, giving rise to enhanced sensitivity [51].

Another electrospray variation that shows great promise for rapid analysis of solid polymers is desorption electrospray ionization (DESI) (Figure 2.12), developed by Cooks and coworkers [52]. DESI is a surface analysis technique requiring little or no sample preparation. A solid polymer sample is simply introduced between the angled emitter and extended nozzle, as shown in Figure 2.12, to obtain a mass spectrum; the ion formation process is similar to that of conventional ESI [53].

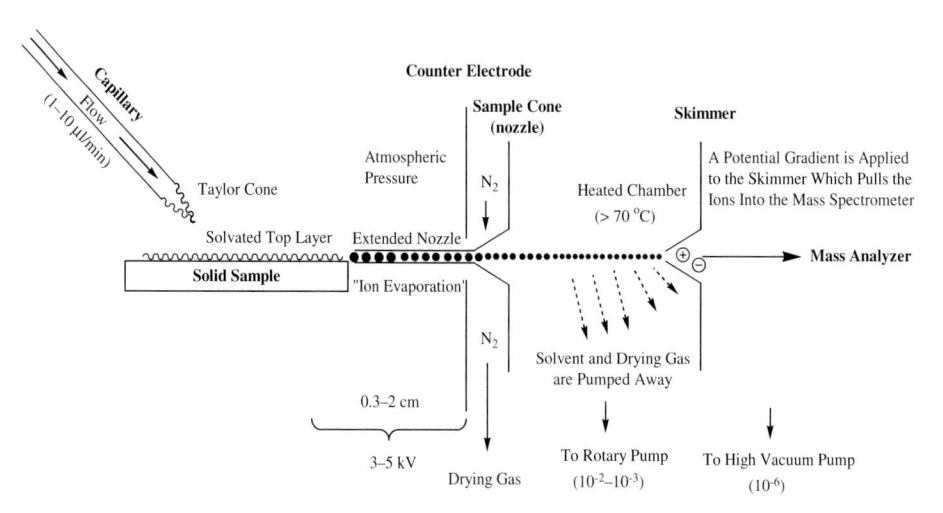

Figure 2.12 Diagram of a desorption electrospray ionization (DESI) source.

2.4.3
Atmospheric Pressure Chemical Ionization (APCI)

In APCI ion–molecule reactions occurring at atmospheric pressure are employed to generate ions – it represents a high-pressure version of conventional CI. The CI plasma is maintained by a corona discharge between a needle and the spray chamber serving as the counter electrode. The ions are transferred into the mass analyzer by use of the same type of vacuum interface that is employed in ESI (Figure 2.11a). Therefore, ESI ion sources can easily be switched to APCI by use of modular source interfaces. In the APCI modular source, the ESI sprayer is replaced with a heated pneumatic nebulizer and a spray chamber with a needle electrode is positioned in front of the orifice [54, 55]. The major advantage of APCI, over conventional electrospray, is that it actively generates ions from neutrals. Therefore, it has the capability of making low to medium polarity analytes, eluting from a liquid chromatograph, accessible for ionization.

Limitations

Early APCI sources, originally developed in the mid-1970s [40, 41], suffered from poor ion formation, due to their ^{63}Ni-based sources; source upgrading to a corona discharge needle has significantly overcome this problem and lead to wider use of this technique. The nature of the APCI plasma varies widely as both solvent and nebulizing gas contribute to the composition of the CI plasma; this explains why APCI conditions suffer from comparatively low reproducibility to that of other ionization methods. In contrast to its development as an ionization method, the application of APCI has seriously lagged behind that of ESI. It was not until the mid-1990s that the use of APCI rapidly grew, primarily due to the vacuum interfaces developed for ESI technology. Nowadays, APCI is used where LC separation is required, but ESI is not applicable to the compound class of interest [56–58].

2.4.4
New Trends

Recently, atmospheric pressure photoionization (APPI) has been introduced [59]. APPI can serve as a complement or alternative to APCI for the analysis of nonpolar samples [57, 60]. In APPI (Figure 2.13), a UV light source replaces the corona discharge-powered plasma, while the pneumatic heated sprayer remains almost unaffected [61–63].

2.4.5
Matrix-Assisted Laser Desorption/Ionization (MALDI)

The introduction of matrix-assisted laser desorption/ionization mass spectrometry (MALDI-MS) was quite possibly the most significant advance in the history of

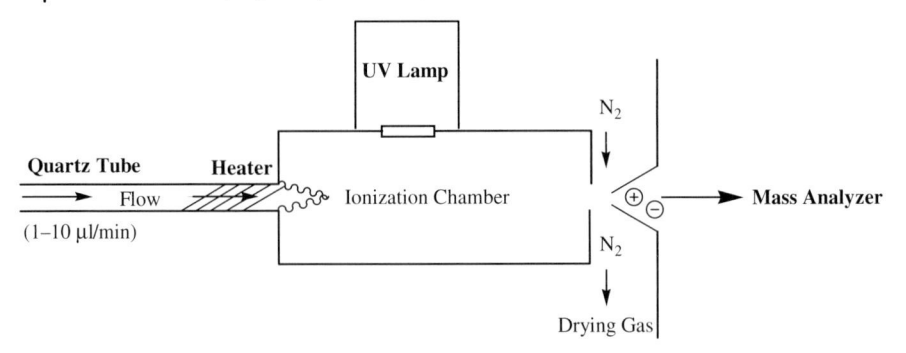

Figure 2.13 Sketch illustrating an atmospheric pressure photoionization (APPI) source.

polymer MS [64]. Two research groups, Karas and Hillenkamp [44] and Tanaka and coworkers [43], concomitantly reported the development of MALDI in 1988. The primary difference between MALDI and LD ionization of the analyte is that, in MALDI sample preparation, a large excess of organic matrix is mixed with a small quantity of analyte. It is this use of an organic matrix (discovered by professors Karas and Hillenkamp) that makes MALDI a "soft" ionization technique [65]. In their initial work, Karas and Hillenkamp used nicotinic acid as the matrix to ionize proteins with masses up to 67 000 Da [44], whereas Tanaka and coworkers used a slurry of analyte, cobalt nanoparticles (10 nm size), and glycerol to examine molecular ions of poly (ethylene glycol) with a mass of 20 000 Da [43]. The matrix performs two functions: (1) it absorbs photon energy from the laser beam and transfers it into excitation energy of the solid system and (2) it serves to dilute the analyte molecules, so that their intermolecular forces are reduced and their aggregation is minimized.

During MALDI sample preparation of synthetic polymers (e.g., via the "dried-droplet method"), a dilute polymer solution is mixed with a more concentrated matrix solution (e.g., dithranol or 2,5-dihydroxybenzoic acid) and a cationization agent (e.g., a metal salt such as sodium iodide or silver trifluoroacetate). Note that the main criterion for choosing a matrix is that it absorbs energy at the wavelength of laser irradiation (e.g., 337 nm). A smaller aliquot of the mixture is applied to a sample plate, and the matrix and polymer cocrystallized as the solvent evaporates, leaving the polymer dispersed in the solid sample-matrix crystals. It should be noted that the more homogeneous and fine-grained the morphology of the crystal formation within the matrix, the more intense is the MALDI spectrum of the analyte [13]. A laser pulse (3–10 ns duration) is focused onto a small portion of the sample-matrix crystals, which is typically 0.05–0.02 mm in diameter, causing the matrix molecules absorb the laser energy, vaporize, and rapidly expand until they explode, carrying the analyte into the gas phase, usually ionized by proton or metal ion addition. A schematic of the MALDI desorption process is shown in Figure 2.14. When performed correctly, very little energy should be transferred to the analyte and MALDI should be considered a "soft" ionization technique because very large molecules can be desorbed into the gas phase without undergoing thermal degradation. Also, unlike in ESI, the ions produced by MALDI are predominantly singly charged.

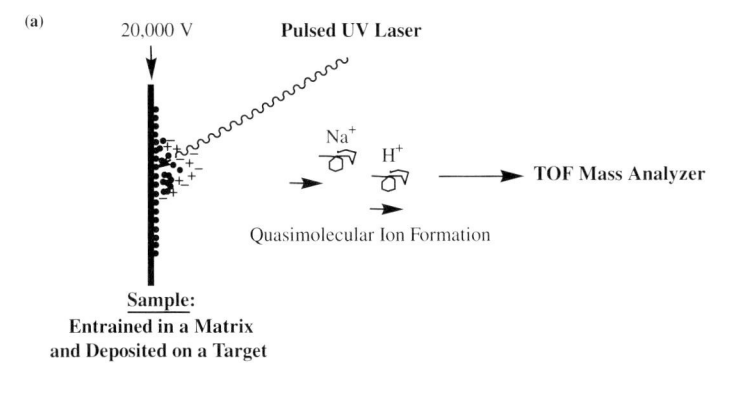

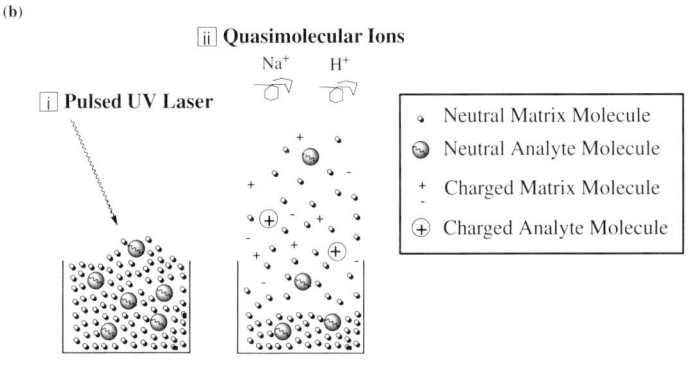

Figure 2.14 Illustration of (a) a typical matrix-assisted laser/desorption ionization (MALDI) source along with (b) a conceptualization of the MALDI process of ion formation.

Limitations

MALDI sample preparation, involving choice of the appropriate polymer/matrix mixture, is critical in obtaining useable MALDI spectra of synthetic polymers, and is considered by many to be an art form. For example, the use of a dithranol matrix to analyze 2,2,6,6-tetramethylpiperidine-*N*-oxyl (TEMPO)-capped polystyrene generates fragment ions, without the presence of a TEMPO end group [66, 67]. By simply changing the MALDI matrix from dithranol to 2,5-dihydroxybenzoic acid (DHB), and repeating the MALDI analysis, polystyrene fragmentation is eliminated and intact TEMPO-capped polystyrene ions will be observed [66]. To assist MALDI users, considerable effort has been made over the past years to identify the optimum matrix-to-polymer "matches" and report these recipes on websites (e.g., the Polymers Division of the National Institute of Standards and Technology homepage – http://polymers.msel.nist.gov/maldirecipes). In addition, it should be noted that direct molecular weight distribution (MWD) determination using MALDI-TOF MS is typically limited to polymers with low polydispersity indices because of mass discrimination effects where high-mass components are generally under-represented compared to low-mass components in the mass spectra. It is often quoted that MALDI MWD measurements are accurate for samples with a polydispersity index of <1.2 [68]. High-mass ions produce a lower detector response than low-mass ions; therefore, molecular weight averages are typically skewed for broadly dispersed oligomers. Nevertheless, Malvagna *et al.* reported accurate MWD determinations for poly(vinylpyrolidone) with masses of 40 000 and 160 000 Da and polydispersity indices of 1.8 and 2.2 using sample preparation techniques involving flash-freezing and freeze-drying of the sample preparations [69].

2.4.6
New Trends

Surface-assisted laser desorption/ionization (SALDI), developed by Sunner and coworkers, uses dry carbon and graphite substrates to generate polymer mass spectra [70]. Similarly, Wallace *et al.* developed "reactive MALDI MS," which uses a molecular solid of fullerene intercalated with cobalt cyclopentadienyl dicarbonyl as the MALDI matrix, to obtain mass spectra of oligomeric polyethylene [71]. Another technique which has found much interest for the analysis of smaller molecules was reported by Siuzdak [72]. This method termed "desorption/ionization on silicon" (DIOS), uses preparations of neat organic samples on porous silicon. Yoctomole (10^{-21} mole) sensitivity has been achieved in this way with a perfluorophenyl-derivatized DIOS system for a small hydrophobic peptide [73].

Recently, MALDI-MS quantification of polymers has been demonstrated. Yan quantified poly(dimethly siloxane) samples with different molecular weights [74]. Murgasova coupled SEC and MALDI-MS to quantify a blend of polystyrene and poly(α-methylstyrene) [75]. MALDI-MS was used to characterize narrow polydispersity SEC fractions because SEC could not differentiate the two polymers.

If an analyte is insoluble, or only soluble in solvents which are not compatible with the standard dried-droplet MALDI sample preparation method, it can alternatively be ground together with a solid matrix. This form of MALDI sample preparation can be accomplished by a variety of "solventless" techniques such as the evaporation-grinding method [76], vortexing method [77], or the mini-ball mill solvent-free method [78]. Presently, the upper mass limit for *insoluble* polymer analysis is 6 kDa; this was accomplished using the evaporation-grinding method to analyze poly(*para*-phenylene terephthalamide) (PPD-T) [79]. Generally, the limiting factor for this type of analysis is the degree of inter-chain entanglement, due to hydrogen bonding or pi-stacking, in the case of PPD-T [79] or poly(*para*-phenylene sulfide) [80, 81], respectively. As with any ionization technique, the MALDI user simply chooses the sample preparation that best solves their problem. It is worth noting that care should be taken when using mechanical grinding methods to mix the matrix and analyte; vigorous mixing generates enormous amounts of sheering force and heat, which can cause macromolecules to undergo main-chain cleavages.

2.5
Conclusions

In conclusion, the preceding sections demonstrate that there are a variety of ion sources available for the practicing polymer mass spectrometrist. It is important to stress that the most effective use of MS in synthetic polymer analysis relies upon matching up the sample to be investigated, along with the information to be obtained, with the ionization technique best suited for solving the problem. However, based upon the inherent limitations of each ionization technique, it should be evident that there is no universal ionization principle. Yet, given the number of review articles [82–89] and books [29, 65, 90] written specifically on MALDI characterization of polymers, one could argue that it is the most versatile technique for synthetic polymer analysis, especially when enabled with ion mobility [91, 92] and CID fragmentation [93, 94] capabilities.

The major challenge of polymer MS is using conventional electrospray and MALDI ionization to analyze the wide variety of polymer classes, which keeps growing each year. In an effort to keep up with the fast-paced development of new polymers, and the rigorous demands for their characterization, a future path for polymer ionization techniques will lie in the developmental modification of the pre-existing electrospray and MALDI sources. Custom tailoring of ionization sources, to address the specific analysis issues associated with a polymer class of interest, shows great potential for the discovery of new characterization information, which would have been previously unavailable using conventional ESI or MALDI techniques.

Acknowledgments

I would like to thank David M. Hercules and Sparkle T. Ellison for all their helpful discussions and contributions to this work.

References

1 Dass, C. (2001) *Principles and Practice of Biological Mass Spectrometry*, John Wiley & Sons, Inc., New York.
2 McLafferty, F.W. and Turecek, F. (1993) *Interpretation of Mass Spectra*, 4th edn, University Science Books, Sausalito.
3 Dempster, A. (1918) *J. Phys. Rev.*, **11**, 316.
4 Bleakney, W. (1929) *Phys. Rev.*, **34**, 157.
5 Nier, A.O. (1947) *Rev. Sci. Instrum.*, **18**, 415.
6 Munson, M.S.B. and Field, F.H. (1965) *J. Am. Chem. Soc.*, **87**, 3294–3299.
7 Munson, M.S.B. (1965) *J. Am. Chem. Soc.*, **87**, 2332–2336.
8 Munson, M.S.B. and Field, F.H. (1966) *J. Am. Chem. Soc.*, **88**, 2621–2630.
9 Munson, M.S.B. (2000) *Int. J. Mass. Spectrom.*, **200**, 243–251.
10 Moldoveanu, S.C. (2005) *Analytical Pyrolysis of Synthetic Organic Polymers*, Elsevier, New York.
11 Wampler, T.P. (2007) *Applied Pyrolysis Handbook*, 2nd edn, CRC Press, Boca Raton, FL.
12 Watanabe, C., Teraishi, K., Tsuge, S., Ohtani, H., and Hashimoto, K. (1991) *J. High Res. Chrom.*, **14**, 269.

13 Watson, J.T. (1997) *Introduction to Mass Spectrometry*, 3rd edn, Lippincott-Raven Publishers, Philadelphia, PA.
14 Beckey, H.D. (1977) *Principles of Field Ionization and Field Desorption Mass Spectrometry*, Pergamon Press, Oxford.
15 Inghram, M.G. and Gomer, R.J. (1954) *J. Chem. Phys.*, **22**, 1279–1280.
16 Matsuo, T., Matsuda, H., and Katakuse, I. (1979) *Anal. Chem.*, **51**, 1329.
17 Lattimer, R.P. and Hansen, G.E. (1981) *Macromolcules*, **14**, 776.
18 Lattimer, R.P. and Schulten, H.-R. (1983) *Int. J. Mass. Spectrom.*, **52**, 105.
19 de Hoffmann, E. and Stroobant, V. (2002) *Mass Spectrometry: Principles and Applications*, 2nd edn, John Wiley & Sons, Ltd, New York.
20 Montaudo, G. and Lattimer, R.P. (2001) *Mass Spectrometry of Polymers*, CRC Press, Boca Raton, FL.
21 Rollins, K., Scrivens, J.H., Taylor, M.J., and Major, H. (1990) *Rapid Commun. Mass Spectrom.*, **4**, 355.
22 Benninghoven, A. (1969) *Z. Angew. Chemie*, **27**, 51–55.
23 Benninghoven, A., Jaspers, D., and Sichtermann, W. (1976) *Appl. Phys. Lett.*, **11**, 35–39.
24 Hittle, L.R., Altland, D.E., Proctor, A., and Hercules, D.M. (1994) *Anal. Chem.*, **66**, 2302.
25 Hercules, D.M. (2002) *Mass Spectrometry of Polymers* (eds G. Montaudo and R.P. Lattimer), CRC Press, Boca Raton, FL, pp. 331.
26 Chairelli, M.P., Proctor, A., Bletsos, I.V., Hercules, D.M., Feld, H., Leute, A., and Benninghoven, A. (1992) *Macromolecules*, **25**, 6970–6976.
27 Barber, M., Bordoli, R.S., Sedgwick, R.D., and Tyler, A.N. (1981) *J. Chem. Soc. Chem. Commun.*, **7**, 325–327.
28 Barber, M., Bordoli, R.D., Elliot, G.J., Sedgwick, R.D., and Tyler, A.N. (1981) *Nature*, **293**, 270.
29 Pasch, H. and Schrepp, W. (2003) *MALDI-TOF Mass Spectrometry of Synthetic Polymers*, Springer, Berlin.
30 Hillenkamp, F., Kaufmann, R., Nitsche, R., and Unsold, E. (1975) *Nature*, **256**, 119–120.
31 Torgerson, D.F., Skrowonski, R.P., and Macfarlane, R.D. (1974) *Biochem. Biophys. Res. Comm.*, **60**, 616–619.
32 Jonsson, G.P., Hedin, A.B., Hakansson, P.L., Sundqvist, B.U.R., Save, B.G., Nielson, P.F., Roepstorff, P., Johansson, K.E., Kamensky, I., and Lindberg, M.S. (1986) *Anal. Chem.*, **58**, 1084–1087.
33 Johnstone, R.A.W. and Rose, M.E. (1996) *Mass Spectrometry for Chemists and Biochemists*, 2nd edn, Cambridge University Press, New York.
34 McNeal, C.J. and McFarlane, R.D. (1981) *J. Am. Chem. Soc.*, **103**, 1609.
35 Daves, G.D. Jr. (1979) *Acc. Chem. Res.*, **12**, 359.
36 Dulcks, T. and Roellgen, F.W. (1995) *J. Mass Spectrom.*, **30**, 324–332.
37 Cook, K.D., Callahan, J.H., and Man, V.F. (1988) *Anal. Chem.*, **60**, 706–713.
38 Cook, K.D. (1986) *Mass Spectrom. Rev.*, **5**, 467–479.
39 Blakley, C.R. and Vestal, M.L. (1983) *Anal. Chem.*, **55**, 750.
40 Carroll, D.I., Dzidic, I., Stillwell, R.N., Haegele, K.D., and Horning, E.C. (1975) *Anal. Chem.*, **47**, 2369–2373.
41 Dzidic, I., Stillwell, R.N., Carroll, D.I., and Horning, E.C. (1976) *Anal. Chem.*, **48**, 1763–1768.
42 Yamashita, M. and Fenn, J.B. (1984) *J. Phys. Chem.*, **88**, 4451–4459.
43 Tanaka, K., Waki, H., Ido, Y., Akita, S., Yoshida, Y., and Yoshida, T. (1988) *Rapid Commun. Mass Spectrom.*, **2**, 151–153.
44 Karas, M. and Hillenkamp, F. (1988) *Anal. Chem.*, **60**, 2299–2301.
45 Schriemer, D.C. and Li, L. (1996) *Anal. Chem.*, **68**, 2721–2725.
46 Whitehouse, C.M., Dreyer, R.N., Yamashita, M., and Fenn, J.B. (1985) *Anal. Chem.*, **57**, 675.
47 Dole, M., Mack, L.L., Hines, R.L., Mobley, R.C., Ferguson, L.D., and Alice, M.B. (1968) *J. Chem. Phys.*, **49**, 2240–2249.
48 Luettgens, U. and Roellgen, F.W. (1991) Electrohydrodynamic disintegration of liquids and ion formation in ESI. in Proceedings of 39th ASMS Conference on Mass Spectrometry and Allied Topics: Nashville, TN, pp. 439–440.

49 Wilm, M.S. and Mann, M. (1994) *Int. J. Mass. Spectrom. Ion Proc.*, **136**, 167.
50 Wilm, M.S., Neubauer, G., and Mann, M. (1996) *Anal. Chem.*, **68**, 527–533.
51 Wilm, M.S. and Mann, M. (1996) *Anal. Chem.*, **68**, 1–8.
52 Takats, Z., Wiseman, J.M., Gologan, B., and Cooks, R.G. (2004) *Science*, **5695**, 471–473.
53 Jackson, A.T., Williams, J.P., and Scrivens, J.H. (2006) *Rapid Commun. Mass Spectrom.*, **20**, 2717–2727.
54 Briuns, A.P. (1991) *Mass Spectrom. Rev.*, **10**, 53–77.
55 Tsuchiya, M. (1995) *Adv. Mass Spectrom.*, **13**, 333–346.
56 Mottram, H., Woodbury, S.E., and Evershed, R.P. (1997) *Rapid Commun. Mass Spectrom.*, **11**, 1240–1252.
57 Hayden, H. and Karst, U. (2003) *J. Chromatogr. A*, **1000**, 549–565.
58 Reemtsma, T. (2003) *J. Chromatogr. A*, **1000**, 477–501.
59 Robb, D.B., Covey, T.R., and Briuns, A.P. (2000) *Anal. Chem.*, **72**, 3653–3659.
60 Keski-Hynnila, H., Kurkela, M., Elovaara, E., Antonio, L., Magdalou, J., Luukkanen, L., Taskinen, J., and Koistianinen, R. (2002) *Anal. Chem.*, **74**, 3449–3457.
61 Yang, C. and Henion, J.D. (2002) *J. Chromatogr. A*, **970**, 155–165.
62 Kauppila, T.J., Kuuranne, T., Meurer, E.C., Eberlin, M.N., Kotiaho, T., and Koistianinen, R. (2002) *Anal. Chem.*, **74**, 5470–5479.
63 Raffaeli, A. and Saba, A. (2003) *Mass Spectrom. Rev.*, **22**, 318–331.
64 Vergne, M.J., Hercules, D.M., and Lattimer, R.P. (2007) *J. Chem. Educ.*, **84**, 81–90.
65 Hillenkamp, F. and Peter-Katalinic, J. (2007) *MALDI MS: A Practical Guide to Instrumentation, Methods and Applications*, Wiley-VCH, Weinheim.
66 Dourges, M.A., Charleux, B., Vairon, J.P., Blais, J.C., Bolbach, G., and Tabet, J.C. (1999) *Macromolecules*, **32**, 2495–2502.
67 Lee, M., Schoonover, D.V., Gies, A.P., Hercules, D.M., and Gibson, H.W. (2009) *Macromolecules*, **42**, 6483–6494.
68 Hanton, S.D. (2001) *Chem. Rev.*, **101**, 527–569.
69 Malvagna, P., Impallomeni, G., Cozzolino, R., Spina, E., and Garozzo, D. (2002) *Rapid Commun. Mass Spectrom.*, **16**, 1599–1603.
70 Sunner, J., Dratz, E., and Chen, Y.-C. (1995) *Anal. Chem.*, **67**, 4335–4342.
71 Wallace, W.E. (2007) *Chem. Commun.*, **43**, 4525–4527.
72 Wei, J., Buriak, J.M., and Siuzdak, G. (1999) *Nature*, **399**, 243–246.
73 Trauger, S.A., Go, E.P., Shen, Z., Apon, J.V., Compton, B.J., Bouvier, E.S.P., Finn, M.G., and Siuzdak, G. (2004) *Anal. Chem.*, **76**, 4484–4489.
74 Yan, W., Gardella, J.A., and Wood, T.D. (2002) *J. Am. Soc. Mass Spectrom.*, **13**, 914–920.
75 Murgasova, R. and Hercules, D.M. (2003) *Anal. Chem.*, **75**, 3744–3750.
76 Gies, A.P., Nonidez, W.K., Anthamatten, M., Cook, R.C., and Mays, J.W. (2002) *Rapid Commun. Mass Spectrom.*, **16**, 1903–1910.
77 Hanton, S.D. and Parees, D.M. (2005) *J. Am. Soc. Mass Spectrom.*, **16**, 90–93.
78 Trimpin, S., Rouhanipour, A., Rader, H.J., and Mullen, K. (2001) *Rapid Commun. Mass Spectrom.*, **15**, 1364–1373.
79 Gies, A.P., Schotman, A., and Hercules, D.M. (2010) *Anal. Bio. Chem.*, **396**, 1481–1490.
80 Gies, A.P., Geibel, J.F., and Hercules, D.M. (2010) *Macromolecules*, **43**, 943–951.
81 Gies, A.P., Geibel, J.F., and Hercules, D.M. (2010) *Macromolecules*, **43**, 952–967.
82 Rader, H.J. and Schrepp, W. (1998) *Acta Polym.*, **49**, 272–293.
83 Nielen, M.W.F. (1999) *Mass Spectrom. Rev.*, **18**, 309–344.
84 Pasch, H. and Ghahary, R. (2000) *Macromol. Symp.*, **152**, 267–278.
85 Murgasova, R. and Hercules, D.M. (2003) *Int. J. Mass. Spectrom.*, **226**, 151–162.
86 Montaudo, G., Carroccio, S., Montaudo, M.S., Puglisi, C., and Samperi, F. (2004) *Macromol. Symp.*, **218**, 101–112.
87 Montaudo, G., Samperi, F., Montaudo, M.S., Carroccio, S., and Puglisi, C. (2005) *Eur. J. Mass Spectrom.*, **11**, 1–14.

88 Jagtap, R.N. and Ambre, A.H. (2005) *Bull. Mater. Sci.*, **28**, 515–528.

89 Montaudo, G., Samperi, F., and Montaudo, M.S. (2006) *Progr. Polym. Sci.*, **31**, 277–357.

90 Li., L. (ed.) (2010) *MALDI Mass Spectrometry for Synthetic Polymer Analysis*, vol. 175, Wiley, Hoboken.

91 Weidner, S.M. and Trimpin, S. (2010) *Anal. Chem.*, **82**, 4811–4829.

92 Gies, A.P., Kliman, M., McLean, J.A., and Hercules, D.M. (2008) *Macromolecules*, **41**, 8299–8301.

93 Crecelius, A.C., Baumgaertel, A., and Schubert, U.S. (2009) *J. Mass Spectrom.*, **44**, 1277–1286.

94 Polce, M.J. and Wesdemiotis, C. (2010) in *MALDI Mass Spectrometry for Synthetic Polymer Analysis* (ed. L. Li), John Wiley & Sons, Inc., Hoboken, pp. 85–127.

3
Tandem Mass Spectrometry Analysis of Polymer Structures and Architectures

Vincenzo Scionti and Chrys Wesdemiotis

3.1
Introduction

The development of soft ionization techniques, in particular electrospray ionization (ESI) [1] and matrix-assisted laser desorption ionization (MALDI) [2–4], set a major cornerstone in mass spectrometry history. These methods enabled the formation of gas-phase ions from nonvolatile molecules, thereby radically expanding mass spectrometry applications to different and diverse fields. Compounds that had been characterized by other platforms started becoming analyzable and more precisely identifiable by mass spectrometry (MS). The potential of this analytical technique was rapidly recognized by polymer scientists, conscious of the great contribution that it can offer to synthetic polymer and materials studies [5, 6]. The growing need for new, inexpensive, and green synthetic routes to both widely used and newly designed polymers has made MS the tool of choice compared to other alternative and less sensitive or selective techniques, such as IR and NMR [7, 8]. MS can rapidly provide the molecular weight distribution (MWD) of the polymer, (co)monomer compositions, and end-group identification. The unique dispersive character of MS allows for the detection of a single component within convoluted polymer distributions arising from different end groups, varied structural arrangements, unpredicted byproducts, or degradation products [9]. Because of these benefits, MS applications to synthetic polymers have increased enormously [10–14].

Single-stage mass spectrometry, also referred to as one-dimensional MS, provides the mass-to-charge ratio (m/z) of each polymer component, from which the corresponding mass can be obtained depending on the size of the polymer and the mass accuracy of the instrumentation used. For new or known polymers prepared via established synthetic procedures, the mass information is often sufficient to derive the elemental composition of the product's constituents and predict their structure. If new synthetic concepts are evaluated or the origin of the sample is unknown, mass data alone may not permit unequivocal compositional or structural assignments. In such cases, two-dimensional or tandem mass spectrometry (MS/MS or MS^2) offers a means to collect additional analytical information, so

Mass Spectrometry in Polymer Chemistry, First Edition.
Edited by Christopher Barner-Kowollik, Till Gruendling, Jana Falkenhagen, and Steffen Weidner
© 2012 Wiley-VCH Verlag GmbH & Co. KGaA. Published 2012 by Wiley-VCH Verlag GmbH & Co. KGaA.

that the problem can be solved. In MS^2, ions of the same m/z ratio corresponding to a specific n-mer are first isolated and then energetically activated in order to undergo structurally diagnostic fragmentations. Detection and interpretation of the resulting fragment ions allows one to reconstruct the primary structure (connectivity) of the selected n-mer. The successive isolation and fragmentation events can take place either in physically separated regions of the mass spectrometer (MS^2 in space) or in the same location at different times (MS^2 in time), depending on the type of mass spectrometer available. MS^2 in space is performed with beam instruments, while the ions travel from the source to the detector; inversely, MS^2 in time is performed in ion traps (ITs), while the ions are stored in the trapping region. The MS^2 in-space experiment requires the coupling of at least two mass analyzers having a collision cell or different excitation section in between. The first analyzer is set to transmit only ions of a specific m/z ratio. After exiting this analyzer, the selected ions (precursor or parent ions) enter the excitation section, where their internal energy is perturbed by collisions with an inert gas, or other reactive species (for example, ions of opposite charge), by collisions with a surface, or by photons. This process increases the ions' internal energy, so that they undergo unimolecular fragmentations. Consecutively, the newly formed fragments and any residual precursor ions travel to the second mass analyzer, which deconvolutes and transmits them to the detector according to their m/z ratio. Instruments performing MS^2 in space may contain quadrupole (Q), time-of-flight (ToF), or IT mass analyzers. Different instrument manufacturers have assembled diverse combinations of these analyzers to achieve enhanced transmission, sensitivity, and resolution of both the selected ions and the fragments generated from them; the most common arrangements have been QqQ, Q-ToF, ToF/ToF, and IT/ToF (q designates a quadrupole used as collision cell).

In the case of mass spectrometers containing trapping analyzers, all ions generated in the source are injected and stored inside the trapping device by electric or magnetic fields. The isolation of a specific precursor ion, its fragmentation, and the dispersion of the resulting fragment ions are executed in time by varying one of these two fields. Examples of such analyzers are the quadrupole (3D) IT, the linear (2D) IT, the orbitrap, and the ion cyclotron resonance (ICR) trap. With quadrupole or linear ITs, the trapped ions are ejected sequentially by m/z value for detection. In contrast, ions stored in an orbitrap or ICR trap are detected nondestructively, based on the image currents induced by the orbiting ions on specific trapping elements; the currents of all ions can be detected simultaneously and converted to mass spectra by Fourier transformation (FT). An advantage of the trapping instruments, compared to the beam instruments, is that the isolation/fragmentation processes can be iterated on the fragment ions in triple-stage MS experiments (MS/MS/MS or MS^3). Theoretically, n stages (MS^n) can be carried out in the same trapping analyzer, provided a sufficient number of ions remain trapped to yield measurable signals. On the other hand, each tandem mass spectrometry stage in beam instruments requires one mass analyzer. To minimize pumping volume and ion losses due to scattering and maximize sensitivity, the vast majority of in-space tandem mass spectrometers contain only two mass-analyzing devices.

3.2
Activation Methods

3.2.1
Collisionally Activated Dissociation (CAD)

The most widely used ionization techniques, namely MALDI and ESI, are characterized as soft, because they generate prevalently molecular ions (M^+ or M^-) or quasi-molecular ions ($[M + H]^+$, $[M - H]^-$, $[M + X]^+$ where X = metal), with little or no fragmentation. These species represent the precursor ions that are mass-selected and energetically activated in MS^2 experiments in order to generate fragments indicative of the precursor ion structure. Since ion fragmentation is the core of tandem mass spectrometry, many different methods for promoting this process have been developed during the past few decades. Collisionally activated dissociation (CAD), also known as collision-induced dissociation (CID) [15], is one of the most common fragmentation methods in use nowadays. Activation of the mass-selected precursor ions is effected by accelerating them, so that they undergo energizing collisions with gaseous targets, commonly argon or helium atoms. During the collision, a fraction of the kinetic (translational) energy of the precursor ion is converted into internal energy, which is redistributed rapidly among the rotational–vibrational degrees of freedom of the ion before it fragments (ergodic process). As a consequence, the weakest bonds in the precursor ion have a higher predisposition for cleavage. Also, unimolecular rearrangement mechanisms are triggered, because they tend to produce stable fragment ions and, hence, have favorable energy requirements.

CAD can be performed in both beam and trapping mass spectrometers. The maximum amount of kinetic energy that can be converted into internal energy after a single collision depends on the masses of both the precursor ion and the collision gas atom or molecule involved, as well as on the kinetic energy of the ion before the collision. This relationship is expressed as follows:

$$E = \frac{N}{m_p + N} E_k \tag{3.1}$$

N identifies the atomic/molar mass of the neutral collision gas, m_p is the molar mass of the selected precursor ion, and E_k is the laboratory-frame kinetic energy of the precursor ion before the collision. E is often termed the center-of-mass kinetic energy (E_{com}). It is the type of instrument and analyzer that set the initial kinetic energy, E_k, according to which CAD can be classified as "high energy" or "low energy."

High-energy CAD is generally performed with ToF-type instruments or older instruments that contain magnetic and electric sectors. The kinetic energies of the ions subjected to CAD are in the order of several keV, and collisions take place at relatively low pressure (10^{-2} mbar). Under these conditions, each precursor ion experiences on average 1–10 collisions [16] that can deposit high enough energy to cause electronic excitation; generally, this process is followed by randomization of the excess energy over the rotational–vibrational degrees of freedom of the precursor ion, which ultimately leads to fragmentation, as mentioned above. With large precursor ions (for example, from polymers), direct excitation of rotational–vibrational modes is more probable at keV collision energies, because the corresponding center-of-mass kinetic energies are too low to access excited electronic states.

Low-energy CAD involves ions with much lower initial kinetic energies (usually ≤ 200 eV) that are required with quadrupole and trapping analyzers for efficient precursor ion selection and fragment ion dispersion. One consequence of using low-energy CAD, compared to high-energy CAD, is that the internal energy transferred during one collision is not enough to cause fragmentation. Multiple collisions are needed, which can be achieved by increasing the pressure of the collision gas (mbar range) or the time of excitation (in ITs). For example, a He pressure of 1 mbar can cause up to 10^6 collisions in a quadrupole IT within tens of milliseconds [16]. Low-energy CAD can only cause rotational–vibrational excitation of the precursor ion.

3.2.2
Surface-Induced Dissociation (SID)

The amount of internal energy that can be transferred to a precursor ion in a collision with a gaseous target depends on the mass of the target atom (or molecule), cf. Eq. (3.1). If the collision gas is substituted with a solid surface, more energetic collisions result because of the much higher mass of the surface compared to that of the gas atom (or molecule). The precursor ion undergoes strictly one collision during SID, gaining a narrower internal energy distribution than upon CAD. Moreover, lower background pressures are maintained, as no collision gas is required, which enhances the reproducibility of the fragmentation process. The first SID experiments were carried out on sector-type instruments [17]. More recently, SID has been adapted to mass spectrometers containing quadrupole, ToF, and ICR analyzers [18]. Since SID deposits higher average internal energies than CAD, but CAD allows for multiple activating collisions, these two activation methods tend to yield comparable fragmentation patterns. SID has not yet been employed to the analysis of synthetic polymers.

3.2.3
Photodissociation Methods

Alternative activation methods that do not require collision(s) of the precursor ion with atomic or larger entities have been considered and developed. The absorption of photons by isolated, mass-selected ions can easily induce excitation that ultimately causes fragmentation. Here, the wavelength of the exciting photons can be varied in the UV, vis, and IR regions, providing different levels of molecular excitement.

IR photons are most widely used to activate mass-selected ions. This radiation has frequencies that fall in the range of molecular vibrational frequencies; hence, the absorption of IR photons increases the vibrational energy. Multiple photon absorption is required to trigger fragmentation due to the low energy transferred by a single photon. Low-power, continuous-wave (cw) CO_2 lasers have served most frequently as the source of radiation, which must intercept the precursor ions for moderately long times (milliseconds or higher) in order to produce fragments with measurable intensities. Trapping instruments are most suitable for such IR multiphoton photodissociation (IRMPD). Note that the background pressure must be maintained low during the absorption window to minimize collisional relaxation. For this reason, IRMPD is performed mainly with FT-ICR mass spectrometers (which operate at ultrahigh vacuum; $\sim 10^{-9}$ bar) and electrostatic trapping instruments that require very low He bath gas pressure inside the trap device. IRMPD applications have so far focused on biological analytes [19].

3.2.4
Electron Capture Dissociation and Electron Transfer Dissociation (ECD/ETD)

Activation methods that exploit the interaction between isolated precursor ions and thermal electrons were first investigated by McLafferty *et al.* [20]. The process, named electron capture dissociation (ECD), involves the capture of thermal electrons (<0.2 eV), emitted from heated tungsten filaments, by isolated multiply charged cations. This reaction gives rise to incipient radical cations that dissociate rapidly via radical-induced decompositions and rearrangements at the site of electron attachment, before energy redistribution occurs (nonergodic process) [21]. Thus, ECD is site-specific, whereas CAD (ergodic process; vide supra) induces bond cleavages that require low activation energy. ECD is almost exclusively performed in FT-ICR instruments, where thermal electrons can be trapped over long periods to produce detectable fragments and where the background pressure is sufficiently low to suppress competitive CAD pathways. The necessity to have multiply charged precursor ions makes ESI the only suitable ionization technique for ECD studies.

The first ECD applications concerned biological analytes, especially proteins and peptides, and documented considerable differences between ECD and CAD fragmentation pathways. The major fragments from ECD of proteins and peptides are c/z-type ions from cleavages at the $N-C^\alpha$ backbone bonds, while CAD on the same species mainly gives rise to b/y-type ions from cleavages at the $C(=O)-N$ backbone bonds. An important feature of ECD is that it generally preserves labile posttranslational modifications at the protein/peptide side chains during fragmentation due to the rapid, nonergodic nature of energy transfer. ECD has been applied on select synthetic polymers, including polyestereamide, poly(alkene glycol), and polyamidoamine samples [22–25].

An alternative, closely related activation method, is electron transfer dissociation (ETD) [26–28]. The inability to store thermal electrons in trapping instruments utilizing strong radiofrequency (rf) fields, such as the 3D and 2D quadrupole IT, has been overcome by employing ion–ion reactions between reagent radical ions (for example, from anthracene or similar substances) and multiply charged precursor ions of opposite charge, both confined in the same trapping device. It is during such reactions that electrons are transferred. Most widely used are ion–ion reactions between reagent anions and multiply charged cations. The negatively charged radical reagents are generated in a separate negative chemical ionization (nCI) source and transmitted to the IT where they are left to react with the mass-selected precursor ions for the necessary time interval. The fragment ions arising from ion–ion reactions are very similar to the ones observed in ECD; an important feature of both methods is that, unlike CAD, they preserve posttranslational modifications in peptides and proteins [29]. Although no ETD application to synthetic polymers has yet been published, preliminary studies on polyethers [30] and polyesters [31] have been reported, based on which increased future use is anticipated.

3.2.5
Post-Source Decay (PSD)

This type of fragmentation takes place in reflectron ToF or ToF/ToF instruments equipped with a MALDI source. A high laser power is utilized to generate ions with enough internal energy to decompose spontaneously after exiting the source but before reaching the detector ("metastable ions"). PSD experiments were first demonstrated on reflectron ToF mass spectrometers, in which the fragments formed from a particular precursor ion in the field-free region between the MALDI source and the reflectron were dispersed by appropriate scanning of the reflecting lens potentials [32]. Precursor and fragment ions have the same velocities due to the principles of conservation of mass and momentum; hence, they travel as a packet to the reflectron. Selection of the desired precursor ion (together with its coherently moving fragments) was achieved with an ion gate that allowed only packets within a specific velocity window, corresponding to a precursor ion resolution of $\sim 20\, m/z$ units, to pass through [33].

The use of the reflectron as mass-analyzing device became obsolete with the introduction of commercial MALDI-ToF/ToF mass spectrometers, in which a short linear ToF analyzer is axially interfaced with a reflectron ToF device. Here, the first (linear) ToF analyzer and an ion gate are used for selection of the precursor ion (and its fragments), while mass analysis of the fragmentation products occurs in the second (reflectron) ToF analyzer. MALDI-ToF/ToF mass spectrometers have significantly improved precursor ion resolution ($<5\, m/z$ units below m/z 3000). Additionally, ToF/ToF instruments may be equipped with collision cells along the beam path from the ion source to the second ToF analyzer to combine PSD with CAD and enhance the fragmentation yield [34].

Mainly biological samples, such as proteins peptides and oligosaccharides [35–37], have been analyzed by MALDI PSD. Investigations on select polymer classes have also been reported [5, 38–41]. MALDI-ToF/ToF instrumentation improved markedly the quality of MALDI MS^2 spectra, as compared to PSD spectra acquired with a simple reflectron ToF mass spectrometer, leading to increased applications of MALDI-ToF/ToF MS^2 to synthetic polymers [42–51].

3.3
Instrumentation

In trapping instruments, the isolation and fragmentation processes take place in the same section (analyzer) of the mass spectrometer but in temporal sequence, whereas in beam instruments, isolation, fragmentation, and deconvolution of the resulting

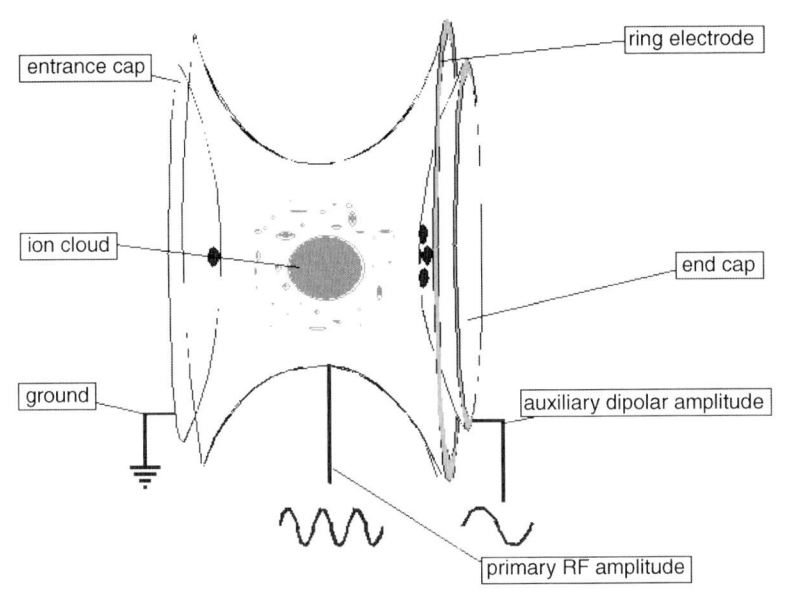

Figure 3.1 Schematic view of a QIT (Bruker HCT ultra). Reproduced from [52] with permission.

fragments occur in different parts of the mass spectrometer, while the ions travel from the source to the detector. This section provides brief descriptions of the trapping and beam instrument setups which, at the present, experience widespread use in synthetic polymer analyses.

3.3.1
Quadrupole Ion Trap (QIT) Mass Spectrometers

Quadrupole ion traps (QITs) are storage devices. They are comprised of a ring electrode and two end caps having small holes for injection of the ions produced in an external ion source or ejection of the stored ions to a detector (Figure 3.1) [52]. Trapping is achieved by grounding the end caps and applying an rf field of fixed frequency (ν) to the ring electrode, $\Phi_0 = V \cos \omega t$, where V is the amplitude of the applied rf potential ($\leq$30 kV) and ω the angular frequency ($\omega = 2\pi\nu$; $\nu \approx 0.75$–1.2 MHz). The trappable m/z range is determined by the rf amplitude. QITs are normally filled with helium bath gas ($\sim 10^{-3}$ mbar), which cools down the ions with collisions and forces them to move to the center of the trap, so that they are not accidentally ejected due to the kinetic energy acquired during the injection step or due to the repulsive forces between ions of the same charge.

The ion motion inside the trap is described by the dimensionless parameter $q_z = 8zeV/m(r_0^2 + 2z_0^2)\omega^2$, which is derived from the Mathieu equation and represents a measure of the rf amplitude V; m and z are the mass and charge of the ion, and r_0 and z_0 the radial and axial dimensions of the trap, respectively [53]. Since modern IT instruments do not utilize dc fields for trapping ($U=0$), the parameter a_z (also derived from the Mathieu equation and representing U) is equal to zero. In rf-only traps, the

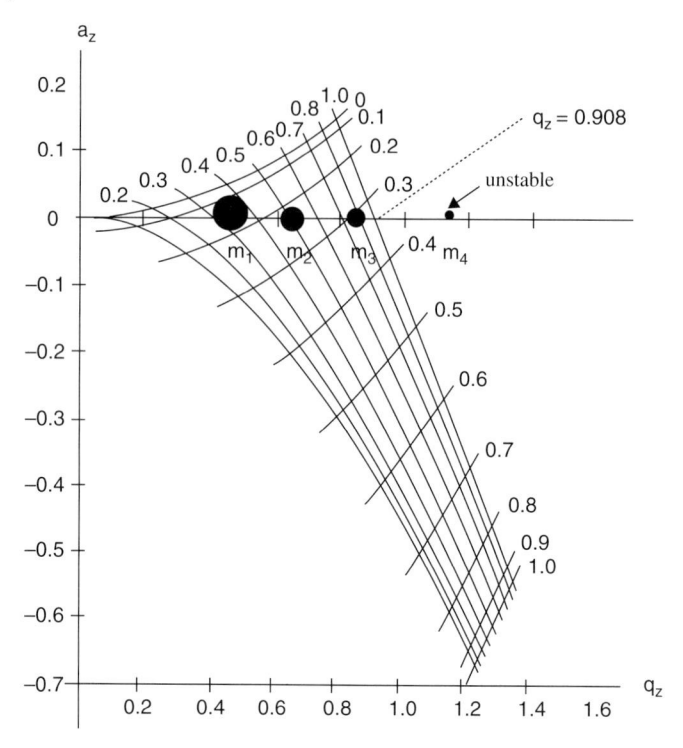

Figure 3.2 Stability diagram for a QIT, showing four ions along the q_z axis; the three ions residing inside the stability region ($q_z \leq 0.908$) are trapped, while the one outside this region ($q_z > 0.908$) is ejected. Adapted from Ref. [53] with permission.

ions are confined in the q_z axis, as shown in Figure 3.2 for four ions differing in mass. Only ions with $q_z \leq 0.908$ have stable trajectories inside the trap and can be stored. Figure 3.2 illustrates the pseudo-potential well that defines the stability region; ions outside this region have unstable trajectories and are ejected. Scanning V successively causes ions of increasing m/z to reach $q_z > 0.908$, at which point they are ejected through the end caps for detection (*mass-selective axial instability* mode) [53].

The limit value $q_z = 0.908$ determines the low-mass cutoff of a QIT; ions below this mass cannot be stored and analyzed. In addition to the mass-selective axial instability mode (vide supra), stored ions can be ejected and detected via resonant ejection. Inside the trap, ions oscillate with a secular frequency (ω_s) that is lower than the main rf field frequency (ω) and inversely proportional to their m/z value. By applying an auxiliary rf voltage to an end cap that has the same angular frequency as a trapped ion (ω_s), this ion will come into resonance and will be ejected from the trap. Scanning the frequency of the auxiliary field allows for the successive ejection (and detection) of all ions stored in the QIT. Instead of a single frequency, an rf signal composed of multiple frequencies can be generated and applied on an end cap to excite and eject many different ions at the same time; this principle is used to isolate specific precursor ions for MS^2 experiments.

Once the precursor ion has been chosen, ions with lower m/z values can be ejected by a rapid scan and ions with higher m/z values by resonance ejection; alternatively, all but the selected ion can be ejected by resonance ejection. The isolated precursor ions are accelerated by an auxiliary rf field, applied to the end caps, which increases their kinetic energy via resonance excitation. The amplitude of the auxiliary field is kept small (~1 V) to avoid ejection of the mass-selected ions from the trap. The excitation time usually is in the range 20–60 ms; during this time, the precursor ions undergo activating collisions with the He bath gas inside the QIT analyzer. Because ion kinetic energies are low in QITs, compared to beam instruments, a higher number of collisions is required to promote efficient CAD. The main fragments arise from dissociations that are associated with low activation energies. Since the auxiliary rf field is in resonance with the selected precursor ions but not their fragments, consecutive fragmentations proceed inefficiently unless they have lower energy requirements than competitive pathways. After the excitation time has elapsed, the CAD products are scanned by resonant ejection to render the corresponding MS^2 spectrum. The individual steps of an MS^2 experiment are generally preprogrammed into a scan function, as shown in Figure 3.3, which sets the order and duration of the events taking place during the acquisition of an MS^2 spectrum.

A specific MS^2 fragment can be isolated and excited to undergo further CAD by repeating the isolation, excitation, and ejection/detection steps outlined above. This procedure leads to the respective MS^3 spectrum; further MS^n cycles are possible, depending on the efficiency of fragmentation and the amount of ions remaining in the trap. The result of a MS^3 experiment, along with the pertinent fragmentation

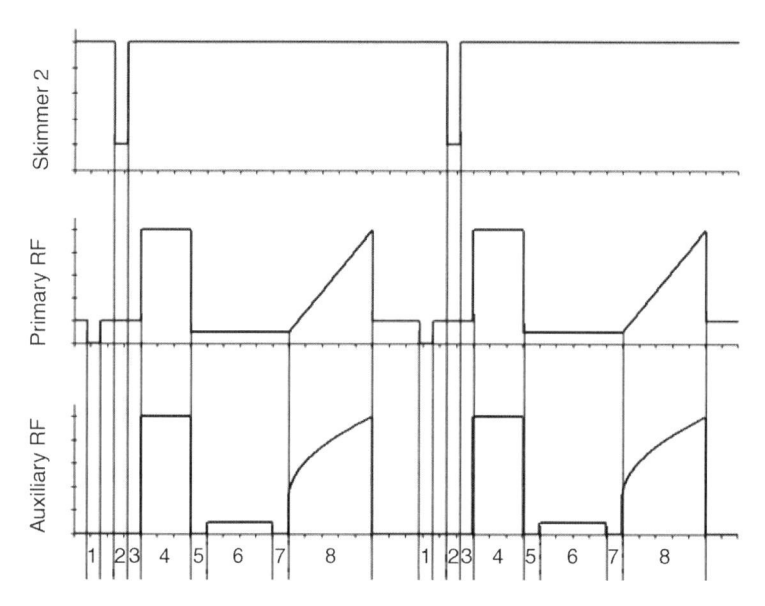

Figure 3.3 QIT MS^2 scan process for precursor ions produced in an external ESI source. 1: clear trap (all stored ions are ejected); 2: accumulation time (ions are injected); 3: isolation delay (cooling time); 4: precursor ion isolation; 5: fragmentation delay (cooling time); 6: fragmentation; 7: scan delay; 8: mass analysis. Reproduced from ref. [52] with permission.

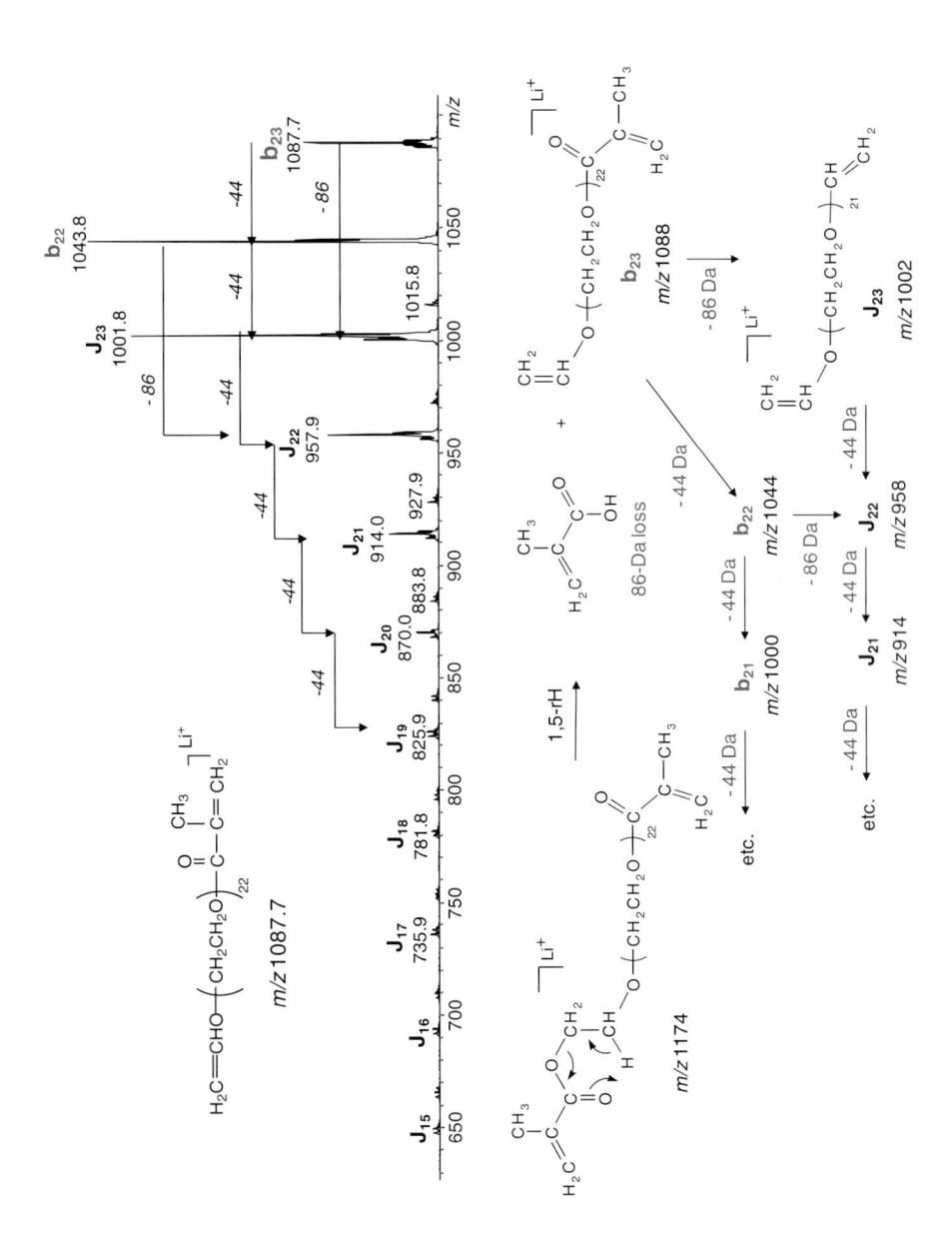

66 | *3 Tandem Mass Spectrometry Analysis of Polymer Structures and Architectures*

mechanism, is provided in Figure 3.4, which shows the MS^3 spectrum of the largest and most abundant fragment from the lithiated 23-mer from poly(ethylene oxide) dimethacrylate (PEO-DMA), namely b$_{23}$ (m/z 1087.7) [54, 55]. The information provided by the combined MS^2 and MS^3 data permits the identification of both end groups and the repeat unit (cf. Figure 3.4).

A shortcoming of CAD experiments in QITs is that fragment ions with m/z ratios smaller than ~1/3 of the precursor ion m/z are not efficiently retained in the trap because the resonance excitation step moves their m/z ratios below the low-mass cutoff. By modulating the QIT electronics, the cutoff can be reduced to <1/4 of the precursor ion m/z, the trade-off being lower sensitivity. Alternatively, a MS^2 method that does not accelerate the precursor ions, such as ETD may be used (see below).

QITs can also be adapted to carry out ETD experiments. Instruments with this capability are equipped with two external ion sources, one ESI source to produce multiply charged precursor ions and one chemical ionization (CI) source to produce reagent ions of opposite charge [26]. In the Bruker HCT model, the CI source is located above the octapole lens used to transfer ESI-generated ions to the IT, cf. Figure 3.5 [56, 57].

In the vast majority of ETD experiments, the CI source is operated in negative mode (nCI), furnishing anions that react with multiply charged cations (as shown in Figure 3.5). Inside the nCI source, 70 eV electrons are emitted from a tungsten filament in a chamber filled with methane at a pressure of 2.0–2.6 bar. The methane gas acts as a mediator that cools down the electrons to thermal energies, so that they attach to the reagent molecules to form intact radical anion species [56, 57]. Different reagents have been tested and utilized for ETD experiments, including 9-anthracenecarboxylic acid, 2-fluoro-5-iodobenzoic acid, 2-(fluoranthene-8-carbonyl)benzoic acid, and fluoranthene [56–58].

Although the ESI and nCI sources operate continuously, the respective ions are transmitted to the QIT alternately. During the transfer of the ions from the ESI source to the IT, the reagent radical anions from the nCI source are blocked at the gate lens by the application of a voltage (cf. Figure 3.5). After a specific ion accumulation time, the ion flow from the ESI source is stopped at the skimmer and the QIT optics start operating partially in the negative mode to move the radical anions formed in the nCI source to the trap. Ion–ion reactions between the two species follow for a specific time interval (ms), which is generally sample-related. Figure 3.6 provides a comparison between the CAD and ETD spectra of polybutylene adipate [31].

Figure 3.4 ESI-QIT MS^3 mass spectrum of the b$_{23}$ fragment (m/z 1088) generated by CAD of the lithiated 23-mer (m/z 1174) from poly (ethylene oxide) dimethacrylate (PEO/DMA). All fragments are generated by 1,5-hydrogen rearrangements (1,5-rH) at the chain ends [54, 55]. MS^2 of m/z 1174 (spectrum not shown) yields predominantly b$_{23}$ by loss of methacrylic acid (86 Da) from one chain end (>75% of total fragment current). Consecutive fragmentation of b$_{23}$, via MS^3 (top), proceeds by a series of 1,5-hydrogen rearrangements from either chain end, which release (bottom) the second methacrylate end group (86 Da) and C$_2$H$_4$O repeat units in the form of acetaldehyde (44 Da), resulting in fragment series b$_n$ (terminal fragments with one methacrylate and one vinyl end group) and series J$_n$ (internal fragments with two vinyl end groups). The combined MS^2 and MS^3 spectra identify both end groups and the repeat unit. Courtesy of Nilüfer Solak [54].

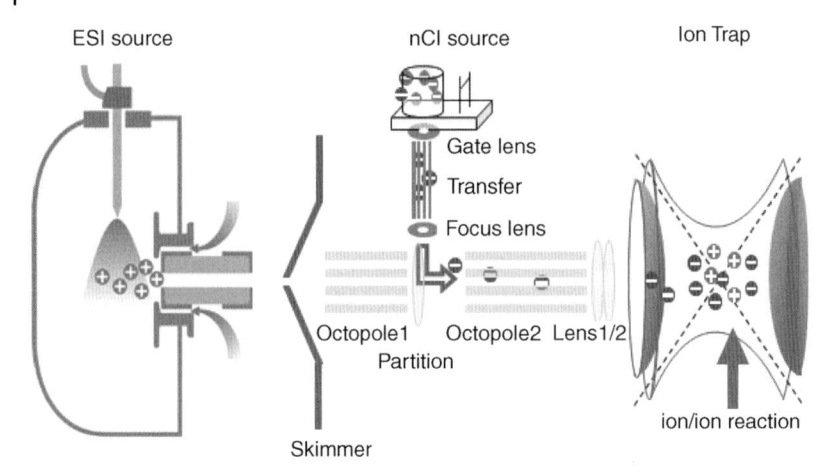

Figure 3.5 QIT with external ESI and CI sources for ETD experiments; the most widespread mode involves ETD between multiply charged cations produced by ESI and reagent anions produced by nCI, as shown. Reproduced from Ref. [57] with permission.

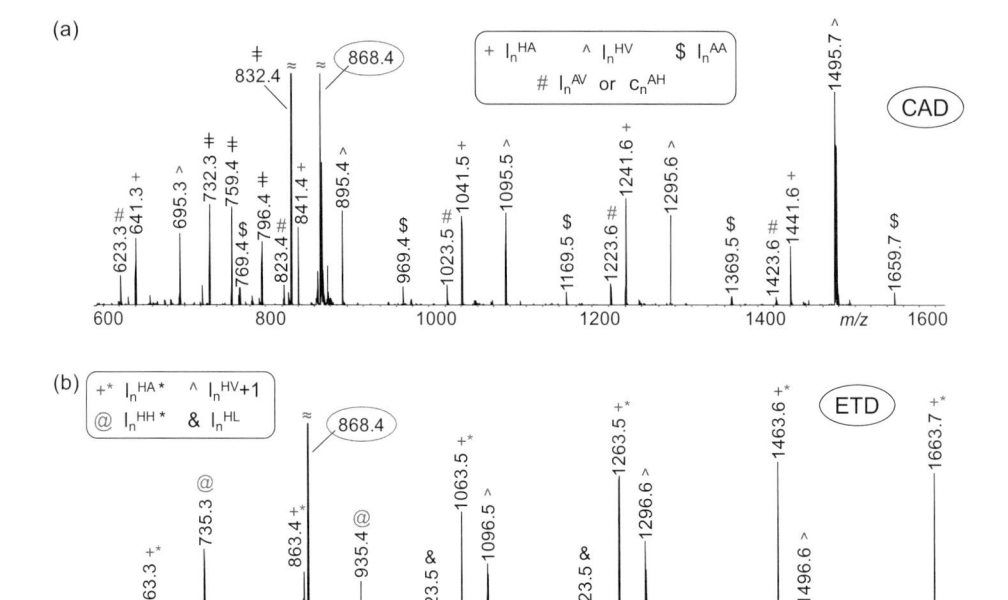

Figure 3.6 ESI-QIT MS^2 spectra of [M + 2Na]$^{2+}$ from the 8-mer of a diol-capped poly(butylene adipate) (m/z 868.4), acquired via (a) CAD and (b) ETD. The letters l and c denote linear and cyclic fragment ions, respectively. The superscripts indicate the corresponding end groups: H, hydroxyl; V, vinyl (butenyl); A, acid (COOH); L, lactone [31]; chains with butenyl and COOH end groups are isomeric with cyclic oligomers. The * and ‡ labels indicate doubly sodiated monocations (sodiated fragments containing COONa groups) and doubly charged fragments, respectively. The notation +1 indicates radical ions [31].

3.3.2
Quadrupole/time-of-flight (Q/ToF) Mass Spectrometers

A tandem mass spectrometer composed of quadrupole and ToF mass analyzers offers the high-mass accuracy and resolving power of double focusing sector mass spectrometers. Moreover, unlike sector instruments, which do not function optimally with pulsed ionization methods (MALDI) or ionization methods requiring high voltages (ESI), Q/ToF instrumentation can be interfaced with almost any ionization technique, including MALDI, ESI, atmospheric pressure chemical ionization (APCI), atmospheric solid analysis probe (ASAP) [59], and desorption electrospray ionization (DESI) [60]. The instrument set up takes advantage of the mass selection and transmission properties of the quadrupole and the resolution and mass accuracy capabilities of the ToF analyzer. Generally, the ToF part is orthogonal to the quadrupole, a geometry imposed by the different features of these devices: a quadrupole transmits ions continuously, whereas a ToF analyzer functions in pulsed mode, by resolving packets of ions having the same initial kinetic energy [33].

A Q/ToF mass spectrometer equipped with a MALDI source and a collision cell between the two mass analyzers is depicted in Figure 3.7 [61]. After passing the quadrupole and the collision cell, precursor and fragment ions are pushed down the ToF tube with an acceleration voltage of $\sim$10 kV, provided by a pusher, which is synchronized with the detector in order to measure accurately the time elapsing after every push until the ions reach the detector. For the acquisition of regular mass spectra, the quadrupole is set in the rf-only mode to transmit all ions generated in the ion source to the ToF segment, where they are accelerated and dispersed by their m/z

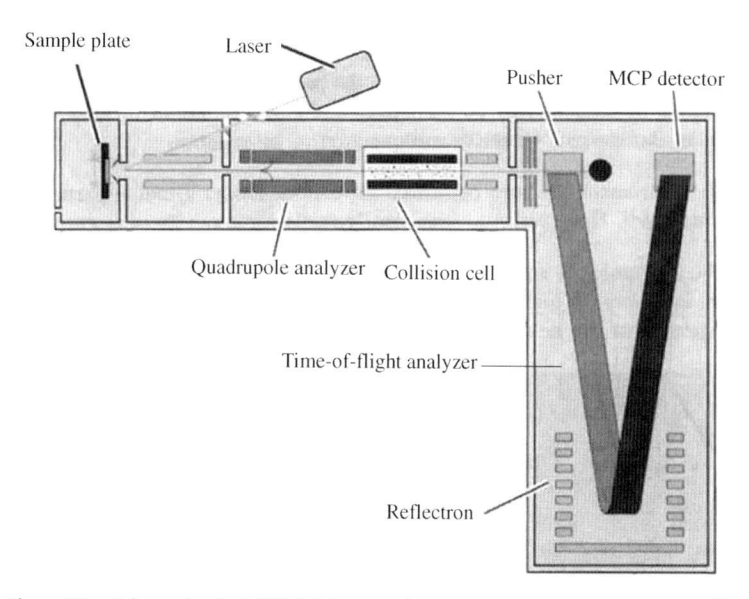

Figure 3.7 Schematic of a MALDI-Q/ToF tandem mass spectrometer (the Waters® Q-ToF Ultima® MALDI). Reproduced from Ref. [61] with permission.

ratios. Conversely, in the MS^2 mode, both dc and rf voltages are applied to the quadrupole and tuned to transmit only the desired precursor ion, which is fragmented in the ensuing collision cell. The collision cell is an rf-only quadrupole or hexapole, filled with a collision gas (typically Ar) to promote CAD. For this, the potential energy of the collision cell, which sets the laboratory-frame collision energy, is raised to several tens of eV (up to ∼200 eV), so that the entering precursor ions undergo energizing collisions and dissociate. The resulting fragments can be mass analyzed in the ToF segment with mass accuracies of <10 ppm if the mass scale is calibrated immediately before mass analysis.

The precursor ion isolation window can be tuned by adjusting the dc and rf voltages applied to the quadrupole rods. This makes it possible to isolate either the complete isotope cluster or a single isotope of the precursor ions. Keeping the isolation window large enough to transmit the entire isotope cluster of the precursor ion is advantageous for the identification of polymers containing elements with unique isotope distributions, such as halogens and sulfur. In these cases, the isotope pattern of the resulting fragment ions unveils not only the presence of a specific atom, but also how many of these atoms are contained in the fragment, which significantly facilitates MS^2 spectral interpretation; Figure 3.8 exemplifies this benefit with the CAD mass spectrum of the 5-mer from poly(dichlorophosphazene) [62]. This polymer carries an alternating backbone of nitrogen and phosphorus atoms, with two chlorine atoms attached at each P atom. The isotope distribution of each fragment ion in the MS^2

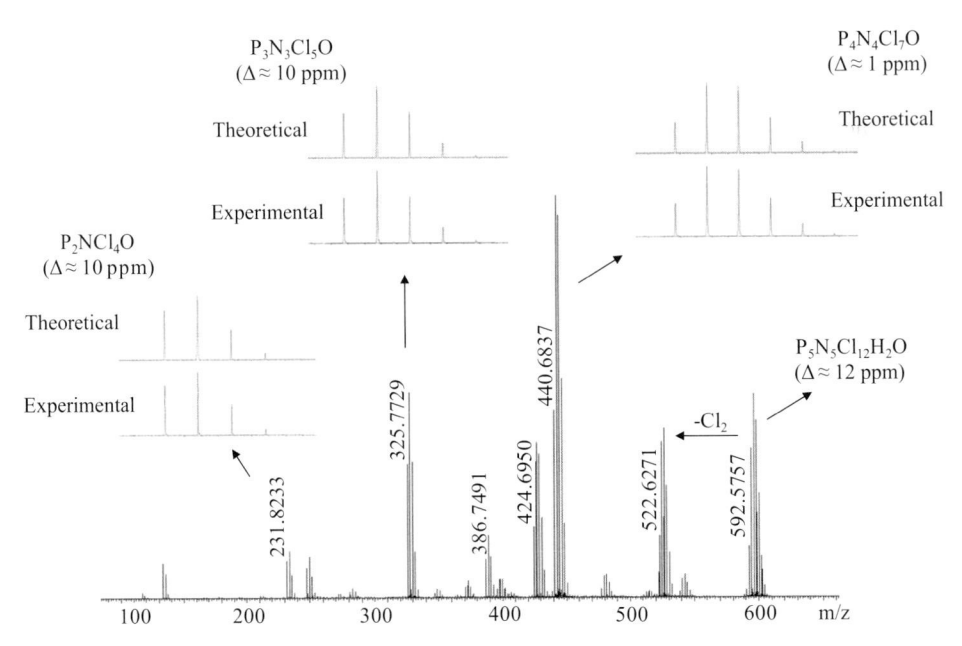

Figure 3.8 MS^2 (CAD) mass spectrum of [M + H$_2$O]$^{\bullet-}$ from the 5-mer of a poly (dichlorophosphazene) with the composition [N = PCl$_2$]$_n$ (m/z 592.6), acquired with an ESI-Q/ToF mass spectrometer [62]. Δ is the difference in ppm between the experimental m/z value and that calculated for the composition shown.

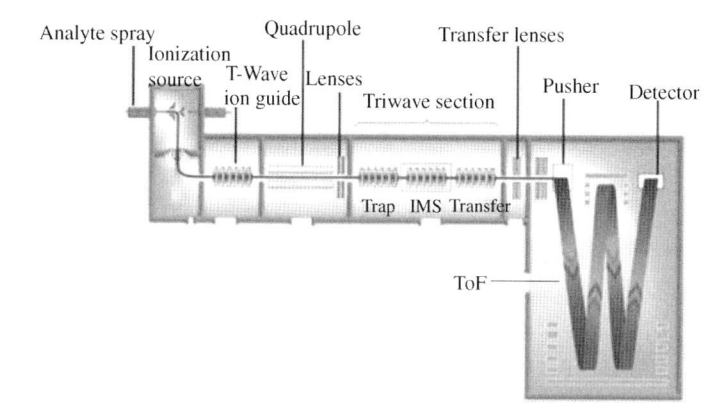

Figure 3.9 ESI-Q/ToF tandem mass spectrometer (the Waters® Synapt HDMS™), equipped with ion mobility spectrometry (IMS) and double-reflectron ToF (W mode) capabilities. Reproduced from Ref. [65] with permission.

spectrum unambiguously provides the number of chlorine atoms. This information is essential for elucidating the fragmentation mechanism and the architecture of the polymer [62].

Recently, Q/ToF instrumentation has become available that enables the combination of MS or MS^2 experiments with ion mobility spectrometry (IMS) [63, 64]. This capability is offered by the Waters Synapt HDMS™ mass spectrometer, which contains a traveling wave (T-wave) section at the interface of the Q and ToF analyzers, consisting of three cells in the order trap cell, ion mobility (IM) cell, and transfer cell (Figure 3.9) [65]. By activating the ion mobility device, ions can be separated first according to their size/shape within the IM chamber and, later, according to their mass-to-charge ratio in the ToF mass analyzer. If a precursor ion is mass-selected by the quadrupole, it can be fragmented in the trap cell, which is operated as a normal collision cell using, typically, Ar as collision gas. When the IM device is also activated, the fragments formed by CAD remain trapped for a short period of time and then are released in the adjacent ion mobility cell. This trapping and release process is synchronized with the pusher located at the entrance of the ToF analyzer, so that the drift time of an ion through the ion mobility cell is matched with the corresponding m/z ratio. Once the ion packet released from the trap cell enters the ion mobility region, the ions move under the influence of a traveling wave electric field in the presence of a drift gas (N_2) which flows in the opposite direction of the ions' motion. Isobaric or isomeric fragments are separated according to their size, shape, and charge state and displayed in a 2-D ion mobility diagram of m/z versus drift time [66]. Alternatively, the mass-selected precursor ions may first be separated in the IM cell before entering the transfer cell for CAD; this way the individual MS^2 spectra of overlapping isobaric or isomeric components can be acquired [54, 67]. An example, documenting the dispersion and characterization of isobaric components of a nonionic surfactant is given in Figure 3.10 [54, 55, 68].

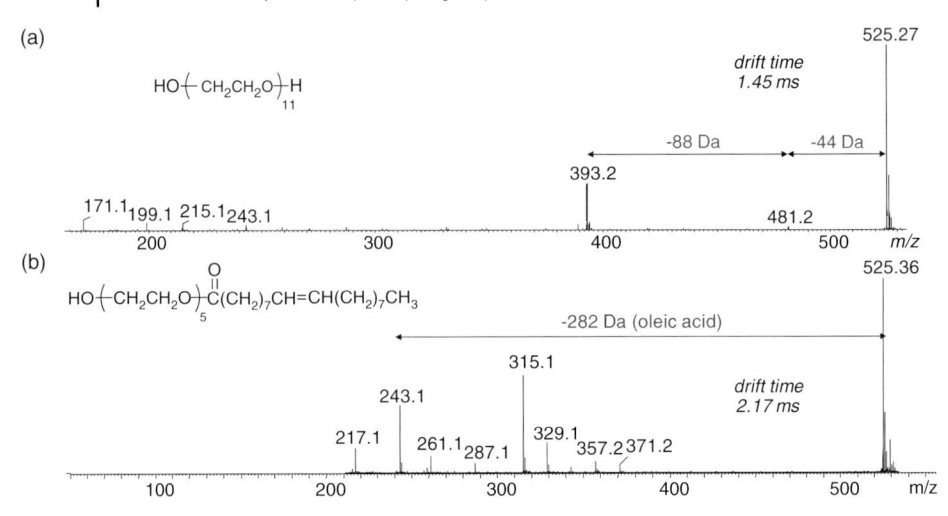

Figure 3.10 MS² (CAD) spectra of the [M + Na]⁺ ions of two isobaric components in sorbitan trioleate, acquired with the Synapt HDMS™ ESI-Q/ToF mass spectrometer. Ions of m/z 525 were mass-selected by Q, dispersed through the ion mobility chamber, and subjected to CAD; ToF analysis of the resulting products gave rise to the spectra shown [54]. The component drifting at 1.45 ms (top) undergoes the 44- and 88-Da losses characteristic for poly(ethylene glycol) [55] and, thus, was assigned to the poly(ethylene glycol) 11-mer (calcd. $m/z = 525.29$). The component drifting at 2.17 ms loses oleic acid (-282 Da) and forms fragments with the repeat unit of 44 Da (m/z 217 and 243), which agree well with the monooleate of the poly(ethylene glycol) 5-mer (calcd. $m/z = 525.38$) [54]. The signals at 315–371 are due to sodiated HO$-$(CH$_2$CH$_2$O)$_5-$C($=$O)$-$(CH$_2$)$_n$CH$=$CH$_2$ ($n = 0$–4), formed via charge-remote pathways at the oleate chain [68].

3.3.3
ToF/ToF Instruments

Tandem ToF mass spectrometers are comprised of two axially interfaced ToF analyzers (*vide supra*). Several models are commercially available; all are equipped with a MALDI source attached to a linear ToF tube, which in turn is interfaced with a reflectron ToF device (cf. Figure 3.11) [69–73]. Mass selection of the desired precursor ion is effected within the linear ToF segment with the help of an ion gate for timed ion selection (TIS). Only fragments formed in the linear drift tube, namely after the precursor ion has left the ion source, are transmitted; since these fragments move with the same velocity as their precursor ion, TIS can take place anywhere within the linear ToF part. The TIS device consists of two deflection gates, which are opened in a synchronized manner for a narrow time window, to allow for passage of only the desired precursor ion and its fragments. Usually the complete isotope distribution is transmitted, as precursor ion selection at higher resolution is accompanied by significant sensitivity losses. ToF/ToF instruments contain dedicated collision cells for CAD, which may be located anywhere in the linear ToF segment [71, 73]. Fragmentation is induced by raising the laser intensity, to cause laser-induced dissociation, as well as by CAD.

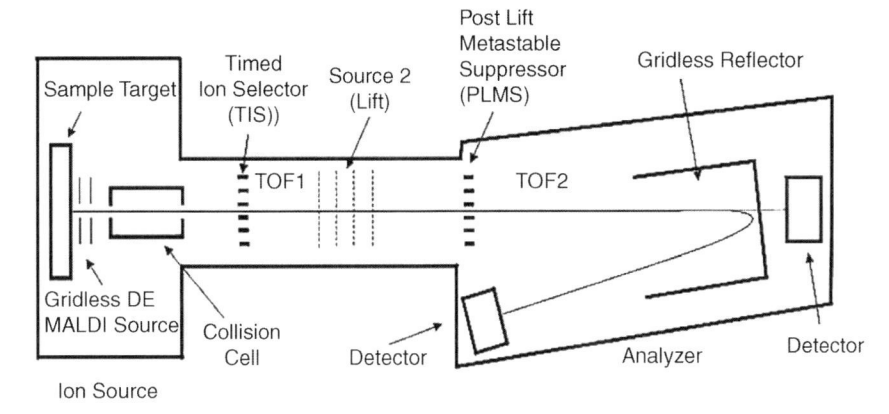

Figure 3.11 A MALDI-ToF/ToF mass spectrometer (the Bruker UltraFlex). Reproduced from [71] with permission.

In the design shown in Figure 3.11, ions formed by MALDI are accelerated to 8 keV and decompose within the linear ToF, primarily due to laser-induced dissociation [71]. Adding CAD increases the yield of low-mass fragments, but also causes scattering losses and, thus, is useful only if the low-mass fragments provide irreplaceable structural insight, as in *de novo* peptide sequencing studies [71, 72]. After the desired precursor ion and its fragments are separated from all other ions by the TIS gates, they enter a reacceleration region that postaccelerates ("lifts") them by $+19$ keV for mass analysis by the reflectron ToF analyzer. Shortly after the "lift" device, a postlift metastable suppressor (PLMS) is located, which is timed to deflect the remaining precursor ions, so that fragmentation is stopped after postacceleration.

In a different design, a time-selected 8-keV precursor ion (or precursor/fragment packet) are decelerated to 1–2 keV as they enter a floated collision cell for CAD [73]. After exiting this cell, precursor and fragments reach a postacceleration region that raises the kinetic energy to 15–20 keV for mass analysis by the reflectron ToF analyzer. In two designs discussed, the reacceleration step reduces the spread of kinetic energies of precursor and fragment ions, so that all can pass the reflectron and reach the detector. In a third design, the ions leave the MALDI source at 20 keV and undergo CAD at this high kinetic energy; the resulting larger range of kinetic energies is accommodated with a curved-field reflectron [70, 74].

The ToF/ToF instruments described can be used to acquire regular mass spectra by grounding the TIS gates and (if present) the deceleration/postacceleration lenses. Further, ToF/ToF equipment can provide information about the absolute MW and MWD of synthetic polymers. Q/ToF or QIT instrumentation is less suitable for such measurements because quadrupolar fields skew the MWD. On the other hand, Q/ToFs and QITs permit the selection of much narrower m/z ranges in MS^2 experiments and offer better control of the energy deposited in CAD than ToF/ToF instruments. Nevertheless, the superior MW and MWD data accessible with ToF/ToF instrumentation and the capability to perform MS and MS^2 analyses within the same experimental setup (*vide supra*) make this type of tandem mass spectrometer essential for the compositional and structural characterization of synthetic polymers.

The MS^2 capabilities of ToF/ToF instrumentation are documented in Figure 3.12, which compares the MALDI MS^2 spectra of the [M + Ag]$^+$ ions from a linear and a

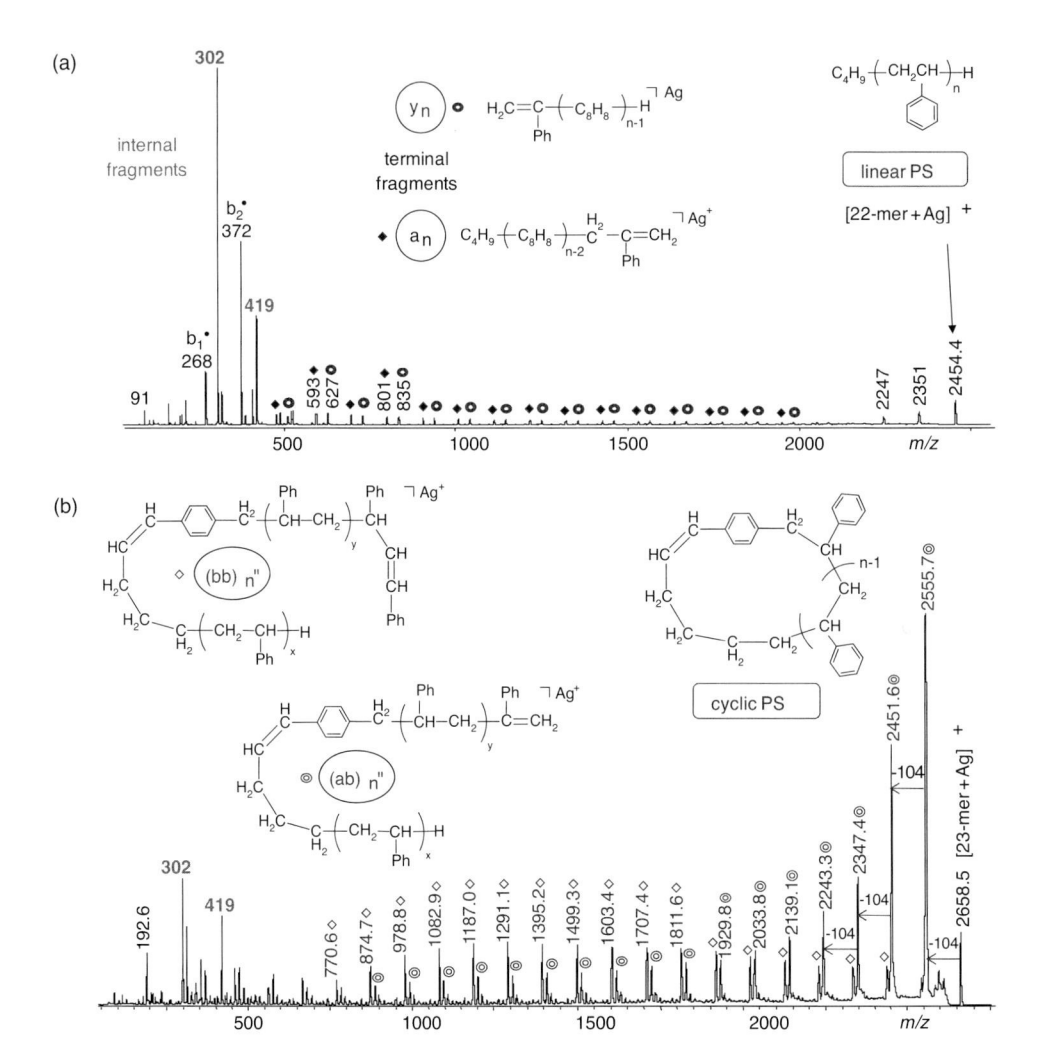

Figure 3.12 MALDI-ToF/ToF MS^2 mass spectra of the [M + Ag]$^+$ ions from a (a) linear and (b) cyclic polystyrene (PS). The linear PS undergoes random C—C bond cleavages, followed by backbiting and β C—C bond scissions to yield mainly the terminal fragments shown along with the internal ions at m/z 302 and 419 [44, 45, 55]. The cyclic oligomer dissociates via ring opening, followed by monomer or monomer + CH$_2$ evaporation, and subsequent intrachain H transfer to mainly yield the series shown [55]. See [55] for a detailed explanation of the fragmentation mechanisms and the nomenclature used.

cyclic polystyrene (PS) oligomer [44, 45, 55]. Silverated polystyrenes dissociate via random homolytic C−C bond cleavages along their backbone, which create charged radicals that decompose further by typical radical site reactions, namely hydrogen rearrangements (backbiting), bond migrations (phenyl shifts), and β scissions. Because of their distinct connectivities, linear and cyclic architectures give rise to unique fragment distributions, based on which they can be conclusively differentiated and identified [55, 75].

3.4
Structural Information from MS^2 Studies

3.4.1
End-Group Analysis and Isomer/Isobar Differentiation

MS analysis of a polymer provides the m/z values of its n-mers. From this information, one can deduce the repeat unit and combined end groups of linear polymers. The addition of MS^2 data makes it possible to elucidate individual end groups [44, 45, 76, 77]. For example, the ESI MS^2 spectrum in Figure 3.10b ascertains the presence of an oleate end group [54]. Similarly, the a_n/y_n fragments in the MALDI MS^2 spectrum of Figure 3.12a reveal that the chain ends must carry C_4H_9 and H substituents [44].

Fragmentation patterns are sensitive to the polymer composition and the substituents present [55]. As a result, isobaric macromolecules are readily distinguished by their MS^2 spectra, as shown in Figure 3.10. If they coexist and overlap, however, prior separation by mass (if the required resolution is available), chromatography, or IMS (which was employed for the data in Figure 3.10) is called for [54].

The characterization of isomers is more challenging and is most facile by interfacing MS^2 with a separation method (see above). If only one isomer is present, the correct structure may be derived by spectral interpretation; for example, amino alcohol and amino methyl ether isomers can be unequivocally identified, because the former lose water and the latter methanol in ESI MS^2 experiments [78].

3.4.2
Polymer Architectures

The MS^2 spectra of the cyclic and linear polystyrene in Figure 3.12 clearly show that the corresponding architectures can be distinguished based on the fragmentation patterns ensuing after CAD [49, 55]. Similarly, linear and branched polyacrylates have been differentiated by the fragments observed in their MS^2 spectra [46]. In these studies, the existence of reference spectra for at least one of the architectures was essential. Expansion of the library of MS^2 spectra, as a result of the tandem MS instrumentation that has become available, will certainly benefit spectral comparisons aiming at the determination of unknown architectures.

3.4.3
Copolymer Sequences

Proteomics studies extensively rely on the sequence analysis of peptides and proteins by tandem mass spectrometry [79–81]. This method could similarly assist in the determination of copolymer sequences, at least for simpler systems containing a limited number of sequence variations [8]. Again, the availability of reference spectra, namely knowledge of the corresponding homopolymer characteristics, is critical for a meaningful analysis of the copolymer fragment distributions. This will be illustrated for a copolymer of styrene and either *meta*- or *para*-dimethylsilylstyrene [82]. Li^+ was used for cationization, because Ag^+, which is commonly utilized with polystyrenes, oxidizes the silane group [83]. The sequences of chains with only one dimethylsilylstyrene unit were examined. The copolymers were produced by mixing the comonomers; hence, the question raised is, where is the substituted styrene incorporated?

The MS^2 spectrum of the homopolymer (Figure 3.13a) shows a series of a_n and y_n fragments, which result from radical-induced decompositions (vide supra) and carry

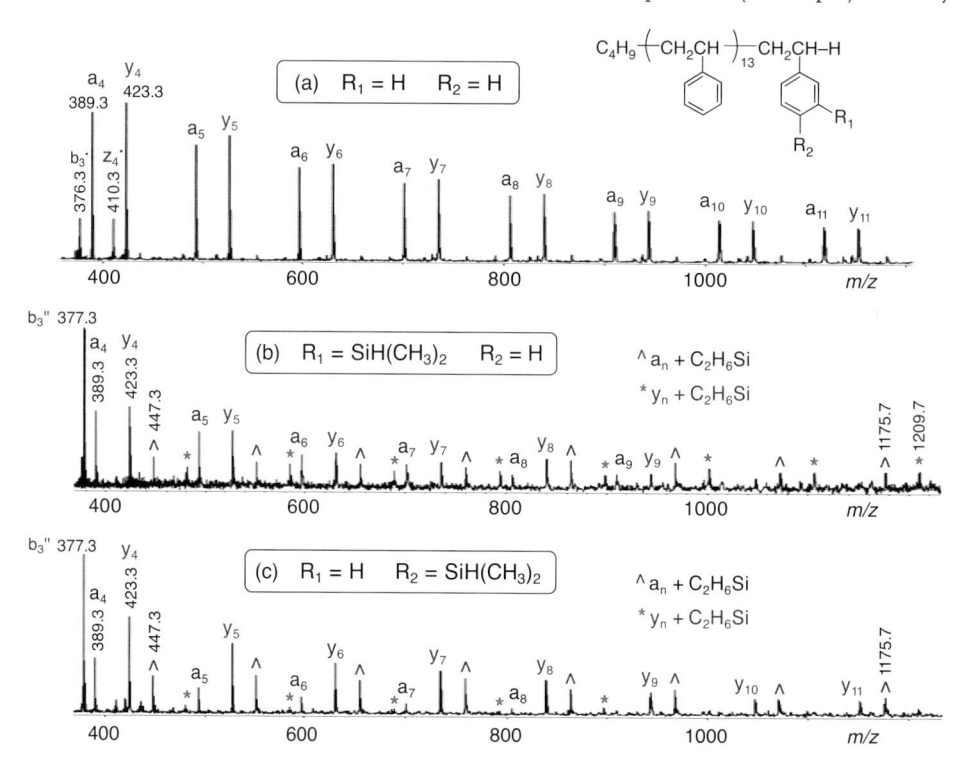

Figure 3.13 MALDI-ToF/ToF MS^2 spectra of the $[M + Li]^+$ ions from poly(styrene-*co*-dimethylsilylstyrene) prepared by living anionic polymerization using mixtures of styrene and either *meta*- or *para*-dimethylsilylstyrene in the molar ratio 8:1; (a) homopolymeric 14-mer (*m/z* 1521.0); (b) copolymeric 14-mer with one *meta*-dimethylsilyl-substituted unit (*m/z* 1579.0); (c) copolymeric 14-mer with one *para*-dimethylsilyl-substituted unit (*m/z* 1579.0). Courtesy of Aleer Yol [84]. See Ref. [55] for a detailed explanation of fragmentation mechanisms and nomenclature.

one original end group (C_4H_9 or H, respectively) as well as a methylene end group produced during fragmentation (cf. top spectrum of Figure 3.12) [44, 55]. Copolymeric oligomers containing the same number of styrene units (14) plus one dimethylsilyl substituent, either in *meta* or in *para* position, show the same series of fragments, some with and some without the dimethylsilyl group. For the oligomer containing the *meta*-substituted styrene (Figure 3.13b), the a_n/y_n fragments increasingly incorporate the comonomer as their mass increases; the higher the mass, the more likely that the substituted comonomer is present in both a_n and y_n. In sharp contrast, for the oligomer containing the *para*-substituted styrene (Figure 3.13c), only copolymeric a_n fragments are observed; the y_n fragments ($n = 4$–11) remain essentially homopolymeric [84]. These results are reconciled if the *meta*-substituted styrene is incorporated randomly in the polymer chain, but the *para*-substituted styrene preferably near the initiating C_4H_9 group and, consequently, is missing from the majority of y_n fragments. The system discussed underscores that copolymer sequence analysis by tandem mass spectrometry presupposes that the fragmentation characteristics of the corresponding homopolymer(s) are well understood. With this information at hand, one can not only distinguish isomeric sequences, but also determine specific sequence motifs.

3.4.4
Assessment of Intrinsic Stabilities and Binding Energies

Using QIT or Q/ToF mass spectrometers, MS^2 (CAD) spectra can be acquired as a function of the center-of-mass collision energy (E_{com}). Plotting relative fragment intensities versus E_{com} leads to break-down graphs [85], which reveal insight about the energy requirements of the various fragmentation pathways and, hence, also about the relative bond strengths in the polymer molecule. This strategy could be used to elucidate the intrinsic stabilities of polymers with different functionalities or different backbones.

An alternative means to probe the stability of a polymer system is to monitor how the intensity of the corresponding precursor ion changes with collision energy. A less stable precursor is depleted more readily and vice versa. Such experiments are most tractable with supramolecular polymers and noncovalent polymer–biomolecule conjugates, which tend to fragment more easily than covalently bonded materials [86, 87]. For a quantitative evaluation, the relative intensity of the precursor ion is plotted versus E_{com} to obtain a fragmentation efficiency curve [88]. The E_{com} value causing 50% dissociation (called E_{50}) can be used as a measure of the stability of the polymer system being analyzed [89]. As an example, Figure 3.14 shows the fragmentation efficiency curves for the complexes between poly(ethylene imine) 400 (PEI) and three different oligodeoxynucleotides (ODNs), constructed from the MS^2 spectra of the corresponding doubly protonated PEI/ODN assemblies [90]. The ODNs included in the plot are d(AAAAA), d(TTTTT), and d(GCGAT), where d symbolizes "deoxy" and A, C, G, and T are nucleotides bearing adenine, cytosine, guanine, and thymine, respectively. The E_{50} values measured reveal the stability order PEI-d(TTTTT) > PEI-d(GCGAT) > PEI-d(AAAAA) (cf. Figure 3.14). PEI has been explored as a delivery vehicle for oligonucleotides or genes to cells [91]; the information provided by the E_{50} data can help to understand the delivery profiles of different ODNs.

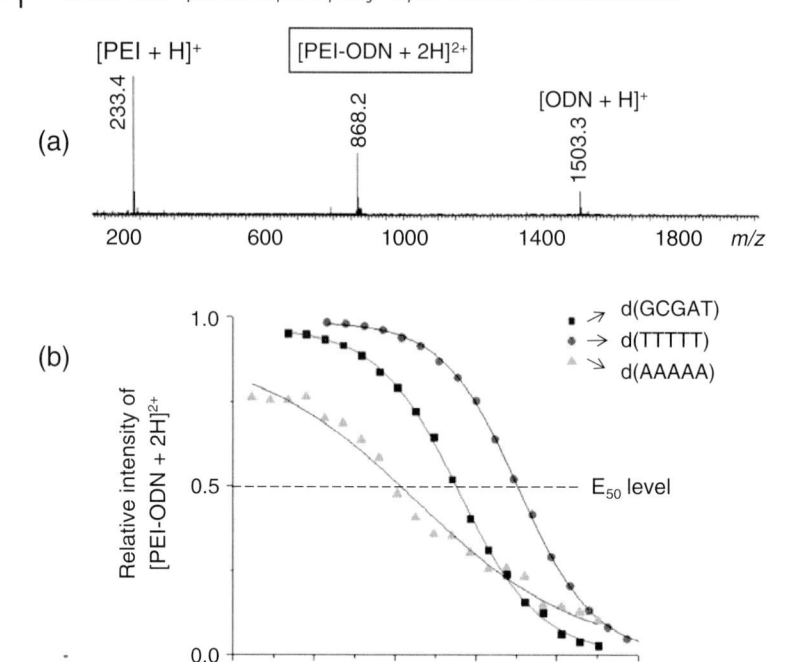

Figure 3.14 (a) MS^2 (CAD) spectrum of the [M + 2H]$^{2+}$ ions generated by ESI from the noncovalent complex of poly(ethylene imine) 400 and the oligodeoxynucleotide d(GCGAT), acquired with a Q/ToF mass spectrometer at a laboratory-frame collision energy of 14 eV using Ar as collision gas (E_{com} = 0.32 eV). The relative precursor ion intensity was calculated from the intensities of [PEI-ODN + 2H]$^{2+}$ (I_1), [PEI + H]$^+$ (I_2), and [ODN + H]$^+$ (I_3) via: relative intensity = $I_1/[I_1 + {}^1/_2(I_2 + I_3)]$. (b) Fragmentation efficiency curves obtained by carrying out analogous MS^2 experiments at different collision energies and for different PEI–ODN complexes. The E_{50} values extracted from the curves are 0.23, 0.29, and 0.35 eV for the complexes of PEI with d(AAAAA), d(GCGAT), and d(TTTTT), respectively. Adapted from [90] with permission.

Similar experiments could benefit many more drug delivery studies using synthetic polymers as the transport medium.

3.5
Summary and Outlook

Tandem mass spectrometry studies of synthetic polymers can be conducted on a variety of mass spectrometers and provide a wealth of information about the macromolecular structures examined, in particular about their end groups, architectures, and sequences. MS^2 (and MS^n) experiments probe the degradation behavior of ionized polymers. Most synthetic polymers ionize most efficiently by metal cationization. Since metal-cationized macromolecules preferentially decompose via charge-remote pathways [55], the fragments observed in tandem mass spectra also unveil insight about the degradation properties of the neutral polymer. Indeed, the MS^2 spectra of silverated polystyrenes were shown to contain abundant low-mass products that have also been detected in pyrolysis studies [44, 47]. Hence, MS^2 experiments can be useful beyond structural characterization, to determine the degradability of synthetic polymers. Dissociations in the mass spectrometer are purely unimolecular, making it easier to elucidate the corresponding degradation mechanisms, which can be used as the basis for understanding the more complicated pathways occurring at ambient conditions, where bimolecular and unimolecular processes compete. Further, evaluation of the dissociation energetics of polymer-biomolecule complexes through MS^2 adds insight that can be beneficial for the planned *in vivo* use of the polymer, as mentioned in Section 3.4.4.

A significant challenge in current MS^2 studies is that they are limited to $m/z < 3000$ because larger ions cannot be effectively activated to dissociate within their residence time in the mass spectrometer (generally ms or less). Unfortunately, a large number of polymers designed for commercial applications have molecular weights much larger than 3000 Da. For such polymers, appropriate ESI conditions must be developed, so that they can be multiply charged to yield ions with $m/z < 3000$ [92]. An alternative and promising approach for ionizing large polymers could be the recently discovered laser spray ionization method [93], which resembles MALDI, can be performed solvent-free, and produces multiply charged ions, all characteristics that might extend the MS^2 applicability to much larger polymers; moreover, multiply charged ions can also be examined by ETD (namely ion–ion reactions).

The CAD fragmentation characteristics of many different classes of polymers have been determined and provide the template on how to interpret tandem mass spectra to deduce the correct structure with confidence [44–49, 54, 55]. Incorporation of these interpretation rules into algorithms for MS^2 spectral interpretation is in its infancy [94], but continued development of such programs would further facilitate and advance tandem mass spectrometry applications to synthetic polymers.

Acknowledgments

The authors are grateful to the National Science Foundation for generous financial support (CHE-0517909, DMR-0821313, CHE-1012636). We thank Dr. N. Solak, A. Yol, and D. Smiljanic for experimental assistance and helpful discussions.

References

1 Fenn, J.B., Mann, N., Meng, C.K., Wong, S.F., and Whitehouse, C.M. (1989) Electrospray ionization for mass spectrometry of large biomolecules. *Science*, **246**, 64–71.

2 Karas, M., Bachmann, U., Bahr, U., and Hillenkamp, F. (1987) Matrix-assisted ultraviolet desorption of non-volatile compounds. *Int. J. Mass Spectrom. Ion Processe*, **78**, 53–68.

3 Karas, M. and Hillenkamp, F. (1988) Laser desorption ionization of proteins with molecular masses exceeding 10,000 daltons. *Anal. Chem.*, **60**, 2299–2301.

4 Tanaka, K., Waki, H., Ido, S., Akita, Y., and Yoshida, Y. (1988) Protein and polymer analyses up to m/z 100,000 by laser ionization time-of-flight mass spectrometry. *Rapid Commun. Mass Spectrom.*, **2**, 151–153.

5 Montaudo, G. and Lattimer, R.P. (eds) (2002) *Mass Spectrometry of Polymers*, CRC Press, Boca Raton, FL.

6 Pasch, H. and Schrepp, W. (2003) *MALDI-TOF Mass Spectrometry of Synthetic Polymers*, Springer, Berlin.

7 Wesdemiotis, C., Arnould, M.A., Lee, Y., and Quirk, R.P. (2000) MALDI TOF mass spectrometry of the products from novel anionic polymerizations. *Polym. Prep.*, **41** (1), 629–630.

8 Wesdemiotis, C., Pingitore, F., Polce, M.J., Russel, V.M., Kim, Y., Kausch, C.M., Connors, T.H., Medsker, R.E., and Thomas, R.R. (2006) Characterization of a poly(fluorooxetane) and poly(fluorooxeetane-co-THF) by MALDI mass spectrometry, size exclusion chromatography, and NMR spectroscopy. *Macromolecules*, **45**, 305–312.

9 Polce, M.J. and Wesdemiotis, C. (2002) Introduction to mass spectrometry of polymers, in *Polymer Mass Spectrometry* (eds G. Montaudo and R.P. Lattimer), CRC Press, Boca Raton, FL, pp. 1–29.

10 Peacock, P.M and McEwen, C.N. (2004) Mass spectrometry of synthetic polymers. *Anal. Chem.*, **76**, 3417–3428.

11 Peacock, P.M. and McEwen, C.N. (2006) Mass spectrometry of synthetic polymers. *Anal. Chem.*, **78**, 3957–3964.

12 Weidner, S.M. and Trimpin, S. (2008) Mass spectrometry of synthetic polymers. *Anal. Chem.*, **80**, 4349–4361.

13 Weidner, S.M. and Trimpin, S. (2010) Mass spectrometry of synthetic polymers. *Anal. Chem.*, **82**, 4811–4829.

14 Gruendling, T., Weidner, S., Falkenhagen, J., and Barner-Kowollik, C. (2010) Mass spectrometry in polymer chemistry: a state-of-the-art up-date. *Polym. Chem.*, **1**, 599–617.

15 Sleno, L. and Volmer, D.A. (2004) Ion activation methods for tandem mass spectrometry. *J. Mass Spectrom.*, **39**, 1091–1112.

16 Håkansson, K. and Klassen, J.S. (2010) Ion activation methods for tandem mass spectrometry, in *Electrospray and MALDI Mass Spectrometry: Fundamentals, Instrumentation, Practicalities, and Biological Applications*, 2nd edn (ed. R.B. Cole), John Wiley & Sons, Inc., New York, pp. 571–630.

17 Cooks, R.G., Ast, T., and Mabud, M.A. (1990) Collision of polyatomic ions with surfaces. *J. Mass Spectrom. Ion Proc.*, **100**, 209–265.

18 Laskin, J., Denisov, E.V., Shukla, A.K., Barlow, S.E., and Futrell, J.H. (2002) Surface-induced dissociation in a Fourier transform ion cyclotron resonance mass spectrometer: instrument design and evaluation. *Anal. Chem.*, **74**, 3255–3261.

19 Laskin, J. and Futrell, J.H. (2005) Activation of large ions in FT-ICR mass spectrometry. *Mass Spectrom. Rev.*, **24**, 135–167.

20 Zubarev, R.A., Kelleher, N.K., and McLafferty, F.W. (1998) Electron capture dissociation of multiply charged proteins cations: a non-ergodic process. *J. Am. Chem. Soc.*, **120**, 3265–3266.

21 Zubarev, R.A., Kruger, N.A., Fridriksson, E.K., Lewis, M.A., Horn, D.M., Carpenter, B.K., and McLafferty, F. (1999) Electron capture dissociation of gaseous multiply-charged proteins is favored at disulfide bonds and other sites of high hydrogen atom affinity. *J. Am. Chem. Soc.*, **121**, 2857–2862.

22 Koster, S., Duursma, M.C., Boon, J.J., Heeren, R.M.A., Ingemann, S., van Benthem, R.A.T.M., and de Koster, C.G. (2003) Electron capture and collisionally activated dissociation mass spectrometry of doubly charged hyperbranched polyesteramides. *J. Am. Soc. Mass Spectrom.*, **14**, 332–341.

23 Cerda, B.A., Breuker, K., Horn, D.M., and McLafferty, F.W. (2001) Charge/radical site initiation versus Coulombic repulsion for cleavage of multiply charged ion: charge salvation in poly(alkene glycol) ions. *J. Am. Soc. Mass Spectrom.*, **12**, 565–570.

24 Lee, S., Han, S.Y., Lee, T.G., Chung, G., Lee, D., and Oh, H.B. (2006) Observation

of pronounced b•, y cleavages in the electron capture dissociation mass spectrometry of polyamidoamine (PAMAM) dendrimer ions with amide functionalities. *J. Am. Soc. Mass Spectrom.*, **17**, 536–543.
25. Cerda, B.A., Horn, D.M., Breuker, K., and McLafferty, F.W. (2002) Sequencing of specific copolymer oligomers by electron-capture dissociation mass spectrometry. *J. Am. Soc. Mass Spectrom.*, **124**, 9287–9291.
26. Pitteri, S.J. and McLuckey, S.A. (2005) Recent developments in the ion/ion chemistry of high-mass multiply charged ions. *Mass Spectrom. Rev.*, **24**, 931–958.
27. Mikesh, L.M., Ueberheide, B., Chi, A., Coon, J.J., Syka, J.E.P., Shabanowitz, J., and Hunt, D.F. (2006) The utility of ETD mass spectrometry in proteomic analysis. *Biochim. Biophys. Acta*, **1764**, 1811–1822.
28. McLuckey, S.A. and Huang, T.-Y. (2009) Ion/ion reactions: new chemistry for analytical MS. *Anal. Chem.*, **81**, 8669–8676.
29. Sobott, F., Watt, S.J., Smith, J., Edelmann, M.J., Kramer, H.B., and Kleeler, B.M. (2009) Comparison of CID versus ETD based MS/MS fragmentation for the analysis of protein ubiquitination. *J. Am. Soc. Mass Spectrom.*, **20**, 1652–1659.
30. Jackson, T., Hilton, G.R., Slade, S.E., and Scrivens, J.H. (2009) Comparison of electron transfer dissociation and collision-induced dissociation fragmentation of multiply charged polyethers. Proceeding of the 57th ASMS Conference on Mass Spectrometry and Allied Topics, May 31–June 4, Philadelphia, PA.
31. Scionti, V. and Wesdemiotis, C. (2010) CAD vs. ETD fragmentation pathways of polyesters. Proceeding of the 58th ASMS Conference on Mass Spectrometry and Allied Topics, May 23–27, Salt Lake City, UT.
32. Spengler, B. (1997) Post-source decay analysis in matrix-assisted laser desorption/ionization mass spectrometry of biomolecules. *J. Mass Spectrom.*, **32**, 1019–1036.
33. De Hoffmann, E. and Stroobant, V. (2001) *Mass Spectrometry. Principles and Applications*, 2nd edn, John Wiley & Sons, Chichester, UK.
34. Jackson, A.T., Yates, H.T., Scrivens, J.H., Critchley, G., Brown, J., Green, M.R., and Bateman, R.H. (1996) The application of matrix-assisted laser desorption/ionization combined with collision induced dissociation to the analysis of synthetic polymers. *Rapid Comm. Mass Spectrom.*, **10**, 1668–1674.
35. Garozzo, D., Nasello, V., Spina, E., and Sturiale, L. (1997) Discrimination of isomeric oligosaccharides and sequencing of unknowns by post-source decay matrix-assisted laser desorption/ionization time-of-flight mass spectrometry. *Rapid Comm. Mass Spectrom.*, **11**, 1561–1566.
36. Kaufmann, R., Chaurand, P., Kirsch, D., and Spengler, B. (1996) Post-source decay and delayed extraction in matrix-assisted laser desorption/ionization-reflectron time-of-flight mass spectrometry. Are there any trade-offs? *Rapid Comm. Mass Spectrom.*, **10**, 1199–1208.
37. Chaurand, P. and Luetzenkirchen, F. (1999) Peptide and protein identification by matrix-assisted laser desorption ionization (MALDI) and MALDI post-source decay time-of-flight mass spectrometry. *J. Am. Soc. Mass Spectrom.*, **10**, 91–103.
38. Przybilla, L., Räder, H.-J., and Müllen, K. (1999) Post-source decay fragment ion analysis of polycarbonates by matrix-assisted laser desorption/ionization time-of-flight mass spectrometry. *Eur. J. Mass Spectrom.*, **5**, 133–143.
39. Goldschimdt, R.J., Wetzel, S.J., Blair, W.R., and Guttman, C.M. (2000) Post-source decay in the analysis of polystyrene by matrix-assisted laser desorption/ionization time-of-flight mass spectrometry. *J. Am. Soc. Mass Spectrom.*, **11**, 1095–1106.
40. Laine, O., Trimpin, S., Räder, H.J., and Müllen, K. (2003) Changes in post-source decay fragmentation behavior of poly (methyl methacrylate) polymers with increasing molecular weight studied by matrix-assisted laser desorption/ionization time-of-flight mass spectrometry. *Eur. J. Mass Spectrom.*, **9**, 195–201.

41 Adhiya, A. and Wesdemiotis, C. (2002) Poly(propylene imine) dendrimer conformations in the gas phase: A tandem mass spectrometry study. *Int. J. Mass Spectrom.*, **214**, 75–88.

42 Rizzarelli, P., Puglisi, C., and Montaudo, G. (2006) Matrix-assisted laser desorption/ ionization time-of-flight/time-of-flight tandem mass spectra of poly(butylenes adipate). *Rapid Commun. Mass Spectrom.*, **20**, 1683–1694.

43 Rizzarelli, P. and Puglisi, C. (2008) Structural characterization of synthetic poly(ester amide) from sebacic acid and 4-amino-1-butanol by matrix-assisted laser desorption ionization time-of-flight/time-of-flight tandem mass spectrometry. *Rapid Commun. Mass Spectrom.*, **22**, 739–754.

44 Polce, M.J., Ocampo, M., Quirk, R.P., and Wesdemiotis, C. (2008) Tandem mass spectrometry characteristics of silver-cationized polystyrenes: backbone degradation via free radical chemistry. *Anal. Chem.*, **80**, 347–354.

45 Polce, M.J., Ocampo, M., Quirk, R.P., Leigh, A.M., and Wesdemiotis, C. (2008) Tandem mass spectrometry characteristics of silver-cationized polystyrenes: internal energy, size, and chain end vs. backbone substituent effects. *Anal. Chem.*, **80**, 355–362.

46 Chaicharoen, K., Polce, M.J., Singh, A., and Pugh, C. (2008) Characterization of linear and branched polyacrylates by tandem mass spectrometry. *Anal. Bioanal. Chem.*, **392**, 595–607.

47 Gies, A.P., Vergne, M.J., Orndorff, R.L., and Hercules, D.M. (2008) MALDI-TOF/ TOF CID study of 4-alkyl-substituted polystyrene fragmentation reactions. *Anal. Bioanal. Chem.*, **392**, 609–626.

48 Ellison, S.T., Gies, A.P., Hercules, D.M., and Morgan, S.L. (2009) Py-GC/MS and MALDI-TOF/TOF CID study of polysulfone fragmentation reactions. *Macromolecules*, **42**, 3005–3013.

49 Dabney, D.E. (2009) Analysis of synthetic polymers by mass spectrometry and tandem mass spectrometry. PhD Dissertation. Akron, OH: The University of Akron.

50 Gies, A.P., Geibel, J.F., and Hercules, D.M. (2010) MALDI-TOF/TOF CID study of poly(*p*-phenylene sulfide) fragmentation reactions. *Macromolecules*, **43**, 952–967.

51 Knop, K., Jahn, B.O., Hager, M.D., Crecelius, A., Gottschaldt, M., and Schubert, U.S. (2010) Systematic MALDI-TOF CID investigation on different substituted mPEG 2000. *Macromol. Chem. Phys.*, **211**, 677–684.

52 Understanding ion trap mass spectrometry, Bruker Esquire/HCT Series User Manual, version 1.3, p. 1–3.

53 March, R.E. (1997) An introduction to quadrupole ion trap mass spectrometry. *J. Mass Spectrom.*, **32**, 351–369.

54 Solak, N. (2010) Structural characterization and quantitative analysis by interfacing liquid chromatography and/or ion mobility separation with multidimensional mass spectrometry. PhD dissertation, The University of Akron: Akron, OH.

55 Wesdemiotis, C., Solak, N., Polce, M.J., Dabney, D.E., Chaicharoen, K., and Katzenmeyer, B.C., Fragmentation pathways of polymer ions. *Mass Spectrom. Rev.*, **2011**, *30*, 523–559.

56 Hartmer, R., Kaplan, D.A., Gebhardt, C.R., Ledertheil, T., and Brekenfeld, A. (2008) Multiple ion/ion reactions in the 3D ion trap: selective reagent anion production for ETD and PTR from a single compound. *Int. J. Mass Spectrom.*, **276**, 82–90.

57 ETD and PTR with the HCT ultra (July 2008) Bruker Compass Application Tutorial, version 1.3, p. 9.

58 Huang, T.-J., Emory, J.F., O'Hair, R.A.J., and McLuckey, S.A. (2006) Electron-transfer reagent anion formation via electrospray ionization and collision-induced dissociation. *Anal. Chem.*, **78**, 7387–7391.

59 McEwen, C.N., McKay, R.G., and Larsen, B.S. (2005) Analysis of solids, liquids, and biological tissues using solids probe introduction at atmospheric pressure on commercial LC/MS instruments. *Anal. Chem.*, **77**, 7826–7831.

60 Takáts, Z., Wiseman, J.M., and Cooks, R.G. (2005) Ambient mass spectrometry using desorption electrospray ionization (DESI): instrumentation, mechanisms

and applications in forensics, chemistry, and biology. *J. Mass Spectrom.*, **40**, 1261–1275.
61 Q-ToF Ultima MALDI User's Guide. Atlas Park, Manchester: Micromass UK Limited M22 5PP. Available at http://www.waters.com.
62 Scionti, V., Tun, Z., Tessier, C.A., Youngs, W.J., and Wesdemiotis, C. (2009) Characterization of poly (organophosphazene)s by mass spectrometry techniques. Proceeding of the 57th ASMS Conference on Mass Spectrometry and Allied Topics, May 31–June 4, Philadelphia, PA.
63 Giles, K., Pringle, S.D., Worthington, K.R., Little, D., Wildgoose, J.L., and Bateman, R.H. (2004) Applications of a traveling wave-based radio-frequency only stacked ring ion guide. *Rapid Commun. Mass Spectrom.*, **18**, 2401–2414.
64 Pringle, S.D., Giles, K., Wildgoose, J.L., Williams, J.P., Slade, S.E., Thalassinos, K., Bateman, R.H., Bowers, M.T., and Scrivens, J.H. (2007) An investigation of the mobility separation of some peptide and protein ions using a new hybrid quadrupole/travelling wave IMS/oa-ToF instrument. *Int. J. Mass Spectrom.*, **261**, 1–12.
65 Waters Synapt High Definition Mass Spectrometry System , Operator's Guide, 71500129902/Revision A, Waters Corporation, 2007.
66 Kanu, A.B., Dwivedi, P., Tam, M., Matz, L., and Hill, H.H. (2008) Ion mobility-mass spectrometry. *J. Mass Spectrom.*, **43**, 1–22.
67 Hilton, G.R., Jackson, A.T., Thalassinos, K., and Scrivens, J.H. (2008) Structural analysis of synthetic polymer mixtures using ion mobility and tandem mass spectrometry. *Anal. Chem.*, **80**, 9720–9725.
68 Cheng, C. and Gross, M.L. (2000) Applications and mechanisms of charge-remote fragmentation. *Mass Spectrom. Rev.*, **19**, 398–420.
69 Medzihradszky, K.F., Campbell, J.M., Baldwin, M.A., Falick, A.M., Juhasz, P., Vestal, M.L., and Burlingame, A.L. (2000) The characteristics of peptide collision-induced dissociation using a high-performance MALDI-TOF/TOF tandem mass spectrometer. *Anal. Chem.*, **72**, 552–558.
70 Loftus, N. (2002) Gold standard: mass spectrometry and chromatography. *Biochemist*, **24**, 25–27.
71 Suckau, D., Resemann, A., Schuerenberg, M., Hufnagel, P., Frantzen, J., and Holle, A.A. (2003) A novel MALDI LIFT-TOF/TOF mass spectrometer for proteomics. *Anal. Bioanal. Chem.*, **376**, 952–965.
72 Macht, M., Asperger, A., and Deininger, S.-O. (2004) Comparison of laser-induced dissociation and high-energy collision-induced dissociation using matrix-assisted laser desorption/ionization tandem time-of-flight (MALDI-TOF/TOF) for peptide and protein identification. *Rapid Commun. Mass Spectrom.*, **18**, 2093–2105.
73 Vestal, M.L. and Campbell, J.M. (2005) Tandem time-of-flight mass spectrometry. *Meth. Enzymol.*, **402**, 79–108.
74 Cotter, R.J., Griffith, W., and Jelinek, C. (2007) Tandem time-of-flight (TOF/TOF) mass spectrometry and the curved-field reflectron. *J. Chromatogr. B*, **855**, 2–13.
75 Dabney, D.E. (2009) Analysis of synthetic polymers by mass spectrometry and tandem mass spectrometry, PhD dissertation, The University of Akron: Akron, OH.
76 Jackson, A.T., Green, M.R., and Bateman, R.H. (2008) Generation of end-group information from polyethers by matrix-assisted laser desorption/ionization collision-induced dissociation mass spectrometry. *Rapid Commun. Mass Spectrom.*, **20**, 3542–3550.
77 Jackson, A.T. and Robertson, D.F. (2008) Chain end characterization. *Compr. Anal. Chem.*, **53**, 171–203.
78 Wollyung, K.M., Wesdemiotis, C., Nagy, A., and Kennedy, J.P. (2004) Synthesis and mass spectrometry characterization of centrally and terminally amine-functionalized polyisobutylenes. *J. Polym. Sci. A: Polym. Chem.*, **43** 946–958.
79 Noga, M., Dylag, T., and Silberring, J. (2009) *Sequencing of peptides and proteins, Mass Spectrometry* (ed. R. Ekman), John Wiley & Sons, Hoboken, NJ, pp. 179–210.
80 Seidler, J., Zinn, N., Boehm, M.E., and Lehmann, W.D. (2010) De novo

sequencing of peptides by MS/MS. *Proteomics*, **10**, 634–649.

81 Huang, T.-Y. and McLuckey, S.A. (2010) Top-down protein characterization facilitated by ion/ion reactions on a quadrupole/time of flight platform. *Proteomics*, **10**, 3577–3588.

82 Janoski, J. (2010) Anionic synthesis of functionalized polymers, PhD dissertation, The University of Akron: Akron, OH.

83 Quirk, R.P., Kim, H., Polce, M.J., and Wesdemiotis, C. (2005) Anionic synthesis of primary amine functionalized polystyrenes via hydrosilation of allylamine with silyl hydride functionalized polystyrenes. *Macromolecules*, **38** 7895–7906.

84 Yol, A., Janoski, J., Quirk, R.P., and Wesdemiotis, C. (2010) Characterization of copolymers by mass and tandem mass spectrometry. Proceeding of the 58th ASMS Conference on Mass Spectrometry and Allied Topics, May 23–27, Salt Lake City, UT.

85 Butcher, C.P.G., Dyson, P.J., Johnson, B.F.G., Langridge-Smith, P.R.R., McIndoe, J.S., and Whyte, C. (2002) On the use of breakdown graphs combined with energy-dependent mass spectrometry to provide a complete picture of fragmentation processes. *Rapid Commun. Mass Spectrom.*, **16**, 1595–1598.

86 Ren, X., Sun, B., Tsai, C.-C., Tu, Y., Leng, S., Li, K., Kang, Z., Van Horn, R.M., Li, X., Zhu, M., Wesdemiotis, C., Zhang, W.-B., and Cheng, S.Z.D. (2010) Synthesis, self-assembly, and crystal structure of a shape-persistent polyhedral-oligosilsesquioxane-nanoparticle-tethered perylene diimide. *J. Phys. Chem. B*, **114**, 4802–4810.

87 Li, X., Chan, Y.-T., Newkome, G.R., and Wesdemiotis, C. (2011) Gradient tandem mass spectrometry interfaced with ion mobility separation for the characterization of supramolecular architectures. *Anal. Chem.*, **83**, 1284–1290.

88 Dongre, A.R., Jones, J.L., Somogyi, A., and Wysocki, V.H. (1996) Influence of peptide composition, gas-phase basicity, and chemical modification on fragmentation efficiency: evidence for the mobile proton model. *J. Am. Chem. Soc.*, **118**, 8365–8374.

89 Rosu, F., Pirotte, S., De Pauw, E., and Gabelica, V. (2006) Positive and negative ion mode ESI-MS and MS/MS for studying drug–DNA complexes. *Int. J. Mass Spectrom.*, **253**, 156–171.

90 Smiljanic, D. and Wesdemiotis, C. (2010) Non-covalent complexes between single-stranded oligodeoxynucleotides and poly(ethylene imine). *Int. J. Mass Spectrom.*, **2011**, *304*, 148–153.

91 Kan, P.L., Schätzlein, A.G., and Uchegbu, L.F. (2006) Polymers used for the delivery of genes in gene therapy, in *Polymers in Drug Delivery* (eds I. Uchegbu and A.G. Schätzlein), CRC Press, Boca Raton, FL, pp. 183–198.

92 Wang, J.-L., Li, X., Lu, X., Chan, Y.-T., Moorefield, C.N., Wesdemiotis, C., and Newkome, G.R. (2011) Dendron functionalized bis(terpyridine)-iron(II) or -cadmium(II) metallomacrocycles: synthesis, traveling-wave ion-mobility mass spectrometry, and photophysical properties. *Chem. Eur. J.*, **17**, 4830–4838.

93 McEwen, C.N. and Trimpin, S. (2011) An alternative ionization paradigm for atmospheric pressure mass spectrometry: flying elephants from Trojan horses. *Int. J. Mass Spectrom.*, **300**, 167–172.

94 Thalassinos, K., Jackson, A.T., Williams, J.P., Hilton, G.R., Slade, S.E., and Scrivens, J.H. (2007) Novel software for the assignment of peaks from tandem mass spectrometry spectra of synthetic polymers. *J. Am. Soc. Mass Spectrom.*, **18**, 1324–1331.

4
Matrix-Assisted Inlet Ionization and Solvent-Free Gas-Phase Separation Using Ion Mobility Spectrometry for Imaging and Electron Transfer Dissociation Mass Spectrometry of Polymers

Christopher B. Lietz, Alicia L. Richards, Darrell D. Marshall, Yue Ren, and Sarah Trimpin

4.1
Overview

This chapter relates to developing analytical technology to probe polymer chemistry and establish more comprehensive analysis methods useful in, for example, consumer product testing. Specifically, methods are presented that have the potential for rapid reliable analyses irrespective of solubility. Total solvent-free analysis (TSA) [1–3] is an approach for which initial results demonstrate ionization [4–16], separation [17–21], and mass analysis of even insoluble compounds. TSA consists of solvent-free ionization and solvent-free gas-phase separation. A newly developed ionization approach, matrix-assisted inlet ionization (MAII) [22, 23] using solvent-free sample preparation methods [3] is combined with solvent-free separation using ion mobility spectrometry (IMS) mass spectrometry (MS) to decouple ionization, separation, and mass analysis from the solubility of materials. For detailed theory and applications of IMS-MS including synthetic polymers, the interested reader is also referred to a recent book on *Ion Mobility Spectrometry–Mass Spectrometry: Theory and Applications* [24]. We make use of the ability to produce singly or multiply charged ions using MAII, the parent classification of laserspray ionization *inlet* (LSII) at atmospheric pressure (AP), previously called laserspray ionization (LSI), and intermediate pressure (IP) and high vacuum conditions laserspray ionization/vacuum (LSIV), to help characterization of polymeric materials. TSA hyphenated with collision-induced dissociation (CID)/electron transfer dissociation (ETD) fragmentation and IMS-MS drift time analysis [3, 25–28] has the potential to provide insight into sequence and structure (shape) of synthetic polymers directly from surfaces. For the reader's convenience, the molecular structures of polymers discussed in this chapter are shown in Scheme 4.1.

Mass Spectrometry in Polymer Chemistry, First Edition.
Edited by Christopher Barner-Kowollik, Till Gruendling, Jana Falkenhagen, and Steffen Weidner.
© 2012 Wiley-VCH Verlag GmbH & Co. KGaA. Published 2012 by Wiley-VCH Verlag GmbH & Co. KGaA.

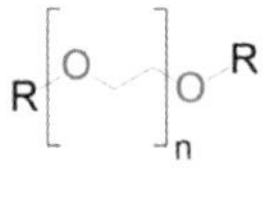

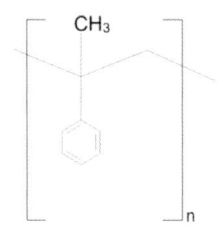

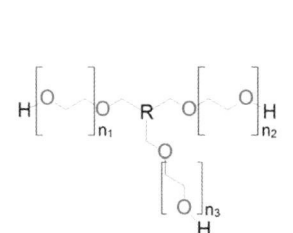

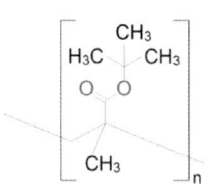

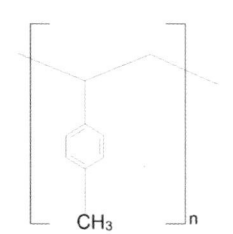

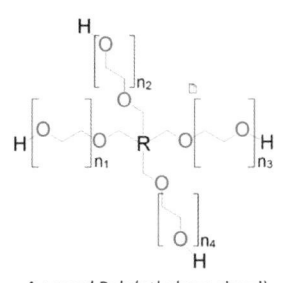

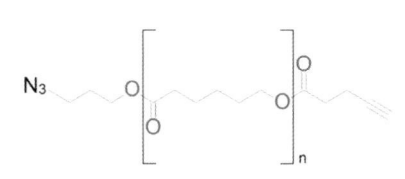

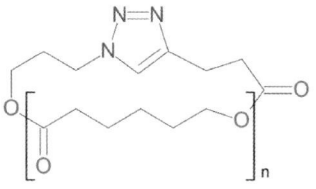

Scheme 4.1 Chemical structures of polymers.

4.2 Introduction

Ideally, to gain a deeper understanding of material composition and to relate these findings to property and function relationships, the entire set of molecules in any polymeric system should be ionized in a quantitative manner. Sample preparation and desorption-ionization in MS is vital for successful comprehensive analyses. The heart of TSA and this book chapter is IMS-MS [1]. We provide detailed descriptions of emerging developments that can improve and expand the utility of IMS-MS for polymer characterization. Solubility-restricted and labile materials, especially when in complex mixtures, are extremely difficult to analyze. No current analytical method is able to master this task. Difficult analytical problems include analysis of cellulose, hemicellulose, lignin, polyelectrolytes, and especially materials directly from surfaces, for which solvent-based separation is not applicable. Our focus is to develop MS techniques to address these daunting materials analyses issues and apply them to mass specific surface imaging with high spatial resolution, specificity, speed of analysis, as well as the potential to characterize solubility restricted and complex materials directly from their native state.

Desorption/ionization and sample preparation/separation methods are currently the most crucial needs for extending analytical capabilities in MS. Solvent-free matrix preparation methods have been extensively used for polymer, organometallic, and asphaltene analyses [29–41]. One important application of interest is analysis of synthetic polymers in electrolyte fuel cells (PEFCs) [42–44]. The poly(electrolyte) [45, 46] commercially used in PEFCs is typically Nafion® (DuPont), a highly fluorinated polymer with sulfonic acid groups having the needed proton conducting properties. Considerable industrial effort is devoted to develop improved polymeric electrolytes. Research aimed at commercially more attractive polymers [47–53] is hampered by lack of suitable analytical technology for such complex and frequently solubility restricted materials. For this kind of analyses, emphasis needs to be placed on developing a set of physical sample preparation tools that operate independent of solubility to allow for the chemical make-up of solubility-restricted materials to be unraveled, and to provide insight for understanding material properties.

Besides the interest in analyzing solubility restricted materials, for which electrospray ionization (ESI) is not suited, solvent-free sample preparation has significant advantages over solvent-based methods with regard to reproducibility of surface imaging and obtaining higher spatial resolution [1, 2, 15, 54, 55]. The solvent matrix preparation method [4–16] is now used extensively for synthetic materials characterization using matrix-assisted laser desorption/ionization (MALDI) [29–41, 56–59], and more recently was extended to other areas such as lipids analysis [60] and imaging MS [61–66]. Matrix crystal sizes, thickness of the matrix, and new means of producing a uniform surface coating of a desired thickness are currently being developed as are methods for relative quantitation using a sample-to-sample comparison along with appropriate standards similar to

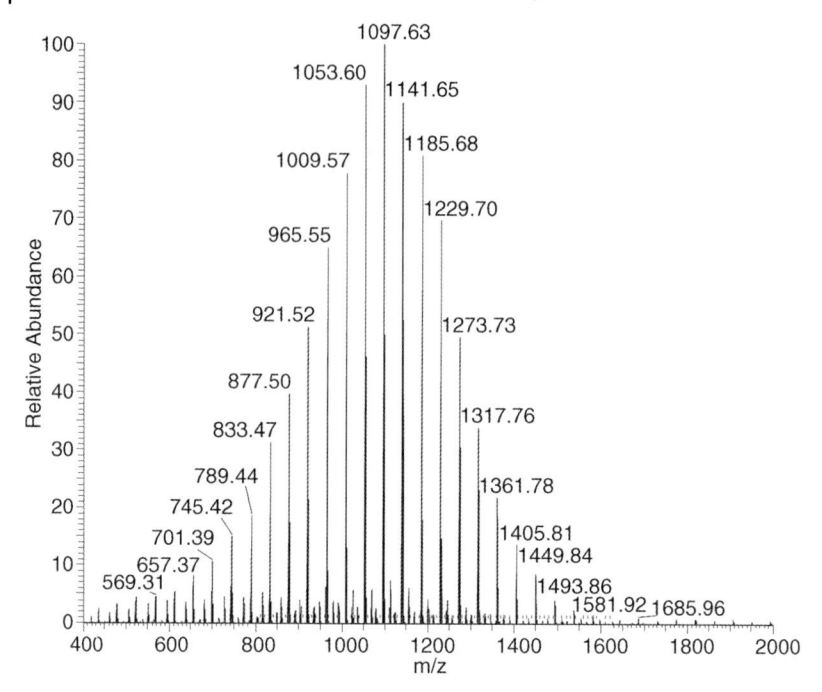

Figure 4.1 LSII-MS analysis of PEG-970 using a dithranol and NaCl matrix. (Reproduced with permission from Analytical Chemistry [67]).

studies of biological surfaces [65, 66]. Solvent-free sample preparation of polymers produces singly charged ions using LSII (Scheme 4.2 I) when matrix and analyte are combined under nonvigorous homogenization conditions using 1 mm metal beads to achieve grinding/homogenization as is shown for poly(ethylene glycol) (PEG)-970 in Figure 4.1 [67], whereas more vigorous homogenization frequencies and longer homogenization times, combined with increased temperature of the atmospheric pressure (AP) to vacuum inlet, produce highly charged ions [3].

IMS separates ions according to the number of charges and the cross section (defined by size and shape). In other words, ions with more charges and smaller cross sections will travel through the drift tube faster [68]. In the studies discussed here, the separation in the ion mobility region is achieved by using an electric field to pull or push ions against a flow of gas held at a constant pressure and temperature. A homebuilt IMS-MS instrument uses a constant field gradient and operates with a helium gas pressure of about 10 Torr [69, 70]. Advantages of this instrument include high-resolution ion mobility separations, multidimensional IMS, and the ability to calculate ion cross sections. The multidimensional character of the instrument is achieved by incorporation of up to three drift regions. Sensitivity is maintained by the use of ion funnels between each drift region. The utility of multidimensional ESI-IMS-MS for the analysis of polymers [17, 18], lipids [19], human plasma [20], and glycans [21] has been demonstrated on this homebuilt instrument. Using this methodology, isomeric polymers having the same size and number of charges [18, 24, 71, 72] have been separated in the gas-phase according to differences as little as their shape.

Two-dimensional (2D) gas-phase separation of drift time versus mass-to-charge (m/z) has been used to produce "snapshot" images of complex polymer systems [17, 18]. The visual pictorial analysis is exemplified in Figure 4.2 for PEG $\sim$ 4250 Da and PEG $\sim$ 3400 Da. The differences between the two very similar polymer samples, differing only in the number of repeating units, are obvious without detailed interpretation. For more comprehensive interpretation, one can extract a wealth of information embedded in the 2D plot, including mass spectra, and drift time distributions of the whole dataset or inset values only. Making use of these interpretation options, the differences in Figure 4.2 c1 and c2 relate to the high-mass component in PEG $\sim$ 3400 Da. Contrary to the ESI-IMS-MS results, MALDI-time-of-flight (TOF)-MS (Figure 4.2b1 and b2), the preferred ionization method for polymer chemists, did not show the differences between these two samples. In agreement with ESI-IMS-MS, size exclusion chromatography (SEC) shows the high mass component of PEG $\sim$ 3400 (Figure 4.2a), exemplifying the need of additional separation power for unequivocal characterization, as is shown here for low-abundance high-mass components present in a sample. IMS-MS relative to SEC is sensitive, fast and does not require calibration standards to obtain separation and mass values.

It is also of interest to note that the analyses of PEG samples show two unique characteristics: charge versus drift time inversions and folding transitions in the IMS dimension. Charge states traditionally fall into families with each charge state forming a straight diagonal line. This, however, is not the case with PEGs ionized with Cs^+ because the higher charge state families travel *slower* rather than faster as expected, but with increasing mass each charge state goes through a folding transition, apparent by a sigmoidal shape of charge state(s) in which the end result is an inversion to the expected drift time versus charge state behavior, common to ions from biological materials and, at this point, most synthetic polymers. This unusual observation relates to drastic changes of the cross section (defined by size and shape) caused by the reorganization of the Cs^+ cation attached to the polymer backbone [18]. Plotting the IMS dimension drift time versus the MS m/z yields a plot that is unique for PEG. Such plots of size and charge versus m/z are frequently unique for each polymer system allowing rapid visual observation of small differences in polymer systems. Further, it is exactly the region of the folding transitions, drift times 40–50 ms and m/z 750–900, where the best solvent-free gas-phase separation enabled by IMS is achieved for the various charge states. IMS-MS is able to obtain the distribution of each charge state. The snapshot approach is considerably enhanced by multiple charging.

Though the drift time resolution of the homebuilt IMS-MS instrument is exceptional, to address possible challenges the broader polymer community faces, this chapter focuses on recent advances of commercially available instrumentation, specifically the Waters Corporation SYNAPT G2. While this instrument has lower IMS resolution than the homebuilt IMS-MS instrument, it has advantages of faster and more sensitive data acquisition and higher resolution in the MS. This instrument uses

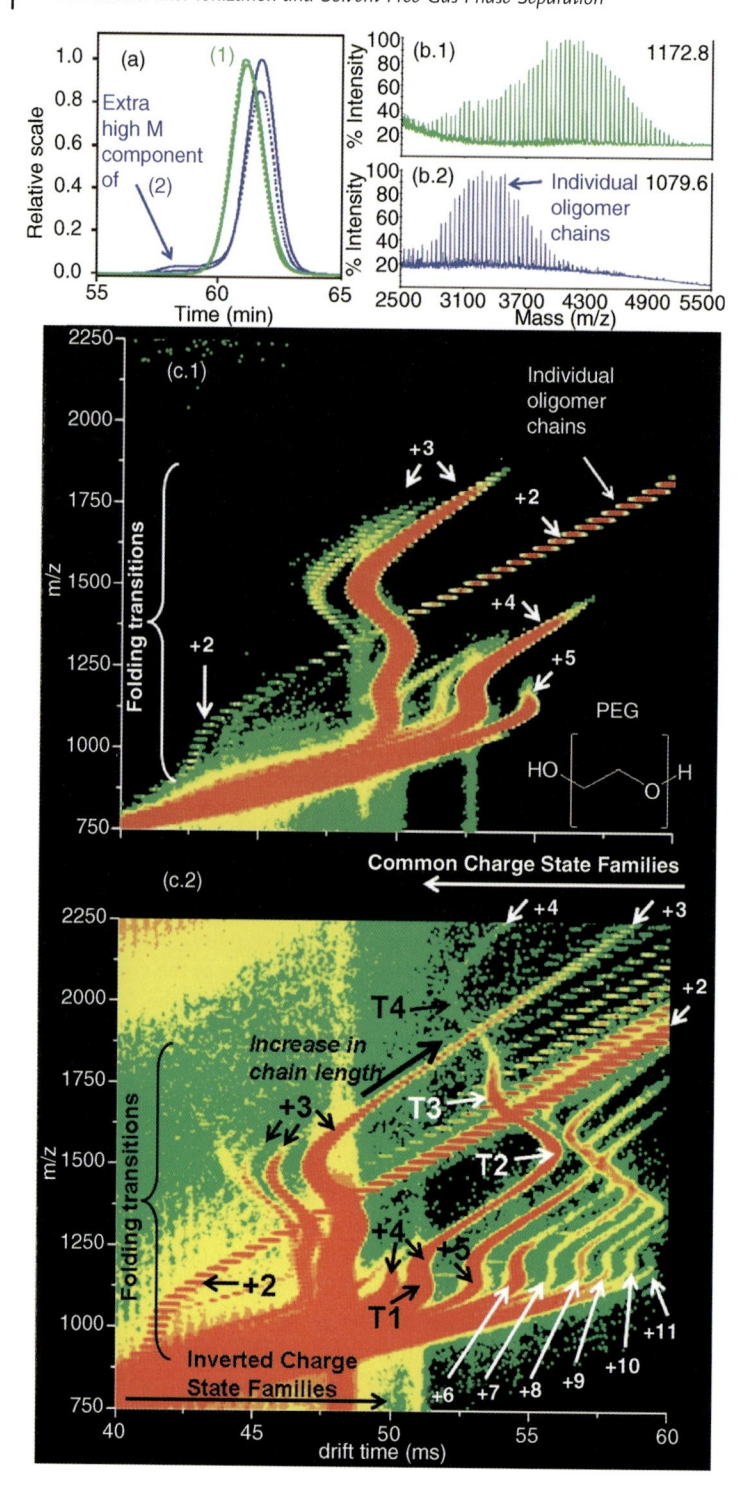

90 | Matrix-Assisted Inlet Ionization and Solvent-Free Gas-Phase Separation

traveling wave-ion mobility that uses a traveling low-voltage pulse to push ions against the countercurrent gas flow. Focusing maintains good ion transmission resulting in high sensitivity. A disadvantage besides the lower, but adequate, IMS resolution is the difficulty calculating ion cross sections relative to drift tube IMS measurements.

Figure 4.3 shows the gas-phase separation of linear poly(ε-caprolactone) (*l*-PCL) $\sim$2200 and cyclic poly(ε-caprolactone) (*c*-PCL) $\sim$2300 using ESI-IMS-MS on the SYNAPT G2 [71, 72]. The pictorial snapshot of the 2D plot rapidly discerns the two different *isomeric* architectures (right panel) whereas the mass spectra show only slight differences in the formation of charge states (left panel). *l*-PCL accommodates the attachment of up to three Li cations, whereas *c*-PCL attaches only two due to size limitations. Similar to linear PEG samples (Figures 4.2 and 4.9), *l*-PCL shows folding transitions (Figure 4.3a2) indicating significant structural rearrangements of the cations bound to the polymer backbone. Extraction of the IMS and m/z data of the +2 charge state for both isomers shows notably different drift time distributions at m/z $\sim$500 to $\sim$900 which converge to nearly identical drift time values at m/z $\sim$1100 to $\sim$1400. Interestingly, for any given m/z between $\sim$500 and $\sim$1300 notable differences in drift-time distributions for charge states +3 between *l*-PCL and *c*-PCL are observed, indicating very different architectural arrangements of these isomeric gas-phase ions. Polymer researchers have started accepting IMS-MS as an analytical tool summarized in review articles [73, 74]. Recent research articles are included for comprehensiveness [75–79].

Using homebuilt or commercial instrumentation, a complete dataset can be obtained and displayed in seconds allowing small changes to be visually obvious nearly instantaneously. However, to increase the ion abundance, IMS-MS experiments are commonly acquired in a 1 to 10 min timescale. The new inlet ionization methods, MAII (Scheme 4.2 II) and LSII, coupled with commercial IMS-MS and Thermo Fisher Scientific LTQ-ETD mass spectrometers provide similar achievements from the solid state. Results can be obtained with high spatial resolution and high throughput from single laser shots [3, 25–28, 54, 55, 67, 80].

Figure 4.2 Analyses of two different samples, (1) PEG $\sim$ 4250 (green) and (2) PEG $\sim$ 3400 Da (blue), which only vary slightly in molecular composition, utilizing a variety of polymer analyses methods. (a) SEC determines the bulk composition, for example, polydispersities of $\sim$ 1.0 were determined for both samples; however, SEC also provides evidence for the high-mass tail of sample 2; dots versus lines indicate different detectors. (b) MALDI-MS of the samples doped with Cs + provides singly charged ions of the oligomer chains of similar ion intensity permitting end-group analysis; however, MALDI-MS does not permit the detection of the high-mass tail of sample (2). (c) ESI-IMS-MS 2D plots (color coding: red most to light blue least abundant) of the samples doped with cesium chloride as charge provider (homebuilt 3 m drift tube instrument by Clemmer and coworkers). Multiple charge-state families are denoted for charge-state families +2 through +5 (sample 1) and +2 through +11 (sample 2), extracted from the mass spectra that are obtained by integration along a diagonal in the [td(m/z)] distribution. (Reproduced with permission from Analytical Chemistry [18]).

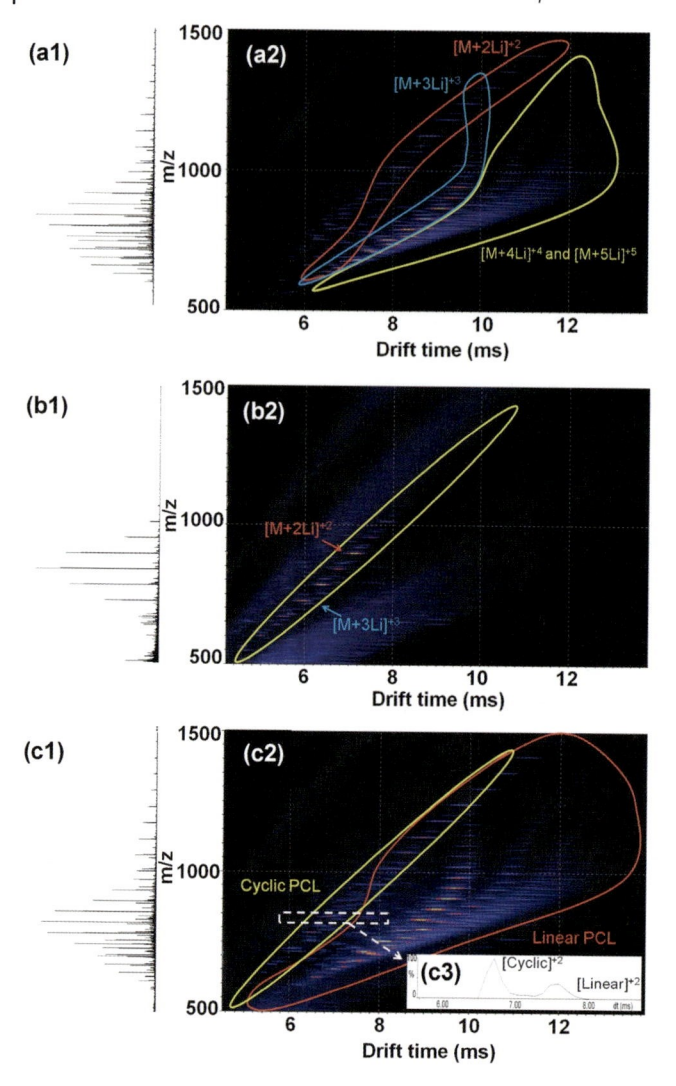

Figure 4.3 ESI-IMS-MS 2D plot of (a) *l*-PCL, (b) *c*-PCL, and (c) an intentional mixture of *l*-PCL and *c*-PCL using a commercial IMS-MS mass spectrometer (SYNAPT G2, Waters). The intensity of the ions detected is incorporated as a false color plot with red as the most abundant and blue as the least abundant ions. The same charge state families are observed for each species in the mixture (c) as shown in (a) and (b). Inset c3 shows the drift time distribution for the mass selected ion at m/z 893.13, corresponding to the doubly charged of both isomeric species. The drift time distribution is distinctly bimodal, indicating that two different isomeric architectures are present at m/z 839.13 and baseline seperated [72]. Reproduced with permission from Macromolecules [72].

I. LSII

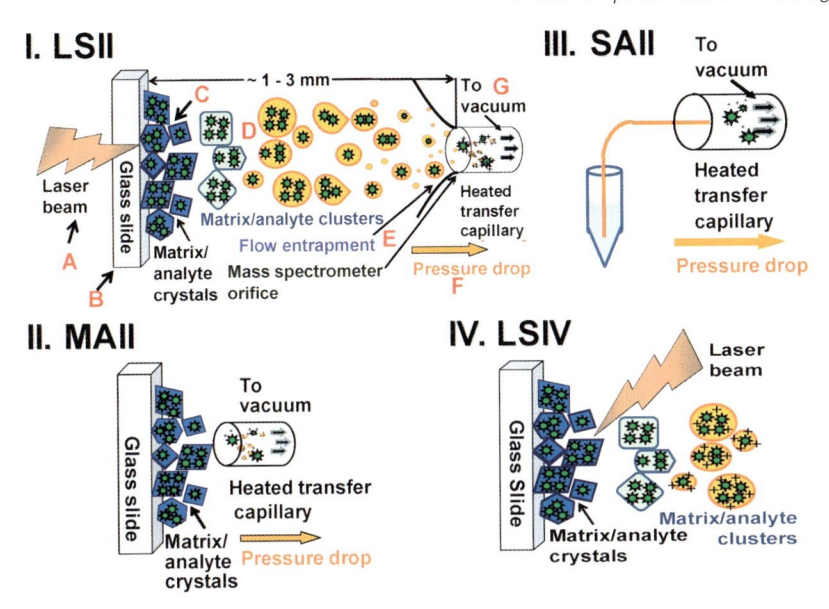

II. MAII

III. SAII

IV. LSIV

Scheme 4.2 Representation of I. Laserspray ionization *inlet* (LSII), former laserspray ionization (LSI) at atmospheric pressure (AP), and in transmission geometry (TG): (A) a laser beam fires through (B) the glass slide sample holder a (C) solid-state matrix/analyte mixture at a 180° angle and (D) ablates clusters to form molten droplets of matrix and analyte. The ablated material is (E) caught in a flow entrapment near the mass spectrometer entrance orifice and travels through a (F) pressure gradient drop from AP-to-vacuum through a (G) heated transfer capillary. Ionization occurs during the heated pressure drop. II. Matrix-assisted inlet ionization (MAII) is similar to LSII but does not use laser ablation as means to transfer sample to the inlet of the ion transfer capillary; thus, the pressure drop/heated transfer region. Matrix/analyte samples are deposited on a surface that is manually "tapped" against the inlet for subsequent ionization. III. Solvent Assisted Inlet Ionization (SAII) uses solvent to transfer sample directly into the heated inlet capillary of the mass spectrometer by a fused silica tube and produces multiply charged ions without the use of a voltage, a laser or a matrix compound. IV. Laserspray Ionization Vacuum (LSIV) is an extension of LSII that produces highly charged LSI ions under vacuum conditions using a commercial MALDI-TOF-MS instrument.

4.3
New Sample Introduction Technologies

More than 30 new direct ambient ionization mass spectrometric approaches were introduced over the past 7 years [81–84]. These include desorption electrospray ionization (DESI) [85–97], direct introduction probe chemical ionization (DAPCI) [98], direct analysis in real time (DART) [99], atmospheric solids analysis probe (ASAP) [15, 100–104], laser ablation electrospray ionization (LAESI) [105–107], electrospray-assisted laser desorption/ionization (ELDI) [108], matrix-assisted laser desorption electrospray ionization (MALDESI) [109], desorption sonic spray ionization (ESSI) [110], desorption atmospheric pressure photoionization (DAPPI) [111], and electrospray ionization (EESI) [112]. Some of these methods are now commercially available. All these methods are based on previously known ionization methods, and those capable of analyzing nonvolatile compounds use ESI. AP-MALDI is also capable of ionizing nonvolatile compounds, but its predominant production of singly charged ions limits the method to compounds that fall within the m/z range of the mass spectrometer being employed. Most mass spectrometers used for AP-MALDI have a limited m/z range.

DESI, LAESI, ELDI, and MALDESI are also capable of chemical analysis and imaging. However, only DESI has been shown capable of obtaining mass spectra of large molecules with similar sensitivity to that currently available with the new ionization methods discussed below. DESI has the advantage of requiring little or no sample preparation, and has a spatial resolution >200 μm. The other ambient techniques that use an ultraviolet (UV) laser to ablate the sample, such as MALDESI [109], are capable of high spatial resolution for imaging as well as analysis of nonvolatile compounds, but have been limited by sensitivity issues. According to literature accounts, these methods require well over an order of magnitude more sample to achieve equivalent results.

A schematic of the LSII method as currently practiced on high-performance instruments, including the Orbitrap Exactive, LTQ/ETD Velos, and SYNAPT G2, is shown in Scheme 4.2 [3, 25–28, 54, 55, 67, 80, 113–118]. This new ionization method for MS uses laser ablation to produce highly charged ions similar to ESI, but directly from the solid state. The m/z range of mass spectrometers is extended and enhanced structural information through improved fragmentation as well as IMS separation is achieved as a direct result of multiply charged ion formation. For example, using ETD combined with LSII, almost complete sequence coverage of ubiquitin (~8.5 kDa), a small protein, was obtained [26]. LSII-ETD has more recently been used to identify myelin basic protein fragment directly from a surface (tissue section) [25]. LSII-MS achieves attomole sensitivity for peptides and low femtomole sensitivity for small proteins making the high spatial resolution provided by the laser practical [26, 55]. Figure 4.4 is the LSII mass spectrum of PEG-6690 obtained on a high-resolution Orbitrap mass spectrometer using m/z of 2000 as the upper limit and demonstrates the utility of this technique for polymer analysis.

An AP-to-vacuum desolvation region was found to be necessary to generate the conditions required for multiply charged ion formation [25, 117]. Further, at 250 °C, few multiply charged protein ions are observed using 2,5-dihydroxybenzoic acid (2,5-DHB) as matrix, but at 425 °C, the ion abundance increased more than 100 fold. It initially seemed reasonable that the generation of matrix/analyte droplets by laser ablation initiated the ionization process in LSII [25, 117]. However, fundamental studies of the LSII process showed that ionization occurs after the neutral matrix/analyte clusters/droplets enter a heated ion transfer region between AP and vacuum [117]. This finding has led to a number of additional ways to transfer matrix/analyte samples to the ion entrance orifice of the mass spectrometer for subsequent ionization. For example, sample can be transferred from a surface using a piezoelectric effect, by laser-induced acoustic desorption from a thin metal surface, by a

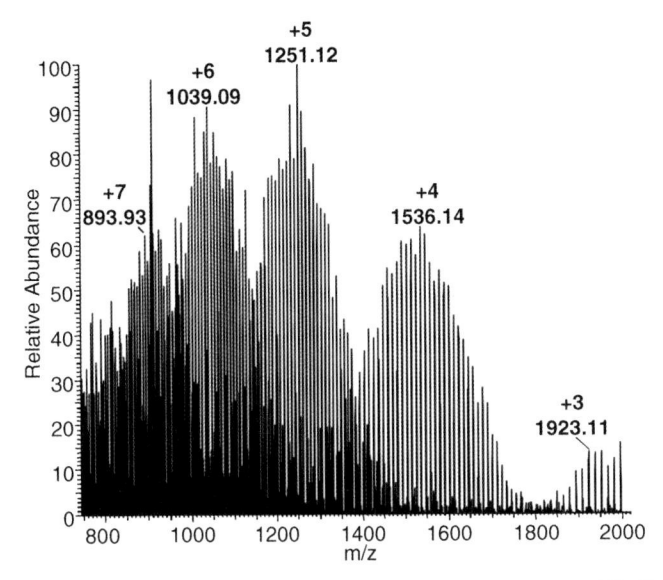

Figure 4.4 LSII mass spectrum of PEG-6690 using a 2,5-DHB and NaCl matrix and the commercially available heated ion transfer capillary of the Orbitrap mass spectrometer. (Reproduced with permission from Analytical Chemistry [67]).

shockwave applied to a metal plate, or by tapping the matrix/analyte sample holder against ion inlet orifice using simply a spatula or glass slide [22]. Further, any laser capable of ablating the surface provides a means of transferring the sample to the ionization region, as was shown for nitrogen and Nd/YAG lasers operated at ultraviolet (UV, 337 and 355 nm), visible (Vis, 532 nm), infrared (IR, 1064 nm) using the same UV matrix (2,5-DHB). The matrix absorption at the laser wavelength is not necessary ionization. These new inlet ionization method (Scheme 4.2), capable of producing highly charged ESI-like ions from large nonvolatile compounds within a heated AP to vacuum transfer region without the use of photons, electrons, or voltage are called MAII [22, 23], and solvent assisted inlet ionization (SAII), Scheme 4.2 III. Multiply charged ions enhance IMS separation [27, 28] and fragment ion yields using CID and ETD [25, 26, 55].

The key to observation of multiply charged ions lies in the mechanism of ionization that appear to be similar to ESI but without the use of a voltage. These clusters are charged within the heated transfer region between AP and vacuum by a mechanism that is unknown but may involve statistical charging upon rapid shearing of clusters or other means [22, 23]. For multiply charged ion formation, evaporation of the matrix (MAII, LSII) or solvent (SAII) but not charge similar to the ionization mechanism ESI results in highly charged unstable surfaces ($>10^9\,V\,m^{-2}$) [119] that create smaller clusters and eventually field evaporation of multiply charged ions. This mechanism, however, does not appear to satisfy all observation, and other, yet to be identified mechanisms may be involved.

4.3.1
Laserspray Ionization – Ion Mobility Spectrometry-Mass Spectrometry

The transmission geometry (TG) ion source used for LSII is field-free (Scheme 4.2). However, it has now been determined that multiply charged ions can be produced with reflective geometry, even using commercial AP-MALDI ion sources [28, 115, 120]. High abundance multiply charged polymer ions are observed on the Orbitrap Exactive (Figure 4.4) and the LTQ Velos (Figure 4.5) using laser ablation of matrix/analyte directly from a surface. In Figure 4.5, we demonstrate the versatility of LSII to ionize polymers (Scheme 4.1) using various matrixes and cations, including 2,5-dihydroxyacetophenone (2,5-DHAP) (Figure 4.5a and c), 2-nitrophloroglucinol (Figure 4.5b), and 2,5-DHB (Figure 4.5d) with NaCl (Figure 4.5a and d) and LiCl cations (Figure 4.5b and c). Similar to what is seen with ESI analysis [18], PEG-1000 and 4-arm PEG-2000 readily form multiply charged LSII ions, while the smaller pentaerythritol ethoxylate (PEEO)-800, a 4-arm branched polymer-2000 with PEG repeating units, and poly(*tert*-butylmethacrylate) (P*t*BMA)-1640 preferentially form singly charged ions. Initial results with synthetic polymers produced low abundant ions using LSII/MAII on the SYNAPT G2, while high abundances were observed with the same sample preparation on a Velos or Orbitrap mass spectrometer, about equal abundance of peptide ions were observed for the various instruments. It is not

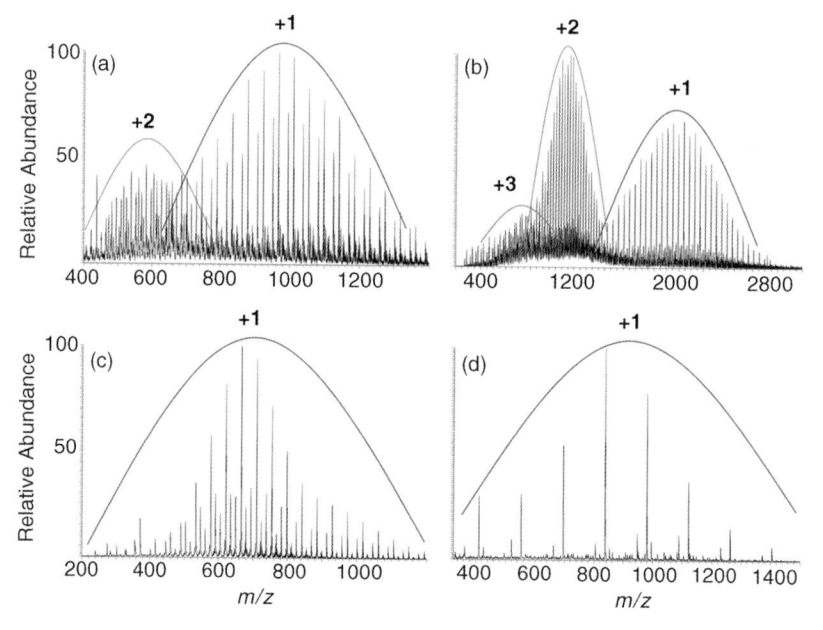

Figure 4.5 LSII-MS analysis of polymers on the LTQ-Velos mass spectrometer with 500:1 salt:analyte molar ratios and a 400 °C ion transfer capillary: (a) PEG-1000 using a 2,5-DHAP and NaCl matrix, (b) 4-arm PEG-2000 using LiCl and 2-nitrophloroglucinol, (c) Pentaethyritol ethoxylate (PEEO) 800 using a 2,5-DHAP and LiCl matrix, and (d) P*t*BMA using a 2,5-DHB and NaCl matrix.

entirely obvious where or how the ions are formed in the new *inlet* ionization methods (LSII, MAII and SAII) in the different instruments. However, clearly instrument dependence is observed with the significant difference being the current inability to provide sufficient heat for ionization on the SYNAPT G2.

Of fundamental and practical interest is the observed temperature dependence of matrixes for the formation of highly charged LSII ions. Two matrixes have produced multiply charged negative ions, and both required substantially higher temperatures (~450 °C) on the LTQ Velos than what was required for their positive ion mode measurements (~300 °C). These findings point to a mechanism that is similar to ESI wherein the solvent needs to evaporate to obtain the highly charged analyte ions. Preliminary work with inlet pressures above and below AP show that pressure, like heat [23, 25, 117], is another important variable. The importance of the heat applied to the inlet capillary was demonstrated using 2,5-DHB on the Thermo mass spectrometers equipped with a commercial heated transfer capillary (room temperature to ~450 °C) [25, 113, 117]. Multiply charged ions are essentially not observed below a transfer capillary temperature of 225 °C using 2,5-DHB but at higher temperature the abundances increase rapidly. Higher molecular weight (MW) analyses (~29 kDa) was obtained at >325 °C using 2,5-DHB as matrix [80]. Subsequent studies show that using a matrix that requires less thermal energy for sublimation or evaporation such as 2,5-DHAP the need for thermal energy in the ion transfer region is reduced [25]. Various desolvation device materials (copper, stainless steel, and glass), shapes, lengths and diameters were employed, all of which provided highly charged LSII ions with 5 pmol bovine insulin applied in 2,5-DHAP [23]. This matrix is less sensitive to synthetic polymers than is 2,5-DHB [25]. However, 2,5-DHAP is capable of producing multiply charged polymer ions when higher salt: analyte molar ratios are used, ranging from 200:1 to 2000:1. Additional improvement in ion abundance occurs using helium gas on the inlet side of the ion transfer capillary.

We attribute insufficient heating abilities (commercially max. 150 °C) on the SYNAPT G2 nano-ESI source, even when modified with a homebuilt desolvation device (~250 °C) [113], as the primary reason for lower sensitivity relative to instruments with desolvation tubes that can be heated to 450 °C (Orbitrap Exactive and XL; LTQ Velos). A systematic study of MAII/LSII matrixes that lower the thermal requirements and carry specificity for ionizing synthetic polymers is in progress. We have discovered a number of compounds useful as matrix materials for MAII and LSII. The best and most versatile discovered for synthetic polymer characterization so far requires much lower AP to vacuum transfer capillary temperature to produce high abundant multiply charged ions. As seen in Figure 4.5b, 2-nitrophloroglucinol produces abundant singly, doubly, and triply charged ions with the capillary at 325 °C. 2,5-DHAP and 2,5-DHB produce optimal polymer spectra closer to 400 °C. This new matrix extends the concept of LSII to high vacuum MALDI-TOF/TOF mass spectrometers, laserspray ionization *vacuum* (LSIV), Scheme 4.2 IV [120].

LSII-IMS-MS of a model polymer blend of PEG-1000 and P*t*BMA-1640 using this new matrix and NaCl salt in a 500:1 salt: analyte molar ratio is shown in Figure 4.6 displays the mass spectrum without incorporating the IMS dimension. A notable convolution of charge states of the blend composition is observed causing difficulties

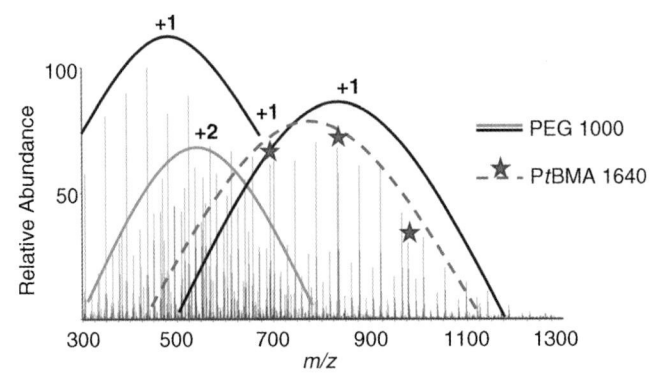

98 | *Matrix-Assisted Inlet Ionization and Solvent-Free Gas-Phase Separation*

Figure 4.6 LSII mass spectrum of a PEG-1000 and P*t*BMA-1640 mixture acquired on the SYNAPT G2 using 2-nitrophloroglucinol and a 500 : 1 salt: analyte ratio and a modified heated desolvation device.

in interpretation. The convoluted and isobaric overlap demonstrates the need for IMS-MS analysis. In Figure 4.7, the same mixture is analyzed by LSII-IMS-MS. Figure 4.7a shows a typical 2D display of drift time versus m/z. The ion intensity is displayed as an arbitrary color plot with yellow being the highest abundance. As is the case with ESI-IMS-MS, the larger PEG oligomers form the doubly charged ions. The deconvolution of blend composition and charge state distribution is enhanced when multiply charged ions are produced as is observed in Inset A for PEG +1 and +2 versus P*t*BMA +1 charge states. Baseline separation is achieved for the +2 charge state, while only half height separation is obtained for the +1 charge state. Better separation of the +1 charge state is attained for the larger PEG and P*t*BMA oligomers

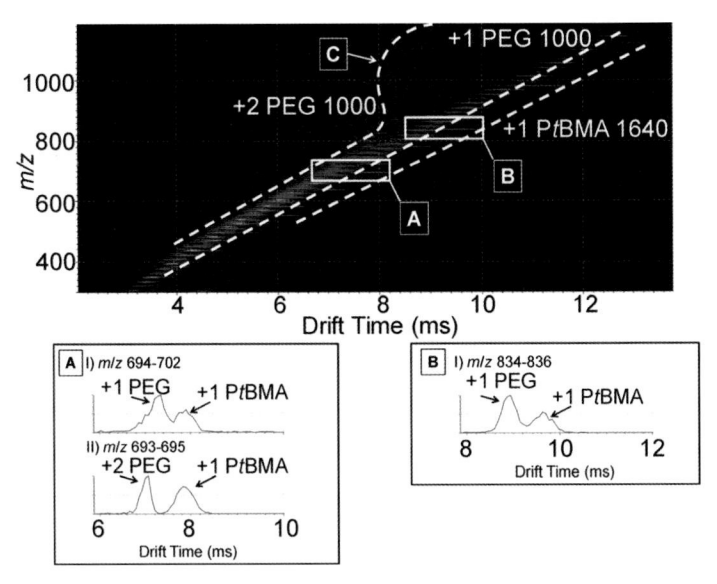

Figure 4.7 LSII-IMS-MS 2D plot of a PEG-1000 and P*t*BMA-1640 mixture with the full spectrum and (a, b) convoluted regions. (C) shows low abundance folding transition of +2 PEG-1000.

(Inset B), although the separation falls short of the baseline separation of the +2 charge state. Multiple charging in MS convolutes mass spectra complicating or sometimes even inhibiting interpretation especially in polymer analyses, however, as shown here, multiple charging enhances the IMS separation. It also provides unique 2D datasets of the blend composition that can be used as pictorial "snapshots" similar to ESI-IMS-MS [17, 18].

MALDI is generally less applicable for small molecule analysis because of the extended background caused by the application of matrix material. LSII (and MAII and LSIV at IP) suffers less from these limitations as the matrix and matrix cluster background can be minimized under proper conditions including sample preparation and deposition, inlet temperature (and low laser fluence at IP and vaccum).

MALDI MS has generally been shown to be a higher energy ionization process than ESI, resulting in more fragmentation. This is especially problematic with ionization of labile compounds and noncovalent complexes. Producing multiply charged ions in LSII, even at ion transfer capillary temperature exceeding 350 °C, no fragmentation is observed. A peptide with a labile c-myc modification and not applicable to MALDI analysis because of fragmentation was shown to produce multiply charged molecular ions very similar to ESI and even allowed ETD fragmentation [115] with identical results to ESI-ETD. Further, IMS-MS of ubiquitin ionized by LSII shows almost identical drift time distributions under similar ion source conditions to ESI; thus, both methods produce similar conformations [118]. This may have important implications for characterization and imaging of labile materials including noncovalent complexes and conformations directly from surfaces [27, 28].

The LSII method appears to be as generally applicable as either ESI or vacuum MALDI, sensitively producing mass spectra of synthetic polymers. Additionally, the LSII method, even in its current early stage of development, is simple to operate; switching between LSII or MAII and ESI, or vice versa, requires less than 2 min. There are numerous reasons why a simple-to-use, high sensitivity AP-ionization method could provide a paradigm shift in materials analysis. The advantages over high-vacuum MALDI include the absence of losses in mass resolution or mass measurement accuracy associated with high laser power, fast analysis as a result of high laser fluence and high single shot ion abundances (versus signal averaging), interfacing with common API instruments which often also have advanced MS/MS capabilities (electron capture dissociation [121–123], and ETD [124, 125]), and the potential to use various atmospheres in studying catalysts and surface activity. Advantages over ESI include speed of analysis, less dependence on solvent, simplicity, and high spatial resolution for imaging analysis.

4.3.2
Matrix Assisted Inlet Ionization (MAII) and Solvent Assisted Inlet Ionization (SAII)

A number of experiments conducted with LSII suggested that ionization might not directly involve the laser energy. At 337 nm, two photons and energy pooling [126] would be necessary to ionize matrix compounds successfully used with LSII. Multiply charged ions of peptides and proteins were observed using a Nd/YAG laser at 1064 nm and frequency doubled at 532 nm with 2,5-DHAP matrix, a compound with very poor absorption at those wavelengths. This experiment eliminated the possibility of photo-induced ionization. As noted above, shockwaves produced with a projectile against a metal plate onto which matrix/analyte sample had been dried on the opposite side, identical to LSII, was shown to transfer sample from the plate to the inlet region of the mass spectrometer and produce MAII mass spectra identical to LSII mass spectra [22, 23]. For more practical use of MAII (Scheme 4.2) we showed a variety of ways to transfer sample, the most simple being tapping the analyte/matrix sample cocrystallized on a surface against the orifice of the mass spectrometer. Mass spectra of peptides, proteins, carbohydrates, lipids, and a variety of small molecules were produced using this field and laser-free ionization method. In Figure 4.8, a MAII mass spectrum of PEG-dimethyl ether (PEGDME)-2000 acquired with a single acquisition with an inlet temperature of 350 °C shows distributions of singly, doubly, and triply charged species very similar to what is obtained by LSII with the use of a laser and ESI with the use of solvent and a voltage. The sample was prepared with 2,5-DHAP matrix and LiCl salt at a salt-to-analyte ratio of 400 : 1. Potential applications may include polymers that are light sensitive and portable mass spectrometers.

4.3.3
LSIV in Reflection Geometry at Intermediate Pressure (IP)

To our knowledge, LSIV is the first example of highly charged ESI-like mass spectra produced under vacuum conditions (subambient ESI has been reported [127]). This discovery is of immense fundamental importance and extremely promising in terms of applications because vacuum conditions eliminate much of ion transmission losses inherent with AP ion sources enhancing sensitivity. LSIV-IMS-MS produces abundant multiply charged ions using the new matrix as is shown for PEGDME-2000 (Figure 4.9); the dataset was acquired using a 500 : 100 : 1 matrix : LiCl : analyte molar ratio. The +2-charge state forms folding transitions as was previously observed for ESI-IMS-MS of PEG (Figure 4.2 and 4.3) and PEGDME [18]. The structural changes are related to conformational changes of the two cations attached to the polymer backbone and dependent on the size of the polymer (see also Figure 4.7, PEG-1000) and the cation. The analytical importance is that the doubly charged ions have a significantly faster drift time than the singly charged ions so that the clean exaction of charge sates become possible. Further, these transitions are easily visualized, speeding data interpretation. These 2D "snapshots" are sufficiently unique that computer programs may help analyses of polymeric materials rapidly and reliably in product control and consumer product safety.

Singly charged MALDI ions are described to be produced close to the sample holder surface ($\sim$15 μm) [128], while the multiply charged LSIV ions, as in LSII, are formed after a delay downstream in the instrument [28]. The results are very similar and MAII matrixes are used so that we can conclude that multiple charging in LSIV is a MAII process. There is a pressure drop of similar magnitude to that occurring between AP and the IP region used in inlet ionization (LSII, MAII and SAII); however, there is currently no means of adding heat to the IP pressure drop zone [28].

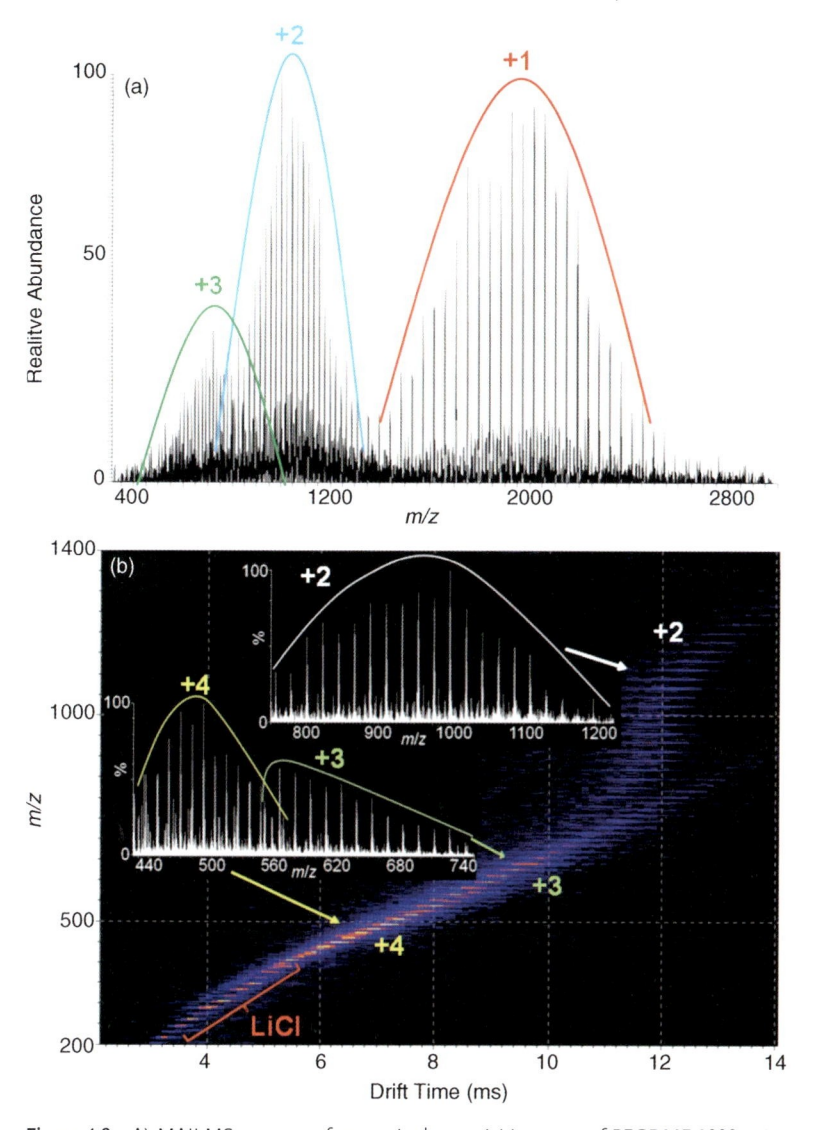

Figure 4.8 A) MAII-MS spectrum from a single acquisition scan of PEGDME-2000 using a 2,5-DHAP and LiCl matrix (400 : 1 salt: polymer molar ratio) on the LTQ-Velos mass spectrometer with an ion transfer capillary temperature of 350 °C. B) SAII-IMS-MS 2-D plot of drift time vs. m/z of PEGDME-2000 (5 pmol μL^{-1}) with LiCl (2 mM). Insets show extracted mass spectra of charge states +2, +3, and +4. LiCl backgroud is apparent below m/z 450.

Dependence on gas and flow conditions as well as laser fluence is observed at AP and IP conditions. Under certain conditions, only multiply charged ions are observed at relative low laser fluence, while only singly charged ions along with metal adduction and chemical background are detected at higher laser fluence [28], common drawbacks in MALDI [9]. Figure 4.10 exemplifies the lack of matrix

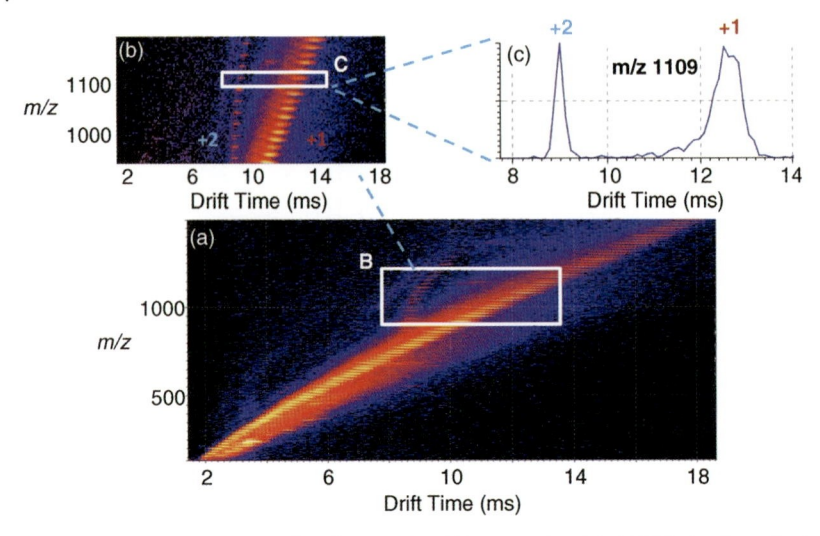

Figure 4.9 LSIV-IMS-MS 2D plot of PEGDME-2000 acquired with LiCl (100 : 1 salt: analyte) and a new matrix on the SYNAPT G2. (a) Full spectrum. (b) Region of overlapping +1 and +2 ions. (c) +2 and +1 ions with baseline separation.

background in LSIV of PEG-400 using 2,5-DHB and LiCl at a 500 : 10 : 1 matrix: salt : analyte molar ratio.

A number of highly charged peptides and small proteins were observed directly from tissue using LSIV, suggesting an entirely new approach to surface analyses and imaging. The highest mass protein observed using LSIV to date is lysozyme (14.3 kDa) but in low abundance [28]. Again, the ionization mechanisms of these unusual observations are unknown including where the ionization events occur, but certainly of immense importance. Critical to this work, however, is the potential to substantially improve the signal for multiply charged ions that we hope to achieve by appropriate matrix choice or more serious instrument modifications.

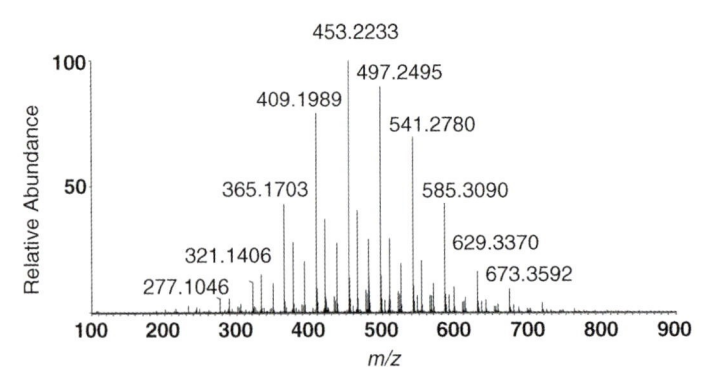

Figure 4.10 LSIV-MS spectrum of PEG-400 using 2,5-DHB and LiCl acquired on the SYNAPT G2 mass spectrometer.

4.4
Fragmentation by ETD and CID

Traditional CID fragmentation provides information on polymeric structures, but is limited in scope, especially with singly charged MALDI ions [129]. Figure 4.11a shows the CID mass spectrum of PEGDME-2000 using 2,5 DHAP and LiCl in a salt:polymer molar ratio of 400:1. The mass spectrum is shown in Figure 4.8. The parent ion of 727.5 (+3) was selected and collisionally activated with helium in a linear ion

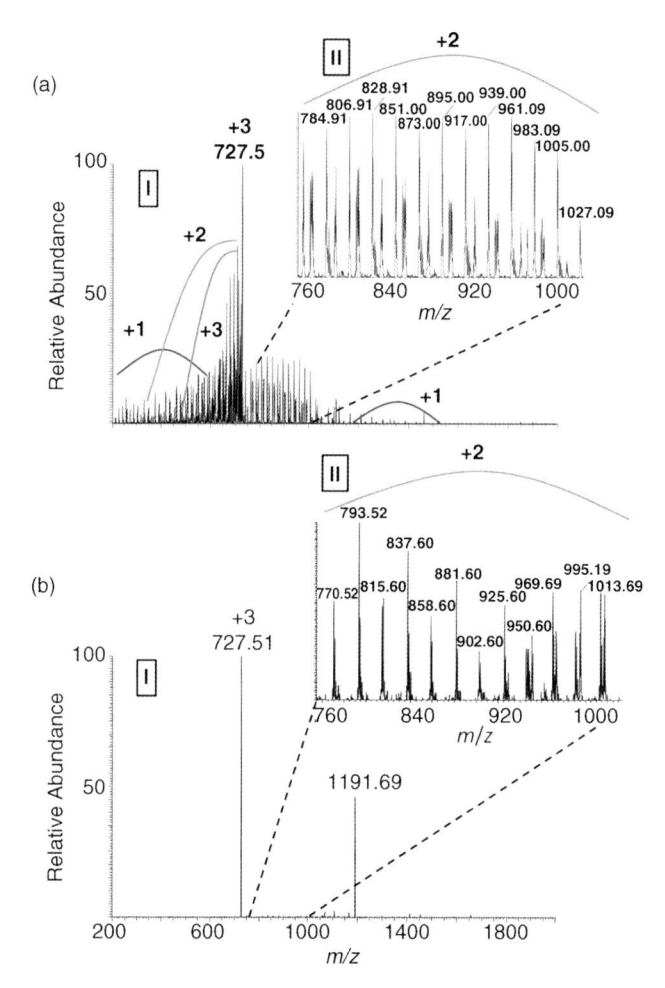

Figure 4.11 A single (a) CID-LSII-MS/MS and (b) ETD-LSII-MS/MS scan of PEGDME-2000 with (I) full and (II) Inset fragment ion mass spectra using a 2,5-DHAP and LiCl matrix (400:1 salt: polymer molar ratio) on an LTQ-Velos mass spectrometer. The triply charged m/z 727.5 was selected with a ± +0.7 Da window. (a) CID fragmentation was induced with collision energy of "50." (b) ETD fragmentation was obtained by permitting the reagent gas fluoranthene to react for 500 ms.

trap (LTQ Velos). Abundant fragment ions are observed for CID (Figure 4.11a) and ETD (Figure 4.11b). Below the selected parent ion, the dominant signals in CID are comprised of a combination of singly, doubly, and triply charged fragment ions, each representing a successive cleavage of the PEG repeating unit (Δ 44, 22, and 14.7, respectively). The fragment ions above the parent ions are doubly and singly charged ions and correspond to successive repeating unit cleavages, as well as loss of a lithium cation. ETD of the same parent ion (Figure 4.11b) produces alternative fragment ions after it receives a single electron from a fluoroanthene radical anion using the same instrument. The interpretation of the fragment ions for structural elucidation of the polymer is currently under way. We also include traditional ESI-MS-MS using CID and ETD on the SYNAPT G2 confirming the radically different fragmentation between CID and ETD (Figure 4.12). The mass spectrum shows charge states +3 to +5 (Figure 4.12I). Two different parent ion selections were performed from the charge state +3, m/z 727 and 624 (Figure 4.12II.1 and II.2). For both selections, CID fragment ion mass spectra (Figure 4.12a) show high abundance fragment ions with a wealth of cleavage sites as was seen for LSII-CID (Figure 4.11a). Similarly, ETD fragment ions (Figure 4.12b) provide only a few cleavage sites but those obtained are complementary to CID as can be best seen in the Insets (Figure 4.12a and b, right panel). The most abundant ETD fragment ions at m/z 1143.8 and 989.6 for the parent ions m/z 727 and 624, respectively, provide the same mass difference of $\sim$114 and are tentatively assigned to the loss of two monomer units including the OMe end group.

Relative to fragments commonly observed from singly charged MALDI ions, a wealth of CID and ETD fragment ions are observed because of the ability to form multiple charging by LSII and ESI. ETD fragment ion spectra may provide a simple means of end-group determination. Just as in protein and peptide MS/MS analysis, fragmentation from both CID and ETD will provide the most complete structural analysis of synthetic polymers. Hopefully, in the future, interpretation programs of CID and ETD data will be available for assignment of polymer structure, similar to biological MS/MS interpretation [73, 74].

4.5
Surface Analyses by Imaging MS

Understanding surfaces, whether geological, environmental, biological, or synthetic, has enormous utility. Spectrometric methods are commonly applied for chemical analysis. However, only mass spectrometry is capable of compound specific analysis of chemically complex surfaces, and until now only mass spectrometric techniques with ionization based on ESI or MALDI were capable of analyzing surfaces for large nonvolatile compounds such as proteins, polymers [130], petroleum, carbohydrates, and ionic materials. ESI-based surface methods such as DESI are promising because of the range of compounds that can be ionized and its use with high performance mass spectrometers, but it currently suffers from low-spatial resolution in imaging. Mass spectrometry in combination with laser ablation offers a direct method of mass-specific surface imaging with high spatial resolution. MALDI, while capable of

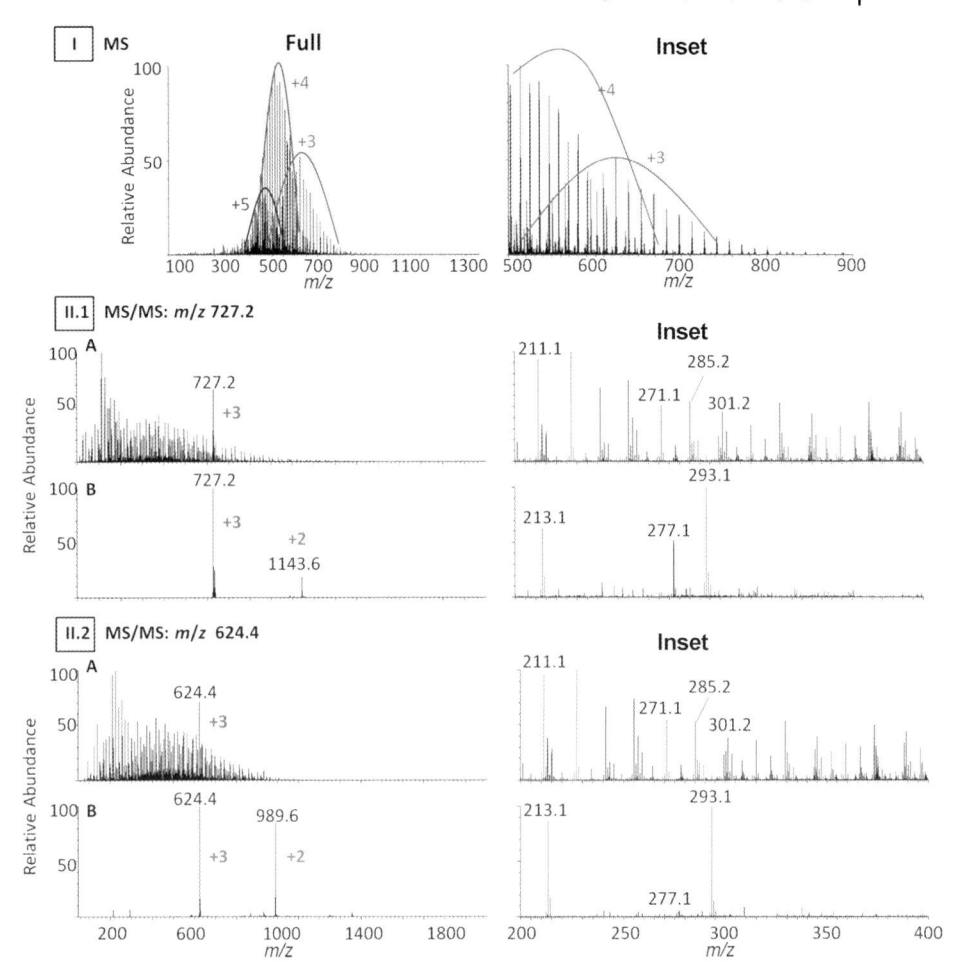

Figure 4.12 An average of 2 min acquisitions of (a) ESI-MS, (b) CID-ESI-MS/MS and (c) ETD-ESI-MS/MS of PEGDME-2000 with (left panel) full and (right panel) Inset mass and fragment ion mass spectra using LiCl and 50/50 ACN/water on an SYNAPT G2 mass spectrometer. Triply charged m/z (I) 727.2 and (II) 624.4 ions were selected, respectively. (b) CID fragmentation was induced with collision energy of "61." (c) ETD fragmentation was obtained with the reagent gas nitrosobenzene.

analysis of a wide range of compounds with good spatial resolution, suffers from the difficulty of identifying the molecular species observed because it is more difficult to obtain structural information by MS/MS from low charge state ions. MALDI is also limited by its inherent need to acquire ∼100 shots per mass spectrum, making imaging mass spectrometry slow (frequently many hours) [131–133]. Improvements to this end have been made by the use of expensive 1000 Hz lasers instead of 20 and 60 Hz lasers [134]. Further, the mass resolution and mass accuracy of the TOF instruments used with vacuum MALDI are often insufficient. High spatial resolution or low sensitivity and the difficulty of analyzing surfaces composed of complex mixtures and solubility-restricted components are shortcomings of current technology.

Surface imaging is an area in which there is high current interest and a method with high spatial resolution, high sensitivity, and the ability to detect high-mass components would be enormously important; initial LSII results have been presented [26, 55]. Typically, mass-specific images are acquired, but for true unknowns, a molecular weight, even if acquired with high-mass accuracy, is insufficient to identify the compound. Further, isomeric compositions cannot be differentiated by MS (Figure 4.3). LSII and LSIV offer the potential for high spatial resolution, postionization separation even of isomers using IMS, high-resolution mass separations, high mass accuracy, and structural analysis using a variety of fragmentation methods, including ETD. As noted earlier, accurate mass measurement and ETD fragmentation allowed database identification of the basic myelin protein end terminal fragment directly from biological surfaces (tissue sections) [26]. Ablations of 15 μm were achieved using a single laser shot per mass spectrum, thus per pixel.

4.5.1
Ultrafast LSII-MS Imaging in Transmission Geometry (TG)

We have found that LSII in TG operation mode, in which the laser beam passes through a glass slide before striking the sample (Scheme 4.2), has favorable implications for imaging including simplicity of laser alignment, higher potential spatial resolution, and speed of analysis enabling use of low-cost lasers. Under AP conditions, mass resolution is not related to laser fluence so that high laser fluence can be used to ablate the entire matrix within the focused area of the laser beam in a single shot, thus potentially drastically reducing the overall imaging acquisition time.

The ability to image the spatial distribution of a model polymer film by LSII in transmission geometry (Scheme 4.2) is demonstrated in Figure 4.13. P*t*BMA-1640

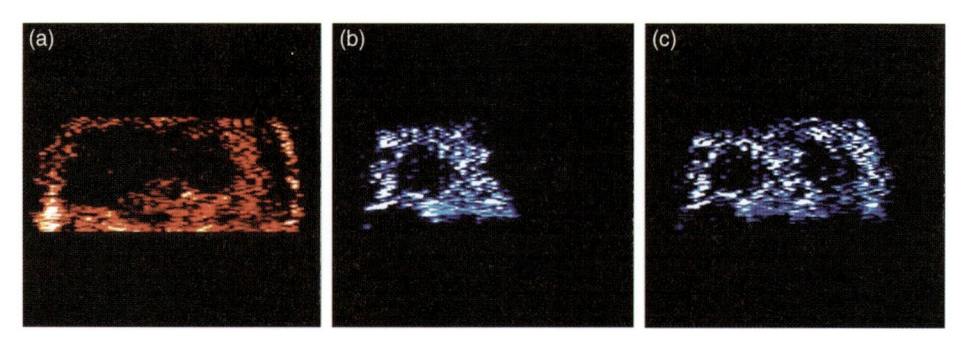

Figure 4.13 TG-AP-LSI-MS images of a crudely prepared mixed-polymer film using a 2,5-DHB and NaCl matrix shows the location of (a) P*t*BMA-1640, (b) 4-arm PEG-2000, and (c) 3-arm PEG-2000.

was dissolved in a 50:50 water: acetonitrile solution containing 10% by volume 0.1 M NaCl, and mixed with a 2,5-DHB solution and subsequently spotted on a glass slide. Individual solutions of 4-arm and 3-arm PEG-2000 were spotted side-by-side on the P*t*BMA model surface. The acquisition time was approximately 50 min. The images display m/z 835.8 for P*t*BMA as the background of the film (Figure 4.13a), m/z 1015.6 for 4-arm PEG (Figure 4.13b), and m/z 1103.9 for 3-arm PEG (Figure 4.13c) showing separate distributions in the center. This technique could possibly be extended to commercial films and surfaces where the properties of the material depend on a homogenous or ordered distribution of its constituent polymers. We are working on ways to apply the matrix without the use of any solvent so that a maximum preservation of a surface is achieved [2].

4.5.2
LSIV-IMS-MS Imaging in Reflection Geometry (RG)

Two isomeric polymers, poly(*alpha*-methylstyrene) (PAMS) 1500 and poly(*para*-methylstyrene) (PPMS) 2480, have a single structural difference in the location of a methyl substituent (Scheme 4.1). Because their molecular weights are identical, MS and imaging MS of a mixture of the two isomeric polymers without prior separation is impossible. MS/MS requires sufficiently different fragmentation pattern which is unlikely for many polymers. However, the application of LSIV-IMS-MS may provide a solution.

The mass spectra obtained for the blend and the individual isomers showed only +1 charge state distributions and were indistinguishable. The molecular weight distribution of the pure and the blend samples show relatively low values in the m/z range of 600 to 1000. The mass spectra were obtained using 2,5-DHB matrix and silver trifluoroacetate (500:10:1 matrix: salt: analyte molar ratio). Use of IMS for the unequivocal characterization of the isomeric blend is demonstrated in Figure 4.14. The analysis is obtained by extracting the drift times of m/z 755. In Figure 4.14a, the PAMS and PPMS blend are displayed, showing near baseline separation in the drift-time dimension allowing unambiguous discrimination of the drift-time distributions of the two isomers. The successful separation of both isomers was verified by measurement of each pure isomeric polymer, respectively. The *alpha* isomer (Figure 4.14c) exhibits the smaller cross section, which is defined by molecular size and shape, and travels faster at an average time of 6.72 ms, while the *para* isomer (Figure 4.14b) exits the IMS after an average of 7.27 ms.

In a final set of experiments, the same isomeric blend was examined for the utility of LSII and LSIV-IMS-MS imaging. By separately dissolving each isomer in tetrahydrofuran (THF), 2,5-DHB, and silver trifluoroacetate solution at a 500:10:1 matrix: salt: analyte molar ratio and then, spotting each solution side by side on a glass microscopy slide, we demonstrate that LSIV-IMS-MS imaging of isomeric polymer blends is possible using the SYNAPT G2 mass spectrometer in the reflection geometry. The resulting images are shown in Figure 4.15. Each colored pixel represents the abundance of a selected signal, here m/z 755, without

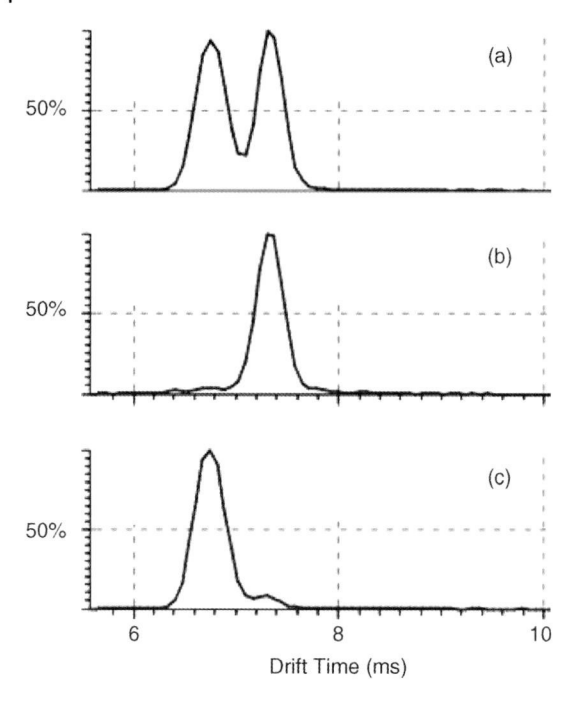

Figure 4.14 LSIV-IMS-MS drift time distributions for (a) the isomeric blend of PPMS-2480 and PAMS-1500 and pure samples (b) PPMS-2480 and (c) PAMS-1500 using a 2,5-DHB and silver trifluoroacetate matrix (50 : 1 salt : polymer molar ratio). Nitrogen gas in a T-Wave ion mobility drift tube was used for separation, applying a 652 m s^{-1} wave velocity and 42-wave height.

(Figure 4.15a) and with (Figure 4.15b and c) the use of the IMS dimension. Results without the incorporation of the IMS separation reflect the ion abundance of both isomers at the signal m/z 755. Distinction between the isomeric blend is achieved employing the drift-time separation. In Figure 4.15b, the image was exclusively created for m/z 755 with a drift time of 6.72 ms, corresponding to the *alpha* isomer. Figure 4.15c shows an image of m/z 755 signals that exited the IMS after 7.27 ms, corresponding to the *para* isomer. Chemical analysis and spatial imaging MS is only achieved for this isomeric polymer blend because of the solvent-free gas-phase separation provided by IMS. Interestingly, the ion abundance of both isomers is observed on the outskirt of the sample spot for the blend and both individual polymers. Undesired inhomogeneities have been reported for polybutyleneglycol (PBG-1000) using a MALDI-TOF imaging MS approach [130].

Solvent-free samples of the isomers were also prepared side-by-side using a ball mill device according a published procedure [15]. The polymers were dissolved in THF and spotted on 1.2 mm stainless steel beads. The samples were dried at 35 °C to ensure the full solvent evaporation and the polymer coated beads were placed in a

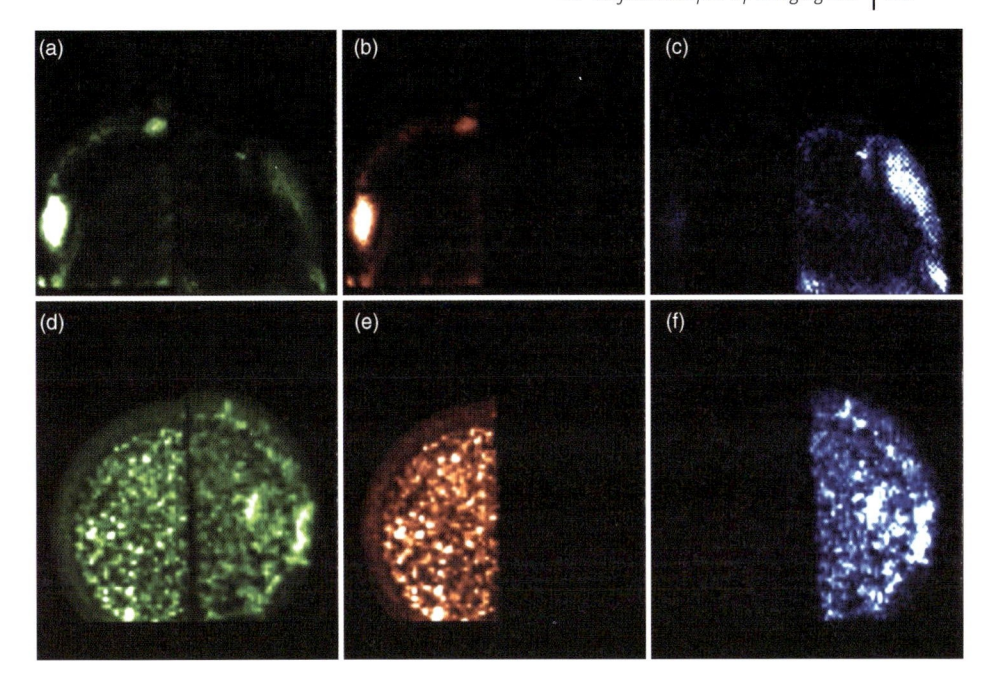

Figure 4.15 LSIV-IMS-MS imaging of a blend of PPMS-2480 and PAMS-1500 isomers. A 2,5-DHB and silver trifluoroacetates matrix (50:1 salt: polymer ratio) was used. The solvent-based preparations were obtained in THF and show the image (a) without respect to drift times, (b) with drift times corresponding to PAMS-1500 drift times, and (c) with drift times corresponding to PPMS-2480. The solvent-free preparations show (d) the image without respect to drift times, (e) the image with PAMS-1500 drift times, and (f) the image with PPMS-2480 drift times.

ball mill device with powdered 2,5-DHB and silver trifluoroacetate (50:1 matrix: salt molar ratio) and ground for 5 min at 25 Hz, homogenizing and transferring the polymer/matrix/salt sample directly to the glass plate surface. Identical to the solvent-based approach, the solvent-free approach was analyzed without (Figure 4.15d) and with the use IMS for which the drift-time distributions of m/z 755 were extracted for the *alpha* (Figure 4.15e) and the *para* isomer (Figure 4.15f). Contrary to the solvent-based LSIV-IMS-MS approach, solvent-free LSIV-IMS-MS, a TSA approach, shows ion abundance throughout the entire sample spot. The solvent-free samples lack the segregation at the edge of the spot seen in Figure 4.15a–c, demonstrating the homogenous shot-to-shot reproducibility inherent to solvent-free ionization [4–16]. As shown here, solvent-free LSIV ionization combined with solvent-free gas-phase separation is a powerful approach for not only characterizing blends, but also blends of isomeric polymers that are impossible to characterize by any sole MS approach.

4.6
Future Outlook

A great many advances in polymer characterization have been achieved using various mass spectrometric approaches [73, 74]. Nevertheless, methods that can be applied to more intractable polymeric compositions still need developing. As noted in the introduction, there are numerous materials, synthetic and natural, that fit this description that are polymeric. In an effort to begin to address these issues, we are using an IMS-MS approach for solvent-free gas phase separation and analysis. Identification of polymeric constituents is enhanced by CID and ETD fragmentation methods. IMS, MS, CID, and ETD are all enhanced in one way or another by formation of multiply charged ions. Preliminary results indicate that SAII permits the use of solvents not amenable to ESI. SAII-MS [135], SAII-IMS-MS and the utility of LC-SAII [136] open entire new ways of solvent-based analysis approaches without the use of voltage (ESI) or a laser (MALDI).

Because solubility issues are common with intractable materials, solvent-free sample preparation suitable for solvent-free ionization needs attention. New matrix-assisted ionization processes which likely have common mechanistic elements that work under AP and IP conditions and produce multiply charged ions directly from the solid state are being developed. Initial work shows that these ionization approaches are applicable to soluble polymers. Further, the multiple charging allows advanced mass spectrometers with high-mass resolution, accurate mass measurement, ion mobility separation, and collision induced and electron transfer dissociation to be applied to polymers. Solvent-free sample preparation methods originally developed for MALDI analysis of polymers is now shown to also work well with the new ionization approaches. New solvent-free methodology is being explored for surface imaging. Combining solvent-free sample preparation with LSII, MAII, and LSIV for ion production and IMS-MS for separation and analysis provides total solvent-free analysis. Directing such technology toward intractable materials and extended to tertiary structural analyses (shape, topology) even obtained directly from surfaces is an important goal. Such studies will be need to be augmented by detailed cross-sectional analysis using experimental data obtained by IMS-MS matched by theoretical cross sections obtained by modeling studies to determine if native structures are sampled directly from complex environments on surfaces so that the polymer chemist ideally not only learns about the MW but the spatial distribution on a surface of a respective polymer topology.

Acknowledgments

We are thankful for the support from NSF (CAREER 0955975 to ST), DuPont (Young Investigator Award to ST and 3 and 4-arm polymer samples), ASMS Research Award (to ST, financially supported by Waters), Wayne State University (Start-Up funds to ST), as well as the continuous support from Waters and Thermo companies.

References

1 Trimpin, S. (2010) A perspective on MALDI alternatives – total solvent-free analysis and electron transfer dissociation of highly charged ions by laserspray ionization. *J. Mass Spectrom.*, **45**, 471–485.

2 Trimpin, S., Herath, T.N., Inutan, E.D., Wager-Miller, J., Kowalski, P., Claude, E., Walker, J.M., and Mackie, K. (2010) Automated solvent-free matrix deposition for tissue imaging by mass spectrometry. *Anal. Chem.*, **82**, 359–367.

3 Wang, B., Lietz,C.B., Inutan, E.D., Leach, S., and Trimpin, S. (2010) Producing highly charged ions without solvent using laserspray ionization: a total solvent-free analysis approach at atmospheric pressure. *Anal. Chem.*, **83**, 4076–4084.

4 Trimpin, S., Rouhanipour, A., Az, R., Räder, H.J., and Müllen, K. (2001) New aspects in matrix-assisted laser desorption/ionization time-of-flight mass spectrometry: a universal solvent-free sample preparation. *Rapid Commun. Mass Spectrom.*, **15**, 1364–1373.

5 Leuninger, J., Trimpin, S., Räder, H.J., and Müllen, K. (2001) Novel approach to ladder-type polymers: polydithiathianthrene via the intramolecular acid-induced cyclization of methylsulfinyl-substituted poly(meta-phenylene sulfide). *Macromolec. Chem. Phys.*, **202**, 2832–2842.

6 Trimpin, S., Grimsdale, A.C., Räder, H.J., and Müllen, K. (2002) Characterization of an insoluble poly(9,9-diphenyl-2,7-fluorene) by solvent-free sample preparation for MALDI-TOF mass spectrometry. *Anal. Chem.*, **74**, 3777–3782.

7 Trimpin, S. and Deinzer, M.L. (2005) Solvent-free MALDI-MS for the analysis of biological samples via a mini-ball mill approach. *J. Am. Soc. Mass Spectrom.*, **16**, 542–547.

8 Trimpin, S., Räder, H.J., and Müllen, K. (2006) Experiments on theoretical principles of matrix-assisted laser desorption/ionization mass spectrometry Part I preorganization. *Int. J. Mass Spectrom.*, **253**, 13–21.

9 Trimpin, S., Keune, S., Räder, H.J., and Müllen, K. (2006) Solvent-free MALDI-MS: developmental improvements in the reliability and the potential of MALDI analysis of synthetic polymers and giant organic molecules. *J. Am. Soc. Mass Spectrom.*, **17**, 661–671.

10 Trimpin, S. and McEwen, C.N. (2007) A multi-sample on-target homogenization/transfer method for solvent-free MALDI-MS analysis of synthetic polymers. *J. Am. Soc. Mass Spectrom.*, **18**, 377–381.

11 Trimpin, S., Weidner, S.M., Falkenhagen, J., and McEwen, C.N. (2007) Fractionation and solvent-free MALDI-MS analysis of polymers using liquid adsorption chromatography at critical conditions in combination with a novel multi-sample on-target homogenization/transfer sample preparation method. *Anal. Chem.*, **79**, 7565–7570.

12 Trimpin, S. and Deinzer, M.L. (2007) Solvent-free MALDI-MS for the analysis of a membrane protein via the mini-ball mill approach: a case study of bacteriorhodopsin. *Anal. Chem.*, **79**, 71–78.

13 Trimpin, S. and Deinzer, M.L. (2007) Solvent-free MALDI-MS for the Analysis of β-amyloid peptides via the mini-ball mill approach: qualitative and quantitative improvements. *J. Am. Soc. Mass Spectrom.*, **18**, 1533–1543.

14 Trimpin, S., Clemmer, D.E., and McEwen, C.N. (2007) Charge-remote fragmentation of lithiated fatty acids on a TOF-TOF instrument using matrix-ionization. *J. Am. Soc. Mass Spectrom.*, **18**, 1967–1972.

15 Trimpin, S., Wijerathne, K., and McEwen, C.N. (2009) Rapid methods of polymer and polymer additives identification: multi-sample solvent-free MALDI, pyrolysis at atmospheric pressure, and atmospheric pressure analysis probe mass spectrometry. *Anal. Chim. Acta*, **654**, 20–25.

16 Trimpin, S. (2007) Solvent-free matrix-assisted laser desorption ionization, in *Encyclopedia of Mass Spectrometry: Molecular Ionization*, vol. 6, Elsevier, The Netherlands, pp. 683–689.

17 Trimpin, S., Plasencia, M.D., Isailovic, D., and Clemmer, D.E. (2007) Resolving oligomers from fully grown polymers with IMS-MS. *Anal. Chem.*, **79**, 7965–7974.

18 Trimpin, S. and Clemmer, D.E. (2008) Ion mobility spectrometry/mass spectrometry snapshots for assessing the molecular compositions of complex polymeric systems. *Anal. Chem.*, **80**, 9073–9083.

19 Trimpin, S., Tan, B., Bohrer, B.C., O'Dell, D.K., Merenbloom, S.I., Pazos, M.X., Clemmer, D.E., and Walker, J.M. (2009) Profiling of phospholipids and related lipid structures using multidimensional ion mobility spectrometry-mass spectrometry. *Int. J. Mass Spectrom.*, **287**, 58–69.

20 Liu, X., Valentine, S.J., Plasencia, M.D., Trimpin, S., Naylor, S., and Clemmer, D.E. (2007) Mapping the human plasma proteome by SCX-LC-IMS-MS. *J. Am. Soc. Mass Spectrom.*, **18**, 1249–1264.

21 Trimpin, S., Plasencia, M., Isailovic, D., Merenbloom, S., Mechref, H., Novotny, M., and Clemmer, D.E. (2007) Analysis of N-linked glycans from human plasma by IMS-MS. 55th American Society for Mass Spectrometry, Indianapolis, IN, June 3–7, IMS Applications-286.

22 McEwen, C.N., Pagnotti, V., Inutan, E.D., and Trimpin, S. (2010) A new paradigm in ionization: multiply charged ion formation from as solid matrix without a laser or voltage. *Anal. Chem.*, **82**, 9164–9168.

23 Trimpin, S. (2010) Fundamentals of laserspray ionization and MAII. Asilomar Conferece in Mass Spectrometry 2010, Pacific Grove, CA, October 8–12 1:35 – 2:10 pm.

24 Trimpin, S., Clemmer, D.E., and Larsen, B.S. (2010) *Ion Mobility Spectrometry-Mass Spectrometry: Theory and Applications* (eds C.L. Wilkins and S. Trimpin), Taylor & Francis Group, Boca Raton, FL, USA, pp. 215–235.

25 Trimpin, S., Inutan, E.D., Herath, T.N., and McEwen, C.N. (2010) Laserspray ionization – a new AP-MALDI method for producing highly charged gas-phase ions of peptides and proteins directly from solid solutions. *Mol. Cell. Proteomics*, **9**, 362–367.

26 Inutan, E.D., Richards, A.L., Wager-Miller, J., Mackie, K., McEwen, C.N., and Trimpin, S. (2011) Laserspray ionization – a new method for protein analysis directly from tissue at atmospheric pressure and with ultra-high mass resolution and electron transfer dissociation sequencing. *Mol. Cell. Proteomics.* doi: 10.1074/mcp.M110.000760

27 Inutan, E.D. and Trimpin, S. (2010) Laserspray ionization-ion mobility spectrometry-mass spectrometry: baseline separation of isomeric amyloids without the use of solvents desorbed and ionized directly from a surface. *J. Proteome Res.*, **11**, 6077–6081.

28 Inutan, E.D., Wang, B., and Trimpin, S. (2011) Intermediate pressure laserspray ionization ion mobility spectrometry mass spectrometry. *Anal. Chem.*, **83**, 678–684.

29 Hanton, S.D. and Parees, D.M. (2005) Extending the solvent-free MALDI sample preparation method. *J. Am. Soc. Mass Spectrom.*, **16**, 90–93.

30 Falkenhagen, J. and Weidner, S.M. (2005) Detection limits of matrix associated laser desorption/ionization mass spectrometry coupled to chromatography – a new application of solvent-free sample preparation. *Rapid Commun. Mass Spectrom.*, **19**, 3724–3730.

31 Hanton, S.D., McEvoy, T.M., and Stets, J.R. (2008) Imaging the morphology of solvent-free prepared MALDI samples. *J. Am. Soc. Mass Spectrom.*, **19**, 874–881.

32 Hanton, S.D. and Stets, J.R. (2009) Determining the time needed for the vortex method for preparing solvent-free MALDI samples of low molecular mass polymers. *J. Am. Soc. Mass Spectrom.*, **20**, 1115–1118.

33 Pizzala, H., Barrere, C., Mazarin, M., Ziarelli, F., and Charles, L. (2009) Solid state nuclear magnetic resonance as a tool to explore solvent-free MALDI samples. *J. Am. Soc. Mass Spectrom.*, **20**, 1906–1911.

34 Sroka-Bartnicka, A., Ciesielski, W., Libiszowski, J., Duda, A., Sochacki, M., and Potrzebowski, M.J. (2010) Complementarity of solvent-free MALDI TOF and solid-state NMR spectroscopy in spectral analysis of polylactides. *Anal. Chem.*, **82**, 323–328.

35 Soltzberg, L.J., Slinker, J.D., Flores-Torres, S., Bernards, D.A., Malliaras, G.G., Abruna, H.D., Kim, J.S., Friend, R.H., Kaplan, M.D., and Goldberg, V. (2006) Identification of a quenching species in ruthenium tris-bipyridine electroluminescent devices. *J. Am. Chem. Soc.*, **128**, 7761–7764.

36 Raeder, H.J., Rouhanipour, A., Talarico, A.M., Palermo, V., Samori, P., and Muellen, K. (2006) Processing of giant grapheme molecules by soft-landing mass spectrometry. *Nat. Mater.*, **5**, 276–280.

37 Kotsiris, S.G., Vasil'ev, Y.V., Streletskii, A.V., Han, M., Mark, L.P., Boltalina, O.V., Chronakis, N., Orfanopoulos, M., Hungerbuhler, H., and Drewello, T. (2006) Application and evaluation of solvent-free matrix assisted laser desorption/ionization mass spectrometry for the analysis of derivatized fullerenes. *Eur. J. Mass Spectrom.*, **12**, 397–408.

38 Hortal, A.R., Hurtado, P., Martinez-Haya, B., and Mullins, O.C. (2007) Molecular weight distributions of coal and petroleum asphaltenes from laser desorption/ionization experiments. *Energ. Fuel.*, **21**, 2863–2868.

39 Cristadoro, A., Raeder, H.J., and Muellen, K. (2008) Quantitative analyses of fullerene and polycyclic aromatic hydrocarbon mixtures via solvent-free matrix-assisted laser desorption/ionization mass spectrometry. *Rapid Commun. Mass Spectrom.*, **22**, 2463–2470.

40 Weidner, S.M. and Falkenhagen, J. (2009) Imaging mass spectrometry for examining localization of polymeric composition in matrix assisted laser desorption/ionization samples. *Rapid Commun. Mass Spectrom.*, **23**, 653–660.

41 Hughes, L., Wyatt, M.F., Stein, B.K., and Brenton, A.G. (2009) Investigation of solvent-free MALDI-TOFMS sample preparation methods for the analysis of organometallic and coordination compounds. *Anal. Chem.*, **81**, 543–550.

42 Gubler, L., Beck, N., Guersel, S.A., Hajbolouri, F., Kramer, D., Reiner, A., Steiger, B., Scherer, G.G., Wokaun, A., Rajesh, B., and Thampi, K.R. (2004) Materials for polymer electrolyte fuel cells. *Chimia*, **58**, 826–836.

43 Deluca, N.W. and Elabd, Y.A. (2006) Polymer electrolyte membranes for the direct methanol fuel cell: A review. *J. Polym. Sci. Part B: Polym. Chem.*, **44**, 2201–2225.

44 Scott, K. and Stamatin, I. (2007) Polymer electrolyte fuel cells, advances in research and development. *J. Optoelectron., Adv. Mater.*, **9**, 1597–1605.

45 Ise, N. and Sogami, I. (eds) (2005) *Structure Formation in Solution: Ionic Polymers and Colloidal Particles*, Springer, Berlin, Germany.

46 Dragan, E.S. (ed.) (2005) *Focus on Ionic Polymers*, Research Signpost, Trivandrum, India.

47 Colby, R.H., Boris, D.C., Krause, W.E., and Tan, J.S. (1997) Polyelectrolyte conductivity. *J. Polym. Sci. Part B: Polym. Phys.*, **35**, 2951–2960.

48 Boris, D.C. and Colby, H.R. (1998) Rheology of sulfonated polystyrene solutions. *Macromolecules*, **31**, 5746–5755.

49 Bae, B. and Kim, D. (2003) Sulfonated polystyrene grafted polypropylene composite electrolyte membranes for direct methanol fuel cells. *J. Membr. Sci.*, **220**, 75–87.

50 Fu, F., Manthiram, A., and Guiver, M.D. (2006) Blend membranes based on sulfonated polyetheretherketone and polysulfone bearing benzimidazole side groups for fuel cells. *Electrochem. Commun.*, **8**, 1386–1390.

51 Wang, D., Pu, H., Jiang, F., Yang, Z., and Tang, L. (2006) Proton-conducting membrane composed of hollow sulfonated polystyrene microspheres,

poly(vinyl alcohol) and imidazole. *Polym. Prepr.*, **41**, 238–239.

52 Goktepe, F., Bozkurt, A., and Gunday, S.T. (2008) Synthesis and proton conductivity of poly(styrene sulfonic acid)/heterocycle-based membranes. *Polym. Intern.*, **57** 133–138.

53 Sahu, A.K., Selvarani, G., Bhat, S.D., Pitchumani, S., Sridhar, P., and Shukla, A.K. (2008) Effect of varying poly (styrene sulfonic acid) content in poly (vinyl alcohol)–poly(styrene sulfonic acid) blend membrane and its ramification in hydrogen–oxygen polymer electrolyte fuel cells. *J. Membr. Sci.*, **319**, 298–305.

54 Trimpin, S., Herath, T.N., Inutan, E.D., Cernat, S.A., Wager-Miller, J., Mackie, K., and Walker, J.M. (2009) Field-free transmission geometry atmospheric pressure matrix-assisted laser desorption/ionization for rapid analysis of unadulterated tissue samples. *Rapid Commun. Mass Spectrom.*, **23**, 3023–3027.

55 Richards, A.L., Lietz, C.B., Wager-Miller, J.B., Mackie, K., and Trimpin, S. (2011) Imaging mass spectrometry in transmission geometry. *Rapid Commun. Mass Spectrom.*, **25**, 815–820.

56 Kulkarni, S.U. and Thies, M.C. (2010) Investigation of solvent-based and solvent-free MALDI-TOF-MS sample preparation methods for the qualitative and quantitative analysis of heavy petroleum macromolecules. 239th ACS National Meeting, San Francisco, CA, March 21–25 Abstracts of Papers.

57 Kulkarni, S.U. and Thies, M.C. (2010) Solvent-based vs. Solvent-free MALDI -TOF-MS sample preparation methods for the quantitative analysis of petroleum macromolecules. *Preprints*, **55**, 108–111.

58 Kudaka, I., Asakawa, D., Mori, K., and Hiraoka, K. (2008) A comparison of EDI with solvent-free MALDI and LDI for the analysis of organic pigments. *J. Mass Spectrom.*, **43**, 436–446.

59 Qiao, H., Piyadasa, G., Spicer, V., and Ens, W. (2009) Analyte distributions in MALDI samples using MALDI imaging mass spectrometry. *Int. J. Mass Spectrom.*, **281**, 41–51.

60 Saraiva, S.A., Cabral, E.C., Eberlin, M.N., and Catharino, R.R. (2009) Amazonian vegetable oils and fats: fast typification and quality control via triacylglycerol (TAG) profiles from dry matrix-assisted laser desorption/ionization time-of-flight (MALDI-TOF) mass spectrometry. *J. Agric. Food Chem.*, **57**, 4030–4034.

61 Hankin, J.A., Barkley, R.M., and Murphy, R.C. (2007) Sublimation as a method of matrix application for mass spectrometric imaging. *J. Am. Soc. Mass Spectrom.*, **18**, 1646–1652.

62 Bouschen, W., Schulz, O., Eikely, D., and Spengler, B. (2010) Matrix vapor deposition/recrystallization and dedicated spray preparation for high-resolution scanning microprobe matrix-assisted laser desorption/ionization imaging mass spectrometry (SMALDI-MS) of tissue and single cells. *Rapid Commun. Mass Spectrom.*, **24**, 355–364.

63 Puolitaival, S.M., Burnum, K.E., Cornett, D.S., and Caprioli, R.M. (2008) Solvent free matrix dry-coating for MALDI imaging of phospholipids. *J. Am. Soc. Mass Spectrom.*, **19**, 882–886.

64 Gholipour, Y., Giudicessi, S.L., Nonami, H., and Erra-Balsells, R. (2010) Diamond, titanium dioxide, titanium silicon oxide, and barium strontium titanium oxide nanoparticles as matrixes for direct matrix-assisted laser desorption/ionization mass spectrometry analysis of carbohydrates in plant tissues. *Anal. Chem.*, **82**, 5518–5526.

65 Goodwin, R.J., MacIntyre, L., Watson, D.G., Scullion, S.P., and Pitt, A.R. (2010) A solvent-free matrix application method for matrix-assisted laser desorption/ionization imaging of small molecules. *Rapid Commun. Mass Spectrom.*, **24**, 1682–1686.

66 Goodwin, R.J.A., Scullion, P., MacIntyre, L., Watson, D.G., and Pitt, A.R. (2010) Use of a solvent-free dry matrix coating for quantitative matrix-assisted laser desorption ionization imaging of 4-bromophenyl-1,4-diazabicyclo(3.2.2)nonane-4-carboxylate in rat brain and quantitative analysis of the drug from laser

67 Trimpin, S., Inutan, E.D., Herath, T.N., and McEwen, C.N. (2010) A matrix-assisted laser desorption/ionization mass spectrometry method for selectively producing either singly or multiply charged molecular ions. *Anal. Chem.*, **82**, 11–15.

68 Wyttenbach, T., Gidden, J., and Bowers, M.T. (2010) *Ion Mobility Spectrometry-Mass Spectrometry: Theory and Applications* (eds C.L. Wilkins and S. Trimpin), Taylor & Francis Group, Boca Raton, FL, USA, pp. 3–30.

69 Koeniger, S.L., Merenbloom, S.I., Valentine, S.J., Jarrold, M.F., Udseth, H., Smith, R., and Clemmer, D.E. (2006) An IMS-IMS analogue of MS-MS. *Anal. Chem.*, **78**, 4161–4174.

70 Merenbloom, S.I., Koeniger, S.L., Valentine, S.J., Plasencia, M.D., and Clemmer, D.E. (2006) IMS-IMS and IMS-IMS-IMS/MS for separating peptide and protein fragment ions. *Anal. Chem.*, **78**, 2802–2809.

71 Trimpin, S., Hoskins, J., Wijerathne, K., Inutan, E.D., and Grayson, S.M. (2010) Library of polymer architectures examined by ion mobility spectrometry-mass spectrometry. Proc. 58th ASMS Conf. Mass Spectrometry and Allied Topics, Salt Lake City, UT, May 23–27 MOCpm 2:50 p.m.

72 Hoskins, J.N., Trimpin, S., and Grayson, S.M. (2011) Architectural differentiation of linear and cyclic polymeric isomers by ion mobility spectrometry-mass spectrometry. *Macromolecules*, **44**, 6915–6918.

73 Weidner, S.M. and Trimpin, S. (2008) Mass spectrometry of synthetic polymers. *Anal. Chem.*, **80**, 4349–4361.

74 Weidner, S.M. and Trimpin, S. (2010) Mass spectrometry of synthetic polymers. *Anal. Chem.*, **82**, 4811–4829.

75 Li, X., Chan, Y.T., Newkome, G.R., and Wesdemiotis, C. (2011) Gradient tandem mass spectrometry interfaced with ion mobility separation for the characterization of supramolecular architectures. *Anal. Chem.*, **83**, 1284–1290.

76 Song, J.K., Grun, C.H., Hereen, R.M.A., Janssen, H.G., and van den Brink, O.F. (2010) High-resolution ion mobility spectrometry-mass spectrometry on poly(methyl methacrylate). *Agnew. Chem. Int. Ed.*, **49**, 10168–10171.

77 Perera, S., Li, X.P., Soler, M., Schultz, A., Wesdemiotis, C., Moorefield, C.N., and Newkome, G.R. (2010) Hexameric Palladium(II) terpyridyl metallomacrocycles: assembly with 4,4′-bipyridine and characterization by TWIM mass spectrometry. *Angew. Chem. Int. Ed.*, **49**, 6539–6544.

78 Angel, L.A., Majors, L.T., Dharmaratne, A.C., and Dass, A. (2010) Ion mobility mass spectrometry of Au-25 (SCH2CH2Ph)(18) nanoclusters. *ACS Nano*, **4**, 4691–4700.

79 Gruendling, T., Weidner, S., Falkenhagen, J., and Barner-Kowolilk, C. (2010) Mass spectrometry in polymer chemistry: a state-of-the-art up-date. *Polym. Chem.*, **1**, 599–617.

80 Richards, A.L., Marshall, D.D., Inutan, E.D., McEwen, C.N., and Trimpin, S. (2011) High throughput analysis of peptides and proteins by laserspray ionization mass spectrometry. *Rapid Commun. Mass Spectrom.*, **26**, 815–829.

81 Van Berkel, G.J., Pasilisi, S.P., and Ovchinnikovia, O. (2008) Established and emerging atmospheric pressure surface sampling/ionization techniques for mass spectrometry. *J. Mass Spectrom.*, **43**, 1161–1180.

82 Venter, A., Nefliu, M., and Cooks, R.G. (2008) Ambient desorption ionization mass spectrometry. *Trends in Anal. Chem.*, **27**, 284–290.

83 Eberlin, L.S., Ifa, D.R., Wu, C., and Cooks, R.G. (2010) Three-dimensional visualization of mouse brain by lipid analysis using ambient ionization mass spectrometry. *Angew. Chem. Int. Edit.*, **49**, 873–876.

84 Van Berkel, G.J., Pasilis, S.P., and Ovchinnikova, O. (2008) Established and emerging atmospheric pressure surface sampling/ionization techniques for mass spectrometry. *J. Mass Spectrom.*, **43**, 1161–1180.

85 Takats, Z., Wiseman, J.M., Gologan, B., and Cooks, R.G. (2004) Mass spectrometry sampling under ambient conditions with desorption electrospray ionization. *Science*, **306**, 471–473.

86 Cooks, R.G., Ouyang, Z., Takats, Z., and Wiseman, J.M. (2006) Ambient mass spectrometry. *Science*, **311**, 1566–1570.

87 Kertesz, V. and Van Berkel, G.J. (2008) Improved imaging resolution in desorption electrospray ionization mass spectrometry. *Rapid Commun. Mass Spectrom.*, **22**, 2639–2644.

88 Kertesz, V. and Van Berkel, G.J. (2008) Scanning and surface alignment considerations in chemical imaging with desorption electrospray mass spectrometry. *Anal. Chem.*, **80**, 1027–1032.

89 Jackson, A.U., Talaty, N., Cooks, R.G., and Van Berkel, G.J. (2007) Salt tolerance of desorption electrospray ionization (DESI). *J. Am. Soc. Mass Spectrom.*, **18**, 2218–2225.

90 Gao, L., Li, G., Cyriac, J., Nie, Z., and Cooks, R.G. (2010) Imaging of surface charge and the mechanism of desorption electrospray ionization mass spectrometry. *J. Phys. Chem. C*, **114**, 5331–5337.

91 Wu, C., Ifa, D.R., Manicke, N.E., and Cooks, R.G. (2009) Rapid, Direct analysis of cholesterol by charge labeling in reactive desorption electrospray ionization. *Anal. Chem*, **81** 7618–7624.

92 Wiseman, J.M., Ifa, D.R., Zhu, Y., Kissinger, C.B., Manicke, N.E., Kisinger, P.T., and Cooks, R.G. (2008) Desorption electrospray ionization mass spectrometry: imaging drugs and metabolites in tissues. *Proc. Natl. Acad. Sci. USA*, **105**, 1–6.

93 Wiseman, J.M., Ifa, D.R., Venter, A., and Cooks, R.G. (2008) Ambient molecular imaging by desorption electrospray ionization mass spectrometry. *Nat. Protoc.*, **3**, 517–524.

94 Ifa, D.R., Gumaelius, L.M., Eberlin, L.S., Manicke, N.E., and Cooks, R.G. (2007) Forensic analysis of inks by imaging desorption electrospray ionization (DESI) mass spectrometry. *Analyst*, **132**, 461–467.

95 Wiseman, J.M., Ifa, D.R., Song, Q., and Cooks, R.G. (2006) Tissue imaging at atmospheric pressure using desorption electrospray ionization (DESI) mass spectrometry. *Angew. Chem. Int. Edit.*, **45**, 7188–7192.

96 Manicke, N.E., Dill, A.L., Ifa, D.R., and Cooks, R.G. (2010) High-resolution tissue imaging on an orbitrap mass spectrometer by desorption electrospray ionization mass spectrometry. *J. Mass Spectrom.*, **45**, 223–226.

97 Roach, P.J., Laskin, J., and Laskin, A. (2010) Nanospray desorption electrospray ionization: an ambient method for liquid-extraction surface sampling in mass spectrometry. *Analyst*, **135**, 2233–2236.

98 Takats, Z., Cotte-Rodriguez, I., Talaty, N., Chen, H.W., and Cooks, R.G. (2005) Direct, trace level detection of explosives on ambient surfaces by desorption electrospray ionization mass spectrometry. *Chem. Commun.*, 1950–1952.

99 Cody, R.B., Laramée, J.A., and Durst, H.D. (2005) Versatile new ion source for the analysis of materials in open air under ambient conditions. *Anal. Chem.*, **77**, 2297–2302.

100 McEwen, C.N., McKay, R.G., and Larsen, B.S. (2005) Analysis of solids, liquids, and biological tissues using solids probe introduction at atmospheric pressure on commercial LC/MS instruments. *Anal. Chem.*, **77**, 7826–7831.

101 McEwen, C.N. and Larsen, B.S. (2009) Ionization mechanisms related to negative ion APPI, APCI, and DART. *J. Am. Soc. Mass Spectrom.*, **20**, 1518–1521.

102 Lloyd, J.A., Harron, A.F., and McEwen, C.N. (2009) Combination atmospheric pressure solids analysis probe and desorption electrospray ionization mass spectrometry ion source. *Anal. Chem.*, **21**, 1889–1892.

103 McEwen, C.N., Major, H., Green, M., Giles, K., and Trimpin, S. (2010) *IMS/MS Applied to Direct Ionization using the Atmospheric Pressure Solids Analysis Probe Method, Ion Mobility Spectrometry: Theory*

104. Pan, H. and Lundin, G. (2011) Rapid detection and identification of impurities in ten 2-naphthalenamines using an atmospheric pressure solids analysis probe in conjunction with ion mobility mass spectrometr. *Eur. J. Mass Spectrom.*, **17**, 217–225.
105. Nemes, P. and Vertes, A. (2007) Laser ablation electrospray ionization for atmospheric pressure, in vivo, and imaging mass spectrometry. *Anal. Chem.*, **79**, 8098–8106.
106. Shrestha, B. and Vertes, A. (2009) In situ metabolic profiling of single cells by laser ablation electrospray ionization mass spectrometry. *Anal. Chem.*, **81**, 8265–8271.
107. Nemes, P. and Vertes, A. (2010) Laser ablation electrospray ionization for atmospheric pressure molecular imaging mass spectrometry. *Method. Mol. Biol.*, **656**, 159–171.
108. Shiea, J., Huang, M.Z., Hsu, C.Y., Lee, C.H., Yuan, I., and Beech, J. (2005) Electrospray-assisted laser desorption/ionization mass spectrometry for direct ambient analysis of solids. *Rapid Commun. Mass Spectrom.*, **19**, 3701–3704.
109. Sampson, J.S., Hawkridge, A.M., and Muddiman, D.C. (2006) Generation and detection of multiply-charged peptides and proteins by matrix-assisted laser desorption electrospray ionization (MALDESI) Fourier transform ion cyclotron resonance mass spectrometry. *J. Am. Soc. Mass Spectrom.*, **17**, 1712–1716.
110. Haddad, R., Sparrapan, R., and Eberlin, M.N. (2006) Easy ambient sonic-spray ionization mass spectrometry. *Rapid Commun. Mass Spectrom.*, **20**, 2901–2905.
111. Haapala, M., Pol, J., Kauppial, T.J., and Kostiainen, R. (2007) Desorption atmospheric pressure photoionization. *Anal. Chem.*, **79**, 7867–7872.
112. Chen, H., Venter, A., and Cooks, R.G. (2006) Extractive electrospray ionization for direct analysis of undiluted urine, milk, and other complex mixtures without sample preparation. *Chem. Commun.*, 2042–2044.
113. McEwen, C.N., Larsen, B.S., and Trimpin, S. (2010) Laserspray ionization on a commercial AP-MALDI mass spectrometer ion source: selecting singly or multiply charged ions. *Anal. Chem.*, **82**, 4998–5001.
114. Inutan, E.D. and Trimpin, S. (2010) Laserspray ionization (LSI) ion mobility spectrometry (IMS) mass spectrometry (MS). *J. Am. Soc. Mass Spectrom.*, **21**, 1260–1264.
115. Herath, T.N., Marshall, D.D., Richards A.L., Inutan, E.D., and Trimpin, S. (2010) Characterization of protein and labile protein modifications using laserspray ionization on a LTQ - electron transfer dissociation mass spectrometer. Proc. 58th ASMS Conf. Mass Spectrometry and Allied Topics, Salt Lake City, UT, May 23–27, WP31.
116. Zydel, F., Trimpin, S., and McEwen, C.N. (2010) Laserspray ionization using an atmospheric solids analysis probe for sample introduction. *J. Am. Soc. Mass Spectrom.*, **21**, 1889–1892.
117. McEwen, C.N. and Trimpin, S. (2011) An alternative ionization paradigm for atmospheric pressure mass spectrometry: flying elephants from trojan horses. *Intern. J. Mass Spectrom.*, **300**, 167–172.
118. Manly, C., Inutan, E.D., Lietz, C., and Trimpin, S, (2011) A comparison of laserspray ionization *inlet* (LSII) and *vacuum* (LSIV), with electrospray ionization (ESI) for structural characterization of ubiquitin protein ions. Proc. 59th ASMS Conf. Mass Spectrometry and Allied Topics, Denver, CO, June 5–9, MPO 64.
119. Vestal, M.L. (1983) Studies of ionization mechanisms involved in thermospray LC-MS. *Int. J. Mass Spectrom. Ion Phys.*, **46**, 193–196.
120. Trimpin, S., Ren, Y., Wang, B., Lietz, C.B., Marshall, D.D., Richards, A.L., and Inutan, E.D. (2011) Extending the laserspray concept to produce highly charged ions at high vacuum on a time-of-flight mass analyzer. *Anal. Chem.*, **83**, 5469–5475.

121 Zubarev, R., Kelleher, N.L., and McLafferty, F.W. (1998) Electron capture dissociation of multiply charged protein cations. A nonergodic process. *J. Am. Chem. Soc.*, **120**, 3265–3266.

122 Yoo, H.J. and Hakansson, K. (2010) Determination of double bond location in fatty acids by manganese adduction and electron induced dissociation. *Anal. Chem.*, **82**, 6940–6946.

123 Voniov, V.G., Deinzer, M.L., Beckman, J.S., and Barofsky, D.F. (2011) Electron capture, collision-induced, and electron-capture collision-induced dissociation in Q-TOF. *J. Am. Soc. Mass Spectrom.* doi: 10.1007/s13361-010-0072-x

124 Syka, J.E.P., Coon, J.J., Schroeder, M.J., Shabanowitz, J., and Hunt, D.F. (2004) Peptide and protein sequence analysis by electron transfer dissociation mass spectrometry. *Proc. Natl. Acad. Sci. USA*, **101**, 9528–9533.

125 Xia, Y., Han, H., and McLuckey, S.A. (2008) Activation of intact electron-transfer products of polypeptides and proteins in cation transmission mode ion/ion reactions. *Anal. Chem.*, **80**, 1111–1117.

126 Knochenmuss, R. (2009) Laser desorption/ionization plumes from capillary-like restricted volumes. *E. J. Mass Spectrom.*, **15**, 189–198.

127 Page, J.S., Tang, K., Kelly, R.T., and Smith, R.D. (2008) Subambient pressure ionization with nanoelectrospray source and interface for improved sensitivity in mass spectrometry. *Anal. Chem.*, **80**, 1800–1805.

128 Knochenmuss, R. and Vertes, A. (2000) Time-delayed 2-pulse studies of MALDI matrix ionization Mechanisms. *J. Phys. Chem. B*, **104**, 5406–5410.

129 Laine, O., Trimpin, S., Räder, H.J., and Müllen, K. (2010) Changes in post-source decay fragmentation behavior of poly(methylmethacrylate) polymers with increasing molecular weight studied by matrix-assisted laser desorption/ionization time-of-flight mass spectrometry. *E. J. Mass Spectrom.*, **9**, 195–201.

130 Weidner, S.M. and Falkenhagen, J. (2009) Imaging mass spectrometry for examining localization of polymeric composition in matrix-assisted laser desorption/ionization samples. *Rapid Commun. Mass Spectrom.*, **23**, 653–660.

131 Chaurand, P., Schwartz, S.A., and Caprioli, R.M. (2004) MALDI imaging MS allows simultaneous mapping of hundreds of peptides and proteins in thin tissue sections with a lateral resolution of ∼30–50 μm. *Anal. Chem.*, **76**, 86A–93A.

132 Chaurand, P., Schriver, K.E., and Caprioli, R.M. (2007) Instrument design and characterization for high resolution MALDI-MS imaging of tissue sections. *J. Mass Spectrom.*, **42**, 476–489.

133 Caldwell, R.L. and Caprioli, R.M. (2005) Tissue profiling by mass spectrometry: a review of methodology and applications. *Mol. Cell. Proteomics*, **4**, 394–401.

134 Brown, J., Murray, P., Claude, E., and Kenny, D., 20 μm resolution MALDI imaging of lipid distributions in tissue using lasers between 1 kHz and 10 kHz. Proc. 58th ASMS Conf. Mass Spectrometry and Allied Topics, Salt Lake City, UT, May 27.

135 Pagnotti, V.S., Chubatyi, N.D., and McEwen, C.N. (2011) Solvent assisted inlet ionization: An ultrasensitive new liquid introduction ionization method for mass spectrometry. *Anal. Chem.* **81**, 3981–3985.

136 Pagnotti V.S., Inutan E.D., Marshall D.D., McEwen C.N., and Trimpin S. (2011) Inlet ionization: A new highly sensitive approach for liquid chromatography/mass spectrometry of small and large molecules. *Anal. Chem.* dx.doi.org/10.1021/ac201982r

5
Polymer MALDI Sample Preparation
Scott D. Hanton and Kevin G. Owens

5.1
Introduction

Since the development of matrix-assisted laser desorption ionization (MALDI) mass spectrometry, the data has been valued for the relative ease of the experiments, the usefulness of the results, the sensitivity of the technique, and the diversity of materials that can be analyzed [1–8]. Early during the development of MALDI, polymer applications were discovered. MALDI experiments can be used to determine the repeat units of polymers, measure the molecular mass distribution, and provide information about the end groups. While the data generated by successful experiments was important in the development of new polymers and the application of existing ones, the key was getting the experiments to work.

With the development of powerful time-of-flight mass spectrometers (TOF MS), the instrumental portion of the MALDI experiment has largely been resolved. Modern mass spectrometers provide high mass accuracy, high mass resolution, a high mass range, and rapid data collection. The key remaining area to investigate for MALDI is sample preparation. In many ways, the expensive TOFMS instrument becomes a detector of the quality of the sample preparation completed at the lab bench.

MALDI sample preparation essentially entails the appropriate mixing of an analyte with a matrix, resulting in a solid sample to be analyzed. For relatively pure, low-molecular-mass molecules with high solubility in common solvents, the sample preparation is easy and straightforward. Unfortunately for the polymer chemist, most polymer samples are not included in that description. A typical polymer sample is a complex combination of closely related molecules with molecular masses that can range over orders of magnitude in mass and with limited solubility in only a few solvents.

Because of the difficulty involved in successfully preparing many polymer materials for MALDI, a misconception that MALDI sample preparation is art or magic has developed. Through the study of the MALDI process, it has been possible to replace

Mass Spectrometry in Polymer Chemistry, First Edition.
Edited by Christopher Barner-Kowollik, Till Gruendling, Jana Falkenhagen, and Steffen Weidner.
© 2012 Wiley-VCH Verlag GmbH & Co. KGaA. Published 2012 by Wiley-VCH Verlag GmbH & Co. KGaA.

the art of sample preparation with the science of sample preparation. It is the intention of this chapter to carefully examine each portion of polymer MALDI sample preparation to shed light on the process. While examined as a whole, polymer sample preparation for MALDI looks complicated; taken in steps, it can be understood and utilized to enable the solution of difficult polymer characterization questions.

5.2
Roles of the Matrix

The matrix really defines the MALDI experiment. In most other analytical techniques, the matrix is the generic background holding the precious analyte. It can be viewed from being an innocent bystander in the experiment, simply taking up space and diluting the analyte, to being an active interference to the experiment, something which must be accounted for to generate accurate results. In MALDI, however, the matrix is a necessary component of the experiment that must be chosen with care to obtain a successful result. Instead of being viewed with either disdain or suspicion, the matrix is now a central part of the experiment.

Before the development of MALDI, plenty of successful laser desorption (LD) experiments had been reported. While LD experiments were effective at analyzing atoms and small molecules, there were relatively few successes in the analysis of higher molecular mass species, like polymers, and those were generally limited to molecular masses below about 2000 Da. There were several issues facing LD experiments on complex, higher mass analytes. Some of the key challenges included efficient absorption of the laser light, nondestructive desorption of the intact analyte, and effective ionization routes for intact molecules. The addition of substantial impurities or matrices to LD experiments was the spark that led to MALDI as it is practiced today.

One interesting aspect of the matrix in the MALDI experiment is that it is present in huge molar excess (discussed below in Section 5.9). The presence of the matrix enables the MALDI experiment to be extremely sensitive to the analyte. Researchers have now published detection limits by MALDI into the low fmole mm^{-2} of sample surface range for some specific biological molecules and synthetic polymers [9, 10]. While it can be counterintuitive, one way to optimize a poorly performing MALDI experiment is to increase the amount of matrix present. The increased number of matrix molecules can enable improved desorption for some analytes and dilute interfering impurities in other experiments. Despite increased dilution for the desired analyte, the exquisite sensitivity of MALDI often enables a successful experiment.

In all MALDI experiments, the matrix must fulfill four different functions to generate a successful result:

- generate intimate contact with the analyte
- absorb the laser light

- desorb into the mass spectrometer
- produce an ionization path for the analyte

5.2.1
Intimate Contact

There is debate in the MALDI community about the absolute nature of the interaction of the matrix and the analyte. The debate centers on the degree of incorporation of the analyte within the matrix, and the degree of crystallinity of the matrix in the final MALDI sample. Despite this debate, it is clear that the matrix must closely interact with the analyte. To be successful, the MALDI sample preparation method must generate intimate contact between the analyte and the matrix. The exact nature of this contact is not yet proven. In cases where the analyte significantly separates from the matrix, or if the analyte is significantly agglomerated, preventing intimate contact with the matrix, then the MALDI experiment typically fails.

As will be discussed in greater depth below, other factors besides the direct choice of the matrix play key roles in developing and maintaining intimate contact between the matrix and the analyte. The choice of solvent and deposition method both strongly contribute to generating the appropriate level of contact for a successful experiment.

5.2.2
Absorption of Laser Light

The MALDI matrix must efficiently absorb the laser light used in the MALDI experiment. The laser is the only source of energy available to convert the solid-phase sample into gas-phase ions ready for mass spectrometric analysis. Most commercial MALDI instruments use ultraviolet (UV) lasers, so most MALDI matrices have functional groups that are strongly absorbing in the UV. Typical matrices have significant unsaturation that results in strong UV absorbance.

Of course, assuming that the presence of an aromatic ring will generate the needed absorbance can be problematic. Early in the history of polymer MALDI, papers were published questioning the usefulness of MALDI to analyze simple polymers [11]. They reported the use of gentisic acid as the matrix. Unfortunately, gentisic acid has different isomers. One of the isomers is the well-used MALDI matrix, 2,5-dihydroxybenzoic acid (DHB). Unfortunately, another isomer is the closely related 2,4-DHB. Figure 5.1 shows MALDI mass spectra of a polymethylmethacrylate (PMMA) 2900 sample prepared with 2,5-DHB (Figure 5.1a) and 2,4-DHB (Figure 5.1b).

The 2,5-DHB produces a typical MALDI mass spectrum and the 2,4-DHB produces noise. While both species are aromatic acids, only the 2,5-DHB absorbs the 337 nm light generated by the nitrogen laser used in the experiment. The 2,4-DHB absorbs very poorly at 337 nm resulting in a failed MALDI experiment. This was the problem faced by those early researchers. It was not a failure of polymer

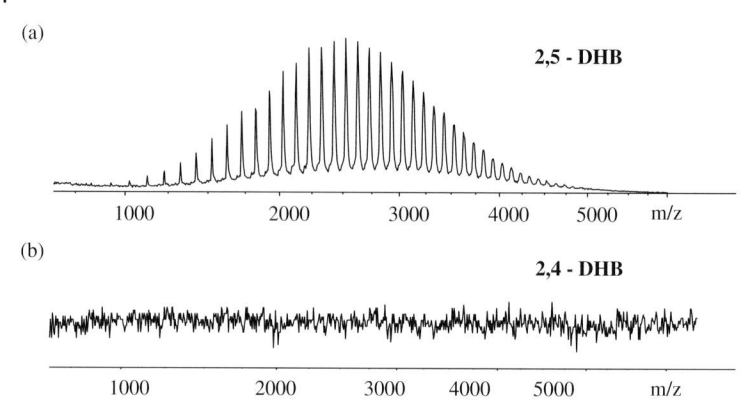

Figure 5.1 MALDI mass spectra of PMMA 2900 prepared with 2,5-DHB (a) and 2,4-DHB (b). The 2,5-DHB is an effective MALDI matrix due to excellent absorbance of the 337 nm laser light. The 2,4-DHB is not effective due to poor absorbance of the laser light.

MALDI. It was a failure of the specific sample preparation method used to explore polymer MALDI.

5.2.3
Efficient Desorption

The energy of the laser that is absorbed by the matrix must efficiently deliver intact analyte molecules into the gas phase inside the mass spectrometer. The interaction of the laser with the solid breaks the relatively weak intermolecular bonds in the matrix crystal. However, there is sufficient energy available to break covalent bonds in the matrix molecules themselves; for example, many of the most popular MALDI matrices are alpha hydroxy acids (like 2,5-DHB), which readily decompose to produce water. Breaking of both the inter- and intramolecular bonds results in the generation of a localized high pressure at the sample surface during the few nanoseconds of the laser pulse. The material in this high-pressure region then expands into the vacuum chamber, resulting in a large number of analyte molecules being ejected into the mass spectrometer. The characteristics of this expansion closely mirror those of a gas beam created by a pulsed valve typically used in supersonic molecular beam experiments [12]. In many ways, the polymer analytes are simply bystanders to the matrix/laser interaction and find themselves transiting to the gas phase along with the volatiles created by the decomposition of the matrix.

The desorption event created by the matrix/laser interaction is a key to the polymer MALDI experiment. This event does not directly involve the analyte. The analyte does not significantly absorb the laser light. The matrix/laser interaction does not impart significant energy to the analyte. The desorption event creates a "cold" process to transition the polymer analyte into the gas phase. One of the key results of this desorption event is intact, individual gas-phase polymer molecules ready for mass spectrometric detection.

(a)

(b)

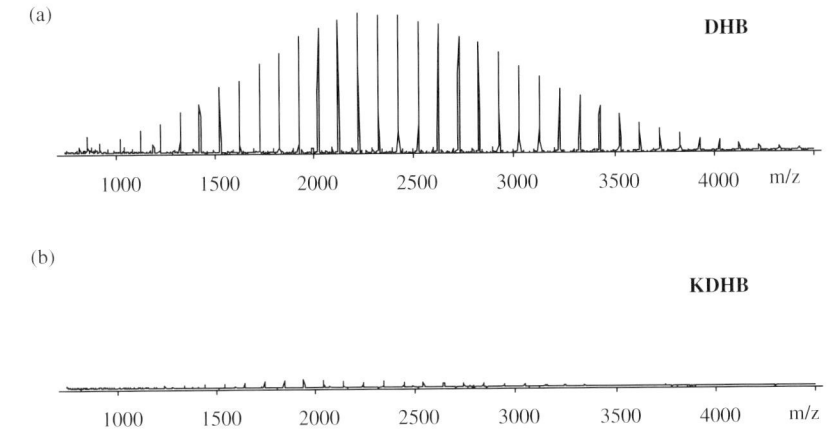

Figure 5.2 MALDI mass spectra of PMMA 2900 prepared with DHB (a) and the potassium salt of DHB (KDHB, b) as the matrix. The DHB effectively desorbs the polymer analyte while the KDHB does not.

If something in the sample preparation method interferes with the desorption process, the MALDI experiment will fail. Figure 5.2 shows MALDI mass spectra of PMMA 2900 prepared with 2,5-DHB (Figure 5.2a), and the potassium salt of 2,5-DHB (KDHB, Figure 5.2b).

Using the acid form of the matrix, a typical MALDI mass spectrum of the PMMA 2900 is observed. If the salt form of the matrix is used, however, vastly reduced signal intensity for the polymer is observed. The salt form of the matrix appears to no longer readily decompose to produce the intense gas pulse which drives the desorption event. Interestingly, the trace of analyte signal observed in the bottom mass spectrum of Figure 5.2 is thought to be derived from the residual acid form matrix present in the sample.

5.2.4
Effective Ionization

The last major role of the matrix is to ensure that the molecules injected into the gas phase have a route to ionization. The mass spectrometer requires charged species to separate. If the analyte is a preformed ion, like a quaternary amine, then this role of the matrix is not needed. Most polymers, however, are naturally neutral and require assistance to ionize.

Typical neutral polymers will readily cationize with different ions, depending on their chemical structure:

- Basic species, like amine functional polymers, will protonate ($+ H^+$).
- Oxygen-containing species, like polyethers and polyesters, will adduct with alkali ions ($+ Na^+$, for example).
- Unsaturated hydrocarbon polymers, like polystyrene or polybutadiene, will adduct with transition metal ions ($+ Ag^+$, for example).

In the MALDI experiment, there is a ready supply of protons available. If the analyte is basic, nothing further is required to ensure a route to ionization. In this case, there is similarity to MALDI of peptides or proteins, which very often have a basic amine site that drives ionization through protonation.

For nonbasic, neutral polymers, however, another route to ionization is required for a successful MALDI experiment. Typically, a metal salt will be added to the organic matrix to provide the ionization. For solvent-based sample preparation methods, trifluoroacetate (TFA) salts are most commonly used due to their solubility in solvents commonly used in polymer MALDI. The most common cationization agents are NaTFA and AgTFA.

The need for a cationization agent for polymer MALDI is illustrated in Figure 5.3.

The figure clearly shows the impact on the MALDI analysis of polystyrene (PS) 5050 prepared with and without the AgTFA cationization agent. With the presence of the metal salt, the PS mass spectrum is clearly observed. Without the metal salt present to provide cationization, the MALDI experiment fails.

There has been significant debate in the MALDI community about the origins of the ions observed. The two main camps in the debate are proponents of preformed ions and post-desorption formed ions. The preformed ion position is that the majority of the ions observed in the MALDI mass spectra are formed in solution during the sample preparation step. These ions are stabilized by the organic matrix and then liberated during the desorption step. The postdesorption ion position claims that the majority of the observed ions are formed from reactive collisions in the transition from the solid phase to the gas phase during the desorption step. This position would require neutral species during the sample preparation process and rely on the thousands of collisions that occur during desorption. We tend toward the postdesorption camp and view the MALDI source as a powerful chemical reactor. The key products of this reactor are the cationized polymers that are observed in the MALDI mass spectra.

In the early days of polymer MALDI, a considerable effort was expended to find the perfect matrix, a single compound that would provide all of the needs of the experiment. By understanding the different roles of the matrix, it has been shown that different molecules can be added to the sample preparation method to account

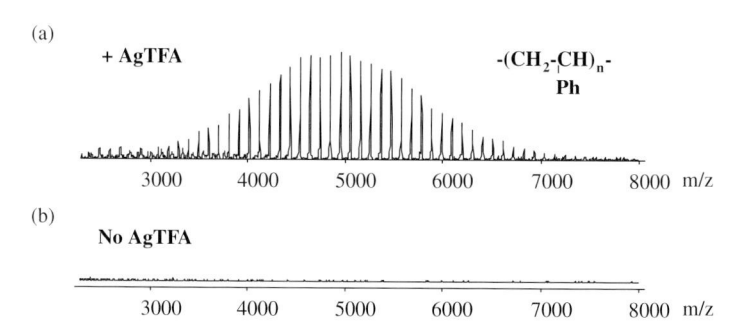

Figure 5.3 MALDI mass spectrum of PS 5050 prepared with (a) and without (b) the silver trifluoroacetate (AgTFA) cationization agent. Without the addition of the Ag, no ionization is observed.

for each of the roles. There is no longer a need or desire for a "silver bullet" matrix compound that does everything.

5.3
Choice of Matrix

With the diversity of small molecule chemistry, there is a considerable list of molecules that will significantly absorb the laser light and decompose into volatile fragments. This list then constitutes the starting point for a decision of what matrix to choose for a specific polymer MALDI experiment [13]. The decision about which matrix to choose for the experiment at hand is then driven by the need to generate intimate contact between the matrix and the analyte. Intimate contact can be generated in different ways. For MALDI experiments, there are two primary pathways to intimate contact, either solution chemistry using appropriate solvents and deposition methods, or solvent-free sample preparation. Each of these approaches will be discussed below.

To better understand the nature of the relationships between the analyte and matrix in the solid samples introduced to the MALDI instrument, a combination of solvent solubility and time-of-flight secondary ion mass spectrometry (ToF-SIMS) experiments were conducted [14, 15]. The results of these experiments enabled the generation of a guide to selecting the matrix for polymer MALDI. Figure 5.4 shows the relative hydrophilicity/hydrophobicity for a series of common polymer MALDI matrices (Figure 5.4a) and common polymer materials (Figure 5.4b).

Figure 5.4 can be a very effective guide to selecting the first best choice for a novel polymer for MALDI analysis. The guide works in two steps. First, understand the dissolution characteristics of the polymer in common solvents and compare it to the common polymers on the right side of the figure. Choose the common polymer with dissolution characteristics closest to the new polymer. Second, move across the figure to the left side and choose the matrix with a similar relative hydrophilicity/hydrophobicity. This guide can be used to dramatically reduce the development time for new sample preparation methods for polymer MALDI [16]. What may have once taken several weeks to develop can now be done in hours.

To help new practitioners of polymer MALDI, there are a couple of listings of previously observed MALDI results [5, 17]. The list of polymer MALDI recipes published online by the polymer group at NIST [17] is particularly helpful.

5.4
Choice of the Solvent

For solution-based MALDI sample preparation, clear consideration must be given to making good solutions of the analyte, the matrix, and any cationization agent needed for the experiment. Partially dissolved materials are a clear sign of trouble for the MALDI experiment. Especially for complex analytes, partial dissolution can leave

(a) Matrices

Hydrophilic ↑

TU
DHB

IAA
CHCA
FA

Dith
DPDB
RA

Hydrophobic ↓

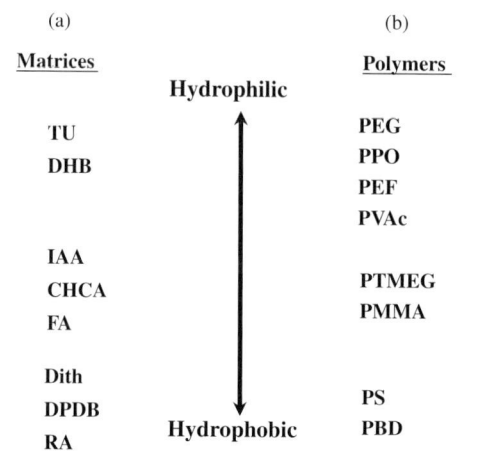

(b) Polymers

PEG
PPO
PEF
PVAc

PTMEG
PMMA

PS
PBD

Figure 5.4 Relative matrix and polymer solubility scale. The more hydrophilic matrices and polymers are at the top of the figure and the more hydrophobic matrices and polymers are at the bottom of the figure. Best MALDI results occur when the matrix is a good match of relative hydrophilicity/hydrophobicity to the polymer analyte. Abbreviations: DPBD = diphenylbutadiene, Dith = dithranol, Ret A = all-trans retinoic acid, FA = ferulic acid, TU = thiourea, DHB = 2,5-dihydroxybenzoic acid, IAA = indole acrylic acid, CHCA = α-cyano-4-hydroxycinnamic acid, PEG = polyethylene glycol, PPO = polypropylene oxide, PEF = polyethynylformamide, PVAc = polyvinylacetate, PTMEG = polytetramethylene glycol, PMMA = polymethylmethacrylate, PS = polystyrene, PBD = polybutadiene. Reproduced from Ref. [15] with permission.

behind a whole set of components and blind the MALDI experiment to the full extent of the material to be analyzed. If using solvent-based sample preparation, it is worth the search for effective solvents that fully dissolve all of the components of the experiment.

Due to the fact that the solvent must be evaporated from the sample prior to analysis, there is significant advantage from an intimate contact perspective to using a single solvent or an azeotrope for the entire experiment. The use of multiple solvents, either as mixtures to dissolve the analyte or from the use of different solvents for the analyte and the matrix, leads to sequential evaporation during the deposition step. This sequential evaporation can lead to sequential precipitation from solution, which can result in clear separation of the key components for the MALDI experiment, and to failure of the experiment. Separation has been observed experimentally by several groups, including by fluorescence microscopy [18], time-of-flight secondary ion mass spectrometry [14], and MALDI imaging [19].

In addition, superior results are obtained by using solvents that evaporate rapidly. As the solvent evaporates, the matrices will start to recrystallize. Recrystallization is a method of purification familiar to most synthetic organic chemists. Purification is exactly the opposite of what is desired for successful MALDI sample preparation. To have the intimate contact required for a good MALDI experiment, the recrystallization process must be prevented. By evaporating the solvent rapidly, the process of going from the liquid phase to the solid phase can be driven by the kinetics of the solvent leaving. If the solvent evaporates slowly, then the process will be driven by thermodynamics, resulting in the most energy optimized end point, which is likely the relatively pure matrix crystal with the analyte excluded. Some very effective solvents which can dissolve a wide range of polymer analytes and rapidly evaporate include methanol, tetrahydrofuran (THF), isopropanol, and acetone.

5.5
Basic Solvent-Based Sample Preparation Recipe

After the specific choices of solvent, matrix, and cationization agent are made, for relatively low mass polymers (up to about 15 000 u), a relatively simple basic recipe works for many polymer samples:

- 5 mg ml^{-1} solution of the analyte;
- 0.25 M solution of matrix;
- 5 mg ml^{-1} solution of the cationization agent;
- mix the three solutions by volume 1 : 10 : 0.5 (analyte: matrix: cation).

This simple recipe assumes success with a number of other variables. Some of the most important will be discussed below, such as matrix-to-analyte (M/A) ratio, salt-to-analyte (S/A) ratio, and deposition method. Understanding these more complex parameters of polymer MALDI sample preparation will enable the analysis of more complex analytes and generate more quantitative results.

5.6
Deposition Methods

Certainly the simplest method of sample deposition is the dried-droplet method originally developed by Karas and Hillenkamp [2], where a small volume (0.5–2 µl) of sample solution is deposited on the MALDI sample target and the solvent is allowed to evaporate at room temperature and pressure. Several modifications to this basic deposition method have been developed, most by those preparing peptide or protein samples for analysis from high water content solutions. The simplest modifications of the dried-droplet method involve accelerating the evaporation of the solvent. Increased solvent evaporation can be obtained by heating the sample, placing the samples under vacuum, or by flowing dry air or nitrogen over the sample surface. The "fast-evaporation method" described by Vorm and coworkers involves predepositing a layer of small matrix crystals from a fast-drying solvent such as acetone [20]. The analyte is then deposited from a high water content solution on top of this previously deposited layer. Analysis of this layered sample yields a more consistent analyte signal across the sample spot. An advantage in polymer MALDI is that most polymer samples require the use of nonaqueous solvents, many of which are significantly more volatile, and thus faster drying than water. Preparing a polymer sample in a dried-droplet fashion from a fast-drying solvent results in the production of smaller matrix crystals and more intimate contact of the various sample components by decreasing the amount of time for sample separation to occur during the drying step. In contrast, while solvents such as water, dimethyl sulfoxide (DMSO), and dimethyl formamide (DMF) can be effective solvents for a wide range of polymer materials, their low volatility and slow evaporation rates make them problematic for dried-droplet MALDI analyses.

Note that due to the higher rate of evaporation, it is generally preferable to premix the sample components, rather than spotting them down on the sample plate sequentially. It is the high rate of evaporation that produces a more uniform sample surface composed of smaller matrix crystals; these crystals are also more impure, as less time is available for thermodynamically driven separation of the different sample components. This is particularly important for those MALDI matrices that were initially developed for the analysis of proteins. One of the advantages of the widely used CHCA matrix is its ability to segregate alkali salts while drying [21], which reduced the quantity of alkali salt adducts in the observed mass spectrum. This ability of CHCA to segregate salts would be detrimental to the analysis of polymer samples which require the alkali cations to ionize. The use of a fast-drying solvent is an effective means to preserving the required intimate contact between the different sample components.

Generally, more reproducible MALDI signals are observed from smaller sample crystals. Smaller crystals can also be produced by the use of smaller droplets of solution, which dry faster due to their correspondingly higher surface area. Faster evaporation of these smaller sample droplets has the additional advantage of allowing even less time for undesired sample segregation to occur. Electrospray deposition is an excellent method to produce highly reproducible MALDI samples. It has been used to analyze samples of low-molecular-weight peptides [22, 23], proteins [24], as well as synthetic polymers [15, 25]. The extremely small droplets of solution produced by the electrospray process (a few μm in diameter) effectively dry in the short distance of travel (~2 cm) between the tip of the electrospray needle and the sample plate. The remarkable uniformity of MALDI samples created by electrospray deposition has been studied using advanced microscopy techniques, such as scanning electron microscopy (SEM), atomic force microscopy (AFM), and imaging secondary ion mass spectrometry (SIMS) [25]. For example, Figure 5.5 shows an SEM image of a MALDI sample prepared by electrospray deposition of PMMA 2900 using DHB as the matrix and methanol as the solvent and spraying from a height of 15 mm. Figure 5.6 is the corresponding AFM image of the same sample surface.

The resulting solid droplets on the surface (which appear as flattened spheres ~250 nm in diameter for DHB) contain the matrix, analyte, and additives (including the cationization agent) in extremely close proximity. Even if the sample components are segregated within these solid droplets, the distance between the materials is insignificant compared to the size of the desorption laser spot, which is generally on the order of 50 μm in diameter. With adequate laser intensity entire solid droplets (which are amorphous solids, not small crystallites) are desorbed from the sample surface. These homogeneous samples often generate much more consistent MALDI

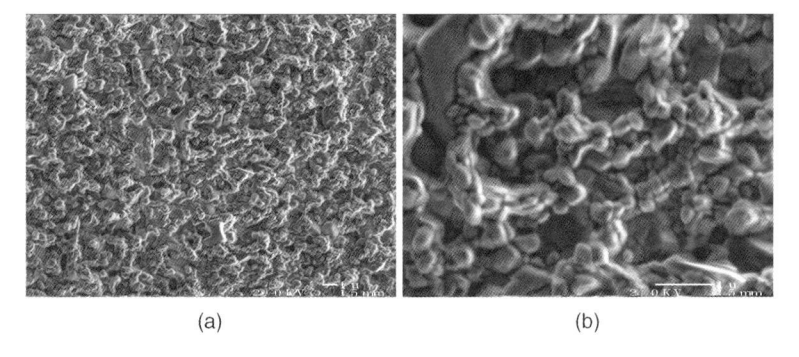

Figure 5.5 SEM images obtained from a PMMA 2900 sample prepared using DHB as the matrix and methanol as the solvent electrosprayed from a height of 15 mm under low magnification (a) and high magnification (b). The images show solid droplets on the surface with an approximate diameter of 250 nm. Reproduced from Ref. [25] with permission.

data than even the best attempts at dried-droplet deposition. The reproducibility of the polymer analyte signal can be significantly improved using electrospray deposition [25]. Calculated values of M_n and M_w were shown to exhibit a 3× improvement (from 0.51 to 0.17% relative standard deviation, RSD), while the reproducibility of the absolute signal area showed an ∼10× improvement from 45 to 4% RSD. The absolute peak area also increased by a factor of 3 for the electrospray deposited samples. In addition to electrospray deposition, there are a number of other techniques that may be employed to produce homogeneous MALDI samples which are sometimes preferred because they do not involve the use of high voltages. These include the aerospray device described by Wilkins and coworkers [26, 27], and the oscillating capillary nebulizer (OCN) device reported by Browner and coworkers [28–30]. In the aerospray device, the flow of gas past the end of a small capillary immersed in an analyte solution is used to entrain small droplets of the

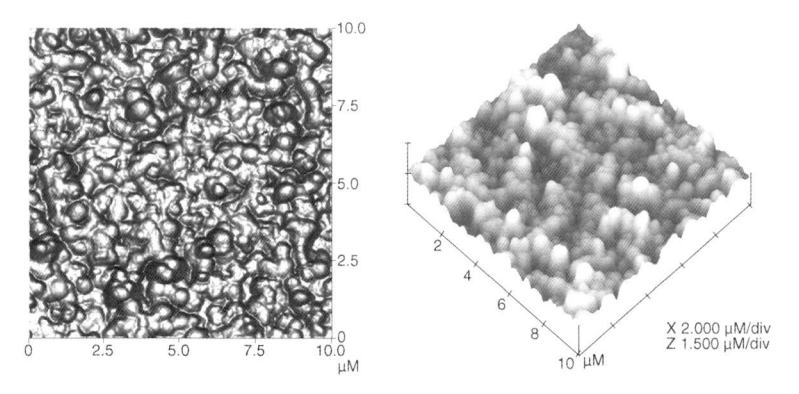

Figure 5.6 AFM images from a sample of PMMA 2900 prepared using DHB as the matrix and methanol as the solvent, electrosprayed from a height of 15 mm. The images show solid droplets on the surface with an approximate diameter of 250 nm. Reproduced from Ref. [25] with permission.

sample in the nebulizing gas through the Venturi effect [26]. In essence, the sample is "painted" on the sample target plate surface in a manner similar to an "air-brush" technique. The OCN was originally developed by Browner to produce small (i.e., low micrometers size) monodisperse aerosol droplets for coupling low-flow rate devices such as HPLC to an inductively coupled plasma for trace metals analysis [28]. In the OCN, a low flow of sample (low μl min^{-1}) is pumped using a syringe pump through a small diameter flexible fused silica capillary which is coaxially mounted in a slightly larger internal diameter (i.d.) fused silica capillary through which a high gas flow is passed. The gas flow induces an oscillation in the internal liquid-filled capillary which acts to break the exiting liquid stream into very small and reproducible droplets which are then entrained in the expanding gas flow. A number of groups have used the OCN to prepare samples for MALDI analysis [31, 32]. Dally recently described a sample preparation device that appears to be a cross between the aerospray and OCN device [33]. The liquid sample is pumped through the inner capillary using a syringe pump; however, no specific attention was paid to the flexibility of the inner capillary and the inner diameter of outer tubing used for the gas sheath was much larger than that used in a traditional OCN. Holcomb and coworkers have investigated the use of this "modified aerospray device" to produce high salt content MALDI samples for analysis and found that excellent reproducibility can be obtained by proper optimization of the experimental variables [34].

Not all samples require the use of one of these spray deposition techniques; highly reproducible results *can* be obtained using the dried-droplet method with a good match of matrix, analyte, cationization agent, and solvent. The basic solvent-based sample preparation recipe described above can be effectively used, for instance, for a low molecular-mass analyte such as polyethylene glycol (PEG) 1500, using DHB as the matrix, sodium trifluoroacetate (NaTFA) as the cationization agent, and THF as the solvent. However, as the match between matrix, analyte, cationization agent, and solvent becomes less, the advantage of use of one of the spray deposition techniques increases.

5.7
Solvent-Free Sample Preparation

As the polymer MALDI community learned about the importance of effective dissolution and intimate contact with the matrix for the success of polymer MALDI experiments, it became clear that to analyze some materials, a solvent-free approach would be needed. Work on some very difficult to dissolve and even insoluble polymers and copolymers could not proceed with the solvent-based methods. Around the same time, several different labs published initial methods to prepare samples for MALDI without the use of solvents [35–38]. These different methods were all developed independently and shared some important aspects. With evaluation by the broader polymer MALDI community, most labs now follow the lead of Trimpin and coworkers [39]. The solvent-free methods show two very significant advantages compared to the more traditional solvent-based methods:

5.7 Solvent-Free Sample Preparation

1) No solvent is used, which leads to no worries about choosing the right solvent, or dealing with analyte solubility issues.
2) The experiments produce high-quality mass spectra.

The solvent-free sample preparation methods opened up a significant variety of new materials to MALDI analysis. Many different kinds of poorly soluble or insoluble polymers could now be analyzed. Figures 5.7 and 5.8 show novel electrolytically polymerized materials that were completely insoluble in any organic solvent tried in previous attempts of analysis.

Figures 5.7 and 5.8 show high-quality MALDI mass spectra of novel materials that were very difficult to analyze due to solubility issues. The solubility issues greatly limited the usefulness of other typical analytical methods like NMR and chromatography. The bulk of the chemical structure information determined for these novel materials came from MALDI, which was only possible through application of the solvent-free sample preparation methods.

One of the interesting side results of the development of solvent-free MALDI sample preparation methods was the observation of excellent mass spectral quality for both sample preparation methods.

Figure 5.9 shows MALDI mass spectra of PS 5050 prepared with both a solvent-free sample preparation method and a typical solvent-based method. The matrix was retinoic acid and the cationization agent was AgTFA in both experiments. The only difference was the use of THF in the solvent-based method, and dry grinding with mortar and pestle in the solvent-free method. Not only does the solvent-free method work very well to analyze this polymer material, it generates mass spectra with just as high mass spectral quality. Both experiments were conducted on the same instrument in back-to-back experiments.

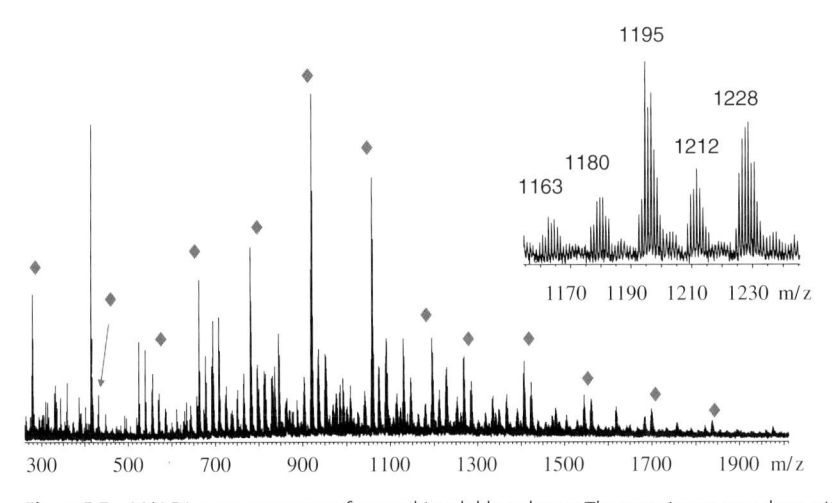

Figure 5.7 MALDI mass spectrum of a novel insoluble polymer. The experiment was done with solvent-free methods from <0.1 mg of material.

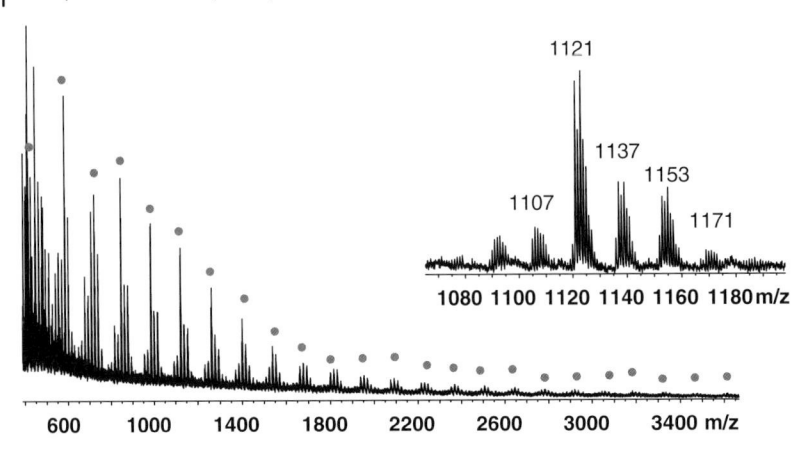

Figure 5.8 MALDI mass spectrum of a novel insoluble polymer produced with solvent-free methods from an emulsion containing <1% solids in water.

5.8
The Vortex Method

While the solvent-free experiments worked very well, two different issues were encountered. Since poorly soluble materials were being analyzed, cleaning the tools became very difficult, and the expensive ball mill specified by Trimpin and coworkers was not accessible to all labs. The problems with clean up manifested as carryover of one material into the next experiment. To address these issues, Parees and Hanton developed the vortex method [40]. The vortex method took the basic tenants of Trimpin's method and used simpler, more available tools, and cheaper components that could be readily discarded, rather than cleaned after use.

The basic recipe for the vortex method is the following:

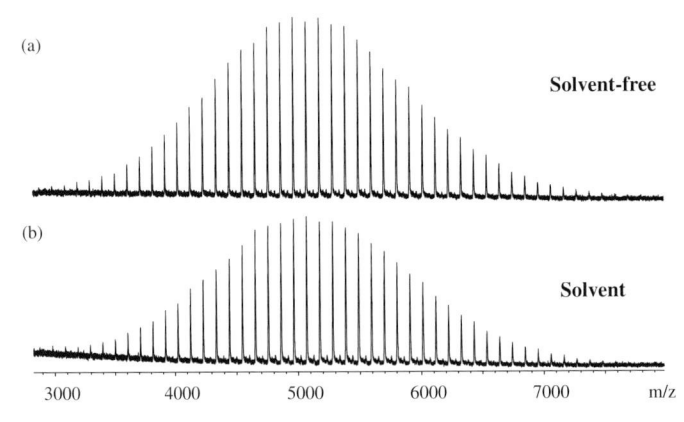

Figure 5.9 MALDI mass spectra of PS 5050 prepared using a solvent-free method (a) and typical solvent-based method (b). The solvent-free method produces mass spectral quality just as high as the traditional solvent method.

- Add <0.1–1.0 mg of analyte to a small glass vial.
- Add 20–100 mg of solid matrix to the vial.
- If needed, add <0.1–0.5 mg of cationization agent to the vial.
- Add 2 BBs to the vial (small round shot used in air rifles, otherwise known as BB guns in the United States).
- Vortex at high speed on a standard lab vortex mixer for ≤1 min.
- Remove a small amount of material with a metal spatula.
- Apply the material to the target plate to form a thin film.

The vortex method has been shown to be a simple method that is applicable to a wide range of analytes. Figure 5.10 shows different PEG analytes analyzed with the vortex method using DHB as the matrix.

The PEG standards shown in Figure 5.10 range from liquids to waxes to partially crystalline solids. The vortex method is sufficiently robust to prepare successful MALDI samples from each of these materials.

The key to the vortex method is the grinding and mixing effect of the two BBs. The solid analyte, matrix, and cationization agent are all mixed together and ground into a very fine powder. During the grinding and mixing, the conditions are created for intimate contact between the analyte and the matrix. Studies of the morphology of samples prepared by the vortex method show that the matrix crystals are ground into remarkably uniform and small particles [41]. This same study also demonstrated that the MALDI samples produced from the vortex method, despite it is apparent simplicity, are remarkably smooth and complete thin films. Figure 5.11 shows a scanning electron microscope (SEM) image of a vortex prepared sample for MALDI at 2000× magnification.

Even at 2000×, it is difficult to measure the particle size of the dithranol matrix in the sample shown in Figure 5.11. One of the benefits of the solvent-free sample

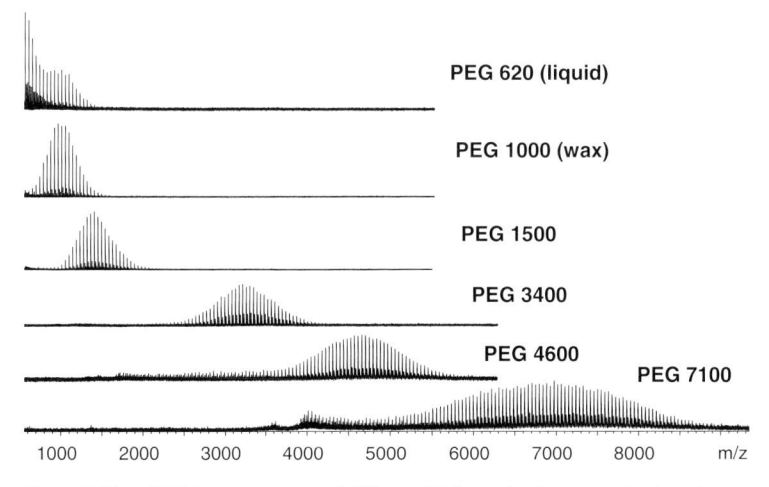

Figure 5.10 MALDI mass spectra of different PEG standards prepared using the vortex method with DHB as the matrix and mixing for about 60 s.

Figure 5.11 SEM image of a vortex prepared sample of PS 2450 with dithranol as the matrix and silver trifluoroacetate (AgTFA) as the cationization agent. The sample was vortexed for 60 s. The sample shows a very diverse morphology. Used by permission from Ref. [41].

preparation methods is that they create a very diverse morphology. This diversity increases the chances that whatever environment will generate the best MALDI experiment for that combination of matrix, analyte, and cationization agent will occur somewhere within the laser volume.

It seemed remarkable that this much work can be accomplished in the vortex method by the two BBs in only 60 s until experiments with the vortexing time were done. Further studies have shown that high-quality MALDI data can be acquired using as little as 10 s of vortex time and that recognizable MALDI spectra can be obtained from as little as 2 s of vortex mixing [42]!

So far, most of the solvent-free MALDI experiments have been done on relatively low-molecular-mass polymers, less than about 10 000 Da. It seems that the Trimpin ball mill method has greater applicability to higher mass analytes. The vortex method may not impart sufficient energy to the system to prepare higher mass analytes.

5.9
Matrix-to-Analyte Ratio

The requirement for a matrix separates the older LD technique, which is generally only successful for low MW analytes, from the MALDI technique. While it was understood early in the development of MALDI that a significant excess of matrix was required, the proper matrix-to-analyte ratio (M/A) was not well understood. Different results can be obtained from the MALDI experiment, even with the same analyte, matrix, and cationization agent if the relative concentrations of the matrix and analyte are significantly different. Early work by Chavez-Eng on the quantitative analysis of peptides and proteins by MALDI showed that the proper choice of the M/A is key to a successful analysis [24]. Figure 5.12 was obtained by creating a series of samples where the quantity of matrix (in this case 4-hydroxy 3-methoxycinnamic acid or ferulic acid) is fixed and the amount of analyte (DDAVP = desamino-8-d-arginine vasopressin, MW = 1069 Da or bovine insulin, MW = 5733 Da) is varied to create samples with varying M/A. The analytes were dissolved in distilled water while the matrix was prepared in methanol; the final sample solutions were mixed in a volume ratio of 5 : 95, respectively, and prepared for analysis via electrospray deposition. The peak areas for the observed protonated peptide/protein species are measured and plotted with M/A increasing while moving to the right in the figure. The M/A curves observed in Figure 5.12 are typical of the results observed for many different analytes, including polymers. Figure 5.13 shows the M/A plot for a PEG 3400 sample obtained using DHB as the matrix, NaTFA as the cationization agent, and methanol as the solvent. Electrospray deposition is used to create the samples. In our experience, the M/A plots for polymer samples are not as "peaked" as those observed for peptides or proteins, but show a similar trend that the peak signal for higher MW polymers is observed at higher M/A values. These M/A plots support the empirical observation that more matrix is required to get the more difficult to analyze, higher molecular mass analytes to generate a significant signal in the MALDI experiment. Understanding the M/A plot is also critical to extending MALDI to successful quantitative experiments. In contrast to a "traditional" analytical calibration curve, smaller quantities of analyte present in the sample occur on the right in these M/A plots. Note in Figure 5.12 that each of the analytes exhibits a linear range for analysis (e.g., at M/A >10 000 for bovine insulin); however, it is also noted that the protein analyte signal "saturates" at the M/A value of ~8000 for bovine insulin and 1000 for DDAVP, and then decreases quickly at lower M/A values. Presumably at low values of M/A, there is not enough matrix present in the MALDI sample to effectively desorb and ionize the analyte. The exact position of the peak in the M/A plot as well as the steepness of the curves before and after the peak is found to be both analyte and matrix dependent. Note that it is difficult to obtain such informative M/A curves without using some type of spray deposition, as the large spatial variability in the

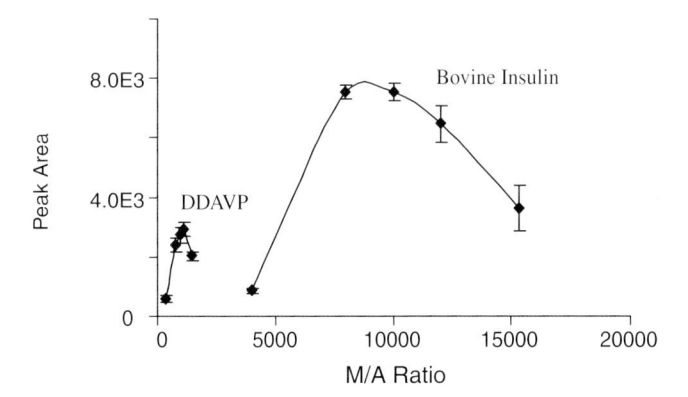

Figure 5.12 Matrix-to-analyte plots for DDAVP and bovine insulin in the ferulic acid matrix using methanol as the solvent. The samples were prepared using electrospray deposition. Reproduced from Ref. [24] with permission.

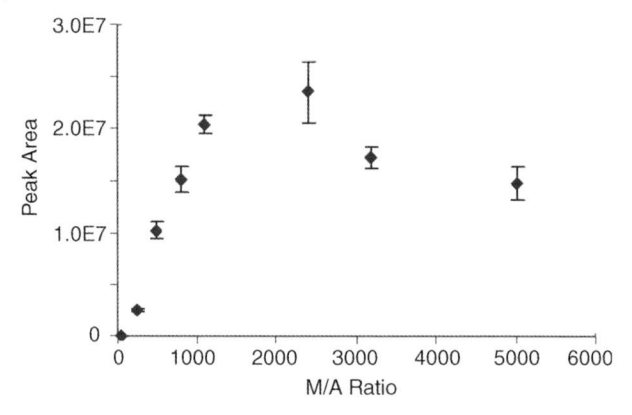

Figure 5.13 Matrix-to-analyte plot for PEG 3400 in the 2,5-dihydroxybenzoic acid matrix, using sodium trifluoracetate as the cationization agent and methanol as the solvent. The samples were prepared using electrospray deposition.

analyte signal obtained from dried-droplet samples leads to extremely large error bars that generally obscure the shape of the observed M/A plots. As will be discussed below concerning the analysis of polymer blends, when wishing to quantitate, it is important to work to the right of the saturation point of the M/A curve where the measured MALDI peak areas change linearly with the amount of analyte present.

5.10
Salt-to-Analyte Ratio

Most polymer analytes are not observed to readily protonate or deprotonate in the MALDI experiment. They require the presence of an alkali or transition metal cation in the MALDI sample to efficiently ionize. For those polymer analytes that have an affinity for alkali cations (e.g., oxygen-containing materials such PEG, PPG, and PMMA), most matrices contain sufficient adventitious sodium that strong signals are readily observed. Other polymers (such as PS and PBD) require the addition of transition metal cations such as Ag or Cu in order for any MALDI signal to be observed. While it is understood which cationization agents to add to efficiently generate MALDI ions from polymer samples, little work had been done to investigate the effect of the amount of cation added to the sample on the MALDI signal observed. Hoteling's initial work in this area involved the study of PMMA 6300 as an analyte with a number of matrices (DHB, dithranol and trans-indole acrylic acid, IAA), using NaTFA as the cationization agent and THF as the solvent [43]. To undertake these studies, a series of MALDI samples are prepared with a constant amount of matrix and analyte, and an increasing amount of the chosen cationization agent. In these studies, the M/A is held constant (note that the M/A plot for the analyte/matrix composition needs to be defined first) while the salt to analyte ratio (S/A) increases from sample to sample. In effect, the analyte is being "titrated" with salt, and the

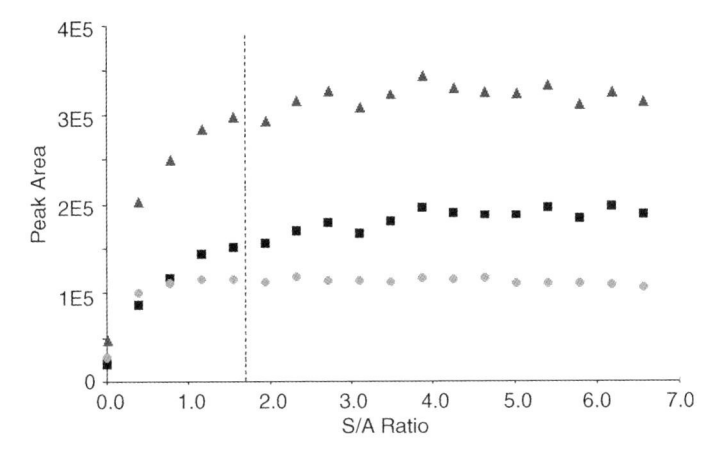

Figure 5.14 Peak area plotted versus salt-to-analyte ratio for three telomer peaks of PMMA 6330 (square: 4430 Da, circle: 6532 Da, triangle: 5531 Da) using 2,5-dihydroxybenzoic acid as the matrix (prepared at a matrix-to-analyte ratio of 1227) and sodium trifluoracetate as the cationization agent. Reproduced with permission from Ref. [43].

expectation is to observe a plot much like a traditional titration curve. An example S/A plot determined using DHB as the matrix (at an M/A of 1227) is shown in Figure 5.14. This is an ideal case, as the peak area of the various PMMA telomers increase linearly to S/A ~1, and then remain constant as excess alkali was added to the sample. Surprisingly, the signal seems to level off at a value very close to the stoichiometric 1 : 1 ratio, indicating a highly sensitive determination.

Unfortunately, not all combinations of analyte, matrix, and cationization agent lead to an ideal case. As shown in Figure 5.15, when titrating a PMMA 6300 sample with

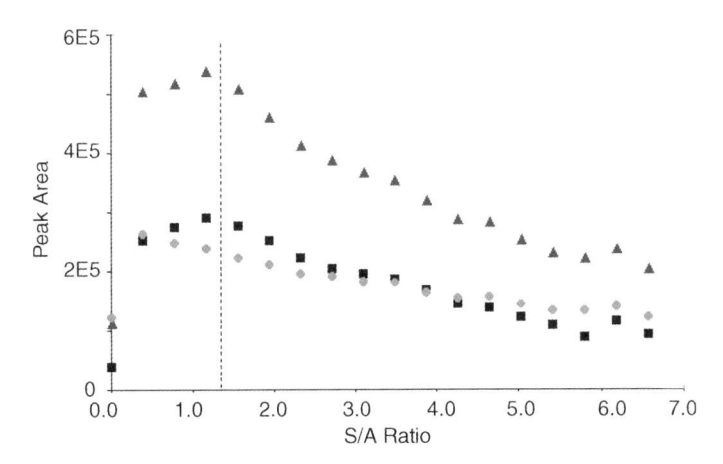

Figure 5.15 Peak area plotted versus salt-to-analyte ratio for three telomer peaks of PMMA 6330 (square: 4430 Da, circle: 6532 Da, triangle: 5531 Da) using indole acrylic acid as the matrix (prepared at a matrix-to-analyte ratio of 1010) and sodium trifluoracetate as the cationization agent. Reproduced with permission from Ref. [43].

IAA as a matrix at M/A = 1010 and NaTFA as the cationization agent, a nonideal case is obtained. In this example, the analyte signal increases as the S/A increases, but after reaching a maximum at S/A ~1, the analyte signal is observed to steadily decrease as additional salt is added. This is significant, as the analyte signal decreases if too much cationization agent is added to the sample. If the amount of analyte present in the sample is not known, it would be difficult to determine the correct amount of cationization agent to add. This is further complicated by the fact that the amount of residual alkali present in the matrix, or extracted from the glassware used in the sample preparation, is often completely unknown. It was found in the case of IAA that this nonideal S/A curve could be converted into an ideal curve by increasing the M/A. Conversely, decreasing the M/A for DHB to ~300 converted the originally ideal S/A curve (Figure 5.14) into a nonideal case. It is proposed that the observation of ideal versus nonideal S/A curves is related to the interactions between the matrix and the salt during the MALDI experiment, such as by the observation of the ease of creation of matrix salt clusters [43].

This initial S/A work was extended by Erb who studied the combination of PMMA 6800 as the analyte with four commonly used MALDI matrices (CHCA, DHB, dithranol, and DCTB) and three alkali cations (Li, Na, and K) added to the sample preparation as the TFA salts [44]. All of these samples were prepared by electrospray deposition and the peak area reproducibility (as measured by the % CV) for most samples was found to be 10% or less. Erb observed both ideal and nonideal S/A curves, as well as a third case, which showed the analyte signal to increase linearly until reaching a plateau at S/A significantly higher than 1. The exact cause for this is still under study, but may be due to competition between the analyte and matrix for the alkali cations available in the MALDI sample. Through this work, it has been found that the MALDI analyte signal is often nonlinearly related to the amount of cationization agent present, suggesting that control of the amount of salt added to a sample is critical for an optimum analysis, and especially for experiments attempting to generate quantitative results.

5.11
Chromatography as Sample Preparation

As described in more detail in the Hyphenated Techniques chapter of this book, chromatography is one of several well-established analytical methods that can be used in combination with MALDI. Early investigations into the analysis of polymers by MALDI showed that the best results were obtained from narrow polydispersity polymers [45]. The rule of thumb for MALDI analysis of polymers of different polydispersity is as follows:

- PD < 1.25: excellent MALDI mass spectra with excellent molecular mass quantitation;
- 1.25 < PD < 1.5: good MALDI mass spectra with reasonable M_n determination, but errors in the M_w measurements;

- $1.5 <$ PD 1.75: OK MALDI mass spectra with errors in both M_n and M_w; molecular mass measurements erring low significantly;
- $1.75 <$ PD < 2.0: poor MALDI mass spectra with no ability to measure molecular masses;
- PD > 2.0: often no recognizable MALDI mass spectra.

One good way to address the issues presented by broad polydispersity polymer samples is to separate the sample into several narrow polydispersity samples first. An effective way to separate polymer samples is to use gel permeation chromatography (GPC). An example of using GPC to separate a relatively broad material into fractions for MALDI analysis is shown in Figures 5.16 and 5.17 [46]. In this example, a polytetramethylene glycol (PTMEG) material having a polydispersity of about 1.7 was separated with GPC and the eluent was captured on a MALDI target in a continuous fashion [47]. Figures 5.16 and 5.17 show the MALDI mass spectra obtained from different spots on the target, which correspond to different elution volumes of the GPC experiment.

Each of the positions analyzed by MALDI shows a robust and effective mass spectrum. The MALDI mass spectra clearly characterize each of the spots analyzed. These data can be inspected to find impurities, like the extra set of peaks observed in the 34.2-min spot, and to calculate average molecular weights that are used to calibrate the GPC for this material. In essence, this creates a novel calibration curve for the GPC instrument specific to PTMEG.

Other chromatography experiments can also be used to help improve MALDI sample preparation for polymers. Any chromatography method that separates the

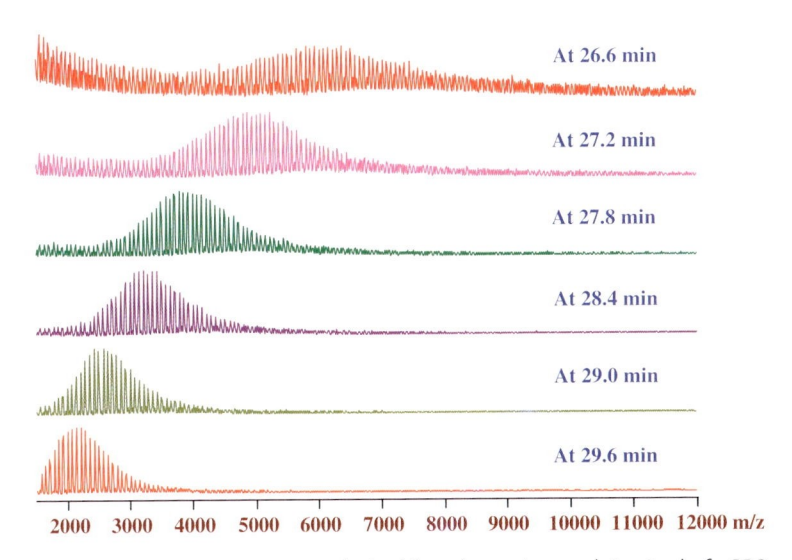

Figure 5.16 MALDI mass spectra obtained from the continuous elution track of a GPC separation of PTMEG 1000. The GPC eluent was deposited directly on a track of 2,5-dihydroxybenzoic acid matrix. Each different elution time produced a narrow distribution of PTMEG telomers.

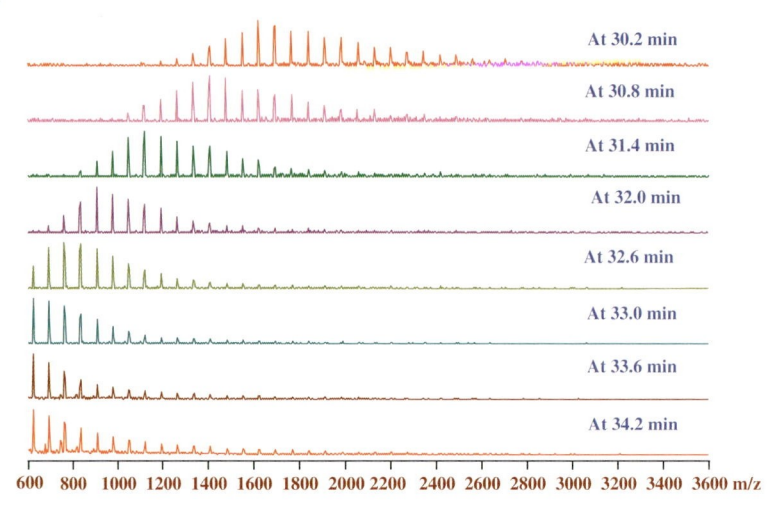

Figure 5.17 MALDI mass spectra obtained from the continuous elution track of a GPC separation of PTMEG 1000. The GPC eluent was deposited directly on a track of 2,5-dihydroxybenzoic acid matrix. Each different elution time produced a narrow distribution of PTMEG telomers.

desired analyte from other species in the sample that might interfere with the MALDI experiment can be applied.

5.12
Problems in MALDI Sample Preparation

To uncover problems in the MALDI sample preparation step, the true composition of the sample must be known. This is often not possible for samples of commercial polymers, so research is often performed with samples created with combinations of lower MW, low polydispersity samples, or through the application of alternate analytical techniques to either simplify (e.g., chromatography) or characterize (e.g., NMR) the samples. Problems that do arise in the MALDI analysis are often termed "discrimination effects" and generally fall into two categories, molar mass discrimination and structural discrimination.

Molar mass discrimination is usually evidenced by a bias in the M_n or M_w values calculated from the observed MALDI spectrum. One reason for this is the differential solubility of different MW oligomers in the solvent chosen for the analysis. Hoteling and coworkers demonstrated this using several low molar mass (<20 000 Da) polyethylene terephthalate (PET) samples analyzed using THF (a poor solvent which dissolves only the lower mass oligomers) and a 70:30 (v/v) methylene chloride (CH_2Cl_2): hexafluorisopropanol (HFIP) azeotrope [48]. For a PET 17.5 kDa sample, no analyte peaks were observed in the MALDI spectrum when using THF as the solvent, whereas peaks out to ~12 kDa were observed using the azeotrope. Even while peaks were observed, the low mass peaks were observed with significantly

5.12 Problems in MALDI Sample Preparation | 141

greater intensity, and the M_n or M_w value calculated from the data is significantly different from those obtained by GPC analysis using HFIP as the solvent.

For samples with a large polydispersity, the molar mass discrimination is generally evidenced by a bias to low mass as the lower mass oligomers are over-represented in the spectrum. As described above, this problem may be overcome by preseparation of the polymer sample using some form of chromatography. Molar mass discrimination may also be due to ionization effects, as higher MW oligomers are often ionized more efficiently than lower MW species. This type of discrimination may also arise from instrumental causes. For example, the microchannel plate detectors found in most TOFMS instruments exhibit mass-dependent sensitivity effects as well as signal intensity-dependent saturation effects which often translate through to the observed polymer MW distribution.

In addition to molar mass discrimination, structural discrimination may also be observed in MALDI samples. Structural discrimination can result from differences in the incorporation of different components in the solid sample (e.g., due to the presence of hydrophilic versus hydrophobic end groups), or from differences in ionization of the components of the mixture. This is particularly important in sample mixtures containing different end groups, as an important result of the analysis may be the quantitative determination of each end group present in the sample. As an example, Figure 5.18 shows the MALDI mass spectrum for an equimolar mixture of four simple, low-molecular-mass homopolymers.

Figure 5.18 shows significant differences in relative sensitivity that can be observed from subtle changes in polymer repeat unit and end group. Interestingly, the use of different matrices yields different relative intensities.

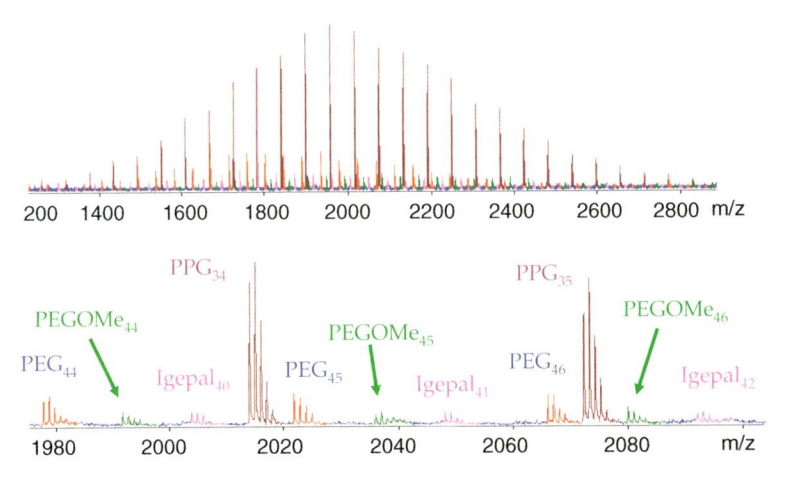

Figure 5.18 MALDI mass spectrum obtained from a 1:1:1:1 mixture of PEG 1900, PPG 2025, methoxy end group polyethylene glycol (PEGOMe) 2200, and Igepal CO-890 prepared with DHB as the matrix, NaTFA as the cationization agent, THF as the solvent and dried-droplet deposition. The spectrum shows very different sensitivity in this experiment for the four analytes.

Szyszka and coworkers have recently combined standard additions with an internal standard method to quantify the amount of residual PEG present in a series of commercially available ethoxylated surfactants [49]. While quantitation was possible, it is clear that the response factors for the various ethoxylated surfactants were significantly affected by the identity of the polymer end group as shown in Figure 5.19 [50].

The different slopes observed in Figure 5.19 demonstrate significantly different response in the MALDI experiment of these analytes, all of which are composed of PEG repeat units with different end groups. The relative values of the slopes do not appear to correlate with the degree of hydrophilicity of the analytes, or their relative strength of surfactancy, as determined by their hydrophilic–lipophilic balance (HLB) value.

Overall, what may be considered "simple" polymer blends can be some of the hardest samples to analyze via MALDI. Without extensive sample knowledge, the intensity of a particular polymer analyte series in the observed mass spectrum cannot be used directly for quantitative analysis.

5.13
Predicting MALDI Sample Preparation

While there is still much work to be done to fully understand the mechanism of the MALDI process, and even on a practical level to be able to rationalize the basic forms

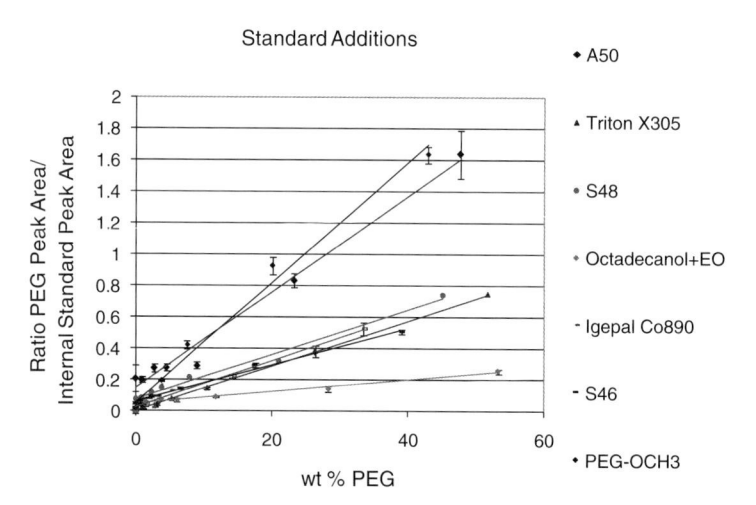

Figure 5.19 Standard additions curves obtained for seven different ethoxylated surfactants investigated. The plot is of the average peak area ratios for PEG to internal standard plotted against weight % of added PEG standard. Error bars represent the 95% confidence intervals. The A50, S46, and S48 are experimental surfactants obtained from Air Products and Chemicals, the Triton and ethoxylated octadecanol (Octadecanol + EO) were obtained from BASF, and the Igepal and methoxy end group polyethylene glycol (PEG-OCH$_3$) were obtained from Aldrich.

of the M/A and S/A plots described above, there are efforts to make the MALDI sample preparation process more predictable for the polymer scientist. A rather comprehensive effort is described in a recent publication by Brandt and coworkers which describes a chemometrics approach (based on a combination of factorial designed experiments and partial least-squares techniques) that can be used to predict the best combination of matrix, cationization agent, and solvent for use with a particular polymer analyte [51]. This follows on earlier work from the same group investigating the choice of solvent for analysis using Hansen solubility parameters for the matrix, analyte, and cationization agent as a guide [52], as well as a means to optimize the selection of what they termed the "mixing ratio" for the sample components [53], which is a combination of the M/A and S/A concepts described here. Their only limitation is a relative dearth of experimental detail, such as the identity of the most useful "molecular descriptors" for use in the sample preparation process. These are not the only active efforts in this area, but show a particularly good application of basic analytical practice toward further understanding the particularly complicated MALDI sample preparation process.

5.14
Conclusions

Sample preparation is the key to a successful polymer MALDI experiment. Research into how successful MALDI experiments work has highlighted several important aspects of the sample preparation process:

- Creating intimate contact between the analyte and the matrix, and preserving that contact through any phase changes occurring in the sample preparation process.
- Balancing the amount of matrix and cationization agent in the sample to provide the optimum ratios for observing maximum analyte signal intensity.
- Effectively depositing the MALDI sample onto the instrument target plate without disrupting the contact and ratios implemented at the bench.
- Sufficient absorption of the laser light by the matrix to deliver energy to the MALDI sample.
- Efficient energy transfer to effectively desorb the analyte from the solid-phase sample to the gas phase.
- Efficient ionization to produce the ions required for mass spectrometry.

If the sample preparation process is not integrated properly into the experiment, or key aspects of sample preparation are ignored, even the most expensive mass spectrometry instrumentation will not yield useful spectra.

Polymer chemistry provides a large structural and chemical diversity of materials for analysis by MALDI, requiring a wide range of matrices and sample preparation methods for successful analysis. Using hydrophilicity/hydrophobicity models, as described in Figure 5.4, a wide range of polymer chemistry can be matched to an appropriate choice of MALDI matrix. With the development of solvent-free sample preparation methods, even completely insoluble polymers can now be analyzed successfully. By using chromatography methods, broad polydispersity polymers and other complex mixtures can be successfully prepared for analysis.

The science of polymer MALDI sample preparation is not yet complete. This is still an area with many interesting and challenging problems to solve, many of which hinge on continuing to develop a deeper understanding of the underlying MALDI mechanism. Areas of continued research interest include improving the quantitative ability of MALDI experiments, the incorporation of ever-improving chromatography instruments into the sample preparation, and the development of new and different methods of sample deposition for high throughput applications.

Acknowledgments

The authors would like to thank Intertek and Drexel University for their support, Renata Szyszka for the data shown as Figure 5.13, and the graduate students from the Owens group, past and present, for their contributions to our understanding of MALDI sample preparation.

References

1 Tanaka, K., Waki, H., Ido, Y., Akita, S., Yoshido, Y., and Yoshido, T. (1988) Protein and polymer analyses up to M/Z 100,000 by laser ionization time-of-flight mass spectrometry. *Rapid Commun. Mass Spectrom.*, **2**, 151.

2 Karas, M. and Hillenkamp, F. (1988) Laser desorption ionization of proteins with molecular masses exceeding 10,000 daltons. *Anal. Chem.*, **60**, 2299.

3 Bahr, U., Deppe, A., Karas, M., Hillenkamp, F., and Giessman, U. (1992) Mass spectrometry of synthetic polymers by UV-matrix-assisted laser desorption/ionization. *Anal. Chem.*, **64**, 2866.

4 Danis, P., Karr, D., Mayer, F., Holle, A., and Watson, C. (1992) The analysis of water-soluble polymers by matrix-assisted laser desorption time-of-flight mass spectrometry. *Org. Mass Spectrom.*, **27**, 843.

5 Hanton, S.D. (2001) Mass spectrometry of polymers and polymer surfaces. *Chem. Rev.*, **101** (2), 527.

6 Nielen, M.W.F. (1999) MALDI time-of-flight mass spectrometry of synthetic polymers. *Mass Spectrom. Rev.*, **18**, 309.

7 Montaudo, G. and Lattimer, R.P. (eds) (2002) *Mass Spectrometry of Polymers*, CRC Press, Boca Raton, FL.

8 Bürger, M., Müller, H., Seebach, D., Börnsen, O., Schär, M., and Widmer, M. (1993) Matrix-assisted laser desorption and ionization as a mass spectrometric tool for the analysis of poly[(R)-3-hydroxybutanoates]. Comparison with gel permeation chromatography. *Macromolecules*, **26**, 4783.

9 Peng, L. and Kinsel, G.R. (2010) Improving the sensitivity of matrix-assisted laser desorption/ionization (MALDI) mass spectrometry by using polyethylene glycol modified polyurethane MALDI target. *Anal. Biochem.*, **400**, 56–60.

10 Falkenhagen, J. and Weidner, S.M. (2005) Detection limits of matrix-associated laser desorption/ionisation mass spectrometry coupled to chromatography – a new application of solvent-free sample preparation. *Rapid Commun. Mass Spectrom.*, **19**, 3724.

11 Lehrle, R.S. and Sarson, D.S. (1995) Polymer molecular weight distribution: results from matrix-assisted laser desorption ionization compared with

those from gel-permeation chromatography. *Rapid Commun. Mass Spectrom.*, **9**, 91–92.

12 Gentry, W.R. (1988) Low energy pulsed beam sources, in *Atomic and Molecular Beam Methods* (ed. G. Scoles), Oxford University Press, New York.

13 Meier, M.A.R., Adams, N., and Schubert, U.S. (2007) Statistical approach to understand MALDI-TOFMS matrices: discovery and evaluation of new MALDI matrices. *Anal. Chem.*, **79**, 863.

14 Hanton, S.D., Cornelio Clark, P.A., and Owens, K.G. (1999) Investigations of matrix-assisted laser desorption/ionization sample preparation by time-of-flight secondary ion mass spectrometry. *J. Am. Soc. Mass Spectrom.*, **10**, 104–111.

15 Hanton, S.D. and Owens, K.G. (2005) Using MESIMS to analyze polymer MALDI matrix solubility. *J. Am. Soc. Mass Spectrom.*, **16**, 1172–1180.

16 Hanton, S.D. (2004) New mass spectrometry techniques for the analysis of polymers for coatings applications: MALDI and ESI. *JCT Coat. Tech.*, **1**, 62–68.

17 http://polymers.msel.nist.gov/maldirecipes/index.cfm, accessed 1/17/2011.

18 King, R.C. (1994) Laser desorption/laser ionization time-of-flight mass spectrometry instrument design & investigation of the desorption and ionization mechanisms of matrix-assisted laser desorption/ionization. Ph.D. Thesis, Drexel University, Philadelphia, PA.

19 Weidner, S.M. and Falkenhagen, J. (2009) Imaging mass spectrometry for examining localization of polymeric composition in matrix-assisted laser desorption/ionization samples. *Rapid Commun. Mass Spectrom.*, **23**, 653–660.

20 Vorm, O., Roepstorff, P., and Mann, M. (1994) Improved resolution and very high sensitivity in MALDI TOF of matrix surfaces made by fast evaporation. *Anal. Chem.*, **66**, 3281.

21 Smirnov, I.P., Zhu, X., Taylor, T., Huang, Y., Ross, P., Papayanopoulos, I.A., Martin, S.A., and Pappin, D.J. (2004) Suppression of α-cyano-4-hydroxycinnamic acid matrix clusters and reduction of chemical noise in MALDI-TOF mass spectrometry. *Anal. Chem.*, **76** (10), 2958.

22 Hensel, R.R. (1996) The quantitative aspects of matrix-assisted laser desorption/ionization time-of-flight mass spectrometry. Ph.D. Thesis, Drexel University, Philadelphia, PA.

23 Hensel, R.R., King, R.C., and Owens, K.G. (1997) Electrospray sample preparation for improved quantitation in matrix-assisted laser desorption/ionization time-of-flight mass spectrometry. *Rapid Commun. Mass Spectrom.*, **11** (16), 1785–1793.

24 Chavez-Eng, C.M. (2002) Quantitative aspects of matrix-assisted laser desorption/ionization using electrospray deposition. Ph.D. Thesis, Drexel University, Philadelphia, PA.

25 Hanton, S.D., Owens, K.G., Blair, W., Hyder, I.Z., Stets, J.R., Guttman, C.M., and Giuseppetti, A. (2004) Investigations of electrospray sample deposition for polymer MALDI. *J. Am. Soc. Mass Spectrom.*, **15** (2), 168.

26 Yao, J., Dey, M., Pastor, S.J., and Wilkins, C.L. (1995) Analysis of high mass biomolecules using electrostatic fields and matrix-assisted laser desorption/ionization in a Fourier-transform mass spectrometer. *Anal. Chem.*, **67**, 3638.

27 Yao, J., Scott, J.R., Young, M.K., and Wilkins, C.L. (1998) Importance of matrix: analyte ratio for buffer tolerance using 2, 5-dihydroxybenzoic acid as a matrix in matrix-assisted laser desorption/ionization-Fourier transform mass spectrometry and matrix-assisted laser desorption/ionization-time of flight. *J. Am. Soc. Mass Spectrom.*, **9**, 805.

28 Wang, L., May, S.W., Browner, R.F., and Pollack, S.H. (1996) Low flow interface for liquid chromatography-inductively coupled plasma mass spectrometry speciation using an oscillating capillary nebulizer. *J. Anal. At. Spectrom.*, **11**, 1137.

29 Wang, L. and Browner, R.F.(March 10 1998) Oscillating Capillary Nebulizer, US Patent #5725153.

30 Hoang, T.T., May, S.W., and Browner, R.F. (2002) Developments with the oscillating capillary nebulizer – effects of spray chamber design, droplet size and turbulence on analytical signals and analytical transport efficiency of selected biochemically important organoselenium compounds. *J. Anal. At. Spectrom*, **17**, 1575.

31 Perez, J., Petzold, C.J., Watkins, M.A., Vaughn, W.E., and Kenttamaa, H.I. (1999) Laser desorption in transmission geometry inside a Fourier-transform ion cyclotron resonance mass spectrometer. *J. Am. Soc. Mass Spectrom.*, **10**, 1105.

32 Lake, D.A., Johnston, M.V., McEwen, C.N., and Larsen, B.S. (2000) Sample preparation for high throughput accurate mass analysis by matrix-assisted laser desorption/ionization time-of-flight mass spectrometry. *Rapid Commun. Mass Spectrom.*, **14**, 1008.

33 Dally, J.E. (2006) Investigation of biopharmaceutical cell culture and fermentation using matrix-assisted laser desorption ionization time-of-flight mass spectrometry. Ph.D. Thesis, University of the Sciences in Philadelphia.

34 Holcomb, A. and Owens, K.G. (2010) Optimization of a modified aerospray deposition device for the preparation of samples for quantitative analysis by MALDI TOFMS. *Anal. Chim. Acta*, **658**, 49.

35 Skelton, R., Dubois, F., and Zenobi, R. (2000) A MALDI sample preparation method suitable for insoluble polymers. *Anal. Chem.*, **72**, 1707–1710.

36 Przybilla, L., Brand, J.-D., Yoshimura, K., Räder, H.J., and Müllen, K. (2000) MALDI-TOF mass spectrometry of insoluble giant polycyclic aromatic hydrocarbons by a new method of sample preparation. *Anal. Chem.*, **72** 4591–4597.

37 Marie, A., Fournier, F., and Tabet, J.C. (2000) Characterization of synthetic polymers by MALDI-TOF/MS: investigation into new methods of sample target preparation and consequence on mass spectrum finger print. *Anal. Chem.*, **72**, 5106–5114.

38 Trimpin, S., Rouhanipour, A., Räder, H.J., and Müllen, K. (2001) New aspects in matrix-assisted laser desorption/ionization time-of-flight mass spectrometry: a universal solvent-free sample preparation. *Rapid Commun. Mass Spectrom.*, **15**, 1364–1373.

39 Trimpin, S. (2010) Solvent free MALDI sample preparation, in *MALDI Mass Spectrometry for Synthetic Polymer Analysis* (ed. L. Li), John Wiley & Sons, Inc., Hoboken, NJ.

40 Hanton, S.D and Parees, D.M. (2005) Extending the solvent-free MALDI sample preparation method. *J. Am. Soc. Mass Spectrom.*, **16**, 90–93.

41 Hanton, S.D., McEvoy, T.M., and Stets, J.R. (2008) Imaging the morphology of solvent-free prepared MALDI samples. *JASMS*, **19**, 874–881.

42 Hanton, S.D. and Stets, J.R. (2009) Determining the time needed for the vortex method for preparing solvent-free MALDI samples of low molecular mass polymers. *JASMS*, **20**, 1115–1118.

43 Hoteling, A.J. (2004) MALDI TOF PSD and CID: Understanding precision, resolution, and mass accuracy and MALDI TOFMS: investigation of discrimination issues related to solubility. Ph.D. Thesis, Drexel University, Philadelphia, PA.

44 Erb, W.J. (2007) Exploration of the fundamentals of matrix assisted laser desorption/ionization time-of-flight mass spectrometry. Ph.D. Thesis, Drexel University, Philadelphia, PA.

45 Montaudo, G., Montaudo, M.S., Puglisi, C., and Samperi, F. (1995) Characterization of polymers by matrix-assisted laser desorption/ionization time-of-flight mass spectrometry: molecular weight estimates in samples of varying polydispersity. *Rapid Commun. Mass Spectrom.*, **9**, 453–460.

46 Hanton, S.D. and Liu, X.M. (2000) GPC separation of polymer samples for MALDI analysis. *Anal. Chem.*, **72**, 4550–4554.

47 Dwyer, J. and Botten, D. (1996) A novel sample preparation device for MALDI-MS. *Am. Lab.*, **28** (17), 51–54.

48 Hoteling, A.J., Mourey, T.H., and Owens, K.G. (2005) The importance of solubility in the sample preparation of poly (ethylene terephthalate) for MALD/I TOFMS. *Anal. Chem.*, **77** (3), 750.

49 Szyszka, R., Hanton, S.D., Henning, D., and Owens, K.G. (2011) Development of a combined standard additions/internal standards method to quantify residual PEG in ethoxylated surfactants by MALDI TOFMS. *J. Am. Soc. Mass Spectrom.*, **22** (4), 633–640.

50 Szyszka, R., Henning, D., Hanton, S.D., and Owens, K.G. (2009) Quantitation of PEG contaminants in ethoxylated surfactant samples by MALDI TOFMS using standard additions and internal standards methods, ASMS poster.

51 Brandt, H., Ehmann, T., and Otto, M. (2010) Toward prediction: using chemometrics for the optimization of sample preparation in MALDI-TOF MS of synthetic polymers. *Anal. Chem.*, **82** (19), 8169–8175.

52 Brandt, H., Ehmann, T., and Otto, M. (2010) Solvent selection for matrix-assisted laser desorption/ionization time-of-flight mass spectrometric analysis of synthetic polymers employing solubility parameters. *Rapid Commun. Mass Spectrom.*, **24**, 2439–2444.

53 Brandt, H., Ehmann, T., and Otto, M. (2010) Investigating the effect of mixing ratio on molar mass distributions of synthetic polymers determined by MALDI-TOF mass spectrometry using design of experiments. *J. Am. Soc. Mass Spectrom.*, **21**, 1870–1875.

6
Surface Analysis and Imaging Techniques
Christine M. Mahoney and Steffen M. Weidner

6.1
Imaging Mass Spectrometry

Imaging mass spectrometry utilizes a probe (ion, laser, plasma, or charged droplet) to desorb or sputter species directly from a solid sample. The ions generated are extracted into a mass spectrometer (time-of-flight (ToF), quadrupole, or magnetic sector), yielding information about the spatial location (and in-depth location in some cases) of various elemental, isotopic, and molecular components contained within the sample. Mass spectrometric imaging can be achieved either by rastering the probe over a selected area as is typically done with a secondary electron microscope (SEM), or by using secondary ion beam focusing elements (as is the case with some magnetic sector based secondary ion mass spectrometers).

More recently, there have been remarkable advancements in the field of imaging mass spectrometry for the direct characterization of soft surfaces, including organic, polymeric, and biological materials. With the advent of several new desorption ionization techniques, including cluster secondary ion mass spectrometry (cluster SIMS), desorption electrospray ionization (DESI) mass spectrometry, plasma desorption ionization mass spectrometry (PDI-MS), and imaging matrix-assisted laser desorption ionization (imaging MALDI), low damage, high mass, and high spatial resolution imaging in soft samples is becoming a reality. Furthermore, 3D analyses are becoming much more commonplace in such systems; a feat that was unheard of a little over a decade ago. Finally, the possibility of analyzing samples at atmospheric pressure allows for direct analysis of real-world samples, and the possibility of characterizing biological samples *in vivo*.

In this chapter we will explore several different methods of surface mass spectrometry, describing in detail the benefits and deficiencies of each technique, as well as reviewing some of the most recent advancements in the area of polymer analysis with SIMS, MALDI, and atmospheric pressure based mass spectrometry techniques. The chapter will start off with an in-depth discussion of SIMS and its utility for surface and in-depth characterization of polymers. Because of the number of recent publications in this area, this particular topic will be discussed in greater detail.

Mass Spectrometry in Polymer Chemistry, First Edition.
Edited by Christopher Barner-Kowollik, Till Gruendling, Jana Falkenhagen, and Steffen Weidner.
© 2012 Wiley-VCH Verlag GmbH & Co. KGaA. Published 2012 by Wiley-VCH Verlag GmbH & Co. KGaA.

Cluster ion beam technology and its relevance to polymer characterization will be a key focal point. This will be followed by a short discussion on MALDI imaging for characterization of polymers. Finally, atmospheric pressure based techniques and other new methods for surface mass spectrometry will be discussed with reference to their utility for polymer analysis.

6.2
Secondary Ion Mass Spectrometry

Secondary ion mass spectrometry (SIMS) utilizes an energetic primary ion beam to sputter material from polymer surfaces. The benefits of SIMS include its extreme surface sensitivity (top one to three monolayers), low detection limits (ppm to ppb), and excellent spatial resolution for analysis of both inorganic and organic materials (typically ~100 nm, but can be as low as 50 nm) [1] as compared to MALDI (typically ~25 µm, but can be as low as 10 µm) [2]. Furthermore, a matrix is not required for SIMS analysis making it an ideal tool for the direct analysis of organic materials. While these benefits are useful, the ionization process is much harsher than MALDI, and therefore the method is typically only useful for molecules below 10 kDa. This of course makes high mass range analysis of most polymers difficult with SIMS, whereas MALDI is useful for molecules and proteins up to 100 kDa [3]. Direct comparisons of MALDI and SIMS can be found in the literature [4, 5].

This section presents an overview of the SIMS method for surface and in-depth characterization of synthetic polymers and includes detailed discussions of its uses and limitations. A fundamental review of static SIMS of polymers will be given, with detailed discussions regarding the fingerprint information from the low-mass region, as well as molecular weight distributions obtained from the high-mass region. Recent advancements in SIMS technologies, such as the advent of the polyatomic primary ion source for polymer depth profiling applications, will be discussed, and recent examples of 3D imaging SIMS will be presented. Some of the pertinent information regarding optimization of experimental conditions for acquiring 3D information from polymers will be addressed. Finally, a basic introduction of the concept of principal components analysis (PCA) will be included, because multivariate statistical analysis approaches are being utilized more and more regularly for SIMS analysis (and other surface mass spectrometry methods) of organic materials.

6.2.1
Static SIMS of Polymers

One of the inherent problems with SIMS for organic samples, particularly when employing atomic ion beam sources, such as Ar^+ or Cs^+, is that the ion beam creates extensive damage to the sample in the form of molecular bond breaking and/or crosslinking, particularly at the site of ion beam track [6]. Therefore molecular information rapidly decays with increasing primary ion dose. In order to minimize this damage, primary ion doses are typically kept low, such that the probability of a primary ion hitting the same spot twice is low; a term dubbed the "static SIMS limit" by Benninghoven and coworkers [7–10]. Mass spectral data acquired below this static SIMS limit are more representative of the pristine undamaged surface of the material, and therefore can contain useful information regarding the molecular structure of the material [9, 11]. In contrast, mass spectra acquired above this limit will usually have increased evidence of damage in the form of aromatic hydrocarbon peaks (initially) [12–15], and amorphous and/or graphitic C peaks after more extensive damage has been done [16–22].

Time-of-flight secondary ion mass spectrometers (ToF-SIMSs) are commonly used for low damage surface analysis of polymers. Unlike dynamic SIMS instruments, which require the use of a continuous primary ion beam current and a scanning of masses, the ToF-SIMS configuration requires beam pulsing, which inherently lowers the dose [23]. Furthermore, the ToF analyzer allows for simultaneous detection of all emitted species with excellent mass resolution and sensitivity, making it an ideal tool for surface characterization of materials that are sensitive to ion irradiation damage, as are polymers. This molecular damage can be further decreased, at least in part, by utilization of cluster sources such as C_{60}^+ or SF_5^+. However, static limits still need to be taken into consideration, particularly for surface sensitive experiments.

One of the first studies employing static SIMS for analysis of polymers was performed by Gardella and Hercules, who employed static SIMS to obtain mass spectra from a series of poly(alkyl methacrylates) [24]. Although this was the first reported study of its kind, the spectra presented in this work contained several peaks characteristic of damage (overwhelming number of C peaks), indicating that static SIMS conditions were not entirely met. Later, Campana, DeCorpo, and Colton published similar spectra under lower damage conditions, yielding mass spectral data more representative of the polymers structure and decreased C-rich fragments [25]. Briggs and coworkers continued with this line of work with a series of systematic investigations of SIMS spectra acquired from various polymers [13, 26–49].

6.2.1.1 The Fingerprint Region

Discussions of static SIMS of polymers will be broken down into two sections: (I) static SIMS in the low mass range (m/z 0 to m/z 500), and (II) static SIMS in the high-mass region (m/z 500 to m/z 10000). The low-mass region, ranging from approximately m/z 0 to m/z 500, is commonly referred to as the "fingerprint" region, for the mass spectral data in this region provide a unique fingerprint for each polymer, depending on its molecular structure [11]. Unique fragmentation patterns are observed, where fragmentation mechanisms for different classes of polymers have been investigated in detail [42, 50–58]. Figure 6.1 shows an example of this, showing representative mass spectra of the structurally similar polymers, poly(hydroxy-ethyl methacrylate) (PHEMA) and poly(propionyl-ethyl methacrylate) (PPEMA) [59]. As can be seen, the two mass spectra displayed in the figure are very different from one another, where the peaks observed are specific for the corresponding molecular structure of the side chain functionalities. This example also shows the utility of SIMS for studying chemistry at surfaces, as this surface was created by reacting PHEMA with gaseous propinylchloride.

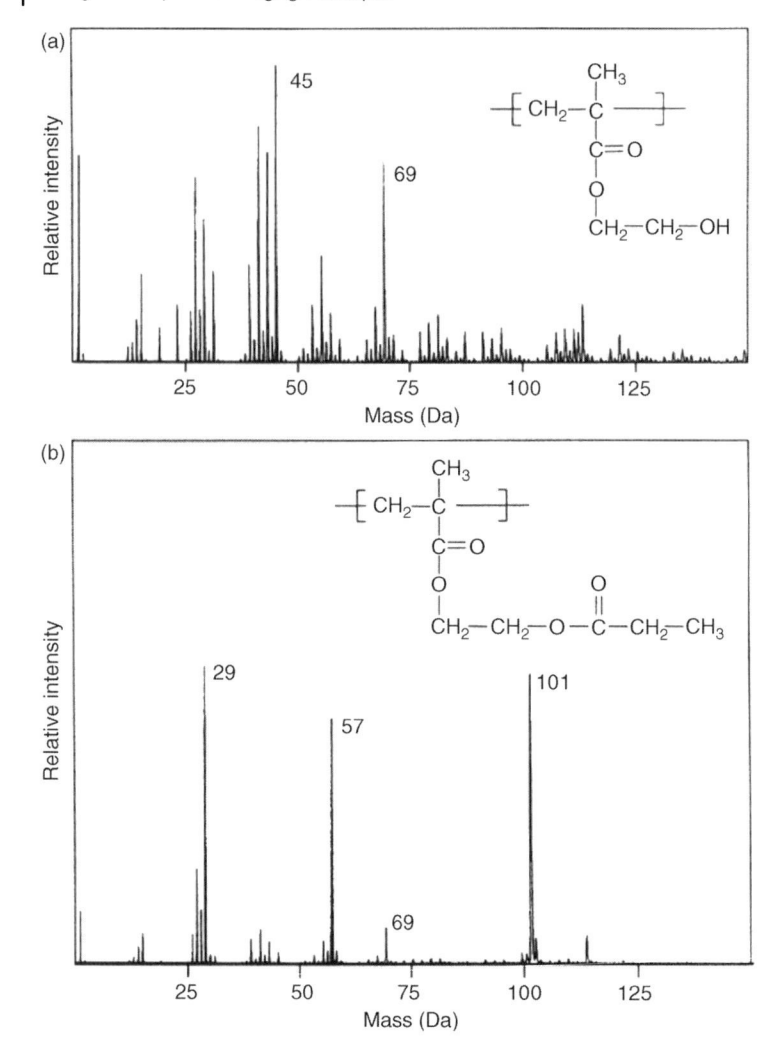

Figure 6.1 Positive secondary ion fingerprint spectrum of a bulk sample of poly(hydroxylethylmethacrylate) (PHEMA) before (a) and after (lower) treatment with gaseous propionylchloride (CH₃CH₂COCl), creating poly(propionylethylmethacrylate) PPEMA). Figure reprinted from Benninghoven, Hagenhoff, and Niehuis (1993), with permission [59].

In addition to obtaining information about the polymers molecular structure, the fingerprint region of the mass spectrum can be extremely useful for the determination of polymer molecular weights [44, 45, 52, 60–67], stereochemistries [68–70], isomeric structures [39, 71–73], and polymer orientations [74, 75].

The low mass range can be particularly useful for surface characterization of multicomponent systems, such as polymer blends, copolymers, and additive containing polymer systems. There have been countless examples of utilizing SIMS for characterizing surface segregation in copolymer and/or polymer blend systems [60, 76–99], where the segregation has been proven to heavily rely on factors including copolymer sequencing [78, 100, 101] (i.e., block vs. Random), copolymer block lengths [41, 77, 78, 81, 101], bulk compositions [41, 77, 78, 81, 87, 93–95, 98, 99, 101], end-group compositions [102, 103], polymer molecular weights [60, 92, 99], and polymer crystallinities [92, 104–109].

SIMS has also proven to be useful in extracting quantitative information from multicomponent polymer systems [32, 37, 43, 80, 90, 93–95, 101, 110–118], as long as a robust set of calibration standards and verification procedures are used. Figure 6.2 depicts an early example of utilizing SIMS for quantitative analysis in a series of nylon-6-co-nylon-6,6 random copolymers of varying bulk compositions [30]. The normalized intensity ratios (to m/z 55) of peaks characteristic of the nylon-6 monomer repeat unit (m/z 114) [signal associated with the $[nM + H]^+$ ion, where n is the number of monomer repeat units (here $n = 1$), and M is the mass of the nylon-6 monomer ($M = 113\,\text{g}\,\text{mol}^{-1}$)], were monitored as a function of the known mass

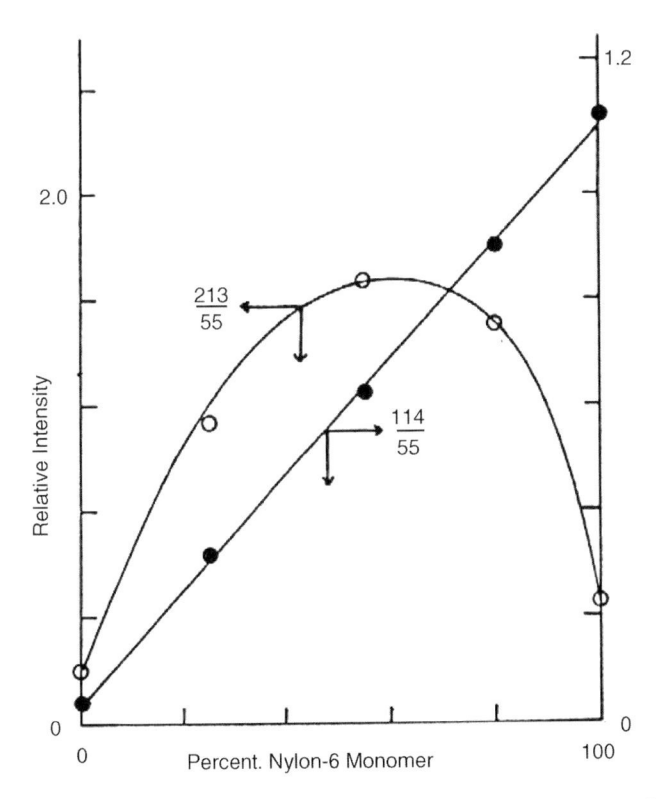

Figure 6.2 Correlation of SIMS relative peak intensities with nylon-6/6,6 copolymer composition in 1 μm thick copolymer films, where m/z 114 is characteristic of nylon-6, m/z 55 is a nonspecific hydrocarbon fragment observed in both polymer mass spectra, and m/z 213 is characteristic of nylon-6/6,6 linkages. Figure reprinted from Briggs (1986), with permission [30].

fraction of nylon-6 (•). Obviously, there is a linear dependence of the nylon-6 ion (m/z 114) with increasing copolymer content, indicating that SIMS is an ideal tool for quantitative analysis in these systems.

The second set of data plotted in Figure 6.2 at m/z 213 (○) are characteristic of nylon-6 (A)/nylon-6,6 (B) monomer linkages (A–B linkages, as opposed to A–A or B–B linkages). This peak is present only in the copolymers and is not present in the mass spectra of the corresponding homopolymers. A maximum is observed in the plot of the normalized m/z 213 curve at a nylon-6 mass fraction of 50%. This is consistent with the fact that the number of A–B linkages will be greatest toward nylon-6 molar contents approaching 50%. It should be noted, however, that the plot shown here is of mass fraction, and not of mole fraction which is published in a subsequent work [37]. The plots of mole fractions show a shift in this maximum toward higher nylon-6 contents – an effect that was attributed to greater ion yields of m/z 213 from ABA linkages. Finally, it should be noted that the contribution from A–B linkages in a random copolymer of A-co-B (ABABAB) will be much greater than in the corresponding block copolymer of A-b-B (AAAA-BBBB), illustrating the ability of SIMS to determine copolymer sequencing.

Similar quantitative studies have been performed on a series of poly(ethyl methacrylate) (PEMA)/PHEMA and PS/poly(4-vinyl phenol) (P4VP) copolymers [35, 101]. In the latter example, the authors were particularly creative and utilized random copolymers of PS-co-P4VP (which were determined not to exhibit preferential segregation or matrix effects) to create calibration curves for use in the quantification of the corresponding block copolymer (PS-b-P4VP) surfaces [101]. While there was no evidence for segregation of either component to the surface in the random copolymers, the block copolymers showed a P4VP-enriched surface. This content was measured quantitatively by SIMS and verified with X-ray photoelectron spectroscopy (XPS).

Once again, the authors were able to differentiate between block copolymers of PS/P4VP and random copolymers of PS/P4VP based solely on the presence or absence of certain characteristic fragment ions indicative of A–B linkages [101] (similar to what was observed in the nylon copolymer work) [30]. This direct measure of copolymer architecture cannot be obtained using other surface analytical methods such as XPS.

The fingerprint region of the mass spectra can be utilized to determine molecular weights of polymers at surfaces, where end-group-specific fragment ions are used to calculate molecular weights, based on the increased molar fractions of these components at lower molecular weights (the number of end groups increases at lower molecular weights) [44, 45, 52, 61–63, 65, 67, 119, 120].

A practical example of this procedure is demonstrated in Figure 6.3, which depicts the process whereby SIMS can be used as an indirect measure of the molecular weights of polycarbonates (PCs) at the surface of a compact disc (CD) [121]. First, a series of PC samples of varying molecular weights were prepared and analyzed using SIMS. Figure 6.3a shows a representative negative ion mass spectrum obtained from the surface of one of the PC samples analyzed in this work. The peak at m/z 149 is characteristic of the *t*-butyl phenyl end group of the PC, while the peak at m/z 211 originates from PC main chain (structures of both fragments are shown in the

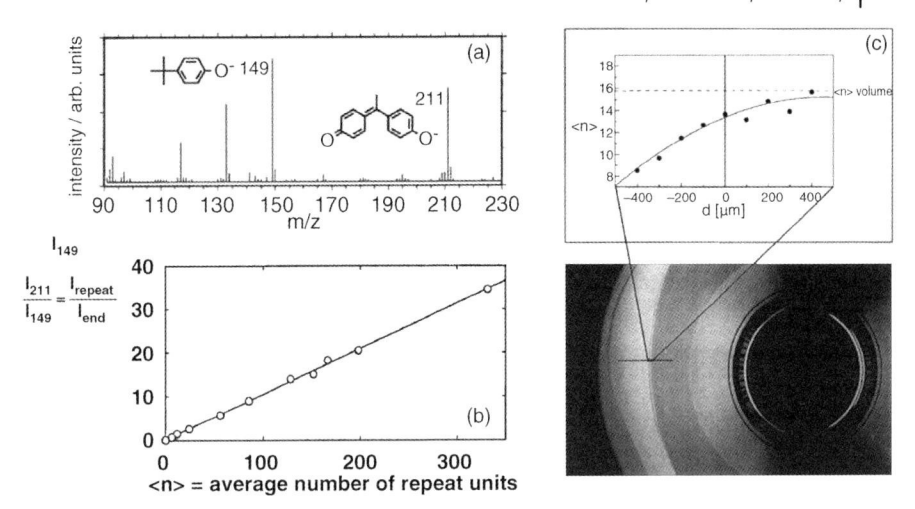

Figure 6.3 Quantitative analysis of PC molecular weight on surface of CD using static SIMS. (a) Negative ion mass spectrum of bisphenol A PC, where m/z 149 represents end-group fragments and m/z 211 originates from the polymer repeat units in the main chain. (b) Intensity ratio of fragments of m/z 211 to m/z 149, plotted as a function of the degree of polymerization, $<n>$ (number of monomer units in polymer chain), and (c) spatially resolved molecular weight determination at the surface of a polycarbonate CD core as a technical application. Figure reprinted from Reihs et al. (1997), with permission [121].

figure). Next, the area ratios of m/z 211 to m/z 149 are plotted in Figure 6.3b as a function of increasing degree of polymerization n or the average number of monomer units contained in the polymer chain. Note that there is a linear increase in this ratio with increasing n. This means that the peaks selected are ideal for quantification of molecular weights.

The calibration curve in Figure 6.3b was used for quantitative analysis of the molecular weight distributions of PC directly on the surface of a CD. The results of this process are shown in Figure 6.3d, which show the spatially resolved molecular weights located at the surface of the disc. These studies, utilizing end-group-specific fragments of polymers for quantitative analysis of molecular weights, are oftentimes more accurate than using high-mass-range molecular weight distributions for quantification of molecular weights (at least for SIMS) due to the lack of problems associated with these higher masses, such as mass bias and chain entanglement issues, which will be discussed later.

A summary of static SIMS mass spectra that have been published in the literature for specific polymers and for different mass regions is given in Table 6.1. Also included in this table is the corresponding dynamic SIMS (depth profiling) papers that are available. This table is broken down into five separate sections: (i) papers dealing with the low-mass region or fingerprint region of the spectrum ($<m/z$ 500) in homopolymers, (ii) papers dealing with the high-mass region of the spectrum, including high-mass fragments (m/z 500 to m/z 10000) in homopolymers, (iii) papers dealing with multicomponent systems at all mass ranges (polymer blends,

Table 6.1 List of reference citations for specific polymers characterized by SIMS: (I) fingerprint region (<m/z 500), (II) high-mass region (m/z 500 to m/z 10000), (III) polymer blends, copolymers, and additive-containing polymers/all mass ranges, (IV) depth profiling/homopolymers, and (V) depth profiling/polymer blends, copolymers, additive containing polymers, and polymer multilayers.

Polymer	(I) Low mass	(II) High mass	(III) Multi-Component	(IV) Depth Profiling (Homo-Polymers)	(V) Depth Profiling (Multi-Component)
Poly(alkyl methacrylate)s					
Poly(methyl methacrylate)	[13, 14, 24–26, 28, 39, 44, 45, 50, 68, 89, 127–133]	[50, 69, 70, 129, 131, 134, 135]	[35, 66, 80, 83, 87, 94, 127, 136–140]	[141–157]	[156, 158–165][a]
Poly(ethyl methacrylate)	[24, 25, 28, 39, 127]	[134]	[35, 127, 140]		
Poly(n-propyl methacrylate)		[134]	[166]	[166][a]	
Poly(n-butyl methacrylate)	[24, 25, 39, 71, 72, 131]		[72, 140]	[94, 167]	
Poly(isobutyl methacrylate)	[24, 25, 39, 71, 72]		[72]		
Poly(t-butyl methacrylate)	[39, 71, 72]		[72]		
Poly(sec-butyl methacrylate)	[71]				
Poly (hexyl methacrylate)	[39]				
Poly(octyl methacrylate)					
Poly(dodecyl methacrylate)				[167]	
Poly(lauryl methacrylate)	[24, 25]			[167]	
Poly(phenyl methacrylate)	[24, 39, 131]				
Poly(cyclohexyl methacrylate)	[24]	[134]			
Poly (hydroxyethyl methacryylate)	[28, 39, 128]		[35, 87, 91, 160]	[40, 147]	
Poly(2-hydroxypropyl methacrylate)	[28]		[87]		
Poly(acetoacetoxyethyl methacrylate)	[28]				

Polymer				
Poly(epoxypropyl methacrylate)	[28]			
Poly(ethylene glycol methacrylate)		[111]		
Poly(ethylene glycol dimethacrylate)		[40]		[162]
Other	[168]			
Poly(alkyl acrylate)s				
Poly(methyl acrylate)	[28, 39, 169]	[171]	[170]	
Poly(ethyl acrylate)	[28]	[87]		
Poly(glycidyl acrylate)		[138]		
Poly[(2-perfluorohexyl-ethyl) acrylate]				
Poly(acrylic acid)s				
Poly(methacrylic acid)	[26, 28, 39]	[88, 171]	[170]	
Poly(acrylic acid)	[28]			
Poly(acrylamide)s				
Poly(acrylamide)		[87]		
Poly(olefin)s				
Poly(propylene)	[26, 28, 29, 64, 73, 129, 172–174]	[129, 175, 176]		[178]
Poly(ethylene)	[26, 28, 29, 64, 73, 172, 174, 177]	[80, 86, 109, 175]		[180]
Poly(iso butylene)	[28, 29, 64]	[179]		
Poly(styrene)s				
Poly(styrene)	[13, 26, 27, 52, 57, 61–64, 68, 93, 114, 129, 133, 172, 174, 181–187]	[12, 52, 57, 118, 122, 126, 133, 174, 185, 188–192]	[60, 66, 88, 90, 91, 93, 101, 112–117, 139, 166, 179, 193, 194]	[92, 142, 150, 155, 156, 166, 195–202][a]	[165, 180, 203–205][a]
Poly(α-methyl styrene)	[181]		[160]	
Poly(4-methyl styrene)	[174, 181]	[174]		
Poly(4-hydroxystyrene)	[54, 115]			
Poly(vinyl naphthalene)		[57]	[115]	

(Continued)

Table 6.1 (Continued)

Polymer	(I) Low mass	(II) High mass	(III) Multi-Component	(IV) Depth Profiling (Homo-Polymers)	(V) Depth Profiling (Multi-Component)
Poly(4-chlorostyrene)	[57]	[57]			
Poly(styrene sulfonate)			[112]		
Poly(pentafluorostyrene)			[101, 116]	[206][a]	
Poly(4-vinyl phenol)			[116, 118]		
Poly(4-vinylpyridine)	[87]			[92][a]	[159]
Poly(vinyl pyrrolidone)					[159]
Poly(hexafluorohydroxyisopropyl-α-methyl styrene)			[117, 118]		
Poly(amide)s					
Nylons	[26, 207]				
Poly(amidoamines)	[208]	[123]			
Other	[168]				
Poly(carbonate)s					
Bisphenol-A poly(carbonate)s	[121, 129, 132, 209–211]	[209]	[97, 105, 107, 142]	[142, 155, 156, 197, 212, 213]	
Poly(tetramethylene carbonate)			[78]		
Other poly(carbonate)s	[209, 214]				
Poly(alkene)s					
Poly(1-butene)	[28, 29, 64]				
Poly(4-methyl-1-pentene)	[28, 29]				
Poly(butadiene)s	[28, 29, 64, 67, 114, 172, 188, 215, 216]		[113, 114, 193, 217]		
Poly(isoprene)s	[28, 29, 172, 218]	[218]			

Poly(ester)s				
Poly(ethylene terephthalate)	[15, 26, 28, 38, 47, 128, 129]		[174, 176]	
Poly(ethylene isophthalate)			[174]	
Poly(butylene terephthalate)		[60, 77]		
Poly(diallyl phthalate)	[28]			
Poly(caprolactone)	[28, 220, 221]		[84]	
Poly(ethylene adipate)	[28, 221, 223]	[223]		
Poly(butylene adipate)	[221]	[119]		
Poly(glycolic acid)	[224–227]	[227–230]		
Poly(lactic acid)	[58, 224, 225, 231, 235]	[229, 230, 235, 236]	[78, 226, 231–233] [99, 231–233]	[234] [95, 96, 98, 130, 178, 234, 237, 239–243]
Poly(hyroxybutyrate)	[220, 224]		[130, 237, 238]	[95, 96, 98, 130, 178, 237]
Poly(orthoesters)	[244]			
Poly(ether)s				
Poly(ethylene oxide)	[28, 129, 245]	[120, 192, 246]	[76]	[146]
Poly(propylene oxide)	[28]	[70, 120]	[82, 247]	[146]
Poly(tetramethylene oxide)	[28, 43, 226]		[43, 76, 77, 226]	[95, 96]
Poly(bisphenol A)	[108]		[105, 107, 248]	
Poly(ether octane)			[105, 107]	
Poly(ether ether ketone)	[133, 249]			
Poly(2,6-dimethyl-1,4-phenylene oxide)	[93]		[68]	[204][a]
Poly(vinyl ether)s				
Poly(vinyl methyl ether)	[28]		[60]	
Poly(vinyl ethyl ether)	[28]			
Poly(vinyl isobutyl ether)	[28]			
Poly(vinyl ketone)s				
Poly(vinyl methyl ketone)	[28]			

(*Continued*)

Table 6.1 (Continued)

Polymer	(I) Low mass	(II) High mass	(III) Multi-Component	(IV) Depth Profiling (Homo-Polymers)	(V) Depth Profiling (Multi-Component)
Poly(vinyl ethyl ketone)	[28]				
Poly(methyl isopropenyl ketone)	[28]				
Poly(vinyl esters)s					
Poly(vinyl acetate)	[28]		[79, 86]		[178]
Poly(vinyl propionate)	[28]				[178]
Poly(vinyl butyrate)	[28]				
Fluropolymers					
Poly(tetrafluoroethylene)	[15, 129]	[250]	[109, 251]	[146]	
Poly(hexafluoropropylene)			[251]	[248]	
Poly(vinylidene difluoride)	[100]				[164]
Perfluoropolyethers	[65, 252, 253]	[65, 252–255]		[256][a]	
other vinyl polymers					
Poly(vinyl chloride)	[13, 14]		[83, 84, 137]	[145]	[163, 164, 205]
Poly(vinyl alcohol)	[28]		[79]		
Poly(acrylonitrile)			[217]		
Conducting polymers					
Polyaniline	[257]				
Polyacetylene	[177]				
Poly(phenyl acetylene)	[182]				
Poly(1-phenyl-1-propyne)	[182]				
Poly(1-chloro-2-phenyl acetylene)	[182]				
Poly(o-trimethyl-silyl phenyl acetylene)	[182]				
Poly(t-butyl acetylene)	[182]				

Poly(pyrrole)	[258, 259]			
Poly(phenylenevinylene)				[206][a]
Poly(ethylene dioxy thiophene)				[206][a]
Poly(thiophene)	[260]			
Poly(anhydride)s				
Poly(adipic anhydride)	[261]			
Poly(suberic anhydride)	[261]			
Poly(azelaic anhydride)	[261]			
Poly(sebacic anhydride)	[261]			
Miscellaneous				
Poly(imide)s			[99]	[263]
Poly(urethanes)	[41, 43, 76, 264]	[119, 122, 265, 266]	[81, 262] [267]	[178]
Poly(urea)s	[43, 264]			
Poly(dimethyl siloxane)s	[122]	[122, 250, 268–270]	[81, 262, 267]	[203, 271][a]
Poly(phosphazene)s	[272]			
Poly(merhacryloyloxy)ethyl phosphate			[74]	
Poly(acetoacetoxy) ethyl methacrylate			[74]	
Celluloses	[140, 171]		[171]	
Tentagel (PS derivative)		[253]	[222]	

[a] Atomic ion beam depth profiling, elemental analysis only.

copolymers, or additive containing polymers), (iv) papers dealing with dynamic SIMS of homopolymers (depth profiling), and (v) papers dealing with dynamic SIMS of copolymers, polymer blends, additive containing copolymers, or polymer multilayers. The latter two sections will be dealt with later on. Now we will move on to discussions of the high-mass region.

6.2.1.2 High-Mass Region

Toward the mid- to late-1980s, Bletsos et al. worked toward utilizing SIMS for high-mass range analysis of polymers and were relatively successful in characterizing the molecular weight distributions in polymers up to m/z 10 000 [122, 123]. The authors employed the method of Ag cationization [124, 125], whereby a very dilute solution or "submonolayer" was cast onto an etched Ag substrate resulting in the formation of intense Ag-cationized polymer oligomers.

An example of this is depicted in Figure 6.4, which shows the positive ion mass spectrum in the range of m/z 600 to m/z 6500 from a sample of low molecular weight (a), and high molecular weight (b) polystyrene. In this work, a series of PS standards are characterized with molecular weights ranging from $M_n = 9000$ to $M_n = 858\,000$ g mol^{-1}, where M_n is defined as the number average molecular weight [126].

Figure 6.4a shows the positive ion mass spectrum acquired from the lower molecular weight PS standard ($M_n = 3770$ g mol^{-1}). Note that the mass spectrum has three distinctive regions, and three overlapping molecular weight distributions:

1) Intact oligomeric ions cationized by silver are observed at higher masses (between m/z 2200 and m/z 6000). The spacing between the peaks is consistent with one polymer repeat unit of PS (m/z 104).
2) Below this region (between m/z 1000 and m/z 3000) lies a doubly charged $(M + 2Ag)^+$ oligomeric distribution of PS, whereby the peaks are separated by 52 mass units. This is not always observed during SIMS experimentation, but is often observed in PS samples.
3) A non-Gaussian Ag-cationized fragment ion distribution is observed throughout the spectrum (m/z 600 to m/z 6000). The spacing between the peaks is consistent with one polymer repeat unit of PS (m/z 104). These peaks arise from fragmentation of the polymer backbone.

Both the number average (M_n) and weight average (M_w) molecular weights can be calculated directly from the intact oligomeric distribution shown in Figure 6.4a. In this particular example, the M_n measured by SIMS was m/z 3550, which is comparable to the value determined by gel permeation chromatography (GPC) (m/z 3770). The values for the lower molecular weight PS standards agreed well with the GPC measurements in the range of m/z 600 to m/z 7400 for this study. However, the M_n determined by SIMS tended to be systematically biased toward lower values than GPC, particularly in the higher molecular weight samples – a problem commonly encountered when employing SIMS for these types of measurements. There are several reasons for this. The primary reason is that the calculation of absolute M_w distributions from mass spectral data assumes that factors such as the sputtering yield, ionization efficiency, and mass spectrometer transmission and

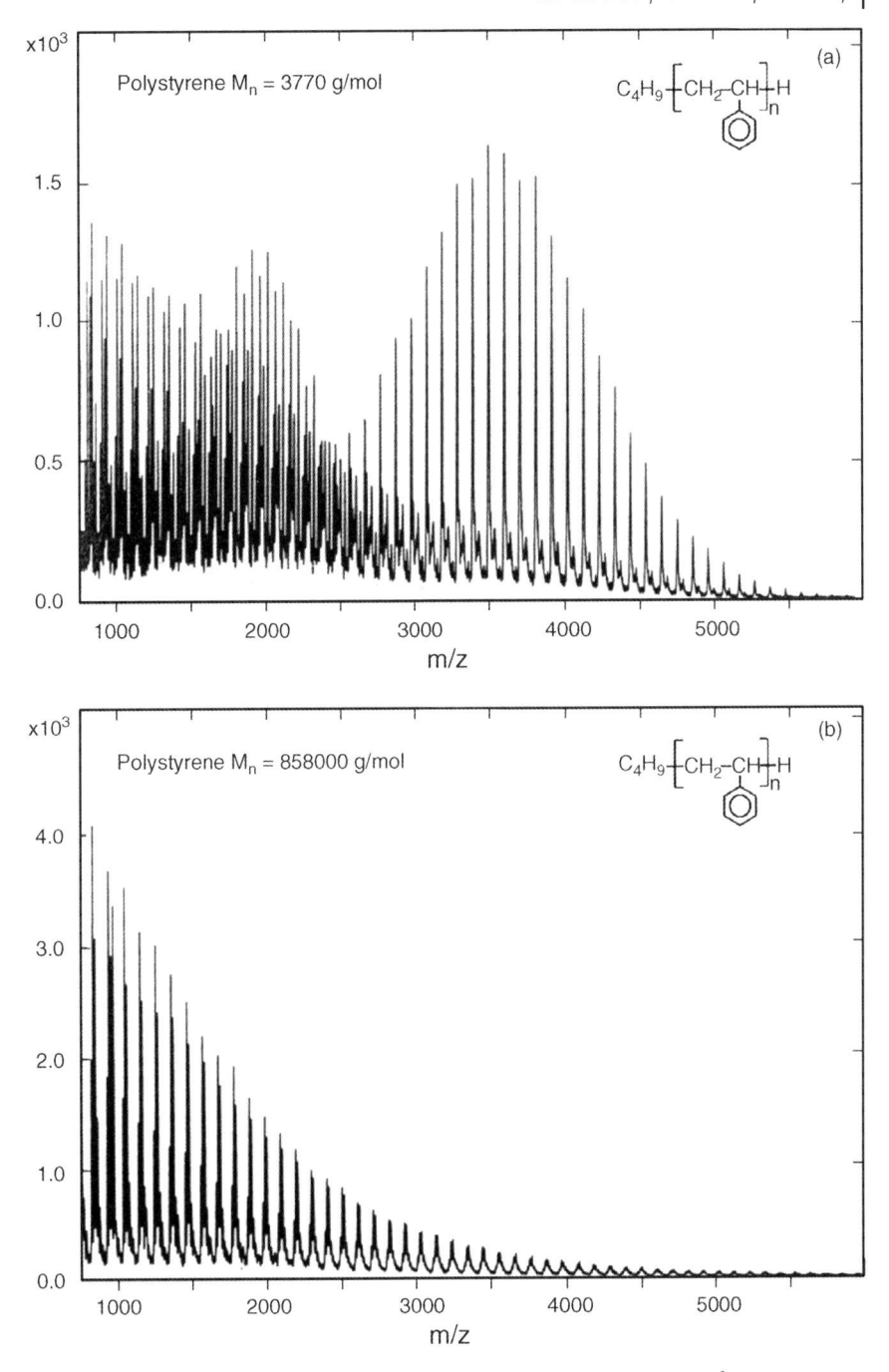

Figure 6.4 Positive secondary ion spectra of polystyrene molecular weight standards prepared as a monolayer/submonolayer film on a silver substrate. Acquired from: (a) PS-3770 ($M_n = 3770$ g mol^{-1}), and (b) PS-858000 ($M_n = 858000$ g mol^{-1}). Figure reprinted from van Leyen et al. (1989) with permission [126].

detection efficiencies are constant as a function of mass, which is not the case. Since both ionization efficiencies and detection efficiencies are expected to decrease with increasing mass, the M_n values calculated by SIMS will be systematically low, unless a correction factor is used. Furthermore it is expected that the effects of increased chain entanglements and fragmentation at higher molecular weights is significant, particularly in SIMS experiments, where the energy density from the ion beam is large and fragmentation is probable. Overall, SIMS measurements in the high-mass range are most accurate for low molecular weight polymers that have narrow molecular weight distributions [126, 188].

Figure 6.4b shows an attempt to obtain a molecular weight distribution from a much higher molecular weight polystyrene sample ($M_n = 858\,000\,\mathrm{g\,mol^{-1}}$) (submonolayer cast on Ag). As expected, only the fragment ion distribution is observed.

One of the greatest benefits of using mass spectrometry methods for direct analysis of polymer molecular weight is that information regarding the polymers structure and chemistry of the end groups are obtained simultaneously [52, 58, 61, 120, 188, 209, 245, 273, 274]. Figure 6.5 illustrates this point rather nicely, showing positive ion SIMS spectra acquired from polystyrene samples having different end-group functionalities [274].

In this work, the authors synthesized PS samples using anionic ("living") polymerization, whereby calculated contents of two separate terminating agents, methanol and chlorodimethylphenylsilane, were added in the appropriate molar ratios. Living polymer chain ends that interacted with a methanol molecule were end-capped with H, while polymer chain ends that interacted with the chlorodimethylphenylsilane were terminated by a dimethylphenyl silane (DMPS) end group. As can be seen from the figure, two separate molecular weight distributions are clearly visible for the DMPS-terminated and H-terminated oligomers, respectively. The authors effectively demonstrate that one can quantify the extent of end-group functionalization in polymers, as well as determine the composition of the end groups in general, through the direct comparison of the signal intensities of the two types of oligomers.

This ability to characterize chain-end compositions is important as the composition of the end groups will affect the thermodynamic properties of the polymer. Therefore mass spectrometry can play a vital role in the verification of end-group chemistry, even in commercial polymer systems. Hittle et al., for example, were able to use SIMS mass spectra to determine that a commercial poly(ethylene glycol) (PEG) distearate polymer – a PEG terminated with stearate groups on both sides – was in actuality terminated with a mixture of palmitate and stearate functionalities [120].

Overall, the development of cationization methods for analysis of molecular weight distributions in SIMS has been extremely useful in polymer analysis. However, the requirement of submonolayer film preparation severely limits the method for analysis of real-world samples. First, only soluble polymers can be analyzed using this method. Second, this method is a means of characterizing the bulk properties of the polymer and not molecular weight distributions at the surface of thick polymer films. Therefore, scientists have been focusing their efforts

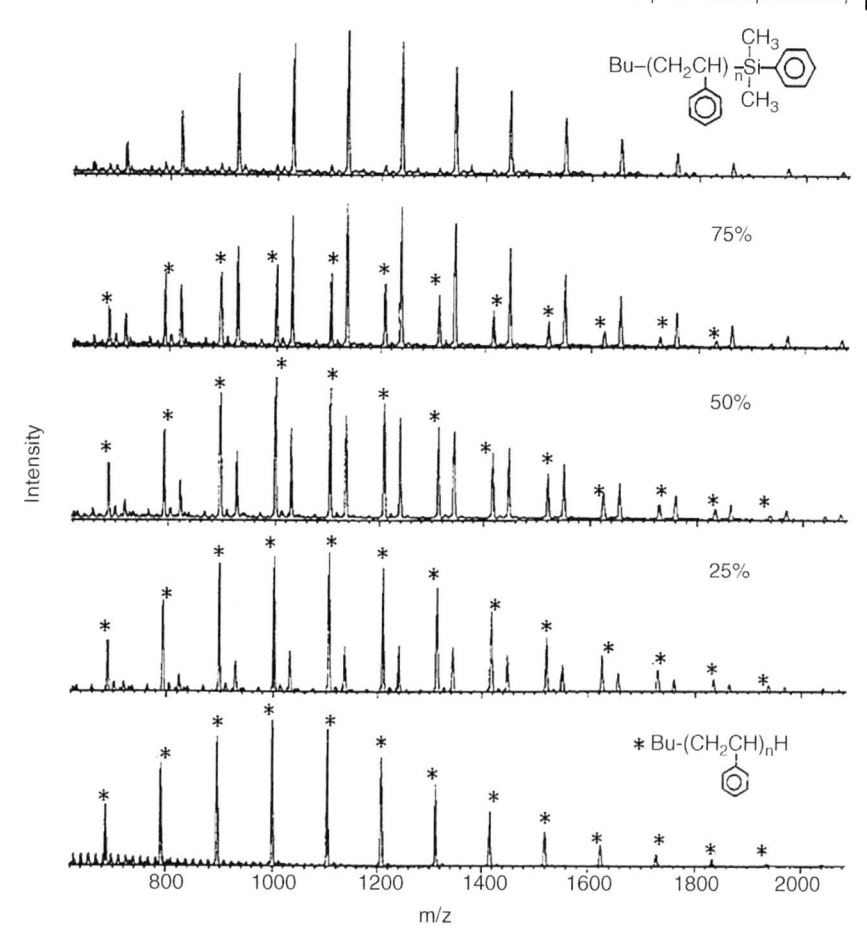

Figure 6.5 ToF-SIMS spectra of polystyrene, functionalized with different molar ratios of dimethyl-phenylsilyl (DMPS) end groups. Peaks labeled with * represent oligomeric distributions obtained from the PS having protonated end groups. Unlabeled peaks represent oligomeric distributions obtained from PS with DMPS end groups. The mole fraction of DMPS end groups relative to H end groups is indicated. The bottom panel represents a DMPS mole fraction of 0%, while the top panel represents a DMPS mole fraction of 100%. Figure reprinted from Belu et al. (1994) with permission [274].

more recently, on developing SIMS methods for the direct characterization of thick polymer films in the high mass range [174, 228, 229, 236, 250, 253, 254, 267, 275].

Although it is extremely difficult to obtain this information from polymer surfaces, primarily due to increased charging and chain entanglements, it is feasible, particularly with the aid of cationizing agents. For example, Bletsos and coworkers were able to characterize the surface of thick poly(dimethyl siloxane) (PDMS) and poly (tetrafluoroethylene) (PTFE) samples (200 μm and 2 mm for PDMS and PTFE, respectively) by covering the sample with an Ni grid. This helped to alleviate problems

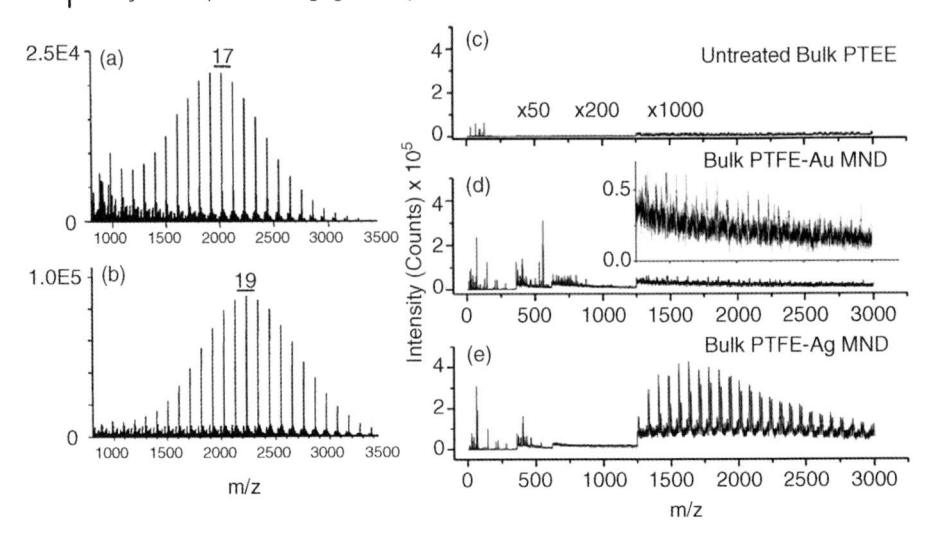

Figure 6.6 Molecular weight distributions measured by meta-SIMS, beam (a,b) [174] and metal nanoparticle SIMS (c,d,e) [253], both using a 15-keV Ga$^+$ primary ion: (a) polystyrene (PS) film (>10 nm) with Au metal evaporated overlayer (20 nmol/cm^2), (b) PS submonolayer film prepared on a Au substrate, (c) PTFE bulk, (d) PTFE with Au nanoparticle deposition, and (c) PTFE with Ag nanoparticle deposition. Figure recreated from Delcorte et al. [174] and Marcus and Winograd (2006) [253], with permission.

arising from charging, which is a common occurrence in thicker polymer samples, but also likely provided a source of Ni cations [250]. When using the Ni grid, the authors were able to observe an Ni-cationized molecular weight distribution on the surface of the PDMS. The cationized molecular weight distribution measured at the surface of the thick PDMS film was in the same general mass range as that observed from the corresponding submonolayer film of the same PDMS.

Later, evaporated metal overlayers were utilized as a means of obtaining high-mass range information from polymeric surfaces, a process dubbed "meta-SIMS" or metal-assisted SIMS [113, 174, 188, 276–279]. In this process, metal overylayers are evaporated onto an organic surface and serve as a cationizing agent. An example of meta-SIMS for analysis of polymers is given in Figure 6.6a and b, which shows representative positive secondary ion mass spectra acquired from a narrow molecular weight polystyrene standard ($M_n = 2180$ g mol^{-1}) in the high-mass range. In Figure 6.6a, the mass spectrum acquired from the Au-metallized PS surface is shown, where the surface of the PS is coated with ∼20 nmol/cm^2 Au. The distribution observed in the Figure is representative of Au-cationized PS molecules. Figure 6.6b shows the Au-cationized PS oligomeric distribution obtained from the corresponding submonolayer film deposited on an Au substrate. The molecular weight distributions observed are clearly very similar for the two methods, although it is apparent that the spectrum in Figure 6.6a is slightly shifted toward lower masses, as would be expected in cases where increased chain entanglements and fragmentations are observed.

Figure 6.6c–e illustrates a related method, whereby metal nanoparticles are deposited from a solution onto the surface of the polymer [253]. Figure 6.6c shows the mass spectrum taken from the surface of a bulk PTFE film (no nanoparticles). No high-mass range peaks are observed here. However, with the deposition of Au (Figure 6.6d) and Ag (Figure 6.15e) nanoparticles a high-mass oligomeric molecular weight distribution characteristic of PTFE appears.

The mechanism of enhancement in nanoparticle systems has been described recently [280]. When an atomic ion impinges upon a gold or silver nanoparticle, it creates a collision cascade in the nanoparticle. The resulting energy of the bombarded nanoparticle is then transferred directly into the polymer surface, creating a dense collision cascade in the polymer surface region, enhancing the sputtering yield of the polymer.

Other methods of enhancing ionization in SIMS experiments have also been studied by scientists. Matrix-enhanced SIMS (MA-SIMS) and related methods, whereby MALDI matrices and glycerol (liquid SIMS) are utilized for SIMS applications, show great promise for characterization of polymeric and protein samples [281–289]. Although these methods are promising, much work needs to be done in order to optimize them for analysis of polymer samples.

The ability to characterize molecular weight distributions from thick films, if optimized, will be a useful tool for the determination of surface molecular weight distributions (as compared to bulk distributions). However, it could also be useful in determining surface molecular weight distributions and block lengths from copolymer surfaces. Zhuang et al. [267], for example, were able to determine the molecular weight distributions of a PDMS block component from the surface of a thick copolymer film of polyurethane-co-dimethylsiloxane (PU-PDMS). In this work, ToF-SIMS spectra were acquired from the surface of thick PU-DMS films, whereby molecular weight distributions, characteristic of the full PDMS block component, were observed. The number average (M_n) and weight average (M_w) molecular weights of the PDMS block were calculated based on these distributions and compared to the amino-terminated PDMS ($M_n = 1000 \,\mathrm{g\,mol^{-1}}$) utilized in the synthesis of the copolymers. The distributions at the surface of the copolymer were found to be very similar to the PDMS homopolymer precursor. Although the authors did not use a conducting grid or metallization in this case, they attribute the success of this particular work to the fact that the copolymer film contained large mass fractions of linear, low molecular weight PDMS (~82%).

Gardella and coworkers later published a series of papers employing ToF-SIMS to study the degradation processes in biodegradable polyesters, such as poly(lactic acid) (PLA) and poly(glycolic acid) (PGA) [228–230, 236, 275]. It was found that the appearance of degradation products in the high-mass range could be used as a means to measure the kinetics of degradation during hydrolysis experiments.

Another way to improve SIMS for polymer analysis in the high-mass region is to use cluster primary ion sources. Cluster sources are extremely beneficial for the analysis of any organic material, and are particularly useful for surface characterization of polymers [129, 135, 169, 185, 221, 235, 246], where significant sputter and

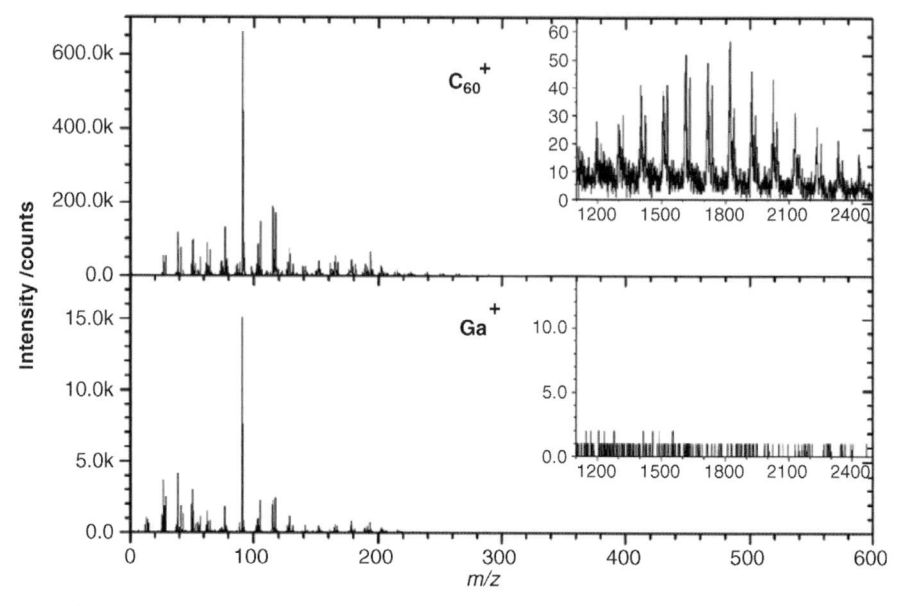

Figure 6.7 Positive ToF-SIMS spectrum of polystyrene (PS) m/z 2000 using 10 keV impact C_{60}^+ and Ga^+. Inset shows the high-mass region, whereby the PS molecular ions are observed only when employing C_{60}^+. Figure reprinted from Weibel et al. (2003), with permission [185].

ion yield enhancements and increased high-mass information from polymers have been observed [129, 185, 235, 246]. An example of this is shown in Figure 6.7, which shows mass spectral data acquired from polystyrene ($M_n = 2000$ g mol^{-1}) using a C_{60}^+ primary ion source as compared to a more conventional Ga^+ primary ion source. The mechanisms behind this have been studied in detail and will be discussed in brevity later on in this chapter [141, 290–294].

6.2.2
Imaging in Polymer Blends and Multicomponent Systems

With ToF-SIMS, one can obtain high spatial resolution molecular images by scanning a focused primary ion beam over the surface, similar to what is done with electron beams during secondary electron microscopy (SEM) imaging. This allows for imaging of the various components in the mass spectrum with spatial resolutions as low as 50 nm in some cases. These modes require the use of a highly focussed primary ion beam source such as a liquid metal ion gun (LMIG). Magnetic sector based mass spectrometers allow for an additional mode of imaging, whereby the secondary ions created are subsequently focused using secondary ion optics.

An example of imaging in a polymer blend system using a magnetic-sector-based configuration (nano-SIMS) is depicted in Figure 6.8, where the authors utilized nano-SIMS for imaging of various elemental components in a copolymer blend of PS and PMMA [66]. The ^{12}C image is from the PS component in the blend, while the ^{16}O

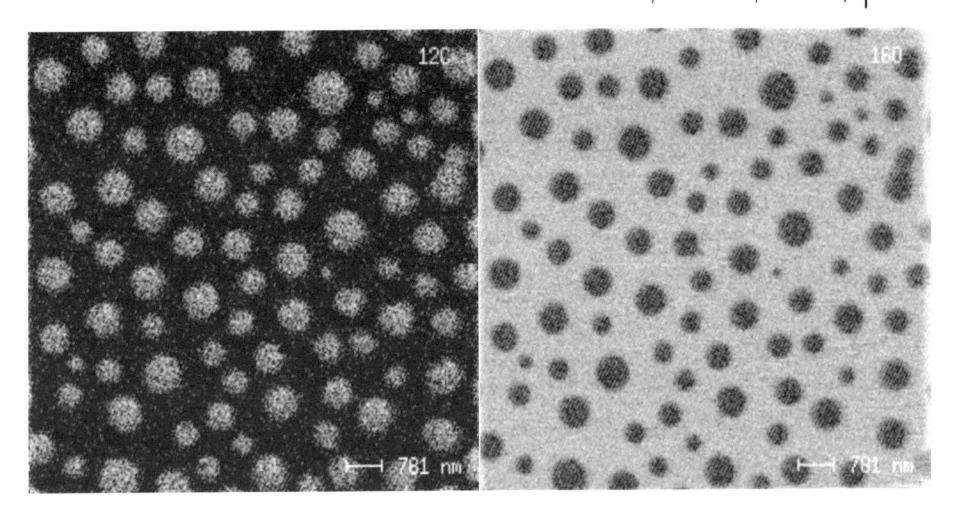

Figure 6.8 Nano-SIMS imaging of C and O components in a polystyrene/poly(methylmethacrylate) (PS/PMMA) polymer blend film. Figure reprinted from Kailas et al. (2004), with permission [66]. Image acquired with Cs^+ ion beam in the microprobe mode (ion beam rastering).

image is from the PMMA component of the blend. The corresponding copolymers were also imaged, but results are not shown here. This image shows domains of PS as little as 100 nm in size – quite an achievement for the field of organic SIMS. Although the spatial resolution is exquisite, however it is difficult to obtain molecular information using the nano-SIMS due to the increased beam damage associated with higher currents.

The spatial resolution observed by the nano-SIMS can be closely matched, however, by using ToF-SIMS with an LMIG source. A recent article published on the microstructure and elution of tetracycline drug from a poly(styrene-co-isobutylene) triblock copolymer matrix illustrates this well [179]. In this work, the authors were able to demonstrate imaging of tetracycline particles of a similar size as that shown in Figure 6.8, whereby the tetracycline molecular ion was imaged as well as the fragments unique to the styrene and isobutylene components in the copolymer. The SIMS results from this work are illustrated in Figure 6.9 [179]. The total secondary ion images are depicted on the left at two separate magnifications, while the corresponding secondary ion image overlays for the drug (red) and copolymer (blue and yellow) components are shown on the right. The small red tetracycline domains range from well below 500 nm to 2 μm in size.

Some other examples of SIMS imaging in multicomponent polymer systems have been published, including images of PVC/PMMA blends [83, 133, 137], PMMA/poly(perfluorohexyl-ethylmethacrylate) PFHEMA copolymers [138], PS/PMMA blends [66, 139], pluronic/PLA/protein blends [96], PP/PET blends [176], PCL/PVC blends [84], ETFE/PMMA blends [295], and PS/PBD blends [193].

An area where SIMS imaging has played a critical role is in the coatings and films industry, where the adhesion properties of complex polymeric coatings need to be understood. An example of SIMS for molecular imaging in paint coatings is given in

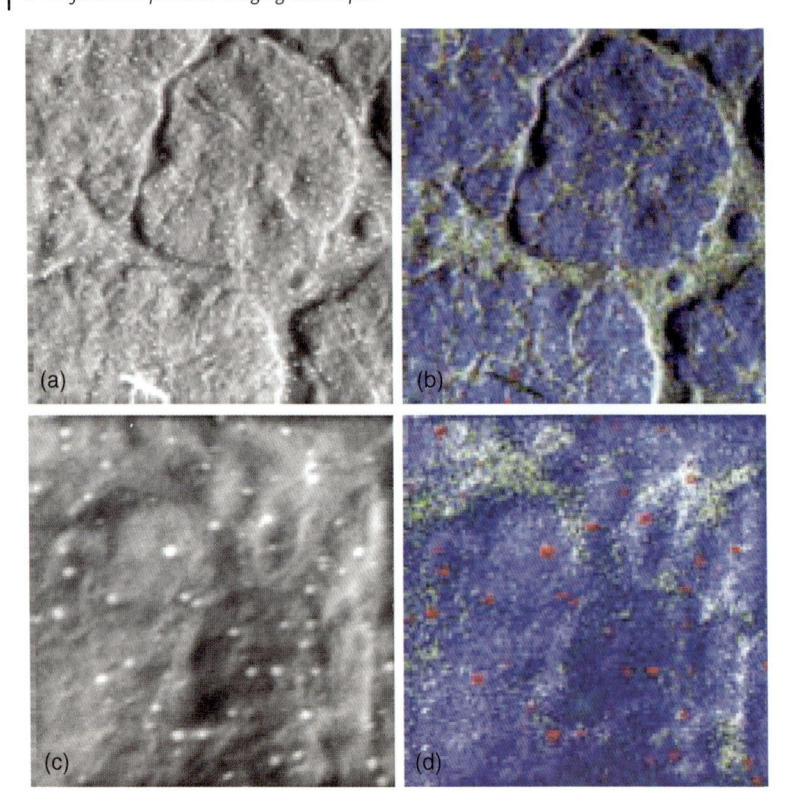

Figure 6.9 Positive secondary ion mass spectral images acquired from a multicomponent sample of poly(styrene-co-isobutylene) containing a tetracycline mass fraction of 8%. (a) Image of total secondary ions (500 × 500 μm), (b) corresponding overlay of secondary ions characteristic of the tetracycline molecular ion at m/z 446 (red), polystyrene block at m/z 91 (yellow), and polyisobutylene block at m/z 97 (blue), (c) image of total secondary ions (100 × 100 μm), and (d) corresponding overlay of secondary ions characteristic of tetracycline (red), polystyrene (yellow), and polyisobutylene (blue). Figure reprinted from McDermott et al. (2010), with permission [179].

Figure 6.10, in which the authors utilized ultralow angle microtomy to expose and image a buried polymer/polymer interface [296]. The slice is created at an angle, and subsequently imaged using SIMS in order to obtain in-depth information from the sample. The negative ion images displayed in the figure were acquired using an Au_3^+ cluster source and show intense molecular and atomic ion images from three different polymeric components in the film: (1) the poly(urethane) primer coat located ∼20 μm below the surface (m/z 66), (b) the poly(vinylidene difluoride) (PVdF) topcoat (m/z 19), and (c) acrylic PMMA/PEMA copolymers which are shown in the figure to segregate to the interface region (m/z 85) [296]. The data is summarized in an overlay of the three regions, created subsequently by the authors of this book chapter (Figure 6.10b). The green areas are the acrylic rich regions, the blue areas are the PU primer, and the red is the PVdF topcoat.

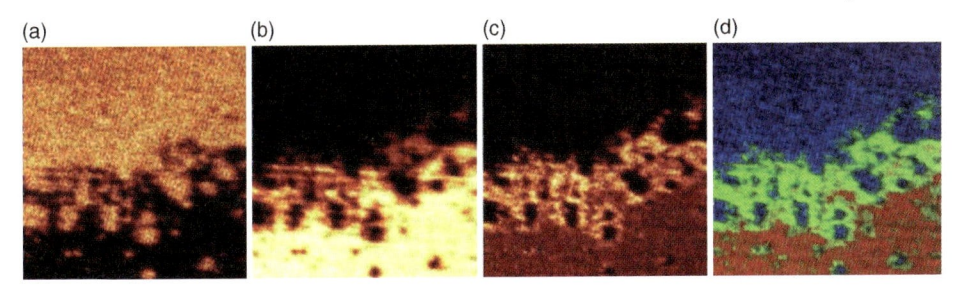

Figure 6.10 Negative ion ToF-SIMS images (500 μm × 500 μm) of a buried polymer/polymer interface exposed by ultra low angle microtomy. (a) m/z 66, PU primer (located 20 μm below surface), (b) m/z 19, poly (vinylidene difluoride) (PVDF) coating, (c) m/z 85, acrylic copolymers, and (d) overlay of PU (blue), PVDF (red), and acrylics (green). Findings confirm that the acrylic copolymers segregate to the topcoat/primer interface where they enhance the adhesive properties. Figure recreated from Hinder et al. (2004), with permission [296].

6.2.3
Data Analysis Methods

More recently, multivariate statistical analysis methods such as PCA and multivariate curve resolution (MCR) have been developed for SIMS [297], and have been used extensively for characterization of polymer surfaces [63, 74, 109, 132, 153, 193, 194, 298–302]. Although there are several other statistical approaches to characterizing data, the discussion here will be limited to PCA [303].

PCA is a method that is used to find variances in complex mass spectral data. The first step in the process is to measure and tabulate the areas of the peaks of interest in the mass spectra. The variance–covariance matrix is calculated from this dataset (X), and the singular value decomposition is calculated, reducing the dimensionality of the data to several key principal components (PCs), PC1, PC2, PC3, and so on, where PC1 describes the greatest variance between the mass spectra, with each consecutively higher PC number describing decreasing amounts of variance in the data. The results of PCA are described by scores and loadings, where the scores (P) indicate how the mass spectra acquired from each sample are correlated with one another, and the loadings (T) indicate what peaks are correlated with the scores data. The process is defined by Eq. (6.1), where X is the mean-centered data matrix, P is the matrix of sample scores, and T is the matrix of loadings. The cross-product of PT^T contains most of the original variance in X with the remaining variance (mostly noise) relegated to the residual matrix E

$$X = PT^T + E \tag{6.1}$$

In the example given in Figure 6.11a and b, Choi et al. wanted to understand the differences in the mass spectra of plasma-polymerized polyethyleneglycol (pPEG) created at different plasma powers. Three different SIMS spectra were acquired at each plasma power, and are associated with a single score, which is given in Figure 6.11a. The first three measurements represent the scores from individual

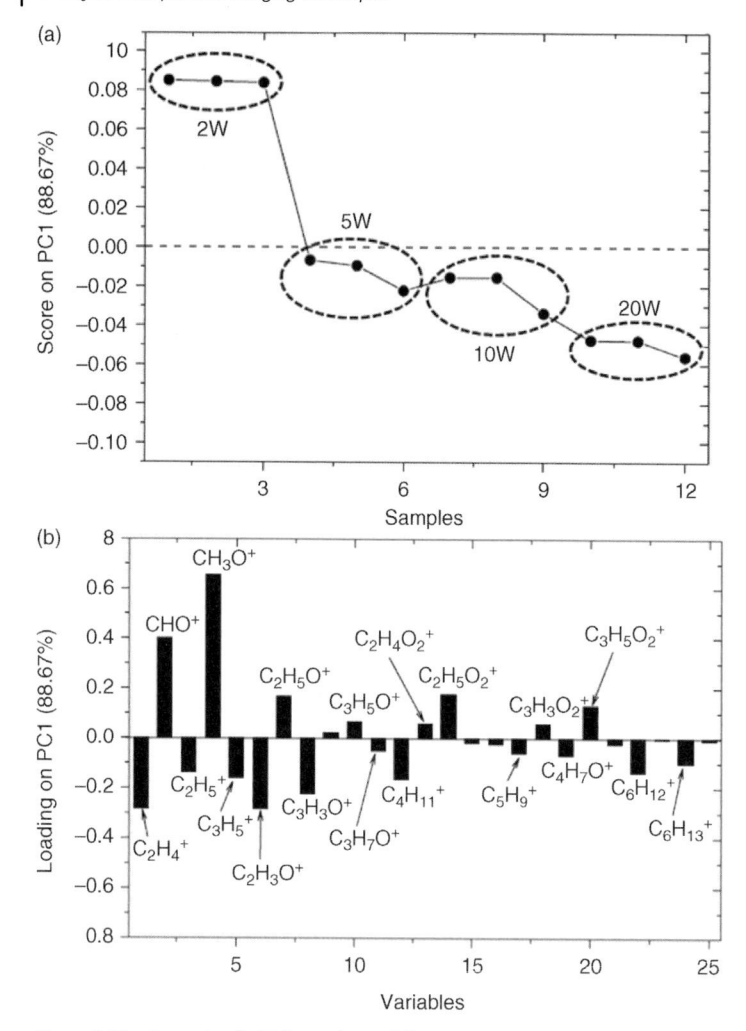

Figure 6.11 Example of PCA for analysis of plasma-polymerized polyethyleneglycol (pPEG). (a) PC1 scores plot for pPEG created at different plasma powers, and (b) corresponding PC1 loadings plots for the pPEG films in (a). Data reprinted from Choi et al. (2010), with permission [300].

mass spectra acquired from pPEG created at 2 W, and the next nine measurements are the resultant scores from pPEG created at 5, 10, and 20 W, respectively. While the first three measurements have positive scores, the remaining spectra in the scores plot have negative scores. This is an indication that the mass spectra are different from each other.

Figure 6.11b shows the loadings plot associated with Figure 6.11a. The positively loaded peaks in this example are associated with O-rich peaks, while the negatively loaded peaks are indicative of hydrocarbons. Therefore, the O-rich peaks are correlated with the pPEG created at 2 W, while the hydrocarbon peaks are correlated with the pPEG created at higher plasma powers. The conclusion from this study is

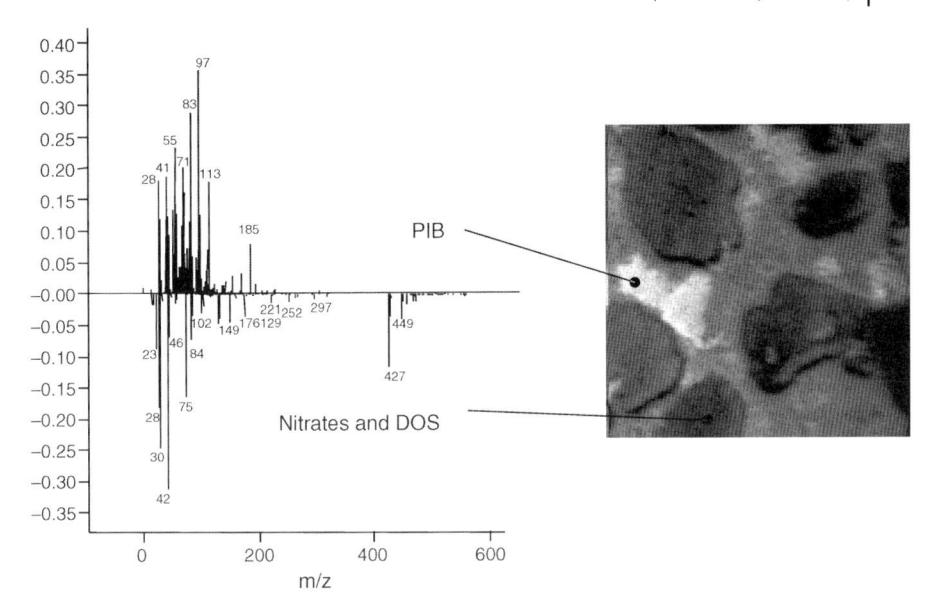

Figure 6.12 Example of image analysis by PCA in a polyisobytylene (PIB) containing plastic explosive (C-4). The image on the right represents a scores image (200 × 200 μm), where the bright regions represent positive scores, and the dark regions represent negative scores. The corresponding loadings plot indicates that the bright regions are correlated with PIB-rich domains, while the dark regions are correlated with the explosives (nitrates) and a di-isooctyl sebacate (DOS) binder. Figure reprinted from Mahoney et al. (2010), with permission [304].

that the structure of the pPEG created at the lower plasma powers is more consistent with the actual structure of PEG.

PCA is also useful for imaging applications, and a recent example of this is given in Figure 6.12, which depicts the PCA image analysis from a solution cast sample of a composition C-4 explosive. The primary constituents in C-4 consist of RDX explosive, a poly(isobutylene) (PIB) binder, and plasticizers such as diisooctylsebacate (DOS) or dicaprylcapryl adipate (DCA). The image on the right-hand side of the figure shows the image of PC2 scores calculated from a series of mass spectral images of the C-4. In this image, the bright regions represent areas with positive scores, while the dark regions represent areas with negative scores. The corresponding loadings plot is shown on the left-hand side, where the positively loaded peaks represent poly(isobutylene), and the negatively loaded peaks represent nitrates (from the RDX) and DOS. The white areas in the image, therefore, are domains that are rich in PIB. This example demonstrates quite nicely the utility of PCA for characterization of complex mixtures. The mass spectrum in this PIB-rich region is complex, containing several additional peaks characteristic of other components in the C-4. These peaks can make it much more difficult to identify peaks specific to the fingerprint spectrum of PIB. The loadings plot in the PCA, however, shows a mass spectrum that is close to identical to that of the pure PIB.

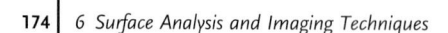

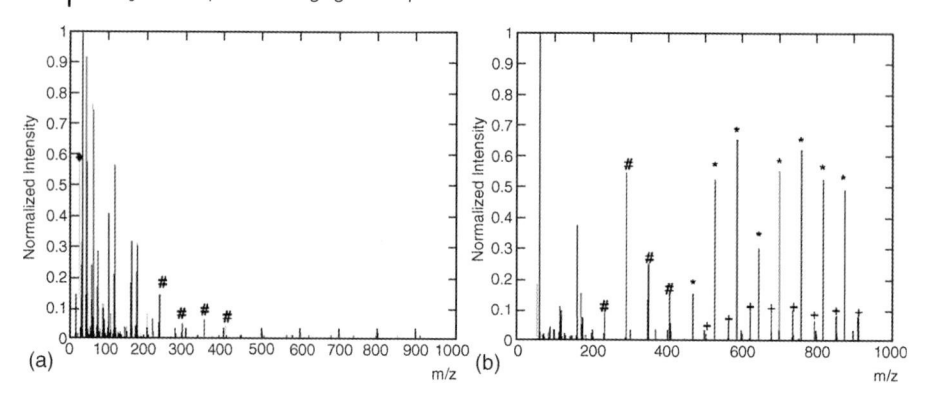

Figure 6.13 Representative positive ion SIMS (a) and G-SIMS (b) mass spectra from a sample of poly(glycolic acid) (PGA). Peak series of $[nM + H]^+$ (#), $[nM - OH]^+$ (+), $[nM - H_2O] + Na^+$ (★), and Na^+ (◆) are labeled accordingly. Figure reprinted from Ogaki et al. (2000), with permission [227].

Another analytical method that is worth mentioning here is a method developed by scientists at the National Physical Laboratory (NPL) in the UK. This method, called gentle SIMS, or G-SIMS, has shown great promise as a tool for decreasing the influence of fragmentation in SIMS spectra of polymers [58, 227, 232, 305, 306]. The method takes advantage of variable ion beam damage conditions to extrapolate information into low damage regimes. During the process of ion bombardment, a high-density, localized plasma is created, where the temperature of this plasma (T_p) varies with changing ion bombardment conditions (such as changing beam energies or using a different primary ion). The G-SIMS process takes into account the differences in mass spectra created at different T_ps, (e.g., high vs. low ion beam energies) and uses these differences to extrapolate into even lower T_p regions, yielding a mass spectrum with significantly reduced fragmentation. An example representing the utility of G-SIMS for analysis of polymers is shown in Figure 6.13, which shows the representative positive ion SIMS spectrum (Figure 6.13a) and corresponding G-SIMS spectrum (Figure 6.13b) acquired from a PGA sample cast on Si [227]. As can be seen, the G-SIMS spectrum has significant reductions in fragmentation allowing for visualization of high-mass fragmentation patterns (★, +).

6.2.4
Polymer Depth Profiling with Cluster Ion Beams

In order to introduce the topic of polymer depth profiling, it is necessary to first introduce the concept of cluster SIMS. Earlier in the chapter, it was shown that polyatomic primary ion sources, such as C_{60}^+, can cause significant increases in the sputtering and ion yields from polymer samples. This is associated with the nonlinear sputtering yield enhancements that occur when several atoms strike the surface at the same time [307, 308]. On the one hand, because the energy of each constituent atom in the cluster will strike the surface with only a fraction of the initial beam energy, cluster ion sources will deposit all their energy close to the surface. On the other hand, high-energy atomic beams will penetrate deep below the surface causing extensive damage to the molecules in the subsurface region.

Recent molecular dynamics (MD) simulations nicely illustrate the differences between atomic ion and cluster ion bombardment [291, 292]. Some examples are shown in Figure 6.14, where Figure 6.14a and b shows a snapshot obtained from a

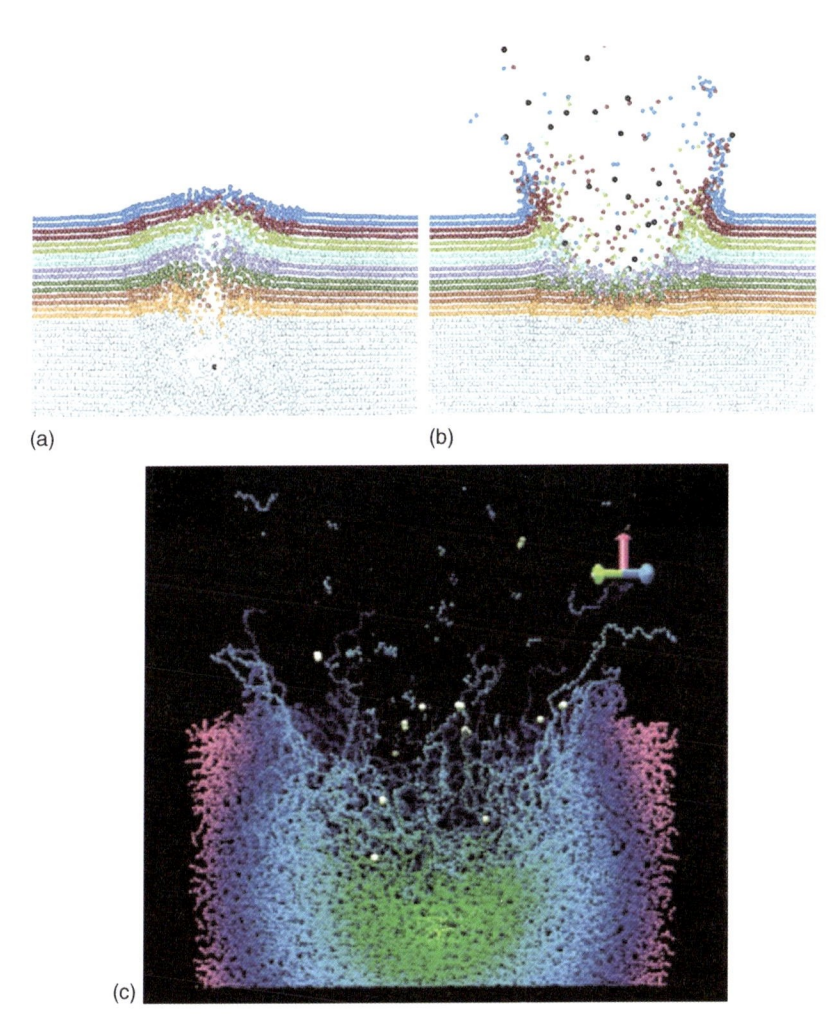

Figure 6.14 Side-view snapshots during a molecular dynamics simulation of sputtering in Ag (a, b) [291] and polyethylene (PE14) (c) [292] substrates. (a) 15 keV Ga ion bombardment of Ag(111) after 1 ps, (b) 15 keV C60 ion bombardment of Ag(111) after 1 ps, and (c) 5 keV C_{60} bombardment of PE with 14 monomer units (PE14) after 5.6 ps. Results reprinted from Postawa et al. (2004) (a, b) [291], and Delcorte and Garrison (2007) (c) [292], with permission.

MD simulation of Ga (a) and C_{60} (b) projectiles impinging on an Ag surface, and Figure 6.14c shows a snapshot from an MD simulation of a C_{60} molecule impinging upon a polyethylene (PE) surface.

The increased subsurface damage created by atomic ion beams coupled with low sputtering yields results in the rapid accumulation of beam damage with increasing ion dose, precluding any possibility of molecular depth profiling. In contrast, polyatomic or cluster ion beams have a combination of increased sputter yields and surface-localized damage, resulting in a system at equilibrium, where most of the damage created by the impact is essentially removed, allowing for molecular depth profiling in organic materials [309, 310]. This represents a major paradigm shift for the method of SIMS.

Figure 6.15 illustrates this by showing molecular depth profiles from a glutamate film (180 nm) vapor deposited on Si, using an Ar^+ (a) and SF_5^+ (b) primary ion source, where molecular and fragment ion intensities characteristic of this amino acid were monitored as a function of increasing sputter time [141]. As can be seen, all characteristic molecular signal rapidly decays, when employing the Ar^+ source. However, the SF_5^+ source yields constant glutamate molecular (and fragment) ion signals as a function of increasing sputtering time. Furthermore, after ~600 s of sputtering, the entire film was etched through, as evidenced by the rise in Si substrate signal commensurate with diminishing glutamate signal.

Depth profiling of polymers was also demonstrated in this same work, where the authors discovered relatively stable signal characteristic of PMMA at m/z 69 ($C_4H_5O^+$ or $[M-CH_3O]^+$) obtained under higher primary ion doses [141]. Shortly

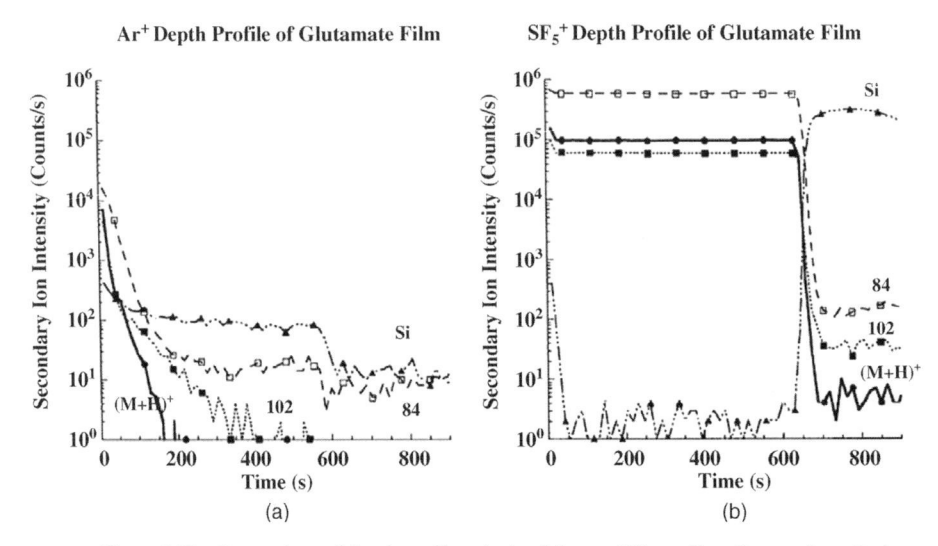

Figure 6.15 Comparison of depth profiles obtained from a 180-nm film of vapor-deposited glutamate using Ar^+ (a) and SF_5^+ (b) primary ions under dynamic SIMS conditions. Figure reprinted from Gillen et al. (1998), with permission [141].

thereafter, Brox *et al.* demonstrated successful attempts to depth profile through thin films of PEG, PPG, PMMA, and PTFE [146]. Although these polymers were proven to be amenable to depth profiling, however, there still were several other polymers that were not. These polymers (i.e., PC, PS, PE, and polyisoprene (PIs)) exhibited the same rapid molecular decays as is typically observed with atomic beams (Figure 6.15, left panel). Furthermore, no bulk polymers, other than PTFE, were amenable to being depth profiled. The authors concluded, therefore, that polymeric depth profiling was more of an exception than a rule [146].

This is not the case today. With the development of advanced cluster sources, such as gas cluster ion beams (GCIB) ($Ar_{n>500}$)$^+$ [156, 196, 311], and a better understanding of optimized beam conditions for molecular depth profiling [142, 290, 312–315], the limitations for polymer depth profiling have essentially been removed. It should be noted that although *molecular* depth profiling is not possible with higher energy atomic ion beams (>300 eV), depth profiling of *elemental* components is feasible using any ion beam, even high-energy atomic ion beams, and has been utilized regularly for this purpose. For more information, see Refs. [166, 198, 203–206, 256, 271].

A particularly successful group of materials in terms of molecular depth profiling are biodegradable polyesters, which are prone to undergo main-chain scission and depolymerization processes upon ion bombardment and are therefore more amenable to molecular depth profiling [95, 96, 98, 130, 178, 237, 240, 241, 290, 312]. Figure 6.16 shows a representative depth profile acquired from a PLA thin film (200 nm) doped with a drug molecule at a mass fraction of 50% [98, 240, 243].

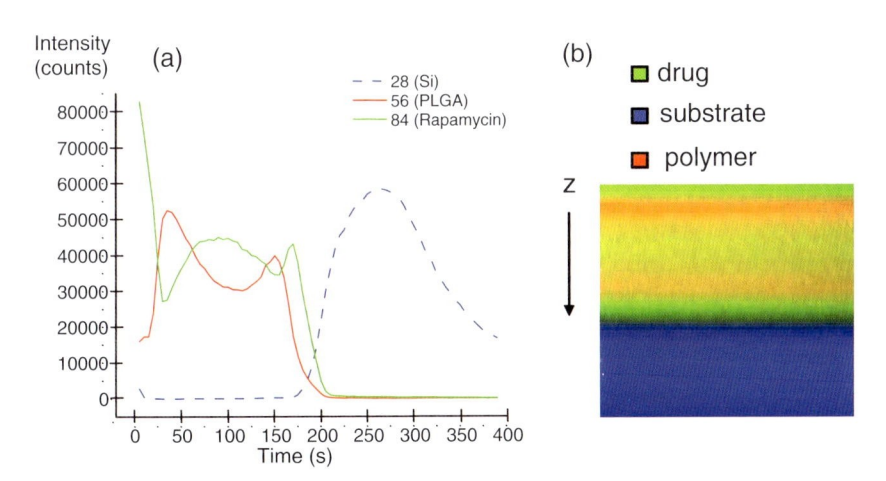

Figure 6.16 Representative depth profile of a thin film (200 nm), containing rapamycin (drug) in a poly(lactic acid) (PLA) matrix (drug mass fraction = 50%); a polymer/drug system employed in drug-eluting stent coatings. (a) Average intensity profiles of m/z 56 (PLA), m/z 84 (rapamycin), and m/z 28 (Si substrate), and (b) z-scale image profile. Figure reprinted from Mahoney *et al.* (2011), with permission [243].

This system represents a model for a drug-eluting coronary stent system, where the drug, rapamycin, serves to prevent restenosis or the renarrowing of the artery after an angioplasty procedure. Figure 6.16a shows the intensities (averaged over 5 s) characteristic of the rapamycin (m/z 84, green), polymer (m/z 56, red), and Si substrate (m/z 28, blue) plotted as a function of increasing sputter time with an SF_5^+ polyatomic primary ion source. The corresponding side-view image of the film is depicted in Figure 6.16b, where the colors shown are commensurate with those described for Figure 6.16a. Clearly, the drug is segregated to the surface and interfacial regions in this example, and a very distinctive diffusion profile is observed at both interfaces. Immediately below these drug-enriched regions are drug-depleted regions, which are marked by a maximum in PLGA signal (red) and a corresponding minimum for the rapamycin signal (green). This example illustrates the utility of cluster SIMS for characterizing the surface and in-depth compositions of additives in polymers, as well as the potential for relating drug release characteristics to the chemical structure within the film.

Quantitative depth profiling in polymers is also possible using cluster ion beams, although there have been very few examples of this in the literature [95, 158]. Figure 6.17 illustrates the process used by Mahoney, Yu, and Gardella, to obtain quantitative depth profiles from a series of PLA/Pluronic P104 blends (~1 µm thick, cast on Si), which have potential as protein drug delivery vehicles [95]. A typical depth profile acquired from a PLA/P104 blend system is displayed in Figure 6.17a, where the known mass fraction of P104 in the blend was 10%. Intensity characteristic of the P104 molecule (m/z 59; solid gray line), PLA (m/z 56, solid black line) and Si (m/z 28, dashed line), are displayed as a function of SF_5^+ sputter time. A very similar diffusion profile is observed in the P104/PLA blend system, as was discussed previously with the model drug delivery system in Figure 6.16, where the P104 appears to be segregating to the surface and interfacial regions.

The data is plotted as a ratio of P104 (m/z 59) to PLA (m/z 56) signal in Figure 6.17b, giving a qualitative assessment of the P104 content relative to PLA. Also note that the data is converted to a depth scale through use of measured sputter rates (in nm/s). It is expected that the intensity ratio observed in the steady-state region of the profile will be associated with the known bulk content in the sample (in this case 10%). Therefore, one can create calibration curves by analyzing a series of polymer blends with varying bulk compositions, and plotting the intensity ratios in the steady-state region as a function of known bulk compositions (Figure 6.17c and d). The resultant equations with fitted parameters were utilized to obtain quantitative depth profiles from the intensity ratio profiles (Figure 6.17e). The concentrations of P104 at the surface, in this example, were measured to be ~25%. This is consistent with previous XPS data [95].

Figure 6.17 (*Continued*) of P104 mass fraction (ranging from 10% to 50%), and (e) composition depth profile plotting the mass fraction of P104 as a function of depth, calculated using the calibration curves in (c) and (d). Values plotted in (c) and (d) are the averaged values from at least four measurements with the error bars representing one standard deviation.

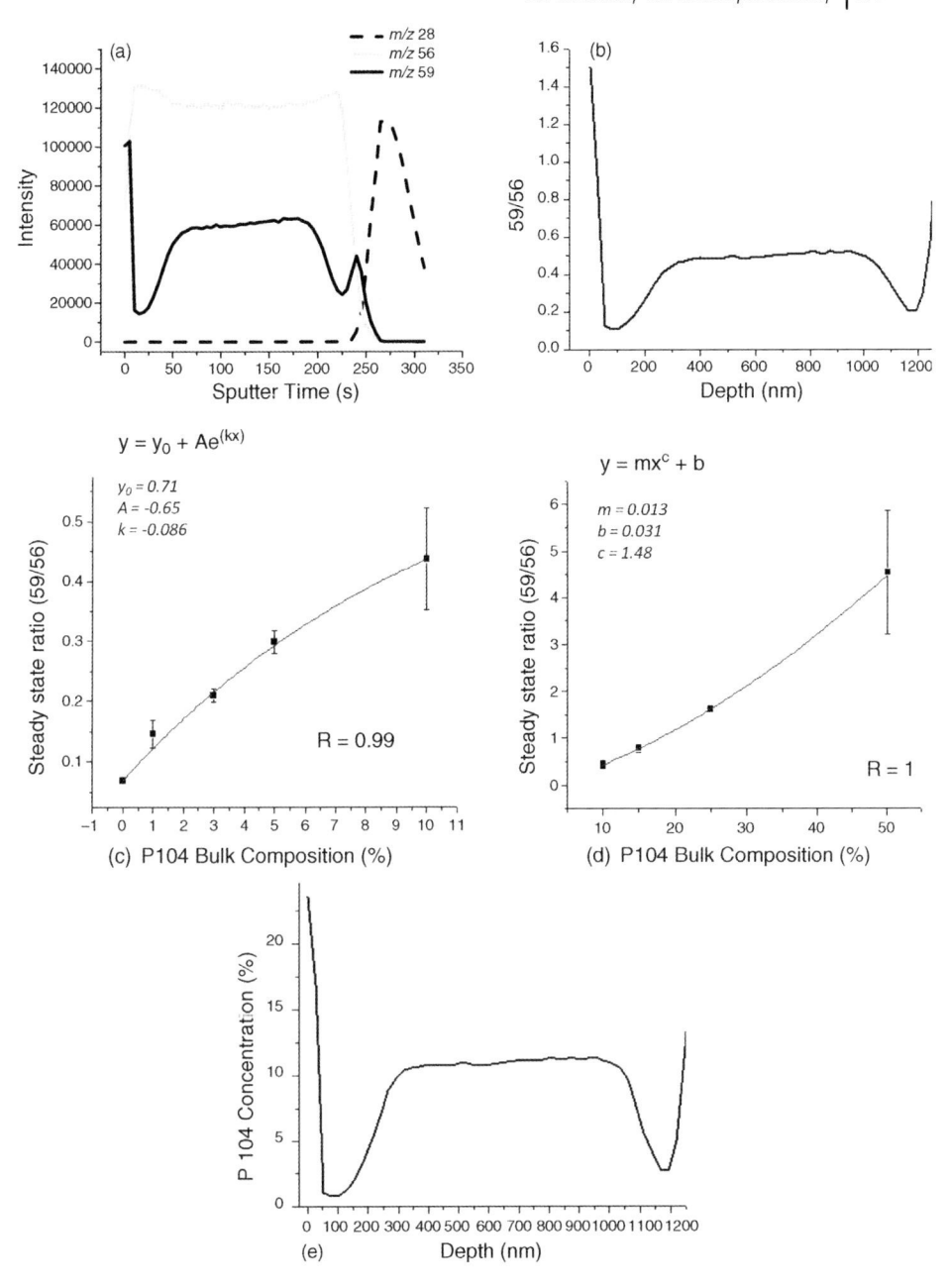

Figure 6.17 Demonstration of quantitative depth profiling in a sample of poly(lactide acid) (PLA) containing 10% by weight of P104. (a) Signal characteristic of P104 (m/z 59), PLA (m/z 56), and Si (m/z 28) plotted as a function of increasing sputter time with a 5-keV SF_5^+ polyatomic primary ion source, (b) ratio of intensities characteristic of P104 (m/z 59) to PLA (m/z 56), (c) average steady-state intensity ratios plotted as a function of P104 mass fraction (ranging from 0% to 10%), (d) average steady-state intensity ratios plotted as a function

6.2.4.2 A Brief Discussion on the Physics and Chemistry of Sputtering and its Role in Optimized Beam Conditions

It is well-known that different polymers will behave differently when irradiated by ion beams or other forms of radiation [162, 167, 170, 290, 316–318]. More specifically, certain polymers will tend to crosslink under ion beam irradiation (e.g., PS, PE, PP, etc.; polymers that have very little branching and high aromaticity), whereas others will undergo extensive degradation via main-chain scission and/or depolymerization processes (e.g., PLA and PMMA; polymers that have increased branching and weakened main chain structure via introduction of chemical functionalities) [290].

Obviously, for sputter depth profiling crosslinking polymers are going to be more difficult to work with, because crosslinked polymers have decreased sputtering yields, which continue to decrease with increasing crosslink concentration [32, 162]. Even in degrading polymers, however, there is often an element of crosslinking that occurs, which will eventually accumulate to a point where depth profiling in these systems also fails. The approach of the "gel point" in the bombarded region will cause an even more rapid decay in molecular signal and corresponding sputter rates until a graphitized surface structure is formed [19, 20, 40, 153, 154, 162, 290, 319–322].

It is also well understood that it is not the initial ion bombardment event that causes the damage, but rather the diffusion of free radicals created by the initial event [6]. One method of decreasing the reactivity of these free radicals is to trap them by decreasing the temperature of the sample during the SIMS depth profile experiment. This significantly decreases the diffusion and reactivity of the free radicals and therefore has often resulted in significant improvements in the depth profilability of organics [98, 149, 154, 160, 178, 323, 324].

Increased temperatures also can play a role, particularly above the glass transition temperature of the polymer of interest, where free radicals have an even greater reactivity and diffusivity resultant from the increased mobility of the polymer chains. While increasing temperature is likely to speed up the crosslinking events in many polymers, it is also expected to increase the chain scission events. This has been proven to be the case in certain polymers, such as PMMA and poly(α-methyl styrene) (PAMS), in which ion-beam-induced depolymerization mechanisms dominate at higher temperatures [152, 160].

Other experimental parameters have proven to be important in polymer depth profiling, including ion-beam angle [197, 312, 325], beam energy [143, 148, 153, 196, 312, 314, 326], sample rotation [312–314], and source design, chemistry, and size [156, 161, 241, 263, 290]. On the one hand, increased beam energies, for example, typically allow for increased maximum achievable erosion depths before the gel point is reached and the molecular signal disappears, likely a result of the increased sputter yields observed at higher beam energies [143, 153, 326]. On the other hand, the higher beam energies will impart increased damage to the sample (increased T_p as described by G-SIMS, see Section 6.2.3), resulting in significant losses in depth resolution and increased fragmentation in the mass spectra.

Organic depth profiling at glancing incidence angles also shows much improved results as compared to normal angles [309, 310, 327, 328]. When the ion impinges upon a surface at $90°$, the energy is distributed downward toward the bulk of the sample, creating a thicker "altered layer" [310], while at a 10° glancing incidence the energy will be distributed at an angle away from the bulk of the sample, resulting in a decreased "altered layer" thickness and therefore decreased overall damage accumulation. Using glancing angles, some have even been able to demonstrate depth profiling in type I crosslinking polymers, such as polystyrene [197].

More recently, sample rotation has been shown to play a very significant role as well, removing the topography formation during the depth profile experiment, and thus allowing for increased sustainability of signal and more uniform sputter properties [312–314].

The parameters discussed above are summarized in Figure 6.18, which shows the results from a recent interlaboratory study performed with a multilayer reference material prepared by NPL [311, 312, 314]. The reference material consists of alternating layers of two UV stabilizers (Irganox 1010 (~55 nm) and Irganox 3114 (3 nm ~ 4 nm)), prepared by evaporation. This type of sample provides an accurate measure of sputter rate nonlinearities and variations in depth resolutions under changing experimental conditions.

A sample depth profile of the NPL reference material using 20 keV C_{60}^{2+} ions with sample rotation is shown in Figure 6.18a. As can be seen, the signal characteristic of the Irganox 3114 is monitored as a function of depth, and four peaks are resolved. Figure 6.18b shows plots of the FWHM measured from each of the four peaks under different beam conditions. The curve on the top (O) shows the control values, acquired using 20 keV C_{60}^{+}, 45° incidence and 25 °C. The depth resolution of the peaks (FWHM) clearly decays with depth. The corresponding sputter rates are also nonlinear with increasing depth (data not shown). The best results for this particular

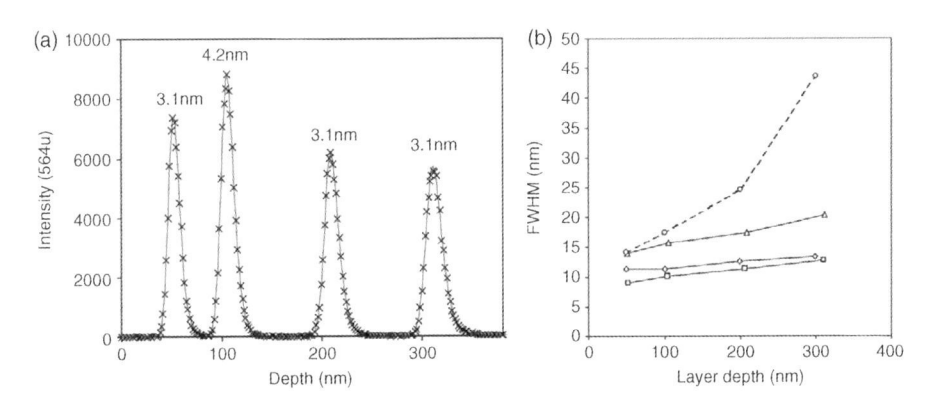

Figure 6.18 Depth profiling in organic multilayer reference material developed by the National Physics Laboratory (NPL) in the UK [329]. Layers consist of Irganox 1010 with very thin layers of Irganox 3114 distributed throughout: (a) example of a depth profile acquired using 20 keV C_{60}^{2+} ions with 14 Hz sample rotation. The thickness of each Irganox 3114 layer is shown and the integrated intensity ratios are in close agreement with thickness ratios. (b) FWHM measured for each layer, using improved parameters such as 20 keV C_{60} at a 76° incidence angle (□), 10 keV C_{60} at −80 °C (◇), and 20 keV C_{60} at 14 Hz rotation (△), compared to mean 20 keV C_{60} values (O). Figure reprinted from Shard et al. (2010), with permission [312].

system are obtained at low temperatures (◇) and with sample rotation (□). This is followed closely by the results obtained at glancing incidence angles (△).

Although these optimized conditions are useful for depth profiling in degrading polymers, they still have not proven to be useful (other than glancing incidence angles) for depth profiling in type I polymers. There are specific chemistries, however, that can be taken advantage of for depth profiling in type I polymers. For example, since crosslinking mechanisms in polymers are largely dependent on free radical chemistry, the use of free radical inhibitors should help quench these reactions. Preliminary results, where a free radical inhibitor gas such as NO was leaked into the main chamber during the experiment, show very promising results for depth profiling in type I polymers such as PS [330]. Another means of characterizing type I polymers is to utilize a low-energy reactive beam, such as Cs^+ (<200 eV), which is superior for analysis of crosslinking polymers (better than most cluster sources) most likely due to the formation of reactive intermediates [142, 155, 212, 213, 331].

The best depth profiles thus far, however, have been obtained from the application of the GCIB in SIMS. GCIBs consist of hundreds to thousands of atoms (such as Ar) or molecules (O_2 or SF_6) of gaseous materials that are formed by a supersonic expansion. These massive gas clusters are capable of profiling through both type I and type II polymers, with superior depth resolutions and greater sampling (etching) depths as compared to more conventional cluster ion sources, such as C_{60}^+ [156, 196, 197, 311]. Although these sources are relatively new to the market, it is expected that they will be the gold standard in sputter sources (particularly for polymers) in the near future.

6.2.5
3-D Analysis in Polymer Systems

One of the most exciting aspects of polymer depth profiling is the potential for visualizing molecular distributions in 3D. This is done by stacking sequential SIMS images acquired at different erosion depths. 3D imaging in polymers was first demonstrated by Gillen et al. [239], where 3D volumetric representations of PLA-based drug delivery systems were presented.

A more recent example is given in Figure 6.19, which shows the 3D volumetric representations of pluronic (P104) in a PLA matrix (same system as was described earlier in Figure 6.17). All samples shown contain a P104 mass fraction of 15%, where Figure 6.19a and b also contains bovine serum albumin (BSA) at a mass fraction of 5%. The BSA was added via an oil-and-water emulsion method [96]. The study presented here is a continuation of a previous work performed by Mahoney et al., where the authors utilized insulin as the model protein [96].

Figure 6.19c and d shows the 3D distribution of pluronic in the control sample containing no BSA, where Figure 6.19c shows a side view of the image (surface toward the bottom of the image), and Figure 6.19d shows a frontal view. P104 crystalline domains are clearly observed in the subsurface region, where the depletion layer is located.

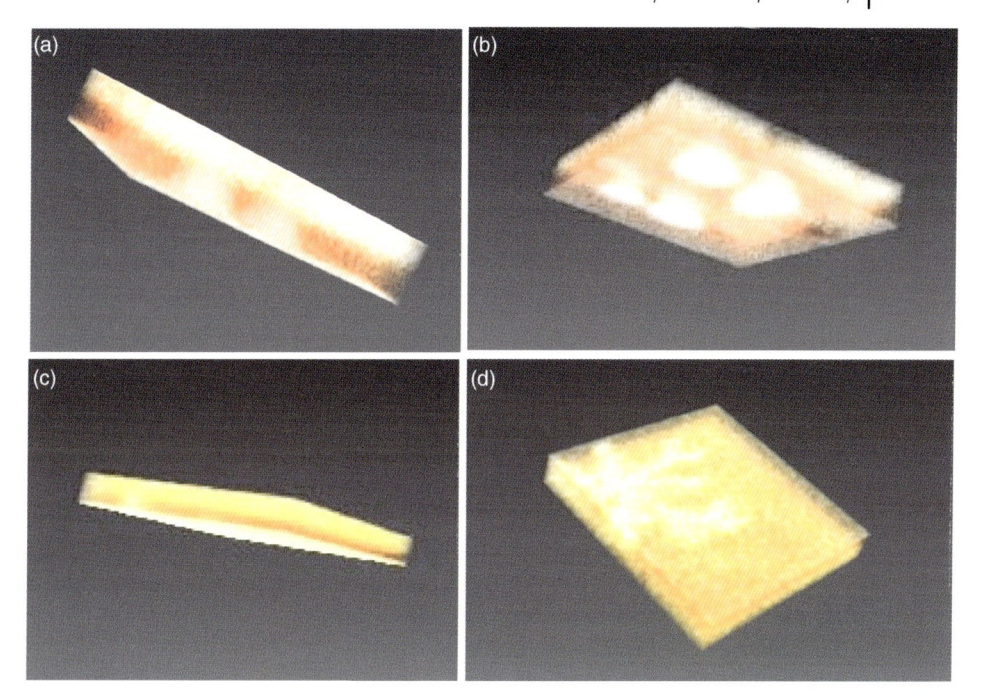

Figure 6.19 3-D volumetric representations of pluronic (P104) in a model protein delivery device (200 × 200 × 0.9 μm). Data were acquired using a Bi_3^+ cluster source for imaging, and an 8-keV SF_5^+ polyatomic source for sputtering. All images contain a P104 mass fraction of 15% in PLA. (a + b) Also contains BSA (mass fraction of 5%); (a) side view, and (b) frontal view. (c + d) Controls sample containing no BSA; (c) side view, and (d) frontal view. Data shown represent all signal acquired prior to the polymer/Si interface region. Therefore, the approximate thickness of the displayed area is approximately 900 nm in both cases. The visual differences in scaling in the z-direction are resultant from differing sputter rates (sputter rates were slower in BSA-containing system).

The pluronic molecule is an amphiphillac triblock copolymer containing PEG and PPG components. The amphiphilic nature of the P104 polymer in combination with its increased biocompatibility means that it is well-suited to encapsulate and protect protein components via micelle encapsulation, and thus can increase their stability in hydrophobic polymer matrices, such as PLA [95, 96, 332].

The images of the BSA-containing system in Figure 6.19a and b show a very different 3D structure. In both the cases, the 2D diffusion profile in the z-direction is clearly observed, whereby the pluronic component is enriched at the surface and depleted in the subsurface region. However, spherical domains are observed in the depletion region in the BSA-containing sample, as opposed to the crystalline P104 domains that were observed in the control. These spherical domains are consistent with the formation of BSA-containing micelles. Indeed, the nitrogen signal is colocated with the P104 signal in the negative ion images (data not shown).

One of the biggest difficulties that arise from analysis of samples of this type is matrix effects. More specifically for the example in Figure 6.19, the proteins are more difficult to ionize relative to the surrounding polymer matrix. Therefore, the protein signal was masked by the overwhelming signal from the polymeric matrix. Although negative ion images do indeed show low mass nitrogen-containing species in this region, the signal was very weak, and there were no molecular signals consistent with the protein. The positive ion mass spectra showed no evidence of any protein anywhere throughout the film. Matrix effects, therefore, remain an elusive problem in the field of SIMS.

Although SIMS 3D imaging does show promise for characterization of polymeric materials, there are certain things that need to be considered when performing any sputter depth profile experiment, including differential sputtering rates in multi-component systems and different damage accumulation rates. This can be corrected for, at least in part by measuring the topography (e.g., AFM) before and after the experiment, and correcting the z-scaling accordingly. Wucher *et al.* defined these and other protocols for obtaining accurate 3D images in soft materials [333].

Another very important area of recent development in cluster SIMS has been in the application of cluster SIMS for 2D and 3D mass spectral analysis in biological systems [2, 334–342]. Although this topic is not the scope of the current work, it is important to mention this work in brevity, as it more than adequately demonstrates the utility of the method for high spatial resolution imaging in 3D. Recent examples in 3D subcellular imaging are very promising, and represent an area where SIMS can potentially make a huge impact [337, 342]. Once again, matrix effects are proving to be an important factor in biological tissue imaging, particularly for pharmaceutical components in tissue [283, 343].

6.3
Matrix-Assisted Laser Desorption Ionization (MALDI)

6.3.1
History of MALDI Imaging Mass Spectrometry

MALDI utilizes a laser probe for desorption of species from surfaces. In comparison to other imaging techniques (e.g., SIMS-ToF), the sample preparation in MALDI-ToF MS requires the deposition of a matrix, which is essential for the soft desorption and ionization of the analyte molecules. Hence, a homogeneous covering of the sample by the matrix represents a basic prerequisite for recording reproducible data. For this purpose, various techniques have been developed and will be discussed in Section 6.3.2.

Soon after its broad introduction in the scientific community in the mid-nineties, the potential of MALDI-ToF mass spectrometry for imaging experiments became obvious. Instrumental developments, especially in the field of lasers, detectors, and digitizers, enabled the fast recording and processing of huge data sets. For example, the laser repetition rate increased from 10 Hz (N_2 laser, 337 nm) to 1 kHz (solid-state lasers, 355 nm), which was especially important for the rapid scanning of larger sample regions.

In 1997, a first example on MALDI imaging of peptides and proteins was presented by the Caprioli group [344]. Although this chapter is dedicated to synthetic macromolecules, a few applications of MALDI imaging of biological samples shall be given. This particular technique was applied, for example, for the determination of pharmaceutical compounds in skin [345], for the direct analysis of drug candidates in tissue [346], and for tissue analysis in general [347]. It can also be applied for determining the location of agrochemical compounds in plants [348]. The coupling with ion mobility mass spectrometry (IMS) was used to determine the distribution of lipids in brain tissue [349]. Combining laser ablation (LA) with inductively coupled plasma mass spectrometry (ICP-MS), a quantitative determination of copper, zinc, lead, and uranium distributions in thin slices of human brain could be performed [350].

There are also examples demonstrating that imaging can be further improved from a lateral 2D method to a 3D technique providing analyte spatial distributions [351, 352]. In this process, MALDI data from a series of consecutive microtomed slices were coregistered with the optical images and superimposed to produce a 3D volumetric representation of the data. This method was shown to provide a better understanding of healthy and pathological brain functions.

In earlier years, the software used for characterization of MALDI data was variable, creating compatibility issues. This has been recently overcome by a new standardized open-source mass spectrometry imaging software that is compatible with all types of instruments [353]. A comprehensive overview on biopolymer imaging can be gained from the following reviews [2, 354, 355]. Meanwhile the number of published MALDI imaging applications increased extraordinarily. More than 450 publications covering the search terms *MALDI* and *imaging* are available from web of science (from August, 2010). Among them are papers dealing with the imaging of the dynamics and gas-phase diagnostics of the desorbed MALDI plume, which is not in the focus of this article [356, 357]. A further limitation of the search by adding the term *synthetic polymers* surprisingly resulted in a very few papers that were published within the last 2 years.

6.3.2
Sample Preparation in MALDI Imaging

The process of desorption/ionization in MALDI requires the use of a low molecular mass compound – the so-called matrix. This matrix must fulfill several requirements. First, the matrix should be miscible with the analyte molecules and must also be easily vaporizable. In addition, it has to absorb the laser energy at the appropriate laser wavelength, and must, in the gas phase, transfer the energy to the analyte molecules. Normally, sample preparation in MALDI proceeds by a mixing of sample and matrix either in solution (solvent-based technique) or in the solid state (solvent-free technique). Neither preparation methods are applicable for imaging of sample surfaces. On the one hand, it is obvious that a simple deposition of dissolved matrix on the sample to be scanned is not very useful, because the analyte distribution in the sample would be disturbed by the solvent. On the other hand, a simple covering of the analyte layer with dry matrix does not enable a sufficient mixing of matrix and analyte molecules, which is essential for achieving good MALDI spectra. Therefore specific methods had to be developed, which guarantee a redissolvation of the sample surface without changing the lateral or spatial analyte distributions. In this regard, several studies have been performed to gain a deeper insight into the mechanism of desorption and ionization.

The influence of sample preparation on the matrix crystal size distribution and its consequence on volatilization in MALDI-mass spectrometry was investigated by Sadeghi *et al.* [358]. Their results indicated that at near-threshold irradiance, smaller crystals are completely volatilized by the laser shot, whereas larger crystals undergo layer-by-layer evaporation (peeling). Later findings showed that intensive mixing of sample and matrix using a solvent-free sample preparation method (e.g., in a mini ball mill) dramatically increases the quality of the spectra, due to the lack of crystallite formation [359, 360].

In MALDI imaging mass spectrometry, where the spatial resolution is determined by the laser spot size, which is normally around 30–100 µm, the importance of depositing small homogeneous crystals across the surface becomes evident. Larger crystals require larger step sizes between laser shots, while smaller crystals allow for more crystals to be ablated in a single laser shot and a decreased laser step size. Moreover, attempts have been undertaken to perform MALDI imaging experiments of features smaller than the size of the laser beam by oversampling with complete sample ablation of the MALDI matrix coating, moving the laser spot a fraction of the laser beam diameter, such that the effective laser spot is reduced [361]. This process yields images with a pixel size as much as four times smaller than the laser spot size.

Practical aspects of sample preparation from initial tissue treatment, matrix application, and MS analysis were comprehensively reviewed [347, 362]. Two different principles involving *spraying* or *spotting* devices were discussed. The spraying method utilizes a glass spray nebulizer to spray small matrix amounts directly onto the tissue surface. In order to avoid a dislocation of sample molecules by redissolving, the spray process has to be repeated several times. Another spray method utilizes the principle of electro spray deposition (ESD) [363]. Charged droplets of dissolved matrix are sprayed on the surface. However, in contrast to the conditions applied in an electro spray ionization (ESI) source (used for soft ionization in mass spectrometry), in ESD a complete drying of the formed droplets before reaching the analyte surface must be avoided to enable a sufficient incorporation of analyte molecules by redissolvation. Once suitable deposition conditions are found, very homogeneous sample spots can be achieved.

In Figure 6.20, an example using 2,5-dihydroxybezoic acid (2,5-DHB) serving as a typical MALDI matrix is presented. In Figure 6.20a, the optical image of a DHB spot is shown, prepared according to the dried droplet method, using methanol as a solvent. Large matrix crystals mainly located at the spot periphery were obtained. The same matrix/solvent system was used for the ESD experiment. The results are shown in Figure 6.20b–d. The optical image (Figure 6.20b) shows a homogeneous covering of the target plate. This could also be confirmed by a successive MALDI imaging experiment (Figure 6.20c). The average size of matrix crystals of ca. 1–2 μm could be determined by means of atomic force microscopy (AFM) (Figure 6.20d). Other versions of the spray technique involve the use of an oscillating capillary nebulizer (OCN) [364] and acoustic nebulization (AN) [365].

As an alternative tool, the deposition of droplets (spotting) can also be used for sample preparation. Starting from manual deposition of matrix solutions, various modifications of this technique have been reported. An acoustic reagent multispotter was developed to provide improved reproducibility for depositing matrix onto a sample surface. The observed matrix spot size was 180–200 μm [365]. The principle of acoustic droplet ejection was used to deposit picoliter-sized matrix droplets onto the surface. Droplets can therefore remain wet longer while minimizing molecule delocalization to within the droplet area [354]. Nano-spotting based on a

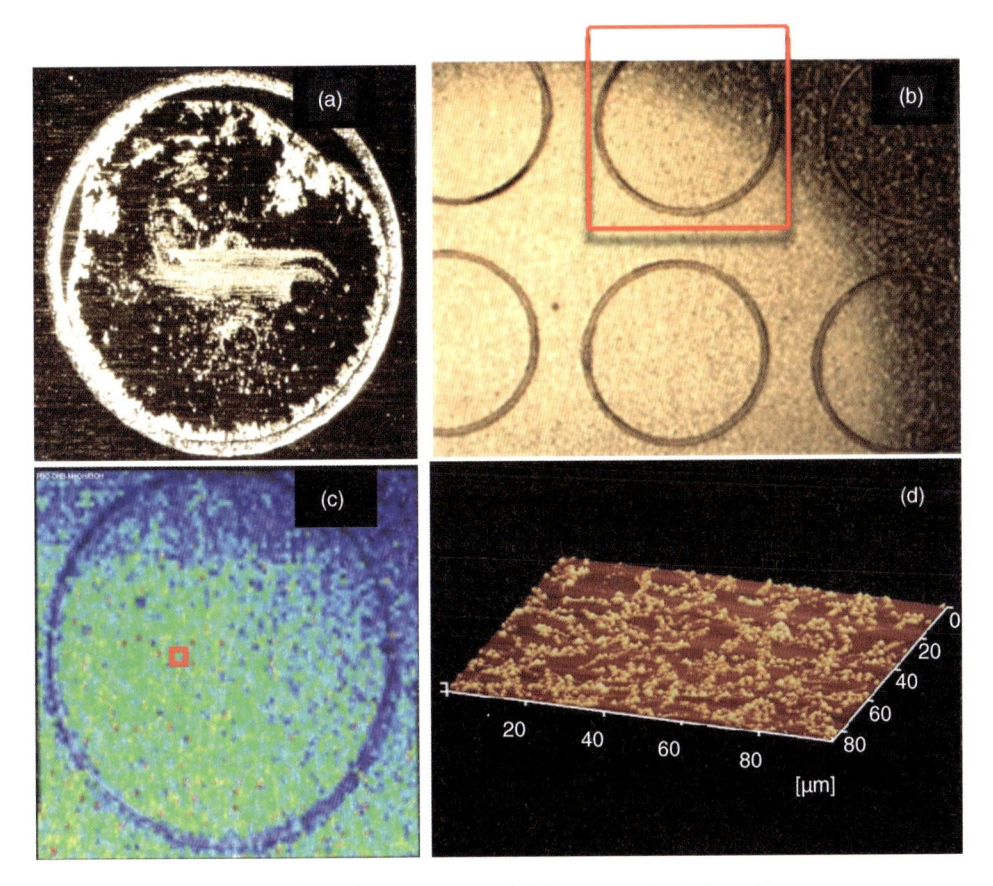

Figure 6.20 Optical images of sample spots of 2,5 DHB dissolved in methanol prepared by "dried droplet" method (a), by ESD (b); MALDI ion intensity image of [DHB]$^+$ recorded from the region indicated by the red frame in (b) and AFM picture taken from the region indicated by the red frame in (c).

microdispenser platform was found to be advantageous compared to a commercially available instrument that applies the principle of matrix sublimation (known to yield small pure crystals with uniform distributions) [366].

In recent years, a variety of other sample deposition techniques for MALDI imaging have been described. Among them are applications using solid and liquid ionic matrices [367–369], graphite- or nanoparticle-assisted MALDI methods [370, 371], and matrix sublimation [372].

An effective analytical resolution of 2 μm in the x- and y-direction was achieved for scanning microprobe matrix-assisted laser desorption/ionization imaging mass spectrometry (SMALDI-MS) by matrix vapor deposition. The preparation procedure involved two steps leading to an improved control of migration and incorporation. In the first step, a dry vapor deposition of matrix onto the investigated sample was performed. In a second step, a saturated water atmosphere is applied which enhanced the incorporation of analyte into the matrix crystal by a controlled recrystallization of matrix [373].

Lately, a solvent-free matrix deposition protocol based on the direct deposition of solid matrix particles after sieving through a 20-μm stainless steel sieve was described [374]. Signals obtained were comparable to those from spray-coated sections. Highly reproducible results were obtained with a simpler and faster sample preparation and virtually no analyte delocalization. Thus, much of the variance caused by operator differences could be eliminated.

6.3.3
MALDI Imaging of Polymers

In contrast to MALDI imaging of biological polymers, which is proving to be quite useful, the utility of MALDI imaging for synthetic polymers is rather limited thus far. This is mainly due to the fact that unlike biological polymers and protein macromolecules, which usually have a single characteristic molar mass associated with them, synthetic polymers are comprised of a distribution of molecular weights as is depicted in Figure 6.21a and b. Moreover, the quality of mass spectra acquired strongly depends on the homogeneity of the matrix/analyte mixture, especially when the so-called dried droplet method is used for sample preparation. A very high spectral intensity is often observed at so-called hot or sweet spots of the preparation. However, the molecular weight distributions observed in these regions are not necessarily correct.

Several efforts have been undertaken to establish routines for sample preparation and, more important, for the recording of spectra. Meanwhile, some of these procedures have been defined in national and international standards [375, 376]. The general advice for recording spectra of such obviously inhomogeneous sample spots recommends the frequent change of spot positions and the accumulation of several series of spectra without changing the laser and instrument settings. Nevertheless, it is well known that even in seemingly homogeneous MALDI sample spots measured average molecular mass values can drastically vary. The deviation of the average molecular masses can be more than 15%.

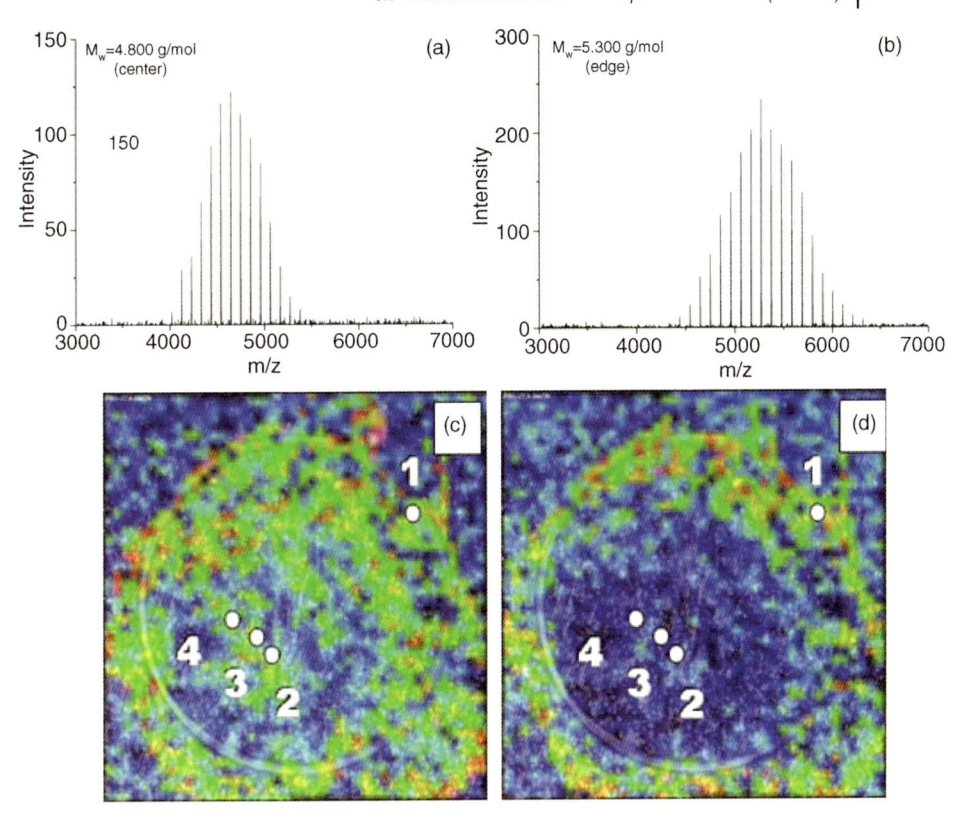

Figure 6.21 Examples of mass segregation effects for imaging MALDI of polymers. (a, b) MALDI-ToF mass spectra of polystyrene (PS) 5100 g mol^{-1} recorded in the middle (a) and at the periphery (b) of a spot prepared according to the dried droplet method (matrix: alpha-retinoic acid, silver trifluoracetate) [377]. (c, d) Ion intensity distribution of PBG 1000 (1.5 mg) in a DHB matrix (10 mg) both dissolved in 1 mL methanol (premixed solutions) using the "dried droplet" technique: (c) [PBG]$_8$Na$^+$ ion at m/z 732.2, (d) [PBG]$_{14}$Na$^+$ ion at m/z 1164.5. Image spot size in (c, d) was 100 µm. Image reprinted from Weidner et al. (2009), with permission [378].

Figure 6.21 shows an example of this, illustrating the differences observed in polymer molecular weights, as a function of changing position within a spot prepared by the dried droplet method. In Figure 6.21a and b, mass spectra of polystyrene PS 5100 were recorded both in the middle and at the outer boundary of a spot in which no visual inhomogeneities were apparent. As can be seen, the number average molecular weight distribution (M_n) measured at the edge of the sample is higher than the values determined in the center of the droplet. It becomes obvious, therefore, that this widely used principle of spot preparation cannot be applied for a reliable molecular mass determination by MALDI-ToF MS.

In the second example (Figure 6.21c and d), various mixtures of poly(butyleneglycol) (PBG) 1000 with alpha-cyanocinnamic acid (CCA) and 2,5-DHB and different solvents were imaged with MALDI [378]. The ion intensity distributions of [PBG]$_8$Na$^+$ (m/z 732.2) (Figure 6.21c) and [PBG]$_{14}$Na$^+$ (m/z 1164.5) are shown. The lower mass ions (Figure 6.21c) are more or less homogeneously distributed over the entire spot. In contrast, the higher mass ions (Figure 6.21d) are predominantly located at the spot periphery. The molecular mass determination performed at four selected spots showed differences of more than 10%.

These segregation effects were also found to be highly dependent upon the solvent employed. It was demonstrated that the use of easy vaporizable solvents (e.g. acetone) led to the formation of smooth layers showing no preferential segregation of matrix and polymer. In contrast to that, the use of methanol and DHB resulted in spots having distinct crystalline rings at the periphery of the droplet (see Figure 6.20a).

Similar results have been reported for polystyrene (PS 5100) [33]. The observed segregation shown in Figure 6.22, for example, shows a chromatography-like separation of single polymer homologues with an enrichment of higher mass species at the spot boundary. A possible explanation has been presented, based on diffusion effects during solvent evaporation, whereby the volatility of the solvents was suggested to play a critical role. The observed mass segregation clearly reveals the necessity to apply alternative spot preparation methods, like spraying, nano-spotting, or solvent-less techniques.

Alternatively, a minimizing of the sample spot size by using micro- or nano-droplet deposition is possible. The MALDI imaging mass spectrum of an array of microspots with a diameter of 900 μm is shown in Figure 6.23. In this particular case 2,5-DHB was spotted on poly(ethylene glycol) (PEG 1000). However, the enlargement (Figure 6.23b) shows that even in these small droplets, a homogeneous distribution of matrix cannot be obtained. A significantly higher concentration of matrix was found at the spot periphery. In addition, a coalescence of matrix spots at those places precoated with PEG 1000 was observed.

Alternatively, a matrix/polymer system consisting of 2,5-DHB and PBG 1000 was investigated using the air spray deposition technique [378]. These results are presented in Figure 6.24. The images clearly show a homogeneous distribution of the DHB matrix as well as of two sample ions, having different masses. Even density differences of the sprayed trace (see the crescent-shaped fronts in the trace) caused by fluctuations of the pump did not affect the ion intensity distribution. The MALDI imaging data illustrate a nearly constant deposition of matrix and polymer, independent of the overall amount that was sprayed on the target. Single MALDI mass spectra exemplarily taken from five positions revealed similar absolute spectra

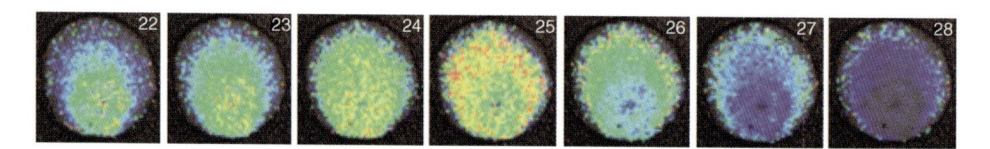

Figure 6.22 Ion intensity distribution of polystyrene [PS]$_{2n+1}$Ag$^+$ ions (where $n = 22$–28) acquired from a PS 5100 sample, mixed with alpha-retinoic acid, DMAA, and AgTFA on a polished steel plate. The single spot size was 100 μm. Image reprinted from Weidner et al. (2010), with permission [377].

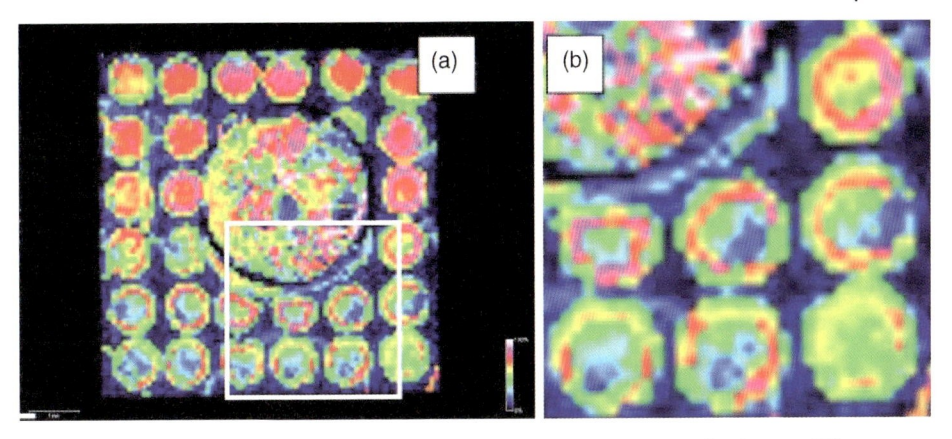

Figure 6.23 (a) Array of droplets of 2,5-DHB matrix on a MALDI target precoated with PEG 1000 (scale bar = 1 mm), (b) enlargement of selected area. Image reprinted from Weidner et al. (2010), with permission [377].

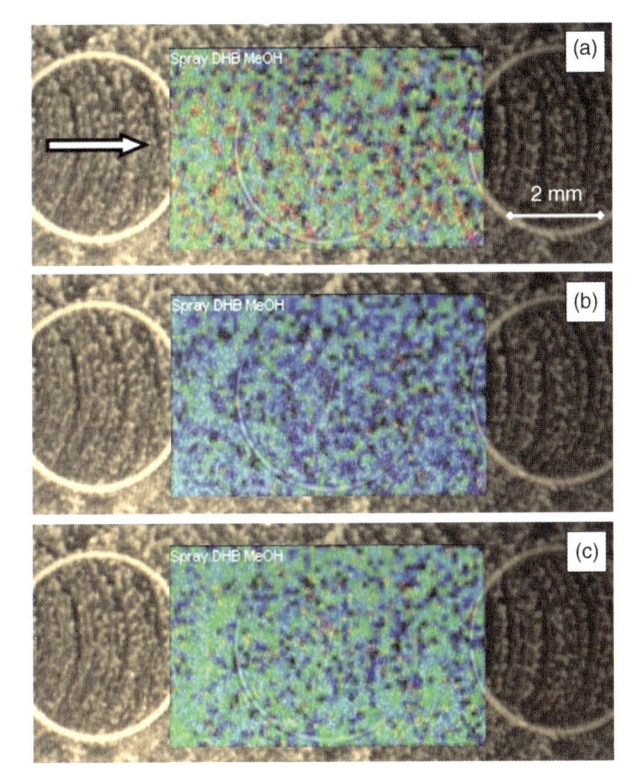

Figure 6.24 Single ion intensity distribution of poly(butylene glycol) PBG 1000 (1.5 mg) in a DHB matrix (10 mg) both dissolved in 1 mL methanol (premixed solutions) using the spray technique, (a) [matrix-H]$^+$ ion at m/z 155.03, (b) [PBG]$_8$Na$^+$ ion at m/z 732.2 g mol^{-1}, and (c) [PBG]$_{14}$Na$^+$ ion at m/z 1164.5 g mol^{-1}. The image spot size was 100 µm. The arrow in (a) indicates direction of spraying. Image reprinted from Weidner et al. (2009), with permission [378].

intensities. The mean deviation between calculated average molecular masses was less than ±1%.

6.3.4
Outlook

In contrast to biopolymers, where MALDI-ToF imaging mass spectrometry has become an important tool for determining local distributions of proteins, peptides, drugs, and pharmaceutical agents in tissues, MALDI imaging is only just now becoming interesting for characterization of synthetic polymers. Matrix application and the formation of homogeneous matrix/polymer layers are challenging. It has been shown that this becomes especially important when solvent-based "dried droplet" methods are used. In those cases, segregation of polymer and matrix was frequently observed. In addition, the use of slow evaporating solvents can result in a distinctive mass segregation of single polymer homologues. This mass segregation of homopolymer species leads one to assume that similar effects could be found for block copolymers as well. In this regard, MALDI-ToF imaging mass spectrometry might be an important tool to localize and visualize the lateral block copolymer composition. Further investigations into this area need to be carried out.

Another interesting field for applying imaging techniques represents the problem of polymer degradation. In contrast to bulk methods, which only provide average data, knowing the lateral distribution of degradation products is highly desirable. Such data would help to understand the mechanism of degradation and to increase polymer stability and lifetime.

6.4
Other Surface Mass Spectrometry Methods

In recent years, several new techniques for surface mass spectrometry at atmospheric pressures have been developed, including atmospheric pressure MALDI (AP-MALDI) [379], DESI and related methods, and PDI methods. The last section of this chapter will be dealing with DESI and PDI. The application of ESI for SIMS will also be explored.

6.4.1
Desorption Electrospray Ionization

Desorption electrospray ionization or DESI is a relatively new method in the surface mass spectrometry community. It works by utilizing fast-moving charged solvent droplets to extract analytes from surfaces and propel the resulting secondary microdroplets toward the mass analyzer [380]. DESI is turning out to be a robust method for characterization of polymer samples, its primary advantage being the ability to characterize samples at atmospheric pressure. It surpasses SIMS in its ability to analyze higher mass species and has even been developed as an imaging tool [380, 381], albeit with significant reductions in the spatial resolution (values ranging from 180 to 220 µm [380], although it has been reported to be as low as 40 µm under optimized conditions) [382]. Direct comparisons of the performance of DESI and other surface mass spectrometry techniques can be found in the literature [383, 384].

DESI has also shown great promise as a tool for characterization of polymers [385, 386]. Accurate in the determinations of molecular weight distributions up to m/z 3000 so far, it is expected to rival MALDI in its ability to characterize large molecules, due to its relatively soft ionization process [385]. An example of DESI for characterization of a PEG sample ($M_n = 3000$ g mol^{-1}) is depicted in Figure 6.25, where multiple molecular weight distributions are observed at different charge-state envelopes. As can be seen, the multiplicity of charge states observed for DESI is much greater than is observed with other methods of surface mass spectrometry – a trait common to ESI methods [385]. Also note that the molecular weight distribution associated with charge state of $+1$ is not shown. This is because the mass spectrometer used in this study was only limited to a mass range of m/z 0 to 2000. Despite this setback, the utility of the method for characterization of polymers is evident.

There are some limitations of the method for polymer analysis as well, in particular with regard to the kind and size of polymers that can be accurately characterized. The selection of solvent can be difficult. Furthermore, it is difficult to discriminate between peaks for higher molecular weight polymers due to the overlapping distributions at different charge states.

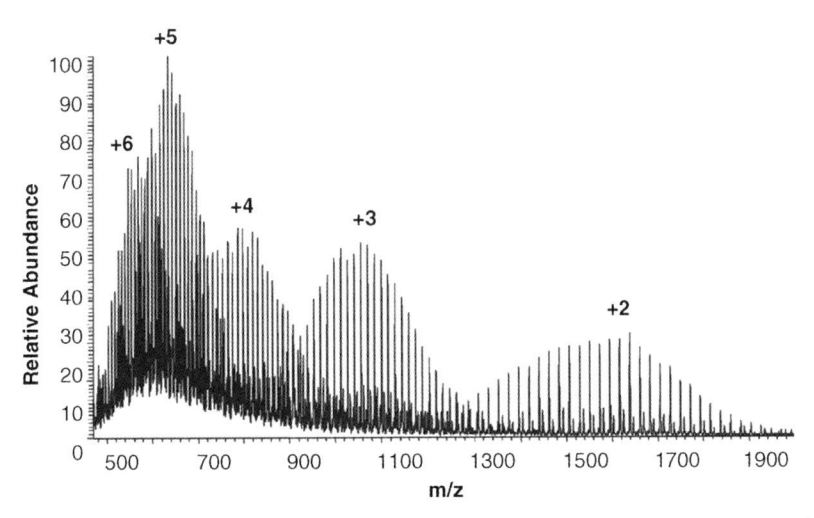

Figure 6.25 Positive ion DESI mass spectrum of poly(ethylene glycol) PEG (M_w 3000 g mol^{-1}): 0.2 µg per sample spot, spray solvent of methanol:water, 1 : 1. Figure reprinted from Nefliu, Venter, and Cooks (2006), with permission [385].

In addition to MADLI [2, 352, 387] and SIMS [2, 342, 388], DESI is making major headway for biological imaging and cancer biomarker applications [380, 381, 389]. The benefit of DESI being that the ionization occurs at atmospheric pressure, making it a versatile technique that can be used to directly analyze tissue samples *in vitro*. Unlike SIMS, which is limited to lower mass regions, DESI is capable of detecting much larger molecules. Although MALDI (thus far) has demonstrated increased ability to characterize high molecular weight proteins, no matrix is required for the DESI method, making it a viable alternative for direct analysis of tissue in real-world samples.

6.4.2
Plasma Desorption Ionization Methods

Plasma desorption ionization techniques, such as direct analysis in real time (DART), atmospheric pressure glow discharge ionization (APGDI), and low-temperature plasma ionization (LTPI), are also promising ambient pressure surface mass spectrometry techniques for characterization of polymeric materials. Similar to DESI, these techniques can be operated at atmospheric pressure and therefore can be used for direct analysis of samples without the requirement of a vacuum [390]. Furthermore, more recent plasma desorption methods, in particular LTP, have enabled a softer ionization process, similar to DESI, whereby a proximate plasma source is used rather than a distal plasma, which requires the use of high temperatures and high voltages.

In general, methods employing plasmas do not extend to the analysis of larger biomolecules [383]. However, plasma desorption ionization methods are showing particular promise as a polymer depth profiling tool. Depth profiling in PMMA and polystyrene, for example, was recently demonstrated by Tuccito *et al.*, where the authors utilized pulsed radiofrequency glow discharge ToF-MS for depth profiling in PMMA and PS films [391]. The results are promising, particularly for type I polymers that tend to crosslink when irradiated.

6.4.3
Electrospray Droplet Impact for SIMS

Electrospray droplet impact can be utilized in conjunction with other methods, such as XPS and SIMS, in order to obtain lower damage mass spectra similar to that observed with MALDI without the matrix requirement. EDI-SIMS and EDI-XPS have recently been developed, whereby charged droplets of 1 M acetic acid are electrosprayed directly into the main chamber vacuum. The difference between DESI and EDI-SIMS is that the former uses electrospray droplets to dissolve/ionize the samples under atmospheric pressure, while the latter uses the electrospray charged droplets accelerated in vacuum and measures the secondary ions produced by the high-momentum collision between the projectile and the solid sample [192]. These large water droplets also serve as a sputter source for XPS and/or SIMS.

EDI-SIMS has been particularly promising as a tool for characterization of polymers. The fragmentation is significantly reduced when using EDI sources as compared to conventional ion beam sources [192, 392]. An example of EDI-SIMS for polymer characterization is depicted in Figure 6.26, which compares the EDI-SIMS mass spectra acquired from PS at two different molecular weights (m/z 2500 and m/z 4000) to MALDI mass spectra of the same polymers [192]. Unlike conventional SIMS, which has been shown to have a systematic shift in the molecular weight distribution toward lower masses relative to MALDI [5], both EDI-SIMS and MALDI yield similar molecular weight distributions that are consistent with the known distributions in the standard. However, while MALDI requires the use of a matrix, no matrix was required when employing EDI-SIMS spectrum (although both methods employed a cationization salt). At lower masses, cationization salts were not necessary for EDI-MS.

ESI has also proven to be capable of low damage etching in polymers, such as polyvinylchloride (PVC), PS, and PMMA, whereby the XPS spectra were found to be independent of the irradiation time when employing EDI as a sputter source [393–396]. The decreased damage processes occurring, particularly with PS and other crosslinking polymers, once again shows the potential of this source for depth profiling of polymers.

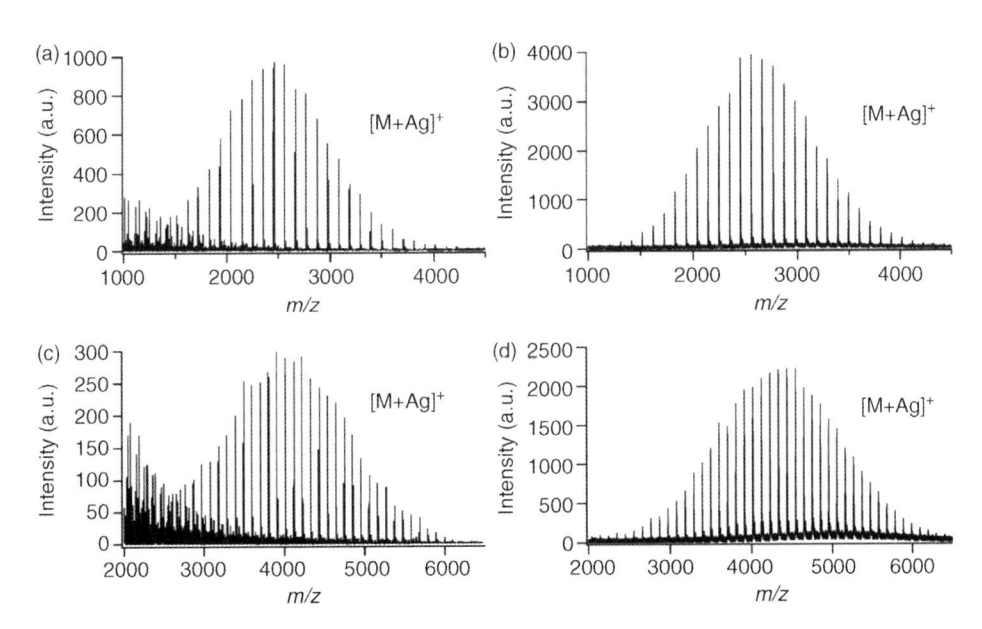

Figure 6.26 Positive ion mass spectra of poly (styrene) (PS) 2500 and 4000 g mol^{-1} using EDI-MS. (a) PS2500 obtained by EDI-SIMS, (b) PS2500 obtained by MALDI, (c) PS4000 obtained by EDI-SIMS, and (d) PS4000 obtained by MALDI. Silver tri-fluoroacetic acid (AgTFA) was added as the cationizati on salt for both EDI-SIMS and MALDI, and dithranol was used as the MALDI matrix. Figure reprinted from Asakawa, Chen, and Hiraoka (2009), with permission [192].

6.5
Outlook

Over the past couple of decades, the field of imaging mass spectrometry has taken a quantum leap in the area of organic, polymeric, and biological analysis. SIMS, originally developed for characterization of inorganic materials and semiconductors, is now used on a regular basis for the surface and in-depth characterization of polymers and other soft samples. The recent development of cluster sources over the past 15 years has resulted in a major paradigm shift in SIMS, moving from 2-D to 3-D analysis in polymeric materials. MALDIs recent development of imaging methodology has made an overwhelming impact in the biological community. The ability to detect high-mass proteins and other organic compounds remains unrivaled. Furthermore, some of these instruments can now be operated under ambient pressure conditions, thus making it a much more versatile tool.

Although this is a time of expansion in the entire field of surface mass spectrometry, DESI and PDI methods are right at the beginning of their growth curve. It is expected, therefore, that significant developments will be observed in the coming years as these techniques mature. It has been said that these methods represent the future of desorption mass spectrometry. This is very likely to be true.

References

1 Kollmer, F. (2004) *Appl. Surf. Sci.*, **231**, 153–158.
2 McDonnell, L.A. and Heeren, R.M.A. (2007) *Mass Spectrom. Rev.*, **26** (4), 606–643.
3 Tanaka, K., Waki, H., Ido, Y., Akita, S., Yoshida, Y., Yoshida, T., and Matsuo, T. (1988) *Rapid Commun. Mass Spectrom.*, **2** (8), 151–153.
4 Parees, D.M., Hanton, S.D., Cornelio Clark, P.A., and Willcox, D.A. (1998) *J. Am. Soc. Mass Spectrom.*, **9** (4), 282–291.
5 Williams, J.B., Gusev, A.I., and Hercules, D.M. (1996) *Macromolecules*, **29** (25), 8144–8150.
6 Licciardello, A., Puglisi, O., Calcagno, L., and Foti, G. (1988) *Nucl. Instrum. Meth B*, **32** (1–4), 131–135.
7 Benninghoven, A. (1969) *Phys. Status Solidi.*, **34**, K169.
8 Benninghoven, A. and Loebach, E. (1972) *J. Radioanal. Chem.*, **12**, 95–99.
9 Briggs, D., Brown, A., and Vickerman, J.C. (1989) *Handbook of Static SIMS*, John Wiley & Sons, Chichester, UK.
10 Benninghoven, A., Jaspers, D., and Sichtermann, W. (1976) *Appl. Phys. A-Mater.*, **11** (1), 35–39.
11 Van Vaeck, L., Adriaens, A., and Gijbels, R. (1999) *Mass Spectrom. Rev.*, **18** (1), 1–47.
12 Licciardello, A., Wenclawiak, B., Boes, C., and Benninghoven, A. (1994) *Surf. Interf. Anal.*, **22** (1–12), 528–531.
13 Briggs, D. and Hearn, M. (1986) *J. Vac.*, **36** (11–12), 1005–1010.
14 Leggett, G.J. and Vickerman, J.C. (1992) *Appl. Surf. Sci.*, **55** (2–3), 105–115.
15 Gilmore, I.S. and Seah, M.P. (1996) *Surf. Interf. Anal.*, **24** (11), 746–762.
16 Marletta, G. (1990) *Nucl. Instrum. Meth. B*, **46** (1–4), 295–305.
17 Marletta, G. and Licciardello Lucia, A. (1989) *Nucl. Instrum. Meth. B*, **37**, 712–715.
18 Marletta, G., Pignataro, S., and Oliveri, C. (1989) *Nucl. Instrum. Meth. B*, **39** (1–4), 792–795.
19 Koval, Y. (2004) *J. Vac. Sci. Technol. B*, **22**, 843.

20 Lazareva, I., Koval, Y., Alam, M., Stromsdorfer, S., and Muller, P. (2009) *Appl. Phys. Lett.*, **90** (26), 262108.
21 Venkatesan, T., Forrest, S.R., Kaplan, M.L., Murray, C.A., Schmidt, P.H., and Wilkens, B.J. (2009) *J. Appl. Phys.*, **54** (6), 3150–3153.
22 Pignataro, S. (1992) *Surf. Interf. Anal.*, **19** (112), 275–285.
23 Steffens, P., Niehuis, E., Friese, T., Greifendorf, D., and Benninghoven, A. (1985) *J. Vac. Sci. Technol. A*, **3** (3), 1322–1325.
24 Gardella, J.A. and Hercules, D.M. (1980) *Anal. Chem.*, **52** (2), 226–232.
25 Campana, J.E., DeCorpo, J.J., and Colton, R.J. (1981) *Appl. Surf. Sci.*, **8** (3), 337–342.
26 Briggs, D. (1982) *Surf. Interf. Anal.*, **4** (4), 151–155.
27 Briggs, D. and Wootton, A.B. (1982) *Surf. Interf. Anal.*, **4** (3), 109–115.
28 Chilkoti, A., Ratner, B.D., and Briggs, D. (1992) *Surf. Interf. Anal.*, **18** (8), 604–618.
29 Briggs, D. (1990) *Surf. Interf. Anal.*, **15** (12), 734–738.
30 Briggs, D. (1986) *Surf. Interf. Anal.*, **9** (1–6), 391–404.
31 Briggs, D., Fletcher, I.W., and Goncalves, N.M. (2000) *Surf. Interf. Anal.*, **29** (5), 303–309.
32 Briggs, D. and Davies, M.C. (1997) *Surf. Interf. Anal.*, **25** (9), 725–733.
33 Briggs, D., Fletcher, I.W., Reichlmaier, S., AguloSanchez, J.L., and Short, R.D. (1996) *Surf. Interf. Anal.*, **24** (6), 419–421.
34 Briggs, D. (1989) *Br. Polym. J.*, **21** (1), 3–15.
35 Briggs, D. and Ratner, B.D. (1988) *Polym. Commun.*, **29** (1), 6–8.
36 Briggs, D. and Munro, H.S. (1987) *Polym. Commun.*, **28** (11), 307–309.
37 Briggs, D. (1987) *Org. Mass Spectrom.*, **22** (2), 91–97.
38 Briggs, D. (1986) *Surf. Interf. Anal.*, **8** (3), 133–136.
39 Briggs, D., Hearn, M.J., and Ratner, B.D. (1984) *Surf. Interf. Anal.*, **6** (4), 184–192.
40 Chilkoti, A., Lopez, G.P., Ratner, B.D., Hearn, M.J., and Briggs, D. (1993) *Macromolecules*, **26** (18), 4825–4832.
41 Hearn, M.J., Ratner, B.D., and Briggs, D. (1988) *Macromolecules*, **21** (10), 2950–2959.
42 Hearn, M.J. and Briggs, D. (1988) *Surf. Interf. Anal.*, **11** (4), 198–213.
43 Hearn, M.J., Briggs, D., Yoon, S.C., and Ratner, B.D. (1987) *Surf. Interf. Anal.*, **10** (8), 384–391.
44 Leeson, A.M., Alexander, M.R., Short, R.D., Hearn, M.J., and Briggs, D. (1997) *Int. J. Polym. Anal. Ch.*, **4** (2), 133–151.
45 Leeson, A.M., Alexander, M.R., Short, R.D., Briggs, D., and Hearn, M.J. (1997) *Surf. Interf. Anal.*, **25** (4), 261–274.
46 Ratner, B.D., Castner, D.G., Lewis, K.B., Edelman, P.G., and Briggs, D. (1990) *Abstr. Pap. Am. Chem. S.*, **199**, 21-MACRO.
47 Reichlmaier, S., Bryan, S.R., and Briggs, D. (2009) *J. Vac. Sci. Technol. A*, **13** (3), 1217–1223.
48 Tyler, B.J., Ratner, B.D., Castner, D.G., and Briggs, D. (1992) *J. Biomed. Mater. Res.*, **26** (3), 273–289.
49 Briggs, D. (1984) *Polymer*, **25** (10), 1379–1391.
50 Lub, J. and Benninghoven, A. (1989) *Org. Mass Spectrom.*, **24** (3), 164–168.
51 Ramsden, W.D. (1991) *Surf. Interf. Anal.*, **17** (11), 793–802.
52 Eynde, X.V., Bertrand, P., and Jerome, R. (1997) *Macromolecules*, **30** (21), 6407–6416.
53 Leggett, G.J., Briggs, D., and Vickerman, J.C. (1991) *Surf. Interf. Anal.*, **17** (10), 737–744.
54 Leggett, G.J., Chilkoti, A., Castner, D.G., Ratner, B.D., and Vickerman, J.C. (1991) *Int. J. Mass Spectrom. Ion Process*, **108** (1), 29–39.
55 Leggett, G.J., Vickerman, J.C., Briggs, D., and Hearn, M.J. (1992) *J. Chem. Soc. Faraday T.*, **88** (3), 297–309.
56 Chilkoti, A., Castner, D.G., and Ratner, B.D. (1991) *Appl. Spectrosc.*, **45** (2), 209–217.
57 Pleul, D., Simon, F., and Jacobasch, H.J. (1997) *Fresen. J. Anal. Chem.*, **357** (6), 684–687.
58 Ogaki, R., Green, F.M., Gilmore, I.S., Shard, A.G., Luk, S., Alexander, M.R.,

and Davies, M.C. (2007) *Surf. Interf. Anal.*, **39** (11), 852–859.
59 Benninghoven, A., Hagenhoff, B., and Niehuis, E. (1993) *Anal. Chem.*, **65** (14), 630–640.
60 Bhatia, Q.S. and Burrell, M.C. (1990) *Surf. Interf. Anal.*, **15** (6), 388–391.
61 Eynde, X.V., Reihs, K., and Bertrand, P. (1999) *Macromolecules*, **32** (9), 2925–2934.
62 Eynde, X.V., Matyjaszewski, K., and Bertrand, P. (1998) *Surf. Interf. Anal.*, **26** (8), 569–578.
63 Eynde, X.V. and Bertrand, P. (1998) *Surf. Interf. Anal.*, **26** (8), 579–589.
64 Galuska, A.A. (1997) *Surf. Interf. Anal.*, **25** (10), 790.
65 Fowler, D.E., Johnson, R.D., VanLeyen, D., and Benninghoven, A. (1991) *Surf. Interf. Anal.*, **17** (3), 125–136.
66 Kailas, L., Audinot, J.N., Migeon, H.N., and Bertrand, P. (2004) *Appl. Surf. Sci.*, **231**, 289–295.
67 Eynde, X.V., Bertrand, P., Dubois, P., and Jérôme, R. (1998) *Macromolecules*, **31** (19), 6409–6416.
68 Eynde, X.V., Weng, L.T., and Bertrand, P. (1997) *Surf. Interf. Anal.*, **25** (1), 41–45.
69 Nowak, R.W., Gardella, J.A. Jr., Wood, T.D., Zimmerman, P.A., and Hercules, D.M. (2000) *Anal. Chem.*, **72** (19), 4585–4590.
70 Zimmerman, P.A. and Hercules, D.M. (1994) *Appl. Spectrosc.*, **48** (5), 620–622.
71 Lub, J. and Van der Wel, H. (1990) *Org. Mass Spectrom.*, **25** (11), 588–591.
72 Castner, D.G. and Ratner, B.D. (1990) *Surf. Interf. Anal.*, **15** (8), 479–486.
73 Lianos, L., Quet, C., and Duc, T.M. (1994) *Surf. Interf. Anal.*, **21** (1), 14–22.
74 Jasieniak, M., Suzuki, S., Monteiro, M., Wentrup-Byrne, E., Griesser, H.J., and Grøndahl, L. (2008) *Langmuir*, **25** (2), 1011–1019.
75 Bertrand, P. (2006) *Appl. Surf. Sci.*, **252** (19), 6986–6991.
76 Deslandes, Y., Pleizier, G., Alexander, D., and Santerre, P. (1998) *Polymer*, **39** (11), 2361–2366.
77 Bhatia, Q.S. and Burrell, M.C. (1991) *Polymer*, **32** (11), 1948–1956.
78 Brinen, J.S., Rosati, L., Chakel, J., and Lindley, P. (1993) *Surf. Interf. Anal.*, **20** (13), 1055–1060.
79 Mazzanti, J.B., Reamey, R.H., Helfand, M.A., and Lindley, P.M. (2009) *J. Vac. Sci. Technol. A*, **10** (4), 2419–2424.
80 Galuska, A. (1996) *Surf. Interf. Anal.*, **24** (6), 380–388.
81 Zhao, J., Rojstaczer, S.R., and Gardella, J.A. (2009) *J. Vac. Sci. Technol. A*, **16** (5), 3046–3051.
82 Burrell, M.C., Bhatia, Q.S., and Michael, R.S. (1994) *Surf. Interf. Anal.*, **21** (8), 553–559.
83 Briggs, D., Fletcher, I.W., Reichlmaier, S., Agulo-Sanchez, J.L., and Short, R.D. (1996) *Surf. Interf. Anal.*, **24** (6), 419–421.
84 Cheung, Z.L., Weng, L.T., Chan, C.M., Hou, W.M., and Li, L. (2005) *Langmuir*, **21** (17), 7968–7970.
85 Affrossman, S., O'Neill, S.A., and Stamm, M. (1998) *Macromolecules*, **31** (18), 6280–6288.
86 Galuska, A.A. (1994) *Surf. Interf. Anal.*, **21** (10), 703–710.
87 Davies, M.C., Lynn, R.A.P., Hearn, J., Paul, A.J., Vickerman, J.C., and Watts, J.F. (1995) *Langmuir*, **11** (11), 4313–4322.
88 Davies, M.C., Lynn, R.A.P., Hearn, J., Paul, A.J., Vickerman, J.C., and Watts, J.F. (1996) *Langmuir*, **12** (16), 3866–3875.
89 Davies, M.C., Lynn, R.A.P., Davis, S.S., Hearn, J., Vickerman, J.C., and Paul, A.J. (1993) *J. Colloid Interf. Sci.*, **161** (1), 83–90.
90 Weng, L.T. and Chan, C.M. (2006) *Appl. Surf. Sci.*, **252** (19), 6570–6574.
91 Castner, D.G., Ratner, B.D., Grainger, D.W., Kim, S.W., Okano, T., Suzuki, K., Briggs, D., and Nakahama, S. (1992) *J. Biomat. Sci. -Polym. E.*, **3** (6), 463–480.
92 Yokoyama, H., Kramer, E.J., Hajduks, D.A., and Bates, F.S. (1999) *Macromolecules*, **32** (10), 3353–3359.
93 Eynde, X.V. and Bertrand, P. (1999) *Surf. Interf. Anal.*, **27** (3), 157–164.
94 Davies, M.C., Lynn, R.A.P., Hearn, J., Watts, J.F., Vickerman, J.C., and Johnson, D. (1994) *Langmuir*, **10** (5), 1399–1409.
95 Mahoney, C.M., Yu, J., and Gardella, J.A. Jr. (2005) *Anal. Chem.*, **77** (11), 3570–3578.

96 Mahoney, C.M., Yu, J.X., Fahey, A., and Gardella, J.A. (2006) *Appl. Surf. Sci.*, **252** (19), 6609–6614.

97 Lhoest, J.B., Bertrand, P., Weng, L.T., and Dewez, J.L. (1995) *Macromolecules*, **28** (13), 4631–4637.

98 Mahoney, C.M., Fahey, A.J., and Belu, A.M. (2008) *Anal. Chem.*, **80** (3), 624–632.

99 Davies, M.C., Shakesheff, K.M., Shard, A.G., Domb, A., Roberts, C.J., Tendler, S.J.B., and Williams, P.M. (1996) *Macromolecules*, **29** (6), 2205–2212.

100 Feng, J., Chan, C.M., and Weng, L.T. (2000) *Polymer*, **41** (7), 2695–2699.

101 Liu, S., Weng, L.T., Chan, C.M., Li, L., Ho, N.K., and Jiang, M. (2001) *Surf. Interf. Anal.*, **31** (8), 745–753.

102 Affrossman, S., Hartshorne, M., Kiff, T., Pethrick, R.A., and Richards, R.W. (1994) *Macromolecules*, **27** (6), 1588–1591.

103 Affrossman, S., Hartshorne, M., Jqróme, R., Pethrick, R.A., Petitjean, S., and Vilar, M.R. (1993) *Macromolecules*, **26** (23), 6251–6254.

104 Brant, P., Karim, A., Douglas, J.F., and Batess, F.S. (1996) *Macromolecules*, **29** (17), 5628–5634.

105 Lau, Y.T.R., Weng, L.T., Ng, K.M., and Chan, C.M. (2008) *Appl. Surf. Sci.*, **255** (4), 1001–1005.

106 Lee, W.-K., Wells, D.W., Goacher, R.E., and Gardella, J.A. Jr. (2011) *Surf. Interf. Anal.*, **43** (1–2), 385–388.

107 Lau, Y.T.R., Weng, L.T., Ng, K.M., and Chan, C.M. (2010) *Surf. Interf. Anal.* **43** (1–2), 340–343.

108 Li, L., Ng, K.M., Chan, C.M., Feng, J.Y., Zeng, X.M., and Weng, L.T. (2000) *Macromolecules*, **33** (15), 5588–5592.

109 Lau, Y.T.R., Weng, L.T., Ng, K.M., and Chan, C.M. (2010) *Surf. Interf. Anal.*, **42** (8), 1445–1451.

110 Briggs, D. and Ratner, B.D. (1988) *Polym. Commun.*, **29** (1), 6–8.

111 Shard, A.G., Davies, M.C., Tendler, S.J.B., Nicholas, C.V., Purbrick, M.D., and Watts, J.F. (1995) *Macromolecules*, **28** (23), 7855–7859.

112 Weng, L.T., Ng, K.M., Cheung, Z.L., Lei, Y., and Chan, C.M. (2006) *Surf. Interf. Anal.*, **38** (1), 32–43.

113 Ruch, D., Muller, J.F., Migeon, H.N., Boes, C., and Zimmer, R. (2003) *J. Mass Spectrom.*, **38** (1), 50–57.

114 Weng, L.T., Bertrand, P., Lauer, W., Zimmer, R., and Busetti, S. (1995) *Surf. Interf. Anal.*, **23** (13), 879–886.

115 Chilkoti, A., Castner, D.G., Ratner, B.D., and Briggs, D. (1990) *J. Vac. Sci. Technol. A*, **8** (3), 2274–2282.

116 Liu, S., Chan, C.M., Weng, L.T., and Jiang, M. (2004) *Polymer*, **45** (14), 4945–4951.

117 Liu, S., Weng, L.T., Chan, C.M., Li, L., Ho, K.C., and Jiang, M. (2000) *Surf. Interf. Anal.*, **29** (8), 500–507.

118 Liu, S., Chan, C.M., Weng, L.T., and Jiang, M. (2004) *Anal. Chem.*, **76** (17), 5165–5171.

119 Cohen, L.R.H., Hercules, D.M., Karakatsanis, C.G., and Rieck, J.N. (1995) *Macromolecules*, **28** (16), 5601–5608.

120 Hittle, L.R., Altland, D.E., Proctor, A., and Hercules, D.M. (1994) *Anal. Chem.*, **66** (14), 2302–2312.

121 Reihs, K., Voetz, M., Kruft, M., Wolany, D., and Benninghoven, A. (1997) *Fresen. J. Anal. Chem.*, **358** (1), 93–95.

122 Bletsos, I.V., Hercules, D.M., VanLeyen, D., and Benninghoven, A. (1987) *Macromolecules*, **20** (2), 407–413.

123 Bletsos, I.V., Hercules, D.M., Greifendorf, D., and Benninghoven, A. (1985) *Anal. Chem.*, **57** (12), 2384–2388.

124 Grade, H., Winograd, N., and Cooks, R.G. (1977) *J. Am. Chem. Soc.*, **99** (23), 7725–7726.

125 Grade, H. and Cooks, R.G. (1978) *J. Am. Chem. Soc.*, **100** (18), 5615–5621.

126 Van Leyen, D., Hagenhoff, B., Niehuis, E., Benninghoven, A., Bletss, I.V., and Hercules, D.M. (1989) *J. Vac. Sci. Technol. A*, **7** (3), 1790–1794.

127 Lub, J., Van Vroonhoven, F., Van Leyen, D., and Benninghoven, A. (1989) *J. Polym. Sci. Pol. Phys.*, **27** (10), 2071–2080.

128 Brown, A. and Vickerman, J.C. (1986) *Surf. Interf. Anal.*, **8** (2), 75–81.

129 Kotter, F. and Benninghoven, A. (1998) *Appl. Surf. Sci.*, **133** (1–2), 47–57.

130 Mahoney, C.M., Roberson, S.V., and Gillen, G. (2004) *Anal. Chem.*, **76** (11), 3199–3207.

131 Lub, J., Van Vroonhoven, F., and Benninghoven, A. (1989) *J. Polym. Sci. Pol. Chem.*, **27** (12), 4035–4049.

132 Baytekin, H.T., Wirth, T., Gross, T., Sahre, M., Unger, W.E.S., Theisen, J., and Schmidt, M. (2010) *Surf. Interf. Anal.*, **42** (8), 1417–1431.

133 Short, R.D., Ameen, A.P., Jackson, S.T., Pawson, D.J., O'toole, L., and Ward, A.J. (1993) *Vacuum*, **44** (11–12), 1143–1160.

134 Zimmerman, P.A., Hercules, D.M., and Benninghoven, A. (1993) *Anal. Chem.*, **65** (8), 983–991.

135 Piwowar, A.M. and Vickerman, J.C. (2010) *Surf. Interf. Anal.*, **42** (8), 1387–1392.

136 Shard, A.G., Davies, M.C., Tendler, S.J.B., Nicholas, C.V., Purbrick, M.D., and Watts, J.F. (1995) *Macromolecules*, **28** (23), 7855–7859.

137 Jackson, S.T. and Short, R.D. (1992) *J. Mater. Chem.*, **2** (2), 259–260.

138 Marien, J., Ghitti, G., Jerome, R., and Teyssio, P. (1993) *Polym. Bull.*, **30** (4), 435–440.

139 Simko, S.J., Bryan, S.R., Griffis, D.P., Murray, R.W., and Linton, R.W. (1985) *Anal. Chem.*, **57** (7), 1198–1202.

140 Davies, M.C., Wilding, I.R., Short, R.D., Melia, C.D., and Rowe, R.C. (1990) *Int. J. Pharm.*, **62** (2–3), 97–103.

141 Gillen, G. and Roberson, S. (1998) *Rapid Commun. Mass Spectrom.*, **12**, 1303–1312.

142 Cramer, H.G., Grehl, T., Kollmer, F., Moellers, R., Niehuis, E., and Rading, D. (2008) *Appl. Surf. Sci.*, **255** (4), 966–969.

143 Wagner, M.S. and Gillen, G. (2004) *Appl. Surf. Sci.*, **231-2**, 169–173.

144 Fuoco, E.R., Gillen, G., Wijesundara, M.B.J., Wallace, W.E., and Hanley, L. (2001) *J. Phys. Chem. B*, **105** (18), 3950–3956.

145 Norrman, K., Haugshoj, K.B., and Larsen, N.B. (2002) *J. Phys. Chem. B*, **106** (51), 13114–13121.

146 Brox, O., Hellweg, S., and Benninghoven, A. (2000) Dynamic SIMS of Polymers? Proc. 12th Int. Conf. Secondary Ion Mass Spectrom, pp. 777–780.

147 Wagner, M.S. (2005) *Anal. Chem.*, **77** (3), 911–922.

148 Fisher, G.L., Dickinson, M., Bryan, S.R., and Moulder, J. (2008) *Appl. Surf. Sci.*, **255** (4), 819–823.

149 Mahoney, C.M., Fahey, A.J., and Gillen, G. (2007) *Anal. Chem.*, **79** (3), 828–836.

150 Nieuwjaer, N., Poleunis, C., Delcorte, A., and Bertrand, P. (2009) *Surf. Interf. Anal.*, **41** (1), 6–10.

151 Mahoney, C.M., Fahey, A.J., Gillen, G., Xu, C., and Batteas, J.D. (2006) *Appl. Surf. Sci.*, **252** (19), 6502–6505.

152 Mahoney, C.M., Fahey, A.J., Gillen, G., Xu, C., and Batteas, J.D. (2007) *Anal. Chem.*, **79** (3), 837–845.

153 Mahoney, C.M., Kushmerick, J.G., and Steffens, K.L., *J. Phys. Chem. C*, **114** (34), 14510–14519.

154 Mahoney, C.M. (2010) *Surf. Interf. Anal.*, **42**, 1393–1401.

155 Houssiau, L. and Mine, N. (2010) *Surf. Interf. Anal.*, **42** (8), 1402–1408.

156 Ninomiya, S., Ichiki, K., Yamada, H., Nakata, Y., Seki, T., Aoki, T., and Matsuo, J. (2009) *Rapid Commun. Mass Spectrom.: RCM*, **23** (11), 1601.

157 Szakal, C., Sun, S., Wucher, A., and Winograd, N. (2004) *Appl. Surf. Sci.*, **231–232**, 183–185.

158 Py, M., Barnes, J.P., Charbonneau, M., Tiron, R., and Buckley, J. (2011) *Surf. Interf. Anal.*, **43** (1–2), 179–182.

159 Mouhib, T., Delcorte, A., Poleunis, C., and Bertrand, P. (2011) *Surf. Interf. Anal.*, **43** (1–2), 175–178.

160 Moellers, R., Tuccitto, N., Torrisi, V., Niehuis, E., and Licciardello, A. (2006) *Appl. Surf. Sci.*, **252** (19), 6509–6512.

161 Ichiki, K., Ninomiya, S., Nakata, Y., Yamada, H., Seki, T., Aoki, T., and Matsuo, J. (2010) *Surf. Interf. Anal.*, **43** (1–2), 340–343.

162 Wagner, M.S., Lenghaus, K., Gillen, G., and Tarlov, M.J. (2006) *Appl. Surf. Sci.*, **253** (5), 2603–2610.

163 Hinder, S.J., Lowe, C., and Watts, J.F. (2007) *Prog. Org. Coat.*, **60** (3), 255–261.

164 Hinder, S.J., Lowe, C., and Watts, J.F. (2007) *Surf. Interf. Anal.*, **39** (6), 467–475.

165 Russell, T.P., Coulon, G., Deline, V.R., and Miller, D.C. (1989) *Macromolecules*, **22** (12), 4600–4606.

166 Lee, J., Yoon, D., Shin, K., Kim, K.J., and Lee, Y. (2010) *Surf. Interf. Anal.*, **42** (8), 1409–1416.

167 Wagner, M.S. (2005) *Surf. Interf. Anal.*, **37** (1), 53–61.
168 Hagenhoff, B., Deimel, M., Benninghoven, A., and Siegmund, H. (1992) *J. Phys. D Appl. Phys.*, **25**, 818.
169 Stapel, D., Thiemann, M., and Benninghoven, A. (2000) *Appl. Surf. Sci.*, **158**, 362–374.
170 Wagner, M.S. (2005) *Surf. Interf. Anal.*, **37** (1), 42–52.
171 Belu, A.M., Davies, M.C., Newton, J.M., and Patel, N. (2000) *Anal. Chem.*, **72** (22), 5625–5638.
172 Van, O. (1989) *Rubber Chem. Technol.*, **62** (4), 656–682.
173 Garbassi, F., Occhiello, E., Polato, F., and Brown, A. (1987) *J. Mater. Sci.*, **22** (4), 1450–1456.
174 Delcorte, A., Médard, N., and Bertrand, P. (2002) *Anal. Chem.*, **74** (19), 4955–4968.
175 Galuska, A.A. (1997) *Surf. Interf. Anal.*, **25** (1), 1–4.
176 Nysten, B., Verfaillie, G., Ferain, E., Legras, R., Lhoest, J.B., Poleunis, C., and Bertrand, P. (1994) *Microsc. Microanal. M.*, **5** (4–6), 373–380.
177 Wee, A.T.S., Huan, C.H.A., Gopalakrishnan, R., Tan, K.L., Kang, E.T., Neoh, K.G., and Shirakawa, H. (1991) *Synthetic Met.*, **45** (2), 227–234.
178 Mahoney, C.M., Patwardhan, D.V., and McDermott, M.K. (2006) *Appl. Surf. Sci.*, **252** (19), 6554–6557.
179 McDermott, M.K., Saylor, D.M., Casas, R., Dair, B.J., Guo, J., Kim, C.S., Mahoney, C.M., Ng, K., Pollack, S.K., Patwardhan, D.V., Sweigart, D.A., Thomas, T., Toy, J., Williams, C.M., and Witkowski, C.N. (2010) *J. Pharm. Sci.*, **99** (6), 2777–2785.
180 Braun, R.M., Cheng, J., Parsonage, E.E., Moeller, J., and Winograd, N. (2006) *Anal. Chem.*, **78** (24), 8347–8353.
181 Chilkoti, A., Castner, D.G., and Ratner, B.D. (1991) *Appl. Spectrosc.*, **45** (2), 209–217.
182 Huan, C.H.A., Wee, A.T.S., Gopalakrishnan, R., Tan, K.L., Kang, E.T., Neoh, K.G., and Liaw, D.J. (1993) *Synthetic Met.*, **53** (2), 193–203.
183 Davies, M.C., Lynn, R.A.P., Davis, S.S., Hearn, J., Watts, J.F., Vickerman, J.C., and Paul, A.J. (1993) *Langmuir*, **9** (7), 1637–1645.
184 Deslandes, Y., Mitchell, D.F., and Paine, A.J. (1993) *Langmuir*, **9** (6), 1468–1472.
185 Weibel, D., Wong, S., Lockyer, N., Blenkinsopp, P., Hill, R., and Vickerman, J.C. (2003) *Anal. Chem.*, **75** (7), 1754–1764.
186 Petrat, F.M., Wolany, D., Schwede, B.C., Wiedmann, L., and Benninghoven, A. (1994) *Surf. Interf. Anal.*, **21** (5), 274–282.
187 Hittle, L.R., Proctor, A., and Hercules, D.M. (1995) *Macromolecules*, **28** (18), 6238–6243.
188 Linton, R.W., Mawn, M.P., Belu, A.M., DeSimone, J.M., Hunt, M.O. Jr., Menceloglu, Y.Z., Cramer, H.G., and Benninghoven, A. (1993) *Surf. Interf. Anal.*, **20** (12), 991–999.
189 Muddiman, D.C., Brockman, A.H., Proctor, A., Houalla, M., and Hercules, D.M. (1994) *J. Phys. Chem.*, **98** (44), 11570–11575.
190 Hittle, L.R., Proctor, A., and Hercules, D.M. (1994) *Anal. Chem.*, **66** (1), 108–114.
191 Lub, J., Van Leyen, D., and Benninghoven, A. (1989) *Polym. Commun.*, **30** (3), 74–77.
192 Asakawa, D., Chen, L.C., and Hiraoka, K. (2009) *J. Mass Spectrom.*, **44** (6), 945–951.
193 Kono, T., Iwase, E., and Kanamori, Y. (2008) *Appl. Surf. Sci.*, **255** (4), 997–1000.
194 Ito, H. and Kono, T. (2008) *Appl. Surf. Sci.*, **255** (4), 1044–1047.
195 Whitlow, S.J. and Wool, R.P. (1989) *Macromolecules*, **22** (6), 2648–2652.
196 Ninomiya, S., Ichiki, K., Yamada, H., Nakata, Y., Seki, T., Aoki, T., and Matsuo, J. (2011) *Surf. Interf. Anal.*, **43** (1–2), 221–224.
197 Iida, S., Miyayama, T., Sanada, N., Suzuki, M., Fisher, G.L., and Bryan, S.R. (2011) *Surf. Interf. Anal.*, **43** (1–2), 214–216.
198 Agrawal, G., Wool, R.P., Dozier, W.D., Felcher, G.P., Zhou, J., Pispas, S., Mays, J.W., and Russell, T.P. (1996) *J. Polym. Sci. Pol. Phys.*, **34** (17), 2919–2940.

199 Zheng, X., Rafailovich, M.H., Sokolov, J., Zhao, X., Briber, R.M., and Schwarz, S.A. (1993) *Macromolecules*, **26** (24), 6431–6435.

200 Whitlow, S.J. and Wool, R.P. (1991) *Macromolecules*, **24** (22), 5926–5938.

201 Pu, Y., White, H., Rafailovich, M.H., Sokolov, J., Schwarz, S.A., Dhinojwala, A., Agra, D.M.G., and Kumar, S. (2001) *Macromolecules*, **34** (14), 4972–4977.

202 Pu, Y., White, H., Rafailovich, M.H., Sokolov, J., Patel, A., White, C., Wu, W.L., Zaitsev, V., and Schwarz, S.A. (2001) *Macromolecules*, **34** (24), 8518–8522.

203 Petitjean, S., Ghitti, G., Jerome, R., Teyssio, P., Marien, J., Riga, J., and Verbist, J. (1994) *Macromolecules*, **27** (15), 4127–4133.

204 Lin, H.C., Tsai, I.F., Yang, A.C.M., Hsu, M.S., and Ling, Y.C. (2003) *Macromolecules*, **36** (7), 2464–2474.

205 Yokoyama, H., Kramer, E.J., Rafailovich, M.H., Sokolov, J., and Schwarz, S.A. (1998) *Macromolecules*, **31** (25), 8826–8830.

206 Bulle-Lieuwma, C.W.T., Van Gennip, W.J.H., Van Duren, J.K.J., Jonkheijm, P., Janssen, R.A.J., and Niemantsverdriet, J.W. (2003) *Appl. Surf. Sci.*, **203**, 547–550.

207 Van Ooij, W.J. and Sabata, A. (1992) *Surf. Interf. Anal.*, **19** (1–12), 101–113.

208 Shard, A.G., Sartore, L., Davies, M.C., Ferruti, P., Paul, A.J., and Beamson, G. (1995) *Macromolecules*, **28** (24), 8259–8271.

209 Lub, J. and Buning, G.H. (1990) *Polymer*, **31** (6), 1009–1017.

210 Van der Wel, H., Van Vroonhoven, F., and Lub, J. (1993) *Polymer*, **34** (10), 2065–2071.

211 Lub, J., Van Vroonhoven, F., Van Leyen, D., and Benninghoven, A. (1988) *Polymer*, **29** (6), 998–1003.

212 Mine, N., Douhard, B., Brison, J., and Houssiau, L., *Rapid Commun. Mass Spectrom.*, **21** (16), 2680–2684.

213 Houssiau, L. and Mine, N. (2011) *Surf. Interf. Anal.*, **43** (1–2), 146–150.

214 Perez-Luna, V.H., Hooper, K.A., Kohn, J., and Ratner, B.D. (1997) *J. Appl. Polym. Sci.*, **63** (11), 1467–1479.

215 Hittle, L.R. and Hercules, D.M. (1994) *Surf. Interf. Anal.*, **21** (4), 217–225.

216 Van, O. (1989) *Rubber Chem. Technol.*, **62** (4), 656–682.

217 Treverton, J.A. and Paul, A.J. (1995) *Int. J. Adhes. Adhes.*, **15** (4), 237–248.

218 Xu, K., Proctor, A., and Hercules, D.M. (1996) *Microchim. Acta*, **122** (1), 1–15.

219 Conlan, X.A., Gilmore, I.S., Henderson, A., Lockyer, N.P., and Vickerman, J.C. (2006) *Appl. Surf. Sci.*, **252** (19), 6562–6565.

220 Davies, M.C., Khan, M.A., Short, R.D., Akhtar, S., Pouton, C., and Watts, J.F. (1990) *Biomaterials*, **11** (4), 228–234.

221 Van Royen, P., Taranu, A., and Van Vaeck, L. (2005) *Rapid Commun. Mass Spectrom.*, **19** (4), 552–560.

222 Fletcher, J.S., Conlan, X.A., Lockyer, N.P., and Vickerman, J.C. (2006) *Appl. Surf. Sci.*, **252** (19), 6513–6516.

223 Kim, Y.L. and Hercules, D.M. (1994) *Macromolecules*, **27** (26), 7855–7871.

224 Davies, M.C., Short, R.D., Khan, M.A., Watts, J.F., Brown, A., Eccles, A.J., Humphrey, P., Vickerman, J.C., and Vert, M. (1989) *Surf. Interf. Anal.*, **14** (3), 115–120.

225 Short, R.D. and Davies, M.C. (1989) *Int. J. Mass Spectrom. Ion Process*, **89** (2–3), 149–155.

226 Brinen, J.S., Greenhouse, S., and Jarrett, P.K. (1991) *Surf. Interf. Anal.*, **17** (5), 259–266.

227 Ogaki, R., Green, F., Li, S., Vert, M., Alexander, M.R., Gilmore, I.S., and Davies, M.C. (2006) *Appl. Surf. Sci.*, **252** (19), 6797–6800.

228 Lee, J.W. and Gardella, J.A. Jr. (2001) *Macromolecules*, **34** (12), 3928–3937.

229 Lee, J.W. and Gardella, J.A. (2002) *J. Am. Soc. Mass Spectrom.*, **13** (9), 1108–1119.

230 Chen, J. and Gardella, J.A. Jr. (1999) *Macromolecules*, **32** (22), 7380–7388.

231 Shard, A.G., Volland, C., Davies, M.C., and Kissel, T. (1996) *Macromolecules*, **29** (2), 748–754.

232 Ogaki, R., Shard, A.G., Li, S.M., Vert, M., Luk, S., Alexander, M.R., Gilmore, I.S., and Davies, M.C. (2008) *Surf. Interf. Anal.*, **40** (8), 1168–1175.

233 Shard, A.G., Clarke, S., and Davies, M.C. (2002) *Surf. Interf. Anal.*, **33** (6), 528–532.
234 Burns, S.A., Hard, R., Hicks, W.L. Jr., Bright, F.V., Cohan, D., Sigurdson, L., and Gardella, J.A. Jr. (2010) *J. Biomed. Mater. Res. Part A*, **94** (1), 27–37.
235 Boschmans, B., Van Royen, P., and Van Vaeck, L. (2005) *Rapid Commun. Mass Spectrom.*, **19** (18), 2517–2527.
236 Lee, J.W. and Gardella, J.A. Jr. (2003) *Anal. Chem.*, **75** (13), 2950–2958.
237 Mahoney, C.M., Roberson, S., and Gillen, G. (2004) *Appl. Surf. Sci.*, **231-2**, 174–178.
238 Shard, A.G., Brewer, P.J., Green, F.M., and Gilmore, I.S. (2007) *Surf. Interf. Anal.*, **39** (4), 294–298.
239 Gillen, G., Fahey, A., Wagner, M., and Mahoney, C. (2006) *Appl. Surf. Sci.*, **252** (19), 6537–6541.
240 Belu, A., Mahoney, C., and Wormuth, K. (2008) *J. Control Release*, **126** (2), 111–121.
241 Fisher, G.L., Belu, A.M., Mahoney, C.M., Wormuth, K., and Sanada, N. (2009) *Anal. Chem.*, **81** (24), 9930–9940.
242 Fisher, G.L., Belu, A.M., Mahoney, C.M., Wormuth, K., and Sanada, N. (2009) *Anal. Chem.*, **81** (24), 9930–9940.
243 Mahoney, C.M., Fahey, A.J., Belu, A.M., and Gardella, J.A. Jr. (2011). *J. Surf. Anal.*, **17** (3), 299–304.
244 Davies, M.C., Leadley, S.R., Paul, A.J., Vickerman, J.C., Heller, J., and Franson, N.M. (1992) *Polym. Advan. Technol.*, **3** (6), 293–301.
245 Shard, A.G., Davies, M.C., and Schacht, E. (1996) *Surf. Interf. Anal.*, **24** (12), 787–793.
246 Aimoto, K., Aoyagi, S., Kato, N., Iida, N., Yamamoto, A., and Kudo, M. (2006) *Appl. Surf. Sci.*, **252** (19), 6547–6549.
247 Burrell, M.C., Bhatia, Q.S., Chera, J.J., and Michael, R.S. (1990) *J. Vac. Sci. Technol. A*, **8** (3), 2300–2305.
248 Li, L., Chan, C.M., Liu, S.Y., An, L.J., Ng, K.M., Weng, L.T., and Ho, K.C. (2000) *Macromolecules*, **33** (21), 8002–8005.
249 Pawson, D.J., Ameen, A.P., Short, R.D., Denison, P., and Jones, F.R. (1992) *Surf. Interf. Anal.*, **18** (1), 13–22.
250 Bletsos, I.V., Hercules, D.M., Magill, J.H., VanLeyen, D., Niehuis, E., and Benninghoven, A. (1988) *Anal. Chem.*, **60** (9), 938–944.
251 Vargo, T.G., Thompson, P.M., Gerenser, L.J., Valentini, R.F., Aebischer, P., Hook, D.J., and Gardella, J.A. Jr. (1992) *Langmuir*, **8** (1), 130–134.
252 Spool, A.M. and Kasai, P.H. (1996) *Macromolecules*, **29** (5), 1691–1697.
253 Marcus, A. and Winograd, N. (2006) *Anal. Chem.*, **78** (1), 141–148.
254 Hues, S.M., Wyatt, J.R., Colton, R.J., and Black, B.H. (1990) *Anal. Chem.*, **62** (10), 1074–1079.
255 Feld, H., Leute, A., Rading, D., Benninghoven, A., Chiarelli, M.P., and Hercules, D.M. (1993) *Anal. Chem.*, **65** (15), 1947–1953.
256 Lorenz, M.R., Novotny, V.J., and Deline, V.R. (1991) *Surf. Sci.*, **250** (1–3), 112–122.
257 Chan, H.S.O., Ang, S.G., Ho, P.K.H., and Johnson, D. (1990) *Synth. Met.*, **36** (1), 103–110.
258 Abel, M.L., Leadley, S.R., Brown, A.M., Petitjean, J., Chehimi, M.M., and Watts, J.F. (1994) *Synthetic Met.*, **66** (1), 85–88.
259 Hearn, M.J., Fletcher, I.W., Church, S.P., and Armes, S.P. (1993) *Polymer*, **34** (2), 262–266.
260 Morea, G., Sabbatini, L., West, R.H., and Vickerman, J.C. (1992) *Surf. Interf. Anal.*, **18** (6), 421–429.
261 Davies, M.C., Khan, M.A., Domb, A., Langer, R., Watts, J.F., and Paul, A.J. (1991) *J. Appl. Polym. Sci.*, **42** (6), 1597–1605.
262 Gardella, J.A. and Mahoney, C.M. (2004) *Appl. Surf. Sci.*, **231-2**, 283–288.
263 Miyayama, T., Sanada, N., Bryan, S.R., Hammond, J.S., and Suzuki, M. (2010) *Surf. Interf. Anal.*, **42** (9), 1453–1457.
264 Shimizu, K., Phanopoulos, C., Loenders, R., Abel, M.L., and Watts, J.F. (2010) *Surf. Interf. Anal.*, **42** (8), 1432–1444.
265 Bletsos, I.V., Hercules, D.M., VanLeyen, D., Benninghoven, A., Karakatsanis, C.G., and Rieck, J.N. (1990) *Macromolecules*, **23** (18), 4157–4163.
266 Bletsos, I.V., Hercules, D.M., VanLeyen, D., Benninghoven, A.,

Karakatsanis, C.G., and Rieck, J.N. (1989) *Anal. Chem.*, **61** (19), 2142–2149.

267 Zhuang, H., Gardella, J.A. Jr., and Hercules, D.M. (1997) *Macromolecules*, **30** (4), 1153–1157.

268 Niehuis, E., Heller, T., Jurgens, U., and Benninghoven, A. (1989) *J. Vac. Sci. Technol. A*, **7** (3), 1823–1828.

269 Elman, J.F., Lee, D.H.T., Koberstein, J.T., and Thompson, P.M. (1995) *Langmuir*, **11** (7), 2761–2767.

270 Hagenhoff, B., Benninghoven, A., Barthel, H., and Zoller, W. (1991) *Anal. Chem.*, **63** (21), 2466–2469.

271 Stein, J., Leonard, T.M., and Smith, G.A. (1991) *J. Appl. Polym. Sci.*, **42** (8), 2355–2360.

272 Groenewold, G.S., Cowan, R.L., Ingram, J.C., Appelhans, A.D., Delmore, J.E., and Olson, J.E. (1996) *Surf. Interf. Anal.*, **24** (12), 794–802.

273 Hunt, M.O. Jr., Belu, A.M., Linton, R.W., and DeSimone, J.M. (1993) *Macromolecules*, **26** (18), 4854–4859.

274 Belu, A.M., Hunt, M.O. Jr., DeSimone, J.M., and Linton, R.W. (1994) *Macromolecules*, **27** (7), 1905–1910.

275 Lee, J.W., Gardella, J.A., Hicks, W., Hard, R., and Bright, F.V. (2003) *Pharmaceutical research*, **20** (2), 149–152.

276 Adriaensen, L., Vangaever, F., and Gijbels, R. (2004) *Anal. Chem.*, **76** (22), 6777–6785.

277 Nittler, L., Delcorte, A., Bertrand, P., and Migeon, H.N. (2011) *Surf. Interf. Anal.*, **43** (1–2), 103–106.

278 Delcorte, A., Bour, J., Aubriet, F., Muller, J.F., and Bertrand, P. (2003) *Anal. Chem.*, **75** (24), 6875–6885.

279 Delcorte, A. and Bertrand, P. (2004) *Appl. Surf. Sci.*, **231**, 250–255.

280 Restrepo, O.A. and Delcorte, A. (2011) *Surf. Interf. Anal.*, **43** (12), 70–73.

281 Nicola, A.J., Muddiman, D.C., and Hercules, D.M. (1996) *J. Am. Soc. Mass Spectrom.*, **7** (5), 467–472.

282 Wu, K.J. and Odom, R.W. (1996) *Anal. Chem.*, **68** (5), 873–882.

283 Jones, E.A., Lockyer, N.P., Kordys, J., and Vickerman, J.C. (2007) *J. Am. Soc. Mass Spectrom.*, **18** (8), 1559–1567.

284 Locklear, J.E., Guillermier, C., Verkhoturov, S.V., and Schweikert, E.A. (2006) *Appl. Surf. Sci.*, **252** (19), 6624–6627.

285 McDonnell, L.A., Heeren, R., de Lange, R.P.J., and Fletcher, I.W. (2006) *J. Am. Soc. Mass Spectrom.*, **17** (9), 1195–1202.

286 Wittmaack, K., Szymczak, W., Hoheisel, G., and Tuszynski, W. (2000) *J. Am. Soc. Mass Spectrom.*, **11** (6), 553–563.

287 Brewer, T.M., Szakal, C., and Gillen, G. (2010) *Rapid Commun. Mass Spectrom.*, **24** (5), 593–598.

288 De Pauw, E. and Pelzer, G. (1984) *Biochem. Biophys. Res. Commun.*, **123** (1), 27–32.

289 Aberth, W., Straub, K.M., and Burlingame, A.L. (1982) *Anal. Chem.*, **54** (12), 2029–2034.

290 Mahoney, C.M. (2010) *Mass Spectrom. Rev.*, **29** (2), 247–293.

291 Postawa, Z., Czerwinski, B., Szewczyk, M., Smiley, E.J., Winograd, N., and Garrison, B.J. (2004) *J. Phys. Chem. B*, **108** (23), 7831–7838.

292 Delcorte, A. and Garrison, B.J. (2007) *J. Phys. Chem. C.* **111** (42), 15312–15324.

293 Delcorte, A. (2005) *Nucl. Instrum. Meth. B*, **236** (1–4), 1–10.

294 Delcorte, A. and Garrison, B.J. (2007) *Nucl. Instrum. Meth. B*, **255** (1), 223–228.

295 Weng, L.T., Smith, T.L., Feng, J., and Chan, C.M. (1998) *Macromolecules*, **31** (3), 928–932.

296 Hinder, S.J., Lowe, C., Maxted, J.T., and Watts, J.F. (2004) *Surf. Interf. Anal.*, **36** (12), 1575–1581.

297 Belu, A.M., Graham, D.J., and Castner, D.G. (2003) *Biomaterials*, **24** (21), 3635–3653.

298 Klerk, L.A., Broersen, A., Fletcher, I.W., van Liere, R., and Heeren, R. (2007) *Int. J. Mass Spectrom.*, **260** (2–3), 222–236.

299 Mahoney, C.M., Fahey, A.J., Gillen, G., Xu, C., and Batteas, J.D. (2006) *Appl. Surf. Sci.*, **252** (19), 6502–6505.

300 Choi, C., Jung, D., Moon, D.W., and Lee, T.G. (2011) *Surf. Interf. Anal.*, **43** (1–2), 331–335.

301 Coullerez, G., Lundmark, S., Malmström, E., Hult, A., and

Mathieu, H.J. (2003) *Surf. Interf. Anal.*, **35** (8), 693–708.
302 Chilkoti, A., Ratner, B.D., and Briggs, D. (1993) *Anal. Chem.*, **65** (13), 1736–1745.
303 NIST/SEMATECH e-Handbook of Statistics.
304 Mahoney, C.M., Fahey, A.J., Steffens, K.L., Benner, B.A. Jr., and Lareau, R.T. (2010) *Anal. Chem.*, **82** (17), 7237–7248.
305 Hawtin, P.N., Abel, M.L., Watts, J.F., and Powell, J. (2006) *Appl. Surf. Sci.*, **252** (19), 6676–6678.
306 Gilmore, I.S. and Seah, M.P. (2000) *Appl. Surf. Sci.*, **161** (3–4), 465–480.
307 Andersen, H.H. and Bay, H.L. (1975) *J. Appl. Phys.*, **46** (6), 2416–2422.
308 Andersen, H.H. and Bay, H.L. (1974) *J. Appl. Phys.*, **45** (2), 953–954.
309 Wucher, A., Cheng, J., and Winograd, N. (2008) *J. Phys. Chem. C. Nanomater. Interf.*, **112** (42), 16550–16555.
310 Wucher, A. (2008) *Surf. Interf. Anal.*, **40** (12), 1545–1551.
311 Lee, J.L.S., Ninomiya, S., Matsuo, J., Gilmore, I.S., Seah, M.P., and Shard, A.G. (2009)
312 Shard, A.G., Goster, I.S., Gilmore, I.S., Lee, L.S., Ray, S., and Yang, L. (2011) *Surf. Interf. Anal.*, **43** (1–2), 510–513.
313 Rading, D., Moellers, R., Kollmer, F., Paul, W., and Niehuis, E. (2011) *Surf. Interf. Anal.*, **43** (1–2), 198–200.
314 Sjovall, P., Rading, D., Ray, S., Yang, L., and Shard, A.G. (2010). *J. Phys. Chem. B*. **114** (2), 769–774.
315 Zheng, L., Wucher, A., and Winograd, N. (2008) *Anal. Chem.*, **80** (19), 7363–7371.
316 Chapiro, A. (1964) *Radiation Res. Suppl.*, **4**, 179–191.
317 Ivanov, V.S. (1992) Radiation chemistry of polymers; Vsp.
318 Wagner, M.S. (2005) *Surf. Interf. Anal.*, **37** (1), 62–70.
319 Puglisi, O., Licciardello, A., Pignataro, S., Calcagno, L., and Foti, G. (1986) *Radiat. Eff. Defect. S.*, **98** (1), 161–170.
320 Puglisi, O., Licciardello, A., Calcagno, L., and Foti, G. (1987) *Nucl. Instrum. Meth. B*, **19**, 865–871.
321 Calcagno, L. (1995) *Nucl. Instrum. Meth. B*, **105** (1–4), 63–70.
322 Calcagno, L. and Foti, G. (1991) *Nucl. Instrum. Meth. B*, **59**, 1153–1158.
323 Mahoney, C.M., Fahey, A.J., Gillen, G., Xu, C., and Batteas, J.D. (2006) *Appl. Surf. Sci.*, **252** (19), 6502–6505.
324 Mahoney, C.M., Fahey, A.J., Gillen, G., Xu, C., and Batteas, J.D. (2006) *Appl. Surf. Sci.*, **252** (19), 6502–6505.
325 Kozole, J., Wucher, A., and Winograd, N. (2008) *Anal. Chem.*, **80** (14), 5293–5301.
326 Shard, A.G., Green, F.M., Brewer, P.J., Seah, M.P., and Gilmore, I.S. (2008) *J. Phys. Chem. B*, **112** (9), 2596–2605.
327 Cheng, J., Wucher, A., and Winograd, N. (2006) *J. Phys. Chem. B*, **110** (16), 8329–8336.
328 Wucher, A. and Winograd, N. (2010) *Anal. Bioanal. Chem.*, **396** (1), 105–114.
329 Shard, A.G., Green, F.M., and Gilmore, I.S. (2008) *Appl. Surf. Sci.*, **255** (4), 962–965.
330 Licciardello, A. (2011) depth profiling in PS using NO backfilling.
331 Houssiau, L., Douhard, B., and Mine, N. (2008) *Appl. Surf. Sci.*, **255** (4), 970–972.
332 Yu, J., Mahoney, C.M., Fahey, A.J., Hicks, W.L. Jr., Hard, R., Bright, F.V., and Gardella, J.A. Jr. (2009) *Langmuir*, **25** (19), 11467–11471.
333 Wucher, A., Cheng, J., and Winograd, N. (2007) *Anal. Chem.*, **79** (15), 5529–5539.
334 Fletcher, J.S. and Vickerman, J.C. (2010) *Anal. Bioanal. Chem.*, **396** (1), 85–104.
335 Fletcher, J.S. (2009) *Analyst*, **134** (11), 2204–2215.
336 Solon, E.G., Schweitzer, A., Stoeckli, M., and Prideaux, B. (2010) *AAPS J*, **12** (1), 11–26.
337 Breitenstein, D., Rommel, C.E., Müllers, R., Wegener, J., and Hagenhoff, B. (2007) *Agnew. Chem. Int. Ed.*, **46** (28), 5332–5335.
338 Jones, E.A., Lockyer, N.P., and Vickerman, J.C. (2008) *Anal. Chem.*, **80** (6), 2125–2132.
339 Yamada, H., Ichiki, K., Nakata, Y., Ninomiya, S., Seki, T., Aoki, T., and Matsuo, J. (2010) *Nucl. Instrum. Meth. B*, **268** (11–12), 1736–1740.
340 Fletcher, J.S., Rabbani, S., Henderson, A., Blenkinsopp, P., Thompson, S.P.,

Lockyer, N.P., and Vickerman, J.C. (2008) *Anal. Chem.*, **80** (23), 9058–9064.

341 Fletcher, J.S., Lockyer, N.P., Vaidyanathan, S., and Vickerman, J.C. (2007) *Anal. Chem.*, **79** (6), 2199–2206.

342 Fletcher, J.S., Lockyer, N.P., and Vickerman, J.C. (2011) *Mass Spectrom. Rev.*, **30** (1), 142–174.

343 Jones, E.A., Lockyer, N.P., and Vickerman, J.C. (2007) *Int. J. Mass Spectrom.*, **260** (2–3), 146–157.

344 Caprioli, R.M., Farmer, T.B., and Gile, J. (1997) *Anal. Chem.*, **69** (23), 4751–4760.

345 Bunch, J., Clench, M.R., and Richards, D.S. (2004) *Rapid Commun. Mass Spectrom.*, **18** (24), 3051–3060.

346 Reyzer, M.L., Hsieh, Y., Ng, K., Korfmacher, W.A., and Caprioli, R.M. (2003) *J. Mass Spectrom.*, **38** (10), 1081–1092.

347 Schwartz, S.A., Reyzer, M.L., and Caprioli, R.M. (2003) *J. Mass Spectrom.*, **38** (7), 699–708.

348 Mullen, A.K., Clench, M.R., Crosland, S., and Sharples, K.R. (2005) *Rapid Commun. Mass Spectrom.*, **19** (18), 2507–2516.

349 Jackson, S.N., Ugarov, M., Egan, T., Post, J.D., Langlais, D., Albert Schultz, J., and Woods, A.S. (2007) *J. Mass Spectrom.*, **42** (8), 1093–1098.

350 Becker, J.S., Becker, J.S., Zoriy, M.V., Dobrowolska, J., and Matusch, A. (2007) *Eur. J. Mass Spectrom.*, **13** (1), 1.

351 Rubakhin, S.S., Greenough, W.T., and Sweedler, J.V. (2003) *Anal. Chem.*, **75** (20), 5374–5380.

352 Crecelius, A.C., Cornett, D.S., Caprioli, R.M., Williams, B., Dawant, B.M., and Bodenheimer, B. (2005) *J. Am. Soc. Mass Spectrom.*, **16** (7), 1093–1099.

353 Jardin-Mathe, O., Bonnel, D., Franck, J., Wisztorski, M., Macagno, E., Fournier, I., and Salzet, M. (2008) *J. Proteomics*, **71** (3), 332–345.

354 Caldwell, R.L., and Caprioli, R.M. (2005) *Mol. Cell. Proteomics*, **4** (4), 394–401.

355 van Hove, E.R.A., Smith, D.F., and Heeren, R.M.A. (2010) *J. Chromatogr. A*, **1217** (25), 3946–3954.

356 Puretzky, A.A. and Geohegan, D.B. (1998) *Chem. Phys. Lett.*, **286** (5–6), 425–432.

357 Leisner, A., Rohlfing, A., Rohling, U., Dreisewerd, K., and Hillenkamp, F. (2005) *J. Phys. Chem. B*, **109** (23), 11661–11666.

358 Sadeghi, M. and Vertes, A. (1998) *Appl. Surf. Sci.*, **127**, 226–234.

359 Trimpin, S., Keune, S., Rader, H.J., and Mnllen, K. (2006) *J. Am. Soc. Mass Spectrom.*, **17** (5), 661–671.

360 Trimpin, S. and Deinzer, M.L. (2007) *J. Am. Soc. Mass Spectrom.*, **18** (8), 1533–1543.

361 Jurchen, J.C., Rubakhin, S.S., and Sweedler, J.V. (2005) *J. Am. Soc. Mass Spectrom.*, **16** (10), 1654–1659.

362 Chaurand, P., Norris, J.L., Cornett, D.S., Mobley, J.A., and Caprioli, R.M. (2006) *J. Proteome Res.*, **5** (11), 2889–2900.

363 Kruse, R. and Sweedler, J.V. (2003) *J. Am. Soc. Mass Spectrom.*, **14** (7), 752–759.

364 Chen, Y.F., Allegood, J., Liu, Y., Wang, E., Cachon-Gonzalez, B., Cox, T.M., Merrill, A.H., and Sullards, M.C. (2008) *Anal. Chem.*, **80** (8), 2780–2788.

365 Aerni, H.R., Cornett, D.S., and Caprioli, R.M. (2006) *Anal. Chem.*, **78** (3), 827–834.

366 Vegvari, A., Fehniger, T.E., Gustavsson, L., Nilsson, A., Andren, P.E., Kenne, K., Nilsson, J., Laurell, T., and Marko-Varga, G. (2010) *J. Proteomics*, **73** (6), 1270–1278.

367 Lemaire, R., Tabet, J.C., Ducoroy, P., Hendra, J.B., Salzet, M., and Fournier, I. (2006) *Anal. Chem.*, **78** (3), 809–819.

368 Tholey, A. and Heinzle, E. (2006) *Anal. Bioanal. Chem.*, **386** (1), 24–37.

369 Liu, Q. and He, L. (2009) *J. Am. Soc. Mass Spectrom.*, **20** (12), 2229–2237.

370 Cha, S. and Yeung, E.S. (2007) *Anal. Chem.*, **79** (6), 2373–2385.

371 Taira, S., Sugiura, Y., Moritake, S., Shimma, S., Ichiyanagi, Y., and Setou, M. (2008) *Anal. Chem.*, **80** (12), 4761–4766.

372 Hankin, J.A., Barkley, R.M., and Murphy, R.C. (2007) *J. Am. Soc. Mass Spectrom.*, **18** (9), 1646–1652.

373 Bouschen, W., Schulz, O., Eikel, D., and Spengler, B. (2010)

Rapid Commun. Mass Spectrom., **24** (3), 355–364.
374 Puolitaival, S.M., Burnum, K.E., Cornett, D.S., and Caprioli, R.M. (2008) *J. Am. Soc. Mass Spectrom.*, **19** (6), 882–886.
375 DIN 55674 (2005) *Plastics – Determination of Molecular Mass and Molecular Mass Distriubtion of Polymers by Matrix Assisted Laser Desorption/Ionization – Time of Flight – Mass Spectrometry*, Beuth Verlag GmbH, Berlin.
376 ASTM D7134 (2005) *Standard Test Method for Molecular Mass Averages and Molecular Mass Distribution of Atactic Polystyrene by Matrix Assisted Laser Desorption Ionization (MaLDI) – Time-of-Flight (ToF) Mass Spectrometry (MS)*, ASTM International, West Conshohocken, PA.
377 Weidner, S., Hoffman, K., Falkenhagen, J., and Thuenemann, A. (2010) Polymer segregation in MALDI spots investigated by MALDI-ToF Imaging MS.
378 Weidner, S.M. and Falkenhagen, J. (2009) *Rapid Commun. Mass Spectrom.*, **23** (5), 653–660.
379 Laiko, V.V., Baldwin, M.A., and Burlingame, A.L. (2000) *Anal. Chem.*, **72** (4), 652–657.
380 Ifa, D.R., Wu, C., Ouyang, Z., and Cooks, R.G. (2010) *The Analyst*, **135** (4), 669–681.
381 Eberlin, L.S., Ifa, D.R., Wu, C., and Cooks, R.G. (2010) *Agnew. Chem. Int. Ed.*, **49** (5), 873–876.
382 Kertesz, V. and Van Berkel, G.J. (2008) *Rapid Commun. Mass Spectrom.*, **22** (17), 2639–2644.
383 Weston, D.J. (2010) *The Analyst*, **135** (4), 661–668.
384 Salter, T.L., Green, F.M., Gilmore, I.S., Seah, M.P., and Stokes, P. (2011) *Surf. Interf. Anal.*, **43** (1–2), 294–297.
385 Nefliu, M. and Venter, A. (2006) *Chem. Commun.* (8), 888–890.
386 Jackson, A.T., Williams, J.P., and Scrivens, J.H. (2006) *Rapid Commun. Mass Spectrom.*, **20** (18), 2717–2727.
387 Chaurand, P., Schwartz, S.A., and Caprioli, R.M. (2004) *Anal. Chem.*, **76** (5), 86–93.
388 Kulp, K.S., Berman, E.S.F., Knize, M.G., Shattuck, D.L., Nelson, E.J., Wu, L., Montgomery, J.L., Felton, J.S., and Wu, K.J. (2006) *Anal. Chem.*, **78** (11), 3651–3658.
389 Dill, A.L., Ifa, D.R., Manicke, N.E., Costa, A.B., Ramos-Vara, J.A., Knapp, D.W., and Cooks, R.G. (2009) *Anal. Chem.*, **81** (21), 8758–8764.
390 Harper, J.D., Charipar, N.A., Mulligan, C.C., Zhang, X., Cooks, R.G., and Ouyang, Z. (2008) *Anal. Chem.*, **80** (23), 9097–9104.
391 Tuccitto, N., Lobo, L., Tempez, A., Delfanti, I., Chapon, P., Canulescu, S., Bordel, N., Michler, J., and Licciardello, A. (2009) *Rapid Commun. Mass Spectrom.*, **23** (5), 549–556.
392 Hiraoka, K., Mori, K., and Asakawa, D. (2006) *J. Mass Spectrom.*, **41** (7), 894–902.
393 Hiraoka, K., Iijima, Y., and Sakai, Y. (2011) *Surf. Interf. Anal.*, **43** (1–2), 236–240.
394 Sakai, Y., Iijima, Y., Takaishi, R., Asakawa, D., and Hiraoka, K. (2009) *J. Vac. Sci. Technol. A*, **27**, 743.
395 Hiraoka, K., Takaishi, R., Asakawa, D., Sakai, Y., and Iijima, Y. (2009) *J. Vac. Sci. Technol. A*, **27**, 748.
396 Sakai, Y., Iijima, Y., Asakawa, D., and Hiraoka, K. (2010) *Surf. Interf. Anal.*, **42** (6–7), 658–661.

7
Hyphenated Techniques

Jana Falkenhagen and Steffen Weidner

7.1
Introduction

The coupling of different analytical techniques results in a significant increase of information and is often able to provide more and reliable data than individual methods used for coupling can do. The combination of liquid chromatographic (LC) separation and mass spectrometric (MS) identification for polymer characterization is of particular interest. The most important reason for this can be seen in reducing the complexity of the analysis of polymers.

In contrast to most other compounds showing distinctive molecular masses, polymer materials are heterogeneous regarding their molecular mass and chemical heterogeneity. First approaches to tackle these problems utilized conventional size exclusion chromatography (SEC) combined with multidetector devices (dual or triple detectors). The combination of structure (e.g., ultraviolet and refractive index) and mass-specific detectors (e.g., multiangle laser light scattering and viscosimetry) was initially used for the universal calibration of SEC instruments. A more sophisticated approach represents the coupling of different modes of LC separation in the so-called two-dimensional (2D) analysis. 2D analysis includes such particular techniques as liquid adsorption chromatography (LAC), liquid chromatography at critical conditions (LC-CCs), gradient polymer elution chromatography, and temperature gradient interaction chromatography. A detailed description of different chromatographic modes used for specific polymer separations is provided in the following section.

It very soon became clear that chromatography alone, even combined in a 2D fashion using various detectors, would not be able to solve important analytical questions completely. This became especially evident in the analysis of complex copolymers. Apart from molecular mass and mass distributions, additional analytical questions had to be answered. Thus, the arrangement and compositional distribution of copolymer segments, the structure of block intermediates in controlled living radical polymerization, the end-group distribution, a differentiation between linear, multiarm, or cyclic structures, and other topological questions became increasingly important. Many of these questions could be readily answered via polymer MS.

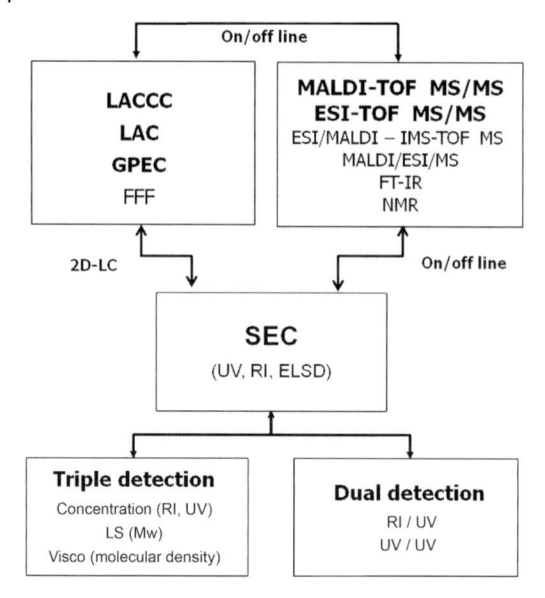

Figure 7.1 Coupling of variable separation techniques with spectroscopic and spectrometric methods for polymer characterization.

Since their introduction in the early 1990s, matrix-assisted laser desorption/ionization (MALDI) and electrospray ionization (ESI) MS have become important tools in polymer analysis. Both techniques *can* (under certain preconditions) enable the simultaneous determination of molecular masses, mass distributions, and end groups. Thus, a combination of chromatographic separation and MS identification can be a powerful tool in polymer analysis. A summary of most applied techniques in polymer characterization is represented in Figure 7.1.

The following chapters are primarily intended to focus on the analytical strengths and opportunities of both methods. Both limitations and drawbacks are discussed, and examples demonstrating their high potential are presented. The development of suitable interfaces enabling routine polymer LC-MS analysis is also discussed.

7.2
Polymer Separation Techniques

Synthetic polymers are heterogeneous in many respects. Since every polymerization reaction involves the statistical processes of initiation, growth, termination, and transfer, polymers always exhibit a molecular mass distribution (MMD). In addition, nearly all synthetic polymers exhibit a concomitant chemical heterogeneity distribution. In principle, one can differentiate between functionality type distributions, copolymer compositions distribution, and topology distribution (e.g., branched polymers, linear or cyclic structures). Moreover, even if the average molecular composition appears to be identical, a sample can consist of a blend, a block, or a

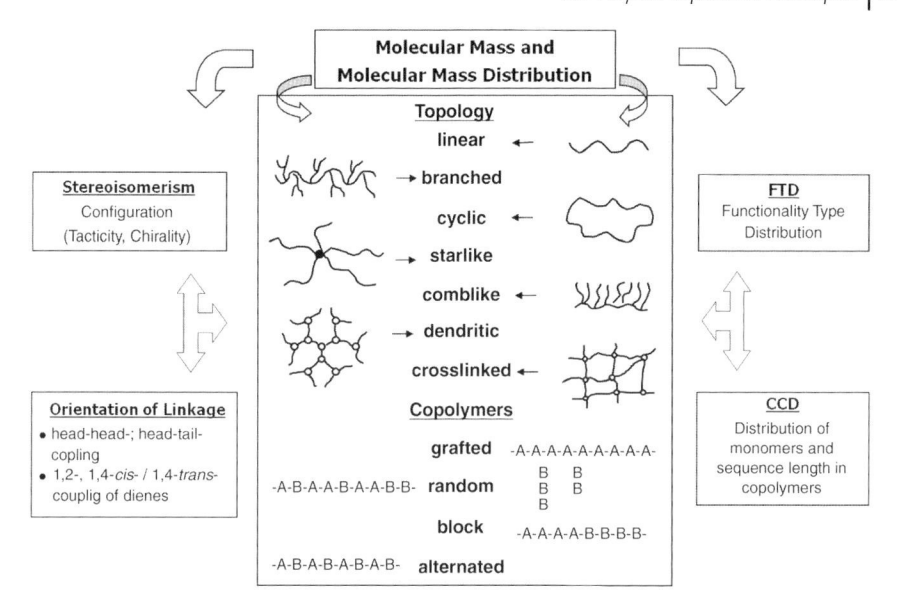

Figure 7.2 Schematic representation of possible heterogeneities of polymers.

statistical copolymer. An overview demonstrating the complexity of polymer structures is shown in Figure 7.2.

The majority of LC investigations of polymers have been performed by means of SEC. Using specific stationary phases with specific pore-size distribution, macromolecules are separated from each other according to the hydrodynamic volume of dissolved molecules. Simply put, larger molecules will be strongly excluded from entering the pores than smaller ones. For these species, the loss of entropy while entering the pores from the free mobile phase is therefore lower. Assuming no enthalpic interactions between the polymer molecule and the pore wall exist, the steric exclusion results in a distribution coefficient $K < 1$. The determination of molecular masses and mass distributions is performed by a previous calibration with polymer standards having similar hydrodynamic volumes.

In contrast to common opinion, chromatography of polymers is much more than SEC only. Another important principle called LAC utilizes interactions between sample molecules and mobile phase on the one hand and the stationary phase on the other hand for separation. This principle is also referred to as high-performance liquid chromatography (HPLC) and has been used for many decades for separating low-molecular compounds. Its usability for separating high mass compounds is limited due to the fact that with increasing number of repeating units in the polymer chain, the number of possible interaction sites increases, too. Thus, the higher the polymer molecular mass ($K > 1$), the stronger the adsorption. All monomer units and heterogeneous groups are able to interact with the surface of the stationary phase, whereas the interaction of polymer segments is reduced. In contrast to the previously described SEC mode with $\Delta S < 0$ and $\Delta H = 0$, the adsorption mode (LAC) is predominantly characterized by enthalpic interactions. This can be expressed through $\Delta H < 0$ and $T\Delta S \ll \Delta H$. The thermodynamic interpretation is given by the Gibbs–Helmholtz equation for the free enthalpy

$$\Delta G = \Delta H - T\Delta S$$

It becomes clear that LC of polymers frequently results in signals caused by a superposition of different separation mechanisms. As an example, the end-group functionalization of a polymer does not necessarily change its hydrodynamic volume. But, due to possible changes of enthalpic interactions, their conventional SEC analysis typically provides incorrect molecular mass values.

The compensation of entropic and enthalpic contributions of a monomer unit ($\Delta G = 0$) enables another very interesting separation mode of polymers. In 1990, Gorshkov and colleagues introduced the principle of LC-CCs [1–4]. At these particular "critical conditions of adsorption," macromolecules are separated according to chemical heterogeneity (functionalities, chemical composition distribution, chirality, tacticity, topology, etc.). The hydrodynamic volume of a macromolecule in the mixed mobile phase is reduced, and its retention is exclusively caused by the adsorption of the characteristic structural unit (e.g. end groups). Consequently, macromolecules with identical repeat units elute independently of their molecular masses. Critical conditions can be adjusted for example, by changing the solvent composition, the stationary phase, and/or the temperature [5]. Typical calibration curves for all three modes of chromatography are presented in Figure 7.3. A more detailed description of the principle of LC-CC can be found in Refs. [6–9]. Alternatively, solvent or temperature gradients can be used for copolymer systems consisting of two structural units strongly differing in their physical properties, such as amphiphilic copolymers. By using gradient systems, several limitations of LAC running at pure isocratic conditions (e.g., limited resolution especially for higher molecular masses, strong peak broadening, or incomplete desorption) can be avoided.

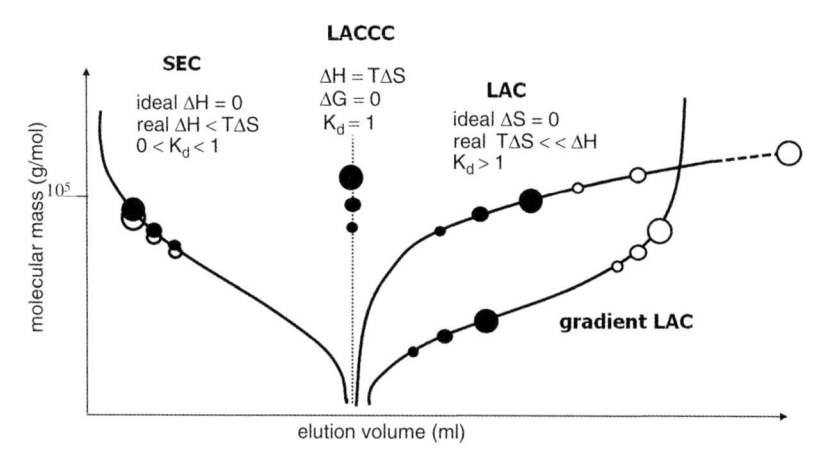

Figure 7.3 Polymer elution curves obtained in different modes of chromatography and contribution of enthalpic and entropic terms of the Gibbs–Helmholtz equation.

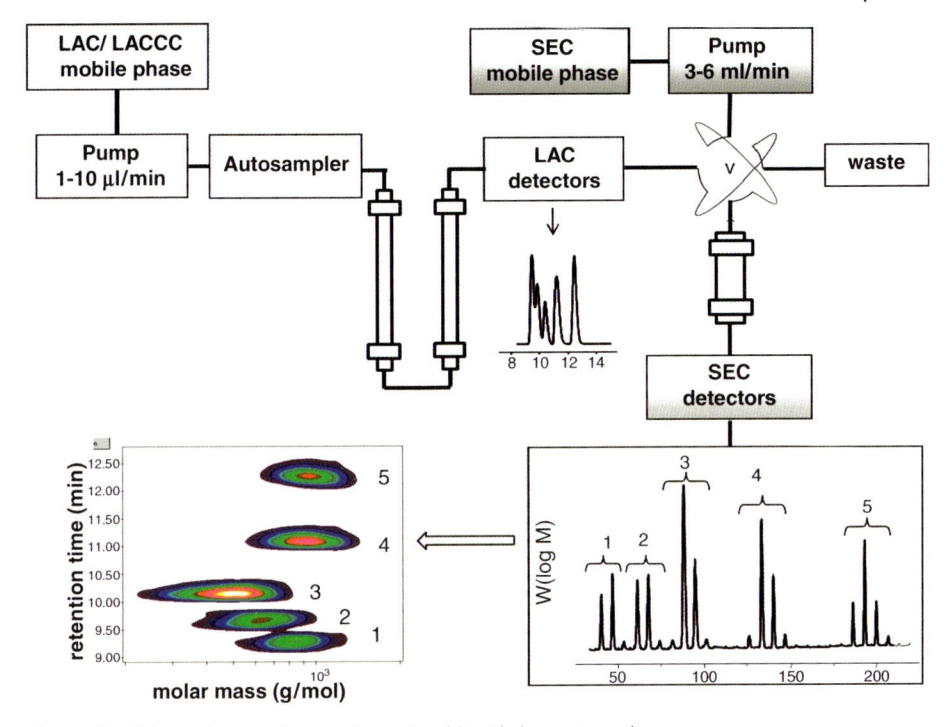

Figure 7.4 Schematic setup for two-dimensional liquid chromatography.

The above remarks clearly indicate that one single chromatographic mode is not sufficient for a comprehensive characterization of polymers. LC-CC leads to a separation of functionalized polymers but lacks information on molecular masses. SEC provides hydrodynamic volumes, which, however, have to be carefully considered when the polymers underwent a chemical modification. LAC seems to be nonsuitable for the separation of polymers with higher masses due to strong enthalpic interactions, frequently causing an irreversible adsorption on the stationary phases. Therefore, a coupling of two specific modes seemed to be mandatory. As a result, the so-called 2D-LC approach was developed. Its schematic principle is depicted in Figure 7.4. Usually, in the first dimension, molecules are separated according to their chemical heterogeneity distribution using "critical," near "critical," or changing (established by a solvent gradient) conditions of adsorption. By means of a switching valve, the collected fractions are transferred online from the first into the second chromatographic dimension. Here, the polymers are separated according to their molecular masses by conventional SEC. In order to obtain a complete sample transfer switching time, loop volume and flow rates of both chromatographic systems have to be carefully adjusted.

However, the coupling of two chromatographic systems is an expensive and time-consuming endeavor. The lack of appropriate standards for SEC calibration can lead to incorrect molecular masses. Often results are reported as polystyrene (PS) equivalents, which causes an erroneous interpretation of mass data. It soon became clear that MALDI- and ESI-MS, initially applied as simple detectors for chromatography, could also be used to replace either chromatographic dimension in a 2D approach. Nevertheless, conventional chromatographic 2D techniques are essential for separating high molecular mass polymers combined with high polydispersities or when ionization in MS could not be achieved.

7.3
Principles of Coupling: Transfer Devices

In contrast to ESI-MS, which can be readily online connected to a chromatographic system, special attention has to be directed to the transfer of dissolved polymer samples into MALDI-TOF-MS instruments (where TOF is time-of-flight). Since this particular MS technique requires the addition of a matrix compound prior the measurement, the online coupling with chromatographic methods still represent a significant challenge. Thus, first attempts were performed manually, that is, simply by taking fractions. Since the current chapter is focused on sophisticated methods, manual fractionation will not be further discussed.

As presented in the following section, several efforts for online connecting chromatographic separation devices with MALDI-MS instruments have been made resulting in mostly inappropriate approaches. The situation has been partially changed after introducing atmospheric pressure (AP) sources in the year 2000. Up to this time, the use of off-line transfer methods was more or less indispensible. Nevertheless, nowadays both online (AP-MALDI and LC-ESI) and off-line devices represent established techniques, constantly undergoing further improvements.

A scheme showing the whole variety of basic LC–MALDI coupling principles is presented in Figure 7.5.

7.3.1
Online Coupling Devices

Various efforts have been made to develop suitable online techniques. Among them were techniques applying rotating wheels, belts, or balls. These interfaces were especially designed for coupling capillary LC or capillary electrophoresis (CE) with MALDI [10–13]. After being separated from each other, samples were deposited on a rotating device at ambient pressure conditions outside the mass spectrometer source. The dried trace, consisting of sample and matrix, was continuously moved in the MS source staying under high vacuum. A laser beam was focused on this trace, and spectra could be continuously recorded. It soon became clear that these interfaces could not be applied for a routine analysis of synthetic polymer samples. In order to enable a co-crystallization of sample and matrix, either the evaporation of the solvent must proceed very fast or the volume of the solution being deposited must be very low. Thus, the liquid flow rate using the rotating wheel approach was in the range of 100–400 nl min^{-1}. Moreover, mass spectra predominantly obtained in the

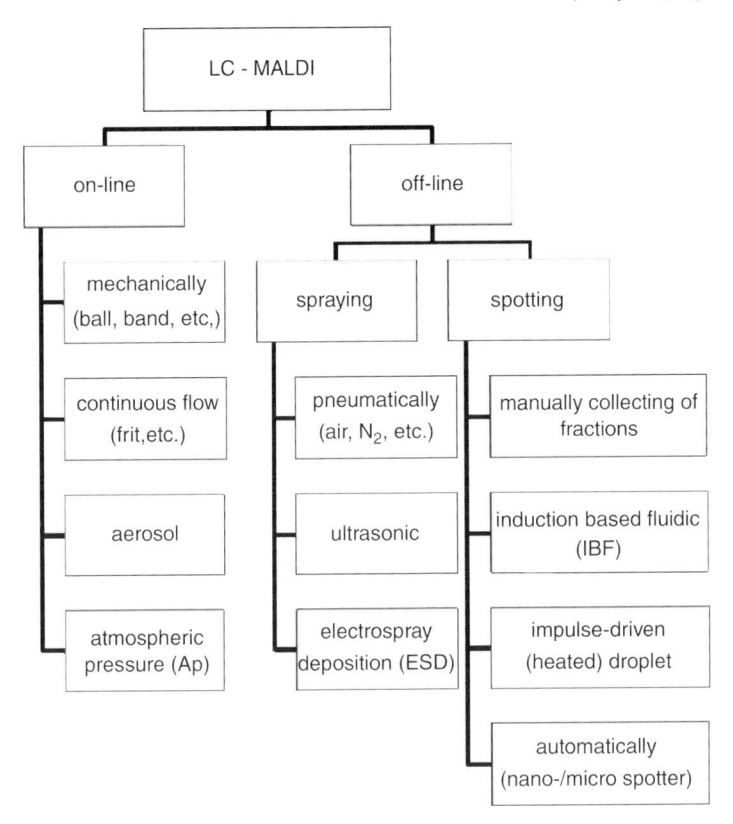

7.3 Principles of Coupling: Transfer Devices | 215

Figure 7.5 Scheme of basic LC–MALDI coupling principles.

low-mass region were characterized by a poor peak resolution. The principle of a rotating ball approach is shown in Figure 7.6 [14].

Other, more applicable approaches utilize aerosol or direct coupling (continuous flow) techniques. The principle of an aerosol interface is based on the direct spaying of the sample/matrix solution into the source of the mass spectrometer [15–19]. In order to keep the high vacuum in the source of the mass spectrometer constant, the flow rate again had to be extremely low. In principle, this desorption step is comparable with those known from thermospray MS and, particularly, from ESI-MS. However, desorption and evaporation proceed in a source held at ambient pressure. Moreover, in an aerosol interface, sample molecules undergo ionization by laser shots after forming the aerosol plume, which is in contrast to ESI, where desorption and ionization of a sample coevally take place. Because of the spatial distribution of the aerosol particles in the MALDI source, only a poor peak resolution can be obtained. This problem could be partially overcome by introducing reflectron MALDI instruments that compensate the spread of ion energies in different regions of the source. Other direct coupling devices are mainly based on the use of frits, filter paper, or steel tips [20]. Here, the sample solution is inserted

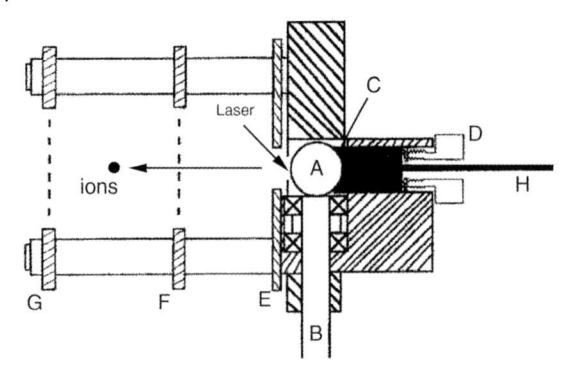

Figure 7.6 Scheme of an online rotating ball MALDI interface. (A) 10 mm in diameter stainless steel ball, (B) drive shaft, (C) gasket, (D) adjustment screw, (E) repeller, (F) extraction grid, (G) ground grid, and (H) capillary. The ball is rotated through the shaft, which is connected to a gear motor positioned outside the vacuum chamber (not shown). (Reprinted from [14] with permission from the American Chemical Society.)

into the MS source using a heated capillary. Matrix solution could be simultaneously added using a T-piece. The one end of the capillary that consists of a frit or filter paper is positioned in the vacuum of the MS source. There, after the evaporation of the solvent, the ionization occurs by a laser, which is focused on the end of the capillary. Several modifications of this technique have been described. Among them are devices working semicontinuously (applying a washing step to clean the frit between two measurements) and approaches using two capillaries (enabling a single delivery of matrix and sample). However, all these modifications have not been overly successful. Again, recorded mass spectra showed a very poor peak resolution and the available mass region was limited to a few thousand Dalton. Both the aerosol and a frit coupling interface are schematically presented in Figure 7.7 [21]. A detailed overview on online coupling techniques can be obtained from Murray [22] and Orsnes [23].

In 2000, first AP-MALDI and AP-MALDI/ion trap interfaces were presented [24, 25]. A scheme of an orthogonal AP–MALDI interface is shown in Figure 7.8. Similar to the ESI, principle desorption/ionization processes are decoupled from the high-vacuum part of the mass spectrometer. In 2004, the first online coupling of liquid sample delivery to AP-MALDI-MS for a variety of matrices and different solvents, polymers (PEG 1000), and peptides was presented [26, 27].

Surprisingly, in this case, a co-crystallization of analyte and matrix was not necessary because analyte and matrix were equally distributed within the droplets. Thus, a segregation of matrix and analyte, which can often be found in conventional crystallized MALDI samples, was not observed. Another advantage of the above method is that in contrast to conventional MALDI, matrices and analytes do not need to be vacuum stable. Depending on the vacuum pump system, flow rates up to 300 µl min^{-1} can be employed. In combination with single-drop microextraction, AP-MALDI was used for a simultaneous determination of cationic

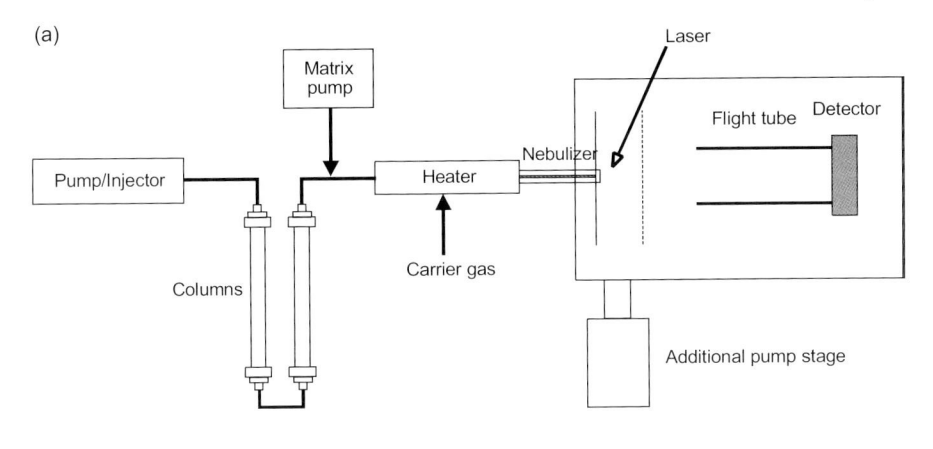

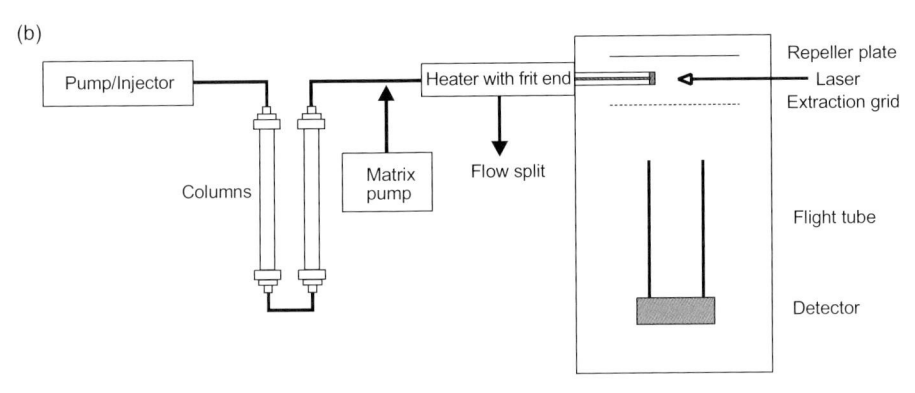

Figure 7.7 Principles of online spray method generating an aerosol (a) and continuous flow coupling method applying a heated frit inlet system (b). (Reprinted from [21] with permission from Wiley.)

surfactants from river and municipal wastewater [28]. Thereby, an additional sample pre- or posttreatment or separation by HPLC, gas chromatography (GC), and CE could be avoided. Recently, a new AP-MALDI-type interface has been presented as an alternative to nano-ESI. This atmospheric pressure free liquid infrared matrix-assisted laser dispersion ionization interface is based on a formation of free liquid microbeam/microdroplets followed by ionization using a mid-infrared optical parametric oscillator. The presented approach is supposed to be well suited for standard HPLC applications, in which flow rates higher than $100\,\mu l\,min^{-1}$ are required [29].

The majority of AP-MALDI applications have been used for biopolymers. Its use for synthetic polymers is still very limited. One main drawback of AP interfaces is the observed in-source decomposition of polymers with masses above 2000–3000 Da. Nevertheless, the AP technique allows detection over a wide range of polarity of sample/solvent and, therefore, appears to be complementary to ESI-MS [30].

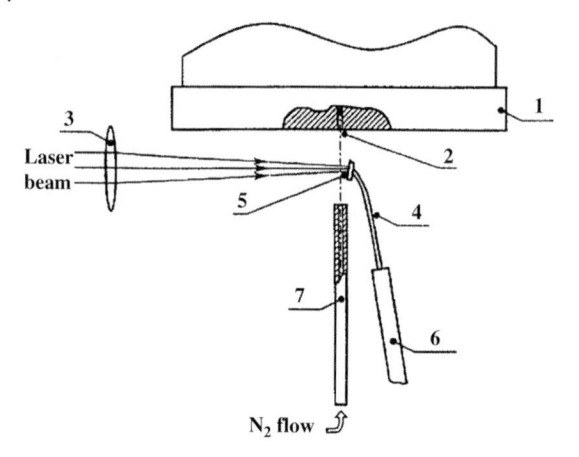

Figure 7.8 Schematic view of the AP-MALDI source: (1) atmospheric pressure interface of the mass spectrometer, (2) inlet nozzle to MS instrument, (3) quartz lens to focus laser beam onto the sample target, (4) replaceable probe tip, (5) stainless steel MALDI target plate, (6) target holder, and (7) stainless steel capillary gas nozzle. The ions are subject to a gas-assisted flow directed toward the nozzle (2). Furthermore, the pressure gradient established in this nozzle orifice drags the ions into the mass spectrometer. (Reprinted from [24] with permission from the American Chemical Society.)

7.3.2
Off-Line Coupling Devices

Off-line transfer devices are realized by spotting or spraying of the chromatographic eluents onto the MALDI target plate. The matrix solution can be added either before the deposition of the sample or together with the sample by premixing via a T-piece. First off-line SEC/MALDI-MS coupling experiments involve fraction collection, solvent evaporation, pipetting, and washing of the transfer needle. This approach, however, was intricate and time-consuming. Modern commercially available spotting devices enable a simultaneous collecting of fractions in vials, which can be used for further investigation and sample preparation (e.g., Probot™ Microfraction Collector, LC Packings USA; DiNa Map MALDI Spotter, Kromatek, UK; MALDILC™ System, Gilson, USA; and SunChrom, Germany) [31, 32].

Several modifications of the spotting technique have been developed, for example, the "heated droplet" or the "impulse-driven heated droplet" deposition [33, 34]. The "heated droplet" interface consists of a transfer tube from the chromatographic system having an outlet being adapted to form continuously replaced, hanging droplets of the liquid stream. A heated MALDI sample plate for collecting the droplets is mounted below. The liquid stream in the transfer tube is heated to a temperature that enables a partial evaporation of the solvent from the hanging droplets. This interface does not show sample loss, and the detection sensitivity of LC/MALDI is comparable to that of standard MALDI-MS. By applying the "impulse-driven heated droplet" deposition, the droplets are actively dislodged from the exit capillary onto the MALDI plate by means of a solenoid plunger.

Direct deposition methods, in which chromatographic fractions and MALDI matrix are directly deposited onto the MALDI target, have become widely used, since they can operate under their respective optimal chromatographic conditions.

One of the first examples of a SEC–MALDI coupling was presented using a commercially available interface (LC transform, LabConnections, USA). This system was originally intended to transfer eluents onto germanium discs for Fourier transform infrared measurements [35, 36]. After passing a heated needle, eluents are sprayed (by a sheet gas or ultrasonically) onto a moving MALDI plate. The matrix could be simultaneously added or precoated target plates could be used. Finally, the resulting polymer "trail" was characterized directly by MALDI [37–41]. The main advantage of these spray devices is that flow rates normally used for SEC experiments (1 ml min^{-1}) could be applied. However, the evaporation of poorly vaporizable solvents requires higher needle temperatures, which can cause polymer degradation and a blocking of the needle tip by matrix crystallization. This could be overcome by using a μ-SEC approach, which reduces the amount of solvent to be evaporated. It could be shown that even though the amount of polymer deposited on the target was drastically reduced, MALDI was still sensitive enough to detect femto mole quantities of poly(ethylene oxides) (PEOs) [42]. A modification of the spray deposition technique represents the oscillating capillary nebulizer (OCN) initially developed for coupling CE with MS [43]. This nebulizer has certain features that make it very suitable for an LC coupling at both microflows (1 μl min^{-1}) and macroflows (1 ml min^{-1}).

Electrospray deposition (ESD) has traditionally been used to prepare thin, uniform samples and to reduce the segregation of analyte from matrix during the sample drying step. Dissolved macromolecules are sprayed through a charged needle. At the tip of the needle, the solvent is dispersed into a fine spray of charged droplets. These droplets migrate in an electric field toward the MALDI target. By applying an additional sheet gas (e.g., nitrogen), the solvent evaporates at elevated temperatures. Thus, during their migration, the evaporating droplets shrink, and finally, Coulomb repulsion forces exceed surface tension of the droplets causing their explosion. Formed smaller droplets undergo the same process again generating even smaller droplets until the ultimate droplet contains only one molecule which contains the residual charge. This resulted in sample spots with significantly improved homogeneity and enables the recording of reproducible spectra characterized by increased signal intensity [44–46]. In 2000, a first interface for the continuous ESD of chromatographic eluents was presented [47]. Its principle is shown in Figure 7.9.

One major drawback of ESD of synthetic polymers is the necessity of using suitable solvents, which are essential for a sufficient spray formation and ionization. Most synthetic polymers are soluble in solvents such as tetrahydrofuran or chloroform, rather than in typical ESI solvents (e.g., water, acetonitrile, and methanol). This can be partially overcome by adding the matrix solution prior to electrospraying to improve the conductivity of the solution. Nevertheless, relatively low flow rates have to be applied, since the efficiency of ion formation is reduced by decreasing the charge per mass ratio at higher chromatographic flow rates. Another disadvantage is the possibility to cause fragmentation of polymers while being sprayed [48]. A new approach represents the use of polycarbonate microfluidic chips with integrated

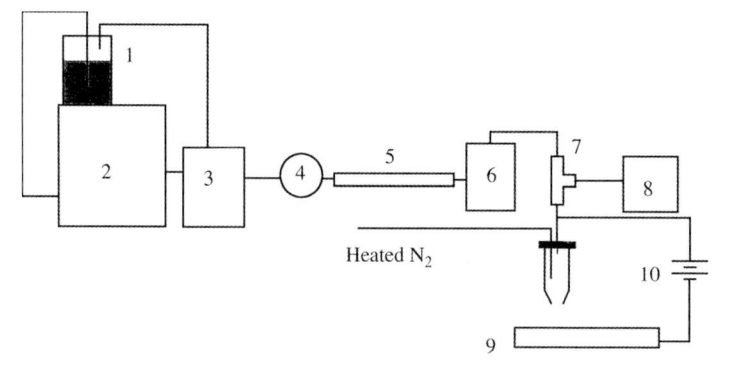

Figure 7.9 Schematic diagram of the experimental setup for direct electrospray sample fraction deposition for SEC/MALDI-TOF-MS. (1) Solvent reservoir, (2) HPLC pump, (3) flow-splitter, (4) autosampler injector, (5) narrow-bore SEC column, (6) UV-vis detector, (7) T-piece, (8) syringe pump for matrix, (9) MALDI plate, and (10) high-voltage supplier. (Reprinted from [47] with permission from Wiley.)

hydrophobic membrane electrospray tips to deposit peptides and proteins onto a stainless steel target followed by MALDI-MS analysis. The arrangement of multiple electrospray tips on a single chip provides the ability to simultaneously elute parallel sample streams onto a MALDI target for high-throughput multiplexed analysis [49]. Although the interface was primarily intended for biopolymer applications, this technology also offers promise for a range of high-throughput analyses in which, simultaneously, several matrices can be checked for its usability for polymer analysis.

Recently, the electric field-enhanced sample preparation for synthetic polymer MALDI-TOF-MS via induction-based fluidics (IBFs) was demonstrated [50, 51]. In contrast to ESD, where a charge is induced on the liquid by passing the fluid through an electric field conductively, an IBF charging proceeds inductively. Therefore, a possible degradation of polymer molecules can be avoided. IBF kinetically launches droplets to targets and, thus, can dynamically direct the liquids to targets in flight. This results in tiny sample spots exhibiting major improvement in both MALDI sensitivity and reproducibility. Although not yet used for coupling with chromatography, IBF could be a valuable tool for future applications in this field.

7.4
Examples

7.4.1
Coupling of SEC with MALDI-/ESI-MS

SEC can be regarded as the "workhorse" technique for the determination of molecular masses and mass distributions of synthetic polymers and has been used for many decades. Nevertheless, due to its principle, which is based on the separation of dissolved macromolecules differing in their hydrodynamic volumes, SEC displays limitations. The main drawback can be seen in the necessity of an accurate calibration using polymer standards having identical (or at least comparable) structure as the polymers to be investigated. Thus, molecular masses of most industrial polymers and copolymers in particular cannot be determined via a simple SEC analysis. As already noted above, the investigation of such polymer systems requires either the use of so-called absolute detectors based on multiangle laser light scattering and/or viscometry or multidetector regimes (dual or triple detectors combining refractive index and ultraviolet detectors, RI/viscometry, RI/viscometry/light scattering, etc.).

However, MALDI- and ESI-TOF-MS provide absolute molecular masses, mass distributions and, moreover, information on end groups without the need of being specifically calibrated. Thus, shortly after the introduction of these MS techniques in the early 1990-ties especially MALDI-TOF-MS was regarded as a complementary analytical method having the potential to replace SEC soon. These high expectations could not be fulfilled since MS of macromolecules also has some limitations. Nowadays, TOF-MS are mainly equipped with multichannel-plate detectors. The principle of signal generation is based on the measurement of cascades of secondary electrons formed after the detector is hit by ions. The response of the detector is characterized by the time it needs to compensate the loss of electrons. If the MMD of the polymer to be investigated is too broad, a fast saturation of a multichannel-plate detector by the first (very fast) ions arriving will be obtained. Unfortunately, this limitation is already fulfilled at a comparatively low polydispersity of 1.2 (M_w/M_n). In these cases, the positive charge of the detector cannot be compensated by electron refilling from the power source until the slower (higher mass) ions arrive. As a result, mass spectra typical for broad distributed polymers will be recorded. They are characterized by a very high intensity in the lower mass range followed by a fast intensity decrease at higher masses. These problems can be overcome by reducing the polydispersity below 1.2 simply by a chromatographic fractionation of the polymer. As the method of choice, SEC can be employed for this purpose. Positively, MALDI provides absolute molecular mass values of such narrowly distributed fractions, which vice versa are useful for calibrating the chromatographic system. This is especially interesting when polymer standards for the calibration of the SEC are not available. Thus, these two techniques can be regarded as complementary to each other.

Because of the lack of suitable devices, first coupling experiments were carried out simply by taking fractions. One of the first applications where MALDI masses of manually taken SEC fractions were used to create, a SEC calibration curve was presented by Krüger et al. [52]. These authors applied various chromatographic techniques, among them SEC, to identify polyester species by spectroscopic methods (NMR and MALDI-TOF-MS) and to calibrate both the detector response and the SEC. Other authors applied this comparatively simple technique for the investigation of polysaccharides [53], during the synthesis of poly(arylether-dendrimers) [54], for the determination of the molecular, functional, and chemical heterogeneities of silsesquioxanes and siloxanes [41] and for coal-derived materials [55]. A first systematic investigation for a wide variety of synthetic polymers (PS, polybutylacrylate, polycarbonate, aromatic polyester resin, and a methyl methacrylate (MMA)-methacrylic acid copolymer) having polydispersities from 1.7 up to 3.0 was presented by Nielen et al. [56]. An impressive example was given by Montaudo et al. [57]. More than 20 fractions of two broadly distributed poly(dimethylsiloxanes) (PDMS) were collected and measured in a linear MALDI-TOF-MS. Finally, SEC calibration in a mass range from a few thousands up to several hundred thousand Dalton (ca. m/z 500 000) could be performed using the MALDI data. Copolymers can also strongly benefit from SEC-MALDI investigations. In order to determine their bivariate distribution and copolymer composition, these data were compared to SEC-NMR experiments [58–62]. The degradation of polyether, polyesters, and polyester polyurethane soft blocks was a matter of investigation, too. Narrow fractions of the degraded and unreacted polyols were collected and analyzed by MALDI, which allowed precise calibration of the SEC chromatograms [63]. Kona et al. investigated the functionalization of polybutadiene double bonds [64]. SEC fractions taken before and after functionalization were used to calibrate the SEC. These authors could show that completely different calibration curves resulted rendering SEC without absolute mass determination worthless. The investigation of synthesis products during the formation of polyalkylcyanoacrylate (PBCA) nanoparticles showed that three polymer series with different end groups were present in the sample, which move through the SEC column as discernable groups [65].

In 1998, manual fractionation was replaced by spotting of chromatographic eluents on the MALDI target using a robotic interface [31]. Its performance was demonstrated by the characterization of a polybisphenol-A-carbonate and a complex dipropoxylated bisphenol-A-/adipic acid/isophthalic acid copolyester resin. A 10-s elution window was used for exact calibration of the micro-SEC, which is less laborious and time-consuming and features very low solvent consumption and reduced solvent emission during the evaporation step.

A first application of air spray-based SEC fractionation combined with MALDI-TOF-MS analysis was demonstrated by Kassis et al. [37] Approximately 15% of the eluted polymer from the SEC was spray deposited onto a rotating matrix-coated substrate. Obtained overall Mn values of a poly(methyl methacrylate) (PMMA) were in fairly good agreement with the manufacturer's estimates. The generation of unwanted by-products (homopolymers) during the synthesis of n-butyl methacrylate/MMA copolymers (PnBMA-b-PMMA) was reported by Esser et al. [39] Moreover, as shown in Figure 7.10, these authors also demonstrated the ability to use SEC-MALDI-MS data for generating 3D fingerprints employing broadly distributed PS. A spray interface was also used for the investigation of silsequioxanes [66], amphiphilic N-vinylpyrrolidone, and vinyl acetate block copolymers [67].

The coupling of thermal field-flow fractionation with a MALDI-TOF-MS applying the OCN was shown by Basile et al. [68]. These studies determined that the OCN was effective in handling liquid compositions from 100% aqueous to 100% organic (reversed-phase LC gradient elution).

In 2000, Lou and van Dongen presented a direct sample fraction deposition of a broad PMMA using an electrospray interface in narrow-bore size-exclusion chromatography/MALDI-TOF-MS [47]. Similar to the OCN device, no substantial effects

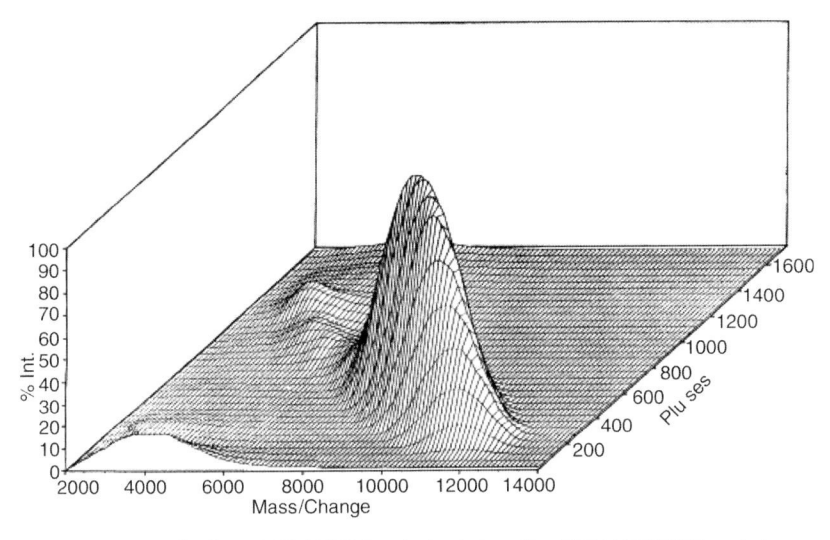

Figure 7.10 3D plot for PMMA 10 900 Da obtained via online SEC-MALDI-TOF analysis. (Reprinted from [39] with permission from Elsevier.)

of composition variation of the SEC effluent on the analytical results using direct fraction deposition were observed.

In contrast to MALDI, the coupling of ESI with LC is far less challenging. Nevertheless, there are some problems which have to be considered. As already mentioned, polymers have to be dissolved in solvents suitable for the ESI process. Often cationization has to be promoted by an addition of salts to the solvent. Furthermore, the quantity of solvent that could be vaporized is limited to only a few microliters per minute, implying that split techniques have to be adapted for ordinary LC apparatus' or micro (or even nano)-LC instruments have to be used [69]. The first polymers investigated in SEC-ESI-MS coupling experiments were lower mass polymers (surfactants and macromonomers) [70–72], complex (methoxymethyl)melamin resins [73], sodium PS sulfonate [74], and methylmethacrylate/butylacrylate [75]. The problem of overlaying charge distributions could be overcome using the superior resolution of FT-ICR mass analyzers. This was demonstrated for polyethylene glycols and glycidylmethacrylate/butyl methacrylate copolymers [76, 77]. The FT-ICR technique enabled the identification of more than 5000 isotopic peaks of 47 oligomers in 10 charge states in the spectrum of a PEG with an average molecular mass of 23 000 Da [76].

Similar to off-line LC-MALDI, the coupling of SEC with ESI-MS is applied for an accurate molecular mass determination of polymers with higher polydispersities [78–80], but also for monitoring the mechanism and structure of products synthesized in photoinitiated, radical, and controlled radical (reversible addition fragmentation chain transfer (RAFT)) polymerizations [79, 81–89]. Recently, a new approach for the determination of accurate mass distributions was presented, which utilizes a computational algorithm based on the maximum entropy principle [84].

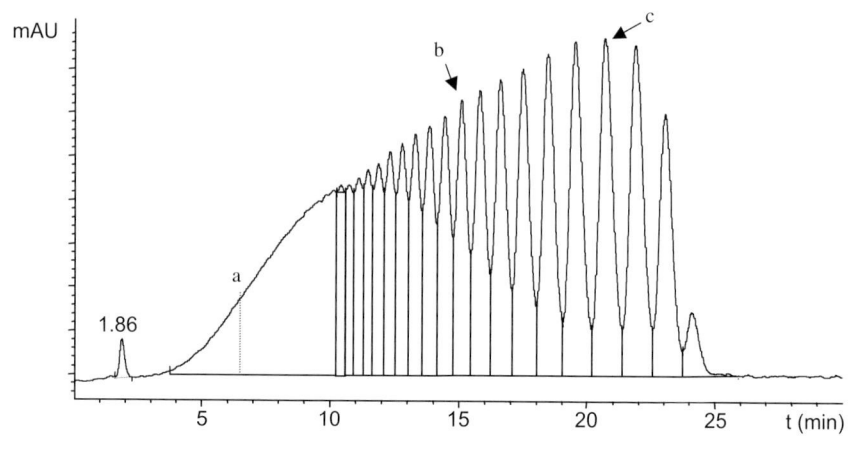

Figure 7.11 Chromatogram of OH-terminated PDMS observed at critical conditions for PDMS. (Reprinted from [21] with permission from Wiley.)

7.4.2
Coupling of LAC/LC-CC with MALDI-/ESI-MS

Compared to a SEC–MS coupling, the combination of MS with adsorption LC offers an additional advantage. Provided that the molecular masses of the sample to be investigated are not too high, macromolecules can be simultaneously separated according their molecular masses *and* chemical heterogeneities by means of LC methods. MS is the perfect complement since it enables simultaneous detection of both properties. First off-line LC-MS experiments were performed on polyethers [90], epoxy resin polyesters [91], and several aliphatic, aromatic, and biodegradable polyesters [92, 93].

A detailed review of applications up to 2002 was given in a monography of Pasch and Schrepp [94].

A nice example that demonstrated the advantage of LAC compared to SEC is exemplarily shown in Figures 7.11 and 7.12. Here the MMD curve of OH-terminated PDMS obtained from a typical SEC experiment (Figure 7.11) is compared to its LC-CC elugram. At these particular conditions, a separation of PDMS according to end groups (methyl and hydroxyl) and molecular masses was observed. The broad separation range in LAC (LC-CC) chromatography resulted in a much higher peak resolution, and, therefore, was shown be beneficial for the ionization process in MS [21].

Poly(ethylene glycols) represent an important class of polymers. They have been widely used in industry, medicine, and cosmetics [40, 95–97]. Their comparatively low molecular masses make these polymers especially suited for MS. LC-CC-MALDI-MS was used to separate various PEOs having different end groups [98]. Moreover, since many chromatographic detectors are quantitative detectors, signal intensities obtained by chromatography can be used to overcome many quantification problems in MS. LC-APCI-MS was also successfully applied for quantification. Mass spectra

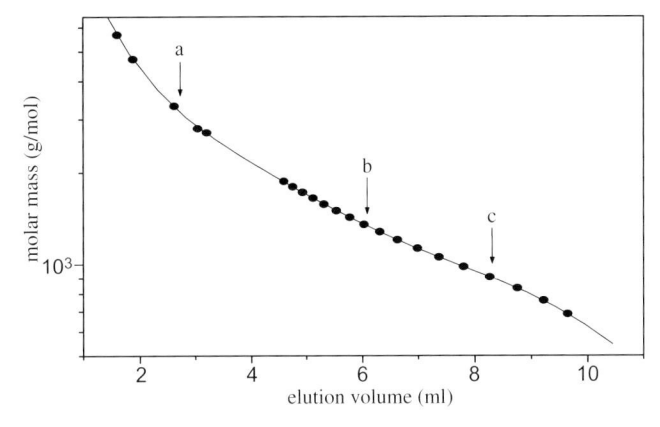

Figure 7.12 Calibration curve of OH-PDMS by means of MALDI-TOF-MS. (Reprinted from [21] with permission from Wiley.)

were recorded using different ratios of functionalized and nonfunctionalized PEO. In addition, evaporative light scattering detection response factors for various PEO standards were compared. It was found that PEO standards with molar masses from m/z 1000 to 8000 show responses that are differing by 10% at maximum, whereas a low molecular mass PEO 400 provides approximately 30% less response compared to its higher molecular mass counterparts [99]. The sensitivity of the MALDI-TOF-MS technique in detecting PEO with different molecular masses after LC separation was investigated. It could be demonstrated that spectra obtained using the spray deposition interface for MALDI sample preparation show a high reproducibility, better homogeneity, and, in particular, a higher sensitivity (see also Chapter 6). By means of a μ-LAC, a surpassing sensitivity of a few femtograms over a broad range of sample-to-matrix ratios was shown [42].

A semionline LC/MALDI-MS system to introduce eluent from monolithic silica capillary chromatographic column directly onto a MALDI sample plate was described [100]. The small elution volume of a monolithic capillary column allows delicate eluents, such as 1,1,1,3,3,3,-hexa-fluoroisopropyl alcohol, to be employed. The mechanism of chromatographic separation of PMMA, nylon-6/6, nylon-11, and linear and cyclic structures of poly(lactic acid) on monolithic columns was discussed and clarified via MALDI-TOF-MS.

Various coupling techniques such as pyrolysis (Py)-GC/MS, ESI-MS, MALDI-TOF-MS, and LAC-MS were studied for usability in the analysis of PDMS for medical applications [101]. Among these techniques, LC/APCI–MS coupling allowed the fastest and most effective analysis. In addition, the complexity of the mass spectra deduced from these LC-MS experiments was simplified compared to the mass spectra obtained by MALDI-TOF-MS. The authors demonstrated how the LC/APCI–MS coupling of this class of polymers permits the complete characterization of end groups being present in very small quantities.

Polyamides have been intensively examined since this class of polymers exhibits a number of different end groups combined with relatively high molecular masses [102, 103]. A first example of a successful substitution of one chromatographic dimension in a 2D approach by MALDI-TOF-MS was presented [15].

ESI-MS in a multidimensional LC approach was applied for studying structure–performance relationships of solid epoxy resins [104]. Several classes of polymers with different functional groups were separated and identified, and their individual molecular mass values were determined.

The application of LAC and LC-CC – in particular for copolymer analysis – has become more and more important. Adjusting critical (or near critical) separation conditions for only one structural unit of a copolymer enables the separation of the second unit in either SEC or LAC mode. MALDI- and ESI-TOF-MS have been used for a fast determination of PEO–PPO copolymer compositions and for elucidating specific separation conditions [105, 106]. The characterization of a block copolymer consisting of methoxy PEO (mPEO), an ε-caprolactone (CL) segment, and linoleic acid (LA), used as surfactant in water-based latex paints, was achieved by LC-APCI and LC-ESI-TOF-MS [107]. Separation was obtained in a reversed phase system based on the number of CL units and the presence of an mPEO and/or LA tail. It can be distinguished between mPEO-pCL$_n$, mPEO-pCL$_n$-LA, pCL$_n$, pCL$_n$-LA, and cyclo- pCL$_n$. The total ion chromatograms and selected mass traces obtained by different ionization methods are given in Figure 7.13.

Gradient chromatography can be problematical since the changing solvent composition can cause problems using conventional detectors. Such problems can be avoided by the use of spray transfer devices followed by MS. A possible transfer between radicals and process solvent in a controlled radical polymerization (RAFT) was observed and main and side products formed in a synthesis of a polyvinylpyrrolidon–polyvinylacetate block copolymers were detected [67]. Especially the development of these living/controlled radical polymerization techniques, such as nitroxide-mediated polymerization (NMP), atom transfer radical polymerization (ATRP), and RAFT, enabled a better control of the structure and architecture of formed polymers. LC-CC-MALDI-TOF-MS was used to monitor chemical transformations of hydroxyl-functional PSs prepared by different pathways and used as macroinitiator for PS-b-PMMA copolymers [108].

LC coupled to tandem MS (MS^n) has been increasingly applied. Several examples demonstrate the usability of this particular MS technique, which can provide additional information on sequence lengths in copolymers. Copolymers consisting of neopentyl diol (NPG), adipic acid, and hexane diol were investigated by LC/MALDI-MS regarding their composition and end-group distribution [109]. The deficiency of MALDI-TOF-MS to distinguish between the number of A and B units and/or species having different end groups and identical chemical structure was overcome by a fragmentation of suitable parent ions. This resulted in typical fragment ion patterns and provided a clear differentiation between cyclic and linear oligomers as well as between isobaric (i.e., identical masses) linear oligomers with different end groups. In addition, this technique could also be used for a detailed examination of longer copolymer sequences. The characteristic fragmentation behavior of PEO-b-PPO copolymers could be used to distinguish between di- and triblock copolymers and to determine the sequence in diblock [110]. Coupled online

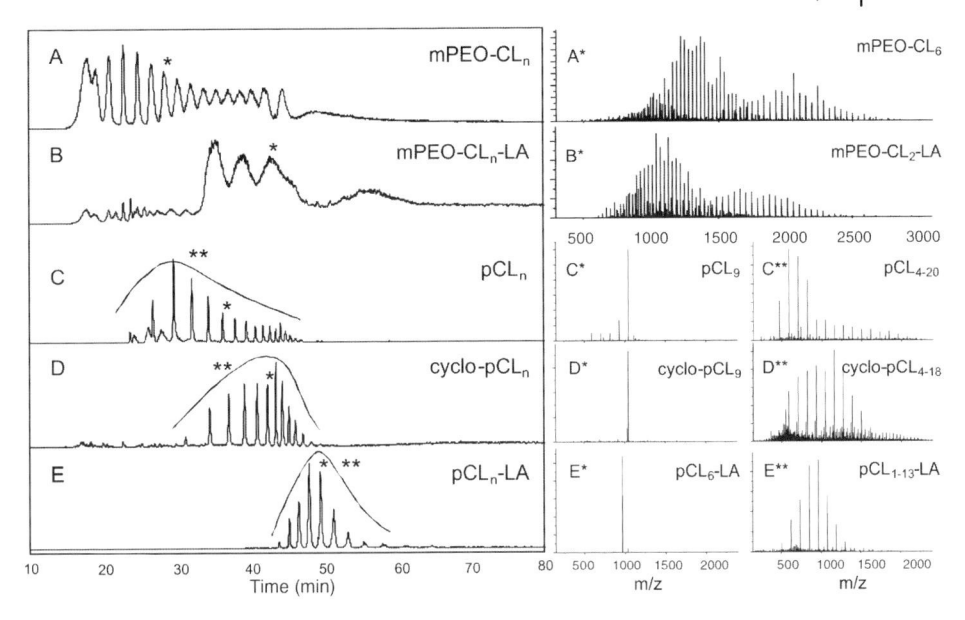

Figure 7.13 LC-MS chromatograms showing the separation of the homologous polymer series contained in the block copolymer samples: (A) total ion chromatogram (TIC) of unfunctionalized polymer, recorded with ESI (+)-MS; (B) TIC of functionalized polymer, recorded with ESI(+)-MS; (C) multiple ion chromatogram (MIC) of 17 selected mass traces of unfunctionalized polymer, recorded with APCI (−)-MS; (D) MIC of 14 selected mass traces of unfunctionalized polymer, recorded with APCI (+)-MS; and (E) MIC of 13 selected mass traces of functionalized polymer, recorded with APCI (−)-MS. Examples of the corresponding mass spectra are given for the marked peaks, as well as for the complete homologous series shown in (C), (D), and (E). (Reprinted from [107] with permission from Wiley.)

with ESI tandem MS, the separation of PEO/PS block copolymer at critical conditions of PEO using a mobile phase that already contained the cationizing agent was demonstrated [111]. Samples were investigated in both the MS and MS/MS mode. Oligomer separation was successfully achieved according to the PS block size. Since the PS block size could be determined by its chromatographic behavior and its characteristic MS/MS fragment patterns, the copolymer microstructure was unambiguously identified.

Ultra performance liquid chromatography (UPLC), recently introduced, provides an ultrafast separation and reduces the sample amount drastically [112]. Its principle is based on the use of column packing materials with smaller particle size (<2 μm), which reduces the time for one analysis without any drawbacks in resolution and sensitivity. Since UPLC systems typically operate at flow rates of 0.05–0.6 ml min^{-1} postcolumn eluents, splitting for online MS is not necessary. A first synthetic polymer application addressed the separation of styrene–acrylonitrile copolymers and epoxy resins [113]. The determination of suitable chromatographic separation conditions, especially of the critical conditions, represents a mostly laborious routine. The online combination of UPLC and ESI-TOF-MS was shown to be extraordinary beneficial for a very fast adjustment of parameters [114]. Because every polymer (even standards with low polydispersities) can be regarded as a mixture of polymer chains with different length, their elution within a single chromatographic peak must occur in a different order depending on the separation mode (SEC, LAC, or LC-CC). Fast data acquisition in UPLC combined with the mass accuracy of an ESI-QTOF instrument revealed the inherent mass distribution of those peaks. This application, representing a fundamentally new approach, was evidenced using PEO and PPO homo- and copolymers. Moreover, these data showed that even with poor chromatographic separation in UPLC structure, composition and molecular weight of these copolymers were obtained in less than 1 min.

Hyphenated coupling methods have also been reported dealing with such particular chromatographic methods such as liquid exclusion adsorption chromatography (LEAC) and temperature gradient interaction chromatography combined with MALDI-TOF-MS identification for the investigation of microwave-assisted polymerization of CL and branched PSs [115, 116].

7.5
Conclusions

Hyphenation of techniques has become a versatile tool in polymer analysis. It has been demonstrated that the complexity and heterogeneity of modern polymers and copolymers, in particular, could only be determined by a combination of several methods. With this regard, a chromatographic separation in various thermodynamic modes combined with mass spectral identification of polymer species provides new insights in polymer synthesis, modification, and degradation. Both methods are widely complementary and can therefore often compensate each other's drawbacks. The development of suitable coupling interfaces, the introduction of AP devices, and an increasing application of ESI pushed polymer analysis toward new regions. Future developments in the field of MS detectors, lasers and software, as well new pressure-stable stationary chromatographic phases will aid in increasing the importance and acceptance of LC–MS coupling methods and will enable new applications in sophisticated polymer analysis.

References

1 Gorshkov, A.V., Much, H., Becker, H., Pasch, H., Evreinov, V.V., and Entelis, S.G. (1990) Chromatographic investigations of macromolecules in the critical range of liquid-chromatography. 1. Functionality type and composition distribution in polyethylene oxide and polypropylene oxide copolymers. *J. Chromatogr.*, **523**, 91–102.

2 Schulz, G., Much, H., Krüger, H., and Wehrstedt, C. (1990) Determination of functionality and molecular-weight distribution by orthogonal chromatography. *J. Liq. Chromatogr.*, **13** (9), 1745–1763.

3 Pasch, H., Much, H., and Schulz, G. (1993) Polymer characterization by liquid chromatography at the critical point of adsorption. *Trends Polym. Sci.*, **3**, 643.

4 Krüger, R.P., Much, H., and Schulz, G. (1996) Determination of Polymer Heterogeneity by Two-Dimensional Orthogonal Liquid Chromatography. *Int. J. Poly. Anal. Charact.*, **2**, 221.

5 Chang, T.Y., Lee, H.C., Lee, W., Park, S., and Ko, C.H. (1999) Polymer characterization by temperature gradient interaction chromatography. *Macromol. Chem. Physic.*, **200** (10), 2188–2204.

6 Belenki, B.G., Gankina, E.S., Tennikov, M.B., and Vilenchik, L.Z. (1976) *Dokl. Acad. Nauk USSR*, **231**, 1147.

7 Belenki, B.G., Gankina, E.S., Tennikov, M.B., and Vilenchik, L.Z. (1978) Fundamental aspects of adsorption chromatography of polymers and their experimental verification by thin-layer chromatography. *J. Chromatogr.*, **147**, 99.

8 Falkenhagen, J. (1998) *Kopplung von Chromatographischen und Spektroskopischen Methoden zur Bestimmung der Heterogenitäten von Polymeren*, Mensch & Buch Verlag, Berlin.

9 Pasch, H. and Trathnigg, B. (1997) *HPLC of Polymers*, Springer, Berlin.

10 Preisler, J., Foret, F., and Karger, B.L. (1998) On-line MALDI-TOF MS using a continuous vacuum deposition interface. *Anal. Chem.*, **70** (24), 5278–5287.

11 Preisler, J., Hu, P., Rejtar, T., and Karger, B.L. (2000) Capillary electrophoresis-matrix-assisted laser desorption/ionization time-of-flight mass spectrometry using a vacuum deposition interface. *Anal. Chem.*, **72** (20), 4785–4795.

12 Musyimi, H.K., Narcisse, D.A., Zhang, X., Stryjewski, W., Soper, S.A., and Murray, K.K. (2004) Online CE-MALDI-TOF MS using a rotating ball interface. *Anal. Chem.*, **76** (19), 5968–5973.

13 Musyimi, H.K., Guy, J., Narcisse, D.A., Soper, S.A., and Murray, K.K. (2005) Direct coupling of polymer-based microchip electrophoresis to online MALDI-MS using a rotating ball inlet. *Electrophoresis*, **26** (24), 4703–4710.

14 Orsnes, H., Graf, T., Degn, H., and Murray, K.K. (2000) A rotating ball inlet for on-line MALDI mass spectrometry. *Anal. Chem.*, **72** (1), 251–254.

15 Beeson, M.D., Murray, K.K., and Russell, D.H. (1995) Aerosol matrix-assisted laser-desorption ionization – effects of analyte concentration and matrix-to-analyte ratio. *Anal. Chem.*, **67** (13), 1981–1986.

16 Fei, X., and Murray, K.K. (1996) On-line coupling of gel permeation chromatography with MALDI mass spectrometry. *Anal. Chem.*, **68** (20), 3555–3560.

17 Fei, X., Wei, G., and Murray, K.K. (1996) Aerosol MALDI with a reflectron time-of-flight mass spectrometer. *Anal. Chem.*, **68** (7), 1143–1147.

18 Murray, K.K., Lewis, T.M., Beeson, M.D., and Russell, D.H. (1993) Matrix-assisted laser-desorption ionization of aerosols for liquid-chromatography TOF mass-spectrometry. *Abstr. Pap. Am. Chem. S.*, **206**, 42–Anyl.

19 Murray, K.K., Lewis, T.M., Beeson, M.D., and Russell, D.H. (1994) Aerosol matrix-assisted laser-desorption ionization for liquid-chromatography time-of-flight mass-spectrometry. *Anal. Chem.*, **66** (10), 1601–1609.

20 Zhan, Q., Gusev, A., and Hercules, D.M. (1999) A novel interface for on-line coupling of liquid capillary chromatography with matrix-assisted laser desorption/ionization detection. *Rapid Commun Mass Sp.*, **13** (22), 2278–2283.

21 Weidner, S. and Falkenhagen, J. (2010) LC-MALDI MS for polymer characterization, in *MALDI Mass Spectrometry for Synthetic Polymer Analysis*, vol. **175** (ed. L. Li), John Wiley & Sons, Hoboken, NJ, pp. 247–265.

22 Murray, K.K. (1997) Coupling matrix-assisted laser desorption/ionization to liquid separations. *Mass Spectrom. Rev.*, **16** (5), 283–299.

23 Orsnes, H. and Zenobi, R. (2001) Interfaces for on-line liquid sample delivery for matrix-assisted laser desorption ionisation mass spectrometry. *Chem. Soc. Rev.*, **30** (2), 104–112.

24 Laiko, V.V., Baldwin, M.A., and Burlingame, A.L. (2000) Atmospheric pressure matrix assisted laser desorption/ionization mass

spectrometry. *Anal. Chem.*, **72** (4), 652–657.

24 Laiko, V.V., Moyer, S.C., and Cotter, R.J. (2000) Atmospheric pressure MALDI/ion trap mass spectrometry. *Anal. Chem.*, **72** (21), 5239–5243.

26 Daniel, J.M., Ehala, S., Friess, S.D., and Zenobi, R. (2004) On-line atmospheric pressure matrix-assisted laser desorption/ionization mass spectrometry. *Analyst*, **129** (7), 574–578.

27 Daniel, J.M., Laiko, V.V., Doroshenko, V.M., and Zenobi, R. (2005) Interfacing liquid chromatography with atmospheric pressure MALDI-MS. *Anal. Bioanal. Chem.*, **383** (6), 895–902.

28 Shrivas, K. and Wu, H.F. (2007) A rapid, sensitive and effective quantitative method for simultaneous determination of cationic surfactant mixtures from river and municipal wastewater by direct combination of single-drop microextraction with AP-MALDI mass spectrometry. *J. Mass Spectrom.*, **42** (12), 1637–1644.

29 Rapp, E., Charvat, A., Beinsen, A., Plessmann, U., Reichl, U., Seidel-Morgenstern, A., Urlaub, H., and Abel, B. (2009) Atmospheric pressure free liquid infrared MALDI mass spectrometry: toward a combined ESI/MALDI-liquid chromatography interface. *Anal. Chem.*, **81** (1), 443–452.

30 Desmazieres, B., Buchmann, W., Terrier, P., and Tortajada, J. (2008) APCI interface for LC- and SEC-MS analysis of synthetic polymers: advantages and limits. *Anal. Chem.*, **80** (3), 783–792.

31 Nielen, M.W.F. (1998) Polymer analysis by micro-scale size exclusion chromatography MALDI time-of-flight mass spectrometry with a robotic interface. *Anal. Chem.*, **70** (8), 1563–1568.

32 Stevenson, T.I. and Loo, J.A. (1998) A simple off-line sample spotter for coupling HPLC with MALDI MS. *Lc Gc-Mag. Sep. Sci.*, **16** (1), 54.

33 Young, J.B. and Li, L. (2007) Impulse-driven heated-droplet deposition interface for capillary and microbore LC-MALDI MS and MS/MS. *Anal. Chem.*, **79** (15), 5927–5934.

34 Zhang, B.Y., McDonald, C., and Li, L. (2004) Combining liquid chromatography with MALDI mass spectrometry using a heated droplet interface. *Anal. Chem.*, **76** (4), 992–1001.

35 Wheeler, L.M. and Willis, J.N. (1993) Gel-permeation chromatography Fourier-transform infrared interface for polymer analysis. *Appl. Spectrosc.*, **47** (8), 1128–1130.

36 Willis, J.N. and Wheeler, L. (1995) Use of a gel-permeation chromatography Fourier-transform infrared spectrometry interface for polymer analysis. *Adv. Chem. Ser.*, **247**, 253–263.

37 Kassis, C.E., DeSimone, J.M., Linton, R.W., Remsen, E.E., Lange, G.W., and Friedman, R.M. (1997) A direct deposition method for coupling matrix-assisted laser desorption/ionization mass spectrometry with gel permeation chromatography for polymer characterization. *Rapid Commun. Mass Sp.*, **11** (10), 1134–1138.

38 Coulier, L., Kaal, E.R., and Hankemeier, T. (2005) Comprehensive two-dimensional liquid chromatography and hyphenated liquid chromatography to study the degradation of poly(bisphenol A)carbonate. *J. Chromatogr. A*, **1070** (1–2), 79–87.

39 Esser, E., Keil, C., Braun, D., Montag, P., and Pasch, H. (2000) Matrix-assisted laser desorption/ionization mass spectrometry of synthetic polymers. 4. Coupling of size exclusion chromatography and MALDI-TOF using a spray-deposition interface. *Polymer*, **41** (11), 4039–4046.

40 Falkenhagen, J., Friedrich, J.F., Schulz, G., Kruger, R.P., Much, H., and Weidner, S. (2000) Liquid adsorption chromatography near critical conditions of adsorption coupled with matrix-assisted laser desorption/ionization mass spectrometry. *Int. J. Polym. Anal. Ch.*, **5** (4–6), 549–562.

41 Kruger, R.P., Much, H., Schulz, G., and Rikowski, E. (1999) Characterization of Si polymers by liquid chromatographic methods in coupling with MALDI-TOF-MS. *Monatsh. Chem.*, **130** (1), 163–174.

42. Falkenhagen, J. and Weidner, S.M. (2005) Detection limits of matrix-assisted laser desorption/ionisation mass spectrometry coupled to chromatography – a new application of solvent-free sample preparation. *Rapid Commun. Mass Sp.*, **19** (24), 3724–3730.

43. Wang, L.Q., May, S.W., Browner, R.F., and Pollock, S.H. (1996) Low-flow interface for liquid chromatography inductively coupled plasma mass spectrometry speciation using an oscillating capillary nebulizer. *J. Anal. Atom. Spectrom.*, **11** (12), 1137–1146.

44. Hensel, R.R., King, R.C., and Owens, K.G. (1997) Electrospray sample preparation for improved quantitation in matrix-assisted laser desorption/ionization time-of-flight mass spectrometry. *Rapid Commun. Mass Sp.*, **11** (16), 1785–1793.

45. Axelsson, J., Hoberg, A.M., Waterson, C., Myatt, P., Shield, G.L., Varney, J., Haddleton, D.M., and Derrick, P.J. (1997) Improved reproducibility and increased signal intensity in matrix-assisted laser desorption/ionization as a result of electrospray sample preparation. *Rapid Commun. Mass Sp.*, **11** (2), 209–213.

46. Hanton, S.D., Hyder, I.Z., Stets, J.R., Owens, K.G., Blair, W.R., Guttman, C.M., and Giuseppetti, A.A. (2004) Investigations of electrospray sample deposition for polymer MALDI mass spectrometry. *J. Am. Soc. Mass Spectr.*, **15** (2), 168–179.

47. Lou, X.W. and van Dongen, J.L.J. (2000) Direct sample fraction deposition using electrospray in narrow-bore size-exclusion chromatography/matrix-assisted laser desorption/ionization time-of-flight mass spectrometry for polymer characterization. *J. Mass Spectrom.*, **35** (11), 1308–1312.

48. Wetzel, S.J., Guttman, C.M., and Flynn, K.M. (2004) The influence of electrospray deposition in matrix-assisted laser desorption/ionization mass spectrometry sample preparation for synthetic polymerst. *Rapid Commun. Mass Sp.*, **18** (10), 1139–1146.

49. Wang, Y.X., Zhou, Y., Balgley, B.M., Cooper, J.W., Lee, C.S., and DeVoe, D.L. (2005) Electrospray interfacing of polymer microfluidics to MALDI-MS. *Electrophoresis*, **26** (19), 3631–3640.

50. Hilker, B., Clifford, K.J., Sauter, A.D., Sauter, A.D., Gauthier, T., and Harmon, J.P. (2009) Electric field enhanced sample preparation for synthetic polymer MALDI-TOF mass spectrometry via induction based fluidics (IBF). *Polymer*, **50** (4), 1015–1024.

51. Hilker, B., Clifford, K.J., Sauter, A.D., Sauter, A.D., and Harmon, J.P. (2009) The measurement of charge for induction-based fluidic MALDI dispense event and nanoliter volume verification in real time. *J. Am. Soc. Mass Spectr.*, **20** (6), 1064–1067.

52. Kruger, R.P., Much, H., and Schulz, G. (1994) Determination of functionality and molar-mass distribution of aliphatic polyesters by orthogonal liquid-chromatography. 1. Off-line investigation of poly(1,6-hexanediol adipates). *J. Liq. Chromatogr.*, **17** (14–15), 3069–3090.

53. Garrozzo, D., Impallomeni, G., Spina, E., Sturiale, L., and Zanetti, F. (1995) Matrix-assisted laser-desorption ionization mass-spectrometry of polysaccharides. *Rapid Commun. Mass Spectrom.*, **9** (10), 937–941.

54. Martinez, C.A. and Hay, A.S. (1997) Synthesis of poly(aryl ether) dendrimers using an aryl carbonate and mixtures of metal carbonates and metal hydroxides. *J. Polym. Sci. Pol. Chem.*, **35** (9), 1781–1798.

55. Johnson, B.R., Bartle, K.D., Cocksedge, M., Herod, A.A., and Kandiyoti, R. (1998) A test of the MALDI-MS calibration of SEC with *N*-methyl-2-pyrrolidinone for coal derived materials. *Fuel*, **77** (14), 1527–1531.

56. Nielen, M.W.F. and Malucha, S. (1997) Characterization of polydisperse synthetic polymers by size-exclusion chromatography matrix-assisted laser desorption/ionization time-of-flight mass spectrometry. *Rapid Commun. Mass Spectrom.*, **11** (11), 1194–1204.

57. Montaudo, M.S., Puglisi, C., Samperi, F., and Montaudo, G. (1998) Application of size exclusion chromatography matrix-assisted laser desorption/ionization

time-of-flight to the determination of molecular masses in polydisperse polymers. *Rapid Commun. Mass Sp.*, **12** (9), 519–528.

58 Montaudo, M. (2000) Bivariate distribution in PMMA/PBA copolymers by combined SEC/NMR and SEC/MALDI measurements. *Abstr. Pap. Am. Chem. S.*, **219**, U392–U392.

59 Montaudo, M.S. (2001) Copolymer characterization by SEC-NMR and SEC-MALDI. *Abstr. Pap. Am. Chem. S.*, **221**, U350–U350.

60 Montaudo, M.S. (2002) Full copolymer characterization by SEC-NMR combined with SEC-MALDI. *Polymer*, **43** (5), 1587–1597.

61 Adamus, G., Rizzarelli, P., Montaudo, M.S., Kowalczuk, M., and Montaudo, G. (2006) Matrix-assisted laser desorption/ionization time-of-flight mass spectrometry with size-exclusion chromatographic fractionation for structural characterization of synthetic aliphatic copolyesters. *Rapid Commun. Mass Sp.*, **20** (5), 804–814.

62 Chikh, L., Tessier, M., and Fradet, A. (2008) Polydispersity of hyperbranched polyesters based on 2,2-bis(hydroxymethyl)propanoic acid: SEC/MALDI-TOF MS and C-13 NMR/kinetic-recursive probability analysis. *Macromolecules*, **41** (23), 9044–9050.

63 Mehl, J.T., Murgasova, R., Dong, X., Hercules, D.M., and Nefzger, H. (2000) Characterization of polyether and polyester polyurethane soft blocks using MALDI mass spectrometry. *Anal. Chem.*, **72** (11), 2490–2498.

64 Kona, B., Weidner, S.M., and Friedrich, J.F. (2005) Epoxidation of polydienes investigated by MALDI-TOF mass spectrometry and GPC–MALDI coupling. *Int. J. Polym. Anal. Ch.*, **10** (1–2), 85–108.

65 Bootz, A., Russ, T., Gores, F., Karas, M., and Kreuter, J. (2005) Molecular weights of poly(butyl cyanoacrylate) nanoparticles determined by mass spectrometry and size exclusion chromatography. *Eur. J. Pharm. Biopharm.*, **60** (3), 391–399.

66 Falkenhagen, J., Jancke, H., Kruger, R.P., Rikowski, E., and Schulz, G. (2003) Characterization of silsesquioxanes by size-exclusion chromatography and matrix-assisted laser desorption/ionization time-of-flight mass spectrometry. *Rapid Commun. Mass Sp.*, **17** (4), 285–290.

67 Fandrich, N., Falkenhagen, J., Weidner, S.M., Staal, B., Thuenemann, A.F., and Laschewsky, A. (2010) Characterization of new amphiphilic block copolymers of N-vinylpyrrolidone and vinyl acetate, 2-chromatographic separation and analysis by MALDI-TOF and FT-IR coupling. *Macromol. Chem. Physic.*, **211** (15), 1678–1688.

68 Basile, F., Kassalainen, G.E., and Williams, S.K.R. (2005) Interface for direct and continuous sample-matrix deposition onto a MALDI probe for polymer analysis by thermal field flow fractionation and off-line MALDI-MS. *Anal. Chem.*, **77** (9), 3008–3012.

69 Prokai, L., Aaserud, D.J., and Simonsick, W.J. (1999) Microcolumn size-exclusion chromatography coupled with electrospray ionization mass spectrometry. *J. Chromatogr. A*, **835** (1–2), 121–126.

70 Prokai, L. and Simonsick, W.J. (1993) Electrospray-ionization mass-spectrometry coupled with size-exclusion chromatography. *Rapid Commun. Mass Sp.*, **7** (9), 853–856.

71 Simonsick, W.J. and Prokai, L. (1993) Size-exclusion chromatography with electrospray mass-spectrometric detection. *Abstr. Pap. Am. Chem. S.*, **206**, 205–PMSE.

72 Simonsick, W.J. and Prokai, L. (1995) Size-exclusion chromatography with electrospray mass-spectrometric detection. *Chromatogr. Characterization Polym.*, **247**, 41–56.

73 Nielen, M.W.F. and vandeVen, H.J.F.M. (1996) Characterization of (methoxymethyl)melamine resin by combined chromatographic mass spectrometric techniques. *Rapid Commun. Mass Sp.*, **10** (1), 74–81.

74 Corless, C., Tetler, L.W., Pam, V., and Wood, D. (1994) Proceedings of the 42nd

ASMS Conference on Mass Spectrometry and Allied Topics, p. 515.

75 Simonsick, W.J. (1994) Proceedings of the 42nd ASMS Conference on Mass Spectrometry and Allied Topics, p. 318.

76 OConnor, P.B. and McLafferty, F.W. (1995) Oligomer characterization of 4-23kDa polymers by electrospray Fourier transform mass spectrometry. *J. Am. Chem. Soc.*, **117** (51), 12826–12831.

77 Aaserud, D.J., Prokai, L., and Simonsick, W.J. (1999) Gel permeation chromatography coupled to Fourier transform mass spectrometry for polymer characterization. *Anal. Chem.*, **71** (21), 4793–4799.

78 Adden, R., Melander, C., Brinkmalm, G., Gorton, L., and Mischnick, P. (2006) New approaches to the analysis of enzymatically hydrolyzed methyl cellulose. Part 1. Investigation of the influence of structural parameters on the extent of degradation. *Biomacromolecules*, **7** (5), 1399–1409.

79 Buback, M., Frauendorf, H., Gunzler, F., and Vana, P. (2007) Initiation of radical polymerization by peroxyacetates: polymer end-group analysis by electrospray ionization mass spectrometry. *J. Polym. Sci. Pol. Chem.*, **45** (12), 2453–2467.

80 Liu, X.M., Maziarz, E.P., and Heiler, D.J. (2004) Characterization of implant device materials using size-exclusion chromatography with mass spectrometry and with triple detection. *J. Chromatogr. A*, **1034** (1–2), 125–131.

81 Barner-Kowollik, C., Buback, M., Charleux, B., Coote, M.L., Drache, M., Fukuda, T., Goto, A., Klumperman, B., Lowe, A.B., Mcleary, J.B., Moad, G., Monteiro, M.J., Sanderson, R.D., Tonge, M.P., and Vana, P. (2006) Mechanism and kinetics of dithiobenzoate-mediated RAFT polymerization. I. The current situation. *J. Polym. Sci. Pol. Chem.*, **44** (20), 5809–5831.

82 Buback, M. and Vana, P. (2006) Mechanism of dithiobenzoate-mediated RAFT polymerization: a missing reaction step. *Macromol. Rapid Comm.*, **27** (16), 1299–1305.

83 Feldermann, A., Toy, A.A., Davis, T.P., Stenzel, M.H., and Barner-Kowollik, C. (2005) An in-depth analytical approach to the mechanism of the RAFT process in acrylate free radical polymerizations via coupled size exclusion chromatography-electrospray ionization mass spectrometry (SEC-ESI-MS). *Polymer*, **46** (19), 8448–8457.

84 Gruendling, T., Guilhaus, M., and Barner-Kowollik, C. (2008) Quantitative LC-MS of polymers: determining accurate molecular weight distributions by combined size exclusion chromatography and electrospray mass spectrometry with maximum entropy data processing. *Anal. Chem.*, **80** (18), 6915–6927.

85 Konkolewicz, D., Hawkett, B.S., Gray-Weale, A., and Perrier, S. (2009) RAFT polymerization kinetics: how long are the cross-terminating oligomers? *J. Polym. Sci. Pol. Chem.*, **47** (14), 3455–3466.

86 Szablan, Z., Junkers, T., Koo, S.P.S., Lovestead, T.M., Davis, T.P., Stenzel, M.H., and Barner-Kowollik, C. (2007) Mapping photolysis product radical reactivities via soft ionization mass spectrometry in acrylate, methacrylate, and itaconate systems. *Macromolecules*, **40** (19), 6820–6833.

87 Guenzler, F., Wong, E.H.H., Koo, S.P.S., Junkers, T., and Barner-Kowollik, C. (2009) Quantifying the efficiency of photoinitiation processes in methyl methacrylate free radical polymerization via electrospray ionization mass spectrometry. *Macromolecules*, **42** (5), 1488–1493.

88 Lovestead, T.M., Hart-Smith, G., Davis, T.P., Stenzel, M.H., and Barner-Kowollik, C. (2007) Electrospray ionization mass spectrometry investigation of reversible addition fragmentation chain transfer mediated acrylate polymerizations initiated via (CO)-C-60 gamma-irradiation: Mapping reaction pathways. *Macromolecules*, **40** (12), 4142–4153.

89 Vana, P., Albertin, L., Barner, L., Davis, T.P., and Barner-Kowollik, C. (2002) Reversible addition-fragmentation chain-transfer polymerization: Unambiguous end-group assignment via electrospray ionization mass spectrometry. *J. Polym. Sci. Pol. Chem.*, **40** (22), 4032–4037.

90 Pasch, H., and Rode, K. (1995) Use of matrix-assisted laser-desorption/ionization mass-spectrometry for molar mass-sensitive detection in liquid-chromatography of polymers. *J. Chromatogr. A*, **699** (1–2), 21–29.

91 Adrian, J., Braun, D., Rode, K., and Pasch, H. (1999) New analytical methods for epoxy resins, 1 – liquid chromatography at the critical point of adsorption and MALDI-TOF mass spectrometry. *Angew. Makromol. Chem.*, **267**, 73–81.

92 Kruger, R.P., Much, H., Schulz, G., and Wachsen, O. (1996) New aspects of determination of polymer heterogeneity by 2-dimensional orthogonal liquid chromatography and MALDI-TOF-MS. *Macromol. Symp.*, **110**, 155–176.

93 Wachsen, O., Reichert, K.H., Kruger, R.P., Much, H., and Schulz, G. (1997) Thermal decomposition of biodegradable polyesters. 3. Studies on the mechanisms of thermal degradation of oligo-L-lactide using SEC, LACCC and MALDI-TOF-MS. *Polym. Degrad. Stabil.*, **55** (2), 225–231.

94 Pasch, H., and Schrepp, W. (2003) *MALDI-TOF mass spectrometry of synthetic polymers*, Springer, Berlin, Heidelberg.

95 Malik, M.I., Trathnigg, B., and Saf, R. (2009) Characterization of ethylene oxide-propylene oxide block copolymers by combination of different chromatographic techniques and matrix-assisted laser desorption ionization time-of-flight mass spectroscopy. *J. Chromatogr. A*, **1216** (38), 6627–6635.

96 Huang, L.H., Gough, P.C., and DeFelippis, M.R. (2009) Characterization of poly(ethylene glycol) and PEGylated products by LC/MS with postcolumn addition of amines. *Anal. Chem.*, **81** (2), 567–577.

97 Himmelsbach, M., Buchberger, W., and Reingruber, E. (2009) Determination of polymer additives by liquid chromatography coupled with mass spectrometry. A comparison of atmospheric pressure photoionization (APPI), atmospheric pressure chemical ionization (APCI), and electrospray ionization (ESI). *Polym. Degrad. Stabil.*, **94** (8), 1213–1219.

98 Weidner, S.M., Falkenhagen, J., Much, H., and Kruger, R.P. (2000) Determination of chemical and molecular-weight distributions of heterogeneous polymers by means of MALDI-MS coupled with chromatography. *Abstr. Pap. Am. Chem. S.*, **219**, 242–POLY.

99 Barman, B.N., Champion, D.H., and Sjoberg, S.L. (2009) Identification and quantification of polyethylene glycol types in polyethylene glycol methyl ether and polyethylene glycol vinyl ether. *J. Chrom. A*, **1216** (40), 6816–6823.

100 Watanabe, T., Nakanishi, K., Ozawa, T., Kawasaki, H., Ute, K., and Arakawa, R. (2010) Semi-online nanoflow liquid chromatography/matrix-assisted laser desorption ionization mass spectrometry of synthetic polymers using an octadecylsilyl-modified monolithic silica capillary column. *Rapid Commun. Mass Sp.*, **24** (13), 1835–1841.

101 Schneider, C., Sablier, M., and Desmazières, B. (2008) Characterization by mass spectrometry of an unknown polysiloxane sample used under uncontrolled medical conditions for cosmetic surgery. *Rapid Commun. Mass Spec.*, **22** (21), 3353–3361.

102 Mengerink, Y., Peters, R., deKoster, C.G., van der Wal, S., Claessen, H.A., and Cramers, C.A. (2001) Separation and quantification of the linear and cyclic structures of polyamide-6 at the critical point of adsorption. *J. Chromatogr. A*, **914** (1–2), 131–145.

103 Weidner, S.M., Just, U., Wittke, W., Rittig, F., Gruber, F., and Friedrich, J.F. (2004) Analysis of modified polyamide 6.6 using coupled liquid chromatography and MALDI-TOF-mass spectrometry. *Int. J. Mass Spectrom.*, **238** (3), 235–244.

104 Julka, S., Cortes, H., Harfmann, R., Bell, B., Schweizer-Theobaldt, A., Pursch, M., Mondello, L., Maynard, S., and West, D. (2009) Quantitative characterization of solid epoxy resins using comprehensive two dimensional liquid chromatography coupled with electrospray ionization-time of flight mass spectrometry. *Anal. Chem.*, **81** (11), 4271–4279.

105 Weidner, S.M., Falkenhagen, J., Maltsev, S., Sauerland, V., and Rinken, M. (2007) A novel software tool for copolymer characterization by coupling of liquid chromatography with matrix-assisted laser desorption/ionization time-of-flight mass spectrometry. *Rapid Commun. Mass Spectrom.*, **21** (16), 2750–2758.

106 Weidner, S., Falkenhagen, J., Krueger, R.P., and Just, U. (2007) Principle of two-dimensional characterization of copolymers. *Anal. Chem.*, **79** (13), 4814–4819.

107 van Leeuwen, S.M., Tan, B.H., Grijpma, D.W., Fejen, J., and Karst, U. (2007) Characterization of the chemical composition of a block copolymer by liquid chromatography/mass spectrometry using atmospheric pressure chemical ionization and electrospray ionization. *Rapid Commun. Mass Sp.*, **21** (16), 2629–2637.

108 Guillaneuf, Y., Dufils, P.E., Autissier, L., Rollet, M., Gigmes, D., and Bertin, D. (2010) Radical chain end chemical transformation of SG1-based polystyrenes. *Macromolecules*, **43** (1), 91–100.

109 Weidner, S.M., Falkenhagen, J., Knop, K., and Thünemann, A. (2009) Structure and end-group analysis of complex hexanediol-neopentylglycol-adipic acid copolyesters by matrix-assisted laser desorption/ionization collision-induced dissociation tandem mass spectrometry. *Rapid Commun. Mass Spec.*, **23** (17), 2768–2774.

110 Maciejczek, A., Mass, V., Rode, K., and Pasch, H. (2010) Analysis of poly (ethylene oxide)-*b*-poly(propylene oxide) block copolymers by MALDI-TOF mass spectrometry using collision induced dissociation. *Polymer*, **51** (26), 6140–6150.

111 Girod, M., Phan, T.N.T., and Charles, L. (2008) On-line coupling of liquid chromatography at critical conditions with electrospray ionization tandem mass spectrometry for the characterization of a nitroxide-mediated poly(ethylene oxide)/polystyrene block copolymer. *Rapid Commun. Mass Spec.*, **22** (23), 3767–3775.

112 Swartz, M.E. (2005) UPLC (TM): an introduction and review. *J. Liq. Chromatogr. R. T.*, **28** (7–8), 1253–1263.

113 Pursch, M., Schweizer-Theobaldt, A., Cortes, H., Gratzfeld-Huesgen, A., Schulenberg-Schell, H., and Hoffmann, B.W. (2008) Fast, ultra-fast and high-resolution LC for separation of small molecules, oligomers and polymers. *Lc. Gc. Eur.*, **21** (3), 152.

114 Falkenhagen, J., and Weidner, S. (2009) *Determ Crit. Cond. Adsorp. Chrom. Polym.*, **81** (1), 282–287.

115 Malik, M.I., Trathnigg, B., Bartl, K., and Saf, R. (2010) Characterization of polyoxyalkylene block copolymers by combination of different chromatographic techniques and MALDI-TOF-MS. *Anal. Chim. Acta*, **658** (2), 217–224.

116 Matsumoto, H., Kawai, T., and Teramachi, S. (2007) MALDI-TOFMS characterization of polystyrene oligomer fractionated by temperature gradient interaction chromatography. *Kobunshi Ronbunshu*, **64** (11), 735–739.

8
Automated Data Processing and Quantification in Polymer Mass Spectrometry

*Till Gruendling, William E. Wallace, Christopher Barner-Kowollik,
Charles M. Guttman, and Anthony J. Kearsley*

8.1
Introduction

The interpretation of synthetic polymer mass spectra is a process that usually requires intricate knowledge of both, the synthetic chemistry of the investigated polymers as well as the measurement process itself and the instrumentation at hand. The growing usage of mass spectrometric tools in the polymer community also leads to an increasing number of nonexpert users of the technique. Sophisticated tools for the automated processing and interpretation of mass spectra have the potential to significantly increase the acceptance of mass spectrometry (MS) as a versatile technique in polymer characterization by these users, while at the same time providing an operator-independent and reproducible outcome of the interpretation process. Especially when extracting quantitative information – for example on the compositional distribution of copolymers or the molecular mass distribution (MMD) of homopolymers – operator-independent approaches for spectral evaluation are highly desirable. Sophisticated data processing and database search tools have the potential to maximize the efficiency and accuracy of investigations by hyphenated techniques, including tandem mass spectrometry (MS/MS), ion mobility spectrometry (IMS), and online liquid chromatography mass spectrometry (LC-MS). The following sections give an overview of the techniques that are currently available to the polymer community and should guide the interested scientist in the selection of suitable tools for spectral interpretation. The reader will realize that – although tools have matured in areas such as copolymer characterization and molecular mass determination – the field of automation in polymer MS is in many areas still in its infancy.

8.2
File and Data Formats

Once the physical process of spectrum acquisition has been carried out, every data analysis effort – automated or manual – requires importing of the mass spectral data

Mass Spectrometry in Polymer Chemistry, First Edition.
Edited by Christopher Barner-Kowollik, Till Gruendling, Jana Falkenhagen, and Steffen Weidner.
© 2012 Wiley-VCH Verlag GmbH & Co. KGaA. Published 2012 by Wiley-VCH Verlag GmbH & Co. KGaA.

into a processing software suite. Here, the generated spectra are either displayed for manual interpretation by the user or further automated data retrieval, preprocessing, interpretation, and quantification steps may be performed. The current situation features a plethora of vendor-specific, proprietary data formats in MS. At times, diversity exists even between instrument generations of the same vendor, and the negative impact on scientific data exchange and efforts to automate spectral interpretation needs to be realized. Polymer mass spectra acquired by direct infusion electrospray ionization mass spectrometry (ESI-MS) or matrix-assisted laser desorption ionization mass spectrometry (MALDI-MS) without chromatographic fractionation of the sample can be stored as a two-column data matrix, and thus are easily handled in the form of simple text files for data export, which is supported by most software.

However, in the last years, a trend is also witnessed within the polymer MS community toward the realization of increasingly more sophisticated scenarios, featuring mass spectrometric analysis in combination with other means of macromolecular separation or fragmentation to maximize information content. Hyphenated techniques such as size-exclusion chromatography (SEC) or chromatography under critical conditions of adsorption (LCCC) coupled online or offline to MS as well as gas phase separation techniques such as IMS may allow more structural data to be gained from the measurements than from a simple MS experiment by itself. Tandem MS in combination with knowledge about the fragmentation pathways of certain polymer species may enhance the information gained on polymer functionality and composition. The multidimensional data obtained from these approaches requires sophisticated means for its compact storage, ideally in openly accessible data formats. In the field of proteomics, the sheer complexity of information from tandem LC/MS^2 experiments necessitates automated processing from the very beginning of the data analysis chain and sophisticated tools have therefore been developed over the last couple of years [2–4]. The requirement of a common open data format to enable the platform and instrument vendor-independent processing and exchange of data has also been realized in recent years within the proteomics community [1, 5, 6]. As a result, two open data formats have emerged, based on the extensible markup language (XML) standard: The mzXML [1] format, developed by the Seattle Proteome Center and later the mzML [5] format developed by the Human Proteome Organization's proteomics standards initiative, which aims to marry the superior elements of mzXML with a third data format, mzData [7]. Figure 8.1 shows the great benefit of this approach in that a common (open source) data analysis pipeline may be used to data analysis which uses one unified open file format as data input. Software developers will be freed from the need to obtain knowledge about the different vendor-specific formats thus enabling the development of truly universal software tools. Furthermore, common file standards will allow public storage and retrieval of data over long periods. A number of open source programs are available today to convert vendor-specific formats to mzXML or mzML files. A list of these converters can be found in Ref. [3]. Increasingly, instrument vendors are also beginning to support these open file standards.

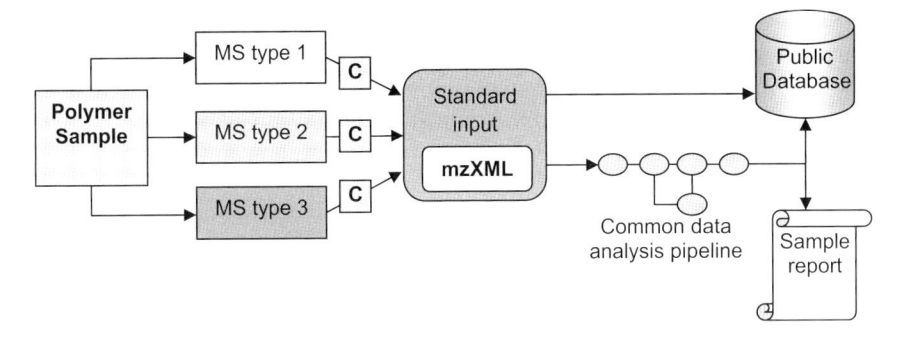

Figure 8.1 The mzXML file acts as a mediator, allowing multiple input formats to be subjected to a common data analysis pipeline. New types of instruments can be integrated into a preexisting analysis framework with only a utility (here represented by C) to convert MS native output to the mzXML format. The open structure of mzXML instance documents makes them suitable for data exchange such that, for example, they may be submitted to a data repository to support the results presented in a publication. Figure adapted from Ref. [1].

8.3 Optimization of Ionization Conditions

Finding the instrumental settings and experimental conditions at which the sensitivity and accuracy of an analytical procedure is at its optimum is one of the main goals in any analytical method development process. In MS, a great number of these chemical and instrument parameters exist that the scientist can tweak. With regards to polymer analysis by MS, depending on what the goal of the mass spectrometric experiment is, the objective to be optimized may vary. The instrumental noise and sensitivity may be important objectives in quantification studies of low abundant polymer species. Minimization of ionization mass bias and optimization of detector linearity or dynamic range are required in MMD determination by MALDI-MS [8]. Recently, the introduction of living/controlled radical polymerization protocols has lead to new kinds of functional polymers carrying end groups such as halogen atoms, dithioesters, or nitroxides which are bound to the polymer terminus by intrinsically weak carbon–chalcogen or carbon–halogen bonds. Especially with nonpolar polymers such as polystyrene [9, 10], which are difficult to ionize in general, but also with poly(methyl methacrylate) (PMMA) [11, 12], loss of these end groups is often observed in MALDI-MS. This leads to information loss and false deductions in the case of mechanistic investigations or when MS is used to verify end-group fidelity. ESI has the advantage that it often provides a much softer ionization of the polymer molecule, with full retention of functionality [12, 13], but in a few cases, end-group losses for polymers synthesized by living/controlled radical polymerization have been observed even with ESI-MS [10]. The retention of functionality is therefore a third important objective in polymer MS requiring optimization, especially when mechanistic and structural studies are to be performed.

The physical processes affecting the performance of the ionization source, mass separating process, and ion detection are often only insufficiently understood and conditions depend critically on the type of mass spectrometer employed. With only little or no *a priori* knowledge of the optimal conditions, the number of parameter settings to be sampled is very large. Often, source optimization in both MALDI-TOF and ESI-MS is performed in a one-factor-at-a-time fashion [14]. This approach, although straightforward to perform, may not yield the best experimental conditions, as interactions between parameters cannot be identified [15, 16]. Design of experiment (DoE) is a useful tool that can be employed to significantly reduce the number of the experiments required in optimizing ionization conditions, while retaining maximum certainty in the effects of the experimental parameters and their interactions on arbitrary objectives to be optimized. Response surface designs allow the statistical and graphical evaluation of the experimental data by regression analysis with suitable model functions [15, 17, 18]. A number of different applications of DoE to LC-MS optimization exist in literature, many covering the optimization of the liquid chromatographic separation [19, 20], but some also covering the optimization of MS source conditions [18, 20, 21]. More recently, Kell and coworkers successfully employed a genetic search method [22] to achieve an operator intervention-free, fully automated numerical optimization of up to 14 instrument settings in polypeptide ESI-MS [23] and gas chromatography-MS [24]. These authors noted that the method could yield optimum conditions by sampling less than 500 of the possible 10^{14} combinations and that relationships between source parameters were identified that accounted for much of the success of the optimization. The hypothesis-generating potential of genetic search processes in which little *a priori* knowledge of the system is available was thus demonstrated.

A selection of articles exist that are directly related to polymer MS. Wetzel *et al.*, for example, employed an orthogonal experimental design to identify parameters that significantly affected signal-to-noise ratio in polystyrene analysis by MALDI-TOF-MS. From a set of five parameters including detector voltage, laser energy, delay time, extraction voltage, and lens voltage, detector voltage and delay time were shown to be the most influential [25]. Later, Wallace *et al.* employed numerical optimization routines to find conditions of minimal instrumental mass bias, which is one goal when employing MALDI-MS to generate absolute MMD standards. Stochastic numerical optimization [26, 27] was employed to this task, and the effects of instrumental noise on the optimization procedure were dealt with by the use of implicit filtering [28]. Optimal values of five instrument parameters were obtained in as few as five iterations and the confidence intervals of the parameters were gained which may serve for a sensitivity analysis of the effects of each parameter.

DoE can be especially useful when optimizing online LC-MS of synthetic polymers. Here, the operator is faced with the challenge of having to find optimum ionization conditions in a system where the concentration of analytes eluting from a chromatographic column is changing rapidly as a function of time. In such cases, parameters may need to be varied between chromatographic runs with the goal to obtain maximum information from a minimum amount of chromatographic runs to save valuable instrument time. Gruendling *et al.* presented a method based on a

D-optimal design, which allows for a modification of the number of experiments included in the design plan [29]. The influence of four ionization source parameters, including cone voltage, spray gas flow rate, and capillary temperature on ionization efficiency and their optimum settings were identified.

In MALDI-MS, the selection of proper conditions for sample preparation including the correct chemical matrix, solvents, and ionization salt in suitable concentrations is a crucial part of the signal optimization process. A number of studies exist in this direction, with one very interesting approach by Schubert and coworkers [30]. These authors used quantitative structure–performance relationships for the rational selection of potentially new well-performing matrices for MALDI of synthetic polymers. Recently, Brandt *et al.* [31–33] employed partial least square regression together with a training set of eight matrices, five cationization reagents and six solvents to predict the performance of untested combinations of matrix, cationization reagent and solvent. Molecular descriptors were used for the matrix and cationization reagent, while Hansen solubility parameters were found to be the most informative for the solvent. The authors concluded that, despite of inconsistencies due to the formation of precipitates with some salts, the established structure–performance relationships may serve as a starting point to predict the performance of matrices in the case of unknown polymers.

8.4
Automated Spectral Analysis and Data Reduction in MS[1]

Numerical spectrum analysis is an often neglected subject in the overall study of MS. It is typically treated as an afterthought to the widely studied subjects of sample preparation, ionization mechanisms, and mass-to-charge separation methods. Yet it is the determination of accurate and precise peak positions that is at the core of chemical identification in MS. Furthermore, it is the determination of peak intensity that underpins any quantitative measurements. This section presents, in brief, two new methods that have been developed at the National Institute of Standards and Technology. These methods make no assumptions about peak shape.

Mass spectral "peaks" are defined as statistically significant excursions in the spectrum intensity from its baseline that are the result of ions of a given mass-to-charge ratio (m/z) being detected by the instrument. Spurious peaks may arise from purely random events of either electronic or chemical origin. Electronic noise arises from the detector, preamplifier, amplifier, or spectrum digitizer. Chemical noise arises from stray ions that have been improperly separated in time, mass, or kinetic energy. Typically spectrum averaging will smooth out such peaks if the noise is truly random and uncorrelated. Spurious peaks may also arise from systematic instrument artifacts, for example, periodic effects such as digitizer jitter (yielding electronic noise) or voltage fluctuations (leading to chemical noise). From a purely

---

[1] Official contribution of the National Institute of Standards and Technology; not subject to copyright in the United States of America.

statistical or numerical point of view, these may be impossible to distinguish from genuine peaks. The analyst needs answers to the following questions:

1) When is a given excursion correctly classified as a genuine peak? (statistical significance)
2) At what m/z is the peak most likely located? (peak location)
3) Does it overlap with other nearby peaks? (peak resolution)
4) Where does a peak begin and end? (integration end points)
5) What is the area of the spectrum underneath the peak? (peak integration).

An answer to the first question is used to separate true peaks from spurious peaks. An answer to the second question is required for species identification and is used predominantly in qualitative analysis. The third question must be answered to determine if two or more peaks overlap as a result of insufficient mass-to-charge resolution. Overlapping peaks may lead to incorrect peak position and intensity determination. Knowledge of the location of the peak beginning and end, the fourth question, is required to determine peak area. Peak area, in turn, is typically required for quantitative analytical results. Succinct answers to these questions will result in a reliable translation between the spectrum and the metrics the analyst wishes to determine. Failure to properly answer these questions renders moot efforts at sample preparation and data collection.

8.4.1
Long-Standing Approaches

Standard approaches to the reduction of mass spectral data have focused on calculating either derivatives or intensity thresholds of the data. A few of the many reviews in the literature can be found in Refs. [34–37]. Typically, excursions from the baseline are found at increases in the first derivative. As the algorithm proceeds sequentially through the data (typically but not necessarily from low m/z to high m/z), an initial excursion of the derivative, or an increase in intensity above a preset threshold, indicates a peak beginning. A peak maximum is found when the derivative after an initial increase flattens out to zero. As the algorithm proceeds sequentially, the derivative will change sign and then flatten out to zero again, or the intensity will drop below the preset threshold value, as the baseline is restored.

Many variations of this basic method exist. For example, second derivatives may be used to find peak maxima. In some cases, third derivatives may also be employed. There are two significant problems that one encounters when using these derivative-based approaches. First, the function whose derivative must be approximated is only available at discrete prescribed points, that is, one has access only to (x, y) pairs of data, not to a continuous function. Second, random noise results in inaccurate derivative estimates. It is well known that the availability and accuracy of derivative approximations decreases as noise in a function increases. The result is that noisy data, when analyzed with algorithms that employ derivative approximations, may fail to find genuine peaks and may identify as peaks features that are purely artifacts. Furthermore, the higher the derivative, the greater its sensitivity to random noise [38].

In this case, smoothing or filtering of the data is one way to ensure existence and computability of needed derivative estimates. Running or windowed averages, Savitsky-Golay smoothing [39–41], Fourier filtering [42], and wavelet decomposition [43] are the most common of the many methods possible and have been extensively discussed in the literature. However, the success of these methods relies on a circular logic in which the type and degree of smoothing determine the effectiveness of the peak finding algorithm, and the effectiveness of the peak finding algorithm determines the amount of smoothing required. The problem is compounded when the noise is variable across the m/z range, or when the noise is not constant between spectra but the analyst wishes to apply to same data analysis methods to all spectra of a series. Different kinds of, or degrees of, smoothing may be required in different parts of the spectrum. Likewise, derivative computation (or other gradient estimates) may be more feasible in one part of the spectrum than another.

8.4.2
Some New Concepts

Many of the new concepts in peak identification and integration attempt to move beyond the purely local approaches of derivatives or thresholds. Furthermore, they attempt to do this without going toward global spectrum smoothing. Here "local" refers to operations on any given mass versus intensity (x, y) data point and its nearest neighbors. "Global" refers to operating on the spectrum as a whole without consideration of any specific local features, such as Fourier filtering. New methods endeavor to treat the spectrum as a series of regions that are larger than a few data points but smaller than the spectrum as a whole. They attempt to isolate peaks into "neighborhoods" or small sets of mass versus intensity data points. By analogy, the spectrum is the city, the peaks are its neighborhoods, and the data points are individual addresses.

8.4.3
Mass Autocorrelation

Signal autocorrelation has an extensive history in the communications field [44]. The mass autocorrelation function, $G(L)$ is defined as

$$G(L) = \frac{\sum_i S(m_i) \cdot S(m_{i+L})}{\sum_i S(m_i)^2} \tag{8.1}$$

where $S(m_i)$ is the signal at mass m_i taken on equal intervals of mass, Δm, and L is the lag which is also measured in units of mass. Equal intervals of mass are used because most correlation algorithms require the signal to be evenly spaced points on the scale of interest. As an aside, remember that in TOF mass separation, the signal, $S(t_i)$, is collected on equal intervals of time. The transformation from this time-base signal

$S(t_i)$ to a mass-base signal $S(m_i)$ involves both an interpolation and a change of the signal itself by a Jacobian transform. The mathematical methods to effect this transformation are discussed in Ref. [45]. Numerical interpolation must be used to convert the spectrum in mass from unequally spaced points to equally spaced points for the application of autocorrelation methods. By choosing reduced (contiguous) sets of the data, information about periodic peaks may be obtained in as local or as global a context as desired. The periodicity of the peaks may arise from the periodic nature of the polymer's structure such as repeat units periodicity or isotopic periodicity.

In addition to verifying the mass of any repeat units found in the sample, there are several important applications of mass autocorrelation to the analysis of polymer mass spectra. By autocorrelating in different regions of the data, and by overlaying these results, subtle changes in polymer architecture can be discovered. This was demonstrated for polysilsesquioxanes where the degree of intermolecular condensation could be quickly and accurately tracked without having to resort to identifying every peak in the spectrum [46, 47]. A second application involves pulling a weak signal out of noisy data [45]. Exploiting the fact that for a polymer there should be a repeating peak sequence at the repeat unit mass, autocorrelation can reveal if the expected polymer ions have been detected in an otherwise noisy spectrum. Figure 8.2 shows a very noisy polystyrene MALDI-TOF spectrum. Identifying the mass difference between pairs of peaks is difficult. Autocorrelation compares intensities at all

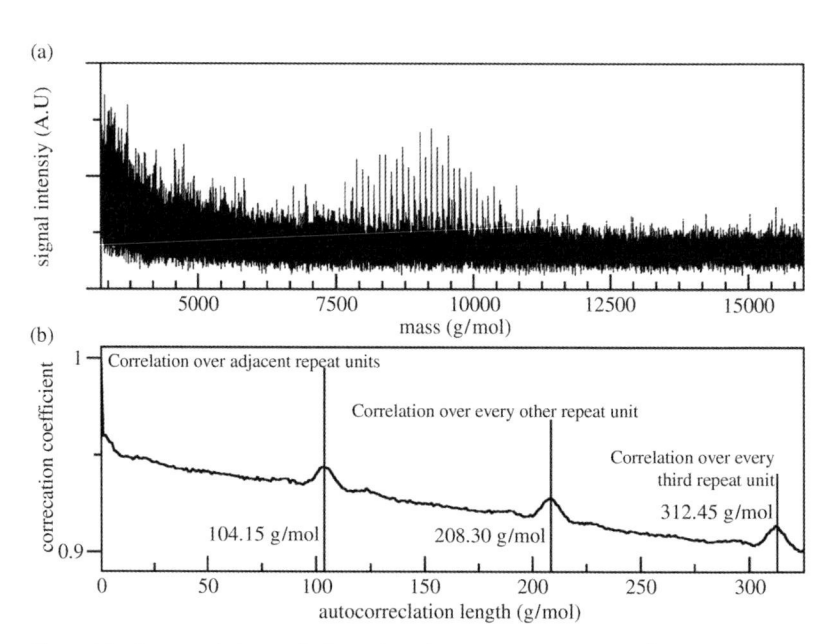

Figure 8.2 Autocorrelation applied to noisy polymer mass spectrometry data. (a) Mass spectrum. (b) Autocorrelation function. Notice the clearly repeating structure at 104.15 g mol^{-1} which is the repeat unit mass of polystyrene.

mass differences across the spectrum. In cases with matching peak distances the autocorrelation coefficient increases. This can be seen in the figure where correlations of peaks one, two, or three repeat unit masses apart are clearly seen and the repeat unit of polystyrene is clearly identified. Careful inspection of Figure 8.2 also shows small peaks on either side of the main autocorrelation series. These are due to either a separate set of end groups or adduct formation. In either case, finding this effect by simple inspection of the original spectrum would be exceedingly difficult. The third application is to mass calibration. If the repeat unit mass of the sample is known, autocorrelation can be used to adjust the slope (but not the offset) of the calibration curve. This serves to improve mass accuracy because if the slope can be corrected, then the peak positions are more accurate. This is important when calculating end group or adduct masses.

A software tool (PolyCalc) which has recently been introduced by Luftmann and Kehr allows the molecular masses of repeating units and end groups, as well as an approximation of the MMD to be obtained from the multiply-charged spectra recorded in direct infusion ESI-MS [48]. The software operates by minimizing the difference between a simulated mass spectrum and the measured ESI spectrum and thus yields the monomer and end-group mass and estimates of the MMD of the polymer, which is assumed to be Gaussian in shape. The software seems to be a useful tool effectively extending the mass range of ESI-MS to around $10-20\,\text{kg mol}^{-1}$ if multiple charging is achieved. Programmatic extensions are needed to allow the analysis of mixtures of multiple end-group-carrying polymers.

8.4.4
Time-Series Segmentation

Another alternative to calculating local derivatives is to consider the spectrum as a whole and to reduce it to a set of concatenated line segments based on its features. As shown in Figure 8.3, by connecting the first (x, y) pair to the last (x, y) pair in the spectrum, a crude baseline for the entire spectrum is created. From this line, the (x, y) pair that is the greatest normal distance from the line is determined. This yields two line segments spanning the spectrum. This procedure is continued until the spectrum is replicated by a series of line segments with each peak determined (at the minimum) by two line segments and the intervening baseline determined (also at a minimum) by a single line segment. After the spectrum has been segmented, least squares or orthogonal distance regression [49] may be used to adjust the line segments to best fit the data; however, caution must be exercised because if the random noise level varies across the spectrum the quality of the fit will also vary across the spectrum. For this reason, the NIST method [50] uses a background spectrum taken at the same instrumental conditions as the spectrum to be analyzed with a sample that is free of analyte (e.g., in the case of MALDI, contains the matrix and the cationizing salt). A background spectrum requires additional experimental effort but yields significant dividends when analyzing the data to determine quantitative measures.

A nonlinear programming algorithm using an L2 (least squares) approximation to an L1 (least absolute-value) fit was employed [51–55]. L1 fits are superior to L2 fits due

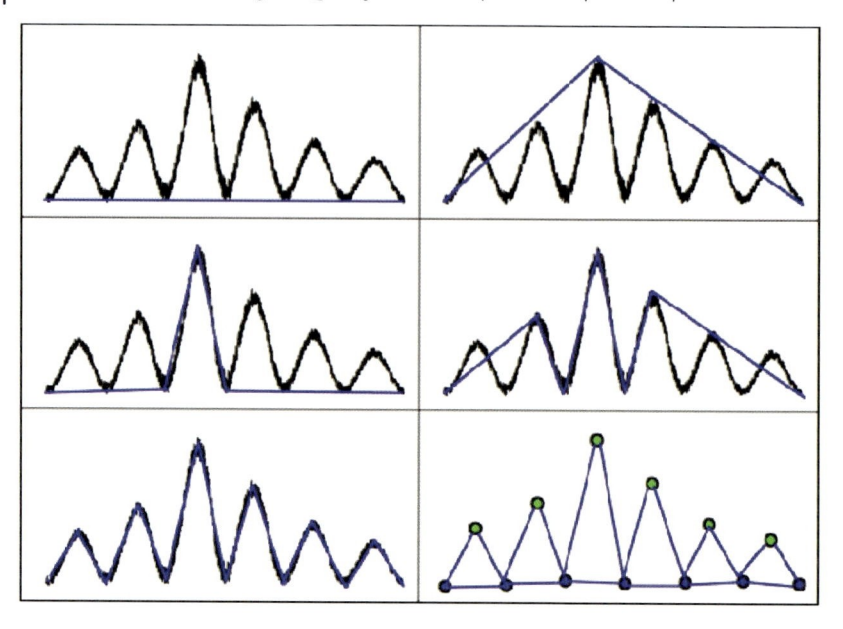

Figure 8.3 Schematic representation of time-series segmentation on a model problem. In the final panel, the green circles represent peak positions, and the blue circles represent peak beginnings/endings. The calculated relative area for each peak is the area of the triangle but many other area-summing routines are possible.

to their increased tolerance for outliers, that is, outlying points do not exert as much control over the final fit. Given a dataset of N points, a collection of strategic points is found and the unique optimal piecewise linear function passes through the x coordinate of each strategic point. This defines a set of function maxima and minima corresponding to the peak maxima and the peak limits, respectively. The original data is then integrated by finding the area of the polygon determined by the strategic points.

Our segmentation method is a two-step algorithm. The first portion requires the selection of strategic points and is derived from the earlier work of Douglas and Peucker [56]. Strategic points are selected based on an iterative procedure that identifies points whose orthogonal distance from the end-point connecting line segment is the greatest. Once a point with greatest orthogonal distance from the mean has been identified, it joins the collection of strategic points and, in turn, becomes an end point for two new line segments from which a point with greatest orthogonal distance is found. This numerical scheme is performed until the greatest orthogonal distance to any end-point connecting line segment drops beneath a prescribed threshold value. This threshold value is the only algorithmic parameter and is based on a statistical analysis of the data and its corresponding analyte-free spectrum. Clearly the selection of these points does not require equally spaced data; therefore, the method is equally well suited for TOF data expressed in either time or mass space. Generally, it is chosen to work in time space with the data in its most basic state and to eliminate for doing a point-by-point correction of intensity using partial integrals [45]. The second phase of the algorithm, developed specifically for this work, requires the solution of an optimization problem, specifically, locating strategic point heights (i.e., adjusting strategic point y-axis values at their associated strategic x-axis values) that minimize the sum of orthogonal distance from raw data. This problem is a nonlinear (and nonquadratic) optimization problem that can be accomplished quickly using a recently developed nonlinear programming algorithm [57].

The algorithm works as shown in Figure 8.3 [50, 54]. Clearly this method requires no knowledge peak shape and no preprocessing of the data (e.g., smoothing), nor does it require equal spacing of data points. Note that the strategic points defining the beginning and end of adjacent peaks are located in the same spot resulting from the choice of $\cos^2(x)$ as the underlying function for this demonstration of the procedure.

Once the data set is fully segmented, strategic points are discarded in accordance with the statistical analysis of the original data set and its corresponding analyte-free data set. This "deflation" of strategic points using statistics-derived thresholds is performed by first analyzing the analyte-free spectrum for peaks and peak areas. Once a collection of peaks and peak areas has been accumulated, the spectrum with sample is then analyzed. Each peak identified from the spectrum with analyte is compared to peaks found in close relative proximity from the analyte-free spectrum algorithm output (i.e., peaks that appear with similar time or mass coordinates). If any peak in the spectrum with analyte has a smaller peak height or smaller peak area than most (about 95%) of the background-spectrum peaks in close proximity, then that peak is ignored. Likewise, any peak that falls outside the statistically significant measure for area and height is also discarded. Thus, no peak is identified from the sample spectrum that could have been identified by height or by area from the background spectrum. This discarding of strategic points also serves to prevent the inadvertent subdivision of larger peaks into a set of smaller peaks. This can sometimes occur if the noise in the analyte spectrum is much greater than the noise in the corresponding background spectrum.

Once the final set of strategic points has been found, the area of the polygon defined by these points is calculated. (The polygon is often, but not always, a triangle. The algorithm will work on polygons of any number of vertices connected by line segments.) The line connecting the first and last strategic points for a given peak determines a "local baseline." The mathematical basis for the polygonal area calculation algorithm is Green's theorem in the plane and can be interpreted as repeated application of the trapezoidal rule for integration [58]. The method returns the exact area of the polygon.

Figure 8.4 shows an example of a MALDI-TOF mass spectrum of polystyrene having three different end groups. Without user intervention, but with the requirement of a background spectrum for statistical deflation of the number of peaks, the algorithm is able to identify and integrate peaks without smoothing or making any assumptions on peak shape. In this case, the areas calculated from the triangular shapes defined by the three strategic points for each peak were calculated. However,

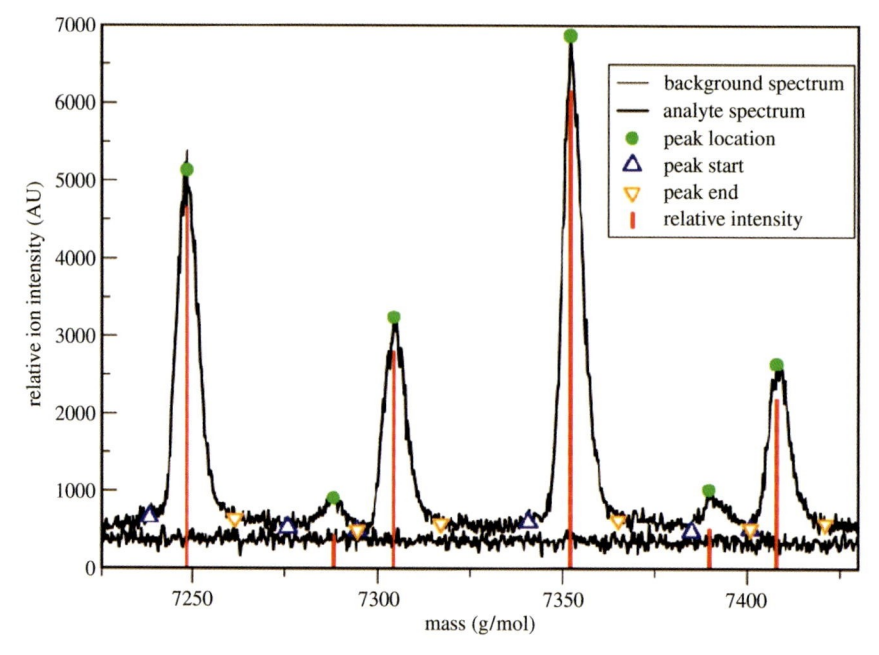

Figure 8.4 Time-series segmentation applied to real polymer MALDI-TOF mass spectral data. Note that the background spectrum has been shifted down slightly for clarity.

any method to determine the area is suitable, for example, simply summing the signal channels between the beginning and ending strategic points defining each peak.

8.5
Copolymer Analysis

The practical details and applications of soft ionization MS to analyze copolymer structure have been discussed in a preceding chapter and shall not be reiterated here. MS can serve two purposes in copolymer characterization; in addition to a determination of the end-group structure, it is possible to glean the copolymer composition from a mass spectrum (a sequence determination may be attempted after partial degradation or via MS/MS). The last two decades have seen the development of a number of mathematical approaches and software tools aimed at a spectral interpretation of copolymer mass spectra (consider also Chapter 9.9 for an alternative viewpoint on the topic). Mass spectra of copolymers are significantly more complex than homopolymer spectra. This is owing to the fact that where the MMD of homopolymers is a one-dimensional function of one repeat unit length, copolymers feature a two-dimensional topology distribution of the chain lengths of two (or multiple) monomer-building blocks.

The monoisotopic m/z of a copolymer ion in charge state z featuring m units of monomer M1 with mass m_{M1} and n repeat units of monomer M2 with mass m_{M2} and with a combined end-group mass m_E is given by the following equation:

$$m/z = \frac{m \cdot m_{M1} + n \cdot m_{M2} + m_E}{z} + m_{M+} \tag{8.2}$$

where m_{M+} is the mass of the adduct metal cation.

Spectral interpretation can be attempted in a generally very straightforward manner: If a hypothesis about the constituent monomers and the end group of the polymer can be made, the resultant copolymer spectrum can be modeled based solely on Eq. (8.2). A problem, however, arises, as the two-dimensional topology distribution is projected onto a one-dimensional mass spectrum, which in most cases leads to loss of information. This is due to the fact that different combinations of m and n can lead to the same or very similar m/z. Consider the example of $m_{M1} = 100$ Da and $m_{M2} = 40$ Da: A polymer with constitution $m = 4$ and $n = 10$ will feature the same mass-to-charge as one with $m = 6$ and $n = 5$. Further complication arises, as each copolymer features a distribution of masses due to the isotope distribution of ^{13}C atoms making up the backbone of the polymer and the limited instrumental resolution. Nevertheless, in the majority of cases, quantitative data may be extracted on the copolymer composition and topology distribution, given that the effects of mass bias on the ionization of copolymers with differing composition can be neglected.

Mathematical tools developed from the early 1990s for the quantitative interpretation of copolymer spectra have been accounted for in an extensive review by Montaudo [59]. This early work focused mainly on the application of chain-statistical models of the copolymerization process. The use of chain statistics to simulate a theoretical copolymer topology distribution can aid spectral interpretation by comparison of a model spectrum with the spectrum measured in reality. It was shown that using MALDI-MS with appropriate chain models, the average monomer composition, c of copolymers could be determined [60–63]. An evaluation by a direct method in which no assumptions about the polymerization process are required is also possible using the following equation [60, 64–66]:

$$c_{M1} = \frac{\sum_m \sum_n m \cdot (I_{m,n})}{\sum_m \sum_n m \cdot n \cdot (I_{m,n})} \tag{8.3}$$

where $I_{m,n}$ is the mass spectral intensity at the mass corresponding to a co-oligomer with m and n repeat units of the respective comonomers. The authors, however, cautioned that inaccuracies result when the tallest peaks in the spectrum are due to chains of less than 10 repeat units in length. Erroneous results will also be obtained in the case of strongly overlapped mass spectral peaks with ambiguous assignments in which case pruning methods have been employed [67]. The naturally occurring isotope distribution and mass-dependence of the instrumental resolution need to be corrected for if peak apices are compared instead of area ratios.

Wilczek-Vera et al. were among the first to demonstrate that results from MALDI-MS can be used to determine the full two-dimensional distribution of copolymer composition and chain lengths [67–69]. These authors also employed random coupling statistics in the case of block-copolymer formation to aid quantitative spectral interpretation. Their work was later followed up by Suddaby et al. [70] and recently by Willemse et al. [63, 71] and Huijser et al. [72, 73] who used the results for the determination of reactivity ratios in free radical copolymerizations [63, 70, 71] as well as for the mechanistic investigation of polycondensation reactions [72, 73]. These authors also derived so-called fingerprint plots (see Figure 8.5) from the mass spectra. These plots depict the two-dimensional contour plots of the probability distribution of both comonomer chain lengths as determined from the mass spectra. They provide a facile means to interpret the polymer mass spectrum and the form of the distribution observed also allows deductions to be made about the type of the analyzed copolymer (block vs. random) [74]. Weidner et al. recently employed the

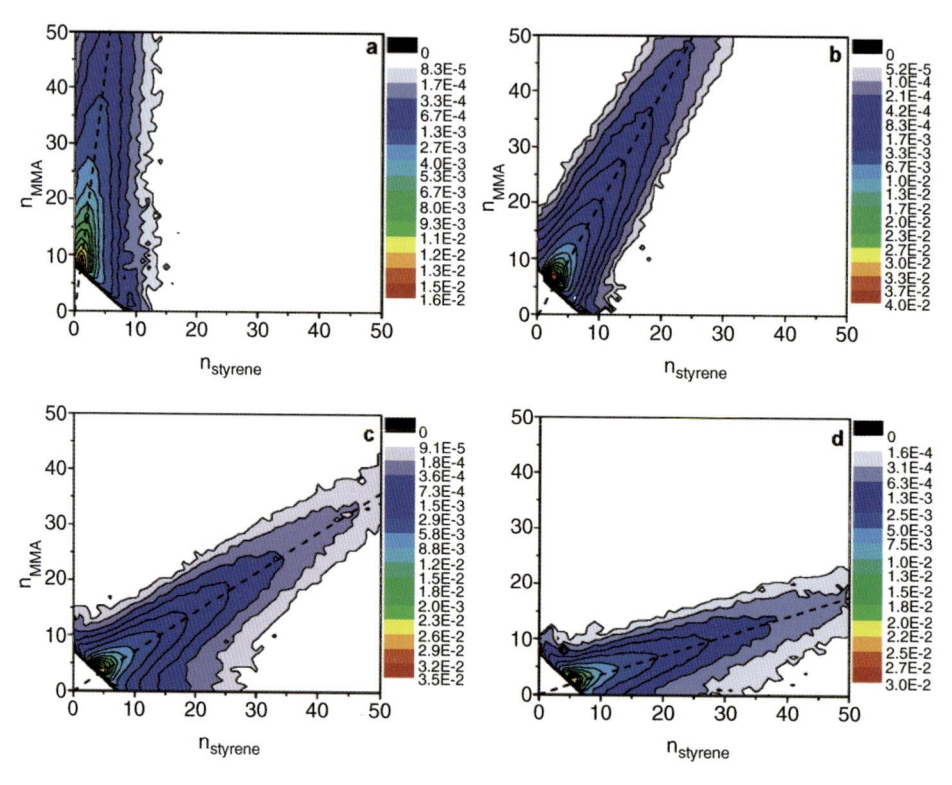

Figure 8.5 Copolymer fingerprint plots obtained from the pulsed laser-initiated radical copolymerization of MMA and styrene at a molar feed ratio (x_{St}) of: $x_{St} = 0.053$ (a), $x_{St} = 0.249$ (b), $x_{St} = 0.600$ (c), and $x_{St} = 0.792$ (d). The dashed lines in the copolymer fingerprint plots are indicative of the average chemical composition of the copolymer. The figure is taken from Ref. [71] with permission from the American Chemical Society.

method to cases in which an online coupling of MALDI and ESI MS to chromatographic separations was required and reported the introduction of a software tool developed in-house (MassChrom2D) [75, 76].

By employing a procedure named strip-based regression, Vivó-Truyols et al. [77] provided for the first time an elegant and statistically sound algorithm for the determination of the topology distribution of copolymers from their highly overlapped spectra in MALDI-MS. The characteristic of the algorithm to use strips whose width is a fraction of the full-acquired mass range for processing allows the data extraction problem to be treated by linear regression methods, whereas changes in the instrumental resolution of the instrument with mass-to-charge and an incorrect calibration do not deteriorate the results. As a further benefit, the application of the regression approach allows the error associated with the relative abundance of each comonomer combination to be determined.

8.6
Data Interpretation in MS/MS

Advanced fragmenting techniques in MS have been around for some while and have been extensively used in the determination of polypeptide sequence by bottom-up proteomics [78, 79]. The unambiguous interpretation of the very information-rich spectra obtained from MS/MS is greatly facilitated by the availability of highly advanced data processing software and database search tools, which are an indispensable part of contemporary proteomics [3, 4]. In recent years, MS/MS has also become a topic of largely increasing popularity in the field of synthetic polymer characterization. The fragmentation pattern of macromolecules can provide detailed information on the structure of the constituent monomer-building blocks as well as on the attached end groups. A number of studies have established the main degradation pathways of common polymers such as PMMA, poly(butyl methacrylate), poly(ethylene glycol), poly(propylene glycol), poly(styrene), poly(2-ethyl-2-oxazoline)s, and poly(α-methyl styrene) [80–94]. Software for the automated interpretation of synthetic polymer tandem mass spectra may prove to be a valuable tool for the determination of the backbone structure and end groups of otherwise uncharacterized polymers as well as for the analysis of the chain structure of copolymers. The advances in software development and the large availability of open source software solutions in biomolecular MS may greatly benefit development of suitable tools for synthetic polymer MS/MS. So far, there is only one tool available developed by Thalassinos et al. [95] that, however, greatly aids interpretation of tandem mass spectra and which is provided free of charge by the author. A screenshot of the software is given in Figure 8.6. Taking user-provided input on the repeating monomer units and the α- and ω-end-groups as well as the type of the attached cation, the software automatically assigns the recorded peaks to fragment ion species, followed by a color coding of the peaks making further spectral interpretation highly intuitive. Tentative assignments of the end groups can be quickly validated, which

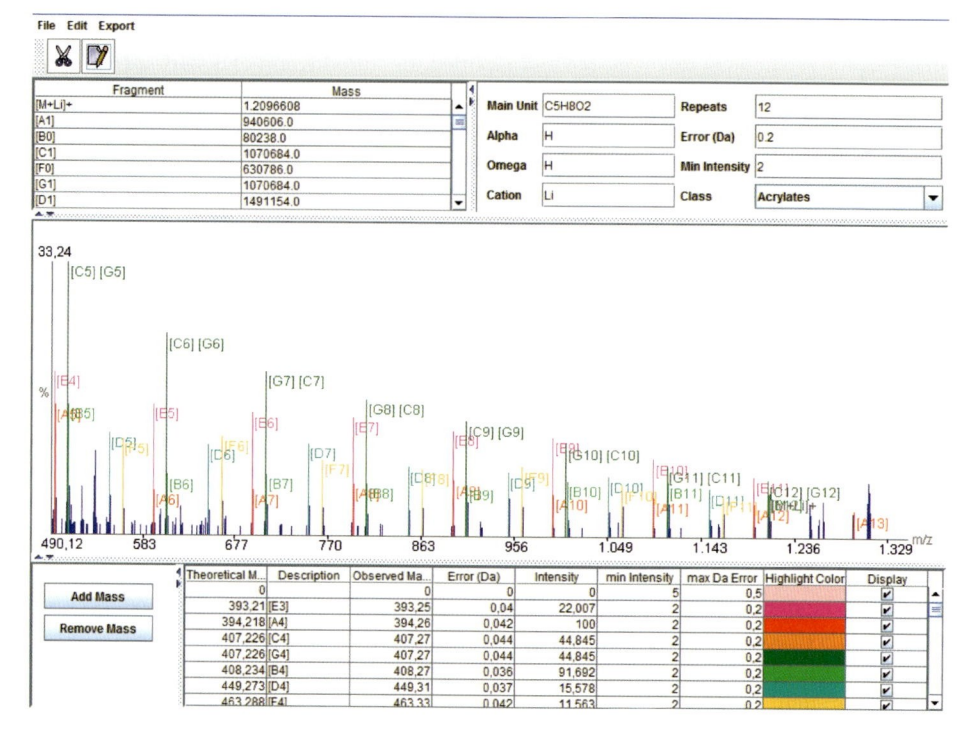

Figure 8.6 Screenshot of the polymerator software showing an annotated expansion (m/z range 490–1350 Th) of the ESI-MS/MS spectrum of the lithiated octadecamer of PMMA. Details of annotated fragment ions are displayed by the software in the table below the spectrum. Predicted fragment ions are also detailed above (left) of the spectrum.

otherwise poses a time-consuming process. The authors have also reported on the application of Polymerator to analyze PPG [96] and poly(hydroxyethyl methacrylate) [97].

8.7
Quantitative MS and the Determination of MMDs by MS

MS with soft ionization has evolved into a powerful analytical tool in macromolecular science within the last two decades. MALDI-MS [98] and ESI-MS [99] are especially versatile tools for the analysis of synthetic polymers. A large field of application of MS in polymer science aims at gaining qualitative information on the chemical identity of the repeat units or end groups of a synthetic macromolecule based on the precise measurement of the molecular weight of individual oligomer molecules. Although MALDI-MS and ESI-MS yield exact molecular weights of individual molecules, accurate MMD of synthetic polymers requires sophisticated spectral processing approaches. This is because, intrinsically, synthetic polymers do not exhibit one uniform chain length but rather a distribution of molecular weights. Although in MS, the molecular weight axis is certain, due to instrumental bias and a dependence of ionization efficiency on molecular weight and charge state, abundances of oligomer ions are not an accurate description of oligomer concentration in the analyzed sample. Classical methods used for the determination of MMDs by SEC yield accurate information about the concentration of the polymer. The molecular weight axis though is uncertain in SEC and existing calibration procedures may introduce errors of up to 30% in the obtained molecular weights [100].

In the following text, two approaches, developed independently at the National Institute of Standards and Technology [101] and at Karlsruhe Institute of Technology [102, 103], respectively, are described and evaluated. The first approach (see Section 8.7.1) has been employed in the context of creating an absolute molecular mass standard (Standard Reference Material™ (SRM) 2881) from MALDI measurements alone, using an internal calibration of the mass spectral intensity axis [101]. The second approach (see Section 8.7.2) relies on the use of SEC coupled online to a quantitative concentration detector (refractive index (RI) detector), whereas molecular mass calibration and band-broadening correction are achieved using peak data obtained from online ESI-MS [102, 103].

8.7.1
Quantitative MMD Measurement by MALDI-MS[2]

The accuracy of a polymer's MMD determined from a well-resolved mass spectrum depends on accounting correctly for the mass bias in the measurement. Here "well-resolved" means having the ability to separate to baseline the individual peaks of two oligomers whose mass differs by one unit of their (typically) periodic mass spacing.

This ability is required for the quantitation methods described in this section. The methods found here have been developed at the National Institute of Standards and Technology and are described in more detail in Refs. [101, 104–106]. Mass bias is the systematic over- or undercounting of specific parts of the MMD by the mass spectrometer. Here "specific parts" can refer to the high-mass or low-mass parts of the spectrum, or to specific types of oligomers as defined by, for example, end group or molecular architecture. Mass bias can occur in any of the three basic functions of the mass spectrometer (sample ionization, separation by m/z, and detection) as well as in the sample preparation or the data analysis. By systematic it is meant that the bias is an inherent aspect of the measurement and how it is conducted and not simply due to imperfect counting statistics. In the latter case, taking more data will resolve the problem; in the former case, taking more data is not a viable solution. For systematic bias, the magnitude of the bias must be found and a correction is applied, otherwise the measured MMD is of little use.

Fundamental metrological principles identify two types of measurement uncertainty, type A and type B. Type A refers to uncertainty that can be evaluated by the

---

[2] Official contribution of the National Institute of Standards and Technology; not subject to copyright in the United States of America.

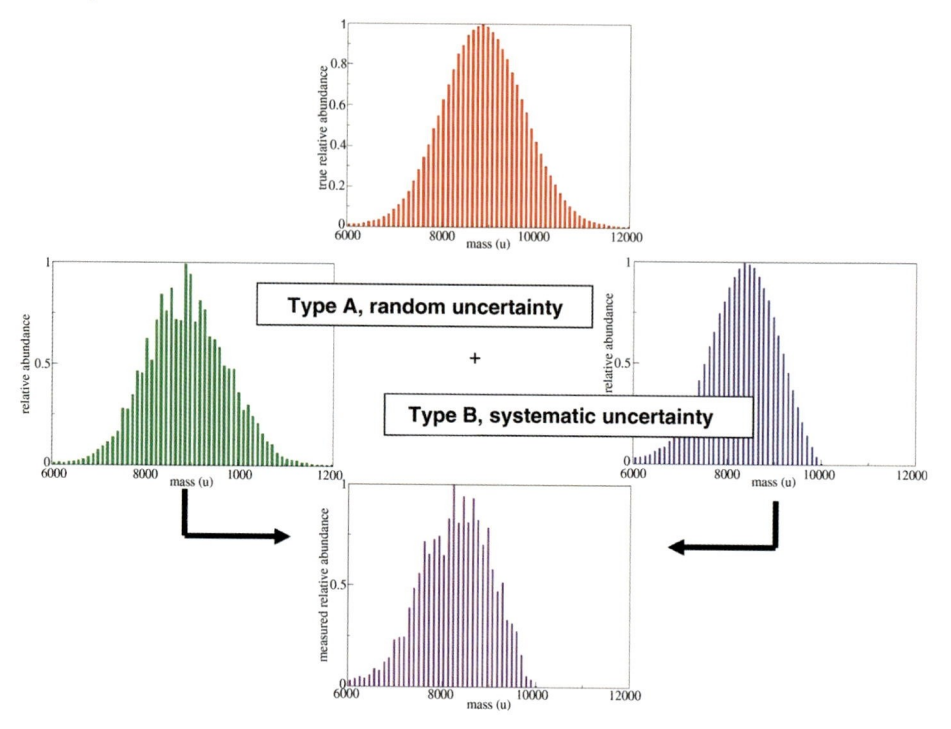

254 | *8 Automated Data Processing and Quantification in Polymer Mass Spectrometry*

Figure 8.7 Schematic illustration of type A ("random") and type B ("systematic") uncertainties in MMD measurement.

statistical analysis of a series of observations, whereas type B refers to uncertainty that cannot be evaluated by statistical methods alone. Generally, type A is spoken of as *statistical* or *random* uncertainty and type B as *systematic* uncertainty. Their differences applied to the MS are shown in Figure 8.7. This section is concerned with the determination of type B. Type A uncertainty that can be determined (and reduced) by repeat measurements is not explicitly discussed here. It is noted that measurement repeatability is critical. If the operator cannot repeat the measurement from run-to-run and from day-to-day, the chances of measuring the correct MMD decrease dramatically. The measurement method must be repeatable and reliable before it can be considered for quantitative, much less for standards, work.

An MMD is a two-dimensional quantity of which the mass spectrum is its (imperfect) representation. Thus, both the mass axis and the signal axis (i.e., the intensity of the ion signal at any given mass) have to be calibrated separately, and their associated type B uncertainties are considered separately.

Mass axis quantification is the most easily performed of the two and is not a significant source of uncertainty in determining the MMD from the mass spectrum. Calibration of most mass spectrometers is usually done with biopolymers of known molecular masses. These biopolymers are selected because they typically provide a single major peak whose mass is known accurately; thus, mass axis quantification is quite straightforward. Calibration must be done using at least two or three of these biopolymers that span the mass range of interest. More calibration points would increase calibration accuracy. Calibration of the mass axis can also be done by combining a single biopolymer with a molecular material calibrant. If this material is close to, or identical to, the material under study then, in general, inaccuracies in mass axis calibration will be minimized. The oligomeric masses, m_i, with n repeat units of mass r and masses of the end group, m_{end}, of the polymer calibrant are given by

$$m_i = nr + m_{end} + m_{adduct} \tag{8.4}$$

where m_{adduct} refers to the mass of any charged or neutral atoms or molecules noncovalently bound to the analyte. This may be, for example, any salts added to the sample preparation to encourage charging of the analyte. Thus, calibration of the mass axis using a homopolymer calibrant (for example) reduces to determining n for one of the peaks. A mass accuracy of better than a few mass units is not necessary since polymer MMDs are not critically dependent on such small mass differences.

Calibration of the signal axis is much more difficult. There are many systematic uncertainties that can arise in the signal axis quantitation. It would be an insurmountable task to try to quantify each of these uncertainties individually. Instead, the systematic bias in the signal axis is best determined heuristically by gravimetric techniques. By mixing together in carefully prepared gravimetric ratios samples having different MMDs, a mixture's MMD can be controlled. By comparing the gravimetric ratios to the signal intensity in the mass spectrum, a calibration curve for the signal axis can be obtained.

Various averages, known as molecular moments, where the entire shape of the distribution is reduced to a single number, serve as useful numerical simplifications of the MMD. Measuring and computing these summary statistics has historically comprised the core of the analysis of molecular materials. The two most common measures of the MMD are the number-average molecular mass, M_n, and the mass-average molecular mass, M_w.

$$M_n = \frac{\sum_i m_i n_i}{\sum_i n_i} \tag{8.5}$$

$$M_w = \frac{\sum_i m_i^2 n_i}{\sum_i m_i n_i} \tag{8.6}$$

$$PD = \frac{M_w}{M_n} \tag{8.7}$$

where m_i is the mass of a discrete oligomer i, n_i is the number of molecules at the given mass m_i, and PD defines the polydispersity (PD) index.

To estimate the level of uncertainty in an instrumental method, a mathematical construct is needed to determine how type B uncertainties affect the final measurand. Assume that there is a point in the experimental parameter space (sample preparation, instrument operation, and data analysis) where the signal intensity, S_i, for an oligomer of mass m_i is linearly proportional to n_i, the number of polymer molecules at that oligomer mass. Mathematically, this is given by

$$S_i = k n_i \tag{8.8}$$

where for a narrow enough range of m_i, it is assumed that k is a constant independent of m_i and the range of linearity, $n_i < n_0$, is about the same for all molecules in the (polydisperse) sample.

If the measurement is performed in the linear region for all the oligomers of the sample, the overall signal from the quantity of analyte introduced into the mass spectrometer is given by

$$\sum_i S_i m_i = k \sum_i n_i m_i \tag{8.9}$$

with $n_i m_i$ summed over all i. From this, it can be derived that

$$\frac{\sum_i S_i m_i}{\sum_i S_i} = \frac{k \sum_i n_i m_i}{k \sum_i n_i} \tag{8.10}$$

The right-hand side of the equation is by definition the exact M_n of the polymer independent of k since k in numerator and denominator cancels out. The same holds for equations for M_w and all higher moments. This is generally true when the measurements are made in the linear range of analyte versus signal strength. However, it is well known that the mass spectra of wide PD analytes give poor representations of the MMD due to large systematic uncertainties in the signal axis. That is, if the values of the m_i span too great a mass range, then the values for k and/or the n_0 saturation limits must change dramatically, otherwise MS would be able to obtain the MMD correctly for very broad distribution analytes which is widely demonstrated not to be the case.

If k is not a constant independent of i then, and if the measurements are made in a linear concentration range for each oligomer i (i.e., $n_i < n_0$), then

$$S_i = k_i n_i \tag{8.11}$$

where k_i is now a function of the oligomer i for a fixed experimental method: sample preparation, instrument operation, and data analysis.

8.7.1.1 Example for Mixtures of Monodisperse Components

The simplest example of gravimetric quantitation to test mass spectrometer response is to create a mixture of two monodisperse compounds: species 1 as a standard and species 2 as the analyte whose concentration is sought. If there is no systematic bias in the measurement, then the ratio of S_2/S_1 is directly proportional to the gravimetric mass ratio G_2/G_1, where G_i is defined as the gravimetric mass of each species.

The signal from such a mixture, call it A, is

$$S_A = k_1 n_1 + k_2 n_2 \tag{8.12}$$

The mass moments would be

$$M_{nA}^{grav\_exp} = \frac{(k_1 m_1 n_1 + k_2 m_2 n_2)}{(k_1 n_1 + k_2 n_2)} \tag{8.13}$$

$$M_{wA}^{grav\_exp} = \frac{(k_1 m_1^2 n_1 + k_2 m_2^2 n_2)}{(k_1 m_1 n_1 + k_2 m_2 n_2)} \tag{8.14}$$

The gravimetric mass of species i is

$$G_i = m_i n_i \tag{8.15}$$

Substituting into Eq. (8.14) we get

$$M_{wA}^{grav\_exp} = \frac{(k_1 m_1 G_1 + k_2 m_2 G_2)}{(k_1 G_1 + k_2 G_2)} \tag{8.16}$$

To simplify this let the mass fraction X be

$$X = \frac{G_1}{G_1 + G_2} \tag{8.17}$$

Substituting Eq. (8.17) into Eq. (8.16) and dividing numerator and denominator by $(G_1 + G_2)$ yields

$$M_{wA} = \frac{(k_1 m_1 X + k_2 m_2 (1-X))}{(k_1 X + k_2 (1-X))} = \frac{(m_1 X + \theta m_2 (1-X))}{(X + \theta (1-X))} \tag{8.18}$$

where

$$\theta = \frac{k_2}{k_1} \tag{8.19}$$

In this way, the mass bias in the mass spectrum is reduced to a single metric, θ. θ equals one for an unbiased system. If species 2 is overcounted with respect to species 1, θ will be greater than one, if species 2 is undercounted, θ will be less than one. The further θ is from one the greater the systematic bias in the mass spectrum.

8.7.1.2 Example for Mixtures of Polydisperse Components

For most mixtures encountered, any given oligomer peak in the mass spectrum cannot be assigned exclusively to one or the other component of the mixture. In fact, a given oligomer peak may have contributions from both components in the mixture.

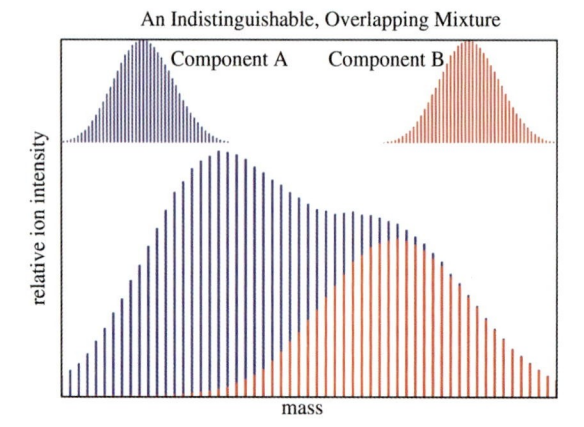

Figure 8.8 Schematic illustration of an indistinguishable and overlapping mixture of two components, where the peak intensities in the mixture are simply sums of component oligomer intensities.

Typically these overlapping MMDs are made up of indistinguishable oligomer components, that is, each component of the mixture has some (but not all) oligomers that are identical to those in the other component as illustrated in Figure 8.8. This means that in this case the mass moments of the mixtures must be calculated and used to create a calibration curve. A full theory for the case of distinguishable oligomer mixtures (shown in Figure 8.9), or nonoverlapping MMDs (shown in Figure 8.10), where each oligomer peak can be assigned to a specific component, is given in Section 8.7.1.5. In this special case, true type B uncertainties can be given for each oligomer in the target material and a true absolute molecular mass standard can be created.

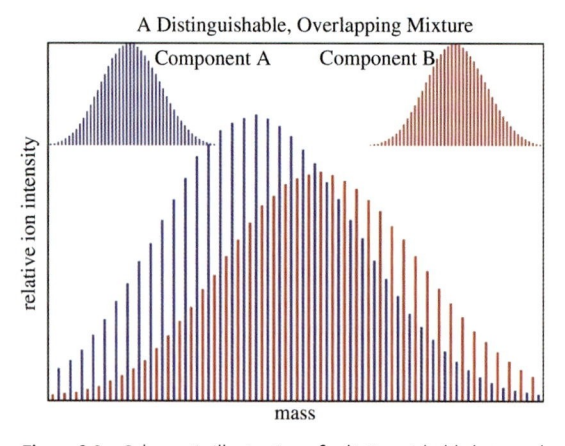

Figure 8.9 Schematic illustration of a distinguishable but overlapping mixture of two components.

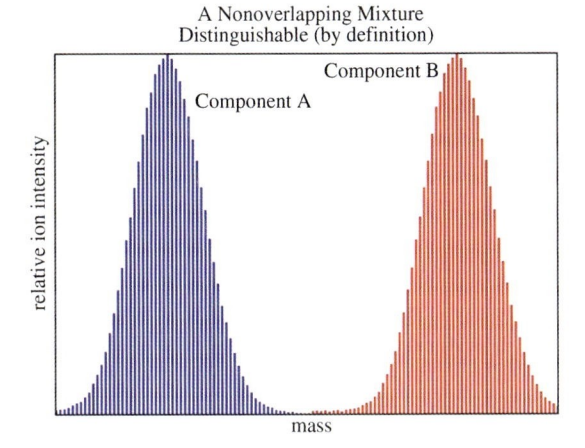

Figure 8.10 Schematic illustration of a nonoverlapping mixture of two components, which is distinguishable by definition.

Equation (8.18) can be extended to a gravimetric mixture of polydisperse components by substituting the experimental average molecular mass of each pure component derived from its mass spectrum. This leads to the mass moments

$$M_{wq}^{\mathrm{grav}} = \frac{(\hat{k}_1 M_{w1}^{\mathrm{exp}} X + \hat{k}_2 M_{w2}^{\mathrm{exp}} (1-X))}{(\hat{k}_1 X + \hat{k}_2 (1-X))} = \frac{(M_{w1}^{\mathrm{exp}} X + \theta M_{w2}^{\mathrm{exp}} (1-X))}{(X + \theta(1-X))} \quad (8.20)$$

where q represents a given gravimetric mixture. In Eq. (8.20), $\hat{k}_1$ and $\hat{k}_2$ replace k_1 and k_2 used in the monodisperse example and are the mass-average means over each component of the mixture which is conceptually similar to the mass-average molecular mass. Likewise, X is now calculated from the gravimetric amounts of each component in the mixture. The mass moments of the pure components are from their mass spectra using Eq. (8.21):

$$M_{wq}^{\mathrm{exp}} = \frac{\sum_i S_{iq} m_i^2}{\sum_i S_{iq} m_i} \quad (8.21)$$

To obtain an estimate of the value of θ, the minimum value of the sum of squares is found. The sum of squares over all mixtures q is expressed as

$$SS_\theta = \sum_q (M_{wq}^{\mathrm{grav}} - M_{wq}^{\mathrm{exp}})^2 \quad (8.22)$$

The simplest way to solve this equation is to insert an arbitrary value for θ (typically $\theta = 1$) and calculate a value for SS_θ then increment θ and recalculate SS_θ. This most basic iterative process will yield an optimal value typically in a few steps and can easily be encoded in spreadsheet software. Recall that values of θ near one indicate systems with little bias in the mass spectrum.

8.7.1.3 Calculating the Correction Factor for Each Oligomer

Once θ has been calculated and found to be near one, the next step in the process is to calculate the various k_i in order to correct the MMD. If the k_i are a smoothly and slowly varying function of i (or m_i), a Taylor expansion on k_i may be made around a mass peak near the center of the MMD, termed M_0. The center is used to assure that the function is changing as little as possible over the entire width of the MMD; however, mathematically, the choice is arbitrary. Thus

$$k_i = k_0 + Q(m_i - M_0) + \text{higher order terms in } m_i \tag{8.23}$$

$$S_i = k_0 n_i + Q(m_i - M_0) n_i + \text{higher order terms in } m_i \tag{8.24}$$

where k_0 and Q are the first two coefficients in the Taylor expansion. They are also functions of all the experimental conditions: the instrument parameters, the sample concentrations, and the sample preparation method. (By k_0, it is not meant the k of the zeroth index oligomer but rather the zeroth derivative of the Taylor expansion). In this way, the entire physics of the experiment is folded into these two coefficients. From these assumptions, and dropping the higher order terms in Eq. (8.24), one can derive the following important relationship:

$$M_{wq}^{\exp} = M_{wq}^0 \left\{ \frac{1 + (Q/k_0)(PD_{wq} M_{wq}^0 - M_0)}{1 + (Q/k_0)(M_{wq}^0 - M_0)} \right\} \tag{8.25}$$

where $M_{wq}^{\exp}$ is the mass spectral mass-average molecular mass for the mixture of analytes given in Eq. (8.21). PD_w is mass average PD (M_z/M_w) and is taken here to be the experimentally measured value ($M_z^{\exp}/M_w^{\exp}$). Equation (8.25) is then solved for M_{wq}^0 for various values of Q/k_0 at a fixed M_0 chosen as described below for the values of the mixtures described by $q = A, B, C$, and so on, and for the initial components of the mixtures described as $j=1$ and $j=2$.

For a gravimetric mixture A, $M_{wA}^{\text{grav}\_0}$ is calculated from the values for the individual components M_{w1}^0 and M_{w2}^0 computed for each Q/k_0 using a simple weighted average:

$$M_{wA}^{\text{grav}\_0} = \frac{G_1}{G_1 + G_2} M_{w1}^0 + \frac{G_2}{G_1 + G_2} M_{w2}^0 \tag{8.26}$$

where G_1 is the gravimetric mass of component 1 in the mix, and G_2 is similarly defined.

For each Q/k_0, the sum of squares, $SS_{(Q/k_0)}$, is computed as

$$SS_{(Q/k_0)} = \sum_q (M_{wq}^{\text{grav}\_0} - M_{wq}^0)^2 \tag{8.27}$$

where the sum is taken over all measured mixtures. The Q/k_0 which gives the minimum value of the $SS_{(Q/k_0)}$ is then taken as the best fit. As with Eq. (8.22), solution of Eq. (8.27) required iteration over incremented values of Q/k_0. Dropping the higher order terms and rearranging Eq. (8.24) yields

$$\frac{S_i}{k_0 n_i} = 1 + \frac{Q}{k_0}(m_i - M_0) \tag{8.28}$$

Equation (8.28) shows us how to apply the correction factor Q/k_0 to each oligomer i to arrive at a more reliable measure of the MMD. If Q/k_0 were equal to zero, then the mass spectrum would show no mass bias and $S_i = k_0 n_i$. This would mean that the peak areas are directly proportional to the oligomer concentrations in the sample. If Q/k_0 is nonzero, then mass bias is present. If M_0 is taken at the middle of the distribution being calibrated, then the sign of Q/k_0 along with where the mass m_i of an oligomer i is greater than or less than M_0 determines whether the correction to the ion intensity is positive or negative.

8.7.1.4 Step by Step Procedure for Quantitation

The steps of the method can be summarized as follows:

1) Obtain at least two samples having different MMDs but with otherwise very similar, if not identical, properties.

    For example, these could be polymers with different degrees of polymerization or nanoparticles with different levels of functionalization. The different samples could be obtained directly by synthesis or by separation of a single broader molecular mass sample. Two samples are required at a minimum, but additional samples will allow for more calibration points. If possible the only difference between the two should be molecular mass. Any other differences, for example, different functional groups may contribute to mass bias in an uncontrolled way.

2) Take mass spectra of each sample endeavoring to keep all experimental conditions constant.

    As much as possible keep all aspects of the measurement constant. This includes sample preparation, instrument settings, and data analysis. Also, measurements should be made contemporaneously to keep any variables that change over time constant. These variables could be sample preparation conditions, for example, water absorption into samples or solvents, or time drift in instrument settings.

3) Use a laboratory balance to make carefully controlled gravimetric mixtures of two samples in several well-spaced ratios.

    The balance needs to be calibrated and accurate to about least 0.1% of the total mass measured. Any gravimetric errors are carried through the entire analysis. Making stock solutions and then mixing solution volumes can be more accurate than repeated weighing of small amounts of material. Generally, as a practical matter, final weights must be at least 25 mg.

4) Take mass spectra of each mixture using the same experimental conditions as used for the pure components.

    The instrument settings may not be optimal for the mixtures, but they must be held constant to satisfy the self consistency of the method. If the experimental conditions are such that some oligomers of the mixture have disappeared (as compared to the pure component measurements), then compromise experimental conditions must be found. If this occurs, then it suggests strong mass bias in the measurement.

5) From the mass spectra, calculate the mass-average molecular masses of the pure components and of the mixtures.

Be careful in the application of "black box" software for this step. Unseen algorithms for data processing can lead to substantial errors in converting the mass spectrum to a MMD. Smoothing can introduce mass bias into a spectrum that is not a product of the measurement itself but of the data analysis method applied.

6) Use Eq. (8.22) to iteratively calculate the minimum value of SS_θ at a given θ.

The most direct way to do this is to set up a simple spread sheet. Start with $\theta = 1$ and change it systematically by small steps until a minimum in SS_θ is found. If θ is between 0.5 and 2 then the possibility exists that the MMD can be corrected. If not, the results should be treated with caution, and the error is too great to be corrected using only the linear term in the Taylor expansion. See Ref. [104] for an example of computer code to make this calculation.

7) Choose M_0, a mass near the center of the average molecular masses of the two components.

The exact choice of M_0 is not critical; however, the correction to the distribution will be more accurate near M_0 and less accurate the farther any given oligomer mass is from M_0. If a certain mass range is more critical, then choose M_0 at the center of that range.

8) Use Eq. (8.27) to iteratively calculate the minimum value of $SS_{(Q/k0)}$ at a given Q/k_0.

Again see Ref. [104] for computational assistance.

9) Use Eq. (8.28) and the value for Q/k_0 to correct the ion intensities S_i in the mass spectrum to arrive at a new MMD.

Individual oligomer intensities may increase or decrease depending on whether they were undercounted or overcounted in the mass spectrum.

At this stage, the analyst should have a good feel for the degree of mass bias in the mass spectra. Furthermore, if this bias is not too large it can be corrected using the methods outlined in this section. If the bias is large, higher order terms in Eqs. (8.23) and (8.24) need to invoked; however, methods to determine the values of the higher order coefficients have not been created. This is a fruitful topic for future research.

8.7.1.5 Determination of the Absolute MMD

The procedures outlined in this section do not provide systematic uncertainties for the corrected values. The corrected mass spectrum is closer to the true MMD, but just how close is it? In order to determine this, the following procedures must be invoked. These procedures require distinguishable or nonoverlapping mixtures as well as numerical instrument optimization to determine the systematic uncertainties inherent in the instrument. This requires extra effort on the part of the analyst, but an

MMD with both type A (random) and type B (systematic) uncertainties is a very useful calibration standard for MS and any other molecular mass measurement technique.

Starting with Eq. (8.11), $S_i = k_i n_i$, if an assumption is made that k_i is a slowly varying function of i (hence also of m_i), then a Taylor expansion around a mass peak near the center of the MMD, termed M_0, can be made. The center of the mass spectrum is used to assure that the function is changing as little as possible over the entire width of the MMD. Then

$$S_i = k_0 n_i + Q(m_i - M_0)n_i + \text{higher order terms in } m_i \tag{8.29}$$

Here Q and k_0 are functions of M_0 as well as of all the experimental conditions: the instrument parameters, the sample concentrations, and the sample preparation method. (By k_0, it is not meant the k of the zeroth index oligomer but rather the zeroth derivative of the Taylor expansion). In the experimental procedures referred to later, once the instrument parameters and experimental preparation methods are optimized, every attempt was made to keep them constant to insure experimental reproducibility. (Later it will be shown how variation in the machine parameters can affect the variation of Q/k_0 and thus the type B uncertainty.)

The implications of the model embodied in Eq. (8.29) will now be explored and it will be shown how small linear shifts of the calibration constant Q over limited mass ranges effect quantities derivable from mass spectral data. First the total signal, the total detected mass, and the mass ratios of mixtures will be considered, and it will be shown how these quantities relate to the true MMD of the analyte.

The total signal, S_T, from the polymer is given by

$$S_T = \sum_i S_i = k_0 \sum_i n_i + Q(M_n^0 - M_0) \sum_i n_i \tag{8.30}$$

while the total mass of polymer detected, G_T^{exp}, is given by

$$G_T^{\text{exp}} = \sum_i m_i S_i = k_0 M_n^0 \sum_i n_i + Q M_n^0 (M_w^0 - M_0) \sum_i n_i \tag{8.31}$$

where M_n^0 and M_w^0 are defined in Eqs. (8.32) and (8.33), respectively, and are the true number average and mass average relative molecular masses.

$$M_n^0 = \frac{\sum_i m_i n_i}{\sum_i n_i} \tag{8.32}$$

$$M_w^0 = \frac{\sum_i m_i^2 n_i}{\sum_i m_i n_i} \tag{8.33}$$

$$\text{PD}_n^0 = \frac{M_w}{M_n} \tag{8.34}$$

$$M_z^0 = \frac{\sum_i m_i^3 n_i}{\sum_i m_i^2 n_i} \tag{8.35}$$

$$PD_w^0 = \frac{M_z}{M_w} \tag{8.36}$$

where m_i is the mass of a discrete oligomer, and n_i is the number of molecules at the given mass m_i. The experimental moments from mass spectrum are defined as M_n, M_w, and M_z, while the true values are given as M_n^0, M_w^0, and M_z^0. PD_n defines the PD index that is a measure of the breadth of the polymer distribution. When PD_n is equal to one (i.e., in statistical terms the variance of MMD is zero), all of the polymer molecules in a sample are of the same molecular mass and the polymer is referred to as monodisperse.

Multiplying Eqs. (8.32) and (8.33) together gives

$$M_w^0 M_n^0 = \frac{\sum_i m_i^2 n_i}{\sum_i n_i} \tag{8.37}$$

Then taking the ratio of Eqs. (8.30) and (8.31), one obtains

$$M_n^{\text{exp}} = \frac{\sum m_i S_i}{\sum S_i} \tag{8.38}$$

with the result that

$$M_n^{\text{exp}} = M_n^0 \left\{ \frac{(1+(Q/k_0)(M_w^0 - M_0)}{(1+(Q/k_0)(M_n^0 - M_0)} \right\} \tag{8.39}$$

where M_n^{exp} is the experimentally measured M_n^0.

For use later in this section by the same algebra is obtained:

$$M_w^{\text{exp}} = \frac{\sum m_i^2 S_i}{\sum m_i S_i} \tag{8.40}$$

with the result that

$$M_w^{\text{exp}} = M_w^0 \left\{ \frac{(1+(Q/k_0)(M_z^0 - M_0))}{(1+(Q/k_0)(M_w^0 - M_0))} \right\} \tag{8.41}$$

All higher moments may be obtained in a similar way and have a similar form.

Equation (8.41) gives by simple division:

$$M_w^0 = M_w^{\text{exp}} \left\{ \frac{1+(Q/k_0)(M_w^0 - M_0)}{1+(Q/k_0)(M_z^0 - M_0)} \right\} \tag{8.42}$$

which yields

$$M_w^0 = M_w^{\text{exp}} \left\{ 1 - \frac{(Q/k_0) M_w^0 (PD_w - 1)}{1+(Q/k_0)(M_z^0 - M_0)} \right\} \tag{8.43}$$

Equation (8.43) states that the deviation of the mass moment measured by mass spectrum from the true mass moment is a function of the PD (arising from that moment) divided by a correction term arising from how far that moment is from the mass M_0 around which the Taylor expansion to obtain k_0 and Q is centered. In Eq. (8.43), the reader should notice that if M_z^0 is close to M_0 the term in $(Q/k_0)(M_z^0-M_0)$ is small compared to 1 and the result depends only on the PD of the polymer.

Since the method depends on gravimetrically mixing analytes to obtain estimates of Q/k_0, it is necessary to consider the equations relating to these mixtures. Equation (8.31) states that the MS-measured total mass, G_T^{exp}, is proportional to the true mass, G_T^0:

$$G_T^0 = M_n \sum n_i \tag{8.44}$$

$$G_T^{\text{exp}} = k_0 G_T^0 \{1 + Q/k_0(M_w^0 - M_0)\} \tag{8.45}$$

Consider now a mixture of the chemically identical analytes with functional groups having different masses, or two different molecular mass analytes having distributions that are well separated, such that each oligomer in the mass spectrum can be assigned to a specific polymer in the mixture. Call them analyte A and analyte B that will make up the components of the gravimetric mixtures. Then the measured ratio of the masses of each is given by

$$\frac{G_{TA}^{\text{exp}}}{G_{TB}^{\text{exp}}} = \frac{k_{0A} G_{TA}^0}{k_{0B} G_{TB}^0} \left\{ \frac{1+(Q_A/k_{0A})(M_{wA}^0-M_0)}{1+(Q_B/k_{0B})(M_{wB}^0-M_0)} \right\} \tag{8.46}$$

Note that the expansions are performed for both polymer distributions A and B around the same M_0. Also note Q_A, Q_B, k_{0A}, and k_{0B} are all functions of M_0. Thus, from Eq. (8.46):

$$\frac{G_{TA}^{\text{exp}}}{G_{TB}^{\text{exp}}} = \frac{G_{TA}^0}{G_{TB}^0} \left\{ \frac{1+(Q/k_0)(M_{wA}^0-M_0)}{1+(Q/k_0)(M_{wB}^0-M_0)} \right\} \tag{8.47}$$

Simple algebra leads us to

$$\frac{G_{TA}^{\text{exp}}}{G_{TB}^{\text{exp}}} = \frac{G_{TA}^0}{G_{TB}^0} \left\{ 1 + \frac{(Q/k_0)(M_{wA}^0-M_{wB}^0)}{1+(Q/k_0)(M_{wB}^0-M_0)} \right\} \tag{8.48}$$

What is measured are $G_{TA}^{\text{exp}}/G_{TB}^{\text{exp}}$ from MS versus G_{TA}^0/G_{TB}^0 gravimetrically determined. The calculated slope is

$$\text{slope} = \left\{ 1 + \frac{(Q/k_0)(M_{wA}^0-M_{wB}^0)}{1+(Q/k_0)(M_{wB}^0-M_0)} \right\} \tag{8.49}$$

As before with Eq. (8.45), the reader should notice if M_{wB}^0 is close to M_0, the term in $(Q/k_0)(M_{wB}^0-M_0)$ is small compared to 1 which means the slope depends only on the difference $(M_{wA}^0-M_{wB}^0)$ and, thus, (Q/k_0) maybe easily calculated. Finally, remember that the gravimetric calibration of the signal axis using chemically identical analytes can avoid the issues pertaining to the uncertainties arising from ablation, ionization, and detection. However, uncertainties in sample preparation as well as data analysis repeatability and consistency still affect the gravimetric calibration techniques.

8.7.2
Quantitative MMD Measurement by SEC/ESI-MS

Today, SEC is used as the method of choice for the determination of MMDs. The method suffers a number of significant drawbacks. First, it requires calibration with polymer standards whose molecular weights need to be determined by independent techniques [107]. For many polymer classes, well-characterized standards are not available. In these cases, universal calibration, heavily relying on the accuracy of Mark-Houwink parameters and the validity of the Flory-Fox equation [108–112] or, alternatively, online calibration by light scattering and viscosimetric detection have to be employed, which can lead to errors in the MMD of up to 30% [100]. Chromatographic band-broadening further deteriorates the SEC results, with an especially strong impact on the apparent MMD of polymers exhibiting sharp peaks or shoulders as in the case of distributions derived by experiments aimed at the determination of kinetic rate constants [113, 114].

8.7.2.1 Exact Measurement of the MMD of Homopolymers
Barner-Kowollik and coworkers have recently shown that by employing SEC with online concentration detection and using ESI-MS for an internal mass calibration, very accurate MMDs of synthetic polymers can be determined [102]. In the employed chromatographic setup [115, 116] (see Figure 8.11), a concentration-sensitive RI detector and the electrospray mass spectrometer are coupled to the chromatographic effluent of a size-exclusion column in parallel.

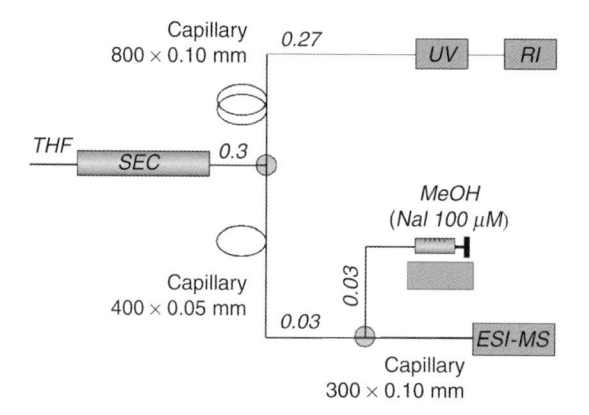

Figure 8.11 Chromatographic setup employed for coupling the concentration-sensitive RI- and/or UV detectors and the ESI-MS to the effluent of an SEC column in parallel. Numbers indicate flow rates in milliliters per minute.

The method accounts for the individual strengths and limitations of both detectors by deriving the absolute polymer concentration solely from the RI-detector trace. The electrospray mass spectrometer is used only in its ability to accurately measure the concentration profiles of the individual oligomers eluting from the chromatographic system for further processing. No use is made of MS to derive absolute concentration data. The elution profiles of the individual oligomers derived by MS contain accurate retention time information. This allows for a precise calibration of the SEC retention time dependence on chain length. A calibration can be derived without additional knowledge of the polymer class or any other physical assumptions as long as the polymer molecule is compatible with ESI. In addition to the position in time, the exact shape of the elution profile can be derived from online ESI-MS, which allows the characterization of the chromatographic band-broadening function as well as corrections to be made for band-broadening effects in the derived MWDs.

The influence of chromatographic band broadening is described mathematically by the discrete form of Tung's convolution equation (8.50) [117]. The SEC-trace, S_{V_R} recorded by a detector with a linear mass concentration response (RI detector) is derived by the convolution of the mass-weighted MMD w_n with the instrumental spread and calibration function, $G_{V_R,n}$. Here V_R and n are the chromatographic retention volume and the polymer repeat unit number, respectively.

$$\mathbf{S} = \mathbf{G} \times \mathbf{w} \quad \text{or} \quad S_{V_R} = \sum_n G_{V_R,n} \cdot w_n \qquad (8.50)$$

Figure 8.12 provides a graphical representation of the process described by Eq. (8.50): The individual elution profiles of each oligomer of a certain chain length – instead of being negligibly narrow spikes or delta peaks – feature a somewhat Gaussian peak shape. The RI-detector trace is obtained by a summation over the elution profiles of all individual oligomers, weighted by their respective concentrations. A band-broadened size-exclusion chromatogram is hence much like a blurred picture, where the detailed patterns are partially hidden, as each pixel is smeared out over a larger area determined by the point spread function of the out-of-focus lens.

If the functional form of $G_{V_R,n}$ is known, a number of deconvolution approaches may be applied to derive the reconstructed MMD, $\hat{w}_n$. Unfortunately inversion of

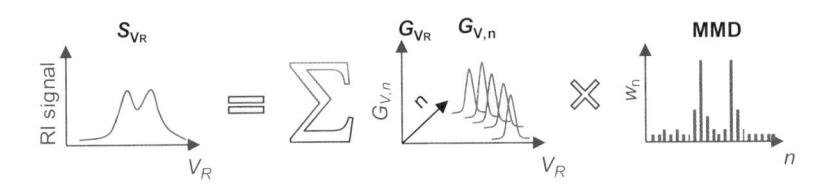

Figure 8.12 Graphical representation of the convolution process in SEC. The concentration of each individual oligomer molecule in the molecular mass distribution is multiplied with its elution profile stored in the instrumental calibration and broadening matrix $G_{V_R,n}$. The individual weighted oligomer profiles are summed to result in the recorded RI-detector trace.

Eq. (8.50) is an "ill-conditioned" problem and direct solution of the convolution equation, for example, by linear regression leads to an amplification of instrumental noise resulting in a highly oscillatory behavior of $\hat{w}_n$ with possibly negative values, lacking physical significance [118, 119]. Sophisticated numerical approaches therefore have to be used for the inversion of (8.50). Today the most widely used and most effective deconvolution approaches are based on singular value filtering and the application of regularization filters [119, 120]. Maximum entropy (MaxEnt) regularization has been successful in a number of related scientific problems in image reconstruction [121, 122] and spectroscopy [123]. It has been argued that use of the Shannon entropy [124] criterion as information constraint is the only consistent way of restoring a probability density function from noisy data [125, 126].

As can be seen in Figure 8.13, online ESI-MS can be used to extract the chromatographic elution profiles of individual oligomers and thus to gain calibration data (retention volume vs. chain length) together with band-broadening information for each individual eluting oligomer. This data can in turn be used to construct $G_{V_R,n}$ without the need for additional information, as long as the extracted ion chromatograms provide a correct representation of the actual elution profiles. Deconvolution of Eq. (8.50) will directly yield the absolutely calibrated MMD, corrected for chromatographic band-broadening effects. In our approach, a MaxEnt-based algorithm was employed in order to compute $\hat{w}_n$. At the heart of this approach lies a constrained nonlinear optimization problem (8.51). The general derivation of the objective function based on Bayesian probability theory can be found elsewhere [127]. The employed algorithm proceeds by calculating the theoretical RI-detector trace from a trial MMD. This concentration trace is then compared against the measured RI-detector trace. The software iteratively manipulates w to obtain the closest possible fit to the measured trace. The total squared sum of error (χ^2) is used to assess the agreement between the measured and the theoretical mass concentration trace. In a typical least squares approach, the single objective would be to minimize χ^2, yielding $\hat{w}_n$ as the maximum likelihood estimator of the MMD. However, as mentioned

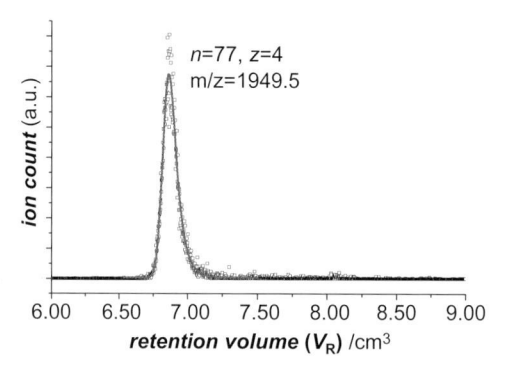

Figure 8.13 Extracted ion chromatogram recorded at $m/z = 1949.5$ Th, effectively reproducing to the elution curve of a hydrogen-functional MMA oligomer of chain length, $n = 77$ in charge state, $z = 4$ (dots), together with a fitted EMG peak profile (grey curve).

before, such an approach would lead to an excessive amplification of noise from the RI-detector trace because in SEC the individual oligomers elute very closely to each other in time, so that many individual elution profiles overlap. This feature of SEC complicates an accurate calculation of the individual contribution of each oligomer to the RI-detector trace and leads to great covariance of the oligomer concentrations in the obtained MMD. The problem is alleviated if a regularization filter – in the current case MaxEnt regularization – is imposed on the estimated MMD [119]. An additional entropy term, S [124–126], introduced in the objective function by means of a Lagrange multiplier, λ ascertains that the MMD $\hat{w}_n$ is as smooth as permitted by some limiting criterion [127]. Nonnegativity constraints in $\hat{w}_n$ and first-order equality constraints of the total area of $\hat{w}_n$ and S_{V_R} can further stabilize deconvolution. Quadratic or Thikonov regularization employing these constraints also yields very good results, while the resulting quadratic optimization problem is less complicated to solve algorithmically.

$$\hat{w}_n = \arg\max_{w}(Q - \lambda \cdot \chi^2) \quad \text{with} \quad w_n > 0 \quad \forall \quad n \tag{8.51}$$

$$\chi^2 = \sum_{V_R} \frac{(S_{V_R,\exp} - \sum_n G_{V_R,n} w_n)^2}{\sigma^2} \tag{8.52}$$

$$Q = \sum_n w_n \cdot \log(w_n) \tag{8.53}$$

Figure 8.14 shows the deconvoluted absolute MMD and the chromatographic peak parameters for a narrow-molecular weight PMMA standard, having a manufacturer-

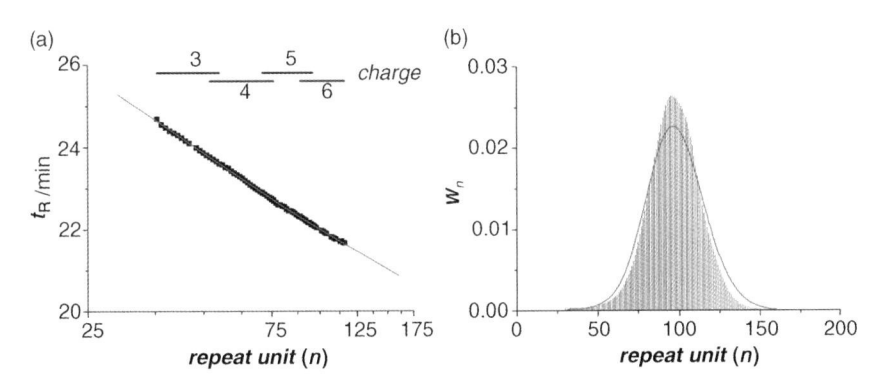

Figure 8.14 (a) SEC retention time (t_R) as a function of the repeat unit length of a narrow PMMA standard with manufacturer-specified $DP_w = 102$, as derived from online ESI-MS (symbols). As can be seen, there is an excellent overlap of the calibration data determined from chromatographic peaks recorded in different charge states. (b) Deconvoluted molecular mass distribution $\hat{w}_n$ obtained from MaxEnt data processing (sticks) and molecular mass distribution obtained from conventional calibration using MS retention time data (curve).

specified weight-averaged degree of polymerization (DP_w) of 102 and polydispersity index (PDI) of 1.03. Linear retention time data free from outliers was obtained for this sample (see Figure 8.14a). In many cases, ions of adjacent charge state appear in the mass spectrum for oligomers of the same chain length. Peak data were extracted for both charge states in these cases and the data is superimposed in Figure 8.14. There was excellent agreement between retention volume data from different charge states as was previously reported by Simonsick et al. [115]. Furthermore, for repeat units corresponding to the concentration apex of the MMD, no concentration-induced bias in the retention times is seen for the analyzed sample.

A comparison of the deconvoluted MMD with that obtained from a conventional calibration approach converting the retention volume axis to a molecular weight axis by using only the retention time information without band-broadening correction (Figure 8.14b, red curve) shows that for these narrow standards, the MaxEnt procedure can effectively account for chromatographic band broadening. As can be taken from, molecular weight moments for this standard obtained from SEC/ESI-MS are about 5% lower than specified by the manufacturer and determined using SEC calibrations which can be traced back to light-scattering measurements as absolute calibration source. The number-averaged degree of polymerization of the $10\,kg\,mol^{-1}$ PMMA standard was calculated independently using ^{1}H-NMR. This technique is an absolute, calibration-free means of determining DP_n. From three consecutive measurements, a $DP_n = 94.5 \pm 1\%$ was calculated, which agrees with the average value obtained by SEC/ESI-MS ($DP_n = 93.4$) within its standard deviation.

8.7.2.2 MMD of the Individual Components in Mixtures of Functional Homopolymers

The great potential of MS in polymer analysis lies in its ability to measure the molecular weight of individual oligomer molecules with an accuracy that allows both the unambiguous identification of the chemical identity of the polymer and its end-groups, as well as an accurate molecular mass calibration of the SEC system. This strength of MS can be used to elucidate the individual MMDs and absolute concentrations of components in mixtures of functional homopolymers. Knowing the identity of the individual polymer species together with their concentrations provides both the synthetic chemist and the polymer kineticist with new toolsets for the characterization and optimization of polymerization processes.

The current approach features a direct extension of the method of absolute molecular weight derivation for pure polymers (described in Section 8.7.2.1) [102] to mixtures of individual polymer species [103]. An assumption is made that ESI-MS can be used to successfully derive the relative concentrations of polymers eluting from the SEC-column, as long as they are of the same chain length. In other words, it is assumed that there is no significant end-group bias on ionization efficiency. Furthermore, it is assumed that influences of the polymer end groups and the chain length of the polymer on the RI increment, dn/dc are negligible. As demonstrated by a number of authors, this assumption is generally valid except for low-molecular weight oligomers of less than about 20 repeat units [128–130].

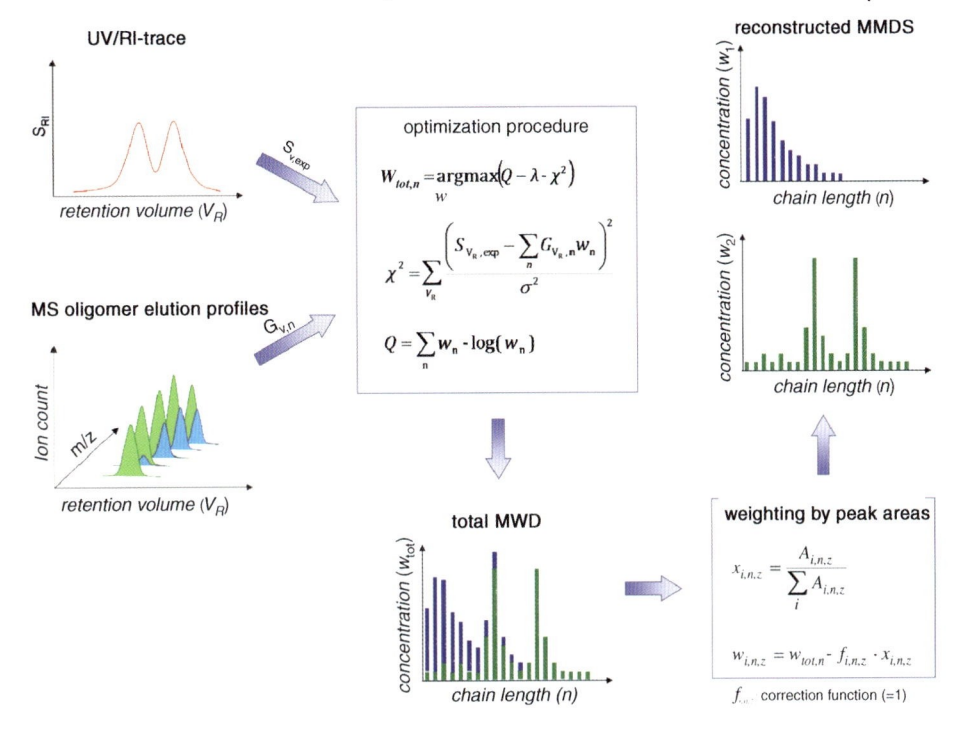

8.7 Quantitative MS and the Determination of MMDs by MS | 271

Figure 8.15 Flow diagram of the principal data processing approach. The instrumental calibration and band-broadening matrix $G_{V,n}$ derived by online mass spectrometry as well as the RI-detector trace $S_{V,\text{exp}}$ are processed to arrive at the deconvoluted total molecular mass distributions w_{tot}. Weighting of w_{tot} by the areas under the peaks of the individual functional oligomers $A_{i,n,z}$ yields w_i.

The MMD of a single species $\hat{w}_{i,n,z}$ is thus calculated by simply weighting the net MMD $\hat{w}_n$ with the ratio $x_{i,n,z}$ of the areas under the individual functional oligomer peaks, $A_{i,n,z}$ of the species to the total area of all functional oligomer elution profiles at the fixed charge state z and repeat unit number n (see Figure 8.15 – "weighting by peak areas").

$$\hat{w}_{i,n,z} = \hat{w}_n \cdot f_{i,n,z} \cdot x_{i,n,z} \quad \text{with} \quad x_{i,n,z} = \frac{A_{i,n,z}}{\sum_i A_{i,n,z}} \tag{8.15}$$

This approach is possible even in the presence of strong molecular mass influences on the ionization efficiency, as a quantitative rationing is carried out only between the abundance of different end-group-carrying polymers of the same repeat unit. Such a methodology is feasible as long as there is only a negligible effect of the end group on ionization efficiency in the electrospray source. Furthermore, all species with the same repeat unit need to arrive at the mass spectrometer at the same retention time, so that ionization occurs in the same chemical background. The latter assumption is valid in most cases, as the influence of the end group on hydrodynamic

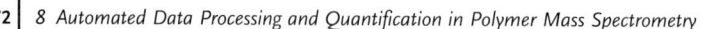

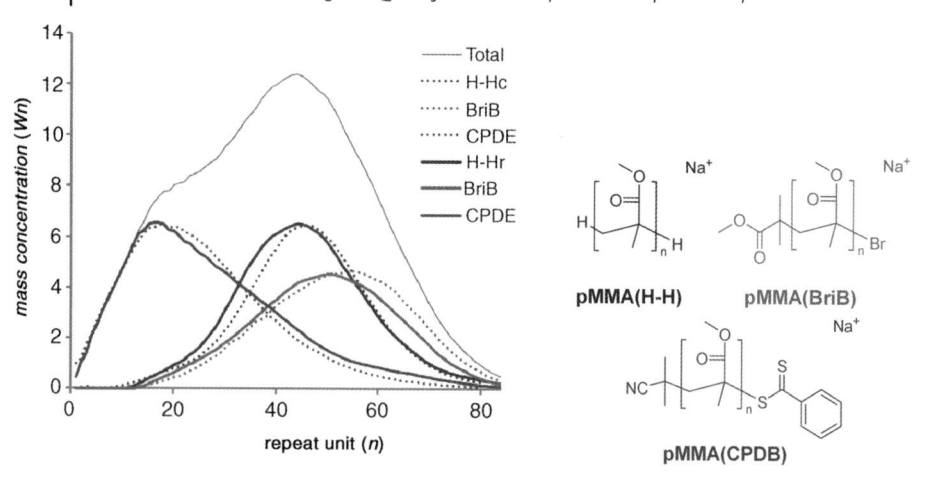

Figure 8.16 Reconstructed and original mass-weighted molecular mass distributions for a ternary 1 : 1 : 1$_w$ mixture of hydrogen-functional (H-H), bromine-functional (BriB), and dithiobenzoate functional (CPDB) PMMA.

volume is typically negligible when compared to the polymer backbone. To account for possible end-group bias, a correction factor $f_{i,n,z}$ may be introduced. Because of a lack of a proper functional description of the ionization process, negligible end-group influences were assumed in the current approach ($f \equiv 1$). A validation of this assumption as well as of the general applicability of the proposed method is given in the following paragraph.

Three functional polymer species were used to validate the developed method (see Figure 8.16): Commercial standards of PMMA synthesized by anionic group transfer polymerization and carrying only hydrogen as end group are denoted by PMMA (H-H). These standards were mixed in different weight ratios with PMMA synthesized by atom transfer radical polymerization carrying a bromine end group denoted by PMMA(BriB) and PMMA(CPDB) synthesized by cyanoisopropyl dithiobenzoate-mediated reversible addition fragmentation chain transfer polymerization.

The original molecular weight distributions of each species together with the reconstructions from a 1 : 1 : 1$_w$ mixture of these species are given in Figure 8.16. Generally, a good agreement was attained between the individual original molecular weight distributions and those that were reconstructed. The average degrees of polymerization of PMMA(CPDB) are overestimated by around 8%, whereas the degrees of polymerization of PMMA(H-H) and PMMA(BriB) are underestimated by around 2% and 5%, respectively. Agreement between the original and reconstructed area under the distribution (total mass concentration) is better, featuring a maximum deviation of 5%.

The roughly 5% lower molecular weight averages of the original distribution of PMMA(H-H) compared to the manufacturer-stated values are in agreement with earlier findings. Any systematic error in the reconstructed molecular weights is likely to be due to resolution limitations of the mass analyzer toward higher molecular weights, as well as to possible unrecognized side products in the polymer standards and adduct formation with salt and solvents. Baseline subtraction in the oligomer peak elution proved to be difficult in some cases, due to background suppression and unresolved peaks, thereby influencing correct calculation of the peak area ratios. The accuracy of the current method especially in cases where there are only minor amounts of one species present in a mixture can be improved. Proper baseline correction and safeguards against mass spectral overlap are thus important issues to be addressed in further investigations, in order to further extent the dynamic range and mass range of the method.

8.7.3
Comparison of the Two Methods for MMD Calculation

The methods presented in Sections 8.7.1 and 8.7.2 – although being based on two conceptually very different approaches – provide for the first time a means to measure the MMD of synthetic polymers with unrivaled accuracy. In the accessible molecular weight range of up to several tens of kilograms per mole, depending on instrumental resolution, these methods therefore pose serious alternatives for the molecular mass measurement of synthetic polymers by MS.

The typical areas of application of each method should be noted: The method based on MALDI-MS with internal calibration together with a very thorough error analysis developed at the NIST provides for the first time a tool to generate truly accurate molecular mass standards where the molecular mass uncertainties can be ultimately traced back to basic gravimetric and volumetric measurements. The polystyrene MMD standard, SRM 2881 is the final product of this effort. The method furthermore allows quantitative assessment of the mass bias observed in polymer analysis not just by MALDI-MS but by chromatographic and other methods as well. The determination of the absolute MMD opens the door for many experiments where the shape of the MMD plays a critical role in polymer behavior as in viscosity and rheology.

In the SEC/ESI-MS method on the other hand, a characterization of the molecular mass bias is not attempted or deemed possible. The method relies on the online internal molecular mass calibration of SEC, using MS only in its potential to determine the molecular weight and elution times of macromolecules with high accuracy and regardless of the polymer chemical identity as long as ionization can be achieved. Any data on the concentration of macromolecules eluting from the SEC column is obtained from a concentration-sensitive detector (an RI detector in the current case). Although relying on a couple of assumptions, this method is deemed to be especially useful when fast but at the same time accurate determinations of MMDs and compositions are required of many polymer species of differing chemical makeup (e.g., in an industrial laboratory setting or in high-throughput experimentation).

The availability of these two methods, which each rely on physically very different approaches to ultimately provide the same molecular mass information, allows an assessment of their accuracy by direct comparison. Figure 8.17 shows the number fraction of styrene oligomers against repeat unit length for SRM 2881, as determined by the NIST MALDI-based method (circles with 95% confidence interval) [101] and by

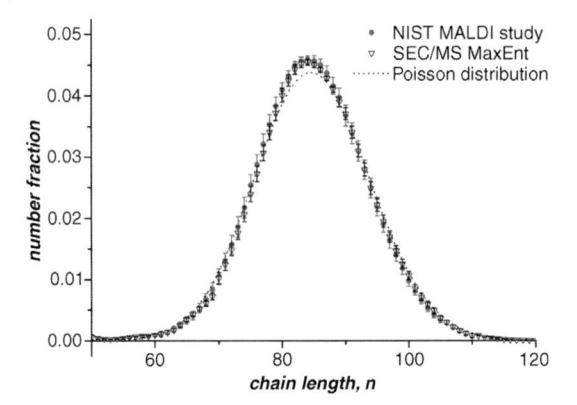

Figure 8.17 The MMD of the NIST polystyrene molecular mass standard SRM 2881. The NIST-certified number fractions (circles) were determined by MALDI-MS using an internal calibration procedure of the mass spectrometric intensity axis [101]. A comparison with the MMD determined by an SEC/ESI-MS analysis with data treatment by the MaxEnt method (triangles) [102] reveals the excellent agreement between the molecular mass data obtained from these two, conceptually very different methods.

a triple repeat measurement of the same standard employing SEC/ESI-MS (triangles) with ionization by $AgBF_4$ [102].[3] The excellent agreement of the MMDs serves as proof of the high accuracy of molecular mass data obtained from each of these two methods. This is further supported by an independent measurement of DP_n by ^{1}H-NMR in the case of the SEC/ESI-MS approach [102].

8.7.4
Simple Methods for the Determination of the Molar Abundance of Functional Polymers in Mixtures

The above described methods have highlighted approaches that can be employed – via the online coupling of SEC and ESI-MS – to arrive at exact mass distributions of components (i.e., individual chain termini differentiated chains) of homo- and copolymers, allowing for the extraction of absolute concentrations of individual chains. While these approaches are – in terms of accuracy – very reliable, they are also associated with a considerable instrumental and mathematical complexity. In the following text, a method is presented that allows to quantitatively evaluate mass spectra with regard to individual chain distributions with minimal computational demand, while providing satisfactory results. In an ideal scenario, one would have

---

[3] Note that the error bars of the MALDI-derived MMD are the 95% confidence intervals, based on a careful assessment of all errors including those in sample preparation (weighting, volumetric measurements, and spotting), whereas the SEC/MS 95% confidence intervals are based on a triple repeat measurement of a single standard solution and therefore only reflect repeatability of the analytical data acquisition process.

mass spectra at hand that display no mass bias at all. In reality, however, there may always be a certain amount of mass bias present. Let's assume a scenario where the ratio of two components in a polymer mixture (i.e., two chain distributions with different yet well-defined end groups) is to be quantified (the presented approach can readily be expanded to more components). In the most simple and straightforward approach for quantitatively evaluating mass data, the height of two peaks that do not show isobaric overlap (i.e., P_1 and P_2, the species of interest), Δh^{P_1}, is evaluated in each repeat unit. As the height of each nonoverlapping single isotopic peak (alternatively the integral could be employed) is proportional to the number of molecules corresponding to the associated mass, the individual peak heights can be employed to arrive at the mole fraction of one of the species, F, via Eq. (8.55). Note that F is given as $F(i)$, as it can be evaluated in every repeat unit, with i being the chain length to which the repeat unit corresponds.

$$F^{P_1}(i) = \frac{\Delta h^{P_1}(i)}{\Delta h^{P_1}(i) + \Delta h^{P_2}(i)} \tag{8.55}$$

While Eq. (8.55) can be a valuable tool as $F(i) = F$ indicates the absence of mass bias [131], it does not take into account any potential chain length-dependent ionization mass biases other than evaluation of every repeat unit. However, a further refinement allows for the elimination of the mass bias. A useful approach was taken by Günzler et al. in previous quantitative mass spectrometric evaluations [132, 133]: Let $G(i)$ be the ratio of the peak heights of P_1 and $P_2 \forall i$. This ratio may directly be plotted against chain length, i, and yields identical information to $F(i)$ as $F(i) = G(i) \cdot (G(i) + 1)^{-1}$. Now define $G'(i, i-1)$, $G''(i, i+1)$ in a similar manner to $G(i)$, however taking the height of $P_2 (\Delta h^{P_2})$ from one repeat unit higher (Eq. (8.57)) or lower (Eq. (8.58)). Thus, the vectors $G(i)$, $G'(i, i-1)$, and $G''(i, i+1)$ are obtained.

$$G(i) = \frac{\Delta h^{P_1}(i)}{\Delta h^{P_2}(i)} \tag{8.56}$$

$$G'(i, i-1) = \frac{\Delta h^{P_1}(i)}{\Delta h^{P_2}(i-1)} \tag{8.57}$$

$$G''(i, i+1) = \frac{\Delta h^{P_1}(i)}{\Delta h^{P_2}(i+1)} \tag{8.58}$$

In a subsequent step, $G(i)$, $G'(i, i-1)$, and $G''(i, i+1)$ are individually averaged $\forall i$, yielding the average values $\langle G \rangle$, $\langle G' \rangle$, and $\langle G'' \rangle$. $\langle G \rangle$ corresponds to the average ratio within the same repeat unit, $\langle G' \rangle$ corresponds to the average ratio within two repeat units with the second repeat unit being at smaller molecular weights, and $\langle G'' \rangle$ corresponds to the average ratio of within two repeat units with the second repeat unit being at larger molecular weights. With $\langle G \rangle$, $\langle G' \rangle$, and $\langle G'' \rangle$ at hand, one can plot these values against $\Delta m/z$, that is, the difference in Th between the positions $(i-1)$, i, and $(i+1)$. The y-intercept $(\Delta m/z = 0)$ of such a plot yields $\langle G \rangle^{\Delta m/z, 0}$, which represents the mass bias free ratio of the two products P_1 and P_2 in the polymer sample. In systems where the mass bias is negligible, all the evaluation procedures, that is, averaging $F(i)$, $G(i)$, and calculating $\langle G \rangle^{\Delta m/z,0}$ should give (near) identical results, providing a guide toward assessing whether more complex evaluation procedures – as detailed in the previous sections – are required. In a range of systems, the above approach has provided satisfactory results [131–133].

8.8
Conclusions and Outlook

The current chapter has provided an account of the contemporary status of the field of automated MS data processing in synthetic polymer chemistry. Especially in the recent couple of years, some important advances have been made in fields including copolymer compositional characterization, the determination of accurate MMDs, and the automated interpretation and integration of polymer MS data. The increasing use of hyphenated techniques including MS/MS [88, 96, 134–136], IMS [137, 138], and LC-MS [76, 103, 139, 140] lead to the generation of large spectroscopic datasets. The necessity of automated techniques for an efficient processing of these large and information-rich data in MS has been realized at an early point in the related fields of biomacromolecular analysis and proteomics. Fragmentation databases and highly sophisticated automated processing tools form an integral part of the MS-based proteomics today. Trends toward the creation of fragmentation databases to aid the interpretation of MS/MS data are seen also in the polymer analysis community [94, 95, 141], but the amount of freely available software tools is presently very limited, with the only example being Polymerator by Thalassinos *et al.* [95]. It is no question that the coming years will see the development of further sophisticated computational tools to aid the polymer science community. The synergies with biomacromolecular MS should be realized and the already (freely) available software tools [3, 4, 78] should be customized to synthetic polymer applications and further exploited.

References

1 Pedrioli, P.G.A., Eng, J.K., Hubley, R., Vogelzang, M., Deutsch, E.W., Raught, B., Pratt, B., Nilsson, E., Angeletti, R.H., Apweiler, R., Cheung, K., Costello, C.E., Hermjakob, H., Huang, S., Julian, R.K., Kapp, E., McComb, M.E., Oliver, S.G., Omenn, G., Paton, N.W., Simpson, R., Smith, R., Taylor, C.F., Zhu, W., and Aebersold, R. (2004) *Nat. Biotech.*, **22** (11), 1459–1466.

2 Listgarten, J. and Emili, A. (2005) *Mol. Cellular Proteomics*, **4** (4), 419–434.

3 Deutsch, E.W., Mendoza, L., Shteynberg, D., Farrah, T., Lam, H., Tasman, N., Sun, Z., Nilsson, E., Pratt, B., Prazen, B., Eng, J.K., Martin, D.B., Nesvizhskii, A.I., and Aebersold, R. (2010) *PROTEOMICS*, **10** (6), 1150–1159.

4 Deutsch, E.W., Lam, H., and Aebersold, R. (2008) *Physiol. Genomics*, **33** (1), 18–25.

5 Deutsch, E. (2008) *PROTEOMICS*, **8** (14), 2776–2777.

6 Shah, A.R., Davidson, J., Monroe, M.E., Mayampurath, A.M., Danielson, W.F., Shi, Y., Robinson, A.C., Clowers, B.H., Belov, M.E., Anderson, G.A., and Smith, R.D. (2010) *J. Am. Soc. Mass Spectrom.*, **21** (10), 1784–1788.

7 Further information can be found on the internet at http://psidev.info and http://tools.proteomecenter.org (last accessed: 12.12.2010).
8 Wallace, W.E., Guttman, C.M., Flynn, K.M., and Kearsley, A.J. (2007) *Anal. Chim. Acta*, **604** (1), 62–68.
9 Gruendling, T., Hart-Smith, G., Davis, T.P., Stenzel, M.H., and Barner-Kowollik, C. (2008) *Macromolecules*, **41** (6), 1966–1971.
10 Ladavière, C., Lacroix-Desmazes, P., and Delolme, F. (2008) *Macromolecules*, **42** (1), 70–84.
11 Jackson, A.T., Bunn, A., Priestnall, I.M., Borman, C.D., and Irvine, D.J. (2006) *Polymer*, **47**, 1044–1054.
12 Jiang, X., Schoenmakers, P.J., van Dongen, J.L.J., Lou, X., Lima, V., and Brokken-Zijp, J. (2003) *Anal. Chem.*, **75**, 5517–5524.
13 Murgasova, R. and Hercules, D.M. (2002) *Anal. Bioanal. Chem.*, **373**, 481–489.
14 Naidong, W., Chen, Y.-L., Shou, W., and Jiang, X. (2001) *J. Pharm. Biomed. Anal.*, **26**, 753–767.
15 Gemperline, P. (ed.) (2006) *Practical Guide to Chemometrics*, 2nd edn, CRC Press, Boca Raton, FL.
16 Asperger, A., Efer, J., Koal, T., and Engewald, W. (2001) *J. Chromatogr. A*, **937**, 65–72.
17 Sautour, M., Rouget, A., Dentigny, P., Divies, C., and Bensoussan, M. (2001) *J. Appl. Microbiol.*, **91**, 900–906.
18 Tak, V., Kanaujia, P.K., Pardasani, D., Kumar, R., Srivastava, R.K., Gupta, A.K., and Dubey, D.K. (2007) *J. Chromatogr. A*, **1161**, 198–206.
19 Rudaz, S., Cherkaoui, S., Gauvrit, J.-Y., Lantéri, P., and Veuthey, J.-L. (2001) *Electrophoresis*, **22**, 3316–3326.
20 Bateman, K., Seto, C., and Gunter, B. (2002) *J. Am. Soc. Mass. Spectrom.*, **13**, 2–9.
21 Riter, L.S., Vitek, O., Gooding, K.M., Hodge, B.D., and Julian, R.K.J. (2005) *J. Mass Spectrom.*, **40**, 565–579.
22 Knowles, J. (2009) *Computational Intelligence Magazine, IEEE*, **4** (3), 77–91.
23 Vaidyanathan, S., Broadhurst, D.I., Kell, D.B., and Goodacre, R. (2003) *Anal. Chem.*, **75** (23), 6679–6686.
24 O'Hagan, S., Dunn, W.B., Brown, M., Knowles, J.D., and Kell, D.B. (2005) *Anal. Chem.*, **77** (1), 290–303.
25 Wetzel, S.J., Guttman, C.M., Flynn, K.M., and Filliben, J.J. (2006) *J. Am. Soc. Mass Spectrom.*, **17** (2), 246–252.
26 Kelley, C.T. (1999) *Iterative Methods for Optimization*, Society for Industrial and Applied Mathematics, Philadelphia.
27 Spall, J.C. (2003) *Introduction to Stochastic Search and Optimization*, Wiley-Interscience, Hoboken, NJ.
28 Gilmore, P. and Kelley, C.T. (1995) *SIAM J. Optim.*, **5** (2), 269–285.
29 Gruendling, T., Guilhaus, M., and Barner-Kowollik, C. (2009) *Macromol. Rapid Commun.*, **30**, 589–597.
30 Meier, M.A.R., Adams, N., and Schubert, U.S. (2007) *Anal. Chem.*, **79**, 863–869.
31 Brandt, H., Ehmann, T., and Otto, M. (2010) *Anal. Chem.*, **82** (19), 8169–8175.
32 Brandt, H., Ehmann, T., and Otto, M. (2010) *Rapid Commun. Mass Spectrom.*, **24** (16), 2439–2444.
33 Brandt, H., Ehmann, T., and Otto, M. (2010) *J. Am. Soc. Mass Spectrom.*, **21** (11), 1870–1875.
34 Armanino, C. and Forina, M. (1987) *Chemometrics and Species Identification*, vol. 141, Springer, Berlin.
35 Zupan, J. (1989) *Algorithms for Chemists*, Wiley, New York.
36 Papas, A.N. (1989) *CRC Crit. Rev. Anal. Chem.*, **20** (6), 359–404.
37 Kateman, G. and Buydens, L. (1993) Data processing, in *Quality Control in Analytical Chemistry*, vol. 60, Wiley, New York.
38 Hippe, Z., Bierowska, A., and Pietryga, T. (1980) *Anal. Chim. Acta*, **122**, 279–290.
39 Savitsky, A. and Golay, M.J.E. (1964) *Anal. Chem.*, **36** (8), 1627–1639.
40 Eilers, P.H.C. (2003) *Anal. Chem.*, **75**, 3299–3304.
41 Vivó-Truyols, G. and Schoenmakers, P.J. (2006) *Anal. Chem.*, **78**, 4598–4608.
42 Kast, J., Gentzel, M., Wilm, M., and Richarson, K. (2003) *J. Am. Soc. Mass Spectrom.*, **14**, 766–776.

43 Zhang, P., Li, H., Zhou, X., and Wong, S. (2008) *Int. J. Hybrid Intel. Sys.*, **5**, 197–208.

44 Wiener, N. (1949) *Extrapolation, Interpolation, and Smoothing of Stationary Time Series with Engineering Applications*, Technology Press of MIT & John Wiley, New York.

45 Wallace, W.E. and Guttman, C.M. (2002) *J. Res. Nat. Inst. Stand. Tech.*, **107**, 1–17.

46 Wallace, W.E., Guttman, C.M., and Antonucci, J.M. (1999) *J. Am. Soc. Mass Spectrom.*, **10**, 224–230.

47 Tecklenburg, R.E., Wallace, W.E., and Chen, H. (2001) *Rapid Commun. Mass Spectrom.*, **15**, 2176–2185.

48 Luftmann, H. and Kehr, S. (2007) *e-polymers*, Art. No. 10.

49 Boggs, P.T., Byrd, R.H., and Schnabel, R.B. (1987) *SIAM J. Sci. Stat. Comput.*, **8** (6), 1052–1078.

50 Wallace, W.E., Kearsley, A.J., and Guttman, C.M. (2004) *Anal. Chem.*, **76**, 2446–2452.

51 Barrondale, I. (1968) *Appl. Stat.*, **17**, 51–57.

52 Barrondale, I. and Roberts, F.D.K. (1973) *SIAM J. Numer. Anal.*, **10** (5), 839–848.

53 Duda, R.O. and Hart, P.E. (1973) *Pattern Classification and Scene Analysis*, John Wiley and Sons, Inc., New York.

54 Kearsley, A.J., Wallace, W.E., and Guttman, C.M. (2005) *Appl. Math. Lett.*, **18**, 1412–1417.

55 Kearsley, A.J. (2006) *J. Res. Nat. Inst. Stand. Tech.*, **111** (2), 121–125.

56 Douglas, D.H. and Peucker, T.K. (1973) *Canadian Cartographer*, **10** (2), 112–122.

57 Boggs, P.T., Kearsley, A.J., and Tolle, J.W. (1999) *SIAM J. Opt.*, **9** (3), 755–778.

58 Beyer, W.H. (1981) *CRC Standard Mathematical Tables*, CRC Press, Boca Raton, FL.

59 Montaudo, M.S. (2002) *Mass Spectrom. Rev.*, **21** (2), 108–144.

60 Montaudo, A.S. and Montaudo, G. (1992) *Macromolecules*, **25**, 4264–4280.

61 Montaudo, M.S. and Montaudo, G. (1993) *Makromol. Chem. Makromol. Symp.*, **65**, 269.

62 Zoller, D.L. and Johnston, M.V. (1997) *Anal. Chem.*, **69**, 3791–3795.

63 Willemse, R.X.E., Staal, B.B.P., Donkers, E.H.D., and van Herk, A.M. (2004) *Macromolecules*, **37** (15), 5717–5723.

64 Simonsick, W.J. and Prokai, L. (1995) *Chromatographic Characterization of Polymers* (eds T. Provder, H., Barth and M. Urban), ACS, Washington, DC, p. 41.

65 Nuwaysir, L.M., Wilkins, C.L., and Simonsick, W.J. (1990) *J. Am. Soc. Mass. Spectrom.*, **1**, 66.

66 Montaudo, G., Montaudo, M.S., Scamporrino, E., and Vitalini, D. (1992) *Macromolecules*, **25**, 5099.

67 Wilczek-Vera, G., Yu, Y., Waddell, K., Danis, P.O., and Eisenberg, A. (1999) *Rapid Commun. Mass Spectrom.*, **13** (9), 764–777.

68 Wilczek-Vera, G., Danis, P.O., and Eisenberg, A. (1996) *Macromolecules*, **29** (11), 4036–4044.

69 Wilczek-Vera, G., Yu, Y., Waddell, K., Danis, P.O., and Eisenberg, A. (1999) *Macromolecules*, **32** (7), 2180–2187.

70 Suddaby, K.G., Hunt, K.H., and Haddleton, D.M. (1996) *Macromolecules*, **29** (27), 8642–8649.

71 Willemse, R.X.E. and van Herk, A.M. (2006) *J. Am. Chem. Soc.*, **128** (13), 4471–4480.

72 Huijser, S., Staal, B.B.P., Huang, J., Duchateau, R., and Koning, C.E. (2006) *Biomacromolecules*, **7** (9), 2465–2469.

73 Huijser, S., Staal, B.B.P., Huang, J., Duchateau, R., and Koning, C.E. (2006) *Angew. Chem. Int. Ed.*, **45** (25), 4104–4108.

74 Staal, B.B.P. (2005) *Characterization of (co)polymers by MALDI-TOF-MS*, Technische Universiteit Eindhoven, Eindhoven.

75 Weidner, S.M., Falkenhagen, J., Maltsev, S., Sauerland, V., and Rinken, M. (2007) *Rapid Commun. Mass Spectrom.*, **21** (16), 2750–2758.

76 Weidner, S., Falkenhagen, J., Krueger, R.-P., and Just, U. (2007) *Anal. Chem.*, **79** (13), 4814–4819.

77 Vivó-Truyols, G., Staal, B., and Schoenmakers, P.J. (2010) *J. Chromatogr. A*, **1217** (25), 4150–4159.

78 Aebersold, R. and Mann, M. (2003) *Nature*, **422** (6928), 198–207.

79 Yates, J.R., Ruse, C.I., and Nakorchevsky, A. (2009) *Annu. Rev. Biomed. Eng.*, **11** (1), 49–79.

80 Selby, T.L., Wesdemiotis, C., and Lattimer, R.P. (1994) *J. Am. Soc. Mass Spectrom.*, **5** (12), 1081–1092.

81 Lattimer, R.P. (1994) *J. Am. Soc. Mass Spectrom.*, **5** (12), 1072–1080.

82 Lattimer, R.P. (1992) *J. Am. Soc. Mass Spectrom.*, **3** (3), 225–234.

83 Lattimer, R.P. (1992) *Int. J. Mass Spec. Ion Proc.*, **116** (1), 23–36.

84 Jackson, A.T., Green, M.R., and Bateman, R.H. (2006) *Rapid Commun. Mass Spectrom.*, **20** (23), 3542–3550.

85 Jackson, A.T., Bunn, A., Priestnall, I.M., Borman, C.D., and Irvine, D.J. (2006) *Polymer*, **47** (4), 1044–1054.

86 Jackson, A.T., Scrivens, J.H., Williams, J.P., Baker, E.S., Gidden, J., and Bowers, M.T. (2004) *Int. J. Mass Spectrom.*, **238** (3), 287–297.

87 Jackson, A.T., Slade, S.E., and Scrivens, J.H. (2004) *Int. J. Mass Spectrom.*, **238** (3), 265–277.

88 Jackson, A.T., Bunn, A., Hutchings, L.R., Kiff, F.T., Richards, R.W., Williams, J., Green, M.R., and Bateman, R.H. (2000) *Polymer*, **41** (20), 7437–7450.

89 Borman, C.D., Jackson, A.T., Bunn, A., Cutter, A.L., and Irvine, D.J. (2000) *Polymer*, **41** (15), 6015–6020.

90 Jackson, A.T., Yates, H.T., Scrivens, J.H., Green, M.R., and Bateman, R.H. (1998) *J. Am. Soc. Mass Spectrom.*, **9** (4), 269–274.

91 Scrivens, J.H., Jackson, A.T., Yates, H.T., Green, M.R., Critchley, G., Brown, J., Bateman, R.H., Bowers, M.T., and Gidden, J. (1997) *Int. J. Mass Spectrom. Ion Proc.*, **165–166**, 363–375.

92 Jackson, A.T., Yates, H.T., Scrivens, J.H., Green, M.R., and Bateman, R.H. (1997) *J. Am. Soc. Mass Spectrom.*, **8** (12), 1206–1213.

93 Jackson, A.T., Yates, H.T., Scrivens, J.H., Critchley, G., Brown, J., Green, M.R., and Bateman, R.H. (1996) *Rapid Commun. Mass Spectrom.*, **10** (13), 1668–1674.

94 Baumgaertel, A., Weber, C., Knop, K., Crecelius, A., and Schubert, U.S. (2009) *Rapid Commun. Mass Spectrom.*, **23** (6), 756–762.

95 Thalassinos, K., Jackson, A.T., Williams, J.P., Hilton, G.R., Slade, S.E., and Scrivens, J.H. (2007) *J. Am. Soc. Mass Spectrom.*, **18** (7), 1324–1331.

96 Jackson, A., Slade, S., Thalassinos, K., and Scrivens, J. (2008) *Anal. Bioanal. Chem.*, **392** (4), 643–650.

97 Jackson, A.T., Thalassinos, K., John, R.O., McGuire, N., Freeman, D., and Scrivens, J.H. (2010) *Polymer*, **51** (6), 1418–1424.

98 Karas, M. and Hillenkamp, F. (1988) *Anal. Chem.*, **60** (20), 2299–2301.

99 Fenn, J.B. (2003) *Angew. Chem. Int. Ed.*, **42** (33), 3871–3894.

100 Barner-Kowollik, C., Davis, T.P., and Stenzel, M.H. (2004) *Polymer*, **45**, 7791–7805.

101 Guttman, C.M., Flynn, K.M., Wallace, W.E., and Kearsley, A.J. (2009) *Macromolecules*, **42** (5), 1695–1702.

102 Gruendling, T., Guilhaus, M., and Barner-Kowollik, C. (2008) *Anal. Chem.*, **80**, 6915–6927.

103 Gruendling, T., Guilhaus, M., and Barner-Kowollik, C. (2009) *Macromolecules*, **42** (17), 6366–6374.

104 Wallace, W.E. and Guttman, C.M. (2010) *Molecular Mass Distribution Measurement by Mass Spectrometry; Special Publication 960-21*, National Institute of Standards and Technology, Gaithersburg, Maryland, www.nist.gov/public_affairs/factsheet/practiceguides.cfm (last accessed: 8/11/2011).

105 Wallace, W.E., Flynn, K.M., Guttman, C.M., VanderHart, D.L., Prabhu, V.M., De Silva, A., Felix, N.M., and Ober, C.K. (2009) *Rapid Commun. Mass Spectrom.*, **23**, 1957–1962.

106 Park, E.S., Wallace, W.E., Guttman, C.M., Flynn, K.M., Richardson, M.C., and Holmes, G.A. (2009) *J. Am. Soc. Mass Spectrom.*, **20**, 1638–1644.

107 Kostanski, L.K., Keller, D.M., and Hamielec, A.E. (2004) *J. Biochem. Biophys. Methods*, **58** (2), 159–186.

108 Grubisic, Z., Rempp, P., and Benoit, H. (1967) *J. Polym. Sci. B Polym. Lett.*, **5**, 753–759.

109 Zammit, M.D. and Davis, T.P. (1997) *Polymer*, **38**, 4455–4468.

110 Zammit, M.D., Coote, M.L., Davis, T.P., and Willetto, G.D. (1998) *Macromolecules*, **31**, 955–963.
111 Tamai, Y., Konishi, T., Einaga, Y., Fujii, M., and Yamakawa, H. (1990) *Macromolecules*, **23** (18), 4067–4075.
112 Flory, P.J. and Fox Jr, T.G. (1951) *J. Am. Chem. Soc.*, **73**, 1904–1908.
113 Buback, M., Busch, M., and Lämmel, R.A. (1996) *Macromol. Theory Simul.*, **5** (5), 845–861.
114 Schnoll-Bitai, I. and Mader, C. (2006) *J. Chrom. A*, **1137** (2), 198–206.
115 Aaserud, D.J., Prokai, L., and Simonsick, W.J. (1999) *Anal. Chem.*, **71** (21), 4793–4799.
116 Prokai, L. and Simonsick, W.J. (1993) *Rapid Comm. Mass Spec.*, **7**, 853–856.
117 Tung, L.H. and Runyon, J.R. (1969) *J. Appl. Polym. Sci.*, **13** (11), 2397–2409.
118 Meira, G.R. and Vega, J.R. (2001) *Dekker Encyclopedia of Chromatography* (ed. J. Cazes), Marcel-Dekker, New York, p. 71.
119 Baumgarten, J.L., Busnel, J.P., and Meira, G.R. (2002) *J. Liq. Chromatogr. Rel. Technol.*, **25** (13–15), 1967–2001.
120 Mendel, J.M. (1995) Least-squares estimation: Singular-value decomposition, in *Lessons in Estimation Theory for Signal Processing, Communications and Control*, Prentice-Hall, New Jersey, pp. 44–57.
121 Reiter, J. (1992) *J. Comput. Phys.*, **103** (1), 169–183.
122 Skilling, J. and Bryan, R.K. (1984) *Mon. Not. R. Astr. Soc.*, **211** (1), 111–124.
123 Splinter, S.J. and McIntyre, N.S. (1998) *Surf. Interface. Anal.*, **26** (3), 195–203.
124 Shannon, C.E. (1948) *Bell System Tech. J.*, **27**, 379 & 623.
125 Shore, J.E. and Johnson, R.W. (1980) *IEEE Trans.*, **IT-26**, 26.
126 Shore, J.E. and Johnson, R.W. (1983) *IEEE Trans.*, **IT-29**, 942.
127 Sivia, D.S. (1996) *Data Analysis: A Bayesian Tutorial*, Oxford University Press, New York.
128 Itakura, M., Sato, K., Lusenkova, M.A., Matsuyama, S., Shimada, K., Saito, T., and Kinugasa, S. (2004) *J. Appl. Polym. Sci.*, **94**, 1101–1106.
129 Gridnev, A.A., Ittel, S.D., and Fryd, M. (1995) *J. Polym. Sci. A Polym. Chem.*, **33**, 1185–1188.
130 Wagner, H.L. and Hoeve, C.A.J. (1971) *J. Polym. Sci. A-2*, **9**, 1763–1776.
131 Koo, S.P.S., Junkers, T., and Barner-Kowollik, C. (2008) *Macromolecules*, **42** (1), 62–69.
132 Günzler, F., Wong, E.H.H., Koo, S.P.S., Junkers, T., and Barner-Kowollik, C. (2009) *Macromolecules*, **42** (5), 1488–1493.
133 Buback, M., Günzler, F., Russell, G.T., and Vana, P. (2009) *Macromolecules*, **42** (3), 652–662.
134 Crecelius, A.C., Baumgaertel, A., and Schubert, U.S. (2009) *J. Mass Spectrom.*, **44** (9), 1277–1286.
135 Giordanengo, R., Viel, S., Allard-Breton, B., Thévand, A., and Charles, L. (2009) *Rapid Commun. Mass Spectrom.*, **23** (11), 1557–1562.
136 Giordanengo, R., Viel, S., Allard-Breton, B., Thévand, A., and Charles, L. (2009) *J. Am. Soc. Mass Spec.*, **20** (1), 25–33.
137 Trimpin, S. and Clemmer, D.E. (2008) *Anal. Chem.*, **80** (23), 9073–9083.
138 Trimpin, S., Plasencia, M., Isailovic, D., and Clemmer, D.E. (2007) *Anal. Chem.*, **79**, 7965–7974.
139 Julka, S., Cortes, H., Harfmann, R., Bell, B., Schweizer-Theobaldt, A., Pursch, M., Mondello, L., Maynard, S., and West, D. (2009) *Anal. Chem.*, **81** (11), 4271–4279.
140 Nielen, M.W.F. and Buijtenhuijs, F.A. (1999) *Anal. Chem.*, **71** (9), 1809–1814.
141 Baumgaertel, A., Becer, C.R., Gottschaldt, M., and Schubert, U.S. (2008) *Macromol. Rapid Commun.*, **29** (15), 1309–1315.

9
Comprehensive Copolymer Characterization
Anna C. Crecelius and Ulrich S. Schubert

9.1
Introduction

Copolymers are polymers with at least two monomeric species. As a consequence, their analysis is considerably more challenging than that of homopolymers. In addition to the determination of the average molar masses and the elucidation of the repeating units and end groups, the topology distribution can be analyzed by mass spectrometry (MS). During the synthesis of copolymers, a variety of structures, such as block, gradient, random, and alternating copolymers can be obtained, as presented in Figure 9.1. Since the structure of copolymers directly affects their physical and chemical properties, this knowledge is essential for scientists working in the field of polymers.

Nowadays soft ionization techniques, such as matrix-assisted laser desorption/ ionization (MALDI) as well as electrospray ionization (ESI), are frequently employed in the field of polymers as a complementary characterization technique to size exclusion chromatography (SEC), nuclear magnetic resonance (NMR) spectroscopy, and light scattering (LS). Nevertheless, MS frequently requires a preseparation for example, by employing liquid chromatography (LC) to reduce the sample complexity. A variety of LC-MS forms have been employed in the field of polymers, such as SEC-MS and LC at critical conditions (LC-CC) combined with MS. All these LC techniques prior to MS analysis have been applied for the analysis of copolymers and will be discussed in the current chapter. Another separation technique, which has been made use of before the detection of the generated ions, is ion mobility spectrometry (IMS). The combination of IMS with MS (IMS-MS) has also been performed on complex copolymers.

The easiest way to obtain more information regarding the structural characteristics of polymers is to study them by tandem MS (MS/MS). Even though the ion selection is not easy in MS/MS experiments since often overlapping peaks are present, several examples of the sequence determination of copolymers have been described in the literature and are discussed in the current chapter. The quantification of monomer ratios in a copolymer sample has rarely been attempted via MS since it is rather

Mass Spectrometry in Polymer Chemistry, First Edition.
Edited by Christopher Barner-Kowollik, Till Gruendling, Jana Falkenhagen, and Steffen Weidner
© 2012 Wiley-VCH Verlag GmbH & Co. KGaA. Published 2012 by Wiley-VCH Verlag GmbH & Co. KGaA.

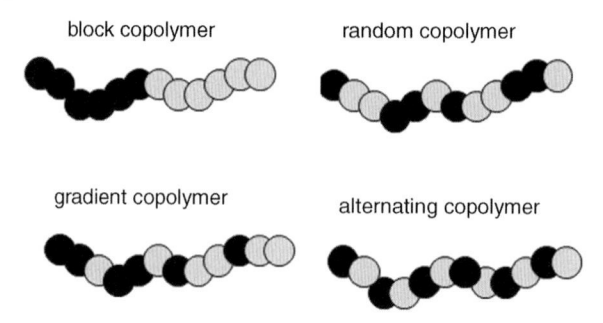

Figure 9.1 Copolymer structures.

difficult and influenced by a variety of factors. However, the few examples which exist will be discussed here.

One of the common applications of copolymers containing poly(ethylene glycol) (PEG) as one block is their use as an active targeted drug-delivery system. Further copolymers, which have been recently described in the literature and which are applicable in the biological and (bio)medical field, will be discussed. Finally, new software developments for enhancing the analysis of copolymers by MALDI-time of flight MS (MALDI-TOF MS) will be presented and evaluated.

9.2
Scope

The aim of this chapter is to provide a compact overview about the literature on synthetic copolymer analysis using MS during the time period of 2002 and the middle of 2010. The year 2002 was selected as starting point because at that time a review written by Montaudo relating to mass spectra of random and block copolymers appeared [1]. We tried to categorize the comprehensive literature in several subheadings to make it easier for the reader to follow. In some cases, the reviewed literature does not only fit into one category; however, repetitions are avoided.

9.3
Reviews

Several reviews dealing with MS of polymers were published approximately every 2 years in the time period considered, covering the mass spectrometric analysis of copolymers as a subitem [2–5]. The characterization of copolymers was presented by Montaudo et al. [6] as one subcategory using the soft ionization technique MALDI. Crecelius et al. [7] described in their review several examples of copolymers, which were analyzed by tandem MS, for example, MALDI-MS/MS or ESI-MS/MS. Moreover, recently an update of the emerging field of MS in polymer chemistry was provided by Gruendling et al. [8].

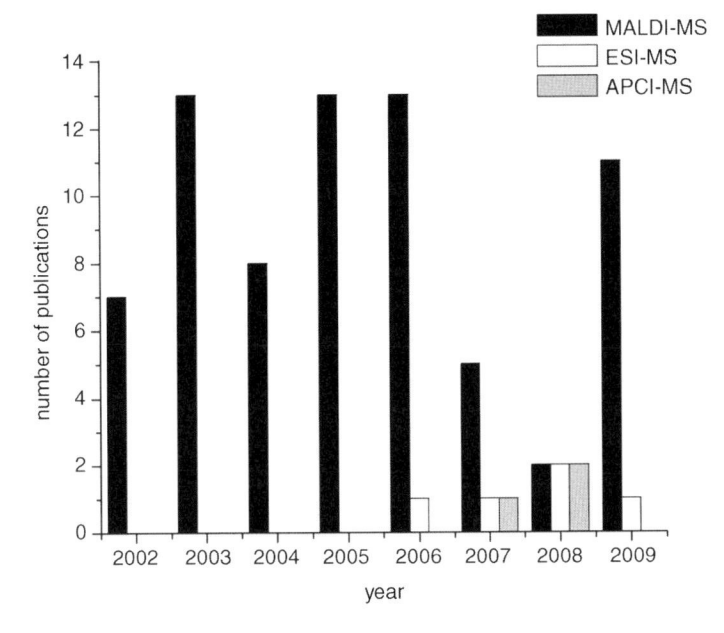

Figure 9.2 Histogram showing the number of papers published between 2002 and 2009, which are discussed in the following three sections in the field of MS of copolymers. The publications are sorted according to the soft ionization technique employed. (Database: SCIFinder, key words: MS and copolymers, date of search: 30.11.2010).

9.4
Soft Ionization Techniques

MALDI seems to be the most promising soft ionization technique employed for the analysis of copolymers, indicated by the number of publications between 2002 and 2009 (see Figure 9.2). The simplicity of the mass spectra, in which typically only singly charged quasi-molecular ions are obtained, makes this ionization technique particularly attractive. Between 2005 and 2006, a maximum can be observed in Figure 9.2 in the number of publications dealing with the analysis of copolymers employing MALDI. Thereafter, ESI and atmospheric pressure chemical ionization (APCI) have started to be investigated more intensively by scientists working in the field of polymers. However, the potential and advantages of using APCI as an ionization technique have not been fully explored so far, possibly due to the thermal decomposition occurring for high-molar-mass polymers.

9.4.1
MALDI

The main application of MALDI in the analysis of copolymers is the characterization of polymers in terms of average molar masses, repeating units, end groups, and architecture. All studies summarized in this section deal with this aspect. Chen *et al.* [9] synthesized conjugated polymers, which have been shown to be the active component in blue light-emitting diodes, through Suzuki polycondensation reaction in the presence of a $Pd(PPh_3)_4$ catalyst. The authors employed MALDI-TOF MS to confirm the proposed end groups by overlaying the simulated isotopic pattern over the observed one. With this technique, they could confirm that hydrogen was the most common end group. The same author synthesized [10] five rod-coil diblock polymers with blue light-emitting properties. The polymers consisted of two monomer units: ethylene oxide, which acted as flexible coils, and *p*-phenylene, which acted as rigid rods. MALDI-TOF MS was used to confirm that the targeted molecules were successfully synthesized by determining the number average molar mass (M_n), the weight average molar mass (M_w), the number of the repeating units, and the end-group masses of the five copolymers. Additionally, by changing the alkali metal ions, used for a better ionization of the polymer, different affinities between the polymer and the alkali metal ions were obtained. The minimum quantity of lithium chloride was determined, which was required for the analysis of the studied polymers.

MALDI-TOF MS was also employed by Chevallier *et al.* [11] to identify the synthesized poly(amide ester)s in terms of their size and cyclic or linear structure created by a polytransesterification or a polycondensation reaction. The polytransesterification yielded only oligomers, and the presence of macrocycles was unambiguously determined by MALDI-TOF MS. The polycondensation led to polymers whatever comonomers were employed and macrocycles were also detected in this case, however with a less relative abundance compared to the polytransesterification. Sen and colleagues [12] demonstrated in their publication that the copolymerization of methyl acrylate with norbornene derivatives by atom transfer radical polymerization (ATRP) could successfully be performed. Besides SEC and NMR spectroscopy, these authors used MALDI-TOF MS as one characterization method to investigate their generated random copolymers. With MALDI-TOF MS, the authors could show that copolymers were formed and not a mixture of homopolymers. The research group of Hercules [13] investigated polyester-based polyurethanes (PURs) among other characterization methods by MALDI-TOF MS. The MALDI-TOF MS analysis of partially acid hydrolyzed PURs confirmed the in the literature proposed degradation mechanism. Viala *et al.* [14] used MALDI-TOF MS to show that the synthesized copolymer consisting of 1,1-diphenylethylene and methyl methacrylate does neither show a statistical nor a block structure, but a composition with a high content of 1,1-diphenylethylene, distributed punctually in the polymer chain.

In 2002, Montaudo [15] proposed a model for the bivariate distribution of the chain size and composition in copolymers. Random copolymers (consisting of methyl methacrylate and butylacrylate and of styrene as well as maleic-anhydride) were fractionated by SEC and analyzed by NMR spectroscopy and MALDI-TOF MS to create this model. In a follow-up study, Montaudo *et al.* [16] replaced this model consisting of one single entity with a new model, which included the sum of a two-bivariate distribution. The new model showed, according to Montaudo *et al.*, satisfactory results by applying it to experimental and theoretical data of block and random copolymers.

Montaudo and colleagues investigated in two successive studies, the exchange reaction occurring during the melt mixing of nylon-6 with either poly(ethylene terephthalate) [17] or poly(butylene terephthalate) [18]. In both cases, NMR spectroscopy and MALDI-TOF MS analysis of the reaction products allowed the identification of the structure, composition, sequence distribution, and copolymer yield of the copolyesteramides produced in the exchange. The essential role of carboxyl end groups in the exchange reaction was shown and a detailed mechanism of both reactions was presented. Puglisi *et al.* [19], from the same research group, showed in a follow-up study that an exchange reaction took place during the melt mixing of nylon-6,10 and nylon-6,6. In this study, NMR spectroscopy and MALDI-TOF MS were also used for the detection and determination of the sequence and composition of the formed copolyamides. An additional study by the same group [20] demonstrated once more the excellent agreement between NMR and MALDI-TOF-MS data for the sequence analysis of copolyamides, synthesized by melt mixing. This time nylon-6 with carboxyl chain ends was melt mixed with high-molar-mass samples of nylon-4,6 and nylon-6,10, respectively. Finally, Montaudo and colleagues [21, 22] summarized in two successive reviews the results obtained in the previously described melt-mixing experiments. Montaudo and coworkers [21] underlined the remarkable progress which was gained in the understanding of the exchange reactions by using MALDI-TOF MS and described [22] the possibility to monitor the yield of the reactive blending reactions by measuring the amount of unreacted homopolymers employing MALDI-TOF MS. The authors also discussed the results obtained using conventional methods, for example, LC and NMR spectroscopy.

The research group of Pasch analyzed for the first time phenol-urea-formaldehyde cocondensates by MALDI-TOF MS using 2,5-dihydroxybenzoic acid or 2,4,6-trihydroxyacetophenone as matrices [23]. The analytical method was established by investigating commercially available phenolic resoles. From the recorded mass spectra, the oligomer peaks could be assigned to a specific chemical structure. As a consequence, the degree of polymerization and the number of reactive methylol groups could be determined. In an additional study, the authors presented in a convincing manner that MALDI-TOF MS can resolve questions at the molecular level, a feat that cannot be accomplished with ^{13}C-NMR spectroscopy [24]. Hong *et al.* [25] constructed a block length distribution map of a poly(ethylene oxide)-*b*-poly-(L-lactide) diblock copolymer ($M_w \cong 2600 \text{ g mol}^{-1}$) by MALDI-TOF MS. Although the resulting MALDI-TOF MS spectra were complicated, most polymer species could be identified by isolating the overlapping isotope patterns and fitting them to the Schulz–Zimm distribution function. The reconstructed spectrum was nearly identical to the measured one highlighting the potential of this procedure as an easy and fast method for characterizing low-molar-mass block copolymers.

Meier *et al.* [26] reported in their contribution the recent advances made for the challenging characterization of supramolecular block copolymers based on bis-terpyridine-ruthenium metal complexes using SEC and MALDI-TOF MS. The MALDI-TOF MS spectra recorded of this kind of block copolymers demonstrated, besides the signals for the intact bis-complexed block copolymer, peaks for both blocks connected via the metal complex. An example of such a MALDI-TOF MS

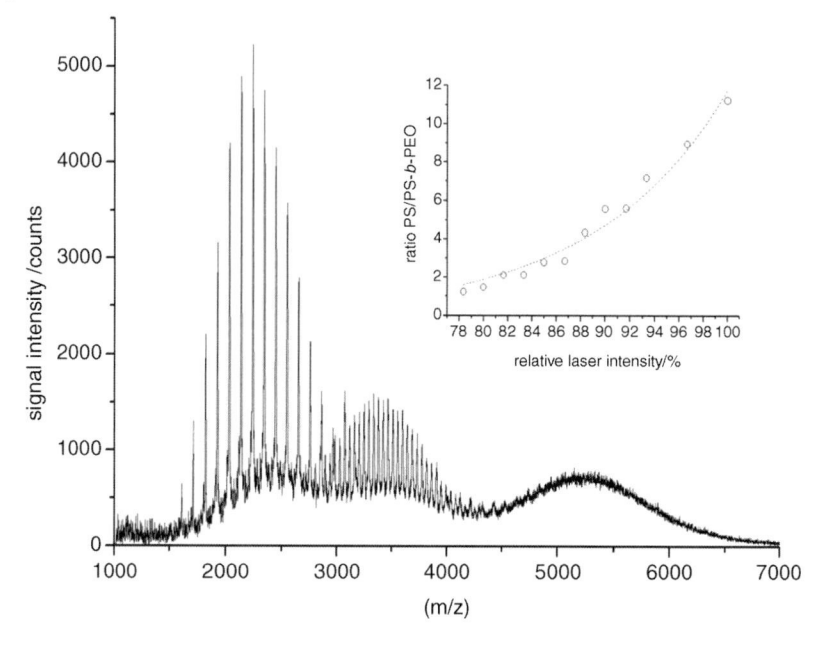

Figure 9.3 MALDI-TOF MS spectrum of PS_{20}-b-PEO_{70}. Inset: Fragmentation behavior of the investigated supramolecular block copolymer in dependence of the applied laser intensity during MALDI analysis. Reproduced from Ref. [26] with kind permission of Wiley-VCH Verlag GmbH & Co. KGaA.

spectrum is presented in Figure 9.3 for the supramolcular copolymer consisting of 20 poly(styrene) (PS) and 70 poly(ethylene oxide) (PEO) repeating units. In the m/z range of 4000–5000, the unfragmented polymer distribution is visible. Both the PS and the PEO building blocks are presented in the m/z range of 1500–3000 and 2800–4200, respectively. From the literature, it is very well-known that such supramolecular assemblies tend to fragment during MALDI analysis. Therefore, by increasing the laser intensity, the copolymer dissociates further to the corresponding individual blocks. This behavior is shown in the inset of Figure 9.3 by plotting the relative laser intensity versus the PS/PS-b-PEO ratio. The ratio increases with increasing relative laser intensity indicating a more pronounced fragmentation behavior of the copolymer sample.

Milani and colleagues [27] tested in their work the application of MALDI-TOF MS for the analysis of CO/styrene/4-methylstyrene terpolymers. For the first time, the authors could show that different amounts of the two olefin monomers in the polymer chain are clearly observable in the recorded mass spectra. A further study demonstrated that an asymmetric alternating copolymerization of cyclohexene oxide and CO_2 takes place when a dimeric zinc complex, which was synthesized in the first step, is added [28]. The MALDI-TOF MS analysis of the obtained copolymer revealed that the copolymerization was initiated by inserting CO_2 into Zn-alkoxide. Using chromium complexes with salalen ligands instead of a dimeric zinc complex resulted in the formation of similar copolymers [29]. In this study, MALDI-TOF MS revealed that the copolymerization was initiated by chloride ions. The preparation of a bisphenol–A carbonate copolymer, containing Cu-diimine units with nonlinear optical properties, was reported by Scamporrino *et al.* [30]. The authors employed MALDI-TOF MS to gain information regarding the composition and structure of the synthesized copolymer. Additionally UV/Vis spectroscopy, SEC, and differential scanning calorimetry (DSC) were employed to estimate the copolymer composition, the average molar masses, and the thermal properties, respectively. The anionic polymerization of ethylene oxide was monitored by Müller and coworkers [31]. In order to obtain more knowledge about the induction period, MALDI-TOF MS measurements were performed on samples taken during the synthesis of a low-molar-mass PS-*b*-PEO diblock copolymer. The conditions for the MALDI-TOF MS analysis were selected in such a way that a resolved copolymer sequence could be obtained in the recorded mass spectra. Williams *et al.* [32] described for the first time the application of synthetic discrete mass oligomers as a quantitative model for the MALDI-TOF MS analysis of macromolecular systems. To the best of our knowledge, this is the only publication, where the influence of sample preparation and instrumental parameters on the molar mass determination of copolymers was investigated in detail.

Cox *et al.* [33] explored the analysis of low-molar-mass ethylene/carbon monoxide copolymers by MALDI-MS. First, a TOF instrument was employed obtaining low-resolution copolymer spectra, even when the carbon monoxide monomer units were reduced to hydroxyl groups. Subsequently, a Fourier transform-ion cyclotron resonance (FT-ICR) instrument was used, capable of a much higher resolving power, allowing the separation between copolymer signals having the same number of monomers, yet different amounts of ethylene and carbon monoxide. Finally, this enabled the determination of the composition and microstructure of the copolymer samples. In a later publication, Cox and colleagues [34] characterized low-molar-mass polyolefin copolymers consisting of isobutylene and *p*-methylstyrene by MALDI-TOF MS. The authors discussed the differences obtained for the average composition by MALDI compared to NMR. In their opinion, the preferential ionization of oligomers with greater *p*-methylstyrene content caused the above discrepancy.

Venkatesh *et al.* [35] used in their investigation MALDI-TOF MS to confirm the formation of copolymers obtained by ATRP and reversible addition-fragmentation chain transfer (RAFT) of acrylates. The authors could demonstrate that several units of the chain-transfer agent allyl butyl ether (ABE) were incorporated in the polymer chain and therefore concluded that ABE behaves as a comonomer under the studied experimental conditions. In a second contribution by Vekatesh *et al.* [36], the copolymers obtained by ATRP and free radical polymerization (FRP) of methyl acrylate with 1-octene were analyzed via MALDI-TOF MS. For the copolymers synthesized by ATRP, only one pair of end groups was observed, in contrast to FRP, where several end groups were recorded. Generally more narrow chemical composition distributions of the copolymer samples were obtained by ATRP as compared to FRP. Müller and coworkers [37] described for the first time the synthesis of the block copolymer poly(*N*-isopropylacrylamide)-*b*-poly(acrylic acid) using the RAFT process.

The determination of the average molar masses was performed, besides SEC and LS, by MALDI-TOF MS. The recorded MALDI-TOF MS spectra indicated not only singly but also multiple-charged ions possibly due to the easy ionization of the acrylic acid blocks. In all cases, the values obtained by SEC were close to one order of magnitude higher than the corresponding MALDI data. Several polystyrene-b-polyisoprene copolymers were investigated by MALDI-TOF MS as presented by van Herk and colleagues [38]. From the resulting MALDI-TOF MS spectra, the authors could calculate the chain length of both blocks, although overlapping isotopic peaks were obtained. The results were further elucidated to random coupling statistics highlighting the accuracy of the applied method. Additionally, Willemse et al. [39] gained insight into the microstructure of two randomly distributed oligoesters synthesized by an esterification of 1,4-butanediol with adipic acid and isophthalic acid or glutaric acid by employing again the soft ionization technique MALDI-TOF MS. The microstructure of the copolymers obtained by MALDI-TOF MS was fitting well to Bernoullian chain statistics, which describes a random distribution of the repeating units. Finally, the same research group [40] successfully applied the combination of pulsed laser polymerization and MALDI-TOF MS on the model copolymerization system methyl methacrylate-styrene.

Krishnan and Srinivasan [41] confirmed the structure of macroinitiatiors and block copolymers of PEO synthesized by ATRP using MALDI-TOF MS, besides other characterization techniques, for example, Fourier transform infrared (FT-IR), SEC, and NMR spectroscopy. The MALDI-TOF MS spectra indicated that triblock copolymers of an ABA structure were produced. The group of Pasch [42] reported in their contribution the analysis of comb-like copolymers of PEO and poly(methacrylic acid) prepared by copolymerization of a macromonomer and methylacrylic acid. The authors summarized their achieved results using the following analytical techniques: SEC, MALDI-TOF MS, 2D-separation, and LC-CC. By using such a variety of analytical techniques, different parameters of the molecular heterogeneity of such hydrophilic copolymers could be explored.

Gies et al. [43] investigated the end groups of several poly(m-phenylene isophthalamide)s by MALDI-TOF MS to clarify the results previously obtained for a Nomex fiber sample [44]. The synthesized oligomers using an excess of diamine or diacid chloride were found to contain either amine or carboxylate end groups, respectively. When equal molar ratios of diamine and diacid were employed for the synthesis, cyclic species were obtained, as previously determined for the commercial product Nomex. Kamitakahara and Nakatsubo [45] described the basic strategies for creating well-defined diblock copolymers with cellulose derivatives. They employed MALDI-TOF MS as the characterization technique. In a follow-up study, the authors synthesized diblock copolymers consisting of cellulose triacetate and oligoamide-15 and employed again MALDI-TOF MS [46]. In this publication, the MS technique was used to determine the molar masses of the synthesized compounds. Scarel et al. [47] could confirm that p-hydroxyphenolic end groups were present in polyketones synthesized via a CO/styrene copolymerization by MALDI-TOF MS analysis. Thus, they were able to explain the major role of 1,4-hydroquinone as molar mass regulator.

Somogyi et al. [48] highlighted in their study the difference between the MALDI-TOF MS spectra of polyesters of 6-hydroxy-2-naptholic acid with either 4-hydroxybenzoic acid or 3-hydroxybenzoic acid. The latter copolymer showed several depolymerization reactions. Furthermore, the distribution of the monomer units was close to the stochiometric ratio, indicating that the monomer units were similar reactive. The study by Sugimoto and coworkers [49] concentrated on the alternating copolymerization of carbon dioxide with cyclohexene oxide. The MALDI-TOF MS analysis revealed that a side-reaction between tetraethylammonium acetate and methylene chloride took place as well as a chain transfer reaction. Therefore, a bimodal distribution of the product was obtained. The careful change of the experimental parameters yielded a unimodal and narrow molar mass distribution of the copolymer.

Zydowicz and colleagues [50] examined the microstructure of the polymer wall of self-formulated polyurethane nanocapsules by MALDI-TOF MS. The authors could prove that the mass range obtained in the recorded mass spectrum correlated with the M_n value achieved by SEC. The contribution by Xia et al. [51] presents a good example of an end-group determination by MALDI-TOF MS of polymers synthesized via a copolymerization of carbon dioxide and cyclohexene oxide. The resulting copolymers were either bearing an ethoxy or a cyclohexyloxyl end group. Endo and coworkers [52] not only determined the end groups (carboxylate or alkoxide end groups, as well as cyclic structures) but also the structure (linear, alternating) of a series of three co-oligomers obtained by anionic copolymerizations of bis(γ-lactone) and epoxide with potassium tert-butoxide. Chatti et al. [53] applied MALDI-TOF MS as a complementary analytical method to characterize their synthesized poly(ether ester)s containing an aliphatic diol based on isosorbide. According to the recorded MALDI-TOF MS spectra, the structure of the synthesized polymers was independent of the applied method of heating (microwave or thermostated oil bath) in the performed polycondensation reactions. Kraft and coworkers [54] also employed MALDI-TOF MS as one possible characterization technique to gain information regarding the purity and the average molar masses of shape-memory polymers containing short aramid hard segments and poly(ε-caprolactone) soft segments. The size of conducting poly[1-(thiophene-2-yl)benzothieno[3,2-b]benzothiophene] films was determined through MALDI-TOF MS, as shown by Aaron and colleagues [55]. Two successive studies by Li et al. describe the MALDI-TOF MS measurements of amphiphilic polymers of poly(ethylene oxide) as the main chain and either poly(methyl acrylate) [56] or PS [57] as the side chain. The intermediates of the synthesized graft polymers were additionally analyzed. The polydispersity indices (PDI) for the intermediates determined by SEC were always slightly higher than the ones obtained by MALDI-TOF MS. Jia et al. [58] also used MALDI-TOF MS to characterize their intermediates and synthesized grafted copolymers. These authors produced amphiphilic macrocyclic copolymers consisting of a PEO ring and multi-PS lateral chains. The composition of two other pairs of amphiphilic copolymers was investigated by MALDI-TOF MS from Park and Kataoka [59, 60]. In the first study, the authors copolymerized poly(2-isopropyl-2-oxazoline)s [59] and in the second one poly(2-alkyl-2-oxazoline)s [60] in a living manner.

The chemical structure of a polymer formed via a spontaneous copolymerization of 1,3-dehydroadamantane with electron-deficient vinyl monomers was determined to be alternating by Matsuoka et al. using MALDI-TOF MS [61]. The MS spectrum of the resulting copolymer showed three series having the same repeating units. The authors easily identified the series in the recorded spectrum and could prove with this that a chain-growth process took place and not a step-growth process. A full characterization of new random copoly(arylen ether sulphone)s was presented by Puglisi et al. [62] employing MALDI-TOF MS besides NMR spectroscopy. Both techniques were used to confirm the chemical structure, the chemical composition, and the end groups of the studied copolymers. Pensec and coworkers [63] synthesized dendritic-linear block copolymers via ring-opening polymerization of lactide using benzyl alcohol dendrons as macroinitiatiors and stannous octoate as catalyst. The MALDI-TOF MS analysis of the obtained products showed well-resolved signals allowing the confirmation of the molecular structure. Howdle and colleagues [64] employed MALDI-TOF MS as a complementary technique to prove that a block copolymer of poly(ε-caprolactone-b-methyl methacrylate) is formed by the one-step chemoenzymatic synthesis of ε-caprolactone and methyl methacrylate and not a mixture of both homopolymers. The analysis of a series of amphiphilic triblock copolymers of poly(ethylene glycol)-b-poly(acrylic acid)-b-poly(n-butyl acrylate) was also performed by MALDI-TOF MS, as presented by Yang et al. [65].

The MALDI-TOF MS measurement of a multiblock copolymer based on poly(2,6-dimethyl-1,4-phenylene oxide) and poly(bisphenol A carbonate) by Samperi et al. [66] revealed that a block-like structure was obtained. Besides the signals for the copolymer, some nonreacted poly(bisphenol A carbonate) oligomers terminated with methyl carbonate groups were present in the recorded spectra, providing information about the purity of the synthesized multiblock copolymer. The study by Faÿ et al. [67] shows once more the successful application of MALDI-TOF MS for the detailed characterization of synthesized copolyesters. The produced copolyesters entailed only one major series in the recorded MALDI-TOF MS spectra proving that exclusively linear species were formed and no cyclic products were present. Pound et al. [68] synthesized copolymers consisting of PEG and poly(vinyl acetate) and determined their end groups by a simple MALDI-TOF MS analysis according to the method described by Willemse et al. [38]. The end-group analysis employing MALDI-TOF MS, besides NMR spectroscopy, was also described by Baéz et al. [69] for poly-(ethylene)-b-poly(ε-caprolactone) diblock copolymer samples. Cyclic species of poly-(ε-caprolactone) were detected in the recorded MALDI-TOF MS spectra, which were produced due to back-biting reactions.

Ihara and coworkers [70] synthesized copolymers of vinyl alcohol and could determine by MALDI-TOF MS that a random structure of the produced polymers was achieved. In the recorded MALDI-TOF MS spectrum of one copolymer sample, the repeating units of styrene and vinyl alcohol could be readily determined. Klumperman and colleagues [71] confirmed with the application of MALDI-TOF MS that they synthesized a low-molar-mass poly(4-vinylpyridine)-b-poly(styrene-co-acrylonitrile) block copolymer. In the case of Cheng et al. [72], MALDI-TOF MS revealed that the decrease in the stability of the produced charged hybrid copolymers can be correlated with the electrostatic effect between the molecular chains. For this purpose, the structure stabilities of five copolymer samples under MALDI-TOF MS conditions were investigated by comparing the occurring fragment ions. The MALDI-TOF MS data obtained by Pfeifer and Lutz [73] indicated that sequence-controlled copolymerizations of styrene with various N-substituted maleimides occurred. As a consequence, functional copolymers with a narrow sequence-distribution were obtained via this synthetic route.

Binder and his research group [74] recently presented an excellent application of MALDI-TOF MS in the polymer field. The authors monitored and evaluated the crossover reactions of four structurally different monomers in a ring-opening metathesis-based block copolymerization. Since MALDI-TOF MS allowed the direct monitoring at the point of the crossover reaction, a detailed evaluation of the polymerization process at this point was possible. The contribution by Dong et al. [75] showed once more that polymer chemists nowadays are more often using MALDI-TOF MS as a complementary technique to characterize their synthesized copolymers. The authors prepared an amphiphilic block copolymer consisting of a PS ring and a PEO tail. The M_n values obtained by MALDI-TOF MS were in good agreement with the estimated values obtained from the NMR spectra. The research group of Charles [76] determined the block size in various nitroxide-capped PEO-b-PS by hydrolysis of the ester bond in the junction moiety of the macromolecules, followed by MALDI-TOF MS analysis of the individual building blocks. However, to determine the block size of copolymers, it is easier and faster to perform tandem MS on the copolymer sample, as presented by Crecelius et al. [77]. This study will be discussed in more detail in Section 9.6.

Schubert and coworkers [78] synthesized well-defined glycopolymers, grafted copolymers and anthracene-containing polymers, and employed MALDI-TOF MS, besides SEC and NMR spectroscopy, to analyze them. Petrova et al. [79] showed the application of MALDI-TOF MS for the intermediates formed by the controlled synthesis of amphiphilic triarm star-shaped block copolymers of an AB_2 type, in which A presented PEO and B poly(ε-caprolactone). Raynaud et al. [80] confirmed the introduction of functional groups at both α and ω positions of PEO chains, which were employed as monomers to form poly(ethylene oxide)-b-poly(ε-caprolactone) block copolymers through a facile procedure. The contribution by Zhang et al. [81] is a typical example that MALDI-TOF MS is nowadays frequently used as a characterization technique to determine the purity of synthesized homopolymers as basis to produce the desired block copolymers. In this paper, in-chain-functionalized poly-(styrene)-b-poly-(dimethylsiloxane) diblock copolymers were formed by anionic polymerization and hydosilylation using dimethyl-[4-(1-phenyl)phenyl]silane, in which the poly(styrene) block was synthesized first and investigated by MALDI-TOF MS before the second comomomer was added.

Penco and colleagues [82] demonstrated via MALDI-TOF MS that the most intense peaks of the synthesized poly(methyl methacrylate)-based copolymers correspond to the expected oligomers, terminated according to the mechanisms of a radical polymerization. The poly(methyl methacrylate)-based copolymers were additionally characterized by means of NMR spectroscopy and SEC. Wu et al. [83] studied systematically the synthesis of all-conjugated copolymers comprising poly(2,5-dihexyloxy-1,4-phenylene) and poly(3-hexylthiophene) blocks using a nickel-based catalyst. The performed MALDI-TOF MS analysis indicated that the addition order of the monomers during the copolymerization can be attributed to the low efficiency of the intramolecular nickel transfer from thiophene to phenylene units. Besides SEC and NMR spectroscopy, MALDI-TOF MS was employed by Weber et al. [84] to characterize their produced comb and graft-shaped poly[oligo(2-ethyl-2-oxazoline) methacrylate]s, which showed a lower critical solution temperature behavior. Huang and coworkers [85] presented the possibility that poly(lactic)-*b*-poly(ethylene glycol) diblock and poly(lactic)-*b*-poly(ethylene glycol)-*b*-poly(lactic) triblock copolymers can also be analyzed by MALDI-TOF MS, besides SEC and ^{1}H-NMR spectroscopy. The M_n values calculated from the SEC data were higher than those obtained from the MALDI-TOF MS spectra and ^{1}H-NMR spectroscopy data.

Since the molar mass determination of random copolymers of poly(methacrylic acid)-poly(methyl methacrylate) (PMAA-PMMA) was not possible, as exemplary presented in Figure 9.4a, Giordanengo et al. [86] derivatized the copolymers by methylation into PMMA homopolymers. The MALDI-TOF MS spectrum of a methylated PMAA-PMMA sample is presented in Figure 9.4b. From this well-resolved MALDI-TOF MS spectrum, the average molar mass values could be readily obtained and were compared to pulsed gradient spin echo NMR spectroscopy data.

Ahmed et al. [87] performed MALDI-TOF MS in addition to different chromatographic techniques to characterize their polymers of poly(ε-caprolactone) generated via microwave-assisted polymerization initiated with PEGs, monodisperse diols, and tin octoate as catalyst. The synthesized block copolymers indicated a much higher content of triblock structures compared to commercially available poly(caprolactone) diols, which contained considerable amounts of diblocks. Ohsawa et al. [88] proved by MALDI-TOF MS measurements of their synthesized alternating copolymers that backbiting reactions were effectively suppressed. Ranimol and colleagues [89] determined the size distribution and purity of the prepared triblock and multiblock copolymers of polysulphone with monodisperse amide segments by MALDI-TOF MS. The recorded MALDI-TOF MS spectra showed signals possibly arising from fragments of the polysulphones. Kumokka [90] classified acrylic- and rubber-based pressure-sensitive adhesives of colorless and transparent oriented polypropylene adhesive tapes distributed in Japan based on MALDI-TOF MS data. The classification according to MALDI was validated by FT-IR spectroscopy.

9.4.2
ESI

ESI-MS, the second soft ionization technique employed in the structural characterization of copolymers, is not as often used as compared to MALDI (see Figure 9.2). One possible reason may be that multiple-charged ions are obtained making the interpretation of the recorded mass spectra more challenging due to overlapping

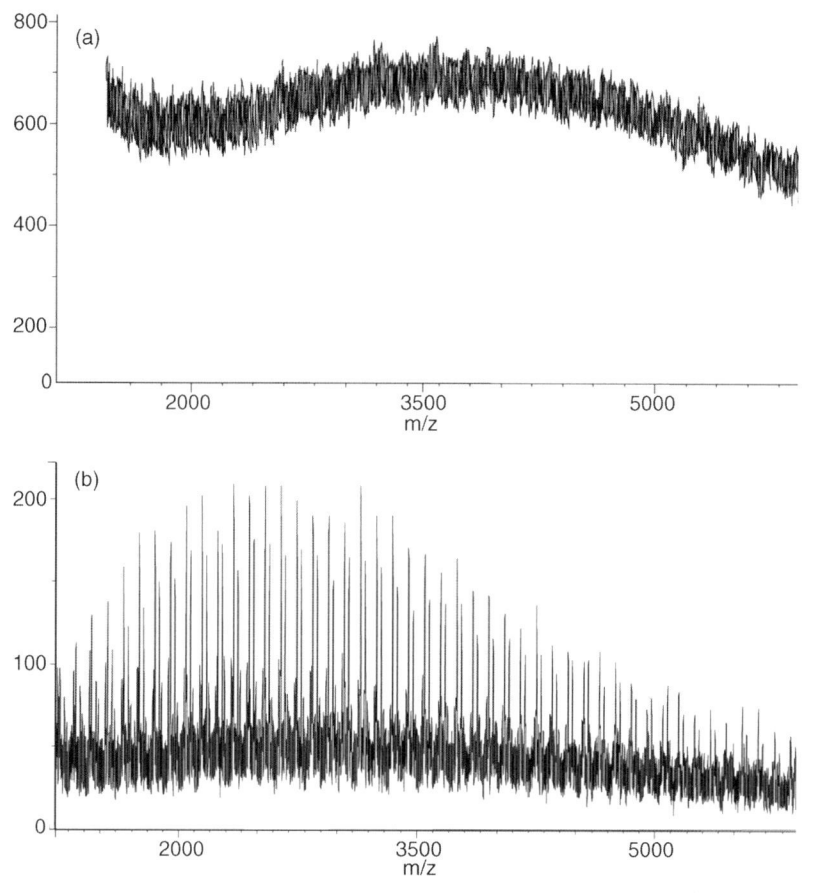

Figure 9.4 MALDI-TOF MS spectra of (a) PMAA-PMMA prior to derivatization and (b) 5.5 mL of derivatization agent (trimethylsilyldiazomethane) was added to obtain the methylated PMAA-PMMA. Reproduced from Ref. [86] with kind permission of Elsevier.

peaks between charge-state distribution and chain length distribution. However, ESI is generally regarded as the softer ionization technique compared to MALDI since the latter one reveals sometimes fragmentation species, as for example, already discussed in Ref. [72] or [89]. However, commonly the addition of the matrix prevents further fragmentation, which typically only occurs when laser desorption/ionization (LDI) is employed. Alhazmi et al. [91] used both soft ionization techniques, MALDI and ESI, to identify the monomer ratios as well as the structure of the end groups of three different copolymers poly(butyl acrylate)-co-poly(vinyl acetate), poly(butyl acrylate)-co-poly(methyl methacrylate), and poly(methyl methacrylate)-co-poly(vinyl acetate), and compared the MS results with ^{1}H-NMR spectroscopy and theoretical predictions. The obtained ESI-MS data of the first two copolymers were in excellent agreement with the results from ^{1}H-NMR spectroscopy, in contrast to the latter one. Comparable distributions of the products were observed in the MALDI and ESI mass spectra of the second copolymer sample. Lu and coworkers [92] reported the synthesis of various poly(propylene carbonate)s with different head-to-tail linkages, which were generated from racemic propylene oxide with chromium salen complexes as catalysts. The analysis of the polymer chain end group (initiating and chain growth species) at various times was achieved by *in situ* ESI-MS. In combination with some control experiments, this enabled an insight into the mechanistic understanding of the copolymerization and the effects of the stereochemistry of the resulting polycarbonates.

The fruitful combination of ESI-MS and ^{1}H-NMR spectroscopy for the characterization of copolymers is also presented in the study by Kasperczyk *et al.* [93]. These authors investigated the hydrolytic degradation of copolymers obtained via the ring-opening polymerization of glycolide and ε-caprolactone. From the recorded ESI-MS spectra, three-dimensional diagrams were constructed, as shown exemplary for the copolymer sample containing 70 mol% glycolidyl units in Figure 9.5. The *x*-axis represents the quantity of the glycolyl units (G), the *y*-axis the quantity of the caproyl units (C), and the *z*-axis the normalized signal intensity of the peaks. After 7 weeks of degradation (see Figure 9.5a), co-oligomers with the type of C_3G_3 to $C_{10}G_{10}$ were observed. The variety of chemical structures of the co-oligomers decreased after 10 weeks (see Figure 9.5b), and homo-oligomers containing only glycolyl units were obtained. At the last stage of degradation, at 26 weeks, co-oligomers with a higher content of caproyl units were visible, besides still remaining homo-oligomers with only glycolyl units.

The application of ESI-MS besides SEC to characterize methacrylic acid oligomers and oligo(methacrylic acid)-*b*-poly(methyl methacrylate) synthesized by RAFT is shown by Nejad *et al.* [94]. The MALDI-TOF MS spectrum of the polymerization of methacrylic acid with 4-cyanopentanoic acid-4-dithiobenzoate (CPADB) as RAFT agent revealed the expected spacing of the peaks corresponding to the repeating unit and as end group the thiocarbonly thio unit. Chen and coworkers [95] employed ESI-FT-ICR MS to analyze their graft copolymers poly(L-glutamic acid)-*g*-poly(*N*-isopropyl acrylamide) as well as their precursors. The ESI-FT-ICR MS spectrum of the precursor amino-semitelechelic poly(*N*-isopropylacrylamide) showed the desired end groups. The research group of Charles combined in the current study [96] NMR spectroscopy with ESI-MS and MS/MS to achieve the complete microstructural characterization of a PMAA–PMMA copolymer synthesized by nitroxide-mediated polymerization (NMP). The structure of the end groups were determined by NMR spectroscopy and further confirmed by ESI-MS. The ESI-MS results obtained for the copolymer composition were not consistent with the NMR data due to the strong mass bias well-known to occur during the ESI process of these kind of polymers. Finally, the random nature of the copolymer was elucidated via ESI-MS/MS.

9.4.3
APCI

Interestingly, APCI has rarely been used as an ionization technique in the analysis of copolymers (see Figure 9.2), although singly charged pseudo-molecular ions are

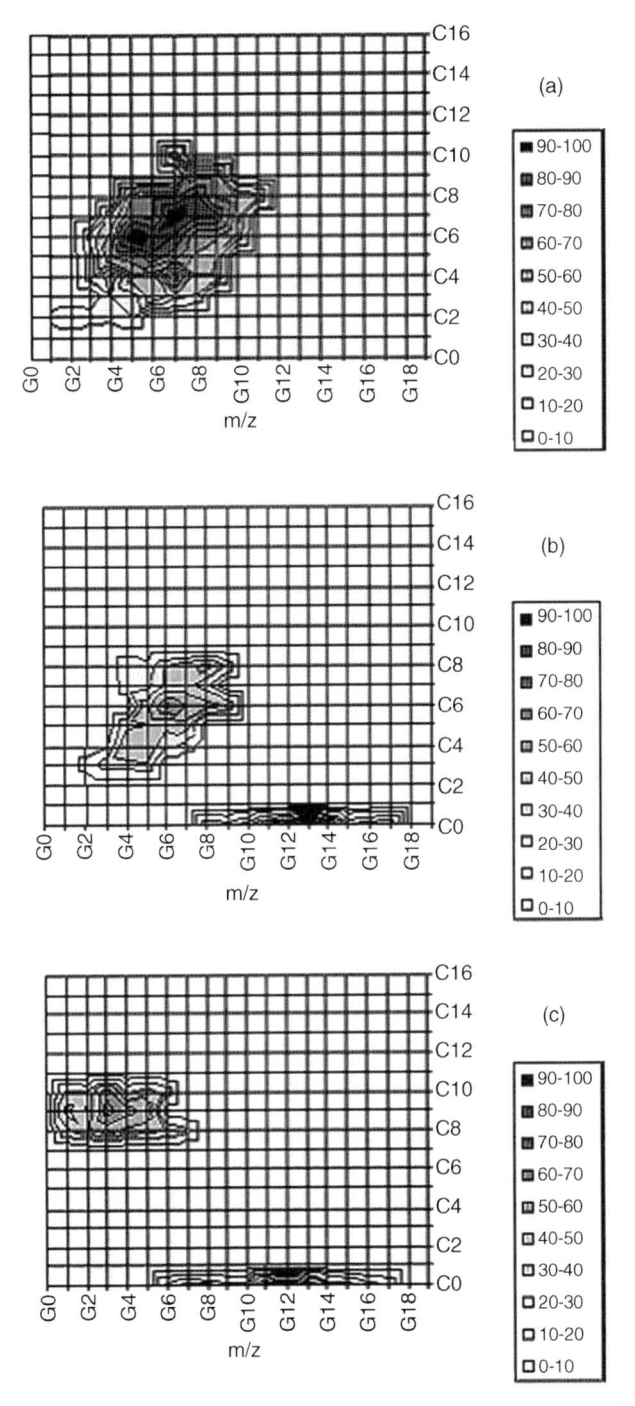

Figure 9.5 Changes in the chemical structure during the degradation of the copolymer containing 70 mol% glycolidyl units: (a) 7 weeks, (b) 10 weeks, and (c) 26 weeks. Reproduced from Ref. [93] with kind permission from Elsevier.

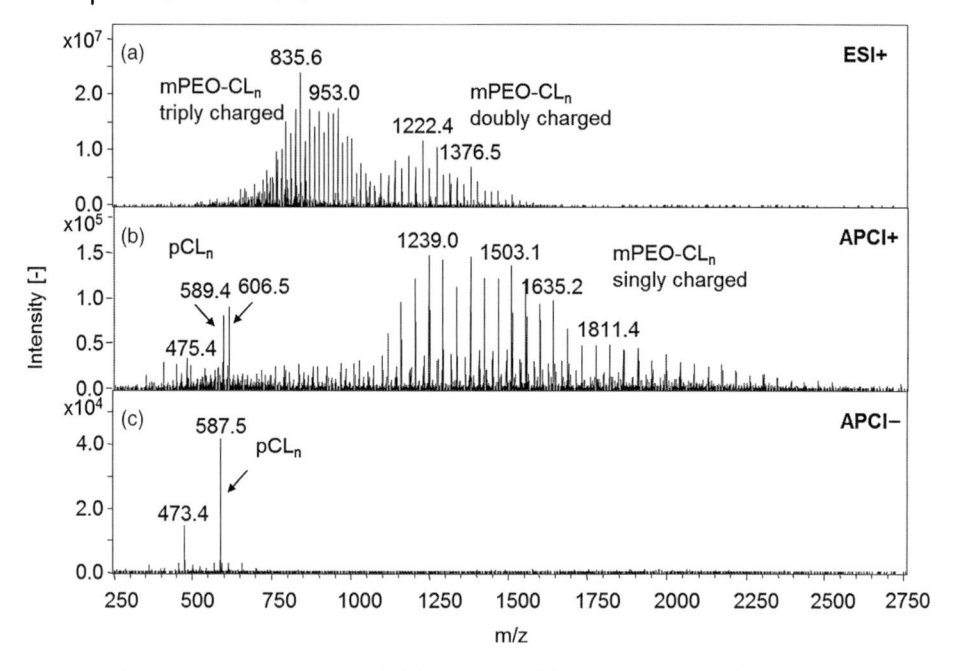

Figure 9.6 Mass spectra recorded from a synthetic block copolymer based on methyl poly(ethylene oxide) (mPEG) and ε-caprolactone (CL) after LC separation using different ionization methods: (a) ESI positive mode, (b) APCI positive mode, and (c) APCI negative mode. Reproduced from Ref. [97] with kind permission of John Wiley & Sons, Ltd.

generally obtained in contrast to ESI. However, APCI is known to be somewhat less mild than ESI due to the use of a heated nebulizer. The complementary character of APCI and ESI, in both the positive and negative ion mode, was highlighted by van Leeuwen et al. [97] for a synthetic block copolymer based on methyl poly(ethylene oxide) (mPEG) and ε-caprolactone (CL), functionalized with linoleic acid. The comparison of the mass spectra recorded for this block copolymer after LC separation by ESI-MS (positive mode), APCI-MS (positive mode), and APCI-MS (negative mode) depicted in Figure 9.6a–c, respectively. Figure 9.6a indicates very well that several charge-state distributions are obtained. This was not the case when the APCI-MS mode was applied (Figure 9.6b). Here only singly charged ions were recorded. Finally, in the negative APCI-MS mode the side product, the homopolymer poly-(ε-caprolactone) (pCl$_n$), is readily observable as a strong peak at the m/z value of 587.5.

Desmazières et al. [98] discussed the benefits and limitations of LC-APCI-MS with a series of polymers, including copolymers of siloxanes. The results were compared to those obtained by SEC- and MALDI-MS. The main drawback of the APCI interface was determined as the in-source decomposition above m/z 2000–3000, which can result in an underestimation of the average molar masses. However, the authors pointed out that APCI allows the detection of a wide range of sample polarities and solvents. Wesdemiotis and coworkers [99] investigated the analysis of complex amphiphilic copolymers by heating them slowly up on a direct probe (DP) inside an APCI source. These so-called DP-APCI measurements provided conclusive information about the nature of the hydrophilic and hydrophobic components of the copolymers, and additionally allowed distinguishing between different comomomer compositions, as well as between crosslinked copolymers and copolymer blends with similar physical properties.

9.5
Separation Prior MS

Copolymer samples are typically complex and if they contain side products formed during the synthesis, a pre-separation step is often required. This step can be performed with a different instrument, for example, a SEC or HPLC instrument in an off or online mode connected to the MS interface, or by using a special MS instrument, for example, in which the ions are separated according to their gas-phase mobility prior to detection. In Sections 9.5.1 and 9.5.2, the applications described in this area found in the literature to analyze copolymer samples are summarized.

9.5.1
LC-MS

LC can be used in different forms. The most widely used technique for copolymers is SEC, although it only provides information regarding the average molar masses and their distributions. More information on the chemical structure of the copolymer is obtained by LC-CC, where separation is based on the number of one comonomer and is independent of the other comonomer. Even higher chromatographic resolution can be obtained with 2D chromatography, which is particular useful for more complex copolymers. A detailed discussion of these hyphenated techniques is presented in Chapter 7; here only the investigations concerning copolymers are outlined.

SEC is typically combined with MALDI-TOF MS in an offline mode, in which SEC fractions are collected, mixed with matrix, a salt additive is added, and applied on a target before introduction into the ion source of a MALDI-TOF mass spectrometer. The earliest application of SEC-MALDI TOF MS for copolymers in the time period considered comes from Gallet *et al.* [100], who compared this technique with SEC-NMR spectroscopy for the analysis of the triblock copolymer poly(ethylene oxide-*b*-propylene oxide-*b*-ethylene oxide) in order to understand the bimodal distribution of poloxamer 407. Montaudo [101] tried to apply the method described by Tatro *et al.* [102], which involves the use of MALDI for the estimation of viscosity parameters of polymers, to polycondensates, without any further modifications. Since this was not successful, he proposed a new methodology based on the universal calibration concept and on SEC-MALDI-TOF MS. Montaudo tested the new method successfully on the terpolymer poly(butylenes succinate-butylenes adipate-butylene sebacate). Approximately 50 fractions were analyzed by SEC-MALDI, and from the recorded MALDI-TOF MS spectra the SEC calibration curve was constructed. From the SEC calibration curve and the calibration of the SEC instrument prior analysis with suitable standards, the parameters describing the viscosity of the studied polymer were obtained and were in good agreement with the calculated ones. Fradet and coworkers [103] studied copolyesters containing aliphatic units in the main chain by coupling SEC with MALDI-TOF MS. This enabled the calibration of the SEC curves against absolute molar masses and the determination of the chemical structure of the copolyesters. The same class of copolymers was investigated by Adamus *et al.* [104]. In their contribution, SEC-MALDI-TOF MS was also employed for the structural elucidation of the copolyesters, including end-group determination. The average molar masses (M_n and M_w) of the copolyester samples were obtained by offline coupling of SEC with MALDI-TOF MS, in which MALDI-TOF MS was used as a detector for the SEC fractions, collected from the polydisperse samples.

Alicata *et al.* [105] analyzed narrowly dispersed molar mass SEC fractions of random and microblock poly[(R)-3-hydroxybutyrate-*co*-ε-caprolatone]s by MALDI-TOF MS. The end groups of the copolymers could be determined with the help of the MALDI-TOF mass spectra. Carboxylic, alcoholic and tosyl end groups were obtained. The ε-caprolactone-rich samples showed mostly tosyl end groups, whereas the 3-hydroxybutyrate-rich samples mainly possessed alcoholic end groups. The ESI-MS spectra of low-molar-mass SEC fractions, which had a higher resolution as compared to the MALDI-TOF MS spectra, enabled the facile identification of the different oligomer species since this was rather difficult in the corresponding MALDI-TOF MS spectra. Weidner and colleagues [106] tried to characterize polyacetal copolymers by a variety of analytical techniques, and among others they employed SEC-MALDI-TOF MS. However, the recorded MALDI-TOF MS spectra from several fractions did not allow the elucidation of the copolymer structure due to the low-mass resolution of the very complex mass spectra; therefore, 2D-separation was applied as well.

Girod *et al.* [107] could detect different ions of the block co-oligomer PEO-*b*-PS by adjusting the salt concentration of the mobile phase of LC-CC combined online to an ESI-MS instrument. Doubly lithiated co-oligomer adducts were obtained at high lithium chloride concentration (1 mM), while both lithiated and protonated species were detected at lower concentrations (down to 0.1 mM). Falkenhagen and Weidner [108] presented for the first time the use of ultra performance liquid chromatography (UPLC) in order to identify the critical conditions for LC-CC coupled to ESI-TOF MS. The authors studied PEOs, poly(propylene oxide)s, and their copolymers by this new technique enabling a faster adjustment of the parameters. Another advantage was that polymer standards were no longer required for the search of the critical conditions.

2D-separation typically involves the combination of LC-CC in the first dimension and SEC in the second dimension. The LC-CC mode is used for the chemical separation and the SEC mode for the molar mass information. For SEC, calibration standards are required, which are often not available. Therefore, Weidner *et al.* [109] suggested the substitution of the second dimension with MALDI-TOF MS. In the above contribution, the authors showed the application of this approach to a complex polyester copolymer sample. The 2D plot, which can be obtained by such an investigation, is presented in Figure 9.7. The structural

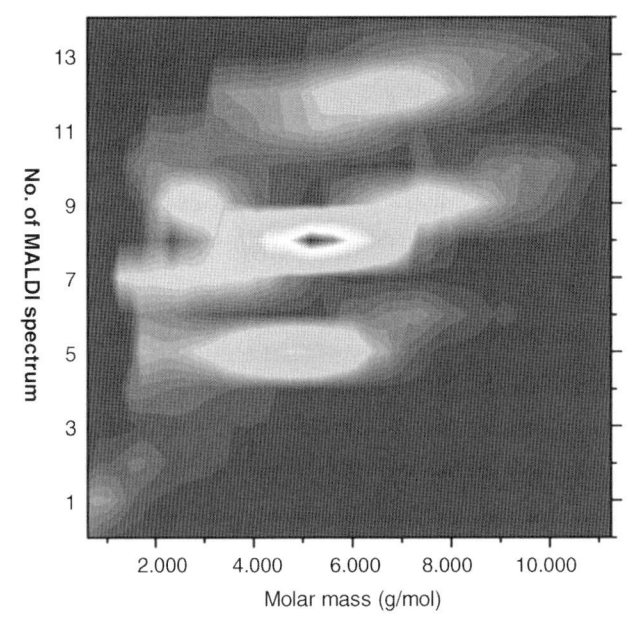

Figure 9.7 2D-plot obtained from a LC-CC run further analyzed by MALDI-TOF MS of a polyester copolymer. Reproduced from Ref. [109] with kind permission of the American Chemical Society.

information obtained in the LC-CC mode and the molar mass information obtained from the following MALDI-TOF MS analysis, by spraying the chromatographic run on the MALDI target, are combined in the figure. This enables the comparison with the 2D-plot obtained from an ordinary 2D-separation (LC-CC combined with SEC). The authors pointed out that the 2D-plots obtained by this procedure offer a more reliable mass scale.

Trathnigg and coworkers applied the above new methodology for the characterization of block copolymers of poly(ethylene oxide)-*b*-poly(propylene oxide) [110] and polyoxyalkylene diblock copolymers [111]. Additionally, they performed offline 2D-separations by combining LC-CC with SEC, LC-CC with LC, and finally LC-CC with LC-CC in the later publication [111]. In both studies, the combination of 2D-chromatography and MALDI-TOF MS enabled the authors to gain information regarding the purity, chemical composition, and molar mass distributions of the synthesized products.

9.5.2
Ion Mobility Spectrometry-Mass Spectrometry (IMS-MS)

Both soft ionization methods, MALDI and ESI, have been used for the ionization of copolymers measured in IMS-MS instruments. Baker *et al.* [112] described the sequence-dependent confirmation of various poly(glycidyl methacrylate-butyl methacrylate)s (PGMA-PBMAs) measured in the gas phase using IMS. PGMA–PBMA

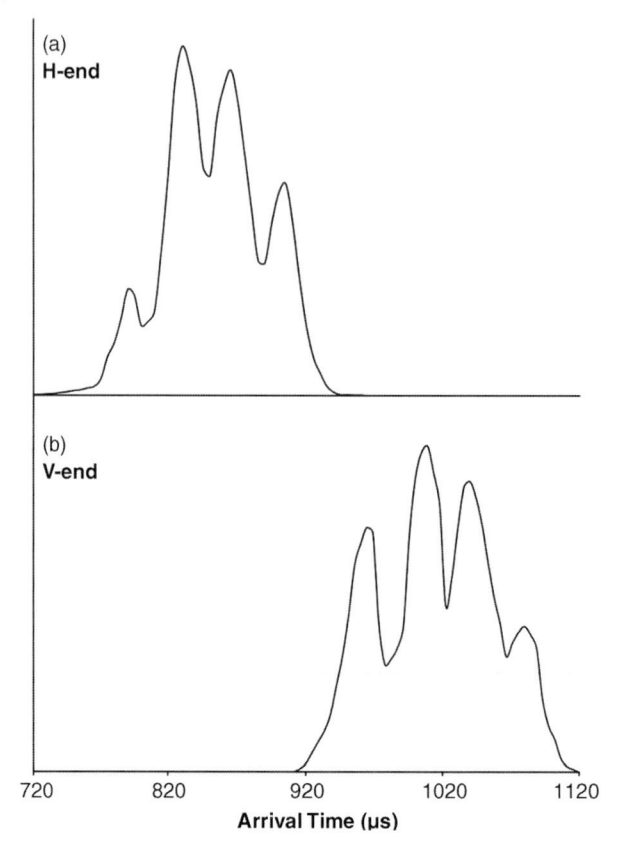

Figure 9.8 Arrival time distributions of the PGMA-PBMA trimer containing (a) a hydrogen and (b) a vaso end group. Multiple peaks indicate more than one sequence with different collision cross-sections. Reproduced from Ref. [112] with kind permission of Elsevier.

copolymers were difficult to characterize by other techniques since both comonomers have the same nominal mass, with a slightly different exact mass (0.036 g mol^{-1}), and thus were investigated by IMS. MALDI was employed to generate sodium adducts of PGMA–PBMA trimer to pentamer, and their different arrival time distributions were collected. An example of such arrival time distributions for the PGMA–PBMA trimer with two different end groups is depicted in Figure 9.8. Four unresolved peaks are visible for each oligomer species since four different sizes of ions were detected, corresponding to four different sequences of each trimer. From these arrival time distributions, the collision cross-section of each composition of the trimer can be calculated and compared to the theoretical one, and finally the most abundant sequence can be determined. This is in the case of the PGMA–PBMA trimer, the combination of BMA–GMA–GMA and GMA–BMA–GMA bearing both a hydrogen end group, and GMA–BMA–GMA possessing a vaso end group (–CNCH$_3$–CCH$_3$C$_2$H$_5$).

A combined structural and conformational study of random and block poly(ethylene oxide-propylene oxide)s has been undertaken by Jackson and coworkers [113]. MALDI has been utilized in combination with IMS, as well as ESI-MS/MS, to establish the fragmentation pathways of the studied copolymers. From the IMS experiments, collision cross-sections were determined and compared to the theoretical ones. The established collision cross-sections were independent of the sequence of the ions in the gas phase; however depending on the size and molar mass of the copolymer species. The resulting IMS data were employed to support the fragmentation routes established by ESI-MS/MS. The most recent application of IMS-MS was reported by Trimpin and Clemmer [114] highlighting its use in the polymer field by providing structural signatures, which allow the effortless recognition of minor differences in blends and copolymers. (A deeper insight of this technique applied in the polymer field is provided in Chapter 4 of this book.)

9.6
Tandem MS (MS/MS)

The analysis of copolymers by tandem MS (MS/MS) is rather difficult since most of the time peak overlap is occurring, which cannot efficiently be avoided. However, if MS/MS experiments can successfully be performed, then the copolymer can be sequenced since each comonomer shows distinct fragmentation pathways along the chain. The prerequisite of such measurements is to understand the fragmentation behavior of homopolymers consisting of each of the comonomers.

Wesdemiotis et al. [115] investigated the copolymer poly(fluorooextane-co-THF) by matrix-assisted laser desorption/ionization-quadrupole-time-of-flight tandem MS (MALDI-Q-TOF MS/MS). The measurements provided structural evidence that the initiator was incorporated in the central position of the polymer chains and that the THF comomomer was located vicinal to the initiator unit by identifying the fragments obtained from the homolytic cleavage at the fluorinated side chains of the copolymer. The sequence of a PUR sample was investigated by Mass et al. [116] employing MALDI-TOF MS/MS. By carefully selecting precursor ions, product ions could be identified in the recorded MS/MS spectra indicating a random or a block-like structure. Weidner and coworkers could clearly differentiate between cyclic and linear oligomers of complex copolyesters by employing MALDI-TOF MS/MS. These authors were also able to determine the end group of isomeric linear oligomers since characteristic product ions could be detected. Crecelius et al. [77] could even determine the block length of a methoxy poly(ethylene oxide)-b-poly(styrene) (mPEG-b-PS) sample by performing MALDI-TOF MS/MS on this rather complex sample. The reason for the possibility to determine the block length was that the MS/MS spectrum revealed characteristic fragments for both blocks and no fragments containing both repeating units. By selecting precursor ions with the same number of PEG repeating units, but an increasing number of PS repeating units, the gap in the recorded mass spectra was increasing, as shown in Figure 9.9, confirming the assumption that a scission between both blocks occurred. From the same group,

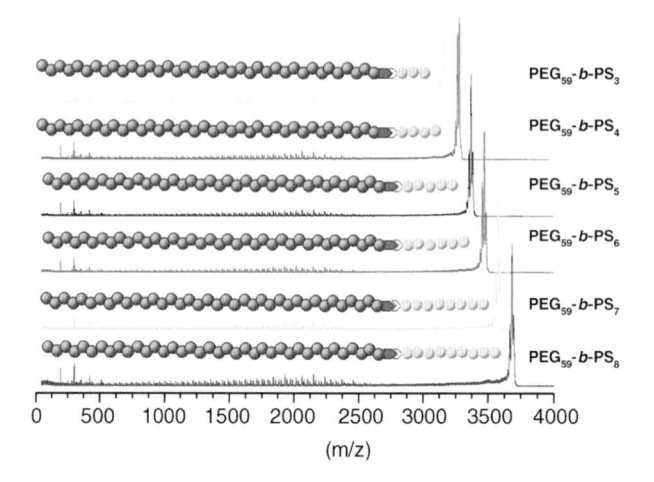

Figure 9.9 MALDI-TOF MS/MS spectra of precursor ions containing 59 repeating units of mPEG and an increasing number of PS repeating units. The selected precursor ions are schematically represented between the spectra. Reproduced from Ref. [77] with kind permission of Wiley Periodicals, Inc.

Schubert and colleagues [117] performed MS/MS using both soft ionization techniques MALDI and ESI for the study of various poly(2-oxazoline) block copolymers to gain more structural information about the block copolymers compositions and their fragmentation behavior. Different mechanisms occurred, such as 1,4-ethylene or hydrogen elimination and McLafferty +1 rearrangement. The copolymers with aryl side-groups showed less fragmentation due to their higher stability compared to copolymers with alkyl side-groups.

Adamus and coworkers [118–121] presented the ESI-MS and MS/MS analysis of various types of copolyesters to identify their molecular architecture, including the chemical structure of their end groups. The MS/MS analysis helped in finding the correct sequence of the copolyesters and to investigate the fragmentation pathways. In the most recent study [121], the author could differentiate by ion trap ESI-MS/MS between random and diblock copolyester samples by comparing the fragmentation patterns obtained by both types of copolyesters. Žagar et al. [122] described in their study, the structural investigations of bacterial poly(3-hydroxybutyrate-co-3-hydroxyvalerate) copolyesters on the molecular level using ESI-MS/MS. The generated data obtained by the tandem MS experiments were compared to NMR spectroscopy data and revealed comparable results. Terrier et al. [123] investigated a series of linear triblock and glycerol derivative diblock copolyethers by ESI-MS/MS. In the first step, homopolymers were analyzed to identify the nature of the product ions and to explore fragmentation pathways. By measuring triblock and diblock copolyethers in the second step, it could be demonstrated that copolyethers with the same composition in each repeating unit, but with inversed block sequence can be distinguished since they form characteristic fragment ions.

Charles and coworkers [124] characterized the microstructure of PEO-*b*-PS, synthesized by a NMP process, using ESI-MS^n. The ESI-MS^2 measurements indicated the homolytic cleavage of the C−ON bond, resulting in the elimination of the terminal SG1 end group as a radical, including a depolymerization process of the PS block from the radical cation. As a consequence, the size of the PS block could be determined. Additionally, an alternative fragmentation route provided structural information on the junction group between both blocks. A specific dissociation of the PEO block did not occur. Hence, information regarding the PEO block and the initiated end group was obtained by ESI-MS^3. The same authors analyzed the PEO-*b*-PS diblock copolymer further by employing LC-CC combined with online ESI-MS and MS/MS [125]. The results further supported the characterization of the PS block by allowing its size determination in both the LC-CC mode combined with the single-stage ESI-MS or in the tandem MS mode. Simonsick and Petkovska [126] highlighted the advantages obtained by combining ultrahigh resolution via ESI-FT-ICR MS with tandem MS analysis to elucidate the structure and sequence of polyesters and acrylates in more detail. (Further information regarding this topic is provided in Chapter 3, in which tandem MS is explored in detail.)

9.7
Quantitative MS

Even though MALDI-TOF MS represents a powerful technique for the analysis of copolymers, the quantification is rather difficult due to variations occurring in the sample preparation and ionization efficiencies. In the short communication by Alhazmi and Mayer [127], the influence of the matrix-to-analyte ratio used in the sample preparation on the measured ratio of two copolymers differing in their end groups is presented. The two copolymers originate from the synthesis of poly(butylacrylate-methyl methacylate). The authors recommend comparing the MALDI results always with ESI-MS and NMR spectroscopy data for confirmation. The question if quantification of monomer ratios in copolymers by ESI-MS and MS/MS is possible was addressed by Prebyl *et al.* in two successive studies [128, 129]. In the first attempt [128], the authors showed for three examples (styrene sulfonic acid, acrylic acid, and 2-acrylamido-2-methylpropane sulfonic acid) that semi-quantitative information can be obtained by ESI-MS/MS experiments in the negative mode. In the second approach [129], they were able to identify suitable conditions for ESI-MS and MS/MS measurements, in which monomers served as standards to assess the composition of copolymers without the need of polymer standards. The authors suggested that once the conditions have been identified, this technique could be utilized for the routine monitoring of monomer ratios.

9.8
Copolymers for Biological or (Bio)medical Application

In the last few years, the application of synthesized copolymers in the biological and biomedical as well as medical fields is strongly increasing. A few examples have been included in this section to provide the reader with an impression of current applications. Impallomeni *et al.* [130] synthesized biodegradable copolymers namely poly(3-hydroxybutyrate-*co*-ε-caprolactone)s. The random and microblock copolymers were analyzed by SEC followed by MALDI-TOF MS to determine their end groups. Copolymers rich in 3-hydroxybutyrate repeating units had hydroxyl and carboxyl end groups, while copolymers rich in ε-caprolactone repeating units had tosyl and carboxyl end groups. A polymeric micelle drug-delivery system was developed by Park and coworkers [131] to enhance the solubility of the poorly soluble drug biphenyl dimethyl dicarboxylate. The produced block copolymers consisting of poly(D,L-lactide) as the hydrophobic segment and *m*PEG as the hydrophilic segment were investigated by MALDI-TOF MS to determine the average molar masses. The recorded mass spectra of the copolymers showed only one broad single peak, without any isotopic pattern, from which the M_n and M_w values were calculated. The polydispercity indices were below 1.05. Rieger *et al.* [132] reported on the synthesis of new PEO macromonomers copolymerizable with ε-caprolactone by a ring-opening polymerization and therefore suitable for the preparation of amphiphilic comblike copolymers with a biodegradable hydrophobic backbone. The created macromonomers were measured via FT-IR spectroscopy, MALDI-TOF MS, and NMR spectroscopy. The MALDI-TOF MS spectra of the synthesized macromonomers showed four distributions: two distributions could be correlated with the desired product ionized as sodium and potassium adducts, and two distributions with a hydroxyl end group and as well ionized as sodium and potassium adducts. The experimental isotopic pattern fitted well to the theoretical one. Gutzler *et al.* [133] developed poly(ethylene oxide)-g-poly(vinyl alcohol) copolymers as an instant-release tablet coating. Besides a 2D-separation, MALDI-TOF MS showed that no free PEO was observed in the created copolymers, which was a prerequisite for the desired application. Fernandez-Megia *et al.* [134] prepared three generations of azido-terminated PEG-dendritic block copolymers, which they reacted with three alkyne-funtionalized unprotected carbohydrates derived from α-D-mannose, α-L-fucose, and β-D-lactose via click chemistry (1,3-dipolar cycloaddition). The azide-containing and the glycosylated block copolymers were analyzed by MALDI-TOF MS. Figure 9.10 presents the MALDI-TOF MS spectra of the glycodentritic copolymers derived from mannose. With these spectra, the authors could prove the purity of the substances and were able to calculate the molar masses. In the case of the second (G2) and third (G3) generation, a series of peaks with a spacing of m/z 44 can be seen, which correlates to monomethyl ether PEG amine, one of the educts, which results due to fragmentation occurring during the MALDI process since high laser power was required for higher generations. The new copolymers were synthesized for the construction of active targeted drug-delivery systems.

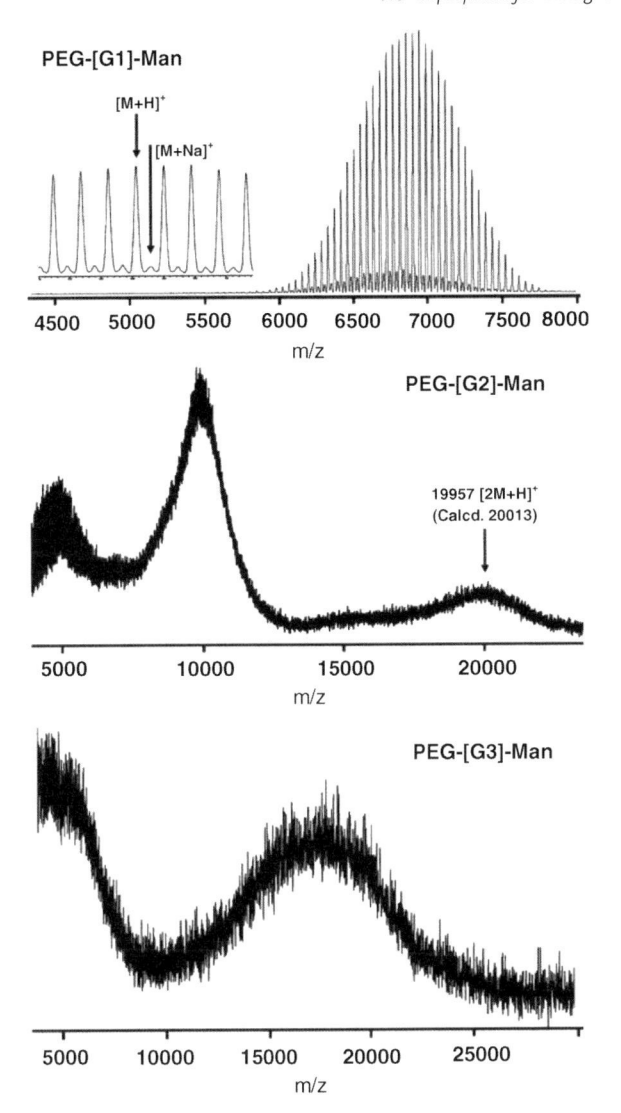

Figure 9.10 MALDI-TOF MS spectra of glycosulated PEG-dendritic copolymers derived from mannose, PEG[G1]-Man, PEG[G2]-Man, and PEG[G3]-Man. Reproduced from Ref. [134] with kind permission of the American Chemical Society.

Potentially biodegradable thermoplastic multiblock copolymers from poly(lactic acid), poly(ε-caprolatone), and/or poly(lactic acid-ε-caprolactone) were produced by Borda et al. [135] employing toluene diisocyanate as chain extender and PEG as intrinsic plasticizer. The obtained copolymers were analyzed besides NMR and IR spectroscopy, by MALDI-TOF MS. To enable the determination of the structure of the generated copolymers by MALDI-TOF MS, only copolymers with short chains were synthesized in model experiments. The performed MALDI-TOF MS analysis unambiguously confirmed the composition of the expected copolymers. Lee et al. [136] also synthesized multiblock copolymers containing poly(ε-caprolactone) as one monomer and claimed that the produced products were biodegradable. These authors compared the average molar masses obtained by SEC with MALDI-TOF MS. The structure of the macroinitiator was also investigated by MALDI-TOF MS. The recorded MALDI-TOF MS spectrum proved that the desired composition of the macroinitiator was obtained. Another synthesis and characterization of multi- and triblock copolymers containing poly(ε-caprolatone), besides poly(1,5-dioxepane-2-one), which has several biomedical applications, such for example, as temporary implants and drug delivery, are described by Adamus and colleagues [137]. The authors could determine the influence of copolymer composition and architecture on the molecular structures at the individual chain level using MALDI-TOF MS. Copolymers of similar structure were already investigated by Henry et al. [138] one year earlier. The authors characterized their synthesized copolymers by NMR and IR spectroscopy and also by MALDI-TOF MS and showed their good biocompatibility. Hakkarainen et al. [139] investigated the hydrolysis of such kind of polymers via ESI-MS. The authors presented with the help of ESI-MS that the hydrophilicity and copolymer architecture both influenced the water-soluble degradation product patterns.

Kaihara and coworkers [140] successfully created biomedical hydrogels based on polyethers consisting of cyclic acetal segments and PEG units. The chemical structures of the hydrogels were confirmed by MALDI-TOF MS. The MALDI-TOF MS analysis proved that alternating copolymers with hydroxyl end groups were synthesized. Nagy et al. [141] studied a series of amphiphilic poly(isobutylene)-b-poly-(vinyl alcohol) copolymers with an increasing block length of the latter monomer. In their contribution, MALDI-TOF MS was used as one characterization technique. The authors further analyzed if the synthesized copolymers can act as drug carrying nanodevices by doping the formed aggregates with indomethacin. Finally, Nagy et al. concluded that the solubility of indomethacin, which is significantly increased in the studied cases, is depending upon the block segment ratios.

Although the current chapter is dedicated to the MS analysis of synthetic copolymers, two examples, where synthetic polymers are combined with biopolymers, will be mentioned here. Alemdaroglu et al. [142] presented the synthesis of DNA block copolymers and their characterization by MALDI-TOF MS and gel electrophoresis. One of the recorded MALDI-TOF MS spectra was used to confirm the desired structure of the synthesized double-stranded DNA pentablock by comparing the obtained molar mass (40 500 Da) with the calculated one (41 000 Da). Hua et al. [143] reported the polymerization of dendronlike polypeptide/linear poly-(ε-caprolactone) block copolymers. The authors also employed MALDI-TOF MS as a characterization tool. One representative MALDI-TOF MS analysis of the precursors was the measurement of a clicked poly(ε-caprolactone)-dendron, which revealed that the expected product was successfully synthesized.

Adamus [144] analyzed the commercially available bipolyesters poly(3-hydroxybutyrate)-co-poly(3-hydroxyvalerate) and poly(3-hydroxybutyrate)-co-poly(3-hydroxyhexanoate) by ESI-MS to obtain information regarding the chemical structures of the end groups and the composition. The drawback of mass spectrometric characterization of copolymers with high molar masses is that with increasing m/z values the resolution decreases, while the number of isobaric structures increases. To overcome this drawback, the author performed a controlled depolymerization and checked with ^{1}H-NMR spectroscopy that oligomers containing carboxylic and olefinic end groups and the same composition as the starting biopolyesters were obtained. Additionally, the arrangement of the repeating units along the copolyester chains was verified by ESI-MS/MS and investigation of the fragmentation route. Impallomeni *et al.* [145] investigated microbial copolymers consisting of poly(3-hydroxybutyric) and poly(3-mercaptoalkanoic acid) by NMR spectroscopy, SEC, and ESI-MS to explore their microstructures. Since ^{13}C-NMR spectroscopy did not enable the determination of the structure of a synthesized triblock copolymer due to the low intensity of the carbon signals of two blocks, ESI-MS analysis was carried out on this sample after partial methanolyzation. The obtained mass spectrum undoubtedly demonstrated the presence of a terpolymer.

9.9
Software Development

A snapshot of the software development improving the characterization of copolymers by MALDI-TOF MS is given in this section. Buchmann and coworkers [146] analyzed triblock copolymers consisting of PEO and poly(propylene oxide) (PPO) via MALDI-TOF MS. To enable the creation of 2D-plots of the monomer composition, a self-made software was used. First, the copolymer sample is measured by MALDI-TOF MS, as presented for example in Figure 9.11a for the triblock PPO-PEO-PPO. The next step is to assign the peaks (Figure 9.11b), for example the ion at m/z 1720 corresponds to $EO_{21}PO_{13}Na^+$. After a de-isotoping process (Figure 9.11c), the software determines the contribution of each monomer for each ion coming from the copolymer and generates a representation of the relative abundance of the ions as a function of the number of PEO and PPO units (Figure 9.11e). The user is only entering the exact masses of the two repeating units, the two end groups, and the mass of the cation involved, as shown in Figure 9.11d. These 2D-plots allowed the rapid qualitative analysis of the copolymers by MALDI-TOF MS measurements.

After Terrier *et al.* [146], Huijser *et al.* showed in two successive studies [147, 148] that with 2D-plots created from MALDI-TOF MS data, the topology of copolymers can be elucidated. The authors demonstrated this very well on the copolymer poly(lactide)-*co*-poly(glycolide), which is extensively used in the medical sector. Finally, the software development for fractionated copolymer samples by LC followed by MALDI-TOF MS analysis was presented by Weidner and colleagues [149]. Various copolymers of PEO-PPO were analyzed and their composition was presented in 2D-plots. The MALDI-TOF MS spectrum of the nonfractionated PEO-PPO copolymer

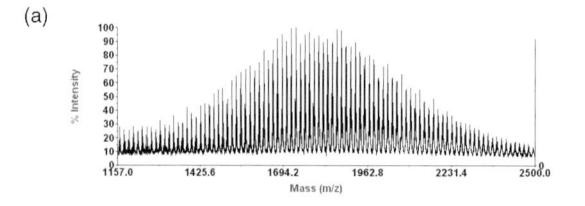

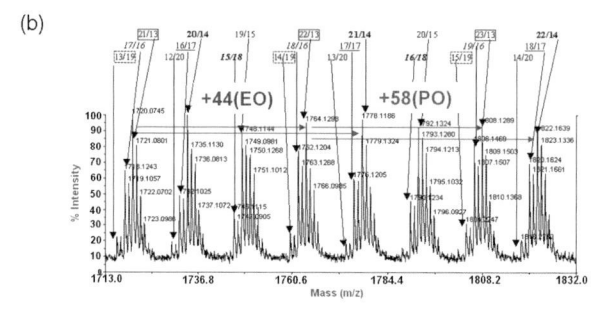

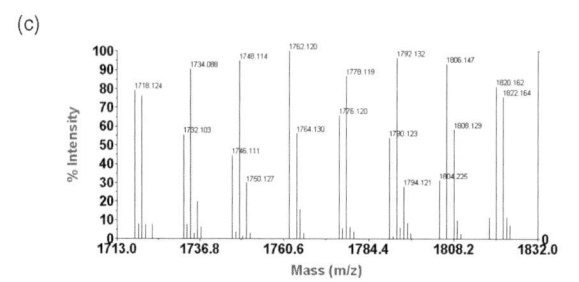

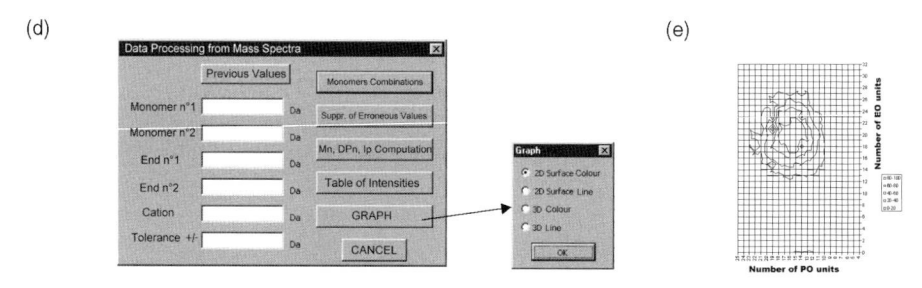

Figure 9.11 Stepwise data treatment of a PPO–PEO–PPO triblock copolymer. Reproduced from Ref. [146] with kind permission of the American Chemical Society.

clearly showed the requirement of LC prior to mass spectrometric analysis. Typical peak distances of m/z 44 (EO unit) and 58 (PO unit) caused overlapping signals. This was overcome by LC near critical conditions, which was employed as an isocratic alteration of the solvent composition or by applying a gradient mode prior to the offline MALDI-TOF MS analysis. With such an approach, structure-dependent separation conditions could be explored.

9.10
Summary and Outlook

A variety of applications in the analysis of copolymers have been presented in this chapter. Most of the applications have been performed with MALDI-TOF MS due to the easy interpretation of the recorded mass spectra. However, the implementation of new ionization techniques, such as desorption electrospray ionization, will presumably appear in the near future since their application for homopolymers has already been successfully demonstrated. Another area, which will be extended in the coming years, is the use of IMS prior MS since only one instrument is used to combine separation with MS detection. Furthermore, a demand on software developments, for example, for ESI-MS measurements, is still not fulfilled and will hopefully be addressed in the near future.

References

1 Montaudo, M.S. (2002) Mass spectra of copolymers. *Mass Spectrom. Rev.*, **21** (2), 108–144.

2 Peacock, P.M. and McEwen, C.N. (2004) Mass spectrometry of synthetic polymers. *Anal. Chem.*, **76** (12), 3417–3428.

3 Peacock, P.M. and McEwen, C.N. (2006) Mass spectrometry of synthetic polymers. *Anal. Chem.*, **78** (12), 3957–3964.

4 Weidner, S.M. and Trimpin, S. (2008) Mass spectrometry of synthetic polymers. *Anal. Chem.*, **80** (12), 4349–4361.

5 Weidner, S.M. and Trimpin, S. (2010) Mass spectrometry of synthetic polymers. *Anal. Chem.*, **82** (12), 4811–4829.

6 Montaudo, G., Samperi, F., and Montaudo, M.S. (2006) Characterization of synthetic polymers by MALDI-MS. *Prog. Polym. Sci.*, **31** (3), 277–357.

7 Crecelius, A.C., Baumgaertel, A., and Schubert, U.S. (2009) Tandem mass spectrometry of synthetic polymers. *J. Mass Spectrom.*, **44** (9), 1277–1286.

8 Gruendling, T., Weidner, S., Falkenhagen, J., and Barner-Kowollik, C. (2010) Mass spectrometry in polymer chemistry: a state-of-the-art up-date. *Polym. Chem.*, **1** (5), 599–617.

9 Chen, H., He, M., Pei, J., and Liu, B. (2002) End-group analysis of blue light-emitting polymers using matrix-assisted laser desorption/ionization time-of-flight mass spectrometry. *Anal. Chem.*, **74** (24), 6252–6258.

10 Chen, H., He, M., Wan, X., Yang, L., and He, H. (2003) Matrix-assisted laser desorption/ionization study of cationization of PEO–PPP rod-coil diblock polymers. *Rapid Commun. Mass Spectrom.*, **17** (3), 177–182.

11 Chevallier, P., Soutif, J.-C., Brosse, J.-C., and Brunelle, A. (2002) Poly(amide ester)s from 2,6-pyridinedicarboxylic acid and ethanolamine derivatives: identification of macrocycles by matrix-assisted laser desorption/ionization mass spectrometry. *Rapid Commun. Mass Spectrom.*, **16** (15), 1476–1484.

12 Elyashiv-Barad, S., Greinert, N., and Sen, A. (2002) Copolymerization of methyl acrylate with norborne derivatives by atom transfer radical polymerization. *Macromolecules*, **35** (19), 7521–7526.

13 Murgasova, R., Brantley, E.L., Hercules, D.M., and Nefzger, H. (2002) Characterization of polyester-polyurethane soft and hard blocks by a combination of MALDI, SEC, and chemical degradation. *Macromolecules*, **35** (22), 8338–8345.

14 Viala, S., Tauer, K., Antonietti, M., Krüger, R.-P., and Bremser, W. (2002)

Structural control in radical polymerization with 1,1-diphenylethylene. 1. Copolymerization of 1,1-diphenylethylene with methyl methacrylate. *Polymer*, **43** (26), 7231–7241.

15 Montaudo, M.S. (2002) Full polymer characterization by SEC-NMR combined with SEC-MALDI. *Polymer*, **43** (5), 1587–1597.

16 Montaudo, M.S., Adamus, G., and Kowalczuk, M. (2002) Bivariate distribution in copolymers: a new model. *J. Polym. Sci., Part A: Polym. Chem.*, **40** (14), 2442–2448.

17 Samperi, F., Puglisi, C., Alicata, R., and Montaudo, G. (2003) Essential role of chain ends in the nylon-6/poly-(ethylene terephthalate) exchange. *J. Polym. Sci., Part A: Polym. Chem.*, **41** (18), 2778–2793.

18 Samperi, F., Montaudo, M., Puglisi, C., Alicata, R., and Montaudo, G. (2003) Essential role of chain ends in the Ny6/PBT exchange. A combined NMR and MALDI approach. *Macromolecules*, **36** (19), 7143–7154.

19 Puglisi, C., Samperi, F., Giorgi, S.D., and Montaudo, G. (2003) Exchange reactions occurring through active chain ends. MALDI-TOF characterization of copolymers from nylon 6,6 and nylon 6,10. *Macromolecules*, **36** (4), 1098–1107.

20 Samperi, F., Montaudo, M.S., Puglisi, C., DiGiorgi, S., and Montaudo, G. (2004) Structural characterization of copolyamides synthesized via the facile blending of polyamides. *Macromolecules*, **37** (17), 6449–6459.

21 Montando, G., Carroccio, S., Montaudo, M.S., Puglisi, C., and Samperi, F. (2004) Recent advances in MALDI mass spectrometry of polymers. *Macromol. Symp.*, **218** (1), 101–112.

22 Montaudo, G., Samperi, F., Montaudo, M.S., Carroccio, S., and Puglisi, C. (2005) Current trends in matrix-assisted laser desorption/ionization of polymeric materials. *Eur. J. Mass Spectrom.*, **11** (1), 1–14.

23 Schrod, M., Rode, K., Braun, D., and Pasch, H. (2003) Matrix-assisted laser desorption/ionization mass spectrometry of synthetic polymers. VI. Analysis of phenol-urea-formaldehyde cocondensates. *J. Appl. Polym. Sci.*, **90** (9), 2540–2548.

24 Du, G., Lei, H., Pizzi, A., and Pasch, H. (2008) Synthesis-structure-performance relationship of cocondensed phenol-urea-formaldehyde resins by MALDI-ToF and ^{13}C NMR. *J. Appl. Polym. Sci.*, **110** (2), 1182–1194.

25 Hong, J., Cho, D., Chang, T., Shim, W.S., and Lee, D.S. (2003) Characterization of poly(ethylene oxide)-*b*-poly(L-lactide) block copolymer by matrix-assisted laser desorption/ionization time-of-flight mass spectrometry. *Macromol. Res.*, **11** (5), 341–346.

26 Meier, M.A.R., Lohmeijer, B.G.G., and Schubert, U.S. (2003) Characterization of defined metal-containing supramolecular block copolymers. *Macromol. Rapid Commun.*, **24** (14), 852–857.

27 Milani, B., Scarel, A., Durand, J., Mestroni, G., Seragila, R., Carfagna, C., and Binotti, B. (2003) MALDI-TOF mass spectrometry in the study of CO/aromatic olefins terpolymers. *Macromolecules*, **36** (17), 6295–6297.

28 Nakano, K., Nozaki, K., and Hiyama, T. (2003) Asymmetric alternating copolymeritzation of cyclohexene oxide and CO_2 with dimeric zinc complexes. *J. Am. Chem. Soc.*, **125** (18), 5501–5510.

29 Nakano, K., Nakamura, M., and Nozaki, K. (2009) Alternating copolymerization of cyclohexene oxide with carbon dioxide catalyzed by (salalen)CrCl complexes. *Macromolecules*, **42** (18), 6972–6980.

30 Scamporrino, E., Bazzano, S., Vitalini, D., and Mineo, P. (2003) Insertion of copper(II)/Schiff-base complexes with NLO properties into commercial polycarbonates by thermal processes. *Macromol. Rapid Commun.*, **24** (3), 236–241.

31 Schmalz, H., Lanzendörfer, M.G., Abetz, V., and Müller, A.H.E. (2003) Anionic polymerization of ethylene oxide in the presence of the phosphazene Bu^{t}P$_4$ - kinetic investigations using *in-situ* FT-NIR spectroscopy and MALDI-TOF MS.

Macromol. Chem. Phys., **204** (8), 1056–1071.

32 Williams, J.B., Chapman, T.M., and Hercules, D.M. (2003) Matrix-assisted laser desorption/ionization mass spectrometry of discrete mass poly(butylene glutarate) oligomers. *Anal. Chem.*, **75** (13), 3092–3100.

33 Cox, F.J., Qian, K., Patil, A.O., and Johnston, M.V. (2003) Microstructure and composition of ethylene-carbon monoxide copolymers by matrix-assisted laser deorption/ionization mass spectrometry. *Macromolecules*, **36** (22), 8544–8550.

34 Cox, F.J., Johnston, M.V., Qian, K., and Pfeifer, D.G. (2004) Compositional analysis of isobutylene/*p*-methylstyrene copolymers by matrix-assisted laser desorption/ionization mass spectrometry. *J. Am. Soc. Mass Spectrom.*, **15** (5), 681–688.

35 Venkatesh, R., Vergouwen, F., and Klumpermann, B. (2004) Copolymerization of allyl butyl ether with acrylates via controlled radical polymerization. *J. Polym. Sci., Part A: Polym. Chem.*, **42** (13), 3271–3284.

36 Venkatesh, R., Harrisson, S., Haddleton, D.M., and Klumperman, B. (2004) Olefin copolymerization via controlled radical polymerization: copolymerization of acrylate and 1-octene. *Macromolecules*, **37** (12), 4406–4416.

37 Schilli, C.M., Zhang, M., Rizzardo, E., Thang, S.H., Chong, B.Y.K., Edwards, K., Karlsson, G., and Müller, A.H.E. (2004) A new double-responsive block copolymer synthesized via RAFT polymerization: poly(*N*-isopropylacrylamide)-*block*-poly-(acrylic acid). *Macromolecules*, **37** (21), 7861–7866.

38 Willemse, R.X.E., Staal, B.B.P., Donkers, E.H.D., and van Herk, A.M. (2004) Copolymer fingerprints of polystyrene-*block*-polyisoprene by MALDI-TOF-MS. *Macromolecules*, **37** (15), 5717–5723.

39 Willemse, R.X.E., Ming, W., and van Herk, A.M. (2005) Solventless liquid oligoesters analyzed by MALDI-TOF-MS. *Macromolecules*, **38** (16), 6876–6881.

40 Willemse, R.X.E. and van Herk, A.M. (2006) Copolymerization kinetics of methyl methacrylate-styrene obtained by PLP-MALDI-TOF-MS. *J. Am. Chem. Soc.*, **128** (13), 4471–4480.

41 Krishnan, R. and Scrinivasan, K.S.V. (2005) Poly(ethylene glycol) block copolymers by atom transfer radical polymerization – synthesis, kinetics and characterization. *J. Macromol. Sci. Pure Appl. Chem.*, **42** (4), 495–508.

42 Adler, M., Rittig, F., Becker, S., and Pasch, H. (2005) Multidimensional chromatographic and hyphenated techniques for hydrophilic copolymers. 1. Analysis of comb-like copolymers of ethylene oxide and methacrylic acid. *Macromol. Chem. Phys.*, **206** (22), 2269–2277.

43 Gies, A.P., Nonidez, W.K., Ellison, S.T., Ji, H., and Mays, J.W. (2005) A MALDI-TOF MS study of oligomeric poly(*m*-phenyleneisophtalamide). *Anal. Chem.*, **77** (3), 780–784.

44 Gies, A.P. and Nonidez, W.K. (2004) A technique for obtaining matrix-assisted laser desorption/ionization time-of-flight mass spectra of poorly soluble and insoluble aromatic polyamides. *Anal. Chem.*, **76** (7), 1991–1997.

45 Kamitakahara, H. and Nakatsubo, F. (2005) Synthesis of diblock copolymers with cellulose derivatives. 1. Model study with azidoalkyl carboxylic acid and cellobiosylamine derivative. *Cellulose*, **12** (2), 209–219.

46 Kamitakahara, H., Enomoto, Y., Hasegawa, C., and Nakatsubo, F. (2005) Synthesis of diblock copolymers with cellulose derivatives. 2. Characterization and thermal properties of cellulose triacetate-*block*-oligoamide-15. *Cellulose*, **12** (5), 527–541.

47 Scarel, A., Durand, J., Franchi, D., Zangrando, E., Mestroni, G., Carfagna, C., Mosca, L., Seraglia, R., Consiglio, G., and Milani, B. (2005) Mono- and dinuclear bioxazoline–palladium complexes for the stereocontrolled synthesis of CO/styrene polyketones. *Chem. Eur. J.*, **11** (20), 6014–6023.

48 Somogyi, A., Bojkova, N., Padias, A.B., and Hall Jr., H.K. (2005) Analysis of all-aromatic polyesters by matrix-assisted laser desorption/ionization time-of-flight mass spectrometry. *Macromolecules*, **38** (10), 4067–4071.

49 Sugimoto, H., Ohtsuka, H., and Inoue, S. (2005) Alternating copolymerization of carbon dioxide and epoxide catalyzed by aluminum Schiff base-ammonium salt system. *J. Polym. Sci., Part A: Polym. Chem.*, **43** (18), 4172–4186.

50 Torini, L., Argillier, J.F., and Zydowicz, N. (2005) Interfacial polycondensation encapsulation in miniemulsion. *Macromolecules*, **38** (8), 3225–3236.

51 Xiao, Y., Wang, Z., and Ding, K. (2005) Copolymerization of cyclohexene oxide with CO_2 by using intramolecular dinuclear zinc catalysts. *Chem. Eur. J.*, **11** (12), 3668–3678.

52 Zhang, C., Ochiai, B., and Endo, T. (2005) Matrix-assisted laser desorption/ionization time-of-flight mass spectrometry study on copolymers obtained by the alternating copolymerization of bis(g-lactone) and epoxide with potassium *tert*-butoxide. *J. Polym. Sci., Part A: Polym. Chem.*, **43** (12), 2643–2649.

53 Chatti, S., Bortolussi, M., Bogdal, D., Blais, J.C., and Loupy, A. (2006) Synthesis and properties of new poly(ether-ester)s containing aliphatic diol based on isosorbide. Effects of the microwave-assisted poycondensation. *Eur. Polym. J.*, **42** (2), 410–424.

54 Rabani, G., Luftmann, H., and Kraft, A. (2006) Synthesis and characterization of two shape-memory polymers containing short aramid hard segments and poly(ε-caprolactone) soft segments. *Polymer*, **47** (12), 4251–4260.

55 Lo, C., Adenier, A., Chane-Ching, K.I., Maurel, F., Aaron, J.J., Kosata, B., and Svoboda, J. (2006) A novel fluorescent, conducting polymer: poly[1-(thiophene-2-yl]benzothieno[3,2-b]benzothiophene] electrosynthesis, characterization and optical properties. *Synth. Met.*, **156** (2–4), 256–269.

56 Li, Z., Li, P., and Huang, J. (2006) Synthesis and characterization of amphiphilic graft copolymer poly (ethylene oxide)-*graft*-poly(methyl acrylate). *Polymer*, **47** (16), 5791–5798.

57 Li, Z., Li, P., and Huang, J. (2006) Synthesis of amphiphilic copolymer brushes: poly(ethylene oxide)-*graft*-polystyrene. *J. Polym. Sci., Part A: Polym. Chem.*, **44** (15), 4361–4371.

58 Jia, Z., Fu, Q., and Huang, J. (2006) Synthesis of amphiphilic macrocyclic graft copolymer consisting of a poly (ethylene oxide) ring and multi-polystyrene lateral chains. *Macromolecules*, **39** (16), 5190–5193.

59 Park, J.-S. and Kataoka, K. (2006) Precise control of lower critical solution temperature of thermosensitive poly (2-isopropyl-2-oxazoline) via gradient copolymerization with 2-ethyl-2-oxazoline as a hydrophilic comonomer. *Macromolecules*, **39** (19), 6622–6630.

60 Park, J.-S. and Kataoka, K. (2007) Comprehensive and accurate control of thermosensitivity of poly(2-alkyl-2-oxazoline)s via well-defined gradient or random copolymerization. *Macromolecules*, **40** (10), 3599–3609.

61 Matsuoka, S.-I., Ogiwara, N., and Ishizone, T. (2006) Formation of alternating copolymers via spontaneous copolymerization of 1,3-dehydroadamantane with electron-deficient vinyl monomers. *J. Am. Chem. Soc.*, **128** (27), 8708–8709.

62 Puglisi, C., Samperi, F., Cicala, G., Recca, A., and Restuccia, C.L. (2006) Combined MALDI–TOF MS and NMR characterization of copoly(arylen ether sulphone)s. *Polymer*, **47** (6), 1861–1874.

63 Richez, A., Belleney, J., Bouteiller, L., and Pensec, S. (2006) Synthesis and MALDI-TOF analysis of dentritic-linear block copolymers of lactides: influence of architecture on stereocomplexation. *J. Polym. Sci., Part A: Polym. Chem.*, **44** (23), 6782–6789.

64 Zhou, J., Villarroya, S., Wang, W., Wyatt, M.F., Duxbury, C.J., Thurecht, K.J., and Howdle, S.M. (2006) One-step chemoenzymatic synthesis of poly(ε-caprolactone-*block*-methyl

methacrylate) in supercritical CO_2. *Macromolecules*, **39** (16), 5352–5358.

65 Yang, H., Su, Y., Zhu, H.Z.H., Xie, B., Zhao, Y., Chen, Y., and Wang, D. (2007) Synthesis of amphiphilic triblock copolymers and application for morphology control of calcium carbonate crystals. *Polymer*, **48** (15), 4344–4351.

66 Samperi, F., Mendichi, R., Sartore, L., Penco, M., and Puglisi, C. (2006) Full characterization of a multiblock copolymer based on poly(2,6-dimethyl-1,4-phenylene oxide) and poly(bisphenol-A carbonate). *Macromolecules*, **39** (26), 9223–9233.

67 Faÿ, F., Renard, E., Langlois, V., Linossier, I., and Vallée-Rehel, K. (2007) Development of poly(ε-caprolactone-*co*-L-lactide) and poly(ε-caprolactone-*co*-δ-valerolactone) as new degradable binder used for antifouling paints. *Eur. Polym. J.*, **43** (11), 4800–4813.

68 Pound, G., Aguesse, F., McLearly, J.B., Lange, R.F.M., and Klumperman, B. (2007) Xanthate-mediated copolymerization of vinyl monomers for amphiphilic and double-hydrophilic block copolymers with poly(ethylene glycol). *Macromolecules*, **40** (25), 8861–8871.

69 Baéz, J.E., Ramirez-Hernández, A., and Marcos-Fernández, Á. (2010) Synthesis, characterization, and degradation of poly(ethylene-*b*-ε-caprolactone) diblock copolymer. *Polymer Adv. Techn.*, **21** (1), 55–64.

70 Ihara, E., Kurokawa, A., Itho, T., and Inoue, K. (2007) A novel synthetic strategy for copoymers of vinyl alcohol: radical copolymerization of alkoxyvinylsilanes with styrene and oxidative transformation of C-Si(OR)$_2$Me into C-OH in the copolymers to afford poly(vinyl alcohol-*ran*-styrene)s. *J. Polym. Sci., Part A: Polym. Chem.*, **45** (16), 3648–3658.

71 Božovic-Vukić, J., Manon, H.T., Meuldijk, J., Koning, C., and Klumperman, B. (2007) SAN-*b*-P4VP block copolymer synthesis by chain extension from RAFT-functional poly(4-vinylpyridine) in solution and in emulsion. *Macromolecules*, **40** (20), 7132–7139.

72 Cheng, X., Zhang, X., Liu, J., and Xu, T. (2008) Novel approaches for the preparation of silica-based zwitterionic hybrid copolymers. *Eur. Polym. J.*, **44** (3), 918–931.

73 Pfeifer, S. and Lutz, J.-F. (2008) Development of a library of *N*-substituted maleimides for the local functionalization of linear polymer chains. *Chem. Eur. J.*, **14** (35), 10949–10957.

74 Binder, W.H., Pulamagatta, B., Kir, O., Kurzhals, S., Barqawi, H., and Tanner, S. (2009) Monitoring block copolymer crossover-chemistry in ROMP: catalyst evaluation via mass-spectrometry (MALDI). *Macromolecules*, **42** (24), 9457–9466.

75 Dong, Y.-Q., Tong, Y.-Y., Dong, B.-T., Du, F.-S., and Li, Z.-C. (2009) Preparation of tadpole-shaped amphiphilic cyclic PS-*b*-linear PEO via ATRP and click chemistry. *Macromolecules*, **42** (8), 2940–2948.

76 Girod, M., Mazarin, M., Phan, T.N.T., Gigmes, D., and Charles, L. (2009) Determination of block size in poly-(ethylene oxide)-*b*-polystyrene block copolymers by matrix-assisted laser desorption/ionization time-of-flight mass spectrometry. *J. Polym. Sci., Part A: Polym. Chem.*, **47** (13), 3380–3390.

77 Crecelius, A.C., Becer, C.R., Knop, K., and Schubert, U.S. (2010) Block length determination of the block copolymer mPEG-*b*-PS using MALDI-TOF MS/MS. *J. Polym. Sci. Part A: Polym. Chem.*, **48** (20), 4375–4384.

78 Krieg, A., Becer, C.R., Hoogenboom, R., and Schubert, U.S. (2009) Tailor made side-chain functionalized macromolecules by combination of controlled radical polymerization and click chemistry. *Macromol. Symp.*, **275–276** (1), 73–81.

79 Petrova, S., Riva, R., Jérôme, C., Lecomte, P., and Mateva, R. (2009) Controlled synthesis of AB$_2$ amphiphilic triarm star-shaped block copolymers by ring-opening polymerization. *Eur. Polym. J.*, **45** (12), 3442–3450.

80 Raynaud, J., Absalon, C., Gnanou, Y., and Taton, D. (2009) N-Heterocyclic carbene-induced zwitterionic ring-opening polymerization of ethylene oxide and direct synthesis of α,ω-difunctionalized poly(ethylene oxide)s and poly(ethylene oxide)-*b*-poly(ε-caprolactone) block copolymers. *J. Am. Chem. Soc.*, **131** (9), 3201–3209.

81 Zhang, W.-B., Sun, B., Li, H., Ren, X., Janoski, J., Sahoo, S., Dabney, D.E., Wesdemiotis, C., Quirk, R.P., and Cheng, S.Z.D. (2009) Synthesis of in-chain-functionalized polystyrene-*block*-poly(dimethylsiloxane) diblock copolymers by anionic polymerization and hydrosilylation using dimethyl-[4-(1-phenylvinyl)phenyl]silane. *Macromolecules*, **42** (19), 7258–7262.

82 Sciucca, S.D., Spagnoli, G., Penco, M., Battiato, S., Samperi, F., and Medichi, R. (2009) Effect on structural relaxation of the poly(methyl methacrylate) copolymers chain flexibility. *J. Polym. Sci., Part B: Polym. Phys.*, **47** (6), 596–607.

83 Wu, S., Bu, L., Huang, L., Yu, X., Han, Y., Geng, Y., and Wang, F. (2009) Synthesis and characterization of phenylene-thiophene all-conjugated diblock copolymers. *Polymer*, **50** (26), 6245–6251.

84 Weber, C., Becer, C.R., Hoogenboom, R., and Schubert, U.S. (2009) Lower critical solution temperature behavior of comb and graft shaped poly[oligo(2-ethyl-2-oxazoline)methacrylate]s. *Macromolecules*, **42** (8), 2965–2971.

85 Siao, S.-Y., Lin, L.-H., Chen, W.-W., Huang, M.-H., and Chong, P. (2009) Characterization and emulsifying properties of block copolymers prepared from lactic acid and poly(ethylene glycol). *J. Appl. Polym. Sci.*, **114** (1), 509–516.

86 Giordaengo, R., Viel, S., Hidalgo, M., Allard-Breton, B., Thévand, A., and Charles, L. (2010) Analytical strategy for the molecular weight determination of random copolymers of poly(methyl methacrylate) and poly(methacrylic acid). *J. Am. Soc. Mass Spectrom.*, **21** (6), 1075–1085.

87 Ahmed, H., Trathnigg, B., Kappe, C.O., and Saf, R. (2010) Synthesis of poly-(ε-caprolactone) diols and EO–CL block copolymers and their characterization by liquid chromatography and MALDI-TOF-MS. *Europ. Polym. J.*, **46** (3), 494–505.

88 Ohsawa, S., Morino, K., Sudo, A., and Endo, T. (2010) Alternating copolymerization of bicyclic bis (γ-butyrolactone) and epoxide through zwitterion process by phosphines. *Macromolecules*, **43** (8), 3585–3588.

89 Ranimol, S., Gibon, C.M., Weber, M., and Gamans, R.J. (2010) Modifying an amorphous polymer to a fast crystallizing semi-crystalline material by copolymerization with monodisperse amide segments. *J. Polym. Sci., Part A: Polym. Chem.*, **48** (1), 63–73.

90 Kumooka, Y. (2010) Classification of OPP adhesive tapes according to MALDI mass spectra of adhesives. *Forensic Science Inter.*, **197** (1–3), 75–79.

91 Alhazmi, A.M., Giguère, M.-S., Dubé, M.A., and Mayer, P.M. (2006) A comparison of electrospray-ionization and matrix-assisted laser desorption/ionization mass spectrometry with nuclear magnetic resonance spectroscopy for the characterization of synthetic copolymers. *Eur. J. Mass Spectrom.*, **12** (5), 301–310.

92 Li, B., Zhang, R., and Lu, X.-B. (2007) Stereochemistry control of the alternating copolymerization of CO_2 and propylene oxide catalyzed by salenCrX complexes. *Macromolecules*, **40** (7), 2303–2307.

93 Kasperczyk, J., Li, S., Jaworska, J., Dobrzyński, P., and Vert, M. (2008) Degradation of copolymers obtained by ring-opening polymerization of glycocide and ε-caprolactone: a high resolution NMR and ESI-MS study. *Polym. Degr. Stab.*, **93** (5), 990–999.

94 Nejad, E.H., Castignolles, P., Gilbert, R.G., and Guillaneuf, Y. (2008) Synthesis of methacrylate derivates oligomers by dithiobenzoate-RAFT-mediated polymerization. *J. Polym. Sci., Part A: Polym. Chem.*, **46** (6), 2277–2289.

95 He, C., Zhao, C., Guo, X., Guo, Z., Chen, X., Zhuang, X., Liu, S., and Jing, X. (2008) Novel temperature- and pH-responsive graft copolymers composed of

poly(L-glutamic acid) and poly(N-isopropylacrylamide). *J. Polym. Sci., Part A: Polym. Chem.*, **46** (12), 4140–4150.

96 Giordaengo, R., Viel, S., Hidalgo, M., Allard-Breton, B., Thévand, A., and Charles, L. (2009) Structural characterization of a poly(methacrylic acid)–poly(methyl methacrylate) copolymer by nuclear magnetic resonance and mass spectrometry. *Anal. Chim. Acta*, **654** (1), 49–58.

97 van Leeuwen, S.M., Tan, B., Grijpma, D.W., Feijen, J., and Karst, U. (2007) Characterization of the chemical composition of a block copolymer by liquid chromatography/mass spectrometry using atmospheric pressure chemical ionization and electrospray ionization. *Rapid Commun. Mass Spectrom.*, **21** (16), 2629–2637.

98 Desmazières, B., Buchmann, W., Terrier, P., and Tortajada, J. (2008) APCI interface for LC- and SEC-MS analysis of synthetic polymers: advantages and limits. *Anal. Chem.*, **80** (3), 783–792.

99 Whitson, S.E., Kennedy, G.E.P., Lattimer, R.P., and Wesdemiotis, C. (2008) Direct probe-atmospheric pressure chemical ionization mass spectrometry of cross-linked copolymers and copolymer blends. *Anal. Chem.*, **80** (20), 7778–7785.

100 Gallet, G., Carroccio, S., Rizzarelli, P., and Karlsson, S. (2002) Thermal degradation of poly(ethylene oxide-propylene oxide-ethylene oxide) triblock copolymer: comparative study by SEC/NMR, SEC/MALDI-TOF-MS and SPME/GC-MS. *Polymer*, **43** (4), 1081–1094.

101 Montaudo, M.S. (2004) MALDI for the estimation of viscosity parameters. A modified method which applies also to polycondensates. *Polymer*, **45** (18), 6291–6298.

102 Tatro, S.R., Baker, G.R., Fleming, R., and Harmon, J.P. (2002) Matrix-assisted laser desorption/ionization (MALDI) mass spectrometry: determining Mark-Houwink-Sakurada parameters and analyzing the breadth of molecular weight distribution. *Polymer*, **43** (8), 2329–2335.

103 Tillier, D., Lefebvre, H., Tessier, M., Blais, J.-C., and Fradet, A. (2004) High temperature bulk reaction between poly(ethylene terephtalate) and lactones: ^{1}H NMR and SEC/MALDI-TOF MS study. *Macromol. Chem. Phys.*, **205** (5), 581–592.

104 Adamus, G., Rizzarelli, P., Montaudo, M.S., Kowalczuk, M., and Montaudo, G. (2006) Matrix-assisted laser desorption/ionization time-of-flight mass spectrometry with size-exclusion chromatographic fractionation for structural characterization of synthetic aliphatic copolyesters. *Rapid Commun. Mass Spectrom.*, **20** (5), 804–814.

105 Alicata, R., Barbuzzi, T., Giuffrida, M., and Ballistreri, A. (2006) Characterization of poly[(R)-3-hydroxybutyrate-co-ε-caprolactone] copolymers by matrix-assisted laser desorption/ionization time-of-flight and electrospray ionization mass spectrometry. *Rapid Commun. Mass Spectrom.*, **20** (4), 568–576.

106 Rittig, F., Fandrich, N., Urtel, M., Schrepp, W., Jost, U., and Weidner, S.M. (2006) Structure determination of polyacetals by liquid chromatography and hyphenated techniques. *Macromol. Chem. Phys.*, **207** (12), 1026–1037.

107 Girod, M., Phan, T.N.T., and Charles, L. (2009) Tuning block copolymer structural information by adjusting salt concentration in liquid chromatography at critical conditions coupled with electrospray tandem mass spectrometry. *Rapid Commun. Mass Spectrom.*, **23** (10), 1476–1482.

108 Falkenhagen, J. and Weidner, S. (2009) Determination of critical conditions of adsorption for chromatography of polymers. *Anal. Chem.*, **81** (1), 282–287.

109 Weidner, S., Falkenhagen, J., Krueger, R.-P., and Just, U. (2007) Principle of two-dimensional characterization of copolymers. *Anal. Chem.*, **79** (13), 4814–4819.

110 Malik, M.I., Trathnigg, B., and Saf, R. (2009) Characterization of ethylene oxide–propylene oxide block copolymers by combination of different chromatographic techniques and matrix-assisted laser desorption ionization time-of-flight mass spectroscopy. *J. Chromatogr. A*, **1216** (38), 6627–6635.

111 Malik, M.I., Trathingg, B., Bartl, K., and Saf, R. (2010) Characterization of polyoxyalkylene block copolymers by combination of different chromatographic techniques and MALDI-TOF-MS. *Anal. Chim. Acta*, **658** (2), 217–224.

112 Baker, E.S., Gidden, J., Simonsick, W.J., Grady, M.C., and Bowers, M.T. (2004) Sequence dependent conformations of glycidyl methacrylate/butyl methacrylate copolymers in the gas phase. *Int. J. Mass Spectrom.*, **238** (3), 279–286.

113 Jackson, A.T., Scrivens, J.H., Williams, J.P., Baker, E.S., Gidden, J., and Bowers, M.T. (2004) Microstructural and conformational studies of polyether copolymers. *Int. J. Mass Spectrom.*, **238** (3), 287–297.

114 Trimpin, S. and Clemmer, D.E. (2008) Ion mobility spectrometry/mass spectrometry snapshots for assessing the molecular compositions of complex polymeric systems. *Anal. Chem.*, **80** (23), 9073–9083.

115 Wesdemiotis, C., Pingitore, F., Polce, M.J., Russell, V.M., Kim, Y., Kausch, C.M., Connors, T.H., Medsker, R.E., and Thomas, R.R. (2006) Characterization of a poly(fluorooxetane) and poly-(fluorooxetane-*co*-THF) by MALDI mass spectrometry, size exclusion chromatography, and NMR spectroscopy. *Macromolecules*, **39** (24), 8369–8378.

116 Mass, V., Schrepp, W., von Vacona, B., and Pasch, H. (2009) Sequence analysis of an isocyanate oligomer by MALDI-TOF mass spectrometry using collision induced dissociation. *Macromol. Chem. Phys.*, **210** (22), 1957–1965.

117 Baumgaertel, A., Altuntaş, E., Kempe, K., Crecelius, A., and Schubert, U.S. (2010) Characterization of different poly(2-oxazoline) block copolymers by MALDI-TOF MS/MS and ESI-Q-TOF MS/MS. *J. Polym. Sci. Part A: Polym. Chem.*, **48** (23), 5533–5540.

118 Adamus, G., Sikorska, W., Kowalczuk, M., Noda, I., and Satkowski, M.M. (2003) Electrospray ion-trap multistage mass spectrometry for characterization of co-monomer compositional distribution of bacterial poly(3-hydroxybutyrate-*co*-3-hydroxyhexanoate) at the molecular level. *Rapid Commun. Mass Spectrom.*, **17** (20), 2260–2266.

119 Adamus, G., Montaudo, M.S., Montaudo, G., and Kowalczuk, M. (2004) Molecular architecture of poly[(R,S)-3-hydroxybutyrate-*co*-6-hydroxyhexanoate] and poly[(R,S)-3-hydroxybutyrate-*co*-(R,S)-2-hydroxyhexanoate] oligomers investigated by electrospray ionization ion-trap multistage mass spectrometry. *Rapid Commun. Mass Spectrom.*, **18** (13), 1436–1446.

120 Adamus, G. (2007) Structural analysis of poly[(R,S)-3-hydroxybutyrate-*co*-L-lactide] copolyester by electrospray ionization ion trap mass spectrometry. *Rapid Commun. Mass Spectrom.*, **21** (15), 2477–2490.

121 Adamus, G. (2009) Molecular level structure of (R,S)-3-hydroxybutyrate/(R,S)-3-hydroxy-4-ethoxybutyrate copolyester with dissimilar architecture. *Macromolecules*, **42** (13), 4547–4557.

122 Žagar, E., Kržan, A., Adamus, G., and Kowalczuk, M. (2006) Sequence distribution in microbial poly(3-hydroxybutyrate-*co*-hydroxyvalerate) co-polyesters determined by NMR and MS. *Biomacromolecules*, **7** (7), 2210–2216.

123 Terrier, P., Buchmann, W., Desmazières, B., and Tortajada, J. (2006) Block length and block sequence of linear triblock and glycerol derivative diblock copolyesters by electrospray ionization-collision-induced dissociation mass spectrometry. *Anal. Chem.*, **78** (6), 1801–1806.

124 Girod, M., Phan, T.N.T., and Charles, L. (2008) Microstructural study of a nitroxide-mediated poly(ethylene oxide)/polystyrene block copolymer (PEO-*b*-PS) by electrospray tandem mass spectrometry. *J. Am. Soc. Mass Spectrom.*, **19** (8), 1163–1175.

125 Girod, M., Phan, T.N.T., and Charles, L. (2008) On-line coupling of liquid chromatography at critical conditions with electrospray ionization tandem mass spectrometry for the characterization of a nitroxide-mediated poly(ethylene oxide)/polystyrene block copoylmer. *Rapid*

Commun. Mass Spectrom., **22** (23), 3767–3775.

126 Simonsick, W.J. and Petkovska, V.I. (2008) Detailed structural elucidation of polyesters and acrylates using Fourier transform mass spectrometry. *Anal. Bioanal. Chem.*, **392** (4), 575–583.

127 Alhazmi, A.M. and Mayer, P.M. (2007) Matrix effects on copolymer quantitation by matrix-assisted laser desorption/ionization mass spectrometry. *Rapid Commun. Mass Spectrom.*, **21** (20), 3392–3394.

128 Prebyl, B.S., Johnson, J.D., Albert, A.Tuinman, Zhou, S., and Cook, K.D. (2002) Qualitative assessment of monomer ratios in putative ionic terpolymer samples by electrospray ionization mass spectrometry with collision-induced dissociation. *J. Am. Soc. Mass Spectrom.*, **13** (8), 921–927.

129 Prebyl, B.S., Johnson, J.D., and Cook, K.D. (2004) Calibration for determining monomer ratios in copolymers by electrospray ionization mass spectrometry. *Int. J. Mass Spectrom.*, **238** (3), 207–214.

130 Impallomeni, G., Giuffrida, M., Barbuzzi, O., Musumarra, G., and Ballistreri, A. (2002) Acid catalyzed transesterification as a route to poly(3-hyroxybutyrate-*co*-(-caprolactone) copolymers from their homopolymers. *Biomacromolecules*, **3** (4), 835–840.

131 Chi, S.-C., Yeom, D.-I., Kim, S.-C., and Park, W.-S. (2003) A polymeric micellar carrier for the solubilization of biphenyl dimethyl dicarboxylate. *Arch. Pharm. Res.*, **26** (2), 173–181.

132 Rieger, J., Bernaerts, K.V., Prez, F.E.D., Jérôme, R., and Jérôme, C. (2004) Lactone end-capped poly(ethylene oxide) as new building block for biomaterials. *Macromolecules*, **37** (26), 9738–9745.

133 Gutzler, R., Smulders, M., and Lange, R.F.M. (2005) The role of synthetic pharmaceutical polymer excipients in oral dosage forms-poly(ethylene oxide)-*graft*-poly(vinyl alcohol) copolymers in tablet coatings. *Macromol. Symp.*, **225** (1), 81–93.

134 Fernandez-Megia, E., Correa, J., and Riguera, R. (2006) "Clickable" PEG-dendritic block copolymers. *Biomacromolecules*, **7** (11), 3104–3111.

135 Borda, J., Kéki, S., Bodnár, I., Németh, N., and Zsuga, M. (2006) New potentially biodegradable polyurethanes. *Polym. Adv. Technol.*, **17** (11–12), 945–953.

136 Lee, R.-S., Lin, Z.-K., Yang, J.-M., and Lin, F.-H. (2006) Synthesis and characterization of biodegradable A-B-A triblock copolymers containing poly((-caprolactone) A blocks and poly(*trans*-4-hydroxy-L-proline) B blocks. *J. Polym. Sci., Part A: Polym. Chem.*, **44** (14), 4268–4280.

137 Adamus, G., Kakkarainen, M., Höglund, A., Kowalczuk, M., and Albertsson, A.-C. (2009) MALDI-TOF MS reveals the molecular level structures of different hydrophilic-hydrophobic polyether-esters. *Biomacromolecules*, **10** (6), 1540–1546.

138 Henry, G.R.P., Heise, A., Bottai, D., Formenti, A., Gorio, A., Giulio, A.M.D., and Koning, C.E. (2008) Acrylate end-capped poly(ester-carbonate) and poly(ether-ester)s for polymer-on-multielectrode array devices: synthesis, photocuring, and biocompatibility. *Biomacromolecules*, **9** (3), 867–878.

139 Hakkarainen, M., Adamus, G., Höglund, A., Kowalczuk, M., and Albertsson, A.-C. (2008) ESI-MS reveals the influence of hydrophilicity and architecture on the water-soluble degradation product patterns of biodegradable homo- and copolyester of 1,5-dixepan-2-one and ε-caprolactone. *Macromolecules*, **41** (10), 3547–3554.

140 Kaihara, S., Matsumura, S., and Fischer, J.P. (2007) Synthesis and properties of poly[poly(ethylene glycol)-*co*-cyclic acetal] based hydrogels. *Macromolecules*, **40** (21), 7625–7632.

141 Nagy, M., Szöllös, L., Kéki, S., Faust, R., and Zsuga, M. (2009) Poly(vinyl alcohol)-based amphiphilic copolymer aggregates as drug carrying nanoparticles. *J. Macromol. Sci. Pure Appl. Chem.*, **46** (4), 331–338.

142 Alemdaroglu, F.E., Safak, M., Wang, J., Berger, R., and Herrmann, A. (2007) DNA multiblock copolymers. *Chem. Comm*, (13), 1358–1359.

143 Hua, C., Dong, C.-M., and Wei, Y. (2009) Versatile strategy for the synthesis of dendronlike polypeptide/linear poly-(ε-caprolactone) block copolymers via click chemistry. *Biomacromolecules*, **10** (5), 1140–1148.

144 Adamus, G. (2006) Aliphatic polyesters for advanced technologies-structural characterization of biopolyesters with the aid of mass spectrometry. *Macromol. Symp.*, **239** (1), 77–83.

145 Impallomeni, G., Steinbüchel, A., Lütke-Eversloh, T., Barbuzzi, T., and Ballsitreri, A. (2007) Sequencing microbial copolymers of 3-hydroxybutyric and 3-mercaptoalkanoic acids by NMR, electrospray ionization mass spectrometry, and size exclusion chromatography. *Biomacromolecules*, **8** (3), 985–991.

146 Terrier, P., Buchmann, W., Cheguillaume, G., Desmazières, B., and Tortajada, J. (2005) Analysis of poly(oxyethylene) and poly(oxypropylene) triblock copolymers by MALDI-TOF mass spectrometry. *Anal. Chem.*, **77** (10), 3292–3300.

147 Huijser, S., Staal, B.B.P., Huang, J., Duchateau, R., and Koning, C.E. (2006) Chemical composition and topology of poly(lactide-*co*-glycolide) revealed by pushing MALDI-TOF MS to its limit. *Angew. Chem. Int. Ed.*, **45** (25), 4104–4108.

148 Huijser, S., Staal, B.B.P., Huang, J., Duchateau, R., and Koning, C.E. (2006) Topology characterization by MALDI-ToF-MS of enzymatically synthesized poly(lactide-*co*-glycolide). *Biomacromolecules*, **7** (9), 2465–2469.

149 Weidner, S.M., Falkenhagen, J., Maltsev, S., Sauerland, V., and Rinken, M. (2007) A novel software tool for copolymer characterization by coupling of liquid chromatography with matrix-assisted laser desorption/ionization time-of-flight mass spectrometry. *Rapid Commun. Mass Spectrom.*, **21** (16), 2750–2758.

10
Elucidation of Reaction Mechanisms: Conventional Radical Polymerization

Michael Buback, Gregory T. Russell, and Philipp Vana

10.1
Introduction

It is often remarked that the molar mass distribution (MMD) of a polymer contains the entire history of its synthesis. It is because of this truism that those interested in the kinetics of polymerization are driven to understand polymer MMDs. As will be seen in this chapter, in the case of (conventional) radical polymerization (RP), this quest has been enormously advanced by the advent of large-molecule mass spectrometry (MS), for it is a characterization technique that furnishes hitherto unimaginable detail about the MMD of a polymer, and thus it may be used to elucidate or confirm the mechanisms of RP in many ways that simply were not previously possible.

What is it that is so special about an MMD yielded by MS? It is not the distribution *amount*, for size exclusion chromatography (SEC) already makes a superior fist of measuring that, mostly through the use of refractive-index (RI) detection. Admittedly there can be issues with this, for example, that RI is molar mass dependent for oligomers [1]. However, such issues apply only in particular circumstances, and they pale beside the general uncertainty one must attach to polymer amounts as returned via MS: there is no doubt that SEC is a vastly superior technique in this respect. Rather, what is so special about MS is its delivery of the MMD *variable*, namely molar mass, M. The two important aspects here are *mass accuracy* and *mass resolving power* [2].

For many large-molecule studies, it is the mass accuracy of MS that is revolutionary. Indeed, the knee-jerk reaction of most RP workers would be that this is also the case in their field. In fact, the matter is not so straightforward. A good illustrative example of this is the pulsed-laser polymerization (PLP) method for measuring propagation rate coefficients, k_p [3, 4], which requires knowledge of absolute values of M. By now an enormous quantity of accurate k_p data has been yielded by this method, almost all of it through the use of SEC [5, 6]. This evidences that for the study of RP kinetics and mechanisms, accurate SEC calibration is commonly possible, either directly through the use of standards, or indirectly through the additional use of Mark–Houwink parameters (so-called universal calibration). So, it is not this aspect alone that makes MS so useful.

Mass Spectrometry in Polymer Chemistry, First Edition.
Edited by Christopher Barner-Kowollik, Till Gruendling, Jana Falkenhagen, and Steffen Weidner
© 2012 Wiley-VCH Verlag GmbH & Co. KGaA. Published 2012 by Wiley-VCH Verlag GmbH & Co. KGaA.

In fact what is so groundbreaking about MS for the study of RP mechanisms is its mass resolving power, that is, the ability of mass analyzers to separate and measure small differences in M, by now to well under 1-Dalton level with most instruments. As Barner-Kowollik *et al.* expressed it, this gives the power "to *visualize* the individual polymer chains present in a given sample" [7]. Thus, for example, polymer chains of identical degree of polymerization but with a different end group can be resolved, something of which SEC will never (in general) be capable. Because of the accurate determination of the M of the resolved species, their precise chemical composition may be deduced, and from this the mechanism of polymer formation can be inferred. Numerous examples of this general principle will be presented in the course of this chapter.

A propos visualization, we will regularly use chemical schemes in this chapter, as we feel that this is the most intuitive way of grasping how MS is such a powerful tool for the elucidation of polymer structure, and therefore (conventional) RP mechanisms.

10.2
Basic Principles and General Considerations

Consider polymerization of a monomer M in which chain growth is started by a species A and chain stoppage results in an end group Z. The polymerization therefore produces chains with structure $A–(M)_i–Z$, where $i = 1, 2, \ldots$ is degree of polymerization. MS of such a sample results in a series of peaks of molar mass $M(A) + i \times M(M) + M(Z)$. Turning this around, if MS of a polymer sample results in a series of peaks with precisely these molar masses, then one may take as likely that initiation produced species A and chain stoppage left behind group Z. Thus, one learns about the chain-starting and -stopping reactions. This is the basic and widely employed principle by which MS is used to shed light on the *mechanism* of polymerization. In the case of RP it was evoked at least as early as 1995, when Haddleton *et al.* [8] used matrix-assisted laser-desorption ionization time-of-flight (MALDI TOF) MS to confirm in this way that the mechanism of catalytic-chain-transfer polymerization (CCTP) (see Chapter 11 of methyl methacrylate (MMA) is as expected. Numerous examples of this principle in action will be seen in this chapter, and indeed elsewhere in this book.

Of course the spectrum from MS of a sample of synthetic polymer is normally more complicated than a single series of peaks. This is because there is usually more than one possible chain-initiating species and/or more than one channel of chain stoppage, with the result that there are multiple series of peaks, each with different intensities that depend on the relative extents to which the various possible reactions occur. Further, frequently it is desirable to identify as many such reactions as possible. This raises the issue of which type of instrument is best for observing different species. An important choice is that between the MALDI and electrospray ionization (ESI) techniques. MALDI is considerably more expensive and has the advantage of early domination of the market. It is therefore of interest that recent studies of products from reversible-deactivation RP (see Chapter 11 of styrene [9] and methyl acrylate (MA) [10] have found that ESI MS detects more species than MALDI MS.

Specifically, the study of MA star formation detected 18 products by both techniques, 42 by ESI only, and 6 by MALDI only [10]. Certainly as far as investigation of RP mechanisms is concerned, this speaks very much in favor of ESI.

It is appropriate also to say something about the possibility of *accurate* MMD determination from MS. This is because one can learn about mechanism not just from species identification, but also from species amount, as this is determined by kinetics, which of course is an important window onto mechanism. As already explained in Section 10.1, SEC measures MMD *amount* accurately, whereas the strength of MS is species separation and determination of M. Thus, the idea arises to couple SEC with MS in order to deliver MMDs of ultimate accuracy. This is an active area of development [11, 12] that may in future years become a standard technique (see Chapter 8).

For TOF detection, there is an important nuance to be aware of with regard to MMD amount. It is that, contrary to what most workers naturally assume, signal intensity is *not* proportional to $n(M)$, the relative number of chains. Rather, this intensity is proportional to $M^{0.5} \times n(M)$ [13]. This can lead to the distribution of intensities from TOF MS being quite different to the underlying $n(M)$, for example, this is the case for the so-called most-probable distribution [13]. It is essential to be aware of this if attempting to obtain MMD from TOF MS results.

Another interesting general result is from the use of MS to shed light on SEC broadening. Polymer standards of different M were analyzed by both MS and SEC. It was found that as M increased, the SEC distribution became increasingly broader than the distribution from MS [14]. This suggests that SEC broadening is greater at large M, as opposed to being constant with M.

The above is a selection of general considerations from the employment of MS to study MMDs. They are merely ones that have piqued our interest. The rest of this book provides exhaustive coverage.

10.3
Initiation

This section is about reactions that generate *new* free radicals. This is as opposed to chain-transfer reactions (Section 10.6), which merely transfer the site of radical activity. Section 10.3.1.1 will be especially in-depth, thereby providing the most detailed case study of this chapter on how MS may be used to learn about RP mechanisms. The principles involved are ones that may be applied for the study of all RP reaction mechanisms by MS.

10.3.1
Radical Generation

10.3.1.1 Thermally Induced Initiator Decomposition
The adequate understanding of RP kinetics and processes requires detailed knowledge about initiation kinetics and mechanism. Of particular interest are the type of initiator-derived radicals and in-cage and out-of-cage reactions of these primary species.

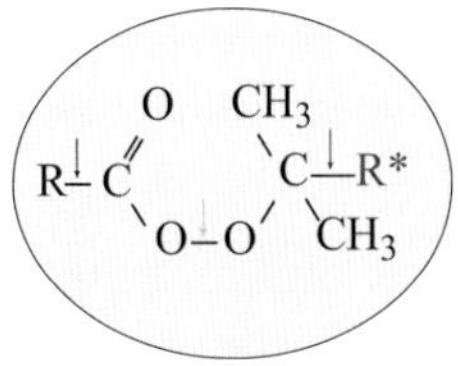

$$R\overset{\downarrow}{-}\overset{O}{\underset{O-O}{C}}\overset{CH_3}{\underset{CH_3}{C}}\overset{\downarrow}{-}R^*$$

Figure 10.1 General structure of a peroxyester, with the arrows indicating the positions where bonds may break during the initiation process.

Figure 10.1 illustrates the situation for alkyl peroxyesters (POEs), $RC(O)OOC(CH_3)_2R^*$. A multitude of such POEs have been investigated with the R and R^* moieties being methyl, ethyl, n-propyl, iso-propyl, tert-butyl, and neo-pentyl. Peroxy-bond scission is the primary decomposition event, which may be (almost) instantaneously followed by β-scission, in particular by decarboxylation accompanied by formation of the radical R. As a second β-scission step, acetone may be released, accompanied by the formation of radical R^*. In the event that the primary oxygen-centered species are fairly stable, the follow-up β-scission processes may occur at such low rate that the primary radical fragments do not undergo bond scission before adding to a monomer molecule and thus starting chain growth.

As decarboxylation of alkylcarbonyloxy radicals, RC(O)O, is exothermic, the simultaneous occurrence of this reaction enhances peroxyester decomposition rate. Thus, higher radical concentrations may be produced at lower temperature. However, there is also a disadvantage associated with (almost) concerted two-bond scission of POEs: the presence of an alkyl radical, from decarboxylation of RC(O)O, and of an alkoxy radical, $OC(CH_3)_2R^*$, may result in combination or disproportionation reactions, thus giving rise to a loss in radical concentration and so to a lowering of initiator efficiency. This disadvantage is particularly severe in the case of R being a tertiary radical, for example, tert-butyl, which may be attacked by an oxygen-centered radical forming isobutene and an alcohol. Combination of two primary radicals from POE decomposition does not result in radical loss, as the POE is re-formed and may succeed in producing growing radicals after a subsequent decomposition reaction. Disproportionation of the primary oxygen-centered radicals is avoided by selecting the POE structure such that C−H bond scission in a β-position to the radical site cannot occur. In addition to the impact on initiator efficiency, follow-up reactions of the primary oxygen-centered radicals are relevant in that they yield carbon-centered radicals that are less prone to chain transfer than oxygen-centered radicals.

Initiator-derived radicals leave the cage ∼1 ns after the primary decomposition event. The leaving radical may be different from the species that adds to a monomer molecule after a microsecond to millisecond time interval. During this time period, several reactions may take place, such as β-scission, 1,5-hydrogen shift, or intermolecular chain transfer to monomer or to solvent. In addition to a potential effect of end groups on polymer properties, which is of particular relevance for low-molar-mass material, the type and concentration of end groups contain valuable kinetic and mechanistic information. The detailed analysis of polymerization kinetics and polymer properties thus requires end-group analysis via mass spectroscopic techniques.

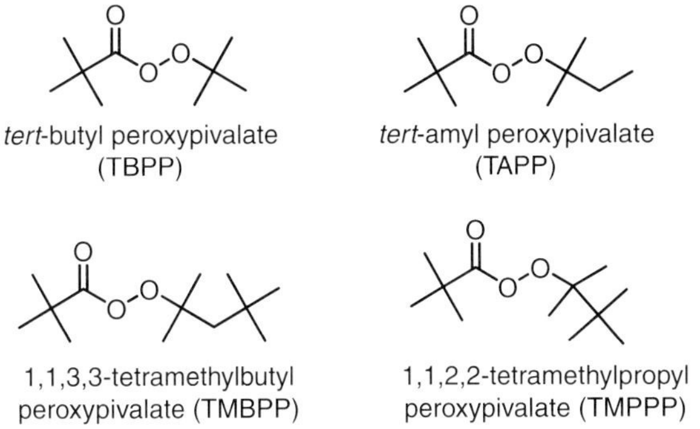

Figure 10.2 Peroxypivalates investigated in Ref. [15], as discussed here.

tert-butyl peroxypivalate (TBPP)

tert-amyl peroxypivalate (TAPP)

1,1,3,3-tetramethylbutyl peroxypivalate (TMBPP)

1,1,2,2-tetramethylpropyl peroxypivalate (TMPPP)

ESI MS is an excellent tool for end-group characterization. ESI is extremely soft, thus preventing the analyte from fragmentation. An extended series of POE-initiated MMA polymerizations has been carried out in toluene solution, and also in benzene in order to reduce the occurrence of chain transfer to solvent. The initiator concentrations were relatively high so as to produce a high fraction of oligomeric material that is suitable for ESI-MS detection.

Alkyl Peroxypivalates The first study to be carried out [15] looked at the four peroxypivalates shown in Figure 10.2. In Figure 10.3 the general decomposition scheme for peroxypivalates is illustrated. The oxygen-centered radical species **A** and **B** are the primary radicals produced from initiator decomposition. Throughout this section, primary radicals are denoted by capital letters, as compared to the lower case letters used for species produced by successive reactions. Peroxypivalates exhibit concerted or close-to-concerted decarboxylation. Thus, **A** is extremely short-lived, which simplifies the kinetic situation in that the so-called "acid side" of the four POEs in Figure 10.2 exclusively yields *tert*-butyl radicals **c**. Differences in end-group characteristics of the resulting polymer are thus entirely due to differences in follow-up reactions of the alkoxy radical **B**. The reactivity of alkoxy radicals in various

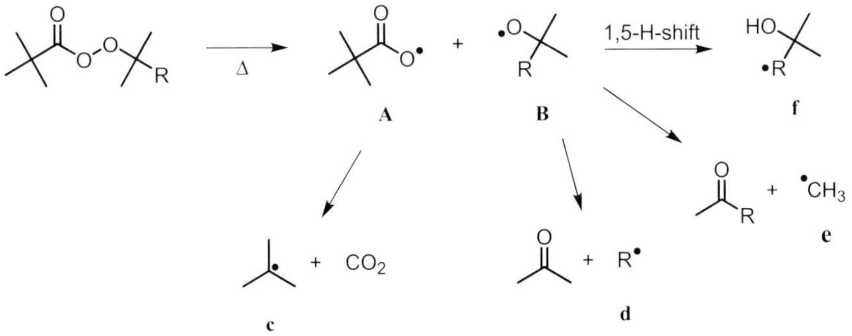

Figure 10.3 Radical species that may be produced from peroxypivalate decomposition.

environments, such as MMA, styrene, or cumene, has been thoroughly investigated by the nitroxide trapping technique [16]. In addition to **c**, the alkoxy radical **B**, the alkyl radicals **d** and **e**, and a hydroxyl radical **f** from a 1,5-H-shift reaction may be produced. The reaction to **f** may only occur in cases where R of Figure 10.3 ($\equiv R^*$ of Figure 10.1) is sufficiently long to enable formation of a six-membered transition-state structure for intramolecular hydrogen transfer.

In the case of radical termination by disproportionation, the polymeric product contains one of the species **B** to **f** as the end group, whereas two such groups occur in the case of termination by combination. There may also be benzyl end groups as a consequence of chain transfer to toluene, the solvent.

In Figure 10.4 the ESI-MS spectrum of MMA oligomers from polymerization in toluene solution at 90 °C with *tert*-amyl peroxypivalate (TAPP) as the initiator [15] is illustrated. Figure 10.4a shows the overall mass spectrum. There is a maximum at around m/z (ratio of relative molar mass to number of unit charges) of 1000. The observed intensity decrease toward high masses reflects the decay of the number distribution for polymer generated via conventional RP, whereas the decrease in relative abundance toward low molar mass is most likely an artifact due to the diminishing ionization probability of small oligomer species. As already

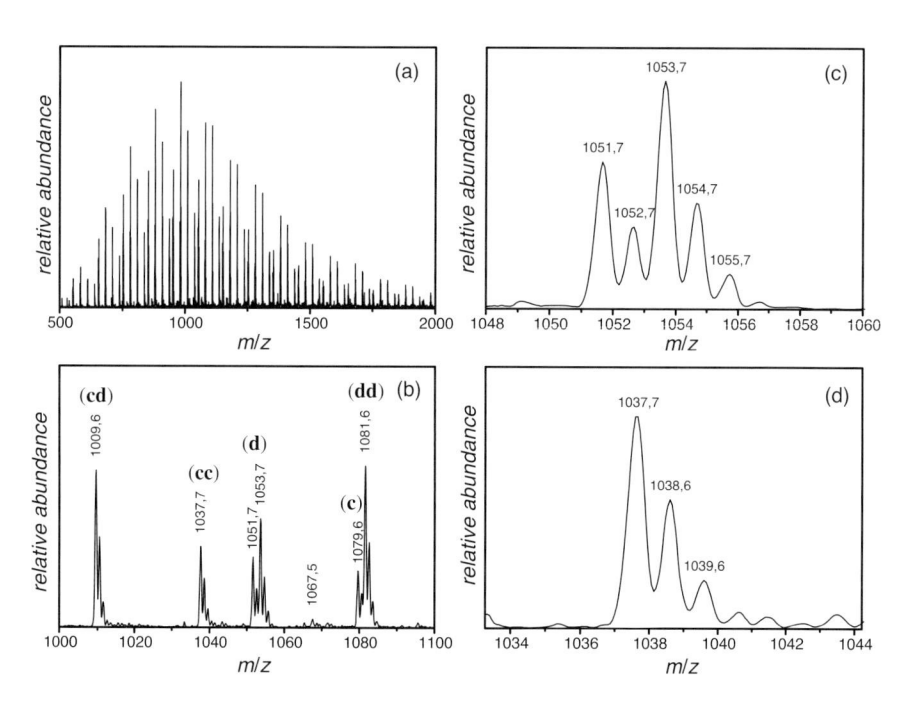

Figure 10.4 ESI-MS spectrum of MMA oligomers from initiation by TAPP in toluene at 90 °C. (a) Full spectrum; (b) zoom into the m/z range of one monomer repeat unit; enlargements of (c) disproportionation peak (d) and (d) combination peak (cc). (Reprinted with permission from Ref. [15]. Copyright 2004 Wiley Periodicals Inc.).

mentioned, the true chain-length distribution (CLD) is not correctly represented by MS intensities.

Figure 10.4b depicts an enlarged section of the spectrum, covering the mass range of one monomer unit (i.e., $M(MMA) = 100.05$ amu). Four major groups of peaks are easily identified for the repeat unit. None of the spectral components is indicative of one of the primary species **A** or **B** being an end group. On the other hand, the observed peaks clearly evidence the end groups *tert*-butyl (**c**) and ethyl (**d**). The masses of the individual peaks are listed elsewhere [15]. Two end groups should give rise to five signals: (**c**), (**d**), (**cd**), (**cc**), and (**dd**), where single and double letters indicate oligomer produced by disproportionation and by combination, respectively. The signals for (**c**) and (**dd**) overlap, explaining the prima facie observation of four groups of peaks rather than five. Polymeric material resulting from initiation by a methyl radical **e** (see Figure 10.3) could not be identified.

Disproportionation occurs via transfer of a hydrogen atom from one radical to the other (see Section 10.5), thus generating two oligomers that differ by 2 amu in molar mass. This is confirmed for the disproportionation peak (**d**) in the close-up that is Figure 10.4c. The additional peaks at spacings one or more amu higher are from one or more ^{13}C atoms being incorporated into the oligomer. The resulting peak pattern is due to the natural abundance of ^{13}C. The same is also observed for combination peaks, as is illustrated for the (**cc**) component in Figure 10.4d.

Taken together these results show how easily and how directly MS reveals the mechanism of initiation with TAPP and confirms that termination occurs via both disproportionation and combination.

Shown in Figure 10.5a is the ESI-MS spectrum for the mass range of one repeat unit of MMA oligomer produced in toluene solution at 90 °C with 1,1,3,3-tetramethylbutyl peroxypivalate (TMBPP) as initiator [15]. In this case, nine ESI-MS components are found. From the m/z values, it may be shown that they are the possible disproportionation and combination products from radicals that have either *tert*-butyl (**c**), 1,1,3,3-tetramethylbutoxyl (**B′**), or *neo*-pentyl (**d′**) as end group. The latter species is produced, together with acetone, by β-scission of **B′**. The occurrence of both the 1,1,3,3-tetramethylbutoxyl (**B′**) and the *neo*-pentyl (**d′**) end groups at first sight suggests that the *tert*-amyloxy radical (species **B** with TAPP) decomposes faster than does **B′**. An alternative explanation is that **B′** undergoes an internal 1,5-H-shift to form 4-hydroxy-2,2,4-trimethylpentyl (**f′**). There is evidence that this is a more likely explanation [16, 17], with the driving force being the lower energy of the carbon-centered radical (**f′**) than that of the oxygen-centered radical (**B′**). Unfortunately, MS is unable to distinguish between an alkoxy radical **B′** and a rearranged hydroxyalkyl radical **f′**, as no mass change is involved in the 1,5-H-shift reaction. This is a limitation of using MS to investigate mechanism, that is, it relies on species having different M. At the same time, if one regards as unlikely that decomposition of **B′** to form **d′** occurs as slowly as addition to MMA, then MS may be said to evidence the occurrence of the 1,5-H-shift reaction, which is a powerful mechanistic insight.

Figure 10.5b presents results from using 1,1,2,2-tetramethylpropyl peroxypivalate (TMPPP) as initiator [15]. The spectrum is far simpler than the previous two examples in that it consists of just two peaks. These are generated by MMA oligomers

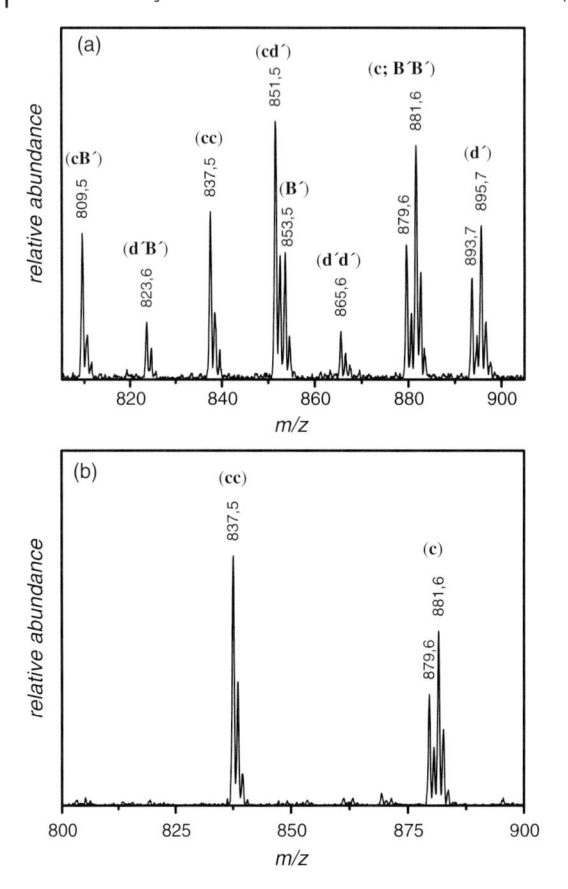

Figure 10.5 ESI-MS spectra of MMA oligomers from initiation by (a) TMBPP and (b) TMPPP in toluene at 90 °C. (Reprinted with permission from Ref. [15]. Copyright 2004 Wiley Periodicals Inc.).

having either one (**c**) or two (**cc**) *tert*-butyl end groups. This is the radical produced both by decarboxylation of the pivaloyloxy radical (**A**) and by β-scission of the 1,1,2,2-tetramethyloxy radical (**B**). The absence of any other group of signals clearly demonstrates both these reactions occur very rapidly in this case, as is fully consistent with data of a different nature [18].

The fourth initiator of Figure 10.2 is *tert*-butyl peroxypivalate (TBPP). Results from using it are given in Figure 10.6 [15]. As with TMBPP (Figure 10.5a), nine groups of peaks occur. They arise from the three radicals *tert*-butyl (**c**), methyl (**d** = **e**), and benzyl (**t**). Methyl radicals originate from β-scission of initiator-derived *tert*-butoxy radicals (**B**) (see Figure 10.3). Benzyl radicals arise from chain transfer to toluene, the solvent. The fact that they do not arise in the other three systems, where toluene was also present, suggests that methyl, the unique and highly reactive radical here, may be the species that reacts with toluene. No *tert*-butoxy (**B**) moieties show up as end groups. This is consistent with results from the other three systems.

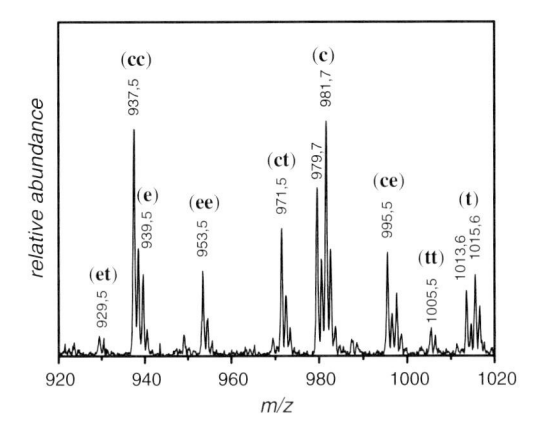

Figure 10.6 ESI-MS spectrum of MMA oligomers from initiation by TBPP in toluene at 90 °C. (Reprinted with permission from Ref. [15]. Copyright 2004 Wiley Periodicals Inc.).

One might suspect that all alkyl peroxypivalates would produce initiating radicals in the same way. Through the use of MS, it has been shown that this is not the case. In fact all four studied initiators have different sets of reactions by which chain-starting radicals are produced [15]. The ease and authority by which MS establishes this interesting mechanistic information should be clear.

Alkyl Peroxyacetates A similar series of ESI-MS investigations as for peroxypivalates has been carried out for members of the alkyl peroxyacetate family [19] (see Figure 10.7). MMA oligomer was produced by polymerization in solution of benzene at temperatures between 125 and 140 °C.

Figure 10.8 shows the section corresponding to one monomer repeat unit of MMA oligomer produced by initiation with *tert*-amyl peroxyacetate (TAPA) [19]. Five groups of peaks are seen. These may be assigned to disproportionation and combination

tert-butyl peroxyacetate (TBPA)

tert-amyl peroxyacetate (TAPA)

1,1,2,2-tetramethylpropyl peroxyacetate (TMPPA)

1,1,3,3-tetramethylbutyl peroxyacetate (TMBPA)

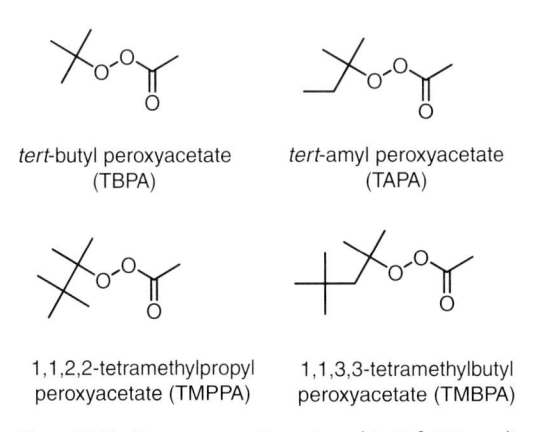

Figure 10.7 Peroxyacetates investigated in Ref. [19], as discussed here.

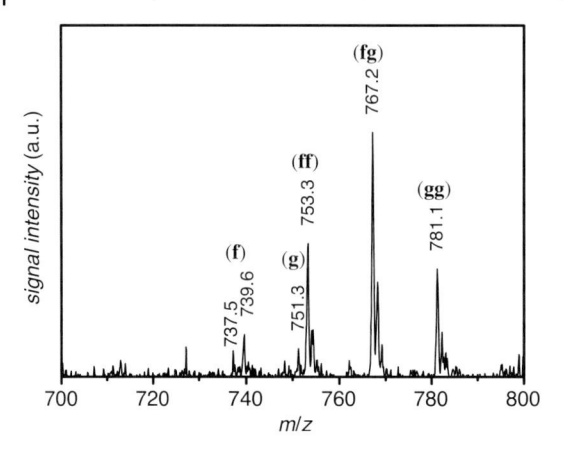

Figure 10.8 ESI-MS spectrum of MMA oligomers from initiation by TAPA in benzene at 135 °C. (Reprinted with permission from Ref. [19]. Copyright 2007 Wiley Periodicals Inc.).

resulting from two types of initiating radical species, as in Figure 10.4b, which are identified as methyl (**f**) and ethyl (**g**). The methyl is from decarboxylation of the primary fragment methylcarbonyloxy, the equivalent of **A** in Figure 10.3. The ethyl is from β-scission of the *tert*-amyloxy radical, **B** in Figure 10.3. Exactly the same was found for initiation by 1,1,2,2-tetramethylpropyl peroxyacetate (TMPPA), for which *tert*-butyl is the analog of ethyl [19].

For initiation by 1,1,3,3-tetramethylbutyl peroxyacetate (TMBPA), on the other hand, nine groups of ESI-MS peaks occur [19]. Chain growth is started by methyl, neopentyl, and 4-hydroxy-2,2,4-trimethylpentyl [19]. There are also nine groups of peaks when *tert*-butyl peroxyacetate (TBPA) is the initiator [19]. One would expect only one initiating radical with TBPA, namely methyl, as produced via both decarboxylation of methylcarbonyloxy (**A**) and β-scission of **B**. Benzyl should not be formed in the present TBPA experiments because benzene rather than toluene was used as solvent. Mass analysis turns out to reveal that the three initiating radicals with TBPA are methyl, as expected; *tert*-butoxy (**B**); and unsaturated MMA radicals formed by hydrogen abstraction from MMA to the highly reactive *tert*-butoxy [19]. Interestingly, there is thus the formation of "naked" MMA oligomer, which consists purely of MMA units, via combination of two growing radicals started from MMA radicals.

Diacyl Peroxides MMA oligomers from RP with initiation by symmetric diacyl peroxides, R(CO)OO(CO)R, have also been analyzed via ESI MS [20]. The initiators studied so far encompass R = *n*-nonyl, *n*-undecyl, 2,4,4-trimethylpentyl, *iso*-butyl, phenyl, and naphthyl. Given in Figure 10.9 is a general representation of the decomposition of these molecules. The primary decomposition step yields the two identical oxygen-centered radicals, **X**. These may decarboxylate to yield the carbon-centered alkyl or aryl radicals, **y**.

Shown in Figure 10.10a is the ESI-MS spectrum of MMA oligomer using di-*n*-dodecanoyl peroxide (DDDP) as the initiator [20]. There are two major components,

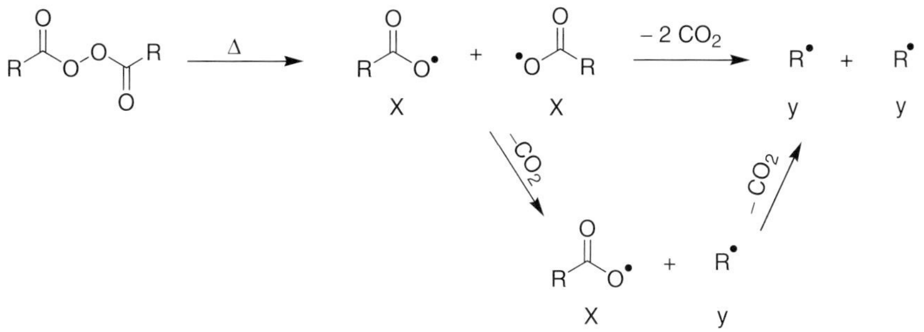

Figure 10.9 Scheme for decomposition of symmetric diacyl peroxides [20].

one of which has twin peaks. As seen with Figure 10.5b, this is the hallmark of there being only one type of end group from initiation, and of radicals terminating by disproportionation (**b**) (the twin peaks) and combination (**bb**). From the masses of these components, one may determine that this end group is undecyl, which is R in the case of DDDP. There are no components in the spectrum suggesting any n-undecyl carbonyloxy moiety as an end group. The third component of the spectrum, the weak signal at 1065.4, has mass of the combination component (**bb**) plus 32. Thus, it is assigned to combination of two radicals, into one of which has been incorporated oxygen (O_2). Such components are known to occur in the early polymerization period as a consequence of imperfect deoxygenation. It is interesting that no such disproportionation species, that is, (**b**) plus 32, are observed, even though the (**b**) signals are the strongest in the spectrum. This suggests that the oligomers containing O_2 are formed from combination involving oxygen-centered radicals that are relatively long-lived. The latter deduction explains the inhibitory role of residual oxygen at the beginning of a polymerization experiment. If it were the case that the incorporated O_2 radical quickly adds to monomer, then the absence of disproportionation product with O_2 cannot be explained.

The MMA oligomer samples prepared with di-n-decanoyl peroxide, bis(3,5,5-trimethylhexanoyl) peroxide and di-iso-butyryl peroxide (DIBP) exhibit the same type of behavior as seen with DDDP [20]. In all four of these cases, the peroxydicarbonate cleavage yields alkyl carbonyloxy radicals. For the first three initiators, these are primary carbon-centered radicals, while with DIBP they are secondary. These are all high in energy. The fact that none are observed as end groups evidences that the subsequent decarboxylation step is fast relative to that of addition to MMA.

Results from using di-benzoyl peroxide (DBP) as initiator are depicted in Figure 10.10b [20]. There are five major spectral components. As in Figure 10.4b, this demonstrates that two different radicals start chain growth. In the present case, they may be identified as R = **f** and RC(O)O = **E**, where the capital letter denotes a primary radical (see Figure 10.3) and R is phenyl. The intensity ratio of signals (**f**) : (**E**) is 3 : 1. If one assumes identical ESI sensitivity of species with these two different end groups, then this suggests that of every four phenyl carbonyloxy radicals that initiate polymerization, three first of all undergo decarboxylation before adding to monomer. Because the concentration of **f** radicals is three times higher than that of **E**, and

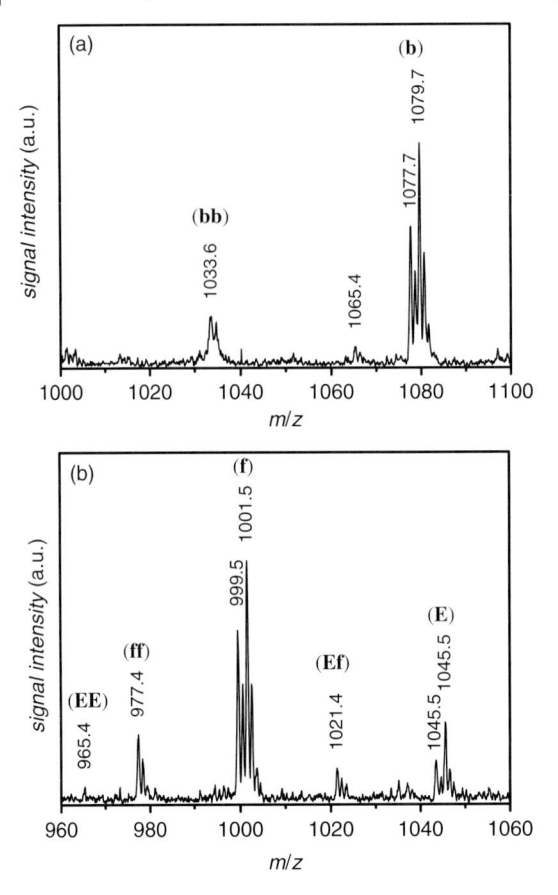

Figure 10.10 ESI-MS spectra of MMA oligomers from initiation by (a) DDDP and (b) DBP in benzene at 95 °C. (Reprinted with permission from Ref. [20]. Copyright 2007 Elsevier Ltd.).

because formation of a combination product involves two specific radicals (cf. disproportionation), the intensity ratio of combination peaks (**ff**) : (**EE**) should be about 3^2 : 1. The experimentally observed ratio of integrated component intensities is 8.3 : 1, which is remarkably close to the prediction.

There is independent evidence for the DBP results here. That the concentration of MMA oligomer species with phenyl end groups is approximately three times higher than that with phenyl carbonyloxy end groups is consistent with NMR results [21]. Quantum-chemical calculations predict that decarboxylation of aryl carbonyloxy radicals is relatively slow, as this process is associated with a reduction of the range of electron delocalization [22, 23]. This explains that primary-radical end groups are not observed with aliphatic but are with aromatic diacyl peroxides. Further evidence of the latter is that investigations into di-2-naphthoyl peroxide reveal that this diacyl peroxide behaves as DBP, except in that the fraction of naphthyl carbonyloxy moieties is even larger [20].

Peroxydicarbonates The ESI-MS method has also been used to study initiation mechanisms for symmetric alkyl peroxydicarbonates [24]. These differ from alkyl diacyl peroxides simply in that there is an oxygen atom between the alkyl and carbonyl groups. Thus, decarboxylation of a primary radical yields an oxygen-centered alkoxy radical.

Di-ethyl peroxydicarbonate (E-PDC), di-n-tetradecyl peroxydicarbonate (TD-PDC), and di-2-ethylhexyl peroxydicarbonate (EH-PDC) were used [24]. From analysis of oligomer end groups, it was found that both alkoxy carbonyloxy primary radicals **A** and the corresponding decarboxylated species **a** add to monomer with all three initiators. The extent to which decarboxylation of **A** into **a** occurs is indicated by the intensity ratio of combination products (**Aa**) : (**AA**). The experimentally determined values are 5.3% for E-PDC, 13.2% for TD-PDC, and 45.5% for EH-PDC [24]. As explained above, these values should reflect the concentration ratio of **a** : **A**.

The given values first of all evidence that only a minority of primary radicals undergo decarboxylation before adding to monomer. At first, this may seem surprising given that for all other peroxide classes considered above, primary radicals either were not observed or were found in only small quantities. The difference here is that decarboxylation does not convert an oxygen-centered radical into a more stable carbon-centered radical, the strong driving force above. Rather, there is only formation of another oxygen centered-radical, and thus the process is relatively slow.

Next one should consider that the three ratios are different. The higher percentages of (**Aa**) : (**AA**) for TD-PDC and EH-PDC are indicative of extra decarboxylation accompanied by a 1,5-H-shift reaction from a CH_2 moiety. The driving force behind this process is the transformation of an oxygen-centered radical into a secondary carbon-centered one. As decarboxylation with EH-PDC is more pronounced than with TD-PDC, a further mechanism probably applies. According to the literature [25], the primary oxygen-centered radical **A** may undergo a 1,5-H-shift reaction with the tertiary carbon atom. The resulting intermediate will rapidly decarboxylate into radical **a**.

In summary, it has been shown in depth in this section that the MS method is widely applicable and extremely powerful for the detailed elaboration of initiation mechanisms. A natural part of this is the identification of polymer end groups, which in itself is of particular relevance for low-molar-mass materials. Of course, the method may also be used for azo and for multifunctional initiators, not just for monofunctional peroxides, as have been discussed here. Indeed, given the tremendous importance of azo initiators it seems remarkable that they have not yet attracted much attention from MS users.

10.3.1.2 Photoinduced Initiator Decomposition

Even if dwarfed by commercial consumption of thermally decomposing initiators, photoinitiator use is nevertheless of high significance and still growing technical relevance. Furthermore, recent decades have seen PLP emerge as an indispensable tool for measurement of RP rate coefficients (see Section 10.4) [6]. For these reasons, it is warranted to study mechanisms of photoinitiation in RP. This section will outline the role that MS has played to date.

Initial investigations were stimulated by a mysterious observation from single-pulse (SP) PLP experiments: that the more 2,2-dimethoxy-2-phenylacetophenone (DMPA) is used as photoinitiator, the *less* conversion of monomer into polymer there is [26]. Of course, under normal circumstances the opposite should be observed. This was explained by postulating that DMPA gives rise to a primary radical that does not initiate polymerization but instead causes termination [27]. The chemical basis of this hypothesis is that the acetal fragment from DMPA photodissociation (Figure 10.11) – should be very stable due to conjugation of the free electron with the adjacent benzene ring. Thus, this radical should be slow to add to monomer, enabling it to be long-lived and therefore have a retardative effect. A test of this hypothesis is to use a photoinitiator that does not generate a similarly stable primary radical. Such an initiator is 2-methyl-4′-(methylthio)-2-morpholinopropiophenone (MMMP) (Figure 10.11). Indeed, SP PLP experiments with it were soon shown to display classical kinetic behavior [28].

None of the above constitutes direct evidence of photoinitiation mechanism. As has been seen above, such can be provided by MS analysis of polymerization product. In a first study using MA and dicyclohexyl itaconate as monomers, it was easily shown that both the primary radicals from MMMP photocleavage are present in near equal quantities as polymer end groups [29]. This confirms that MMMP functions as a photoinitiator that is at least close to ideal.

Next, attention shifted to DMPA. MS analysis of product from PLP experiments with MMA substantiated that chains are started with the benzoyl fragment [30]. The same was found when benzoin was used as initiator [30], as one would expect given its photochemistry and the nature of the primary radicals produced, which are analogous to those with DMPA (see Figure 10.11). However, a problem with this study is that it could find no sign of the inhibiting primary radicals, namely acetal in the case of DMPA and benzyl alcohol with benzoin. This uncertainty was cleared up some years later when a higher quality ESI-MS instrument was used. In the case of benzoin, it was found that the alcohol fragment is indeed present as an end group [31], but it simply had not been clearly observable in the pioneering-days study using a MALDI-MS instrument with poor resolution [30]. Similarly with DMPA, except that there some additional chemistry enters the picture: the acetal radical can undergo secondary cleavage to give methyl radical and methyl benzoate, as shown in the bottom of Figure 10.11. This was evidenced by the finding of MMA oligomers with a third type of end group, namely methyl [31]. In essence, the same was found for DMPA and benzoin in a companion study extending this work to PLP of MA and dimethyl itaconate (DMI) [32].

The question arises as to how these MS investigations can evidence that the stable free radicals from photodissociation are terminating agents, for chain structure is the same if they initiate chain growth, and thus MS cannot tell the difference. There are two complementary ways of deciding this: (1) one looks for the *amount* of material containing two identical initiator-derived end groups. For example, the absence from MS spectra of any oligomers with two methyl end groups gives a strong indication that methyl radicals do not initiate MMA and MA polymerization, because if they did, then one would observe combination product with methyl groups at each end. On the

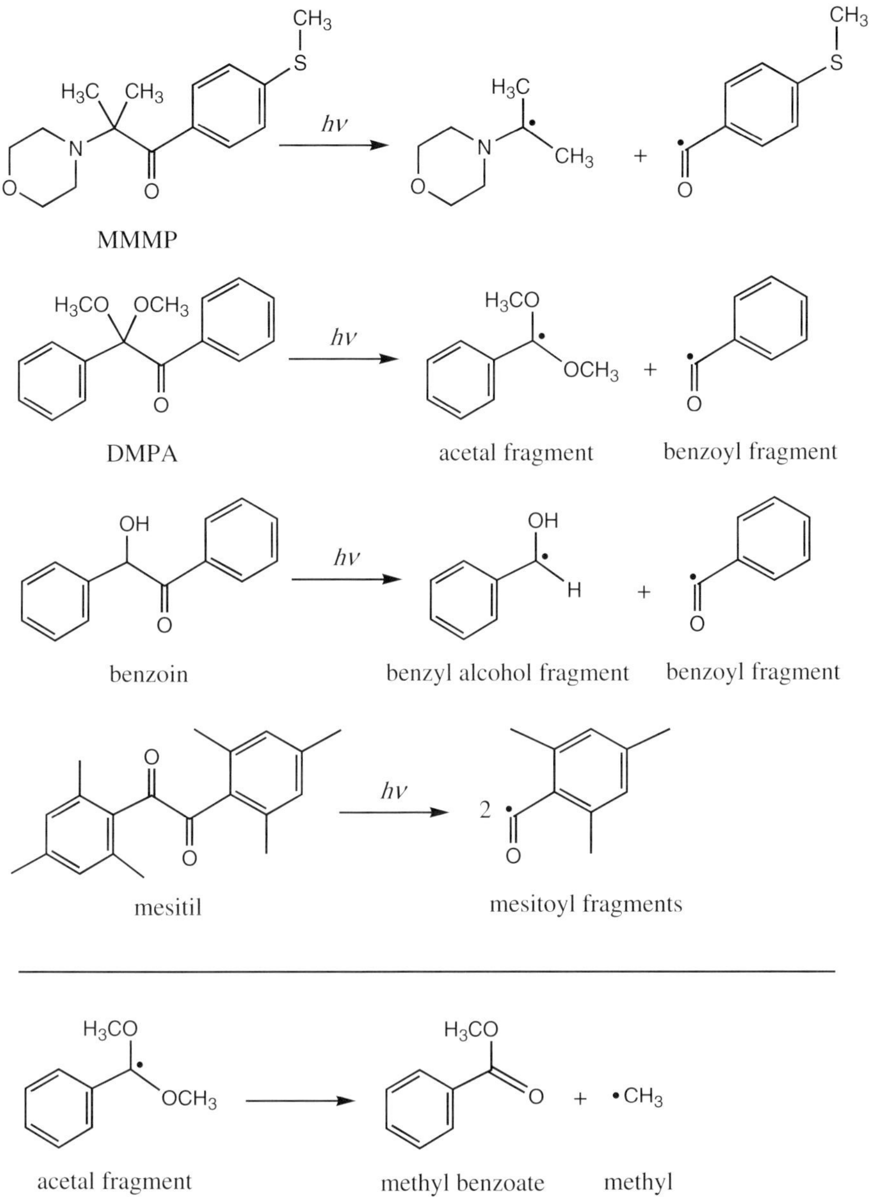

Figure 10.11 Above dividing line: photodissociation of four photoinitiators discussed in this work; below: secondary cleavage of the acetal radical from DMPA photodissociation.

other hand, one does observe much larger amounts of combination product with one methyl and one benzoyl end group. (2) One looks at the relative amounts of disproportionation products, because in such species the initiator-derived end group must arise from chain start. So in essence one uses the disproportionation products to identify chain-starting radicals, and then any *additional* initiator-derived end groups detected in combination products must be terminating agents. By and large it was found in the above studies that the benzyl alcohol, acetal, and methyl radicals predominantly act as terminating moieties, even if in some cases they do seem to play a small role as chain-starting agents [31, 32]. This is all consistent with the original kinetic studies by SP PLP [26, 27].

Barner-Kowollik's most recent MS-based investigation of photoinitiation involved PLP of MMA with cocktails of benzoin and mesitil (Figure 10.11). From the intensities of the resulting MS signals, it was deduced that the benzoyl radical is 8.6 times more likely than the mesitoyl fragment to initiate polymerization [33]. Such analyses are highly relevant where it is of interest – from the viewpoint of material properties – to know what the polymer end groups are.

Another MS study involving photoinitiation utilized a monoacylphosphine oxide with both styrene and MMA [34]. It was found that the resulting phosphorous-centered radical is much more reactive toward monomer than the accompanying benzoyl fragment. This was as expected from independent studies. It is consistent with the picture from the benzoin/mesitil study [33] that primary photoradicals of a different nature should not be assumed to be evenly incorporated into polymer chains.

10.3.1.3 Other Means

In many of the studies in Sections 10.3.1.1 and 10.3.1.2, it may be said that the expected mechanism of initiation was simply confirmed by MS, whereas this technique offers unprecedented and still largely unexplored opportunities in the study of systems where the mechanism of initiation is shrouded in mystery and can only be guessed at.

The most obvious example of this is autoinitiation, that is, the spontaneous polymerization of monomer to which no initiating species has been added. This is a fascinating phenomenon, and in the case of styrene the rate is sufficiently high that commercial polymer is made this way. The mechanism of this process has been debated for over half a century [35, 36], hampered by a lack of concrete information. Therefore, this system would seem to be an obvious candidate for urgent study by MS, yet we are only aware of one such investigation [37], and even there it was peripheral to the primary purpose. In accordance with Mayo's mechanism [35], dimeric styrene end groups were observed [37], although of course this does not discriminate between different chemical structures for this dimer [36], and so the question of mechanism remains open. For other monomers there is also autoinitiation, even if at a lower rate than styrene, for example, MMA [38] and MA [39]. Nevertheless the rate can still be significant, and the mechanism of radical production is more obscure than in the case of styrene. Therefore, these should also be prime candidates for study by MS.

Another important means of initiation in RP is via γ-radiation, usually interacting with water. Again this is an area that is ripe for investigation using MS, because to the best of our knowledge there have only been two such studies to date, both involving RAFT (reversible addition-fragmentation chain-transfer, see Chapter 11 polymerizations [40, 41]. In the aqueous-phase study, it was found that •H and •OH radicals initiate chain growth, as expected. However, it was found that water also gives rise to other radicals, and as well there are end groups from radiolysis of RAFT agent and of monomer [40]. Similar was found in the companion study involving bulk acrylate polymerization [41].

Finally, there is the situation of the truly miscellaneous, where a totally unexpected initiating end group is observed. An example of this comes from MS analysis of poly (butyl methacrylate) formed from initiation by TBPA at high temperature [42]. In addition to the expected MS peaks, there were also ones that could only be rationalized through the sequential occurrence of intermolecular chain transfer to polymer (CTP), propagation, backbiting, and then β-scission, ultimately forming an initiating species very close to butyl methacrylate dimer in structure.

10.3.2
Initiator Efficiency

Initiator efficiency is the fraction, f, of generated primary radicals that go on to initiate polymerization. It is an important quantity because the rate of polymerization and average polymer size depend on the product fk_d, where k_d is the rate coefficient for thermal decomposition of initiator. While k_d is straightforward to measure (which should be done in a solvent medium similar to that of the polymerization of interest), f has proven to be a surprisingly tough nut to crack [43]. The problem is that whereas it is easy to obtain k_d by observing the disappearance of initiator, for f one must measure the appearance of polymer end groups, which can be like searching for needles in a haystack. From the rest of this section, it should be obvious that MS might be helpful in this regard, for it identifies end groups. Nevertheless, there is still the challenge of f being *quantitative*, as opposed to the largely just qualitative identification of initiating species throughout this section. Furthermore, it is not really feasible to use MS intensities to obtain changes in *absolute* (as opposed to relative) amount with time.

Given all this, a relative method has been proposed and successfully used to obtain f via MS [44]. The idea is that an initiator of known f and k_d is used together with one of known k_d but *unknown f*. A polymerization is then carried out with this cocktail and the product is analyzed with MS. Signals from both initiators are observed. The relative heights of these different peaks should reflect the relative rates of initiation, because all else apart from this is equal once the primary radicals add to monomer. From this quantitative information one may thus determine the only unknown, which is the unknown f. This procedure has been successfully used to determine f for two symmetric acyl peroxides [44]. It remains to be seen whether this method receives wider uptake. Of course, it is hostage to the assumption that different initiator-derived end groups have only a negligible effect on ionization tendency in MS.

10.4
Propagation

We have seen in Section 10.3 that the initiation process leaves its footprints in the final polymeric product via the incorporation of initiator-derived fragments as termini. The mass of these end groups can easily be detected via MS by subtracting multiples of the mass of a monomer unit. These units are built into the polymer via the propagation reaction, in which a growing macroradical adds to monomer. This reaction increases the total mass of the growing polymer chain by the molar mass of one monomer unit. That is, features of propagation – that is, the actual polymerization reaction – are conserved in the polymer sample as the typical repetitive MS peak profile. In the case of homopolymers, individual peak series belonging to species that carry identical end groups are thus separated exactly by the molar mass of one monomer unit. Since the molar mass of the monomer unit generally is known and the monomer units are incorporated into the chain without any fragmentation that may alter their individual mass, the information that can be extracted from this spectral feature is rather limited.

What then can then be learned about propagation using MS? Within this context, three aspects regarding the mechanism and kinetics of propagation have been addressed. Firstly, MS has been used as method for determining full MMDs as an alternative to size-exclusion chromatography (SEC) in PLP for measuring the propagation rate coefficient, k_p. Secondly, MS has been exploited for evaluating the dependence of k_p on chain length, both using PLP and trapping experiments. Thirdly, MS has been used to trace the different amounts of comonomer that are incorporated into copolymers. This gives insight into the individual propagation steps of various macroradical types occurring in copolymerization.

10.4.1
Propagation Rate Coefficients

The determination of the CLD of the polymer produced via a pseudo-stationary pulsed-laser experiment allows accurate values for k_p to be obtained [5, 6, 45, 46]. The polymerizable system – containing monomer and photoinitiator, and occasionally solvent – is irradiated by a laser beam that is pulsed periodically. The CLD of the polymer thus formed exhibits a characteristic structure with one or more extra peaks. The determination of the points of inflection on the low-molar-mass side of the n extra peaks gives values for the characteristic chain lengths i_n, given by

$$i_n = nk_p c_M t_0 \qquad (10.1)$$

By measuring MMD and thus determining values of i_n, k_p becomes available via Eq. (10.1) because the monomer concentration, c_M, and the time interval between laser pulses, t_0, are known for an experiment. This method has developed into the IUPAC-recommended method for k_p determination [46].

Typically, the MMD is determined via SEC. Despite its many advantages, it also suffers from limitations including the fact that SEC in its basic form needs molecular weight calibration, either by a monodisperse standard or by "universal calibration" using online viscometry. SEC can be turned into an absolute method, for example, by using online light scattering detection. However, this is restricted to high molar masses. In addition, SEC inherently exhibits considerable instrumental broadening

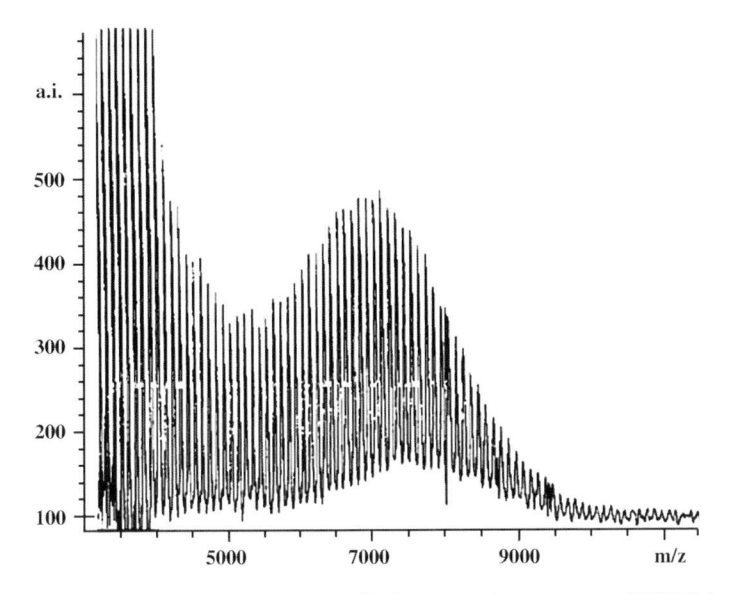

Figure 10.12 MALDI-MS spectrum of poly (MMA) from PLP at $-8\,°C$ [47]. The additional peak due to the pulsing action of the initiation-inducing laser is clearly visible (the feature with maximum at around 7000 Da). It is from the position of this peak that k_p is determined. (Reprinted with permission from Ref. [47]. Copyright 1993 American Chemical Society).

of the experimental CLDs. Danis et al. [47] were the first who proposed MS as an alternative for mapping MMDs in PLP as a route to k_p. It appeared intriguing to overcome the shortcomings of SEC with respect to calibration and peak dispersion, and to use a method that needs only tiny sample quantities. In addition, MS was considered to be superior to SEC in the low-molar-mass regime. Danis et al. [47] found by MALDI MS a distinct additional peak in the MMD of polymer from PLP (see Figure 10.12). This easily allowed the determination of k_p for MMA. However, they found their k_p value from MS to be around 25% smaller than the value from SEC. The differences could not be explained by the fact that different types of MMD are obtained by these two methods, namely number-MMD from MS versus hypermass-MMD from SEC. The authors thus speculated that the calibration standards in SEC are beset with systematic error.

This explanation seems to be unlikely, however, since several well-established techniques such as light scattering and osmometry are usually used to characterize the average molar masses of polymer standards. It is more likely that the mass sensitivity of the electron multiplier detection of MS and preferential ionization were responsible for a distortion of the measured distributions. This effect generally results in low-mass macromolecules being observed in preference to high-mass macromolecules, whereby the inflection point of an additional peak is dragged toward lower M.

Zammit et al. [48] explored this effect systematically and found a significant impact of the power of the ionizing laser on the final kinetic data. By carefully tuning the various MS parameters, these authors arrived at a k_p value for MMA at $0.2\,°C$ that exactly matched the value from SEC analysis of the same sample and was in agreement with k_p from an independent PLP SEC study. It was thus shown that provided the MS operating parameters are judiciously chosen, k_p can indeed be measured by MS. A subsequent study from the same group involving MMA at −34 °C corroborated this finding [30] (see below). Similarly, the long-chain k_p found by Willemse et al. for MMA over a range of temperatures using MALDI MS [49] are in good agreement with the benchmark values obtained using SEC [50]. On the other hand, their styrene (MS) results [49] are consistently *higher* than the benchmark SEC results [46] by about 10%. This trend is opposite to the initial finding of Danis et al. [47], and so is probably just small systematic error of an unknown nature or else an artifact of SEC broadening (see below). Thus, one can say with confidence that MS is perfectly adequate for determining k_p in conjunction with PLP.

In addition to the MS parameters, the MS signal is also dependent on the absolute broadness of a studied CLD. Schweer et al. [51] found that the CLDs of MALDI MS and SEC match much better in the case of low-molar-mass dispersity, D. Polystyrene and poly(MMA) samples of $D = 1.1$ and below showed very similar distributions in MS and SEC; however, peak widths were smaller in the case of MS, which is due to the instrumental broadening of SEC. The signal-to-noise ratio, on the other hand, is better with SEC. Differences were found to be much larger with broader polymer samples. Typically, there are distortions of MS distributions due to the tendency to underestimate the amounts of higher-M species. From these findings, it was concluded that MS as partner for PLP is better suited to the so-called high-termination-rate limit [52], in which PLP is performed under very high radical concentrations leading to relatively narrow additional peaks. Under these conditions the maximum – which can more easily be determined from MS traces than the inflection point – is the better measure for extracting k_p.

Within this context, it should however be noted that there also is evidence that MALDI underestimates the D of narrow CLDs [13]. When investigating broader CLDs, it was found that the larger peak dispersion in SEC shifts the point-of-inflection of the additional peak toward lower M, which leads to smaller k_p values than those obtained from MS. This effect obviously counterbalances the distortion of broad CLDs in MS, which generally reduces the amount of high-M species. This would inherently lead to k_p values from MS that are lower than those from SEC. The impact of these two effects is not easy to separate, and differences of 15% between MS and SEC were found, with no way to decide which values are closer to the true ones.

Interestingly, it was overlooked for quite some time that MS does not yield a clean number distribution; rather, the raw MS signal still needs to be converted by transforming the time-of-flight into a mass axis (see Section 10.2 and Ref. [13]). Differences in the type of the underlying distributions, however, do not change the evaluated k_p value by more than a few percent. With respect to the determination of k_p via PLP, this uncertainty in the nature of the distribution type cannot be responsible for the sometimes observed discrepancies between MS and SEC. In any case, it is difficult to transform the MS signal trace with its highly resolved discrete peaks into a continuous distribution function, as obtained with SEC and as required for determining the location of the inflection point. For instance, Barner-Kowollik et al. [30]

used an approach of integrating the MS trace over 100 amu followed by differentiation. This procedure gave a smooth envelope trace of the discrete MS peaks, resembling all the features of a continuous distribution. The k_p values in MMA polymerization which these authors extracted from such transformed MS distributions – in which even two overtones of the primary additional peak could be identified – were indeed close to the ones that were obtained by SEC. The value from MS was found to be 7% larger than the value from SEC, which is in contrast to earlier findings, where MS data usually gave smaller results. Both values, however, were found to be around 15% higher than the IUPAC-recommended value that was obtained in a temperature regime of -1 to 90 °C [50]. This may be due to the use of a temperature (-34 °C) outside the IUPAC range, or it may be due to relatively small chains coupled with the phenomenon of chain-length-dependent propagation (CLDP) (see Section 10.4.2).

MS appears to be advantageous in comparison with SEC in the case of arbitrarily branched polymer, for which SEC calibration is nearly impossible since the separation in terms of molecular weight is incomplete, that is, a given SEC-elution slice contains a range of molecular weights. Willemse and van Herk [53] exploited the advantage that MALDI-MS truly separates species according to their mass no matter whether they are branched or not and performed the determination of k_p in acrylate polymerization, in which substantial branching occurs due to backbiting reactions [54]. The MS approach revealed small changes in k_p as function of initiator concentration and initiating laser pulse energy. Following the recommendation described above [52], these authors also worked in the limit of high termination rate, using the position of the PLP peak maximum to determine k_p. Values in MA polymerization were found to decrease systematically by up to 6.5% as a function of the initiating radical concentration (i.e., the product of the laser intensity and the photoinitiator concentration). Again, another parameter impacting the k_p values obtained by PLP was hereby identified. By trying to keep this parameter as constant as possible, these authors arrived at a family-type behavior of k_p in acrylate bulk polymerization, studying methyl, ethyl, n-butyl, n-hexyl, and benzyl acrylate. The results demonstrate that the k_p value increases with the size of the acrylate ester moiety, consistent with findings for methacrylates [55]. This is highly important mechanistic information.

In the mid-1990s, MS was widely seen as the successor of SEC for determining CLDs, and hence enhanced speed and accuracy in measuring k_p via PLP was anticipated for the future [51]. However, even now MS has not nearly displaced SEC, which still is the first choice for obtaining polymer distributions. This may partly be due to the fact that SEC is readily available in nearly all polymer laboratories and partly due to the shortcomings of MS, which does not yield results that *a priori* can be regarded as superior. This is basically due to the fact that ion formation, mass spectrometric separation, ion transfer, and ion detection all show a functional dependence on the molecular mass, which impacts the MMD [2, 56].

Recently, however, Gruendling *et al.* [57] have proposed a combination of MS and SEC, an approach that may again boost the role of MS in the field of PLP. These authors coupled SEC with ESI MS, elegantly combining the strengths of these techniques: SEC provides the molar mass separation, refractive index detection provides the polymer concentration, and MS provides the absolute determination of the molar mass. The approach allows for an internal calibration of SEC and for elimination of SEC broadening. A calibration can be derived without knowledge of the polymer class as long as the polymer is amenable to electrospray ionization. In effect the MS equipment is used as an absolute molar mass detector connected to SEC. The polymer sample, however, needs to be of relatively low molecular weight, as the typical mass range in ESI MS is limited to up to a few thousand in m/z, and multiply charged species of higher molar mass would confound the obtained MMD. The authors arrived at k_p values for methyl, ethyl, and butyl methacrylate, again finding a family-type behavior, as also by SEC for these monomers [55] and by MS for acrylates [53]. Whereas the activation energies of propagation remained basically constant for all investigated methacrylates, the frequency factor increased with increasing ester moiety size. They also found that the k_p values from their method are 5–10% higher than the values obtained from calibrated SEC on higher-M samples. This effect was again attributed to chain-length dependent k_p since low-M ensembles were investigated.

10.4.2
Chain-Length Dependence of Propagation

CLDP is the variation with radical chain length, i, of propagation rate coefficients, k_p. Thus, properly one should speak of k_p^i. The energy of a growing radical should barely be affected by the addition of successive monomer units; however, for relatively small species it makes sense that there could be a *steric* effect from the increase in size that comes from addition of a monomer unit. Thus, one expects CLDP to be operative only at small i and for the effect to be *entropic* in origin. As reviewed a few years ago [58], the existing evidence by and large supports this picture. In arriving at this point, MS has played a role in two ways: (i) PLP studies involving measurement of (average) k_p values and (ii) quantitative determination of oligomer amounts. These will now be outlined in turn.

As long ago as 1993, Deady *et al.* pointed out that there was variation of styrene k_p, as obtained from PLP SEC experiments, with laser pulsing time, t_0, and that this variation could be explained by the first few propagation steps being very rapid [59]. The Olaj group started to carry out a systematic study of this in 2000 [60]. They indeed agreed that there is CLDP; however, they felt their PLP SEC data was indicating an effect that is relatively small in magnitude (k_p^1 only 30–40% above the long-chain value of k_p) but which persists out to chain lengths of several hundred. The latter may be considered to be counter-intuitive. Therefore, Willemse *et al.* [49] wondered whether this conclusion was an artifact of using SEC to obtain k_p. Specifically, (i) SEC calibration is most difficult at small chain lengths, which one must probe to investigate CLDP; (ii) SEC broadening might result in k_p being systematically in error (see above); and (iii) RI detection of polymer amount also becomes difficult for oligomers [1]. All these pitfalls may be eliminated by using MS instead of SEC. Therefore, Willemse *et al.* used PLP MS to study CLDP [49]. Some of their results for (average) k_p versus i_1 (see Eq. (10.1))

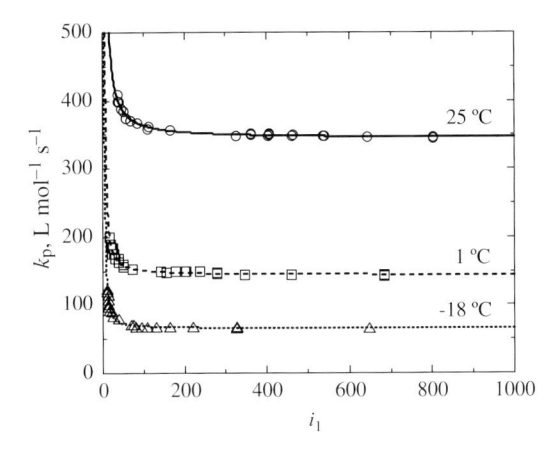

Figure 10.13 Average propagation rate coefficient, k_p, as a function of chain length of the first PLP peak of the MMD, i_1, for MMA at three different temperatures, as indicated. Points: results from PLP-MS experiments [49]; lines: best fits using Eq. (10.2) to calculate average k_p [61].

are shown in Figure 10.13. The evident increase of k_p as i_1 becomes lower is because CLDP exerts a stronger and stronger influence on the average value of k_p as average chain size in a PLP experiment becomes smaller. Willemse et al. showed that their data was consistent with a step function for k_p^i, that is, very high for the first few i, followed by a constant k_p for i thereafter. Soon after it was shown that the following more reasonable equation fitted the PLP MS data flawlessly [61]:

$$k_p^i = k_p^\infty \left\{ 1 + C_1 \exp\left[\frac{-\ln 2}{i_{1/2}}(i-1)\right] \right\} \quad (10.2)$$

The fits from using this empirical equation are also shown in Figure 10.13. The obtained parameter values were $C_1 = 15.8$, meaning k_p^1 is 16.8 times the long-chain value, k_p^∞, and $i_{1/2} = 1.12$, meaning that $k_p^i - k_p^\infty$ halves in value with every increase of chain length by 1.12 [61]. This is completely consistent with theory and other experimental data [58]. Gruendling et al. have recently found in their PLP-MS investigation that Eq. (10.2) with the same parameter values gives a good description of (average) k_p for methyl, ethyl, and n-butyl methacrylate [57], and it explains why their PLP-MS values are 5–10% higher than the benchmark values from SEC of much longer chains [55], as has often been found in PLP-MS studies (see above). In this way, MS has helped to elucidate a more detailed picture of CLDP.

Prior to the above studies, a group at Griffith University employed ESI MS in making what it labeled as "the first realistic estimates of individual rate constants for the early propagation steps in a free RP" [62]. Their method involved carrying out RP in the presence of a nitroxide radical trap. The nitroxide was maintained at a level sufficiently high to prevent the formation of high polymer, yet low enough to allow competitive monomer addition to form the lower members of the propagation series before being trapped to form the oligomeric addition products. These oligomers were identified and quantified by ESI MS operated in the selected ion recording mode in series with HPLC UV to effect oligomer separation. Concentration ratios thus obtained are equal to ratios of addition to trapping rate coefficients for specific chain lengths. Knowledge of trapping rate coefficients as a function of oligomer size thus enables k_p^i to be deduced.

This procedure has been used for acrylonitrile [62], styrene [63], and methacrylonitrile [64]. Typically k_p^i up to about $i = 10$ were reported. Pleasingly, all these studies found k_p^i values greater than the long-chain k_p measured by PLP SEC for the same conditions. However, the reported variations of k_p^i with i do not follow any logical trend. For example, the acrylonitrile results show an initial decrease in k_p^i until $i = 4$, after which an increase is observed [62]. On the other hand, the styrene data suggest that k_p^i increases in going from $i = 1$ to 4, stays constant for a while, and then decreases [63]. Different again is methacrylonitrile, for which k_p^i was found to be constant within experimental error [64]. Not only do these results lack internal consistency, but also each of them defies reasonable explanation even when taken individually. Probably this is due to difficulties in the method, for example, the need to know the total oligomer amount, which is difficult to measure, the need to know absolute values of trapping rate coefficients as a function of i, and of course the often-mentioned (and flawed) assumption that there is no chain-length bias in ionization tendency in MS. If these and other obstacles could be overcome, this method might fulfill its promise. Until then, the PLP MS method must be preferred as a far more reliable method for investigating CLDP, even though it is indirect (measures only average k_p) whereas the trapping method is more direct (uses individual oligomer amounts).

As a final point, it should be stressed that long-chain k_p, the subject of Section 10.4.1, is always the quantity of primary importance to know. The details of CLDP are of mechanistic interest, but they only ever have a perturbing effect on the value of average k_p, and this effect is only of any significance when average chain length is of order 100 or less [58].

10.4.3
Copolymerization

Mechanistic investigations regarding copolymerization mainly addresses reactivity ratios, which are usually obtained by relating the composition of the monomer feed with the composition of the copolymer that is produced by that feed. This can be accomplished by measuring the concentrations of unreacted monomer or by determining the composition of the resulting copolymer (e.g., by NMR or IR). These techniques give average copolymer compositions and only limited information about sequence distributions. Performing MS of copolymers, however, provides chain length information and the composition distribution at each chain length. This allows study not only of the average composition of the copolymer, which is necessary to evaluate the reactivity ratios, but also gives access to microstructure, for example, whether a copolymer is alternating, and to differences in the monomer composition along the main chain, as occurs for example in gradient copolymers.

Typically, the MS spectrum of a copolymer is rather complex, as the mixture within the polymer chain of two monomers having different molar masses disrupts the ordered sequence of individual peaks occurring in MS spectra of synthetic homopolymers. That is, macromolecules of a certain degree of polymerization can be composed of various combinations of monomers A and B, resulting in a series of peaks at each individual chain length. In the case that the molar masses of the two comonomers is rather similar, the spectrum may still be interpretable. With two very different comonomer masses, however, the spectrum may become extremely crowded, making a proper peak assignment nearly impossible. An extreme example of the former is the pair *n*-butyl methacrylate (BMA) and glycidyl methacrylate (GMA), which differ in molar mass by only 36 mDa, that is, the mass difference between CH_4 and O, respectively. This system was studied by Shi *et al.* [65], who resolved the isobaric fine structure of the peaks (separation 36 mDa) from the isotopic distribution (separation $\sim$ 1 Da) via high-resolution Fourier-transform ICR MS. From the isobaric distribution, the authors concluded that GMA is less reactive than BMA in the polymerization process. This hints at how MS information can be useful for mechanistic investigation of copolymerization.

Haddleton and coworkers were the first to explore the possibility of using MS to study copolymers made via RP. They looked at poly(MMA-*co*-styrene), poly(MMA-*co*-methacrylic acid) [66], and poly(MMA-*co*-BMA) [67]. By keeping the degree of polymerization small (around about 10) and having MMA as the dominant component, it remained easy to identify the different peaks, because these were not large in number. In fact the aim of these early papers was simply to show that MS can be used to identify copolymers of different composition. Soon after they presented a more sophisticated analysis of the MMA–BMA system [68]. By evaluating the bivariate distribution of composition and chain length for a series of copolymers, these authors were able to arrive at reactivity ratios obtained purely from MS. The approach rests on the deconvolution of the bivariate distribution into the individual chain length and composition distributions, which were then fitted to calculated distributions using a probability approach. The obtained values were reasonable, even if they deviated to some extent from values obtained by ^{1}H NMR as part of the same study, and from other literature values [68]. This was ascribed to a bias in the MALDI-MS spectra toward chains rich in MMA. This emphasizes that ionization bias, which hampers other quantitative MS approaches, may be even more severe in the case of copolymers, as the chemical composition, and hence ionization probability, changes systematically. It must also be admitted that this MS method requires a laborious evaluation of a multitude of individual MS peaks, making the analysis far more complicated than with traditional methods, which arguably are more accurate. So monomer reactivity ratios in copolymerization are a case where MS can provide mechanistic information but is not needed to.

Willemse *et al.* developed the most advanced MS analysis of copolymers to date by introducing the concept of "copolymer fingerprints." This was first of all presented in a study of polystyrene-block-polyisoprene [69], as made via sequential anionic polymerization. Figure 10.14 shows MALDI-MS spectra from various times during the second-stage polymerization. The formation of the polyisoprene block is clearly

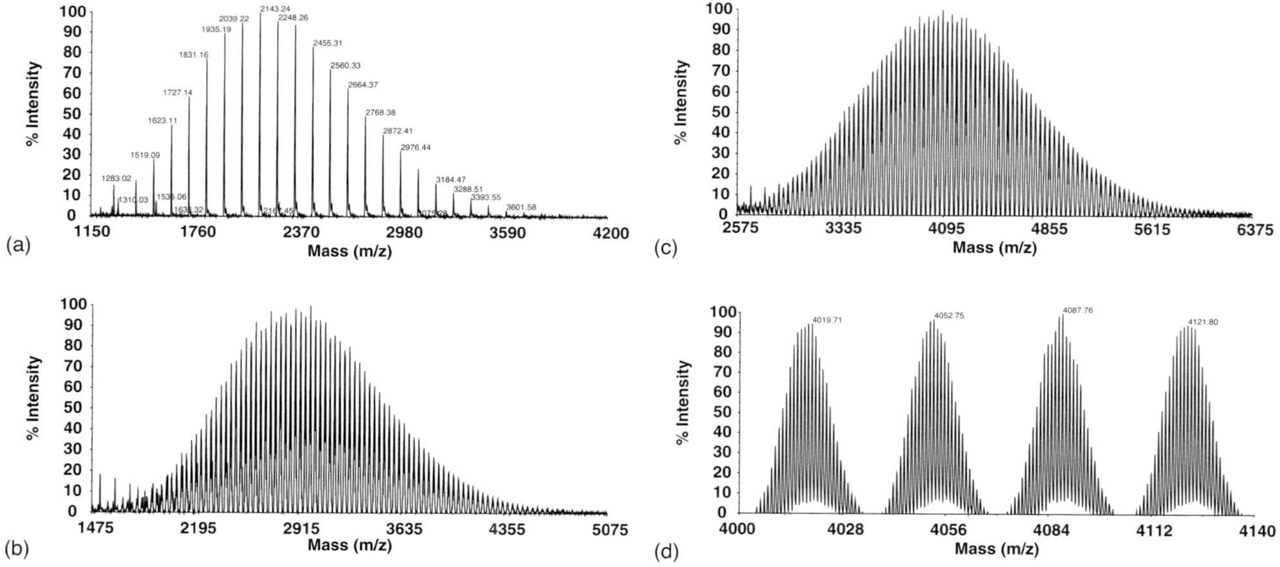

Figure 10.14 MALDI-TOF-MS spectra of the system polystyrene-*block*-polyisoprene after (a) 0%, (b) 50%, and (c) 100% conversion of isoprene, and (d) an enlargement of a portion of (c). (Reprinted with permission from Ref. [69]. Copyright 2004 American Chemical Society).

evident from the way the starting CLD (Figure 10.14a), which is that of the polystyrene, migrates to higher m/z as isoprene conversion increases (50% in Figure 10.14b, 100% in Figure 10.14c). Figure 10.14d is fascinating in that it reveals that each line in the preceding two spectra is in fact an envelope of signals. The shape of this envelope immediately tells that it cannot be from isotopes alone, as isotopic distributions are necessarily peaked toward lower m/z. In fact this complicated pattern arises because three isoprene units and two styrene units are different in mass by only 4 Da [69]. For example, a copolymer with 20 styrene units and 30 isoprene units is nearly isobaric with one consisting of 22 styrenes and 27 isoprenes. Thus, each envelope arises from copolymers of slightly different composition, where each copolymer also has a distribution of isotopes.

The question arises as to how to make use of the tremendous amount of information in the spectra of Figure 10.14 in order to gain insight into the obviously complex situation lying underneath, as described above. The approach of Willemse et al. [69] rests on the development of a matrix of signal intensities consisting of n_A rows and n_B columns, with n_A and n_B being the numbers of units of monomers A and B, respectively, in a chain, that is, the chemical composition. The theoretical molar mass of singly charged copolymer, M_{th}, can be calculated using

$$M_{th} = n_A M_A + n_B M_B + M_I + M_{II} + M_+ \tag{10.3}$$

where M_A and M_B are the molar masses of the monomeric units, M_I and M_{II} are the masses of the end groups, and M_+ is the mass of the cationization agent. Masses obtained from MS spectra, M_{exp}, were compared with M_{th} calculated according to Eq. (10.3). The signal at M_{exp} was identified as corresponding to the copolymer of composition (n_A, n_B) if

$$|M_{exp} - M_{th}| \leq \Delta M/2 \tag{10.4}$$

where ΔM is the accuracy in a value of M measured by MS [69]. The intensity of the signal for copolymer of composition (n_A, n_B) was multiplied by a correction factor that accounts for the fact that only the most abundant isotope is considered and that the peak intensity and not the area beneath an isotope is used, the latter being difficult to obtain. Finally, the resulting matrix of corrected intensities is normalized and represented as a two-dimensional contour plot.

Figure 10.15 presents an example of a contour plot from the polystyrene-*block*-polyisoprene study [69]. It is immediately clear why these plots have been dubbed "copolymer fingerprints." In this case, one sees that the polystyrene blocks remain unchanged during the second-stage polymerization as the polyisoprene block becomes progressively longer with conversion. Of course, the same is implicit in the MS spectra of Figure 10.14, but it is not nearly as vivid.

In follow-up papers, Willemse et al. used this approach to study random [70] and statistical [71] copolymers. These were shown to result in different looking fingerprints to those of Figure 10.15, which are round and vertically moving. By contrast, random oligoester copolymers give elliptical contours that migrate along the plot diagonal [70], while a styrene-MMA statistical copolymer from RP results in a long plume, the

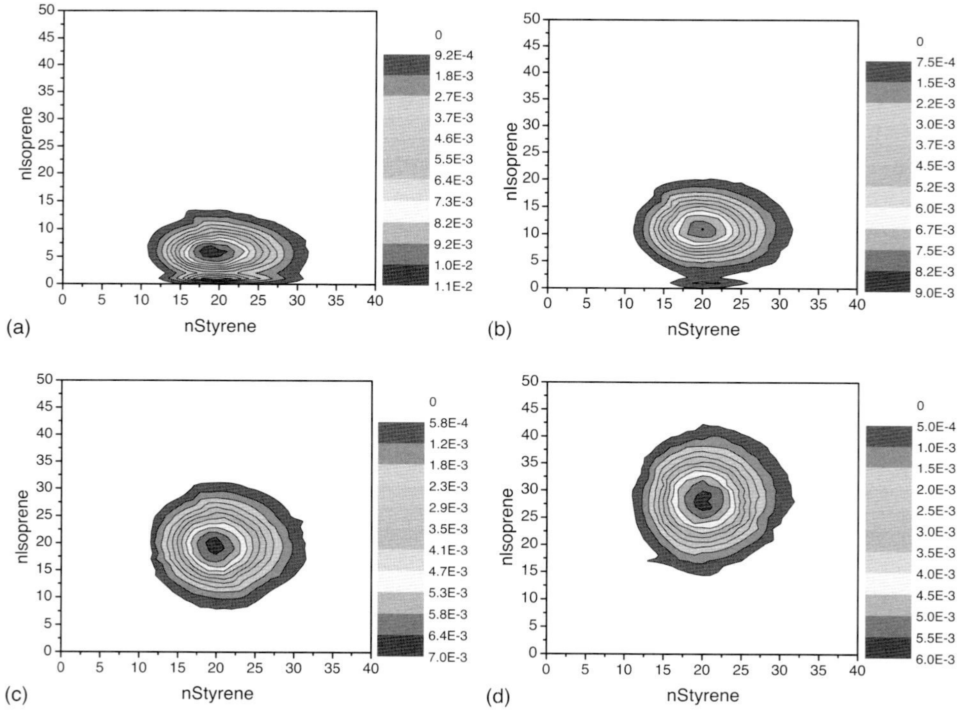

Figure 10.15 Copolymer fingerprints of the system polystyrene-*block*-polyisoprene corresponding to approximately (a) 25%, (b) 50%, (c) 75%, and (d) 100% conversion of isoprene. (Reprinted with permission from Ref. [69]. Copyright 2004 American Chemical Society).

orientation of which is dictated by the average chemical composition [71]. In this way, the fingerprint visually reveals the microstructure. While this approach is elegant, it strikes as being somewhat like a person using a Rolls-Royce to travel 100 m: it executes the function, but is an unnecessarily over-elaborate way of doing so. For one thing, it may be regarded as merely confirming the obvious: what else but a block copolymer could be yielded by sequential anionic polymerization of styrene then isoprene? Even allowing for the need to confirm the highly likely, this is much more easily done by NMR, which is well capable of establishing copolymer microstructure [72].

Where the labors of the fingerprint approach are more enlightening is in how these data can be further processed. For example, Willemse *et al.* showed that the CLD of the polystyrene block did not change as isoprene units were added, and that this distribution exactly matches a Poisson distribution, confirming the living nature of the anionic process [69]. The polyisoprene blocks were also shown to evolve with conversion in a living-like way; however, the CLDs were not perfect Poisson distributions. The most likely cause of this is an irregularity in the CLD at chain length 2; this evidences retardation in the incorporation of the second isoprene unit, for which there is evidence in the literature [69]. Analysis of the distributions of both blocks showed that the so-called random coupling hypothesis is obeyed, which confirms that the first block does not influence the growth of the second block [69]. The follow-up papers are similar in presenting statistical analyses that reveal or confirm mechanistic details about the copolymerizations under study [70, 71], for example styrene/methacrylate reactivity ratios [71] (see above).

10.5
Termination

Consider polymerization of MMA initiated by the primary radical A•. As shown in Figure 10.16, termination by disproportionation gives rise to polymer product with end groups of A at one terminus and monomer (either saturated or unsaturated) at the other. On the other hand, termination by combination results in product with end group A at both termini. Thus, as long as $M(A) \neq j \times M(M)$, where $j = 1, 2, \ldots$, the series of MS peaks from combination products will be distinct from the series due to disproportionation. Thus, with MS one can directly differentiate the products from these two termination pathways. Obviously, this applies for any monomer M, not just MMA. All this was implicit in the examples of Section 10.3.1, but has here been fleshed out in detail. Of course, in practice it is best if $M(A)$ is not too close in value to $j \times M(M)$, so that the two sets of MS peaks are clearly separated. A rough guideline is that there should be a difference of about 10 amu or more, for example, see Figure 10.17, in which the different sets of peaks from disproportionation and combination are clearly visible.

Termination is one of the fundamental reactions of RP, and right from the beginning it was recognized that it may occur by the two avenues shown in Figure 10.16. From the point of view of rate, it is not important to know the extent to which each pathway is taken, because rate depends only on the overall rate

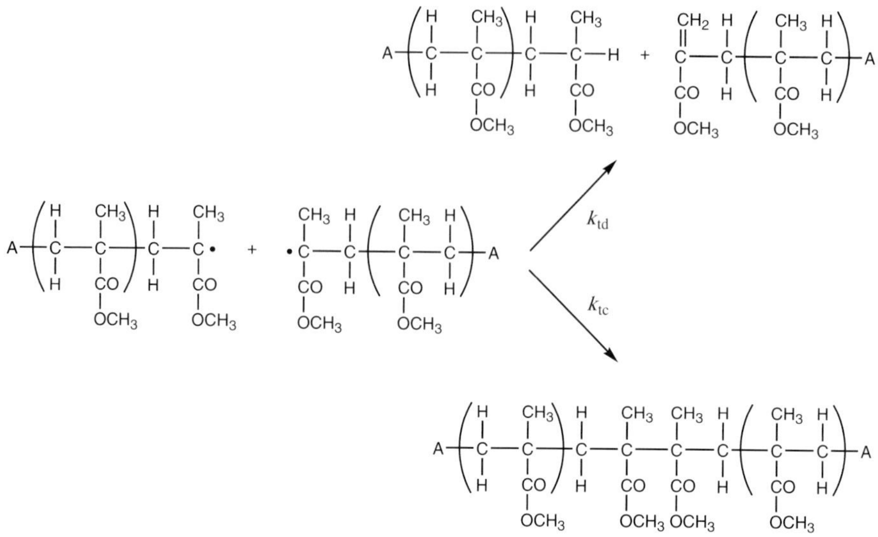

Figure 10.16 Termination by disproportionation, rate coefficient k_{td}, and by combination, k_{tc}, in radical polymerization of MMA.

coefficient for termination, $k_t = k_{td} + k_{tc}$. However, as is obvious from Figure 10.16, the individual values of k_{td} (disproportionation) and k_{tc} (combination) are of importance for average polymer size, because combination gives rise to longer polymer molecules. Furthermore, these values are also crucial where the end-group identity is important, for example, if the end group A gives rise to desirable material properties

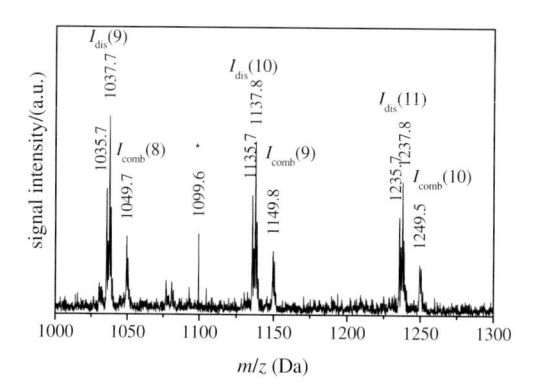

Figure 10.17 ESI-MS spectrum of poly(MMA) from polymerization in benzene at 85 °C with BTMHP as initiator. The m/z of each of the prominent peaks is given. All of these peaks are either combination products ("comb"), disproportionation products ("dis") or internal standard (*), as indicated. The numbers in brackets are the number of MMA residues. For example, the signals labelled $I_{comb}(8)$ are from oligomer made up of 8 MMA units and 2,4,4-trimethylpentyl at each end. (Reprinted with permission from Ref. [73]. Copyright 2009 American Chemical Society).

or contains a functional group to be exploited in a subsequent polymer modification reaction. For these reasons, it is desirable to know the extent to which a monomer terminates by disproportionation and by combination. However, obtaining such information has proven to be surprisingly difficult, because until the advent of large-molecule MS, it was not easy to separate chains according to their end groups.

Workers have realized the exciting potential of MS to improve this situation, in the fashion foreshadowed above: because with MS one can directly see the different chains from each type of termination reaction, it becomes possible to learn about this aspect of mechanism. Mostly this principle has been used in only a *qualitative* way. Zammit *et al.* were the first, noting that disproportionation peaks dominated in their MALDI-MS spectrum from PLP of MMA initiated by 2,2′-azobisisobutyronitrile (AIBN; A≡cyanoisopropyl) [48]. This may also be seen in Figure 10.17 [73]: the disproportionation peaks are much higher in intensity than those from combination. On the other hand, Barner-Kowollik *et al.* used MMMP (see Figure 10.11 for the resulting A) to photoinitiate MA polymerization, and found the situation to be the reverse of that for MMA: there are still disproportionation peaks, but the combination ones prevail in intensity. Mechanistically this makes sense, whereas steric hindrance disfavors combination for monomers like MMA (see Figure 10.16), there is no such barrier for MA, because it lacks an α-methyl group (Figure 10.18). Dicyclohexyl itaconate provides an extreme example of steric hindrance (Figure 10.18), and indeed, the CAMD group found no combination product in their ESI-MS study of its photopolymerization [29]. Copolymerization of styrene and MMA is interesting in that the former monomer gives a secondary propagating radical, and thus terminates predominantly by combination [74], whereas the latter gives a tertiary radical, and so terminates largely by disproportionation, as already seen. This system has been investigated by Willemse and van Herk via MS [71]. As one can imagine, it is

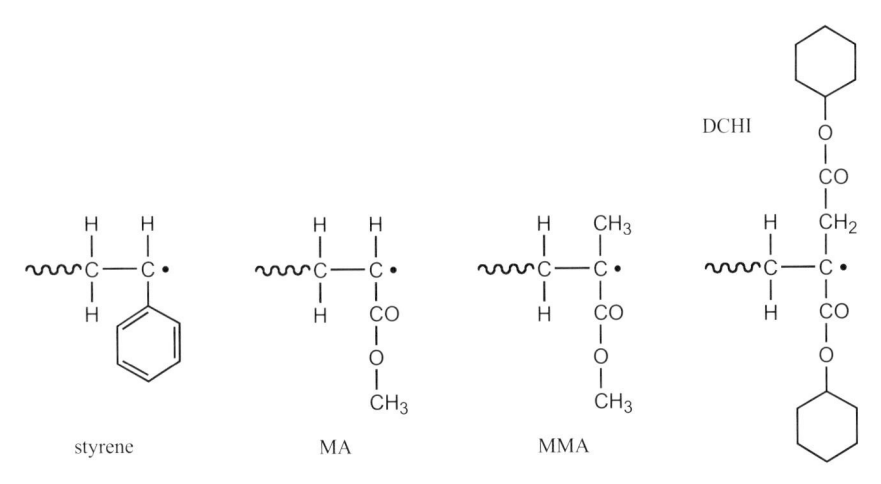

Figure 10.18 Radical structures in the polymerization of (left to right, in order of increasing tendency to terminate by disproportionation) styrene, methyl acrylate, methyl methacrylate, and dicyclohexyl itaconate.

complicated. However, it is clear that qualitative expectations are met: as one adds styrene to the mixture, the proportion of termination by combination increases, and it does so in a way that reflects the relative abundance of each type of radical, as calculated using known values of reactivity ratios [71]. Figure 10.18 shows the structures of all the monomers discussed in this paragraph, thereby assisting in the understanding of these MS findings regarding termination mode.

In order to calculate average polymer size, one needs more than just a qualitative impression about the termination channel: *quantitative* information is needed. Most workers use one of two equivalent indexes: either $\delta = k_{td}/k_{tc}$ or $\lambda = k_{td}/k_t = \delta/(\delta + 1)$. Because disproportionation results in twice as many polymer molecules as combination (see Figure 10.16), it follows that $\delta = (n_d/2)/n_c$, where n_d and n_c are the *total* numbers of molecules formed by disproportionation and combination, respectively. This would seem to open the door to determining δ via MS. However, the problem is that MS does not yield the entire CLD, and thus n_d and n_c cannot confidently be determined, because this requires summing the areas of *all* peaks from each reaction. Therefore, Zammit et al. [75] hit upon the idea of comparing only peaks of the same i. This has the added advantage of reducing any effects of mass-related ionization bias: because only peaks of the same i are being compared, the signal intensity for these peaks should be proportional to the number of chains. Using $\delta = [n_d(i)/2]/n_c(i)$, it was obtained that $\delta = 4.37$ for MMA at 90 °C and $\delta = 0.057$ for styrene at the same temperature [75].

Zammit et al. observed variations of δ with chain length and initiator concentration that they ascribed to experimental error [75]. Recently, Buback et al. [73] realized that these variations in fact evidence an over-simplified kinetic analysis. The idea is actually quite simple: because disproportionation generates an exponential number-CLD whereas combination generates one of the form $i \times \exp(-ki)$, where k is a constant, the ratio $n_d(i)/n_c(i)$ must actually have form $1/i$, that is, it must decrease as chain length increases [73]. This is essentially what Zammit et al. found but were unable to explain, so they did not even recognize it as a real trend. Using Schulz–Flory theory, one may derive the following precise relation for this variation [73]:

$$\frac{\frac{1}{2}n_d(i)}{n_c(i)} = \frac{1}{(i-1)C(\frac{1}{\lambda}-1)} = \frac{\delta}{(i-1)C}, \quad \text{where} \quad C = \frac{(2k_t R_i)^{0.5}}{k_p c_M} \quad (10.5)$$

Here, R_i is rate of initiation and all other parameters are as previously defined. Buback et al. fitted Eq. (10.5) to MS data they obtained from MMA polymerization initiated by bis-3,5,5-trimethylhexanoyl peroxide (BTMHP; A ≡ 2,4,4-trimethylpentyl; see Diacyl peroxides in Section 10.3.1). Figure 10.17 presents an example of such data. The value $\lambda = 0.63$ ($\delta = 1.70$) at 85 °C was found, independent of chain length and initiator concentration [73]. A very important result from Eq. (10.5) is that $n_d(i)/n_c(i)$ depends on parameters other than just δ, meaning that *definitive* statements about termination mode cannot be made from MS peak intensities alone.

There seems little doubt that this MS-based method [73] for determining the value of λ is without peer, and it is to be hoped that it will be widely exploited in years to come, because there is an ignominious lack of good data for this quantity. Even where it is declined to carry out such a quantitative kinetic analysis, at the very least workers presenting MS results for RP should learn the habit of making a qualitative comment on what their data indicates about the mode of termination. This is a case where MS is truly an indispensable tool for studying the mechanism of RP.

10.6
Chain Transfer

In RP, a chain-transfer reaction is one in which radical activity is transferred from one place to another, thereby creating a new radical site and leaving behind a now unreactive one. Because radicals are very reactive molecules, this reaction takes many, many forms in RP, most of which have been looked at in some way or another via MS. Here, we divide these studies into two broad groupings.

10.6.1
Transfer to Small Molecules

In this section, we treat transfer from a growing macroradical to a small molecule. There are many such classes of transfer [74, 76]. The general chemistry is illustrated in Figure 10.19 by considering the example of AIBN-initiated polymerization of MMA in the presence of *tert*-butyl mercaptan (2-methyl-2-propanethiol; tBuS-H) [77], which is a *traditional chain-transfer agent* (TCTA). One sees that the tBuS-H converts a macroradical into a (saturated) dead polymer molecule, and then the resulting small-molecule, sulfur-centered radical adds to monomer so that a new polymerizing chain is formed. It eventually reacts with another tBuS-H molecule and this cycle is repeated. The result is formation of tBuS–(MMA)–H polymer molecules, as shown in Figure 10.19. In other words, the net outcome is a *polymer insertion* reaction into the TCTA.

Commonly the transferred atom in a chain-transfer reaction is H (as with tBuS-H), but it is not always. For example, halogen atoms may also function this way, explaining that alkyl halides, for example, CBr_4 and CCl_4, are used at TCTAs [74, 76]. For this reason, a CTA should be given generic formula B-Z, where Z is the transferred atom and B• (denoting a chain-starting group of atoms, not boron) is the reinitiating radical. In the so-called *transfer limit* [78], there are countless transfer events per termination event, and B–(M)–Z is essentially the only product of the polymerization, where M is monomer. Such conditions are commercially attractive because they afford simple control of molecular weight [76] and they result in polymers of uniform composition. However in the general situation of termination and transfer occurring competitively [78], the situation is far more complex, because there are (at least) two types of end groups from chain start, namely A from symmetric initiator (the situation becomes even more complicated if an asymmetric initiator is used) and B from CTA, and each of these macroradical types may be converted into a dead chain by three different reactions, namely transfer, disproportionation, and combination. The different families of species that one may anticipate observing by

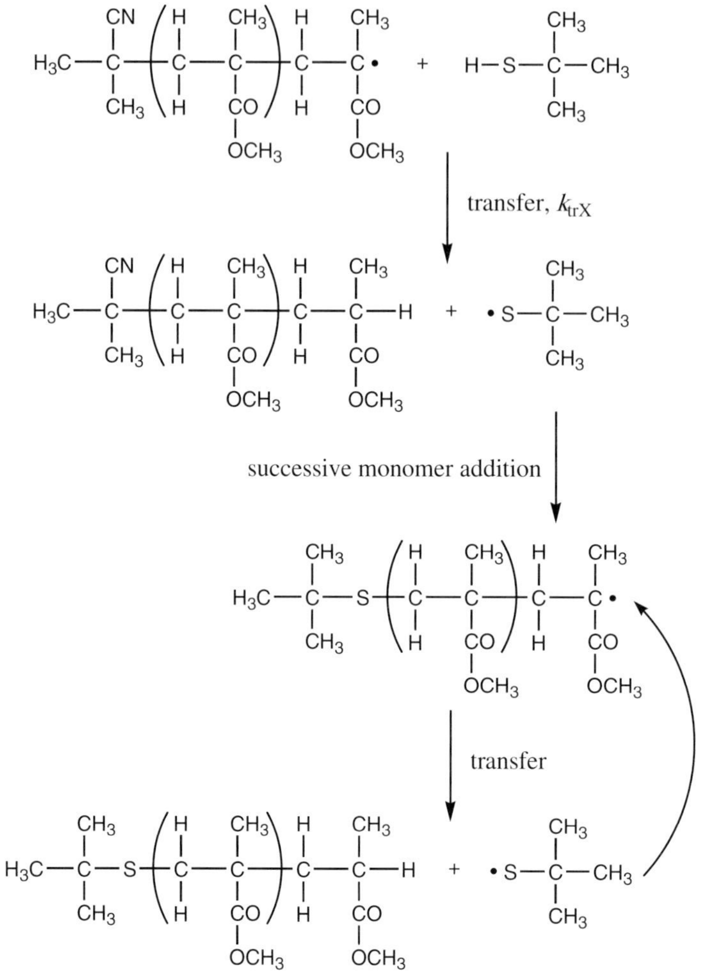

Figure 10.19 Mechanism of chain transfer to *tert*-butyl mercaptan in the AIBN-initiated polymerization of MMA.

MS are shown in Figure 10.20. Thus, in principle, there can be up to seven series of peaks, although where Z = H (e.g., a thiol) this reduces to five, because the products from transfer will be isobaric with the (saturated) products from disproportionation.

Figure 10.21 [77] shows MS results from the example of Figure 10.19, that is, for the case of A = cyanoisopropyl, B = tBuS, and Z = H in Figure 10.20. Knowing that in MMA polymerization, there are significant amounts of termination by both disproportionation and combination (see Section 10.5), there is thus the expectation of either one series of peaks (i.e., transfer control) or five (i.e., competitive termination and transfer). In fact, there are four. This turns out actually to be three, because the two most intense sets of peaks – {898, 998} and {914, 1014} – are from B–(M)–Z with Li^+ and Na^+ respectively. The {877, 977} series is due to A–(M)–Z with Li^+. The final

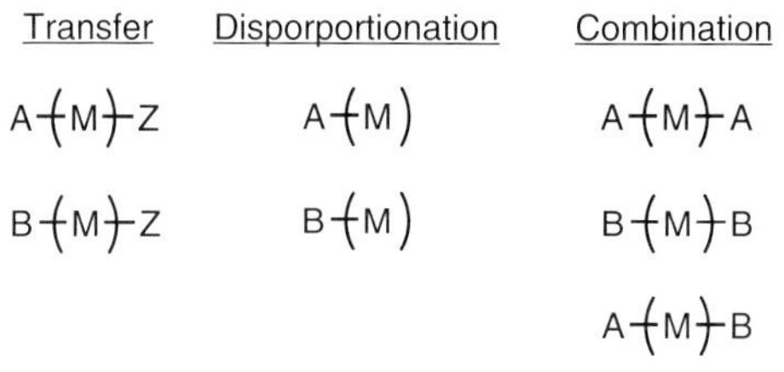

10.6 Chain Transfer

Transfer	Disporportionation	Combination
A−(M)−Z	A−(M)	A−(M)−A
B−(M)−Z	B−(M)	B−(M)−B
		A−(M)−B

Figure 10.20 The different series of dead polymer molecules that arise from transfer, disproportionation and combination in a system comprising monomer M, chain-transfer agent B−Z, and initiator that yields primary radicals A•. For convenience each structural grouping from disproportionation includes both the saturated and unsaturated species (see Figure 10.16); although these are 2 Da apart, in practice they both fall within the one cluster of MS signals (e.g., see Figure 10.17), hence grouping them together.

set of peaks, {860, 960} is ascribed to A−(M)−B [77], but this assignment is implausible for two reasons: (i) the theoretical masses are 5 Da different to the observed masses, a ΔM that way exceeds those for the other three sets of peaks; and (ii) If A−(M)−B species are observed, then so too must be B−(M)−B, because clearly B-ended radicals are the most numerous in the system. But such species are not observed. In fact the origin of the {860, 960} series is not the only mystery here. The steady-state hypothesis dictates that for every initiation event there is one termination event. So, the observation of

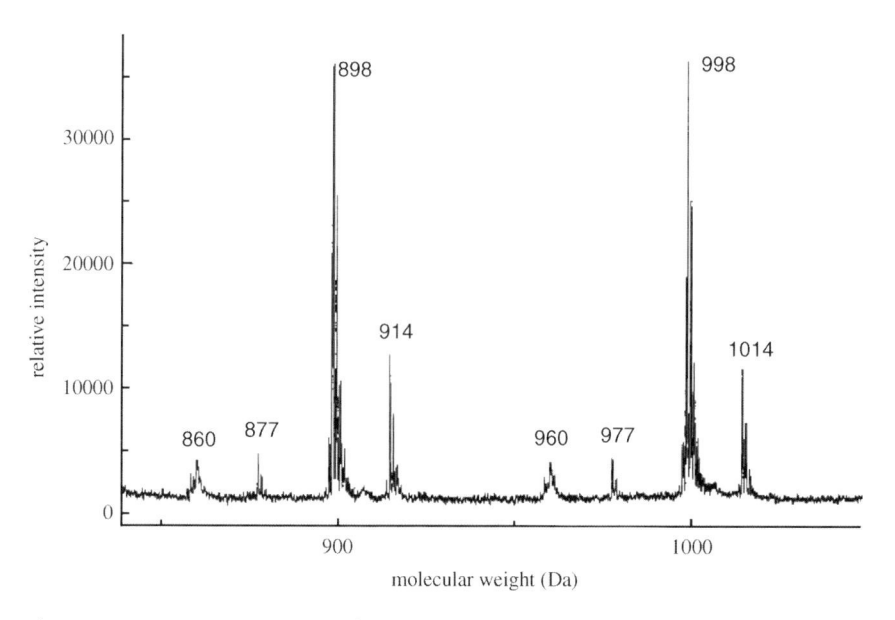

Figure 10.21 MALDI-MS spectrum of poly(MMA) from polymerization at 60 °C with AIBN as initiator and *tert*-butyl mercaptan as chain-transfer agent. The *m/z* of each peak is given, and these values are discussed in the text. (Reprinted with permission from Ref. [77]. Copyright Wiley-VCH Verlag GmbH & Co. KGaA).

A–(M)–Z means that termination product should also be observed. The disproportionation products are obviously buried among the transfer products, but there should be observable combination products; however, there are not.

Remarkably, Kopecek *et al.* found exactly the same in their MS spectra from polymerization of N-(2-hydroxypropyl) methacrylamide (HPMA) in the presence of various functionalized thiols, that is, they generally observed an intense set of peaks from B–(M)–Z, a less intense set from A–(M)–Z, and a set of unknown structure [79, 80]. It may be that the latter signals result from an unsuspected counterion rather than an unexpected reaction. It is certainly clear that further work remains to be done in the study by MS of transfer to thiols.

The main purpose of the work by Kopecek *et al.* was to confirm the synthesis of semitelechelic poly(HPMA) for protein modification, where the functionality at the end of the biopolymer comes from B containing a reactive group [79, 80]. In other words, in the standard way (see Section 10.2) they were using MS to confirm end-group identity. Considering the discussion here, the interesting thing is that where polymerization overwhelmingly results in formation of B–(M)–Z, MS does nothing more than confirm the occurrence of a polymer insertion reaction, not *where* the insertion occurs, that is, the individual identity of each end group. To achieve this, one requires the occurrence of a significant amount of termination, so that B and Z can be observed separate to each other (see Figure 10.20). Of course chemical intuition leaves little doubt as to the transferred atom in most instances, for example with thiols. Nevertheless as a point of principle, one should recognize that confirmation of the structure B–(M)–Z by MS does not unambiguously prove what B and Z are.

There are two more recent studies that are similar in spirit to those of Kopecek *et al.* [79, 80]. In the first [81], α-thioglycerol was used as TCTA in the polymerization of styrene, resulting in the formation of α-dihydroxylated styrene oligomers, as confirmed by MS. Similarly for the second study [82], in which thioglycolic acid was the TCTA and BMA the monomer, yielding α-carboxyl-terminated oligomers. The chemistry in all these cases is easily understood by analogy with Figure 10.19.

A second type of chain-transfer reaction is *transfer to solvent*. Formally this is identical to chain transfer to TCTA, with the only difference being that the CTA is solvent rather than a chemical expressly added for its activity in this regard. A well-known example is toluene [83], where there is a driving force for hydrogen to be extracted from the methyl group, thereby producing a benzyl radical, which of course is relatively stable because the free electron is resonance stabilized by the phenyl ring. In other words, B = benzyl and Z = H in the previously used notation. Needless to say, the transfer rate coefficient, k_{trX} (see Figure 10.19), for a reaction like this is not as favorable as that for the case of a TCTA. But the point is that the frequency of transfer, $k_{trX}c_X$, depends also on the *concentration* of transfer agent X, c_X. Thus, the high value of a solvent concentration can make up for k_{trX} being relatively low, and thereby give a rate of transfer that is significant.

We are aware of surprisingly few MS investigations of transfer to solvent. For example, toluene seems never to have been looked at in this regard. One study was of RP of N-vinylpyrrolidone in 3-methylbutan-2-one [84]. It found extensive dead-polymer formation by transfer to solvent. It is reasonable to speculate that the mechanism of this reaction is transfer of H from the isopropyl position, because this generates a relatively stable tertiary radical. However as explained above, the observation of B–(M)–Z via MS does not prove this hypothesis. In another study, it was found by MS that transfer to xylene occurs when it is used as solvent in acrylate polymerizations [85]. This comes as no surprise given what is known about toluene.

Finally, there is *chain transfer to monomer*, that is, monomer itself acts as CTA. This is the RP reaction in which organic chemists most struggle to believe. However, it is a fact that as initiation rates approach zero, chain sizes reach a limit rather than being ever increasing, as established for MMA [38] and for styrene [86], for example. Such behavior can only be explained by the occurrence of a transfer reaction, and in the absence of any other candidate it is reasonably assumed that the CTA is monomer. As with solvent, the value of k_{trX} might be very low, but to some extent a high value of $c_X = c_M$ compensates for this. Further, it is not the competition with propagation that counts, but rather it is that with other dead-chain-forming events, which by definition are rare; otherwise, polymer molecules would not be long.

As with autoinitiation (Section 10.3.1.3), it is often unclear how transfer to monomer occurs. Styrene contains no labile H, and so it is still debated how it acts as a CTA [74]. In the cases of MMA and α-methylstyrene (AMS), there is an α-methyl group that presumably provides "reactive" hydrogen atoms and thus participates in the chain-transfer process. Indeed, MS from bulk polymerization of AMS without TCTA has confirmed that the dominant species is B–(M)–Z, where B–Z = M [87]. Nevertheless, as already explained in general, this does not prove the mechanism of transfer to monomer, which in principle may involve H transfer either *from* or *to* monomer [88], as shown in Figure 10.22. Of course, it is most likely that transfer is from monomer to the radical, as this is the way transfer occurs with TCTA and solvent, and there is tautomeric stabilization of the product monomeric radical (see Figure 10.22). But the point must be stressed that MS proves only that transfer to monomer occurs, not the mechanism of this reaction.

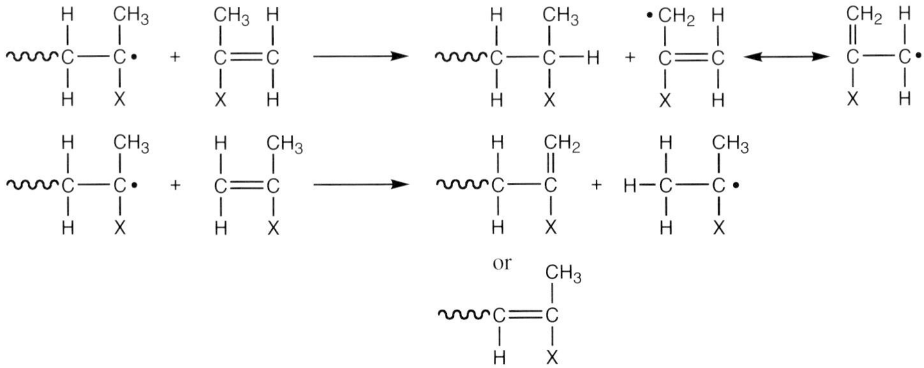

Figure 10.22 Chain transfer to monomer for monomers with an α-methyl group, for example, MMA (X = CO-O-Me) and AMS (X = Ph). Top reaction: transfer of H *from* monomer; bottom: *to* monomer.

In the same way, chain transfer to monomer has also been proven by MS to occur in RP of DMI [32] and several vinyl phosphonates [89]. In the case of DMI, it seems reasonable to speculate that the α-methylene group is involved (see Figure 10.18 for itaconate structure). This is so not just by analogy with Figure 10.22, but also considering that an adjacent carbonyl group activates a C-H bond for H-transfer (see Section 10.6.2). This is also consistent with transfer to monomer in vinyl acetate being held to involve the pendant methyl group of the monomer [88].

Covered in Chapter 11 rather than here is CCTP. Several examples of this have been mentioned in other contexts in the present chapter [8, 66–68]. The essential difference between a catalytic chain-transfer agent (CCTA) and a TCTA is that a CCTA is catalytic, and thus it is regenerated rather than being consumed by the reaction. Because of this, CCTP delivers polymers of schematic structure (M) [8, 90]. This means they are isobaric with those from chain transfer to monomer (see above), even though the mechanism is different.

All the discussion so far has been of how MS has been used for *qualitative* investigation of chain transfer, that is, the confirmation that a molecule B–Z acts as a transfer agent through the presence of a series of MS signals at M_{exp} equal to M_{th} for B–(M)–Z. To the best of our knowledge, there have been only two attempts at *quantitative* analysis, that is, the use of peak intensities to determine transfer constants, $C_X = k_{trX}/k_p$. Kapfenstein and Davis [77] used a relationship between C_X and the average number of initiator end groups per polymer molecule. To determine the latter they had to average peak areas across the entire MS spectrum. The value obtained for C_{tBuS-H} was not consistent with that obtained by traditional methods. This is not surprising given that this process implicitly assumes that MS intensities give the *entire* number-CLD accurately. The same inconsistency problem was experienced by Suddaby et al. [68], who used the bivariate distribution of chain length and composition obtained from MS (see Section 10.4.3) to extract C_{CCTA} for both macroradicals in a CCT copolymerization.

This section has made clear that there is a lot of room for further work in the area of MS investigation of transfer to small molecules. This is especially so with regard to determination of transfer constants, but also applies for the seemingly simpler task of identifying transfer reactions that occur. A difficulty is that transfer to solvent and to monomer usually results in large average sizes. However, it may be anticipated that improving MS technology will overcome this problem [2].

10.6.2
Acrylate Systems

In RP of acrylates, there occur two reactions in addition to those of the fundamental suite of RP reactions. These are CTP and β-scission, and they make acrylate systems fascinating to study [54]. CTP will first of all be considered here, because β-scission depends on its prior occurrence. There are two forms of CTP: *intra*molecular CTP, also known as backbiting (BB), and *inter*molecular CTP (inter-CTP). These reactions are shown in Figure 10.23. It is evident that basically they are the same very simple reaction: radical attack of a C–H bond resulting in transfer of the H atom. All that is

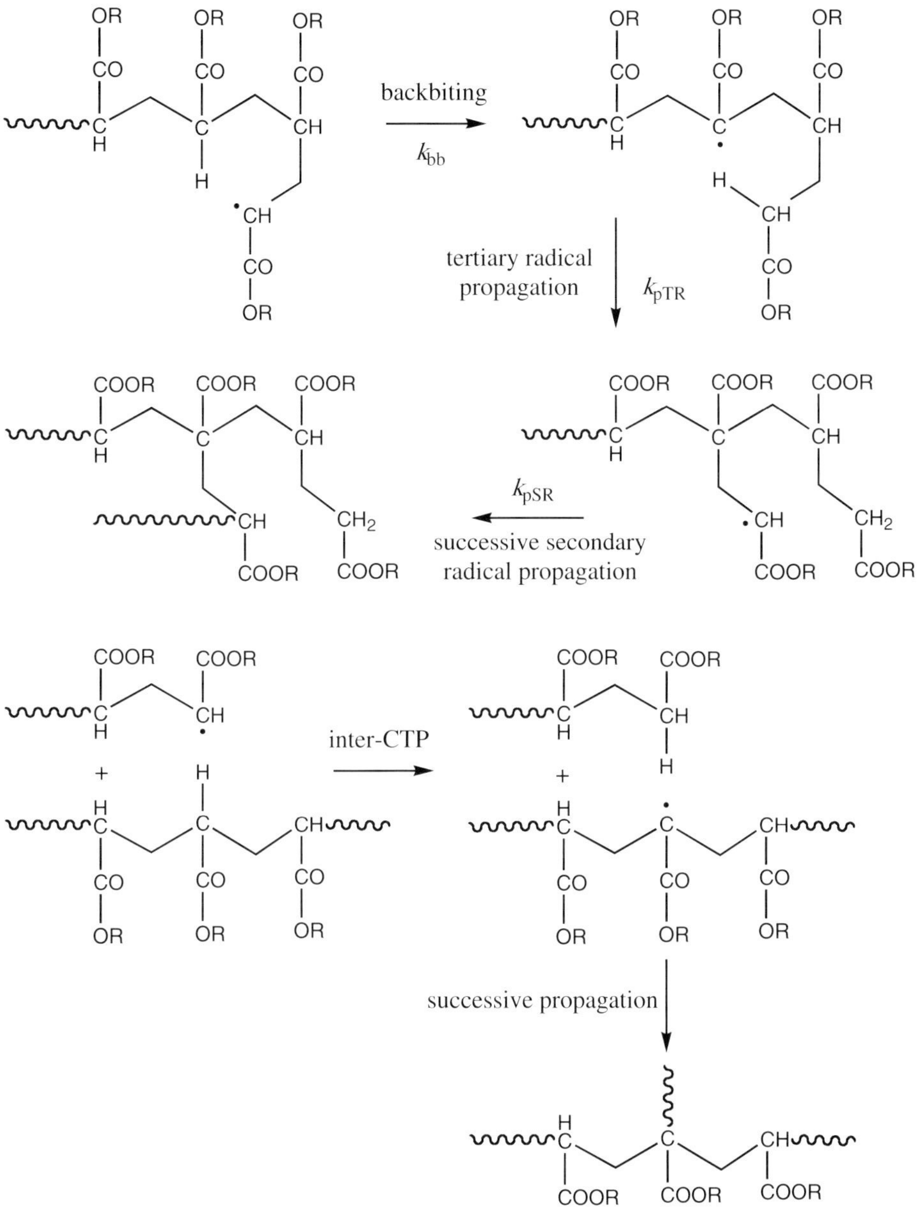

Figure 10.23 Chain transfer to polymer and subsequent reactions in RP of acrylates. Top half of scheme: *intra*molecular CTP; bottom: *inter*molecular CTP.

different is that in one case the transfer is to a site on the same macroradical that is very close to the radical chain-end, while in the other it is to a distant site, usually on a different molecule, which may or (usually) may not be a macroradical. In the former case the result is a *short-chain branch*, while in the latter a *long-chain branch* is formed.

This is a profound difference, which may be unexpected given how similar the two reactions are.

It has long been known that both forms of CTP occur in ethylene polymerization; however, it was considered a major discovery when Ahmad et al. proved that CTP occurs in RP of n-butyl acrylate (BA) [91]. They did this by using ^{13}C NMR to show the existence of quaternary carbons in the product polymer: these can only result from CTP (see Figure 10.23). This occurs because the carbonyl group of the repeat unit of a polyacrylate activates the C–H bond that is α to it, thereby making it amenable to CTP, as shown in Figure 10.23. This is common to all acrylates, and therefore the occurrence of CTP should be expected throughout this family, not just in BA. On the other hand, most other vinyl polymers do not have this propitious arrangement, explaining that CTP is otherwise uncommon.

Nothing about this basic chemistry hints at whether BB or inter-CTP dominates in acrylate systems. The study of Ahmad et al. does not help in this regard, because NMR cannot indicate how a quaternary carbon arises [91]. One factor that BB has on its side is that its transition state involves a six-membered ring, as intimated in Figure 10.23. This of course is energetically favorable. At low conversions, the polymer concentration is sufficiently low that inter-CTP can additionally be considered unlikely: the only C–H bonds a radical chain end will encounter are those adjacent to it. Nevertheless, one should like to see this proven.

It was Farcet et al. who designed an ingenious way of distinguishing between BB and inter-CTP via MS [92]. Their experiment utilized nitroxide-mediated polymerization (NMP), a topic belonging to Chapter 11. Nevertheless, the finding holds also for conventional RP, so this important study is entirely relevant here. If D-Z denotes an alkoxyamine, where D is an alkyl and Z = •O-N(X_1)(X_2) is a stable nitroxide radical, then properly functioning NMP delivers polymers of structure D–(M)–Z [93]. In other words, NMP is a polymer insertion reaction. Farcet et al. realized that the occurrence of BB does not change this structure, although there will be SCBs, every chain will still be started by a D group and ended by a Z. However, inter-CTP does make a difference. Consider the example of Figure 10.24. It shows how inter-CTP can

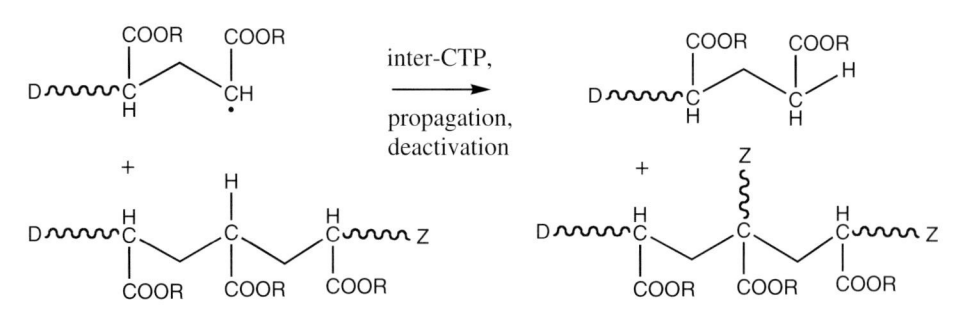

Figure 10.24 Occurrence of intermolecular CTP in nitroxide-mediated polymerization of an acrylate, showing how this can result in polymer molecules with 0 or 2 Z end groups, where D-Z denotes the alkoxyamine used in the polymerization (see the text).

result in one polymer molecule with 2 Z groups and another with 0, instead having an H terminus. If the molecule with 0 Z groups later undergoes inter-CTP, then it can be converted into a molecule containing 1, while that containing 2 can be increased to 3. Thus, it is clear that inter-CTP will result in a *distribution* about the average value of 1 Z per polymer molecule. Since MS is sensitive to the number of Z groups per polymer molecule (because it detects the different M that result), Farcet *et al.* realized that MS could thus indicate whether inter-CTP occurs [92].

Figure 10.25 shows the MALDI-MS spectrum from a properly functioning NMP of BA [92]. These results are from the first 90% conversion of monomer into polymer, by which stage a radical finds itself virtually surrounded by C–H bonds on other polymer molecules. This of course should favor inter-CTP. However, the preponderance of MS signals from molecules with 1 Z group, as opposed to the minor quantities of polymer with 0 and 2 Z groups, proves that BB is the dominant form of CTP over this nearly complete range of conversion. At lower conversions, the MS signals from 0 and 2 Z groups are barely discernible [92]. Evidently, the stabilizing influence of a 6-membered ring in the transition state is a powerful facilitator of CTP in acrylate systems. It would be warranted to repeat this work with a current MS instrument that is capable of individually resolving the signals from molecules with 0, 1 and 2 Z groups (cf. Figure 10.25).

An interesting aspect of the above method is that it relies on end groups rather than monomer residues. Of course, MS is even blinder to monomer arrangements than

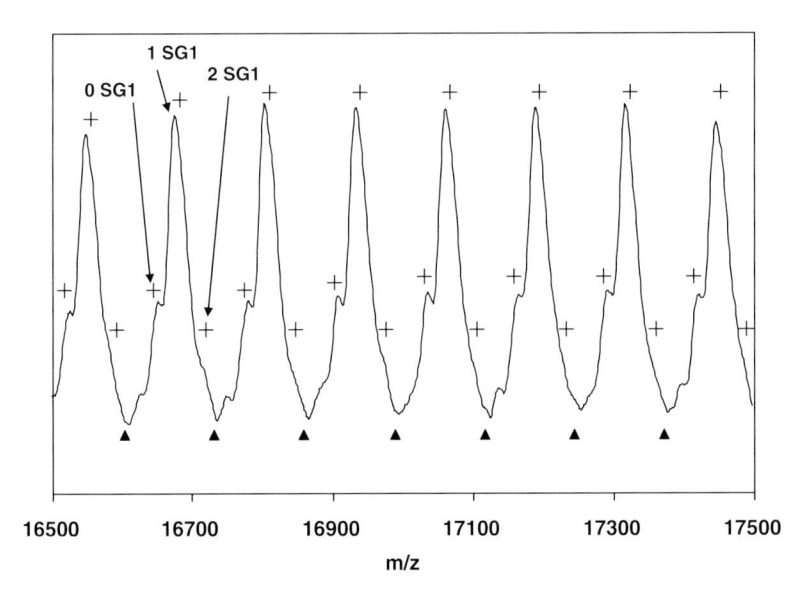

Figure 10.25 MALDI-MS spectrum of poly(BA) from nitroxide-mediated polymerization at 112 °C. The + signify the theoretical m/z for chains with 0, 1, and 2 nitroxide (SG1 = Z) end groups as indicated. (Reprinted with permission from Ref. [92]. Copyright 2002 American Chemical Society).

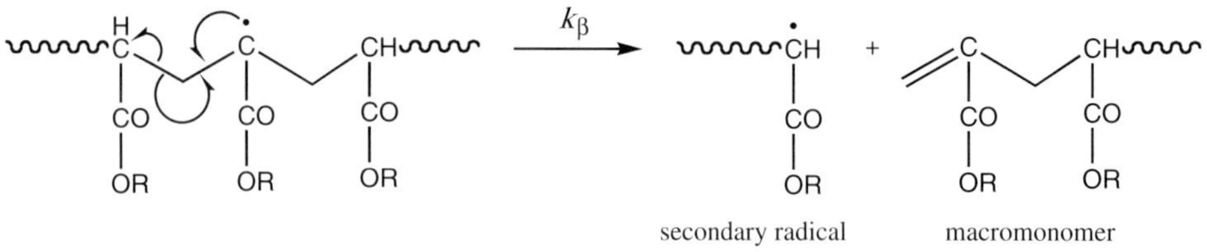

Figure 10.26 β-scission in acrylate polymerization.

NMR, because it cannot say anything about different isobaric molecules, for example, n-pentane versus isopentane versus neopentane (which NMR can distinguish). Given this, the idea occurs that the experiment of Farcet et al. should also work, in principle at least, with conventional RP. For example, if termination of BA is by combination, then all chains will have structure A–(M)–A in the absence of inter-CTP (see Figure 10.20), where A is the end group from initiation. However if inter-CTP occurs, then this situation will change to there being a *distribution* around the average value of 2 A per molecule.

Implicit in Figures 10.23 and 10.24 is the occurrence of tertiary radical propagation, rate coefficient k_{pTR}. It has long been known that acrylate polymerizations have markedly nonclassical order with respect to monomer, and it is now recognized that this is due to very low k_{pTR}: in effect a tertiary radical from CTP acts to retard polymerization [54, 94]. The latest measurements of k_{pTR} have found that $k_{pTR}/k_{pSR} \approx 0.001$ for BA at common polymerization temperatures [95], where k_{pSR} is k_p for a chain-end (secondary) radical, as shown in Figure 10.23. Thus, a tertiary radical is long-lived, meaning that it has much time at its disposal to undergo other reactions. One such reaction is β-scission, as shown in Figure 10.26. In contrast to CTP, which does not change the number of polymer molecules and therefore does not affect number-average degree of polymerization, DP_n, β-scission creates a new polymer molecule, and therefore it is like chain transfer to a small molecule in that it reduces DP_n. Further, it produces a so-called *macromonomer* as nonradical product and it converts a tertiary radical back into a chain-end radical, thereby reducing the retardation effect of CTP.

Chiefari et al. ignited interest in this area by using ^{1}H NMR to show that high-temperature polymerization of various acrylates produces macromonomers in high yield [96]. The first employment of MS to prove the formation of macromonomers in acrylate systems comes from Grady and coworkers [85, 97]. A recent MS study even shows that backbiting followed by β-scission seems to occur in RP of butyl vinyl ether, resulting in a carbonyl group being incorporated into the polymer backbone [98]. However, it is Barner-Kowollik and colleagues who have made far the greatest effort to study the consequences of β-scission in acrylate polymerization, using MS as their scalpel.

First of all Barner-Kowollik et al. pointed out the sheer complexity that is introduced by β-scission: even ignoring tertiary radical propagation, termination, and all forms of small-molecule chain transfer, one still has the reaction scheme shown in Figure 10.27 [99]. As is so often the case in polymerization, one is left amazed at

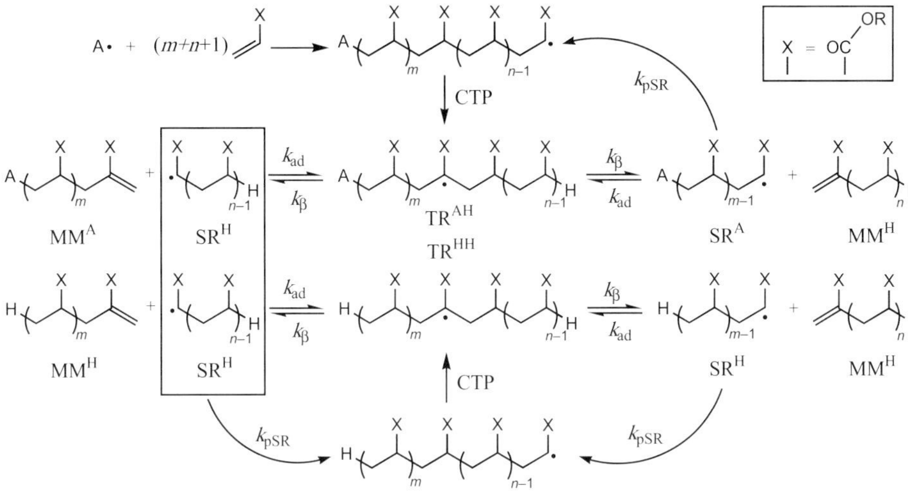

Figure 10.27 Mechanism of macromonomer formation in acrylate RP [99]. Abbreviations: MM: macromonomer; SR: secondary radical; TR: tertiary radical: superscripts: end groups; CTP: chain transfer to polymer; β: β-scission; ad: addition; pSR: SR propagation.

how a simple organic-chemistry reaction introduces such multiplicity by virtue of polymers being involved (e.g., also the different branch structures possible from the elementary CTP reaction). In the present instance, the richness is derived from the different tertiary radicals (TRs) delivered by backbiting ($n = 2$) versus inter-CTP; the reversible nature of β-scission – this is the same addition-fragmentation equilibrium that is at the heart of RAFT polymerization [100]; the fact that β-scission can occur on either side of the radical position in the TR; and the fact that β-scission introduces a different secondary-radical (SR) end group into the system, namely H. And so all this leads to Figure 10.27.

Still, the end result can be remarkably simple. For one thing, there are actually only two types of nonradical polymer in Figure 10.27: macromonomer capped by an initiator fragment (MM^A) and macromonomer capped by hydrogen (MM^H) (one hesitates to call MM a "dead" polymer, because the terminal double bond is obviously reactive). As a radical undergoes cycles of addition, CTP and β-scission, there can only be one MM^A molecule formed, as is clear from Figure 10.27 (and indeed from simple mass balance). Thus, MM^H must be the dominant product, which is why Junkers et al. described the chemistry of Figure 10.27 as a "macromonomer production machine" [99]. It is interesting to note that MM^H is isobaric with the product from chain transfer to monomer (compare Figures 10.22 and 10.27), and in some ways the two processes do resemble each other, as explained: similarly for MM^A and disproportionation product (Figures 10.16 and 10.27). Nevertheless, the structures are not identical, because for MM the two hydrogens at the unsaturation site are geminal, while for acrylate disproportionation products they are vicinal, because H must be abstracted from an α-methylene group in the chain backbone [99].

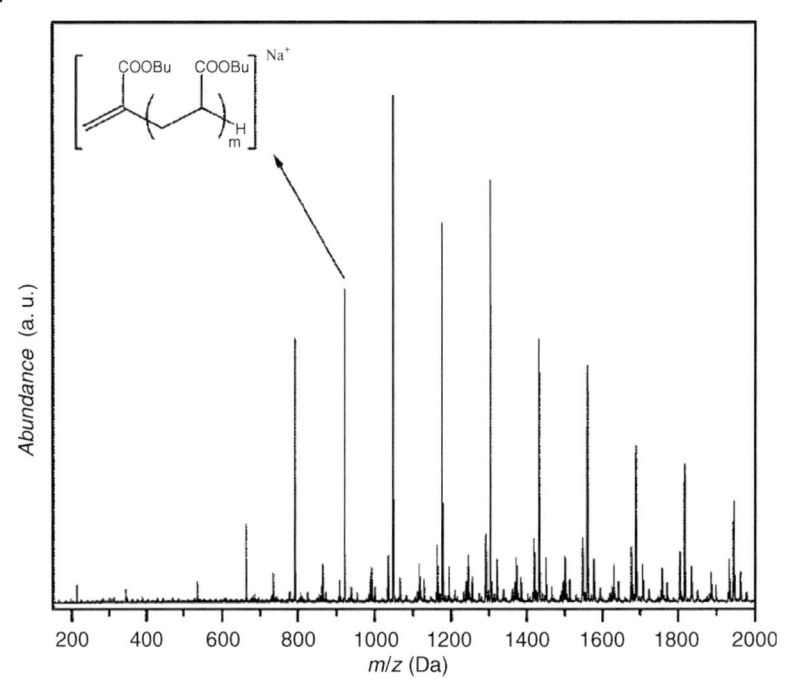

Figure 10.28 ESI-MS spectrum of poly(BA) from AIBN-initiated bulk polymerization at 100 °C in the presence of a low concentration of 1-octanethiol. (Reprinted with permission from Ref. [101]. Copyright 2009 American Chemical Society).

Figure 10.28 shows the MS of product from a high-temperature RP of BA [101]. As is shown, the dominant signals are those from MM^H. In this way, MS confirms the mechanism of Figure 10.27. This chemistry has therefore been used by the Barner–Kowollik group to build up a "macromonomer library" [102] for acrylates, all the time verifying the structure by MS [99, 102].

A very interesting result was obtained when 1-octanethiol (OctS-H), a TCTA, was used at very high concentration in BA polymerization: under conditions that otherwise give MM^H as the dominant product, instead this was B–(M)–H, to use the notation of Figure 10.20, where B = OctS is the chain-starting group from the thiol [101, 103]. A typical MS spectrum is presented in Figure 10.29. This result has been interpreted as meaning that the thiol "patches" a TR [101, 103]. In other words, in the language of Figure 10.27, the species TR^{BH} reacts with thiol, B–H, to produce B–(M)–H and B•. This TR is thereby "patched" by a hydrogen atom and the chain is prevented from becoming a macromonomer (via β-scission) or a branched polymer (via TR propagation). Instead, it is just a standard linear, saturated polymer.

The reasoning behind this conclusion seems to be as follows. Firstly, that B–(M)–H is linear rather than branched (the structures being isobaric) follows from β-scission being dominant in the absence of thiol (see Figure 10.28): adding a thiol will not change that $k_\beta > k_{pTR}c_M$, meaning that branched structures still do not form readily. The fact that MM^H formation is suppressed by the addition of thiol shows that a faster

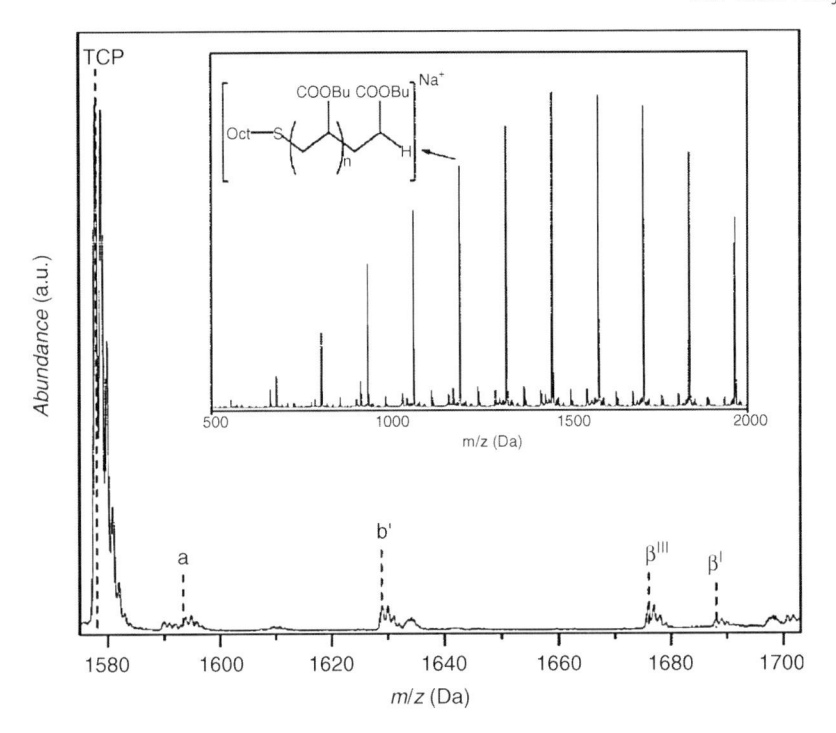

Figure 10.29 ESI-MS spectrum of poly(BA) from AIBN-initiated bulk polymerization at 100 °C in the presence of a very high concentration of 1-octanethiol ("Oct-S-H") [101]. The dominant series of peaks is from "thiol-capped polymer" (TCP) of the shown structure. (Reprinted with permission from Ref. [101]. Copyright 2009 American Chemical Society).

reaction must become available to TR species. The only such candidate is reaction of TR with thiol. Thus most of the reactions in Figure 10.27 are not accessed, and instead one gets TR^{AH} forming A–(M)–H and B•, which then forms TR^{BH} via secondary-radical propagation and CTP, which then forms B–(M)–H and B•, and so on. The fact that only small amounts of A–(M)–H are observed by MS (cluster b′ in Figure 10.29) proves that one has transfer-dominant conditions, that is, many cycles of transfer for each initiation event.

There are two loose ends in all this. One is that the formation of TRs is not proven: the same product distribution would be generated by chain-end radicals (SR^A and SR^B) reacting with thiol in the standard fashion, and TRs never forming simply because the thiol concentration is so high (e.g., $c_X = 0.4$ mol l^{-1} [101]). Secondly, it is not clear why $k_{trX,SR}/k_{pSR}$ should be low enough for polymer to grow, but at the same time $k_{trX,TR}/k_{pTR}$ should be so much higher that no addition of TR to monomer occurs. The fact that recent NMR experiments have shown that thiol reduces the degree of branching in poly(BA) systems [104] does not resolve either of these loose ends.

Where acrylate studies once were full of mysteries, now they are rapidly crystallizing into clarity [54], with MS playing a major role in this mechanistic advancement.

It has been made clear that uncertainties remain, for example, why did Farcet *et al.* [92] not observe macromonomer formation at temperatures where others have? Improved MS will no doubt be to the fore in tidying up these loose ends.

10.7
Emulsion Polymerization

Emulsion polymerization (EP) performs the seeming miracle of taking a water-insoluble reactant and converting it into a water-insoluble product, all in water. Because most monomers and polymers are hydrophobic, and because water is a highly desirable solvent, EP is therefore an important means of carrying out RP. In terms of reagents, the trick to the process is a very simple one: employ a water-soluble initiator. Thus, radicals are generated in the aqueous phase, and a crucially important aspect of the process is the passage of these radicals into the particles, which is where polymerization occurs [105]. It is therefore not surprising that MS has been used to try to reach a better understanding of this aspect of EP, termed *entry* [105]. In many ways these investigations could be classified as ones into initiation, but it seems prudent to give EP its own section rather than to include these works in Section 10.3. It needs to be stated at the outset that by and large these studies were not as successful, in the sense of giving mechanistically clear results, as most of those in this chapter. This is not surprising: chemistry in water is rarely simple, and there are obvious obstacles in trying to analyze molecules from one phase of a heterogeneous system. For these reasons, only a superficial treatment of MS-based investigations of EP will be given here.

Thomson *et al.* were the first on the scene [106]. They used MS to characterize the water-soluble MMA oligomers from EP at 80 °C initiated by ammonium persulfate. MALDI MS gave an average size of 8–9 MMA units, in reasonable agreement with that deduced from NMR. However, the MS spectrum was surprisingly crowded. A lot of these peaks could be identified as species with $A = {}^-OSO_3$ as one end group, as would be expected. However, there was considerable variety for the other end group, *not* including the anticipated ones of H (from disproportionation) or another $^-OSO_3$ (from combination). This suggests a lot of unexpected chemistry occurring in the aqueous phase and influencing the entry process.

Naturally Gilbert has been involved. In one study, he and De Bruyn used MS to investigate whether so-called induced decomposition of initiator (which is chain transfer to initiator, that is, initiator functioning also as a CTA) occurs with vinyl acetate and persulfate, as is widely reputed [107]. They concluded that this reaction does occur, but only at persulfate concentrations much higher than those typically employed in EP. In another study, Lamb *et al.* tackled the topic of radical entry mechanisms in redox-initiated EP [108]. Redox initiators are useful because they give functional entry rates at relatively low temperature. However, the way they produce radicals is something of a black box. Lamb *et al.* were able to shed some light on this for two redox couples, primarily by not finding hypothesized end groups in their MS spectra. At the same time, it did not seem possible to reach definite conclusions about what species did initiate their polymerizations.

In EP one is not limited to chemical initiators. Thus, Pusch and van Herk used MALDI-MS to probe how an electron beam initiates EP [109]. Not surprisingly, hydrogen and hydroxyl radicals – from interaction of electrons with water – were identified as the main initiating species. Also not surprisingly, other initiating species were found, for example monomeric radicals.

γ-initiation has been tremendously important in kinetic investigations into the mechanism of EP [105]. It is held to function in the same way as an electron beam, that is, via production of H• and •OH from interaction with water, and to a lesser extent M• from monomer. As yet this awaits investigation via MS.

10.8
Conclusion

Hart-Smith and Barner-Kowollik recently opined that MS remains of "untapped potential" in the polymer field [2]. They are correct! This may seem a strange claim to make given that it comes at the end of a long chapter on the use of MS. However, a little reflection reveals that this chapter has more been about what *can be done* with MS, not what commonly *is done* in order to study conventional RP mechanisms. For example, Hart-Smith and Barner-Kowollik quote that approximately 15% of papers on polyacrylates from the last decade used NMR for characterization, whereas this figure is only 3% for MS. This makes no sense given what has been exhibited in this chapter: MS is such a useful tool for learning about what takes place in an acrylate polymerization that one can almost argue it should be standard practice to look at the polymer with this technique.

This is not to say that MS provides all answers. For example, as has often been stressed here, MS cannot distinguish between two different mechanisms that yield polymer of the same M. However, the way in which MS reveals the precise M of all polymers in a sample makes it at least as useful a characterization tool as NMR, with the two often being complementary (for example, NMR can directly prove the presence of a particular functional group). Further, NMR has no capacity to deliver MMD, whereas MS already has ability in this regard, and it can be anticipated that this situation will considerably improve in the future. Thus, the utility of MS will continue to grow – an already extremely useful tool for elucidating the mechanisms of RP, conventional and otherwise, will become indispensable.

References

1 Gridnev, A.A., Ittel, S.D., and Fryd, M. (1995) A caveat when determining molecular weight distributions of methacrylate oligomers. *J. Polym. Sci., Polym. Chem. Ed.*, **33**, 1185–1188.

2 Hart-Smith, G. and Barner-Kowollik, C. (2010) Contemporary mass spectrometry and the analysis of synthetic polymers: trends, techniques and untapped potential. *Macromol. Chem. Phys.*, **211**, 1507–1529.

3 Olaj, O.F., Bitai, I., and Hinkelmann, F. (1987) The laser-flash-initiated polymerization as a tool of evaluating

(individual) kinetic constants of free-radical polymerization. 2. The direct determination of the rate constant of chain propagation. *Makromol. Chem.*, **188**, 1689–1702.
4 Buback, M., Gilbert, R.G., Hutchinson, R.A., Klumperman, B., Kuchta, F.-D., Manders, B.G., O'Driscoll, K.F., Russell, G.T., and Schweer, J. (1995) Critically evaluated rate coefficients for free-radical polymerization. 1. Propagation rate coefficient for styrene. *Macromol. Chem. Phys.*, **196**, 3267–3280.
5 van Herk, A.M. (2000) Pulsed initiation polymerization as a means of obtaining propagation rate coefficients in free-radical polymerizations. II. Review up to 2000. *Macromol. Theory Simul.*, **9** (8), 433–441.
6 Beuermann, S. and Buback, M. (2002) Rate coefficients of free-radical polymerization deduced from pulsed laser experiments. *Prog. Polym. Sci.*, **27**, 191–254.
7 Barner-Kowollik, C., Davis, T.P., and Stenzel, M.H. (2004) Probing mechanistic features of conventional, catalytic and living free radical polymerizations using soft ionization mass spectrometric techniques. *Polymer*, **45**, 7791–7805.
8 Maloney, D.R., Hunt, K.H., Lloyd, P.M., Muir, A.V.G., Richards, S.N., Derrick, P.J., and Haddleton, D.M. (1995) Polymethylmethacrylate end-group analysis by matrix-assisted laser desorption ionisation time-of-flight mass spectrometry (MALDI-TOF-MS). *J. Chem. Soc., Chem. Commun* (5), 561–562.
9 Ladavière, C., Lacroix-Desmazes, P., and Delolme, F. (2009) First systematic MALDI/ESI mass spectrometry comparison to characterize polystyrene synthesized by different controlled radical polymerizations. *Macromolecules*, **42**, 70–84.
10 Hart-Smith, G., Lammens, M., Du Prez, F.E., Guilhaus, M., and Barner-Kowollik, C. (2009) ATRP poly(acrylate) star formation: a comparative study between MALDI and ESI mass spectrometry. *Polymer*, **50**, 1986–2000.
11 Gruendling, T., Guilhaus, M., and Barner-Kowollik, C. (2008) Quantitative LC-MS of polymers: determining accurate molecular weight distributions by combined size exclusion chromatography and electrospray mass spectrometry with maximum entropy data processing. *Anal. Chem.*, **80** (18), 6915–6927.
12 Gruendling, T., Guilhaus, M., and Barner-Kowollik, C. (2009) Fast and accurate determination of absolute individual molecular weight distributions from mixtures of polymers via size exclusion chromatography-electrospray ionization mass spectrometry. *Macromolecules*, **42**, 6366–6374.
13 Schnöll-Bitai, I., Hrebicek, T., and Rizzi, A. (2007) Towards a quantitative interpretation of polymer distributions from MALDI-TOF spectra. *Macromol. Chem. Phys.*, **208**, 485–495.
14 van Herk, A.M. (2009) III. Heterogeneous polymerization. Modeling of emulsion polymerization, will it ever be possible? Part-2: determination of basic kinetic data over the last ten years. *Macromol. Symp.*, **275–276**, 120–132.
15 Buback, M., Frauendorf, H., and Vana, P. (2004) Initiation of free-radical polymerization by peroxypivalates studied by electrospray ionization mass spectrometry. *J. Polym. Sci., Polym. Chem. Ed.*, **42**, 4266–4275.
16 Nakamura, T., Busfield, W.K., Jenkins, I.D., Rizzardo, E., Thang, S.H., and Suyama, S. (1997) Free radical initiation mechanisms in the polymerization of methyl methacrylate and styrene with 1,1,3,3-tetramethylbutyl peroxypivalate: addition of neopentyl radicals. *J. Am. Chem. Soc.*, **119**, 10987–10991.
17 Nakamura, T., Watanabe, Y., Suyama, S., and Tezuka, H. (2002) Study of alkyl radicals fragmentation from 2-alkyl-2-propoxyl radicals. *J. Chem. Soc, Perkin Trans.*, **2** (7), 1364–1369.
18 Nakamura, T., Suyama, S., Busfield, W.K., Jenkins, I.D., Rizzardo, E., and Thang, S.H. (1999) Initiation mechanisms for radical polymerization of styrene and

methyl methacrylate with highly substituted peroxypivalate initiators. *Polymer*, **40** (6), 1395–1401.

19 Buback, M., Frauendorf, H., Günzler, F., and Vana, P. (2007) Initiation of radical polymerization by peroxyacetates: polymer end-group analysis by electrospray ionization mass spectrometry. *J. Polym. Sci., Polym. Chem. Ed.*, **45**, 2453–2467.

20 Buback, M., Frauendorf, H., Günzler, F., and Vana, P. (2007) Electrospray ionization mass spectrometric end-group analysis of PMMA produced by radical polymerization using diacyl peroxide initiators. *Polymer*, **48**, 5590–5598.

21 Hatada, K., Kitayama, T., Ute, K., Terawaki, Y., and Yanagida, T. (1997) End-group analysis of poly(methyl methacrylate) prepared with benzoyl peroxide by 750MHz high-resolution ^{1}H NMR spectroscopy. *Macromolecules*, **30** (22), 6754–6759.

22 Abel, B., Assmann, J., Botschwina, P., Buback, M., Kling, M., Oswald, R., Schmatz, S., Schroeder, J., and Witte, T. (2003) Experimental and theoretical investigations of the ultrafast photoinduced decomposition of organic peroxides in solution: formation and decarboxylation of benzoyloxy radicals. *J. Phys. Chem. A*, **107** (26), 5157–5167.

23 Buback, M., Kling, M., Schmatz, S., and Schroeder, J. (2004) Photo-induced decomposition of organic peroxides: ultrafast formation and decarboxylation of carbonyloxy radicals. *Phys. Chem. Chem. Phys.*, **6** (24), 5441–5455.

24 Buback, M., Frauendorf, H., Janssen, O., and Vana, P. (2008) Electrospray ionization mass spectrometric study of end groups in peroxydicarbonate-initiated radical polymerization. *J. Polym. Sci., Polym. Chem. Ed.*, **46**, 6071–6081.

25 Mekarbane, P.G. and Tabner, B.J. (2000) An EPR spin-trap study of the radicals present during the thermolysis of some novel monoperoxycarbonates and peroxydicarbonates. *J. Chem. Soc., J. Chem. Soc, Perkin Trans.* **2** (7), 1465–1470.

26 Buback, M., Kowollik, C., Kurz, C., and Wahl, A. (2000) Termination kinetics of styrene free-radical polymerization studied by time-resolved pulsed laser experiments. *Macromol. Chem. Phys.*, **201**, 464–469.

27 Buback, M., Busch, M., and Kowollik, C. (2000) Chain-length dependence of free-radical termination rate deduced from laser single-pulse experiments. *Macromol. Theory Simul.*, **9**, 442–452.

28 Buback, M. and Kuelpmann, A. (2003) A suitable photoinitiator for pulsed laser-induced free-radical polymerization. *Macromol. Chem. Phys.*, **204** (4), 632–637.

29 Vana, P., Davis, T.P., and Barner-Kowollik, C. (2002) End-group analysis of polymers by electrospray ionization mass spectrometry: 2-methyl-1-[4-(methylthio) phenyl]-2-morpholinopropan-1-one initiated free-radical photopolymerization. *Aust. J. Chem.*, **55**, 315–318.

30 Barner-Kowollik, C., Vana, P., and Davis, T.P. (2002) Laser-induced decomposition of 2,2-dimethoxy-2-phenylacetophenone and benzoin in methyl methacrylate homopolymerization studied via matrix-assisted laser desorption/ionization time-of-flight mass spectrometry. *J. Polym. Sci., Polym. Chem. Ed.*, **40**, 675–681.

31 Szablan, Z., Lovestead, T.M., Davis, T.P., Stenzel, M.H., and Barner-Kowollik, C. (2007) Mapping free radical reactivity: a high-resolution electrospray ionization-mass spectrometry study of photoinitiation processes in methyl methacrylate free radical polymerization. *Macromolecules*, **40**, 26–39.

32 Szablan, Z., Junkers, T., Koo, S.P.S., Lovestead, T.M., Davis, T.P., Stenzel, M.H., and Barner-Kowollik, C. (2007) Mapping photolysis product radical reactivities via soft ionization mass spectrometry in acrylate, methacrylate, and itaconate systems. *Macromolecules*, **40**, 6820–6833.

33 Günzler, F., Wong, E.H.H., Koo, S.P.S., Junkers, T., and Barner-Kowollik, C. (2009) Quantifying the efficiency of photoinitiation processes in methyl methacrylate free radical polymerization via electrospray ionization mass spectrometry. *Macromolecules*, **42**, 1488–1493.

34 Wyzgoski, F.J., Polci, M.J., Wesdemiotis, C., and Arnould, M.A. (2007) Matrix-assisted laser desorption/ionization time-of-flight mass spectrometry investigations of polystyrene and poly(methyl methacrylate) produced by monoacylphosphine oxide photoinitiation. *J. Polym. Sci., Polym. Chem. Ed.*, **45**, 2161–2171.

35 Mayo, F.R. (1953) Chain transfer in the polymerization of styrene. VIII. Chain transfer with bromobenzene and mechanism of thermal initiation. *J. Am. Chem. Soc.*, **75** (24), 6133–6141.

36 Khuong, K.S., Jones, W.H., Pryor, W.A., and Houk, K.N. (2005) The mechanism of the self-initiated thermal polymerization of styrene. Theoretical solution of a classic problem. *J. Am. Chem. Soc.*, **127**, 1265–1277.

37 Dourges, M.-A., Charleux, B., Vairon, J.-P., Blais, J.-C., Bolbach, G., and Tabet, J.-C. (1999) MALDI-TOF mass spectrometry analysis of tempo-capped polystyrene. *Macromolecules*, **32**, 2495–2502.

38 Stickler, M. and Meyerhoff, G. (1978) Die thermische Polymerisation von Methylmethacrylat. 1. Polymerisation in Substanz. *Makromol. Chem.*, **179**, 2729–2745.

39 Srinivasan, S., Lee, M.W., Grady, M.C., Soroush, M., and Rappe, A.M. (2009) Computational study of the self-initiation mechanism in thermal polymerization of methyl acrylate. *J. Phys. Chem. A*, **113**, 10787–10794.

40 Hart-Smith, G., Lovestead, T.M., Davis, T.P., Stenzel, M.H., and Barner-Kowollik, C. (2007) Mapping formation pathways and end group patterns of stimuli-responsive polymer systems via high-resolution electrospray ionization mass spectrometry. *Biomacromolecules*, **8**, 2404–2415.

41 Lovestead, T.M., Hart-Smith, G., Davis, T.P., Stenzel, M.H., and Barner-Kowollik, C. (2007) Electrospray ionization mass spectrometry investigation of reversible addition fragmentation chain transfer mediated acrylate polymerizations initiated via ^{60}Co χ-irradiation: mapping reaction pathways. *Macromolecules*, **40**, 4142–4153.

42 Wang, W. and Hutchinson, R.A. (2009) Evidence of scission products from peroxide-initiated higher temperature polymerization of alkyl methacrylates. *Macromolecules*, **42**, 4910–4913.

43 Buback, M., Huckestein, B., Kuchta, F.-D., Russell, G.T., and Schmid, E. (1994) Initiator efficiencies in 2,2′-azoisobutyronitrile-initiated free-radical polymerizations of styrene. *Macromol. Chem. Phys.*, **195**, 2117–2140.

44 Buback, M., Frauendorf, H., Günzler, F., Huff, F., and Vana, P. (2009) Determining initiator efficiency in radical polymerization by electrospray-ionization mass spectrometry. *Macromol. Chem. Phys.*, **210**, 1591–1599.

45 Olaj, O.F., Bitai, I., and Hinkelmann, F. (1987) The laser-flash-initiated polymerization as a tool of evaluating (individual) kinetic constants of free-radical polymerization. 2. The direct determination of the rate constant of chain propagation. *Makromol. Chem.*, **188**, 1689–1702.

46 Buback, M., Gilbert, R.G., Hutchinson, R.A., Klumperman, B., Kuchta, F.-D., Manders, B.G., O'Driscoll, K.F., Russell, G.T., and Schweer, J. (1995) Critically evaluated rate coefficients for free-radical polymerization. 1. Propagation rate coefficient for styrene. *Macromol. Chem. Phys.*, **196**, 3267–3280.

47 Danis, P.O., Karr, D.E., Westmoreland, D.G., Piton, M.C., Christie, D.I., Clay, P.A., Kable, S.H., and Gilbert, R.G. (1993) Measurement of propagation rate coefficients using pulsed laser polymerization and matrix-assisted laser desorption/ionization (MALDI) mass spectrometry. *Macromolecules*, **26**, 6684–6685.

48 Zammit, M.D., Davis, T.P., and Haddleton, D.M. (1996) Determination of the propagation rate coefficient (k_p) and termination mode in the free-radical polymerization of methyl methacrylate, employing matrix assisted laser desorption ionization time-of-flight mass spectrometry for molecular

weight distribution analysis. *Macromolecules*, **29**, 492–494.

49 Willemse, R.X.E., Staal, B.B.P., van Herk, A.M., Pierik, S.C.J., and Klumperman, B. (2003) Application of matrix-assisted laser desorption ionization time-of-flight mass spectrometry in pulsed laser polymerization. Chain-length-dependent propagation rate coefficients at high molecular weight: an artifact caused by band broadening in size exclusion chromatography? *Macromolecules*, **36**, 9797–9803.

50 Beuermann, S., Buback, M., Davis, T.P., Gilbert, R.G., Hutchinson, R.A., Olaj, O.F., Russell, G.T., Schweer, J., and van Herk, A.M. (1997) Critically evaluated rate coefficients for free-radical polymerization. 2. Propagation rate coefficients for methyl methacrylate. *Macromol. Chem. Phys.*, **198**, 1545–1560.

51 Schweer, J., Sarnecki, J., Mayer-Posner, F., Müllen, K., Räder, H.J., and Spickermann, J. (1996) Pulsed-laser polymerization/matrix-assisted laser desorption/ionization mass spectrometry: an approach toward free-radical propagation rate coefficients of ultimate accuracy? *Macromolecules*, **29**, 4536–4543.

52 Sarnecki, J. and Schweer, J. (1995) Conditions for the determination of precise and accurate free-radical propagation rate coefficients from pulsed-laser-made polymer. *Macromolecules*, **28**, 4080–4088.

53 Willemse, R.X.E. and van Herk, A.M. (2010) Determination of propagation rate coefficients of a family of acrylates with PLP-MALDI-ToF-MS. *Macromol. Chem. Phys.*, **211**, 539–545.

54 Junkers, T. and Barner-Kowollik, C. (2008) The role of mid-chain radicals in acrylate free radical polymerization: branching and scission. *J. Polym. Sci., Polym. Chem. Ed.*, **46**, 7585–7605.

55 Beuermann, S., Buback, M., Davis, T.P., Gilbert, R.G., Hutchinson, R.A., Kajiwara, A., Klumperman, B., and Russell, G.T. (2000) Critically evaluated rate coefficients for free-radical polymerization. 3. Propagation rate coefficients for alkyl methacrylates. *Macromol. Chem. Phys.*, **201**, 1355–1364.

56 Gruendling, T., Weidner, S., Falkenhagen, J., and Barner-Kowollik, C. (2010) Mass spectrometry in polymer chemistry: a state-of-the-art up-date. *Polym. Chem.*, **1** (5), 599–617.

57 Gruendling, T., Voll, D., Guilhaus, M., and Barner-Kowollik, C. (2010) A perfect couple: PLP/SEC/ESI-MS for the accurate determination of propagation rate coefficients in free radical polymerization. *Macromol. Chem. Phys.*, **211**, 80–90.

58 Heuts, J.P.A. and Russell, G.T. (2006) The nature of the chain-length dependence of the propagation rate coefficient and its effect on the kinetics of free-radical polymerization. 1. Small-molecule studies. *Eur. Polym. J.*, **42** (1), 3–20.

59 Deady, M., Mau, A.W.H., Moad, G., and Spurling, T.H. (1993) Evaluation of the kinetic parameters for styrene polymerization and their chain length dependence by kinetic simulation and pulsed laser photolysis. *Makromol. Chem.*, **194**, 1691–1705.

60 Olaj, O.F., Vana, P., Zoder, M., Kornherr, A., and Zifferer, G. (2000) Is the rate constant of chain propagation k_p in radical polymerization really chain-length independent? *Macromol. Rapid Commun.*, **21**, 913–920.

61 Smith, G.B., Russell, G.T., Yin, M., and Heuts, J.P.A. (2005) The effects of chain length dependent propagation and termination on the kinetics of free-radical polymerization at low chain lengths. *Eur. Polym. J.*, **41** (2), 225–230.

62 Zetterlund, P.B., Busfield, W.K., and Jenkins, I.D. (1999) Free radical polymerization of acrylonitrile: mass spectrometric identification of the nitroxide-trapped oligomers formed in and estimated rate constants for each of the first eight propagation steps. *Macromolecules*, **32**, 8041–8045.

63 Zetterlund, P.B., Busfield, W.K., and Jenkins, I.D. (2002) Free radical polymerization of styrene: mass spectrometric identification of the first 15 nitroxide-trapped oligomers and

estimated propagation rate coefficients. *Macromolecules*, **35**, 7232–7237.

64 Ewing, K., Busfield, W.K., Jenkins, I.D., and Zetterlund, P.B. (2003) Early propagation kinetics in the free radical polymerization of methacrylonitrile investigated by a nitroxide trapping technique. *Polym. Int.*, **52** (11), 1671–1675.

65 Shi, S.D.-H., Hendrickson, C.L., Marshall, A.G., Simonsick, W.J. Jr., and Aaserud, D.J. (1998) Identification, composition, and asymmetric formation mechanism of glycidyl methacrylate/butyl methacrylate copolymers up to 7000 Da from electrospray ionization ultrahigh-resolution Fourier transform ion cyclotron resonance mass spectrometry. *Anal. Chem.*, **70**, 3220–3226.

66 Haddleton, D.M., Feeney, E., Buzy, A., Jasieczek, C.B., and Jennings, K.R. (1996) Electrospray ionisation mass spectrometry (ESI MS) of poly(methy methacrylate) and acrylic statistical copolymers. *Chem. Comm.*, (10), 1157–1158.

67 Kukulj, D., Davis, T.P., Suddaby, K.G., Haddleton, D.M., and Gilbert, R.G. (1997) Catalytic chain transfer for molecular weight control in the emulsion homo- and copolymerizations of methyl methacrylate and butyl methacrylate. *J. Polym. Sci., Polym. Chem. Ed.*, **35** (5), 859–878.

68 Suddaby, K.G., Hunt, K.H., and Haddleton, D.M. (1996) MALDI-TOF mass spectrometry in the study of statistical copolymerizations and its application in examining the free radical copolymerization of methyl methacrylate and *n*-butyl methacrylate. *Macromolecules*, **29** (27), 8642–8649.

69 Willemse, R.X.E., Staal, B.B.P., Donkers, E.H.D., and van Herk, A.M. (2004) Copolymer fingerprints of polystyrene-block-polyisoprene by MALDI-ToF-MS. *Macromolecules*, **37**, 5717–5723.

70 Willemse, R.X.E., Ming, W., and van Herk, A.M. (2005) Solventless liquid oligoesters analyzed by MALDI-ToF-MS. *Macromolecules*, **38**, 6876–6881.

71 Willemse, R.X.E. and van Herk, A.M. (2006) Copolymerization kinetics of methyl methacrylate–styrene obtained by PLP-MALDI-ToF-MS. *J. Am. Chem. Soc.*, **128** (13), 4471–4480.

72 Maxwell, I.A., Aerdts, A.M., and German, A.L. (1993) Free radical copolymerization: an NMR investigation of current kinetic models. *Macromolecules*, **26**, 1956–1964.

73 Buback, M., Günzler, F., Russell, G.T., and Vana, P. (2009) Determination of the mode of termination in radical polymerization via mass spectrometry. *Macromolecules*, **42** (3), 652–662.

74 Moad, G. and Solomon, D.H. (2006) *The Chemistry of Radical Polymerization*, Elsevier, Oxford, UK.

75 Zammit, M.D., Davis, T.P., Haddleton, D.M., and Suddaby, K.G. (1997) Evaluation of the mode of termination for a thermally-initiated free-radical polymerization via matrix-assisted-laser-desorption-ionization time-of-flight mass spectrometry. *Macromolecules*, **30**, 1915–1920.

76 Barson, C.A. (1989) Chain transfer, in *Comprehensive Polymer Science: The Synthesis, Characterization, Reactions and Applications of Polymers*, vol. **3** (eds G.A. Allen and J.C. Bevington), Pergamon, Oxford, pp. 171–183.

77 Kapfenstein, H.M. and Davis, T.P. (1998) Studies on the application of matrix-assisted-laser-desorptionionisation time-of-flight mass spectrometry to the determination of chain transfer coefficients in free radical polymerization. *Macromol. Chem. Phys.*, **199**, 2403–2408.

78 Smith, G.B. and Russell, G.T. (2007) The cutthroat competition between termination and transfer to shape the kinetics of radical polymerization. *Macromol. Symp.*, **248**, 1–11.

79 Lu, Z.-R., Kopečková, P., Wu, Z., and Kopeček, J. (1998) Functionalized semitelechelic poly[*n*-(2-hydroxypropyl) methacrylamide] for protein modification. *Bioconjugate Chem.*, **9**, 793–804.

80 Lu, Z.-R., Kopečková, P., Wu, Z., and Kopeček, J. (1999) Synthesis of

semitelechelic poly[*n*-(2-hydroxypropyl) methacrylamide] by radical polymerization in the presence of alkyl mercaptans. *Macromol. Chem. Phys.*, **200** 2022–2030.

81 Stumbé, J.F., Limal, D., Kessler, M., and Riess, G. (2001) Kinetic study of the oligomerization of styrene in the presence of α-thioglycerol – comparison between batch and semi-batch processes. *Eur. Polym. J.*, **37**, 1519–1526.

82 Liu, P., Ding, H., Liu, J., and Xiaosu, Y. (2002) Synthesis of telechelic methacrylate oligomers with carboxyl ends through radical polymerization–chemical modification. *Eur. Polym. J*, **38** 1783–1789.

83 Lonsdale, D.E., Johnston-Hall, G., Fawcett, A., Bell, C.A., Urbani, C.N., Whittaker, M.R., and Monteiro, M.J. (2007) Degradative chain transfer in vinyl acetate polymerizations using toluene as solvent. *J. Polym. Sci., Polym. Chem. Ed.*, **45**, 3620–3625.

84 Liu, Z. and Rimmer, S. (2002) Studies on the free radical polymerization of *N*-vinylpyrrolidinone in 3-methylbutan-2-one. *Macromolecules*, **35**, 1200–1207.

85 Quan, C., Soroush, M., Grady, M.C., Hansen, J.E., and Simonsick, W.J.Jr. (2005) High-temperature homopolymerization of ethyl acrylate and *n*-butyl acrylate: polymer characterization. *Macromolecules*, **38** (18), 7619–7628.

86 Tobolsky, A.V. and Offenbach, J. (1955) Kinetic constants for styrene polymerization. *J. Polym. Sci.*, **XV**, 311–314.

87 Kukulj, D., Davis, T.P., and Gilbert, R.G. (1998) Chain transfer to monomer in the free-radical polymerizations of methyl methacrylate, styrene and α-methylstyrene. *Macromolecules*, **31**, 994–999.

88 Rudin, A. (1982) *The Elements of Polymer Science and Engineering*, Academic Press, Orlando and London.

89 Bingöl, B., Hart-Smith, G., Barner-Kowollik, C., and Wegner, G. (2008) Characterization of oligo(vinyl phosphonate)s by high-resolution electrospray ionization mass spectrometry: implications for the mechanism of polymerization. *Macromolecules*, **41**, 1634–1639.

90 Heuts, J.P.A., Roberts, G.E., and Biasutti, J.D. (2002) Catalytic chain transfer polymerization: an overview. *Aust. J. Chem.*, **55**, 381–398.

91 Ahmad, N.M., Heatley, F., and Lovell, P.A. (1998) Chain transfer to polymer in free-radical solution polymerization of *n*-butyl acrylate studied by NMR spectroscopy. *Macromolecules*, **31**, 2822–2827.

92 Farcet, C., Belleney, J., Charleux, B., and Pirri, R. (2002) Structural characterization of nitroxide-terminated poly(*n*-butyl acrylate) prepared in bulk and miniemulsion polymerizations. *Macromolecules*, **35**, 4912–4918.

93 Hawker, C.J., Bosman, A.W., and Harth, E. (2001) New polymer synthesis by nitroxide mediated living radical polymerizations. *Chem. Rev.*, **101** 3661–3688.

94 Nikitin, A.N. and Hutchinson, R.A. (2005) The effect of intramolecular transfer to polymer on stationary free radical polymerization of alkyl acrylates. *Macromolecules*, **38**, 1581–1590.

95 Barth, J., Buback, M., Hesse, P., and Sergeeva, T. (2010) Termination and transfer kinetics of butyl acrylate radical polymerization studied via SP–PLP–EPR. *Macromolecules*, **43**, 4023–4031.

96 Chiefari, J., Jeffery, J., Mayadunne, R.T.A., Moad, G., Rizzardo, E., and Thang, S.H. (1999) Chain transfer to polymer: a convenient route to macromonomers. *Macromolecules*, **32**, 7700–7702.

97 Grady, M.C., Simonsick, W.J., and Hutchinson, R.A. (2002) Studies of higher temperature polymerization of *n*-butyl methacrylate and *n*-butyl acrylate. *Macromol. Symp.*, **182**, 149–168.

98 Kumagai, T., Kagawa, C., Aota, H., Takeda, Y., Kawasaki, H., Arakawa, R., and Matsumoto, A. (2008) Specific polymerization mechanism involving β-scission of mid-chain radical yielding oligomers in the free-radical

polymerization of vinyl ethers. *Macromolecules*, **41**, 7347–7351.

99 Junkers, T., Bennet, F., Koo, S.P.S., and Barner-Kowollik, C. (2008) Self-directed formation of uniform unsaturated macromolecules from acrylate monomers at high temperatures. *J. Polym. Sci., Polym. Chem. Ed.*, **46**, 3433–3437.

100 Barner-Kowollik, C. (ed.) (2008) *Handbook of RAFT Polymerization*, Wiley-VCH, Weinheim.

101 Koo, S.P.S., Junkers, T., and Barner-Kowollik, C. (2009) Quantitative product spectrum analysis of poly(butyl acrylate) via electrospray ionization mass spectrometry. *Macromolecules*, **42**, 62–69.

102 Zorn, A.-M., Junkers, T., and Barner-Kowollik, C. (2009) Synthesis of a macromonomer library from high-temperature acrylate polymerization. *Macromol. Rapid Commun*, **30**, 2028–2035.

103 Junkers, T., Koo, S.P.S., Davis, T.P., Stenzel, M.H., and Barner-Kowollik, C. (2007) Mapping poly(butyl acrylate) product distributions by mass spectrometry in a wide temperature range: suppression of midchain radical side reactions. *Macromolecules*, **40**, 8906–8912.

104 Gaborieau, M., Koo, S.P.S., Castignolles, P., Junkers, T., and Barner-Kowollik, C. (2010) Reducing the degree of branching in polyacrylates via midchain radical patching: a quantitative melt-state NMR study. *Macromolecules*, **43**, 5492–5495.

105 Gilbert, R.G. (1995) *Emulsion Polymerization: A Mechanistic Approach*, Academic Press, London.

106 Thomson, B., Wang, Z., Paine, A., Lajoie, G., and Rudin, A. (1995) A mass spectrometric investigation of the water-soluble oligomers remaining after the emulsion polymerization of methyl methacrylate. *J. Polym. Sci., Polym. Chem. Ed.*, **33**, 2297–2304.

107 De Bruyn, H. and Gilbert, R.G. (2001) Induced decomposition of persulfate by vinyl acetate. *Polymer*, **42**, 7999–8005.

108 Lamb, D.J., Fellows, C.M., and Gilbert, R.G. (2005) Radical entry mechanisms in redox-initiated emulsion polymerizations. *Polymer*, **46**, 7874–7895.

109 Pusch, J. and van Herk, A.M. (2005) Pulsed electron beam initiation in emulsion polymerization. *Macromolecules*, **38**, 8694–8700.

11
Elucidation of Reaction Mechanisms and Polymer Structure: Living/Controlled Radical Polymerization

Christopher Barner-Kowollik, Guillaume Delaittre, Till Gründling, and Thomas Paulöhrl

The current chapter will in depth collate, describe, and critically evaluate studies in which soft-ionization mass spectrometry (MS) techniques have been employed to study polymeric materials generated through processes where control is exerted over the radical polymerization process via the addition of a specific agent. The mass spectrometric investigation of polymers prepared via free radical polymerization in the absence of controlling agents is the focus of the Chapter 10. However, one exception is made: the regulation of the molecular weight via the addition of conventional chain transfer agents (such as thiols) is also included in the previous chapter.

It seems apt in the context of the current chapter to distinguish – if only for descriptive purposes – between *radical polymerizations with living characteristics* (i.e., those where there is a linear relationship between the monomer conversion and the chain length of the individual macromolecules, recently been given the IUPAC label of "reversible deactivation radical polymerization") [1] and *controlled radical polymerizations*, where the addition of a reagent merely regulates the molecular weight as a function of its concentration yielding broad distributions, yet does not introduce a linear growth of each macromolecule with conversion under retention of the end-group functionality. Note that there is no strict or even IUPAC-recommended definition of *controlled polymerization*.

Living and controlled radical polymerizations (following the above definition) have been a constant topic of mass spectrometric investigations, particularly protocols that induce living characteristics via reversible activation/deactivation mechanisms. Control is achieved either by reversible termination or reversible chain transfer, especially in nitroxide-mediated polymerization (NMP), atom transfer radical polymerization (ATRP), and reversible addition-fragmentation chain transfer (RAFT) polymerization (Scheme 11.1). In addition, soft-ionization MS has been employed to map the products of several minor (or not as prominent) living radical protocols as well as mechanisms only inducing molecular weight control and no living character onto the polymerization such as catalytic chain transfer (CCT) or enhanced spin-capturing polymerization (ESCP). All of the above polymerization techniques have

Mass Spectrometry in Polymer Chemistry, First Edition.
Edited by Christopher Barner-Kowollik, Till Gruendling, Jana Falkenhagen, and Steffen Weidner.
© 2012 Wiley-VCH Verlag GmbH & Co. KGaA. Published 2012 by Wiley-VCH Verlag GmbH & Co. KGaA.

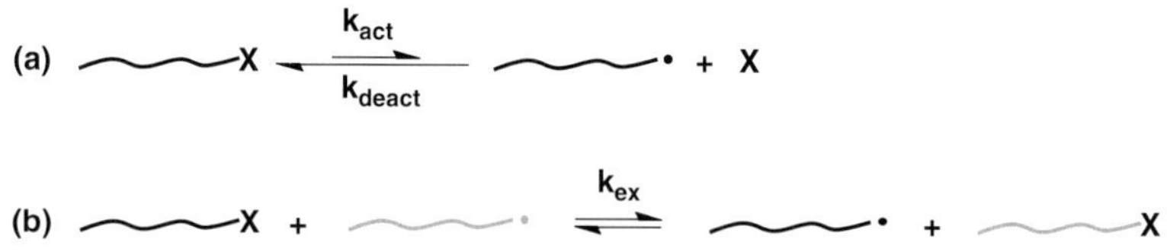

(a) $\sim\sim\sim\text{X} \underset{k_{deact}}{\overset{k_{act}}{\rightleftarrows}} \sim\sim\sim\cdot + \text{X}$

(b) $\sim\sim\sim\text{X} + \sim\sim\sim\cdot \overset{k_{ex}}{\rightleftarrows} \sim\sim\sim\cdot + \sim\sim\sim\text{X}$

Scheme 11.1 Simplified mechanisms of controlled/living radical polymerization based on either (a) reversible termination or (b) reversible chain transfer.

been extensively subjected to matrix-assisted laser desorption/ionization-time of flight (MALDI-TOF) as well as size exclusion chromatography electrospray ionization mass spectrometry ((SEC/)ESI-MS) analysis to assess end-group fidelity as well as to clarify complex mechanistic questions. Since our last assessments of the state-of-the-art on the application of soft-ionization MS toward polymers were prepared via living/controlled radical processes [2, 3], research in this area has been steady. It is notable that the investigations have advanced to more complex systems (such as stars, see below). Importantly, two systematic studies were carried out by Barner-Kowollik, Du Prez and colleagues as well as Ladavière and coworkers, who assessed the differences in the results obtained by different ionization protocols (MALDI vs. ESI) for polymers prepared via NMP, ATRP as well as RAFT [4, 5]. Such comparative assessments are highly important – and should be carried out for further systems (whether living radical or controlled radical), as the two ionization protocols can yield complementary information.

11.1
Protocols Based on a Persistent Radical Effect (NMP, ATRP, and Related)

It seems appropriate to open the exploration of protocols that induce living characteristics onto a free radical polymerization via a persistent radical effect with the two above-mentioned notable comparative ESI and MALDI studies [3, 4]. Both studies provide significant evidence that ESI-MS methodologies can identify a greater number of species than MALDI techniques. For example, during the analysis of polymers generated via a typical ATRP, ESI-MS was able to identify 46 species more than MALDI-TOF, whereas MALDI-TOF only identified 6 species which did not appear in the ESI-MS spectra. The number of commonly identified species was 18. It is very important to note that there exists significant evidence that polymers carrying labile end groups, that is, those typically obtained via living radical processes, can be subject to considerable fragmentation during MALDI. Thus, great care is advised when such polymers are analyzed via MALDI-MS techniques, especially when mechanistic questions are to be discussed.

Our first survey into the use of soft-ionization mass spectrometry is concerned with polymers that have been generated via protocols based on a persistent radical effect with NMP [6], which is one of the earliest living/controlled protocols. Jasieczek et al. [7] were among the first investigators to probe – via MALDI-TOF, ESI-MS, and liquid secondary-ion mass spectrometry (L-SIMS) – polystyrenes prepared via TEMPO (2,2,6,6-tetramethylpiperidine)-mediated living radical polymerization. Employing a thermally decomposing initiator (dibenzoyl peroxide) in the presence of TEMPO several ion distributions were detected, of which only two could be assigned with certainty to polystyrene chains capped by benzoyloxy fragments on both chain termini. In addition, a population of chains terminated by a benzoyloxy fragment and an unsaturated styryl moiety were identified. As noted in the comparative MALDI-TOF-MS and ESI-MS studies discussed above [4, 5], ESI-MS is often capable of imaging a greater variety of structures than MALDI-TOF-MS. Thus, it comes at little surprise that no polymer chains carrying TEMPO-based alkoxyamine end-groups were detected via MALDI-TOF-MS, while ESI-MS and L-SIMS techniques were able to detect TEMPO-capped chains. It remains unclear why certain structures preferentially ionize in a given protocol, yet the above study provides further support to the fact that ESI-MS can detect more structures and leads to less (potential) polymer degradation processes as MALDI. An additional early investigation into nitroxide-mediated polymerization was provided by Dourges et al. – also employing MALDI-TOF-MS – who were able to map TEMPO-end-capped macromolecules by tuning the matrix and salt system [8]. Similar to Jasieczek et al., these authors also found ample evidence for end-group fragmentation during the ionization process (or in the gas phase). Importantly, the obtained fragmentation patterns were strongly dependent on the choice of counterion and matrix. Nevertheless, these early studies provided additional evidence that a reversible activation/deactivation mechanism is operative in NMP. Expanding the range of NMP-generated polymers under investigation, Schmidt-Naake and colleagues subjected chlorine-terminated and TEMPO-capped polystyrenes to MALDI-TOF-MS [9]. These authors showed that under conditions of self-protonation, chlorine-, acrylate- and amine-end-functionalized as well as mono- and bis-TEMPO-capped polystyrenes can be readily imaged via MALDI-TOF-MS. In subsequent studies on NMP-prepared polystyrenes, the same authors expanded their analysis to variable nitroxide-capped polystyrenes and investigated their behavior in MALDI-TOF experiments [10, 11]. Ionization conditions were identified where TEMPO-capped polymers could be ionized without significant fragmentation, while polymers capped by 2,2,5-trimethyl-4-phenyl-3-azahexane-3-oxyl (TIPNO) and 2,2,5-trimethyl-4-(isopropyl)-3-azahexane-3-oxyl (BIPNO) displayed a fragmentation of the N-oxyl group, evidenced by the loss of a tert-butyl or a phenylisobutyl group in the case of these α-hydrogen nitroxides. In an extension on the studies of NMP of styrene, several authors studied block copolymers prepared via this process. For example, Burguière et al. have analyzed block copolymers of n-butyl acrylate (BA) and styrene prepared via NMP [12]. These authors reacted BA at elevated temperature ($T = 130\,°C$) in the presence of TEMPO, initiated by either a low-molecular-weight alkoxyamine or a TEMPO-capped polystyrene. The MALDI-TOF-MS results indicated that the initiating radical fragment could always be identified at the chain termini. In addition, it was demonstrated that a block copolymer was produced with styrene. Interestingly, the opposing chain-end functionality was not the expected alkoxyamine, but rather a vinyl function. The authors hypothesized that the main chain-stopping event is likely to be a β-hydrogen transfer from a propagating radical to free TEMPO radicals, that is, an undesired side reaction has occurred. Nevertheless, it can also not be excluded that a fragmentation during the ionization process may occur. The above investigation demonstrates that molecular information obtained via mass spectrometry can be readily employed to improve the reaction conditions: To avoid the elimination reaction at the chain termini (a rather stable alkoxyamine ω-end-group is required) the controlling parameter that needed to be changed was the chemical structure of the nitroxide. Such a structural change should decrease the rate coefficient for disproportionation relative to the rate coefficient governing the recombination of propagating radicals with TEMPO radicals. In a similar approach to block copolymers, Baumann et al. synthesized polystyrene-block-poly(styrene-co-acrylonitrile) copolymers by chain extension of TEMPO-capped polystyrene [13]. These authors subsequently investigated the thermal stability of the resulting block copolymers by pyrolysis gas chromatography coupled with MS, indicating that up to a temperature of 200 °C the carbon–oxygen bond between the TEMPO end group and the main polymer chain was stable. More recently, it has been demonstrated that hyphenated techniques, where chromatographic separations are coupled with a mass analyzer as detector, can be employed to obtain detailed structural information on block copolymers prepared via NMP. For example, Charles and colleagues coupled liquid adsorption chromatography under critical conditions (LACCCs) with ESI-MS to characterize poly(ethylene oxide)/polystyrene block copolymers [14]. Via additional MS/MS experiments, the copolymer structure was successfully elucidated. In related work on poly(ethylene oxide)-b-poly(styrene) copolymers by Phan and colleagues, both ESI-MS and MALDI-TOF were employed to characterize in depth the difunctional poly(ethylene oxide) precursors [15], which were based on an alkoxyamine derived from the well-known SG1 mediator (N-tert-butyl-N-(1-diethylphosphono-2,2-dimethylpropyl nitroxide)). These authors confirmed the purity of the macroinitiators via the above techniques, clearly observing the desired main species. The study also demonstrated that at low-molecular-weight ESI-MS-based approaches allow for the elucidation of finer details of the polymeric materials, even up to molecular weights higher than 2000 Da under exploitation of the occurrence of higher charge states (remember that mass spectrometry measures mass-to-charge ratio, m/z). MALDI-TOF spectra of the poly(ethylene oxide) precursors were of a poorer signal-to-noise quality, but still allowed the observation of singly charged species at these masses.

Apart from polystyrene, a wide range of other homopolymers is also accessible via NMP and some of the prepared structures were analyzed by both MALDI-TOF and ESI-MS. The polymerization of BA mediated by SG1 [16] produced polymer for which the mass spectrometric analysis demonstrated that the overwhelming majority of the chains carry an initiator fragment at one end and the nitroxide on the other, confirming the suitability of SG1 nitroxide to induce strong living characteristics onto BA radical polymerizations. A further example of the analysis of a nonstyrenic homopolymer by Dire et al. [17] indicated that all chains of a poly(methyl methacrylate) (PMMA) prepared in the presence of a large excess of free SG1 were terminated by an alkene function. This implied that β-hydrogen transfer from propagating radicals to the nitroxide was the predominant chain-stopping event.

However, at low SG1 concentrations the authors could demonstrate that two termination modes existed. Studer and coworkers investigated the 2,2,6,6-tetraethylpiperidin-4-on-N-oxyl-mediated polymerization of N-isopropylacrylamide (NIPAM) via MALDI-TOF and provided evidence that chain-end degradation occurred during the MALDI process [18]. Yet, these authors also noted that the degree of chain-end degradation varied with the applied laser intensity and could be suppressed by employing lower pulse energies. Charleux and coworkers observed the same chain end and polymer degradations during the analysis by MALDI-TOF of functional PS prepared by SG1-mediated polymerization [19]. They used the carboxylic acid functionality of the alkoxyamine BlocBuilder® (N-(2-methylpropyl)-N-(1-diethylphosphono-2,2-dimethylpropyl)-O-(2-carboxylprop-2-yl)hydroxylamine) to introduce a nonactivated alkene at the α-end of polystyrene chains. The presence of the alkene moiety could be observed, but the technique failed to reveal SG1-capped species although the polymerization was rather well controlled. Very recently, Delaittre et al. showed that the polymerization of another acrylamide derivative, namely N,N-diethylacrylamide (DEAM), was well controlled by the alkoxyamine BlocBuilder®, as proven by the reported kinetic data [20]. A clean SEC/ESI-MS spectrum of a DEAM oligomer evidenced the presence of a single population of chains carrying both the initiating fragment of the alkoxyamine and the nitroxide (see Figure 11.1).

Cyclic polymers are also accessible via NMP. An approach toward such structures has been taken by Hémery and coworkers, who characterized the obtained structures via MALDI-TOF-MS [21]. Heterotelechelic polystyrene chains having α-hydroxyl and ω-carboxyl end groups were cyclized via an intramolecular reaction. NMP using 4-hydroxy-TEMPO and a thermally decaying azo initiator was employed to generate difunctional macromolecules. The success of the cyclization reaction was evidenced

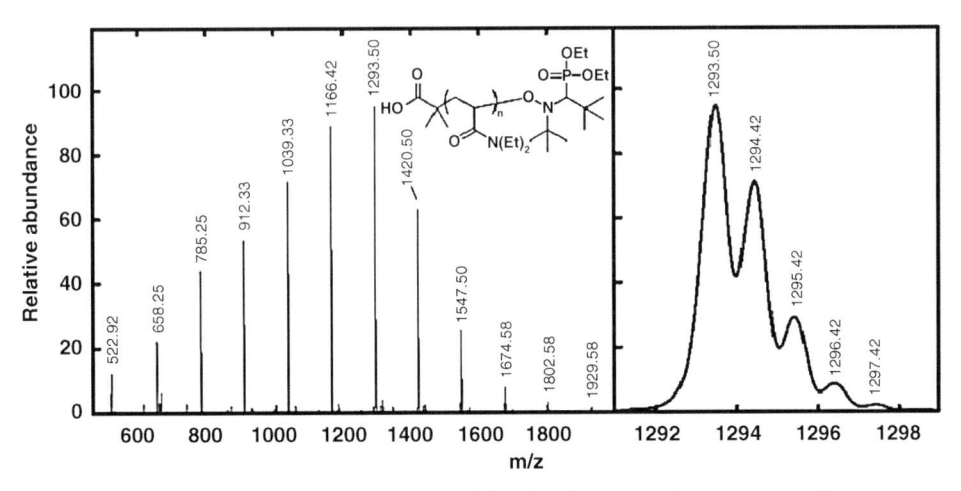

Figure 11.1 (a) SEC/ESI-MS spectrum of a N,N-diethylacrylamide (DEAM) oligomer prepared by BlocBuilder®-mediated polymerization. (b) Magnification of the $m/z = 1291–1299$ region. Reproduced with kind permission from Ref. [20]. Copyright 2011 American Chemical Society.

by FT-NIR spectroscopy, SEC, and MALDI-TOF-MS. The MALDI spectrum of the linear precursor corresponds to the structure [HOOC–(styrene)$_n$–OH + H]$^+$, whereas the cyclic product can be assigned to [–OC–(styrene)$_n$–O– + H]$^+$. The success of the cyclization was clearly supported by a molecular weight gap of 18 Da between the two series, corresponding to the water molecule expelled during the esterification. MALDI-TOF-MS has also been employed to evaluate the success of modification reactions on nitroxide-terminated polymers. For example, Beyou et al. have functionalized nitroxide-terminated polymers by a combination reaction with disulfide compounds such as tetraethyl thiuram disulfide [22]. The obtained polymers were analyzed via MALDI-TOF-MS, ESI-MS, and L-SIMS. Importantly, these authors confirmed the observation of earlier studies [7] that data from MALDI-TOF-MS have to be interpreted with care, as fragmentation reactions during the ionization process can readily occur. This phenomenon has been particularly prominent for polymer chains carrying thiocarbonylthio end groups (see the below section on RAFT). It is thus highly advisable to record a UV spectrum of the polymer to be analyzed and adjust the MALDI laser wavelength accordingly, if possible. More recently, Braslau and colleagues studied via MALDI-TOF-MS the preparation of ω-functionalized polymers, by converting nitroxide-capped polystyrene to a keto-terminated macromolecule. Although the obtained mass spectra were of relatively low resolution, the transformation could nevertheless be confirmed [23].

A further highly important technique relying on a reversible termination mechanism is ATRP. ATRP is arguably one of the most developed and frequently used living radical processes used for controlling radical polymerization and to prepare complex macromolecular architectures [24–27]. ATRP was first reported by Matyjaszewski and Sawamoto [24] and has since advanced into several subvariants [28–32]. Thus, it comes with no surprise that MS has frequently been employed to assess the structure of ATRP-generated polymers and to elucidate chain-end populations that reveal important information about the polymerization mechanism. In the following, an overview will be given of examples where MS was employed to study the ATRP process, going from fundamental mechanistic questions to the analysis of polymers of increasing complexity. Over the period that ATRP has been in use, both ESI and MALDI have been employed and – depending on the specific analytical problem – both methods can be viable option for ATRP polymer analysis. From the very beginning of ATRP developments, MALDI-TOF-MS was used to evidence the presence of halides within the polymeric material. For example, Matyjaszewski and colleagues employed this technique to study the chain-end structure of polyacrylonitrile generated via ATRP using 2-bromopropionitrile as an initiator and CuBr/2,2′-bipyridine (bpy) as a catalyst [33]. The main chain population in this specific case was a polymer with a halide end group, with small additional amounts of unsaturated chains as well as products from bimolecular termination events being present in the polymeric material. Decomposition of the chain end during ionization, that is, loss of HX (X = Br, Cl), can contribute to unsaturated chain ends. At high monomer-to-polymer conversions, evidence for additional side reactions was detected, i.e., the loss of the halide and its replacement by a hydrogen atom [34]. Indeed, a subsequent study into the kinetics of chain-end formation in

ATRP evidenced that loss of HX is clearly possible (thus limiting the functionality of the final polymer), underpinning the results of the MALDI-TOF-MS findings [35]. Intrigued by the observation that in ATRP of *tert*-butyl acrylate (*t*-BA) the loss of terminal bromine occurred as a significant side reaction [36], Kubisa and colleagues undertook a systematic study of the (low molecular weight, $M_n \cong 2000$ Da) product spectrum generated in ATRP of acrylates via MALDI-TOF-MS [37]. At least under the reaction conditions of the above study, the authors found that for all three acrylate ATRP systems studied (i.e., methyl acrylate (MA), BA, and *t*-BA), macromolecules without a terminal bromine group were clearly detected together with bromine-capped species (see Figure 11.2). In addition, the fraction of dead polymer in the reaction mixture increased significantly as the reaction progressed to higher monomer-to-polymer conversions to the extent that, close to complete conversion, the majority of macromolecules were irreversibly terminated. Surprisingly, the polymerizations proceeded in a living fashion to high conversions, raising the possibility that the alteration of the chain ends occurred during the ionization process. For the formation of macromolecules devoid of bromine groups, the authors suggested a transfer mechanism involving the amine ligand. However, they were careful enough to stress that mechanistic generalizations to other ATRP systems should be avoided.

Nevertheless, the observation that the reaction conditions (including the choice of the ligand) can have a dramatic impact on the polymer end groups was also recently demonstrated by Singha and colleagues via MALDI-TOF [38]. Analyzing poly(ethyl acrylate) (PEA) prepared using bpy as ligand, these authors demonstrated – via albeit relatively low resolution mass spectra – the presence of $-$Br as the end group. Interestingly, when PMDETA (N,N,N',N',N-pentamethyldiethylenetriamine) was used as a ligand, the mass spectral analysis evidenced almost exclusively terminal-hydrogen material. Some loss of a halogen end group using bpy as a ligand was also observed in the polymerization of hexyl acrylate by the same group [39]. Earlier, Matyjazsewski and colleagues observed the well-resolved MALDI-TOF-MS spectra of ATRP-prepared PBAs, with the expected structure (i.e., [CH$_3$–CHO$_2$CH$_3$)–BA$_n$–Br + Na]$^+$) [40]. A minor peak series was detected indicating the elimination of HBr, yet, since no unsaturated end groups were detected by ^{1}H-NMR spectroscopy, the authors concluded that the elimination had most likely occurred during the MALDI process. A similar situation could have occurred in a study by Dervaux *et al.* where the ATRP of isobornyl acrylate employing PMDETA as a ligand in a CuBr system was carefully investigated [41]. The main distribution could be clearly assigned to the expected terminal-bromine material (confirmed by the comparison of the theoretical and experimental isotopic pattern distributions), yet also two minor distributions occurred which could not be assigned. The question regarding whether these distributions originated during the ATRP process or during MALDI analysis remained open. Sometimes, however, it is possible to differentiate between material that is formed during the polymerization and during the MALDI process. Jackson and colleagues have prepared low-molecular-weight PMMAs (using ethyl-2-bromoisobutyrate as initiator and CuBr/bpy as catalytic system), which were subsequently analyzed via a host of techniques, including MALDI-TOF-MS [42]. Besides finding evidence for the presence of Br elimination, these authors found terminal-lactone

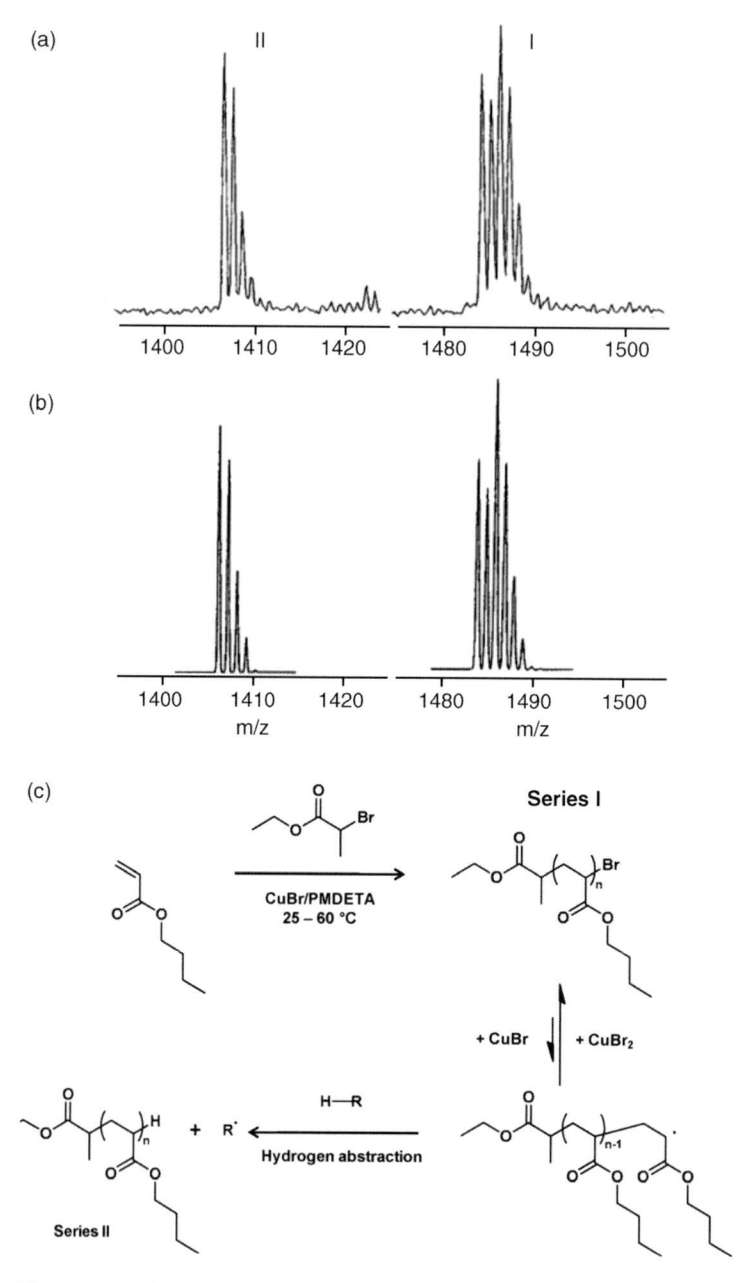

Figure 11.2 Observed (a) and calculated (b) isotope distributions for MALDI-TOF mass spectrum signals of PBA with $DP_n = 10$ containing the initiator fragment as a head group and a bromine atom as a terminal group (I) or devoid of bromine(II). (c) Mechanism of formation of species corresponding to the two series of peaks. Reproduced with kind permission from Ref. [37]. Copyright 2001 Wiley InterScience.

polymers, which were likely formed during the polymerization process. A differentiation with regard to their formation time (polymerization vs. ionization) was possible via the evaluation of not well-resolved peaks in combination with CID experiments. A single distribution of bromine-capped PMMA without any side products was found in a study employing 1-adamantyl-2-bromoisobutyrate as the initiator and PMDETA as the ligand [43], as well as in a study involving the interesting monomer ethyl-3-(acryloyloxy)methyloxetane [44]. In an investigation testing the ability of the catalyst to undergo halogen exchange, Singha et al. have prepared PMMA employing copper(I) thiocyanate (CuSCN) as a catalyst with N-n-pentyl-2-pyridylmethanimine as a ligand [45]. The MALDI-TOF spectra of the obtained polymer demonstrated that halogen exchange during ATRP in cases where the copper catalyst carries a pseudohalide (thiocyanate) counterion is well possible. The introduction of a thiocyanate group at the polymer terminus may have some interesting applications. However, pseudo-halide initiators are not as efficient in ATRP as classical halide ones [46].

The above studies all have in common that the efficiency of the ATRP process is established by a postmortem analysis of the generated polymeric material. However, mass spectrometric investigations also open an opportunity to assess the potential suitability of ATRP catalysts without conducting a polymerization experiment. It has recently been demonstrated that the activity of ATRP catalysts could be established via consecutive competitive experiments, evaluating both the relative binding affinities of several ligands and the relative halidophilicities of the resulting complexes [47]. These studies were subsequently extended – also employing ESI coupled with a quadrupole ion trap (QIT) mass analyzer – to include the identification of the most active copper catalysts and the assessment of the effects of the reaction medium on the relative stabilities of the catalyst complexes. The influence of the chemical nature of the ligand on both the complex halogenophilicity and the metal–ligand stabilities was evaluated as well [48]. A similar use of ESI-MS in the context of studying the catalytic activity of ATRP initiators has been followed by Shen and colleagues [49].

While the above studies all employed copper-based catalytic systems, a significant effort has been directed to understanding the mechanism and product spectrum in ruthenium-mediated living free radical polymerizations, which proceed also via the same basic mechanism. Most importantly, Sawamoto and coworkers applied MALDI-TOF-MS to study the polymers generated in ruthenium-mediated free radical polymerization of MMA, MA, and styrene [50]. The observed mass spectra of the generated polymers were very clean and only displayed a single series of peaks associated with the theoretically expected molecular structure, that is, having an initiator fragment at the α-end and a halide (i.e., chlorine) group at the opposite end. The authors were also able to demonstrate that the theoretical isotopic pattern distribution of the living polymers matched excellently with the experimental one. The study compared the living polymers with polymers derived from the same monomers, yet prepared via conventional (i.e., AIBN (2,2'-azobisisobutyronitrile)-initiated) free radical polymerization, indicating the presence of a high degree of bimolecular termination. In a rare example of employing a molybdenum-based catalyst, Kubisa and colleagues studied ATRP of MA in the ionic liquid 1-butyl-3-methylimidazolium hexafluorophosphate [51]. A combination of a bromine-containing initiator (ethyl-2-bromopropionate) with the catalyst MoBr$_3$(PMe$_3$)$_3$/MoBr$_4$(PMe$_3$)$_3$ led to a controlled polymerization, with a high terminal −Br group fidelity of the polymer evidenced by MALDI-TOF-MS. An even more intriguing application of the MALDI-TOF-MS for the elucidation of reaction mechanisms was carried out by Le Grognec et al., who reported on the free radical polymerization of styrene mediated by molybdenum(III)/(IV) couples [52]. CpMo(PMe$_3$)$_2$Cl$_2$ and CpMo(1,2-bis(diphenylphosphino)-ethane)Cl$_2$ species were used to induce a successful living ATRP process in styrene polymerizations. MALDI-TOF-MS was subsequently applied to analyze the polymers and four major peak series were distinguished. The first series of peaks corresponded to vinyl-terminated polymers; however, the authors concluded by comparison with ^{1}H-NMR spectroscopic data that these species corresponded to bromine-terminated species that have undergone a dehydrobromination sequence under MALDI conditions. The second series of peaks matched the expected dormant chains, with the third series of peaks corresponding to a vinyl-terminated series (as above), but carrying sodium instead of silver as counterion. Finally, the fourth series is associated with dormant chains carrying a −Cl chain terminus. Even though the Mo−Br bond is significantly weaker than the Mo−Cl bond, radical selectivity is not entirely in favor of bromine abstraction. Due to the low population of the Cl-terminated chains, no evidence of proton resonances associated with this series was detected by ^{1}H-NMR spectroscopy.

Statistical copolymers and block copolymers have been prepared in an extensive variety by ATRP since its initial conception. Some of the studies on copolymer formation processes employed soft-ionization mass spectrometric analysis. For example, Klumperman and colleagues investigated olefin copolymerizations under ATRP control, specifically MMA and 1-octene [53]. These authors observed a relatively complex spectrum of overlapping distributions for molecular weights lower than 15 000 Da. Identified distributions could be assigned to chains all having the *p*-toluenesulfonyl initiator fragment at the α-end but carrying different ω-end-groups: (i) a halide corresponding to the dormant chains, (ii) a hydrogen arising from the loss of the aforementioned halide, and (iii) a lactone [54]. The authors attributed the lactone end group to the cyclization of the terminal two repeat units and the subsequent loss of chloroform. Most observed copolymer chains contained at least one 1-octene group.

A multitude of di- and triblock copolymers – too numerous to exhaustively cover them all here – have been prepared via the ATRP process and their structural features have to some extent been characterized via mass spectroscopic techniques. For example, Bednarek et al. demonstrated that ABA-type block copolymers with poly(oxyethylene) and PMMA segments can be analyzed via MALDI-TOF-MS via a combination of mass-centered peak assignments and isotopic-pattern recognition [36]. Kuckling and colleagues prepared block copolymers of poly(glycidol) and PNIPAM via a switch from anionic polymerization to ATRP. The required change in end group was monitored via MALDI-TOF-MS, evidencing that a relatively clean ATRP macroinitiator had been generated [55]. A further elegant dual mechanism polymerization was recently reported by Frey and coworkers, who prepared poly(lactide)-*block*-poly

(2-hydroxyethyl methacrylate) polymers employing an -OH functional ATRP initiator for the ROP process [56]. MALDI-TOF-MS was used to evidence the high end-group fidelity of the ATRP poly(lactide) macroinitiator. An example of using ESI-MS to study the structure of ATRP macroinitiators has been reported by Matyjaszeski and coworkers in their successful attempt to incorporate vinyl acetate (VAc) into block copolymer structures [57]. Their approach in generating block copolymers of VAc followed two principal approaches, that is, the use of a difunctional initiator and the redox initiation of a halogen-terminated (macro)initiator. The provided data confirmed the presence of PVAc-Br (see Figure 11.3) and PBA-Br. It is an interesting observation – not limited to ATRP – that most studies into block copolymers often rely on the characterization of the precursor blocks via MALDI-TOF, yet the generated block copolymers are rarely investigated themselves. This is presumably due to the complexity of the resulting mass spectra, which are difficult to interpret and in which peak overlap often occurs. Thus, it comes as no surprise that when Vidts et al. carefully investigated the poly(ethylhexyl acrylate) prepared via ATRP with MALDI-TOF, noting only a minor side distribution caused by a bromine/hydrogen interchange, they refrained from analyzing the block copolymers of poly(ethylhexyl acrylate) and poly(acrylic acid) (PAA) which were subsequently prepared [58]. One of the fewer recent examples of employing ESI-MS for the analysis of block copolymer precursors comes from Zhang et al., who characterized hydroxyl-functionalized PMMA carrying an opposite −Br terminus, which they subsequently employed to generate PMMA-b-PBA copolymers [59]. As relatively typical for ESI-MS, these authors observed several series of multiply charged signals as well as some minor species, to which they did not assign a chemical structure. Nevertheless, they noted that ESI-MS may be better suited for the analysis of ATRP-prepared polymers and may detect more species – a notion that has since been confirmed [5]. An example came very recently to contradict the aforementioned observed trend when Schubert employed MALDI-TOF MS/MS to determine the block chain lengths of poly(ethylene glycol)-b-PS block copolymers made by ATRP from a bromine-capped PEG initiator [60]. During the ionization, scission between the two blocks occurred and helped to accurately determine the length of the PS blocks.

On a higher level of architectural complexity, the degree of functionalization of star-shaped ATRP macroinitiators can be accessed via the analysis of the isotopic pattern distribution of individual polymer peaks. Typically, the number of initiating groups (and thus the molecular weight of the polymer) is determined via ^{1}H-NMR spectroscopy. However, such an approach is limited since it only provides average numbers and does not differentiate between polymer chains with different molar masses. Kubisa and Bednarek, for example, have applied isotopic-pattern identification via MALDI-TOF-MS on poly(3-ethyl-3-hydroxy-methyloxetane)-derived macroinitiators and successfully determined the degree of end-group functionalization [61]. A clear fine splitting signal due to the isotope distribution was observed, largely caused by the ^{79}Br and ^{81}Br isotopes. Such an approach is applicable to macromolecular architectures which feature halogens with large masses. Although in principle possible, the applicability of the above technology to NMP- and RAFT-generated star initiators is significantly more challenging, if not impossible

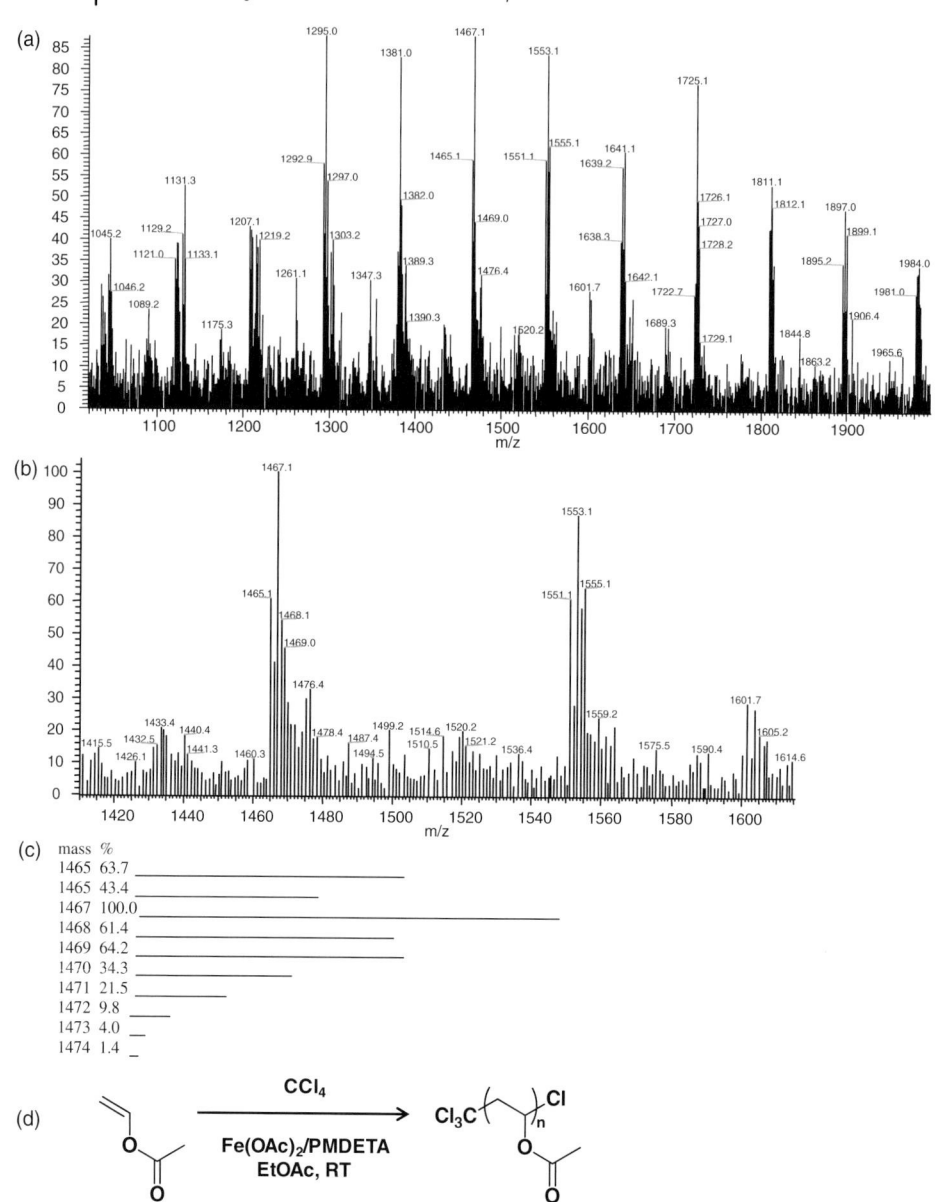

Figure 11.3 (a) ESI mass spectrum, (b) expanded ESI mass spectrum, and (c) simulated isotope pattern of pVA-Cl. Reproduced with kind permission from Ref. [57]. Copyright 2004 American Chemical Society. (d) Corresponding reaction scheme.

due to the lack of a distinctive isotope distribution such as that of ^{79}Br and ^{81}Br. Over the past years ATRP has been frequently employed to generate more complex macromolecular architectures, which have subsequently been characterized via MALDI-TOF-MS and listing all of these here is not within the scope of the present chapter. Nevertheless, a recent example by Li and colleagues showcased the preparation of tadpole-shaped polymers (circular polystyrene attached to a PEG tail) via a combination of ATRP and two modular ligations based on Cu-catalyzed azide/alkyne reactions [62]. The spectra of the precursor polymers (PEG strand, molecular weight close to 2000 Da) were recorded with relatively high resolution, while the final tadpole-shaped macromolecules of higher molecular weight yielded MALDI-TOF spectra featuring a broad molecular weight distribution with nondiscernable and isotopically nonresolved peaks. Nevertheless, a MALDI-TOF measurement was employed to estimate the molecular weight (in this case close to 9500 Da), which agreed well with the value deduced from NMR. Such a molecular weight match indicates that in some cases soft-ionization MS can yield relatively accurate number average molecular weights without the need for sophisticated techniques for data treatment.

ESI-MS was also employed to great extent to study the end-group transformation of ATRP made polymers, particularly for the characterization of nucleophilic substitutions of the bromide end-group functionality. For example, our team was able to equip ATRP made polymers with the highly reactive cyclopentadiene (Cp)-moity [63]. The mild procedure used nickelocene as the source of the Cp unit, tributylphosphine as ligand and sodium iodide as metathesis reagent at ambient temperature. Although the Cp moiety is clearly visible in NMR, the quantitative integration of the corresponding multiplett remained problematic due to the presence of different nickel-Cp species. Again it was ESI-MS that provided detailed information on the end-group fidelity of the macromolecular building block and demonstrated the high efficiency of the transformation (see Figure 11.4).

Further interesting examples of applying soft-ionization MS techniques to ATRP-made polymers come from the field of telechelic polymers. For example, Shen *et al.* employed MALDI-TOF-MS to characterize C_{60}-capped polystyrenes prepared by ATRP [64]. These authors studied both the polystyrene precursor molecules and the C_{60}-functionalized final products via MS. Concomitantly with the findings of Kubisa and coworkers [37], these authors found no polymer with bromine end groups. It can thus not be excluded that bromine is easily cleaved off from the main polymer chain under MALDI conditions. The analysis of the C_{60}-monosubstituted polymers clearly showed that the reaction was successful although the polymer underwent some rearrangement during the ionization process again leading to the loss of the bromine functionality. In a further example, the efficiency of an end-group modification of ATRP polymers was recently monitored via MALDI-TOF by Monge *et al.* [65] These authors transformed bromine-terminated poly(*t*-BA) via the Gabriel reaction into amine-terminated species. MALDI-TOF confirmed that the reaction was successful, yet a side distribution was observed caused by a transesterification of the ATRP initiator fragment at the opposite chain terminus.

Finally, both MALDI-TOF and a novel matrix-free ionization method, desorption ionization on silicon (DIOS), were employed to study PMMA chains tethered to porous silicon and anodic aluminum oxide surfaces via surface-initiated ATRP by Gorman *et al.* [14]. The study effectively demonstrated that ATRP surface-tethered chains can be efficiently analyzed via MS techniques. Matrix-free DIOS-MS was effective for the direct analysis of the polymers up to a molecular weight of

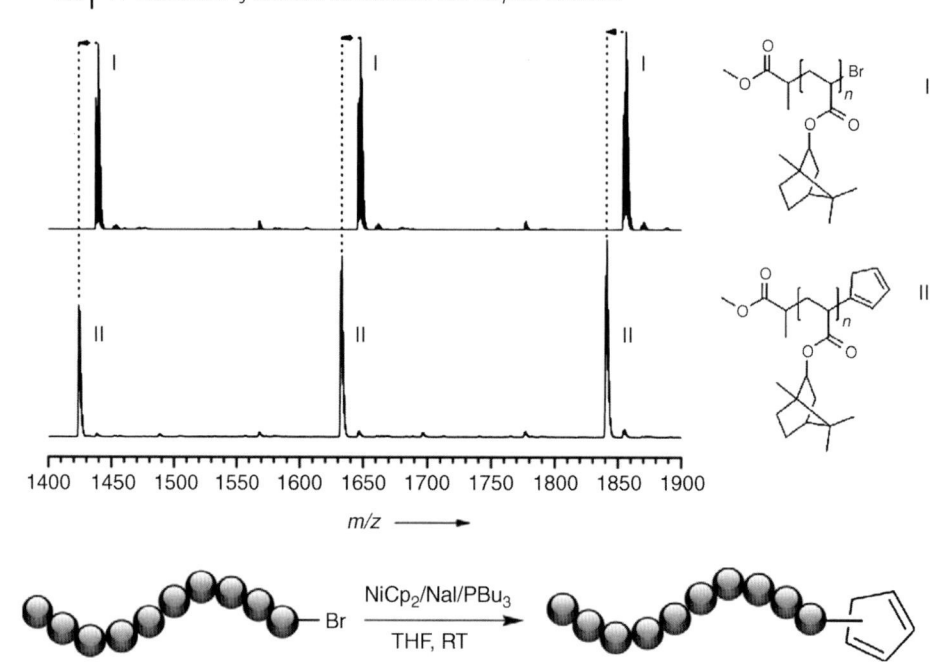

Figure 11.4 Electrospray ionization mass spectra of the end-group transformation of P*i*BoA-Br into cyclopentadienyl-capped P*i*BoA-Cp in the charge state $z = 1$. The isotopic pattern clearly changes from bromide terminated species to Cp-terminated species. Reproduced with kind permission from Ref. [63]. Copyright 2011 American Chemical Society.

approximately $6\,\mathrm{kg\,mol^{-1}}$. Beyond this molecular-weight threshold, the signal-to-noise ratio rapidly decreased. Based on the MS analysis, the study concluded that under the same polymerization conditions, PMMA grown on both substrates had a significantly lower molecular weight and a broader molecular weight distribution than the polymer formed in solution. In a study investigating Co(II)-mediated living radical polymerization Langlotz et al. [66] employed liquid injection field desorption/ionization mass spectrometry (LIFDI-MS) to monitor the reaction intermediates online during the polymerization process. The LIFDI methodology is a very soft-ionization method and especially suited for systems where moisture and air must be rigorously excluded from the reaction system.

11.2
Protocols Based on Degenerative Chain Transfer (RAFT, MADIX)

Perhaps the liveliest application of MS techniques to elucidate a macromolecular reaction mechanism via a postmortem inspection of the products has been in the field of reversible addition fragmentation chain transfer (RAFT) polymerization [67–72] and the mechanistically equivalent technique of macromolecular design via the interchange of xanthates (MADIX). It is not within the scope of the current book chapter to review the entirety of the mechanistic debate that has been part of the development of RAFT chemistry almost from its beginning. Rather, the current chapter will compile the most important conclusions drawn from MS data regarding the RAFT/MADIX mechanism as well as demonstrate that MS has aided greatly in the design of novel materials based on RAFT chemistry. To this day, a debate about the details of the mechanism of the RAFT process is ongoing although a preliminary summary (not solution) was published in 2006 in an article by an IUPAC working party as well as other overview articles [73–80]. The first investigations into the microstructure of RAFT polymers have been conducted via MALDI-TOF-MS [67] to demonstrate that the fundamental mechanism suggested by the CSIRO group is indeed plausible, that is, generating polymers that feature the R and Z groups on opposite ends of the polymer chain. However, polymers made by RAFT/MADIX are colored and feature UV absorptions over a wide wavelength range. Thus, care has to be taken to ensure that the laser pulse during a MALDI experiment does not lead to a fragmentation of the thiocarbonylthio end groups. Indeed, such fragmentation reactions have been observed and can lead to difficulties in spectral interpretation or even incorrectly assigned peaks. The early MALDI-TOF-MS data on RAFT polymers were later unambiguously confirmed using ESI-MS coupled to a quadrupole mass analyzer on PMAs made via cumyl dithiobenzoate-, cumyl *p*-fluorodithiobenzoate- and 1-phenylethyl dithiobenzoate-mediated polymerizations, where no danger of UV-induced fragmentation exists [81]. These initial ESI-MS studies included collision-induced decay (CID) experiments (i.e., MS/MS) to underpin the structure of the polymeric material. After the introduction of the process and in a quest to elucidate the structure of RAFT-made polymers constituted of variable monomers, Ganachaud *et al.* employed MALDI-TOF-MS of RAFT-generated PNIPAM [82]. Although all the expected R- and Z-group carrying chain ends could be clearly identified, the relative low level of mass spectroscopic resolution coupled with a significant noise level did not allow the detection of polymer chains carrying initiator (AIBN) fragments. The analysis further indicated that significant amounts of vinyl-terminated polymeric materials were present, suggesting that either disproportionation events were operative or that the RAFT end group was eliminated, perhaps via reactions akin to a Chugayev elimination. In a further study concerning RAFT-generated PNIPAMs [83] Müller and colleagues not only identified the expected chains with thiocarbonylthio groups on one end and the R group on the other chain terminus, but also detected evidence of initiator fragment-capped chains, which must be present – even if in low proportion – in polymeric material generated via the RAFT process. The propensity of RAFT-generated polymers to undergo fragmentation during the MALDI process may have, however, also interfered in the above analysis, as the authors observed series of peaks that could not be assigned to a specific product (see Figure 11.5). Similar to the findings of Ganachaud *et al.* [82], a significant number of vinyl-capped polymeric materials were found. In a postsource decay analysis (PSD) [84], these authors showed that large numbers of vinyl-terminated polymers are generated, underpinning the notion that thiocarbonylthio end groups can indeed be readily eliminated under formation of a terminal vinyl moiety.

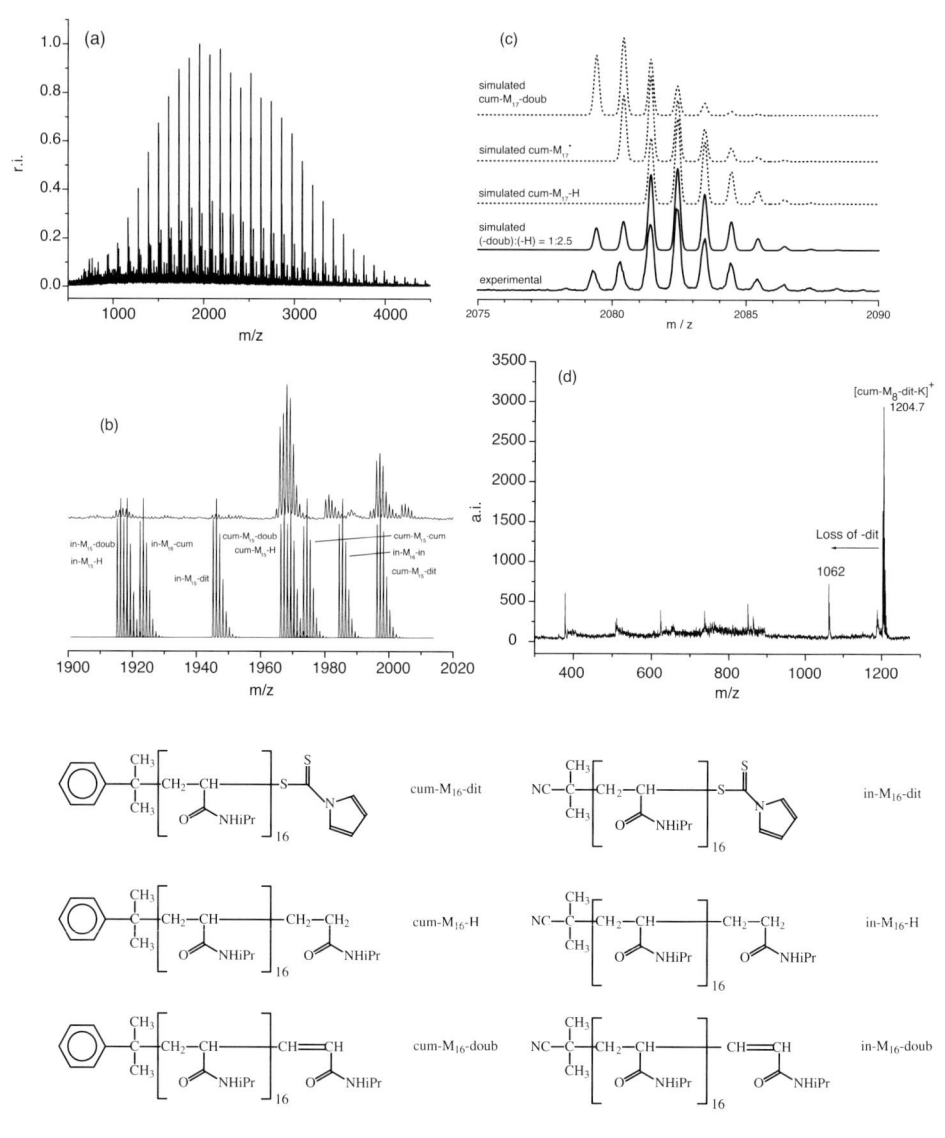

Figure 11.5 MALDI-TOF mass spectrum of poly(N-isopropylacrylamide) prepared in presence of cumyl 1-pyrrolecarbodithioate (sample taken at 13% conversion). (a) Complete spectrum of K^+-ionized sample. (b) Determination of the chain-end structures, with experimental (top) and simulated (bottom) data. (c) Simulation of signal overlap of assumed disproportionation/transfer signals. (d) Post-source decay MALDI-TOF mass spectrum of poly(N-isopropylacrylamide) with 8 monomer units. The most intense fragment peak corresponds to the loss of the dithioester residue. Reproduced with kind permission from Ref. [83]. Copyright 2002 American Chemical Society.

In a further MALDI-TOF-MS study, Favier et al. investigated the products generated in a RAFT-mediated polymerization of *N*-acryloylmorpholine (NAM) employing (*tert*-butyl dithiobenzoate) as a RAFT agent [85]. It is noteworthy that these authors carried out a systematic variation of the counterion to maximize the quality and information depth of the MS spectra. Similar to the studies mentioned above, these authors also found circumstantial evidence that a series of (proton-terminated species) peaks may have been generated during the ionization process, rather than during the polymerization. In addition, the study identified the expected dithioester-capped material as well as oxidation products of the same. The relevance of oxidation products generated from thiocarbonyl thio compounds as well as the identification of these products via MS will be discussed in depth below. In the same study the authors also noted products that may have been generated via termination reactions of the intermediate radicals (resulting from the addition of a propagating radical onto the thiocarbonyl sulfur atom, before fragmentation), although the agreement with the theoretically expected molecular weights of those structures is relatively poor. The search for intermediate termination products via the MS techniques will be discussed in greater detail below. In a related study employing NAM as a monomer, D'Agosto et al. generated amphiphilic poly(NAM)-*block*-polystyrene copolymers [86], which were also characterized via MALDI-TOF-MS. A series of five chain populations was identified; however, only two of these populations could be identified with certainty, that is, the expected population of RAFT-prepared chains carrying sodium as a counterion as well as potassium, and a series of peaks assigned to a population of macromolecules featuring the (acid-containing) R group on one side and a proton at the other end. A study by Loiseau et al. investigated the microstructure of RAFT-prepared PAA via MALDI-TOF-MS as negative-ion spectra of neutralized PAA and identified peak series corresponding to proton- and thiol-terminated species, yet could not identify the (main) thiocarbonylthio-capped species [87]. In a MALDI-TOF-MS study on thiocarbonylthio polymers generated in dispersed systems, Sanderson and colleagues clearly identified a dominant population congruent with the RAFT polymer, yet also identified distributions associated with bimolecular termination events of secondary propagating radicals, as well as species carrying both the initiator fragment and the thiocarbonylthio group. Again, non-assignable peak series were observed, which may further support the notion that the MALDI-TOF mass spectra of RAFT polymers can be problematic to interpret. One of the first examples of a MALDI-TOF-MS investigation into polymers generated by the MADIX/RAFT process was reported by Zard and colleagues on the example of styrene, vinyl acetate and acrylates. Again, besides the expected thiocarbonylthio-capped polymer species, chains carrying hydrogen caps as well as initiator fragments were reported [88].

Given the above findings employing MALDI-TOF-MS, it appears that despite the undoubted advantages of the technology such as access to higher molecular weight regimes, the study of RAFT/MADIX-generated polymers is problematic. As already noted above, ESI-MS (coupled to either QIT detectors or higher resolving mass analyzers) is an attractive alternative for RAFT polymer analysis. Consequently, ESI-MS has found an increasing use in this domain, especially for the clarification of mechanistic questions. Among the earliest investigations on the complex mechanism [72] of the RAFT process via ESI-MS are two complimentary studies, where size-exclusion chromatography (SEC) was interfaced online to a QIT detector equipped with an ESI source [89, 90]. Via such an approach, the cumyl dithiobenzoate as well as the cumyl phenyldithioacetate-mediated polymerization of MA and BA were studied. No termination products of the intermediate radicals – neither with themselves nor with secondary propagating radicals – could be identified. Such an observation is quite surprising, as conventional bimolecular termination products could be readily observed although the concentration of these species should be considerably lower than those of the intermediate termination products (based on the relative radical concentrations). Either three- and four-armed star polymers cannot be mapped by ESI-MS – a notion that, however, can be discounted as intentionally made star polymers readily ionize in ESI-MS processes – or the potential star polymer species form in much lower concentrations (and thus with a much reduced termination rate coefficient) as suggested by some of the early mechanistic models [74, 91, 92]. In contrast to the above findings, Geelen et al. [93] found evidence of three-armed star structures using MALDI-TOF-MS. However, the approaches of Feldermann et al. [89] as well as Ah Toy et al. [90] were significantly different to that of Geleen et al., while the former applied no fractionation to their samples measured via ESI-MS, the latter carried out prefractionation before MALDI-TOF-MS analysis. Both studies were based on polyacrylates, which can give rise to structures through the formation of mid-chain radicals that are isobaric to terminated RAFT intermediates. In a further study on styrene-based systems, some evidence for intermediate-radical termination could be identified by Zhou et al. [94] Thus, on the balance of all evidence it seems likely that the RAFT intermediates may terminate to a certain extent, however, not in the quantities predicted by some kinetic models [74, 91, 92]. A highly interesting study on the same topic was provided by Bathfield et al. [95] who found in the polymerization of N-acryloylmorpholine (also via a MALDI-TOF investigation) that the intermediate RAFT radicals only terminate with very short radicals, that is, initiator-derived fragments. It thus seems probable that RAFT intermediate radicals only terminate to significant extents – due to their sterically hindered nature – with small radical species and largely refrain from termination with oligomeric or polymeric radicals. Interestingly, this notion has been followed by Perrier and coworkers with success to provide a compromise mechanistic model of the RAFT mechanism, incorporating both elements of slow fragmentation and intermediate-radical termination [96, 97]. Since the above mechanistic investigations, ESI-MS has become a standard tool – if not the preferred tool – for the mass spectrometric investigation of RAFT polymers. In a very recent example, our group demonstrated that even synthetic rubbers (acrylonitrile/butadiene (NBR)) generated via the RAFT process are readily ionizable via ESI-MS coupled to linear quadrupole ion trap detectors, confirming high degrees of end-group fidelity of RAFT-made NBRs [98]. When Vana and coworkers used a trimethoxysilane-containing RAFT agent to control the polymerization of vinylic monomers at the surface of silica particles, they first carried out polymerizations in solution [99]. They subsequently analyzed the products by ESI-MS and observed two series of peaks corresponding to single and double charged species of the expected structure.

While the above studies concentrated on the elucidation of the structure of linear chains, the formation of star polymers via RAFT polymerization has also been studied via ESI-MS methods. Barner-Kowollik and colleagues have investigated in two studies the formation of star polymers via the RAFT process, in the so-called R-group approach-RAFT polymerizations, where the core itself carries one or multiple radical functionalities during macromolecular growth. These authors could demonstrate that star RAFT polymers readily ionize during ESI-MS, thus suggesting that the above discussed 3- and 4-armed star polymers formed during linear RAFT processes should also ionize. Both acrylate and styrene systems were investigated [100, 101]. The occurrence of coupling of the core with propagating chains as well as star–star coupling was proven. In the case of acrylate systems, additional species originating from the generation of mid-chain radicals and their follow-on reactions (i.e., termination and β-scission) were identified with high certainty. Similarly, the Z-group-approach RAFT polymerization was investigated by Vana and coworkers, who reported well-resolved ESI-MS spectra of PVAc and poly(vinyl propionate) star polymers [102], which displayed the expected structure (i.e., no star–star coupling reaction was observed). Directly probing the ability of three- and four-armed polymers to ionize during MALDI-TOF-MS, Monteiro and colleagues synthesized those structures directly and also found them to be readily ionizable [91]. Recently, Maynard reported the synthesis of 4-armed ω-functionalized PNIPAM stars by the core-first method [103]. MALDI-TOF spectra of both the star polymer and its arms after cleavage from the core confirmed the expected structures.

ESI-MS was also employed to great extent to study the end-group transformations of RAFT polymers, particularly oxidation processes due to the peroxides formed in cyclic ethers. The first report on this matter described how PMA prepared via cumyl dithiobenzoate-mediated polymerization, could be readily and quantitatively transformed into thioester-capped material via simple stirring with *tert*-butyl peroxide at ambient temperature for 12 h [104]. Circumstantial evidence that storing RAFT polymers in cyclic ethers such as tetrahydrofuran (THF) leads to the loss of end groups was recently substantiated via an in-depth ESI-MS investigation employing linear quadrupole ion trap as well as FT-ICR mass analyzers. These studies evidenced that the peroxy radicals generated in the presence of air in cyclic ethers could remove the thiocarbonylthio end group and lead to the (quantitative) formation of hydroperoxy-capped macromolecules [105–107]. Interestingly, the above oxidation process can be efficiently employed to quantitatively convert RAFT-made polymers into terminal-hydroxyl entities, via coupling a transformation step in THF with an *in situ* one-pot reduction. The efficiency of the process has been evaluated for a series of RAFT agent and monomer combinations via ESI-MS (see Figure 11.6).

In 2010 Theato reported the synthesis of several polymethacrylates by RAFT polymerization using either an activated ester-containing RAFT agent or its subsequently alkyne functionalized counterpart, followed by removal of the dithioester end group by radical addition with AIBN [108]. MALDI-TOF MS analysis confirmed the efficiency of the reaction scheme.

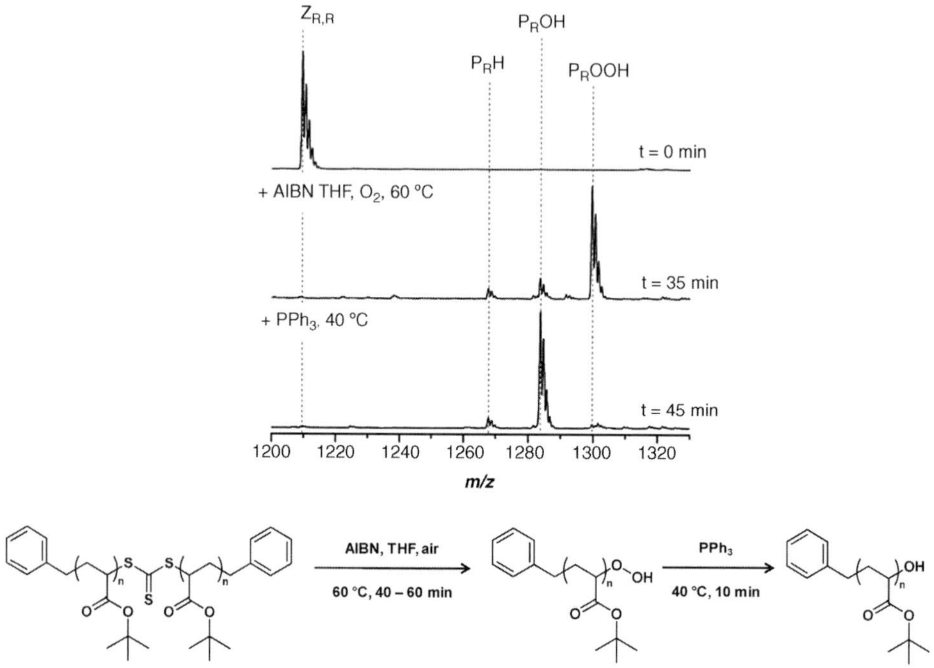

Figure 11.6 Electrospray ionization mass spectra of the transformation of P(t-BA) with a trithiocarbonate moiety in the middle of the chain into hydroxyl functional ptBA in the charge state $z = 1$. The reagents AIBN/THF and PPh$_3$ were added sequentially at $t = 0$ and 35 min. Full conversion was reached after 45 min. Reproduced with kind permission from Ref. [107]. Copyright 2010 Royal Society of Chemistry.

To note another example for the end-group modification of polymers synthesized by RAFT polymerization, ESI-MS was employed to provide an in-depth analysis of the photo-induced incorporation of alkenes into the polymer chain end [109]. Specifically, irradiation of a solution of 1-pentene and benzyl dithioacetic acid ester functional poly(alkyl acrylate) at the absorption wavelength of the thiocarbonyl group (315 nm) resulted in an efficient incorporation of the alkene into the polymer chain end (see Figure 11.7).

A mechanistic study revealed that conjugation proceeded by the UV-induced radical β-cleavage of the dithioester followed by an addition of the alkene and subsequent recapping by the RAFT-end-group. An integration of the mass spectral abundances of the photoconjugation product allowed for the derivation of quantitative data on the concentrations and concentration ratios of the species formed. Other minor products could be assigned to the incorporation of two 1-pentene units, Norrish type II fragmentation or radical termination by disproportionation. It would clearly exceed the scope of this chapter to illustrate more examples of end-group modifications of RAFT polymers. An excellent and comprehensive overview on this topic was published in 2010 by O'Reilly and Willcock [110].

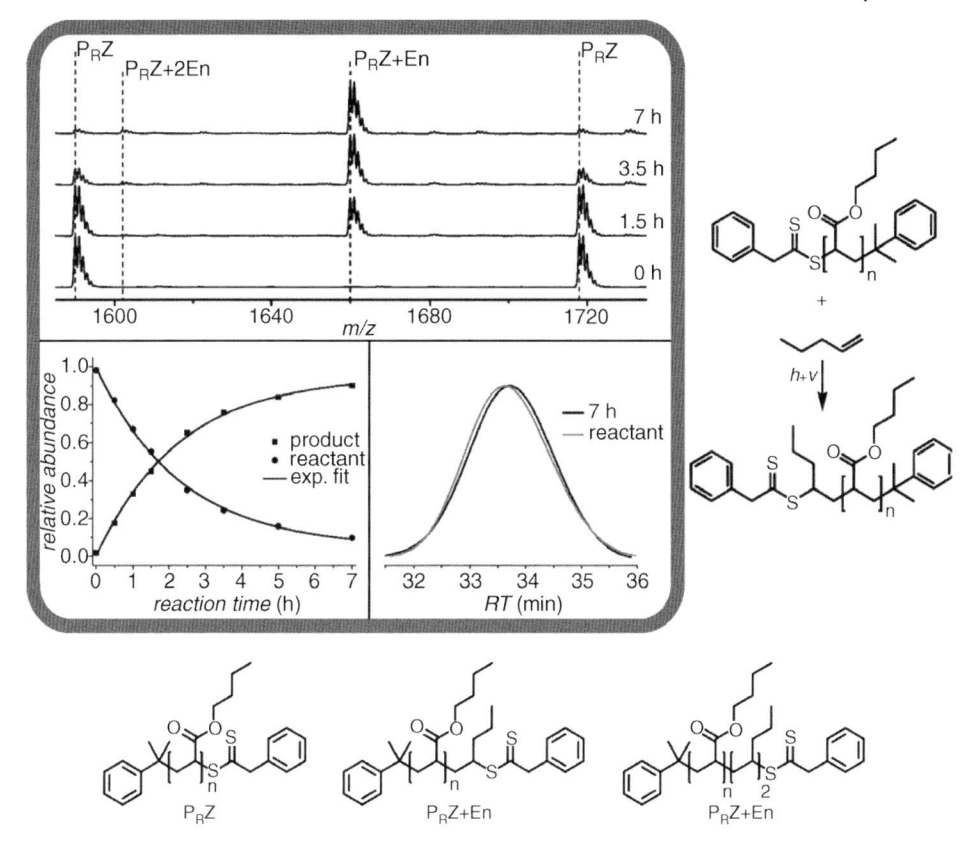

Figure 11.7 SEC/ESI-MS investigation of the species formed during the photoinduced conjugation reaction between pentene and poly(butyl acrylate) carrying a dithioester end group from cumyl phenyl dithioacetate-(CPDA)-mediated polymerization at 315 nm. The development of the relative abundance of the major components during the reaction and the SEC-traces before and after the conjugation is shown in the lower left and lower right graphs of each inset, respectively. Reproduced with kind permission from Ref. [109]. Copyright 2011 American Chemical Society.

11.3
Protocols based on CCT

CCT polymerization is certainly the most efficient and simple method to obtain macromonomers via a free-radical pathway [111–115]. It is based on the transfer of a hydrogen atom from a radical species to an olefin, which is catalyzed by certain transition metal complexes. The most employed class of catalyst is undoubtedly based on low-spin Co(II) complexes such as cobaloximes. This technique additionally allows one to obtain low-molecular-weight polymers (oligomers), whereas in classical free radical polymerization high-molecular-weight species are formed at the very beginning of the reaction. The catalytic cycle is believed to rely on two consecutive reactions: (i) the abstraction of a hydrogen atom in α-position of a macroradical giving rise to a dead unsaturated polymer chain and a metal complex at one higher oxidation state and (ii) the back transfer of the hydrogen atom onto a monomeric double bond producing a monomeric radical able to reinitiate the polymerization and the regeneration of the complex in its lower oxidation state (Scheme 11.2). Since these events have an impact exclusively on the chain end groups, MS is a method of choice to validate the proposed mechanism and identify the products which should ideally correspond to the mass formula $m = n \times M + M_{ion}$, where n is the degree of polymerization, M is the mass of the repeat unit, and M_{ion} is the mass of the counterion [116, 117].

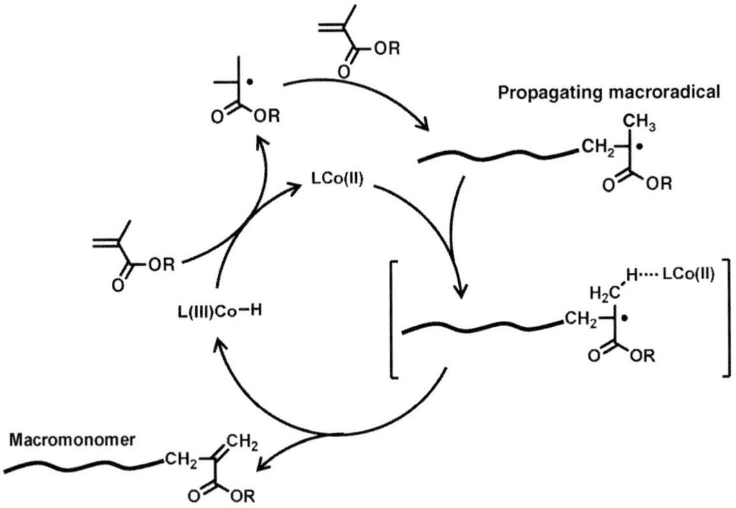

Scheme 11.2 Catalytic cycle operative in catalytic chain transfer (CCT) polymerizations in the presence of a methacrylate and a Co(II) species.

In practice, there is always a portion of chains which have been initiated by a conventional initiator fragment although they are sometimes rather difficult to observe due to their relatively low concentration. In methacrylate-based systems, the occurrence of Co–C bonds is very limited and does not affect the general mechanism. However, in systems consisting of secondary radical species, this phenomenon is more pronounced – very predominant in the case of acrylates – and can have a dramatic effect leading to a delay in the establishment of the CCT polymerization mechanism. Typically, in these cases an induction period is observed: up to a critical conversion – which depends on the Co concentration – an increase in molecular weight occurs before the transfer mechanism takes over [118, 119]. For instance, Roberts et al. isolated the oligomers formed during the induction period of the polymerization of MA in the presence of a cobaloxime and analyzed them by

MALDI-TOF [120]. Two distinct series of peaks were observed; one corroborating the CCT mechanism, that is, no initiator fragment was present on those chains; the second one carrying the cobalt catalyst. To obtain further evidence for these conclusions, the authors reacted a sample with α-methylstyrene at elevated temperatures and under oxygen-free conditions. The product was again analyzed by MALDI-TOF and the formation of α-methylstyrene-capped oligoMA was evidenced, originating from the catalyst-capped chains, while the CCT-formed chains were intact. These experiments proved that the Co—C bond can be cleaved at high temperature. Due to its low propagation rate and its tertiary radical structure, the α-methylstyrene oligoradical exclusively underwent CCT to yield unsaturated species.

Some research groups were interested in the possibility of *in situ* copolymerization of the macromonomers generated during CCT polymerization with the starting monomer [121, 122]. Thus, Davis and coworkers isolated dimers, trimers, and tetramers obtained from the CCT polymerization of MMA and introduced them in the polymerization of *n*-butyl methacrylate (BMA) and deuterated MMA in the presence of a cobaloxime. Based on MALDI-TOF analysis, they concluded that copolymerization may occur at high conversions (high relative concentration of unsaturated oligomers in the medium) in the case of MMA, leading to a small amount of branching, yet that for BMA the steric hindrance around the radical helped to maintain an undisturbed CCT mechanism throughout the whole reaction. Different types of macromonomers have been synthesized and MALDI-TOF or ESI-MS were often employed to evidence their structure. In a further example, Muratore *et al.* synthesized vinyl-capped PMMA by CCT polymerization in order to obtain xerogels by copolymerization with either *N*,*N*-dimethylacrylamide or HEA and *N*-vinyl-2-pyrrolidone [123, 124]. To obtain acrylate-based macromonomers, a method based on copolymerization with CCT-effective monomers can be used. For instance, Kukulj *et al.* demonstrated that it is possible to obtain a large majority of vinyl-capped polystyrene by incorporating only 10 mol% of α-methylstyrene in the reaction mixture [125]. Hence, Heuts, Davis, Barner-Kowollik, and coworkers obtained unsaturated oligoBA also using a small amount of α-methylstyrene [126]. However, when they used benzyl methacrylate as a comonomer at high catalyst concentration MALDI peaks corresponding to pure PBA were found together with CCT product peaks. An alternative way of obtaining macromonomers using CCT was recently introduced by Soeriyadi *et al.* by taking advantage of the RAFT process [127]. In this approach low-polydispersity end-functionalized PMMA and PBMA were obtained in one-pot. By action of a cobaloxime catalyst vinyl-capped oligomers of MMA were generated as proven by the absence of saturated species by ESI-MS and MALDI-TOF (see Figure 11.8). Although no further experimental data is provided, the synthesis of α-carboxylic acid ω-vinyl heterotelechelic counterparts has been accomplished. Other groups previously focused their attention on the synthesis of telechelic oligomers using a CCT process. For example, Haddleton and colleagues used a dimer of hydroxyethyl methacrylate as an addition fragmentation chain transfer agent [128] to obtain dihydroxy telechelic PMMA also bearing a double bond at its ω-end [129]. A few years later, the same authors employed an identical process with benzyl methacrylate as a transfer agent in the polymerization of MMA

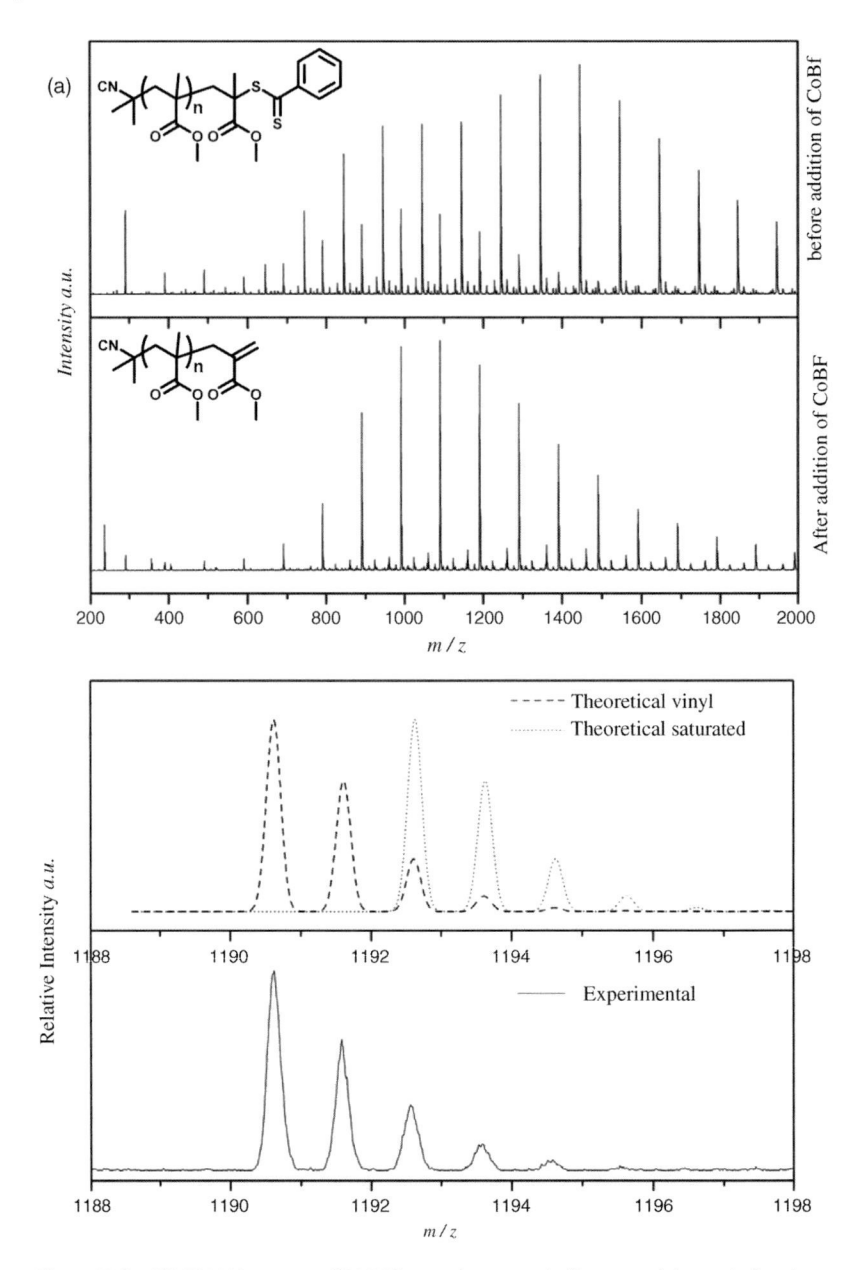

Figure 11.8 (A) ESI-MS spectra of PMMA prepared by RAFT polymerization using 2-(2-cyanopropyl)dithiobenzoate as a transfer agent ($M_n = 1000\,\text{g}\,\text{mol}^{-1}$) before (above) and after (below) addition of CoBF; (B) comparison between vinyl/saturated theoretical and experimental spectra for $DP_n = 11$. Reproduced with kind permission from Ref. [127]. Copyright 2010 Royal Society of Chemistry.

followed by catalytic hydrogenation to produce saturated dicarboxyl telechelic PMMA [130].

Gridnev *et al.* obtained ω-vinyl heterotelechelic oligomers by using the same MMA dimer approach but surprisingly did not invoke an addition fragmentation mechanism that would give in this case species indiscernable in a mass spectrum from species obtained by the simple dimer reinitiation mechanism they claimed [131]. In the same study, these authors also showed that it is possible to use unreactive olefins which are only able to initiate, thus giving an α-functional ω-vinyl oligomer by CCT polymerization, as proven by K^+ IDS MS (potassium ionization of desorbed species). A special case of CCT polymerization was reported by Morrison *et al.* with the use of 2-phenylallyl alcohol (PhAA) [132]. PhAA is actually not able to homopolymerize by CCT polymerization since it undergoes an isomerization to 2-phenylpropanal catalyzed by cobaloxime. The authors used this particularity to produce ω-functionalized PMMA: as soon as a PMMA macroradical adds a unit of PhAA, the chain end isomerizes to an aldehyde unable to propagate further

11.4
Novel and Minor Protocols

Recent years have seen the development of alternative protocols to control radical polymerization. Some of these have been assessed via mass spectrometric methods. Our group recently introduced the thioketone-mediated polymerization (TKMP) process [133]. TKMP is based on the idea that thioketones carrying substituents which will stabilize an adduct radical can impart living characteristics to a radical polymerization process. The working principle is in essence identical to that of NMP, with the difference that in TKMP the dormant species is a stabilized radical, whereas in NMP it is a nonradical species [134]. A TKMP process will only function if the adduct radical displays a sufficiently high lifetime to serve as an effective sink for the propagating radicals, which are thus reduced in their concentration. The TKMP process is thus related to the retardation phenomenon in some RAFT systems, where an additional level of control is induced via the longevity of the RAFT adduct radical [135, 136]. Soft ionization MS can contribute to unraveling the mechanism of TKMP systems via an analysis of the product spectrum. In TKMP systems that induce living behavior, products associated with bimolecular termination of the adduct radicals with propagating free radicals can be clearly identified [137]. Such an observation explains why typical TKMP systems show living behavior that is coupled with continuously broadening molecular weight distributions. However, ESI-MS on TKMP-prepared polymers clearly evidences the adduct radical – a testament to its stability [138]. Interestingly, an early SEC/ESI-MS study on RAFT polymers also provided evidence for the observation of RAFT adduct radicals as directly observable (and stable) species in mass spectra [90].

11.5
Conclusions

The present collation of mass spectrometric studies in living/controlled radical polymerization processes has demonstrated that soft ionization techniques coupled to suitable modern mass analyzers can provide in-depth molecular insight into the generated polymeric materials. In particular, these techniques are not only capable of providing information on the end-group fidelity of macromolecular building blocks prepared by radical techniques, but they also substantially contribute to an enhanced understanding of the underpinning polymerization mechanisms. Outstanding challenges – including an accurate detection of absolute concentrations of polymer chains – are most likely overcome by the emergence of hyphenated techniques with high-resolution chromatographic separations, as already demonstrated on selected examples. In particular, the further exploitation of the hyphenation of size-exclusion chromatography (SEC) with ESI-MS (online) as well as MALDI (offline) is envisaged to bring further advances in polymer chain quantification and in counteracting ionization suppression effects. Via the coupling of SEC to ESI-MS – especially in combination with high-resolution mass analyzers such as the Orbitrap – the multiply charged species occurring during ESI ionization may be further exploited to access molecular weight regimes exceeding 10–20 kDa. Additionally, physicists and developers of MS instrumentation are called upon to work on technical solutions to extend the mass ranges of mass analyzers, that is, m/z ratios, beyond the currently accessible 3000 to 4000 m/z – in collaboration with the polymer chemistry community. On the part of the ionization processes – especially MALDI – the use of less energetic laser irradiation or IR-MALDI may reduce the impact of in-source fragmentation on the more labile end-group chemistries often encountered in polymers prepared via living/controlled radical polymerization processes.

List of Abbreviations

AIBN	2,2'-azobisisobutyronitrile
ATRP	atom transfer radical polymerization
BA	*n*-butyl acrylate
BlocBuilder	*N*-(2-methylpropyl)-*N*-(1-diethylphosphono-2, 2-dimethylpropyl)-*O*-(2-carboxylprop-2-yl)hydroxylamine
BIPNO	2,2,5-trimethyl-4-(isopropyl)-3-azahexane-3-oxyl
CCT	catalytic chain transfer
CID	collision-induced decay
Cp	cyclopentadiene
DIOS	desorption ionization on silicon
DEAM	*N*,*N*-diethylacrylamide
ESCP	enhanced spin-capturing polymerization
ESI	electrospray ionization
FT-NIR	Fourier transform near-infrared

LACCC	liquid adsorption chromatography at critical conditions
LIFDI	liquid injection field desorption/ionization
L-SIMS	liquid secondary-ion mass spectrometry
MADIX	macromolecular design via the interchange of xanthates
MALDI	matrix-assisted laser desorption ionization
MMA	methyl methacrylate
MS	mass spectrometry
NAM	*N*-acryloylmorpholine
NIPAM	*N*-isopropylacrylamide
NMP	nitroxide-mediated polymerization
PAA	poly(acrylic acid)
PhAA	2-phenylallyl alcohol
PMDETA	N,N,N',N',N-pentamethyldiethylenetriamine
QIT	quadrupole ion trap
RAFT	reversible addition fragmentation chain transfer
SEC	size exclusion chromatography
SG1	*N-tert*-butyl-*N*-(1-diethylphosphono-2,2-dimethylpropyl nitroxide
t-BA	*t*-butyl acrylate
TEMPO	2,2,6,6-tetramethylpiperidine
TIPNO	2,2,5-trimethyl-4-phenyl-3-azahexane-3-oxyl
TOF	time of flight
UV	ultraviolet
VAc	vinyl acetate

References

1 Jenkins, A.D., Jones, R.G., and Moad, G. (2010) *Pure Appl. Chem.*, **82**, 483.
2 Gruendling, T., Weidner, S., Falkenhagen, J., and Barner-Kowollik, C. (2010) *Polym. Chem.*, **1**, 599–617.
3 Barner-Kowollik, C., Davis, T.P., and Stenzel, M.H. (2004) *Polymer*, **45**, 7791–7805.
4 Ladavière, C., Lacroix-Desmazes, P., and Delolme, F. (2008) *Macromolecules*, **42**, 70–84.
5 Hart-Smith, G., Lammens, M., Du Prez, F.E., Guilhaus, M., and Barner-Kowollik, C. (2009) *Polymer*, **50**, 1986–2000.
6 Hawker, C.J., Bosman, A.W., and Harth, E. (2001) *Chem. Rev.*, **101**, 3661–3688.
7 Jasieczek, C.B., Haddleton, D.M., Shooter, A.J., Buzy, A., Jennings, K.R., and Gallagher, R.T. (1996) *Am. Chem. Soc. Polym. Prepr.*, **37**, 451–452.
8 Dourges, M.-A., Charleux, B., Vairon, J.-P., Blais, J.-C., Bolbach, G., and Tabet, J.-C. (1999) *Macromolecules*, **32**, 2495–2502.
9 Bartch, A., Dempwolf, W., Bothe, M., Flakus, S., and Schmidt-Naake, G. (2003) *Macromol. Rapid Commun.*, **24**, 614–619.
10 Dempwolf, W., Flakus, S., and Schmidt-Naake, G. (2008) *Macromol. Symp.*, **275**, 166–172.
11 Dempwolf, W., Flakus, S., and Schmidt-Naake, G. (2007) *Macromol. Symp.*, **259**, 416–420.
12 Burguière, C., Dourges, M.-A., Charleux, B., and Vairon, J.-P. (1999) *Macromolecules*, **32**, 3883–3890.

13 Baumann, M., Roland, A.I., Schmidt-Naake, G., and Fischer, H. (2000) *Macromol. Mater. Eng.*, **280**, 1–6.

14 Girod, M., Phan, T.N.T., and Charles, L. (2008) *Rapid Commun. Mass. Spectrom.*, **22**, 3767–3775.

15 Perrin, L., Phan, T.N.T., Querelle, S., Deratani, A., and Bertin, D. (2008) *Macromolecules*, **41**, 6942–6951.

16 Farcet, C., Belleney, J., Charleux, B., and Pirri, R. (2002) *Macromolecules*, **35**, 4912–4918.

17 Dire, C., Belleney, J., Nicolas, J., Bertin, D., Magnet, S., and Charleux, B. (2008) *J. Polym. Sci., Part A: Polym. Chem.*, **46**, 6333–6345.

18 Schulte, T., Siegenthaler, K.O., Luftmann, H., Letzel, M., and Studer, A. (2005) *Macromolecules*, **38**, 6833–6840.

19 Bernhardt, C., Stoffelbach, F., and Charleux, B. (2010) *Polym. Chem.*, **2**, 229–235.

20 Delaittre, G., Rieger, J., and Charleux, B. (2010) *Macromolecules*, **44**, 462–470.

21 Lepoittevin, B., Perrot, X., Masure, M., and Hémery, P. (2001) *Macromolecules*, **34**, 425–429.

22 Beyou, E., Chaumont, P., Chauvin, F., Devaux, C., and Zydowicz, N. (1998) *Macromolecules*, **31**, 6828–6835.

23 Chau, W., Turner, R., and Braslau, R. (2008) *React. Funct. Polym.*, **68**, 396–405.

24 Wang, J.S. and Matyjaszewski, K. (1995) *J. Am. Chem. Soc.*, **117**, 5614–5615.

25 Kato, M., Kamigaito, M., Sawamoto, M., and Higashimura, T. (1995) *Macromolecules*, **28**, 1721–1723.

26 Sawamoto, M. and Kamigaito, M. (2001) *Chem. Rev.*, **101**, 3689–3745.

27 Matyjaszewski, K. and Xia, J. (2001) *Chem. Rev.*, **101**, 2921–2990.

28 Xia, J. and Matyjaszewski, K. (1997) *Macromolecules*, **30**, 7692–7696.

29 Gromada, J. and Matyjaszewski, K. (2001) *Macromolecules*, **34**, 7664–7671.

30 Jakubowski, W. and Matyjaszewski, K. (2005) *Macromolecules*, **38**, 4139–4146.

31 Matyjaszewski, K., Jakubowski, W., Min, K., Tang, W., Huang, J., Braunecker, W.A., and Tsarevsky, N.V. (2006) *Proc. Natl Acad. Sci. USA*, **103**, 15309–15314.

32 Jakubowski, W. and Matyjaszewski, K. (2006) *Angew. Chem. Int. Ed.*, **45**, 4482–4486.

33 Matyjaszewski, K., Jo, S.M., Paik, H.J., and Gaynor, G. (1997) *Macromolecules*, **30**, 6398–6400.

34 Matyjaszewski, K., Jo, S.M., Paik, H.J., and Shipp, D.A. (1999) *Macromolecules*, **32**, 6431–6438.

35 Lutz, J.-F. and Matyjaszewski, K. (2002) *Macromol. Chem. Phys.*, **203**, 1385–1395.

36 Bednarek, M., Biedron, T., and Kubisa, P. (1999) *Macromol. Rapid Commun.*, **20**, 59–65.

37 Bednarek, M., Biedron, T., and Kubisa, P. (2000) *Macromol. Chem. Phys.*, **201**, 58–66.

38 Datta, H., Bhowmick, A.K., and Singha, N.K.J. (2007) *J. Polym. Sci., Part A: Polym. Chem.*, **45**, 1661–1669.

39 Datta, H. and Singha, N.K.J. (2008) *J. Polym. Sci., Part A: Polym. Chem.*, **46**, 3499–3511.

40 Matyjaszewski, K., Nakagawa, Y., and Jasieczek, C.B. (1998) *Macromolecules*, **31**, 1535–1541.

41 Dervaux, B., van Camp, W., van Renterghem, L., and Du Prez, F. (2008) *J. Polym. Sci., Part A: Polym. Chem.*, **46**, 1649–1661.

42 Jackson, A.T., Bunn, A., Priestnall, I.M., Borman, C.D., and Irvine, D.J. (2008) *Polymer*, **49**, 5254–5261.

43 Kavitha, A.A. and Singha, N.K.J. (2008) *J. Polym. Sci., Part A: Polym. Chem.*, **46**, 7101–7113.

44 Singha, N.K. and de Ruiter, B. (2005) U.S. Schubert, *Macromolecules*, **38**, 3596–3600.

45 Singha, N.K., Rimmer, S., and Klumperman, B. (2004) *Eur. Polym. J.*, **40**, 159–163.

46 Singha, N.K. and German, A.L. (2007) *J. Appl. Polym. Sci.*, **103**, 3857–3864.

47 di Lena, F. and Matyjaszewski, K. (2008) *Chem. Commun.*, 6306–6308.

48 di Lena, F. and Matyjaszewski, K. (2009) *Dalton Trans.*, 8878–8884.

49 Zhang, H.Q., Tang, H.D., Tang, J.B., Shen, Y.Q., Meng, L.Z., Radosz, M., and Arulsamy, N. (2009) *Macromolecules*, **42**, 4531–4538.

50 Nonaka, H., Ouchi, M., Kamigaito, M., and Sawamoto, M. (2001) *Macromolecules*, **34**, 2083–2088.
51 Maria, S., Biedron, T., Poli, R., and Kubisa, P. (2007) *J. Appl. Polym. Sci.*, **105**, 278–281.
52 Le Grognec, E., Claverie, R., and Poli, R. (2001) *J. Am. Chem. Soc.*, **123**, 9513–9524.
53 Venkatesh, R. and Klumperman, B. (2004) *Macromolecules*, **37**, 1226–1233.
54 Borman, C.D., Jackson, A.T., Bunn, A., Cutter, A.L., and Irvine, D.J. (2000) *Polymer*, **41**, 6015–6020.
55 Mendrek, S., Mendrek, A., Adler, H.J., Walach, W., Dworak, A., and Kuckling, D. (2008) *J. Polym. Sci., Part A: Polym. Chem.*, **46**, 2488–2499.
56 Wolf, F.F., Friedemann, N., and Frey, H. (2009) *Macromolecules*, **442**, 5622–5628.
57 Pai, H.J., Teodorescu, M., Xia, J., and Matyjaszewski, K. (1999) *Macromolecules*, **32**, 7023–7031.
58 Vidts, K.R.M., Dervaux, B., and Du Prez, F.E. (2006) *Polymer*, **47**, 6028–6037.
59 Zhang, H.G., Jiang, X.L., and van der Linde, R. (2004) *Polymer*, **45**, 1455–1466.
60 Crecelius, A.C., Remzi Becer, C., Knop., K., and Schubert, U.S. (2010) *J. Polym. Sci., Part A: Polym. Chem.*, **48**, 4375–4384.
61 Bednarek, M. and Kubisa, P. (2004) *J. Polym. Sci., Part A: Polym. Chem.*, **42**, 608–614.
62 Dong, Y.Q., Tong, Y.Y., Dong, B.T., Du, F.S., and Li, Z.C. (2009) *Macromolecules*, **42**, 2940–2948.
63 Inglis, A.J., Paulöhrl, T., and Barner-Kowollik, C. (2010) *Macromolecules*, **43**, 33–36.
64 Shen, X., He, X., Chen, G., Zhou, P., and Huang, L. (2000) *Macromol. Rapid Commun.*, **21**, 1162–1165.
65 Monge, S., Giani, O., Ruiz, E., Cavalier, M., and Robin, J.-J. (2007) *Macromol. Rapid Commun.*, **28**, 2272–2276.
66 Langlotz, B.K., Fillol, J.L., Gross, J.H., Wadepohl, H., and Gade, L.H. (2008) *Chem. Eur. J.*, **14**, 10267–10279.
67 Mayadunne, R.T.A., Rizzardo, E., Chiefari, J., Chong, Y.K., Moad, G., and Thang, S.H. (1999) *Macromolecules*, **32**, 6977–6980.
68 Donovan, M.S., Lowe, A.B., Sumerlin, B.S., and McCormick, C.L. (2002) *Macromolecules*, **35**, 4123–4132.
69 Barner, L., Barner-Kowollik, C., Davis, T.P., and Stenzel, M.H. (2003) *Aust. J. Chem.*, **57**, 19–24.
70 Barner-Kowollik, C., Davis, T.P., Heuts, J.P.A., Stenzel, M.H., Vana, P., and Whittaker, M. (2003) *J. Polym. Sci., Part A: Polym. Chem.*, **41**, 365–375.
71 Barner-Kowollik, C. and Perrier, S. (2008) *J. Polym. Sci., Part A: Polym. Chem.*, **46**, 5715–5723.
72 Barner-Kowollik, C. (ed.) (2008) *Handbook of RAFT Polymerization*, Wiley-VCH GmbH & Co. KGaA, Weinheim.
73 Barner-Kowollik, C., Coote, M.L., Davis, T.P., Radom, L., and Vana, P. (2003) *J. Polym. Sci., Part A: Polym. Chem.*, **41**, 2828–2832.
74 Wang, A.R., Zhu, S., Kwak, Y., Goto, A., Fukuda, T., and Monteiro, M.S. (2003) *J. Polym. Sci., Part A: Polym. Chem.*, **41**, 2833–2839.
75 Perrier, S. and Takolpuckdee, P. (2005) *J. Polym. Sci., Part A: Polym. Chem.*, **43**, 5347–5393.
76 Favier, A. and Charreyre, M.-T. (2006) *Macromol. Rapid Commun.*, **27**, 653–692.
77 Moad, G., Rizzardo, E., and Thang, S.H. (2005) *Aust. J. Chem.*, **58**, 379–410.
78 Moad, G., Rizzardo, E., and Thang, S.H. (2006) *Aust. J. Chem.*, **59**, 669–692.
79 Moad, G., Rizzardo, E., and Thang, S.H. (2009) *Aust. J. Chem.*, **62**, 1402–1472.
80 Klumperman, B., van den Dungen, E.T.A., Heuts, J.P.A., and Monteiro, M.J. (2010) *Macromol. Rapid Commun.*, **31**, 1846–1862.
81 Vana, P., Albertin, L., Barner, L., Davis, T.P., and Barner-Kowollik, C. (2002) *J. Polym. Sci., Part A: Polym. Chem.*, **40**, 4032–4037.
82 Ganachaud, F., Monteiro, M.J., Gilbert, R.G., Dourges, M.-A., Thang, S.H., and Rizzardo, E. (2000) *Macromolecules*, **33**, 6738–6745.

83 Schilli, C., Lanzendoerfer, M., and Müller, A.H.E. (2002) *Macromolecules*, **35**, 6819–6827.

84 Kaufmann, R., Spengler, B., and Lutzenkirchen, F. (1993) *Rapid Commun. Mass Spectrom.*, **7**, 902–910.

85 Favier, A., Ladavière, C., Charreyre, M.-T., and Pichot, C. (2004) *Macromolecules*, **37**, 2026–2034.

86 D'Agosto, F., Hughes, R., Charreyre, M.-T., Pichot, C., and Gilbert, R.G. (2003) *Macromolecules*, **36**, 621–629.

87 Loiseau, J., Doerr, N., Suau, J.M., Egraz, J.B., Llauro, M.F., Ladavière, C., and Claverie, J. (2003) *Macromolecules*, **36**, 3066–3077.

88 Destarac, M., Charmot, D., Franck, X., and Zard, S.Z. (2000) *Macromol. Rapid Commun.*, **21**, 1035–1039.

89 Feldermann, A., Ah Toy, A., Davis, T.P., Stenzel, M.H., and Barner-Kowollik, C. (2005) *Polymer*, **46**, 8448–8457.

90 Ah Toy, A., Vana, P., Davis, T.P., and Barner-Kowollik, C. (2004) *Macromolecules*, **37**, 744–751.

91 Monteiro, M.J. and de Brouwer, H. (2001) *Macromolecules*, **34**, 349–352.

92 Venkatesh, R., Staal, B.B.P., Klumperman, B., and Monteiro, M.J. (2004) *Macromolecules*, **37**, 7906–7917.

93 Geelen, P. and Klumperman, B. (2007) *Macromolecules*, **40**, 3914–3920.

94 Zhou, G. and Harruna, I.I. (2007) *Anal. Chem.*, **79**, 2722–2727.

95 Bathfield, M., D'Agosto, F., Spitz, R., Ladavière, C., Charreyre, M.T., and Delair, T. (2007) *Macromol. Rapid Commun.*, **28**, 856–862.

96 Konkolewicz, D., Hawkett, B.S., Gray-Weale, A., and Perrier, S. (2008) *Macromolecules*, **41**, 6400–6412.

97 Konkolewicz, D., Hawkett, B.S., Gray-Weale, A., and Perrier, S. (2009) *J. Polym. Sci., Part A: Polym. Chem.*, **47**, 3455–3466.

98 Gruendling, T., Kaupp, M., Blinco, J.P., and Barner-Kowollik, C. (2011) *Macromolecules*, **44**, 166–174.

99 Rotzoll, R., Nguyen, D.H., and Vana, P. (2009) *Macromol. Symp.*, **275–276**, 1–12.

100 Chaffey-Millar, H., Hart-Smith, G., and Barner-Kowollik, C. (2008) *J. Polym. Sci., Part A: Polym. Chem.*, **46**, 1873–1892.

101 Hart-Smith, G., Chaffey-Millar, H., and Barner-Kowollik, C. (2008) *Macromolecules*, **41**, 3023–3041.

102 Boschmann, D. and Vana, P. (2005) *Polym. Bull.*, **53**, 231–242.

103 Tao, L., Kaddis, C.S., Ogorzalek Loo, R.R., Grover, G.N., Loo, J.A., and Maynard, H.D. (2009) *Macromolecules*, **42**, 8028–8033.

104 Vana, P., Albertin, L., Barner, L., Davis, T.P., and Barner-Kowollik, C. (2002) *J. Polym. Sci., Part A: Polym. Chem.*, **40**, 4032–4037.

105 Gruendling, T., Pickford, R., Guilhaus, M., and Barner-Kowollik, C. (2008) *J. Polym. Sci., Part A: Polym. Chem.*, **46**, 7447–7461.

106 Gruendling, T., Dietrich, M., and Barner-Kowollik, C. (2009) *Aust. J. Chem.*, **62**, 806–812.

107 Dietrich, M., Glassner, M., Gruendling, T., Schmid, C., Falkenhagen, J., and Barner-Kowollik, C. (2010) *Polym. Chem.*, **1**, 634–644.

108 Wiss, K.T. and Theato, P. (2010) *J. Polym. Sci., Part A: Polym. Chem.*, **48**, 4758–4767.

109 Gründling, T., Kaupp, M., Blinco, J.P., and Barner-Kowollik, C. (2011) *Macromolecules* **44**, 166–174.

110 Willcock, H. and O'Reilly, R.K. (2010) *Polym. Chem.*, **1**, 149–157.

111 Enikolopyan, N.S., Smirnov, B.R., Ponomarev, G.V., and Bel'govskii, I.M. (1981) *J. Polym. Sci. Polym. Chem. Ed.*, **19**, 879–889.

112 Davis, T.P., Haddleton, D.M., and Richards, S.N. (1994) *J. Macromol. Sci. Rev. Macromol. Chem. Phys.*, **C34**, 243–324.

113 Gridnev, A. (2000) *J. Polym. Sci., Part A: Polym. Chem.*, **38**, 1753–1766.

114 Heuts, J.P.A., Roberts, G.E., and Biasutti, J.D. (2002) *Aust. J. Chem.*, **55**, 381–398.

115 Gridnev, A.A. and Ittel, S.D. (2001) *Chem. Rev.*, **101**, 3611–3659.

116 Davis, T.P., Kukulj, D., Haddleton, D.M., and Maloney, D.R. (1995) *Trends Polym. Sci.*, **3**, 365–373.

117 Kukulj, D., Davis, T.P., Suddaby, K.G., Haddleton, D.M., and Gilbert, R.G. (1997) *J. Polym. Sci., Part A: Polym. Chem.*, **35**, 859–878.

118 Heuts, J.P.A., Forster, D.J., Davis, T.P., Yamada, B., Yamazoe, H., and Azukizawa, M. (1999) *Macromolecules*, **32**, 2511–2519.

119 Roberts, G.E., Barner-Kowollik, C., Heuts, J.P.A., and Davis, T.P. (2003) *Macromolecules*, **36**, 1054–1062.

120 Roberts, G.E., Davis, T.P., and Heuts, J.P.A. (2000) *Macromolecules*, **33**, 7765–7768.

121 Haddleton, D.M., Maloney, D.R., Suddaby, K.G., Clarke, A., and Richards, S.N. (1997) *Polymer*, **38**, 6207–6217.

122 Pierik, S.C.J. and van Herk, A.M. (2003) *Macromol. Chem. Phys.*, **204**, 1406–1418.

123 Muratore, L.M. and Davis, T.P. (2000) *J. Polym. Sci., Part A: Polym. Chem.*, **38**, 810–817.

124 Muratore, L.M., Steinhoff, K., and Davis, T.P. (1999) *J. Mater. Chem.*, **9**, 1687–1691.

125 Kukulj, D., Heuts, J.P.A., and Davis, T.P. (1998) *Macromolecules*, **31**, 6034–6041.

126 Chiu, T.Y.J., Heuts, J.P.A., Davis, T.P., Stenzel, M.H., and Barner-Kowollik, C. (2004) *Macromol. Chem. Phys.*, **205**, 752–761.

127 Soeriyadi, A.H., Boyer, C., Burns, J., Remzi Becer, C., Whittaker, M.R., Haddleton, D.M., and Davis, T.P. (2010) *Chem. Commun.*, **46**, 6338–6340.

128 Chiefari, J. and Rizzardo, E. (2002) *The Handbook of Radical Polymerization* (eds K. Matyjaszewski and T.P. Davis), Wiley and Sons, New York, NY, pp. 629–689.

129 Haddleton, D.M., Topping, C., Hastings, J.J., and Suddaby, K.G. (1996) *Macromol. Chem. Phys.*, **197**, 3027–3042.

130 Haddleton, D.M., Topping, C., Kukulj, D., and Irvine, D. (1998) *Polymer*, **39**, 3119–3128.

131 Gridnev, A.A., Simonsick, W.J., and Ittel, S.D. (2000) *J. Polym. Sci., Part A: Polym. Chem.*, **38**, 1911–1918.

132 Morrison, D.A., Eadie, L., and Davis, T.P. (2001) *Macromolecules*, **34**, 7967–7972.

133 Ah Toy, A., Chaffey-Millar, H., Davis, T.P., Stenzel, M.H., Izgorodina, E.I., Coote, M.L., and Barner-Kowollik, C. (2006) *Chem. Commun.*, 835–837.

134 Junkers, T., Stenzel, M.H., Davis, T.P., and Barner-Kowollik, C. (2007) *Macromol. Rapid Commun.*, **28**, 746–753.

135 Feldermann, A., Coote, M.L., Davis, T.P., Stenzel, M.H., and Barner-Kowollik, C. (2004) *J. Am. Chem. Soc.*, **126**, 15915–15923.

136 Chernikova, E., Golubev, V., Filippov, A., Lin, C.Y., and Coote, M.L. (2010) *Polym. Chem.*, **1**, 1437–1440.

137 Günzler, F., Junkers, T., and Barner-Kowollik, C. (2009) *J. Polym. Sci., Part A: Polym. Chem.*, **47**, 1864–1876.

12
Elucidation of Reaction Mechanisms: Other Polymerization Mechanisms

Grażyna Adamus and Marek Kowalczuk

12.1
Introduction

Understanding of polymerization mechanisms and a detailed description of the individual reactions taking place at the initiation, propagation, and termination of the polymer chain growth are essential from the point of view of the relations between polymer structure, properties, and function. Mass information alone may be insufficient to elucidate macromolecular structures, especially in the case of new polymerization methods and/or polymerization processes proceeding through complex and unknown mechanisms. In these cases, multistage tandem mass spectrometry (MS^n) can be utilized to determine how a polymer's constituents are connected to each other and to characterize individual end groups of polymers as well as to identify macromolecular sequences and architectures especially of copolymers [1]. Recent studies demonstrated that the collision energy necessary to drive fragmentation of polymers decreased in the order of polyethers > polymethacrylates > polyesters > polysaccharides. The characteristic collision energy to obtain 50% fragmentation, expressed as the characteristic collision voltage (CCV), was employed as a tool to compare different classes of polymers. These results suggest, that among the polymers studied polyesters fragment relatively readily (low CCV), while polyethers require the highest collision energy. [2] In the current chapter recent reports on the application of multistage MS methods for the elucidation of ionic and coordinative ring-opening polymerization (ROP) mechanisms of cyclic ethers and esters are presented. Furthermore, application of MS techniques for the evaluation of ring-opening metathesis polymerization (ROMP) and step-growth polymerization mechanisms, in which bifunctional or multifunctional monomers react to form oligomers and eventually long chain polymers, are discussed. In such cases, the situation is more complex since macromolecules with undefined initiating and terminating substituents are frequently formed especially in condensation processes.

12.2
Ring-Opening Polymerization Mechanisms of Cyclic Ethers

Among oxirane monomers, the most important ones for potential applications are ethylene oxide (EO) and propylene oxide (PO). Their polymerization can proceed via radical, ionic coordinate, and ionic mechanisms, yet only the latter ones have wide practical application. Anionic polymerization of EO may be initiated by bases, such as alcoholates, hydroxides, or other compounds of alkali or alkaline earth metals; their reaction mechanism was previously described [3]. Based on the principles of an ionic mechanism, microstructural, and conformational studies of polyether copolymers derived from EO and PO have been performed employing ESI-MS and MALDI-TOF-MS methods together with ESI-MS/MS. Since the properties of the of EO/PO copolymers depend on initiator, molecular weight, molecular weight distribution, and sequence, analysis of mass spectra, including accurate mass measurements, was performed together with ESI-MS/MS experiments which enable to establish individual fragmentation pathways. By utilizing the acquired data, including accurate mass information, the incorporated initiator molecule was identified and the random or block nature of the copolymer molecules was recognized. Moreover, a complete sequence of the comonomers was derived for the block EO/PO copolymer oligomers studied [4].

Remarkable progress in the EO polymerization has been recently observed when N-heterocyclic carbene-organocatalyzed ROP of this monomer has been developed [5].

Using N-heterocyclic carbene catalysts (NHC, Scheme 12.1) in the presence of alcohols or trimethylsilyl nucleophiles (NuE) as chain moderators, polyethylene oxide (PEO) of dispersities lower than 1.2 and molar masses matching the [EO]/[NuE] ratio were obtained. The controlled/living character of these NHC-catalyzed polymerizations has been proved via MALDI-TOF mass spectrometry among other techniques. The molecular structure of the synthesized PEOs revealed a quantitative introduction of Nu and OH groups at the chain ends as evidenced by MALDI-TOF MS analysis (Figure 12.1). Moreover, simulations of the theoretical isotope distributions of signals being attributed to the cationized adduct (with Na^+) of the targeted α-benzyl,ω-hydroxyl PEO were in agreement with the experimental ones (Figure 12.1b).

Cationic ROP of cyclic ethers may proceed via an activated monomer (AM) mechanism or via an active chain end (ACE) mechanism [6]. MALDI-TOF MS was employed for an evaluation of the AM mechanism of cationic ROP of ethylene oxide, initiated with ethylene glycol in the presence of acid-exchanged montmorillonite clay. Polymers with narrow dispersity were obtained and the molecular weights were controlled with the feed ratio of the monomer to the initiator. The MS studied confirmed the chemical structure of the polymer end groups (HO-$(EO)_n$-OH) and no cyclic byproducts were observed [7].

Polymerization of oxiranes may be initiated by ionic coordination catalysts with a metal, M, such as Al, Ca, Fe, Li, Sn, or Zn and ligands, Y, such as Br, Cl, NH_2, OH, OSnR′, or OR. These catalysts ionically coordinate with the oxirane oxygen, after

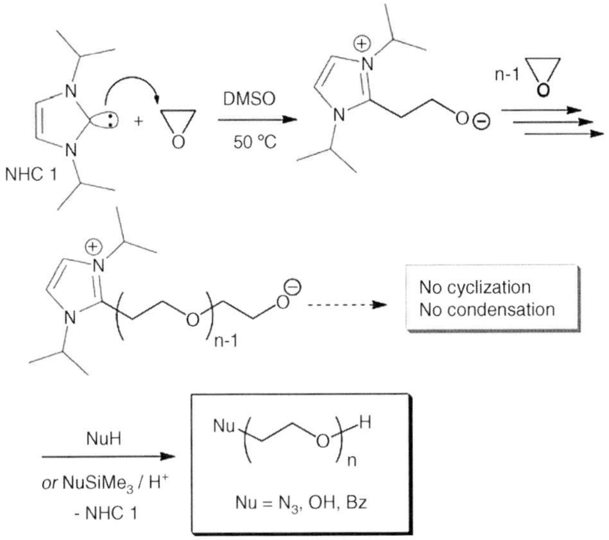

Scheme 12.1 Proposed mechanism of zwitterionic ring-opening polymerization of ethylene oxide. Reproduced with permission from Ref. [5].

which there is a nucleophilic attack to open the ring and form the species that propagates and forms high-molecular-weight polymer. Recently, mixtures of hindered (poly)phenols and alkylaluminum compounds have been used as catalysts for the polymerization of ethylene oxide (EO) and propylene oxide (PO). It was found that certain ligands bearing four phenol groups form particularly active catalysts when combined with triisobutylaluminum and triethylamine as initiator. Repeated chain-exchange reactions among aluminum complexes have been postulated and PEO of low M_w (produced under conditions of low EO/Al by the bisphenol based system) was analyzed by MALDI-TOF mass spectroscopy (MS). The resulting spectrum (Figure 12.2) indicates that indeed ring-opening by trialkylamine is the predominant initiation reaction pathway [8].

Hyperbranched macromolecules have become of increasing interest as a potential alternative for dendrimers since their synthesis do not require tedious, stepwise synthetic approaches. Such polymers are generally prepared by polymerization of ABm-type monomers, which leads to randomly branched structures. Glycidol is frequently used in the synthesis of branched polyethers and it has been polymerized cationically to branched polymers by Penczek and coworkers as well as Dworak and colleagues [9, 10]. The controlled anionic ROP of glycidol has been studied by Frey et al. using MALDI-TOF MS [11]. The hyperbranched polyols with polyether structure were obtained in a controlled manner and the degree of branching of the hyperbranched polyethers was determined. Partially deprotonated (10%) 1,1,1-tris(hydroxymethyl)propane (TMP) was used as an initiator for the anionic polymerization. MALDI-TOF MS studies evidenced a complete attachment of the hyperbranched structures to the TMP initiator and the absence of macrocyclics as well as hyperbranched macromolecules without initiator. In the further studies MALDI-TOF-MS

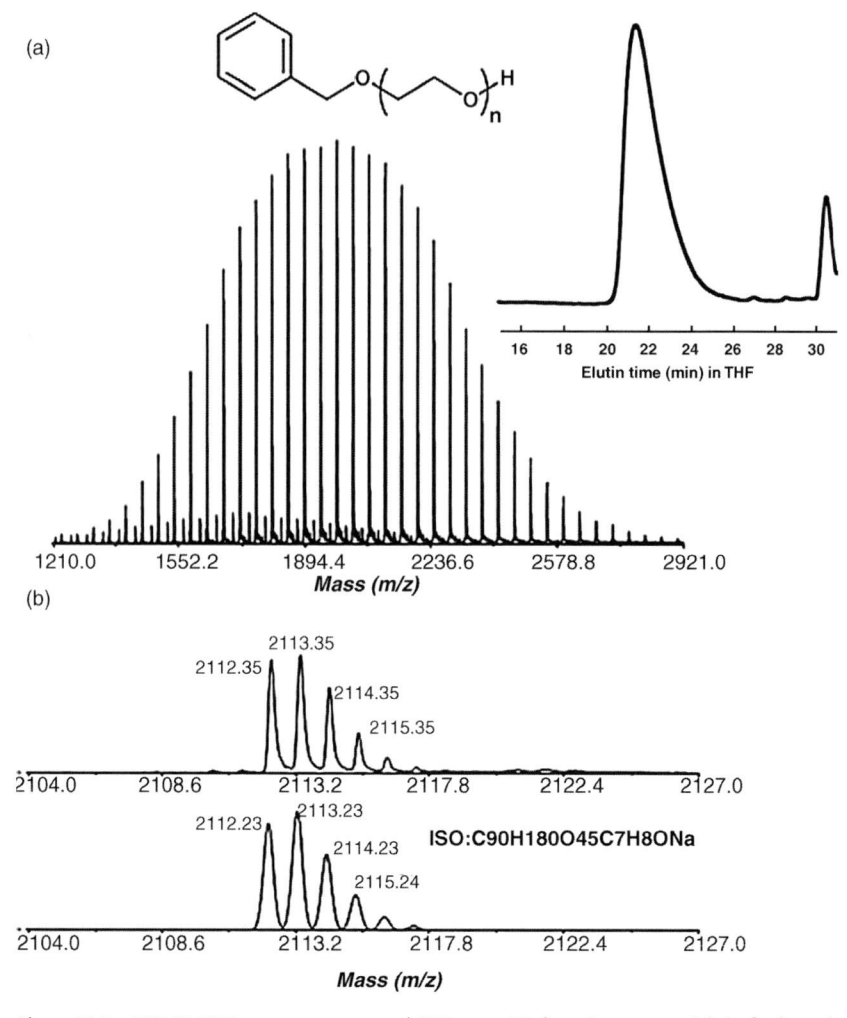

Figure 12.1 MALDI-TOF mass spectrum and SEC trace (RI detection on top right) of α-benzyl, ω-hydroxy polyethylene oxide. Reproduced with permission from Ref. [5].

was employed for the molecular characterization of poly(L-lactide) multiarm star polymers [12] as well as hyperbranched polyglycerol-based lipids [13].

12.3
Ring-Opening Polymerization Mechanisms of Cyclic Esters and Carbonates

Aliphatic polyesters have a leading position among the family of biodegradable polymers, since their hydrolytic and/or enzymatic degradation yields hydroxy acids which in most cases are metabolized. Regardless of the variety of biodegradable

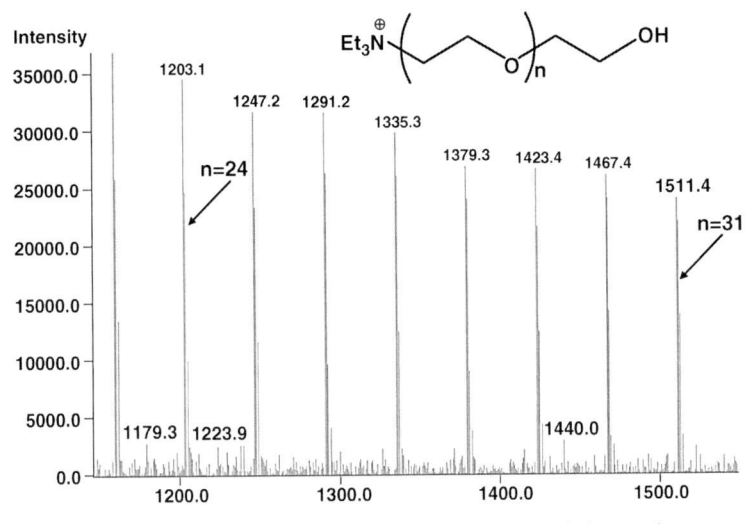

Figure 12.2 Portion of the MALDI-TOF mass spectrum of polyethylene oxide using a tetraphenol ligand/Al(i-Bu)$_3$/NEt$_3$ initiator system. Reproduced with permission from Ref. [8].

polyester materials available on the market, till now, there is no unique biodegradable polymer which can fulfill all requirements needed for specific applications in, for example, health care. Therefore, there is a continuous demand for the development of new monomers and catalytic systems as well as a detailed characterization of new polymer materials with well-defined architecture and properties. Polyesters are currently synthesized via two ways: by polycondensation, which will be described later, or by ROP of cyclic esters and their related compounds.

The chemistry of β-lactones has been of ongoing interest due to their usability as monomers for the preparation of biomimetic polymers. A suitable synthetic route to poly(β-hydroxyalkanoate)s (PHAs) is the ROP of β-alkylsubstituted four-membered β-lactones, which enable better control of molecular parameters (M_n; M_w/M_n) than step growth or biotechnological processes. Moreover, various racemic as well as enantiopure β-substituted-β-lactones have recently become available [14–17]. Depending on the type of initiator the ROP of four-membered β-lactones proceeds according to anionic, cationic, and coordination–insertion mechanisms.

It is known that in ROP of β-lactones, depending on the anionic initiator used, the anion can attack the lactone ring at the carbonyl carbon or carbon atom adjacent to the ether oxygen atom position leading to the acyl-oxygen or alkyl-oxygen bond cleavage with the formation of alkoxide or carboxylate propagation active species, respectively. However, during both initiation and propagation steps of anionic ROP of β-lactones, side-reaction such as transesterification and chain-transfer reaction can occur, which influence the chemical structure of chain-end-groups and molecular weight [17, 18]. The extent of these undesirable side-reactions strongly depends on the kind of the initiating system employed as well as the reaction conditions (temperature, solvent, concentration of monomer) under which the ROP of β-lactones is conducted. In the presence of typical anionic initiators, the mode of β-lactones ring scission depends strongly on the initiators nucleophilicity and the structure of the lactone monomer. The mechanism of anionic ROP of β-propiolactone and α,α-dialkylsubstitutet-β-lactones initiated by weak nucleophiles such as alkali metal carboxylates is well-known. These types of initiators open the alkyl-oxygen bonds of β-lactones via the formation of carboxylate active species. Moreover, as confirmed by NMR, these polyesters contain end groups derived from the initiator used. Further studies revealed that when the carboxylate initiators are activated with the crown ethers or cryptands β-butyrolactone polymerizes according to the above process [19]. The electrospray ionization tandem mass spectrometry technique (ESI-MS) allowed fast and reliable identification of the various macromolecules of poly(3-hydroxybutyrate) synthesized via ROP of β-butyrolactone and thus enabled verification of the polymerization mechanism. The correctness of the mechanism of anionic ROP of β-lactones initiated by weak nucleophiles was provided by ESI-MS structural studies of poly(3-hydroxybutyrate) samples prepared via polymerization of β-butyrolactone using carboxylate salts (e.g., potassium crotonate and sodium 3-hydroxybutyrate) as initiators. Evaluation of the structure of the resulting polyesters at the molecular level clearly evidenced that the propagation proceeds on the carboxylate anions and the polyesters obtained this way contain end groups derived from the initiator [20].

Adamus and Kowalczuk applied ESI-MS^n techniques (in positive- and negative-ion mode) for the study of the ROP of (R,S)-β-butyrolactone initiated by complexes of penicillin G potassium salt with 18-crown-6 [21]. Based on the MS results, these authors demonstrated that carboxylate anions of penicillin G open the alkyl-oxygen bonds of the (R,S)-β-butyrolactone via the formation of poly[(R,S)-3-hydroxybutyrate] macromolecules bearing penicyllin G and carboxylic end groups. In addition, MS^n fragmentation experiments performed in positive- and negative-ion mode confirmed the chemical structure of the end groups and showed some differences in the fragmentation pathway of the penicillin G covalently bonded to the poly[(R,S)-3-hydroxybutyrate] polymer chains in contrast to those observed previously by other authors for pure penicillin G. In further studies, it was demonstrated and confirmed by ESI-MS that the oligomeric (R,S)-3-hydroxbutyrate can be obtained via anionic ROP of (R,S)-β-butyrolactone when highly polar DMSO was employed as a solvent, activating the anionic polymerization species [22]. This finding was applied to the synthesis of ibuprofen-oligo(R,S)-3-hydroxybutyrate and aspirin-oligo(R,S)-3-hydroxbutyrate conjugates. In both cases, the structure of respective conjugates was evaluated with the aid of multistage MS ESI-MS^n. The ESI-MS^n fragmentation experiments of selected conjugate macromolecules confirmed that the respective initiator moieties (2-(4-isobutylphenyl) propionate or acetylsalicylate, respectively) are covalently bonded with polyester chains [23]. The versatility of mass spectrometry in the structural studies of dipeptide-based oligoconiugates, obtained via anionic ROP of β-butyrolactone with a dipeptide bearing two carboxylate groups as potassium salt, has been demonstrated by Buruiana et al. [24]. The application of multistage electrospray MS enabled the determination of the chemical structure of the individual conjugate macromolecules composed of 3-hydroxybutyrates oligomers covalently conjugated to dipeptide. Moreover, the results obtained indicated that the above-mentioned reaction is accompanied by formation of 3-hydroxybutyrate oligomers with crotonate and carboxyl end groups.

For a long time, discrepancies concerning the mechanism of the reactions of β-lactones with strong nucleophiles in aprotic solvents existed in the literature. Different initiation mechanisms were proposed when strong nucleophiles were used as initiators in anionic polymerization of these monomers [25–27]. The mechanistic studies performed by Jedliński and coworkers revealed that in contrast to the polymerization of β-lactones initiated by weak nucleophiles, in the case of the polymerization of these lactones initiated with strong nucleophiles (e.g., alkali metal alkoxides) initially the alkoxide initiator attacks the carbonyl carbon atom with a subsequent acyl-oxygen bond scission.

Subsequently, an elimination of the alkali metal hydroxide occurs with formation of an unsaturated ester, unreactive in further polymerization. According to the authors, the alkali metal hydroxide is an actual initiator of the polymerization and propagation proceeds via a carboxylate active species with alkyl-oxygen bond cleavage with a total inversion of the configuration (Scheme 12.2). Polymers with hydroxyl and unsaturated end groups are formed due to the elimination of water from the potassium salt of β-hydroxy acid formed during the initial step [28, 29]. The direct evidence of ROP, the mechanism of β-lactones, initiated with strong nucleophiles was provided by ESI-MS^n [20].

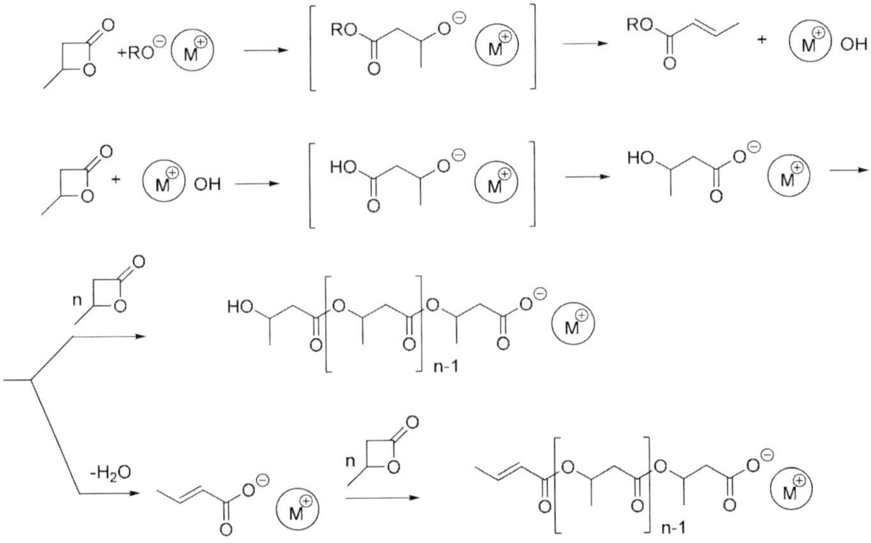

Scheme 12.2 Proposed mechanism for anionic ring-opening polymerization of β-butyrolactone initiated by alkali metal alkoxides.

The ESI-MS analysis of low-molecular-weight poly(3-hydroxybutyrate) obtained with potassium metoxide/18-crown-6 complex as an initiator revealed the presence (at each step of polymerization) of two series of molecular ions with a mass difference

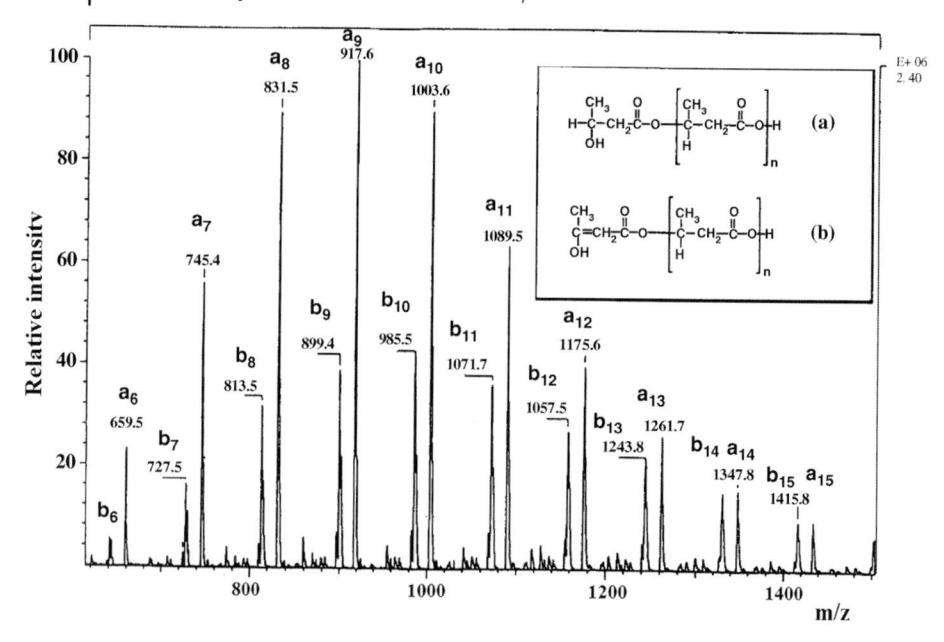

Figure 12.3 ESI-mass spectrum of poly(3-hydroxybutanoic acid) obtained via polymerization of β-butyrolactone initiated with a potassium methoxide/18-crown-6 complex. Two types of polymer chains, (a) and (b), are visible, showing different end groups. Reproduced with permission from Ref. [20].

of 18 Da (Figure 12.3). The respective series of ions represents two kind of poly(3-hydroxybutyrate) chains possessing the same degree of polymerization but containing different end groups that is, crotonate and hydroxybutyrate, respectively. Moreover, the molecular ions corresponding to polymer chains with end groups derived from the alkoxide initiator were not observed. The obtained results were in perfect agreement with previous findings based on NMR analysis [28, 30] and model reactions [31].

It was therefore demonstrated that poly(3-hydroxybutyrate) prepared via anionic polymerization of β-butyrolactone using potassium alkoxide/18-crown-6 as an initiator does not contain end groups derived from the initiator. Thus, an evaluation of the structure of poly(3-hydroxybutyrate) at the molecular level supports the proposed addition-elimination mechanism of the ROP of β-butyrolactone initiated by strong nucleophiles, since clearly macromolecules with either hydroxy or crotonate end groups only are formed.

Electrospray multistage MS (as a technique supplementary to NMR analysis) was employed in a structural study of polyesters obtained via anionic ROP of racemic α-methyl-β-pentyl-β-propiolactone initiated by supramolecular complexes of potassium methoxide and potassium hydroxide with 18-crown-6, respectively [32]. In this case, evaluation of the end groups of the resulting polymer (possessing a relatively long alkyl side chain in β-position) by NMR was difficult, due to the overlapping of the respective resonance signals, and no detailed information was provided. Structural studies with the aid of ESI-MS^n indicated that regardless of the anionic initiator used (potassium methoxide or potassium hydroxide) the mass spectra of the resulting poly (2-methyl-3-hydroxyoctanoate) contained two kind of polyester macromolecules terminated by 2-methyl-3-hydroxyoctanoate and carboxylate as well as (2-methyl-2-octenoate) and carboxylate end groups. On the basis of the end-groups analysis as well as the distribution of polyester macromolecules with hydroxyl and unsaturated end groups, the authors postulated that ROP of α,β-dialkyl-substituted-β-lactones follows the addition–elimination mechanisms previously proposed for simple β-lactones and potassium hydroxide acts as a real initiator in both systems. In further studies, Adamus and Kowalczuk applied ESI-MS^n to the studies of ROP of β-alkoxy-substituted-β-lactones initiated by a potassium acetate supramolecular complex, that is, tetrabutylammonium acetate (Bu_4N^+Ac), as well as by tetrabutylammonium hydroxide (Figures 12.4 and 12.5) [33].

Application of ESI-MS^n fragmentation experiments for structural studies enable the determination of molecular masses and structures of fragment ions of mass-selected macromolecules, thus showing the chemical nature of the polyesters studied and their end groups. Similar behavior of β-alkoxy-substituted-β-lactone with respect to β-alkyl-substituted-β-lactone and α,β-dialkyl-substituted-β-lactone (including side reactions leading to unsaturated end groups) was observed under conditions of anionic ROP [33]. Moreover, the authors observed that the distribution of the signals of macromolecules with unsaturated (4-alkoxy-2-butenoate) and carboxylate end groups in the ESI-mass spectra of polyesters obtained by ROP of β-alkoxy-substituted-β-lactones depends on the nucleophilicity of the initiator system employed. In the case of the ROP of β-alkoxy-substituted-β-lactones initiated via tetrabutylammonium acetate or supramolecular complexes of potassium acetate, the chains containing unsaturated end groups were distributed at the lower mass range of the polyester samples studied (Figure 12.4, series B). However, in the case when tetrabutylammonium hydroxide was used as a initiator, the macromolecules terminated with unsaturated end groups were observed for each step of polymerization and their distribution was similar to that of the hydroxyl ones (Figure 12.5, series B).

The authors concluded that formation of unsaturated end groups in the anionic ROP of β-alkoxy-β-lactones initiated by activated acetate is caused either by a chain transfer reaction to the monomer and/or by intermolecular carboxylate-induced α-deprotonation [34]. However, when tetrabutylammonium hydroxide is employed as an initiator, the unsaturated end groups are formed preferentially due to an elimination reaction and therefore their distribution is different to that observed in the case of polymerizations initiated by acetates (compare distribution of series B in Figures 12.4 and 12.5). The above findings were also observed in ring-opening copolymerization of β-alkoxy-substituted-β-lactone with β-butyrolactone [35].

Recently, Hendrick and coworkers and subsequently Coulembier and colleagues reported the application of N-heterocyclic carbenes as a reactive organic catalyst for the controlled ROP of cyclic esters such as lactides, ε-caprolactone, and β-butyrolactone. The mechanistic studies performed by Coulembier *et al.* suggested that the alcohol adducts of N-heterocyclic carbenes due to their reversible dissociation with

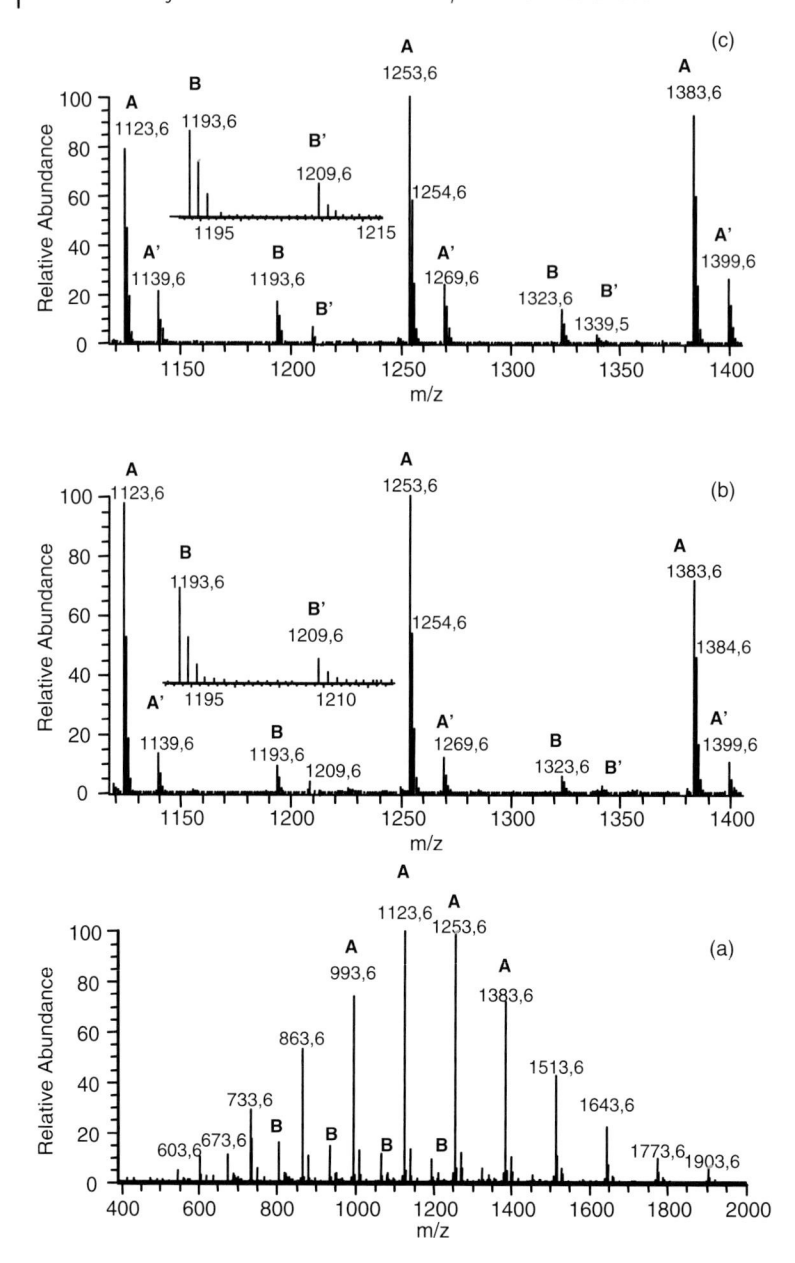

Figure 12.4 (a) ESI-mass spectrum (positive ion-mode) of poly(3-hydroxy-4-etoxybutyrate) synthesized via ROP of β-etoxymethyl-β-propiolactone initiated by a supramolecular complex of potassium acetate; (b) expansion of spectrum a) in the mass range m/z 1100–1420, (c) spectral expansion in the mass range m/z 1100–1420 of the ESI-mass spectrum of poly(3-hydroxy-4-etoxybutyrate) synthesized using tetreabutylamonium acetate. Series A represents polyester macromolecules terminated by acetate and carboxyl end groups; series B represents polyester macromolecules terminated by 4-etoxy-2-butenoate and carboxyl end groups. Reproduced with permission from Ref. [33].

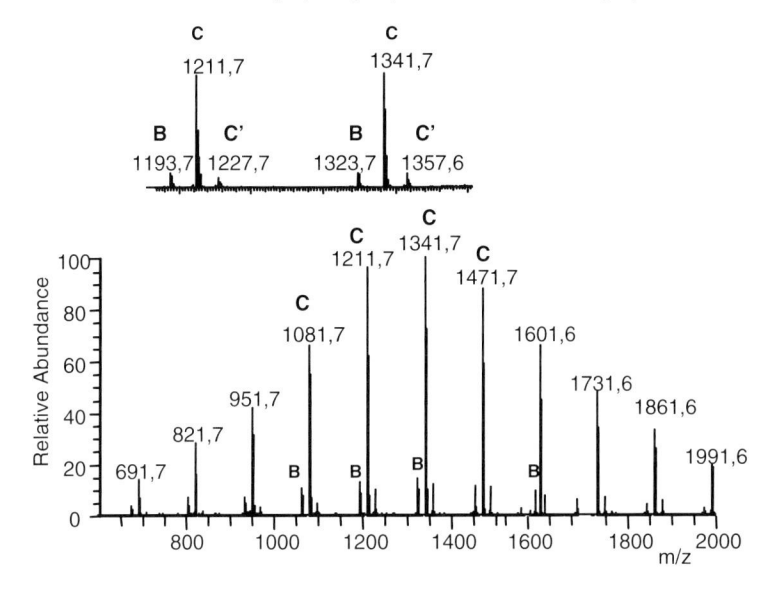

Figure 12.5 ESI-mass spectrum (positive ion-mode) of poly(3-hydroxy-4-etoxybutyrate) synthesized using tetrabutylammonium hydroxide. Series C represents polyester macromolecules terminated by hydroxyl (3-hydroxy-4-etoxybutyrate) and carboxyl end groups; series B represents polyester macromolecules terminated by 4-etoxy-2-butenoate and carboxyl end groups. Reproduced with permission from Ref. [33].

the formation of alcohol and triazolidene carbenes enable the ROP of cyclic esters which proceeds via the monomer activation mechanism.

According to the above mechanism, the free carbene opened the acyl-oxygen bond of the cyclic ester with the formation of intermediates of the hydroxyl-terminated polymer and free carbene (Scheme 12.3) [36, 37]. The formation of a

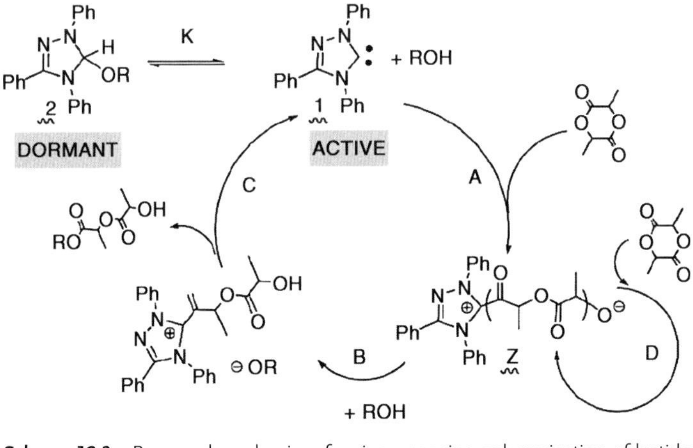

Scheme 12.3 Proposed mechanism for ring-openning polymerization of lactide with alkohol adduct of N-heterocyclic carbene. Reproduced from Ref. [37].

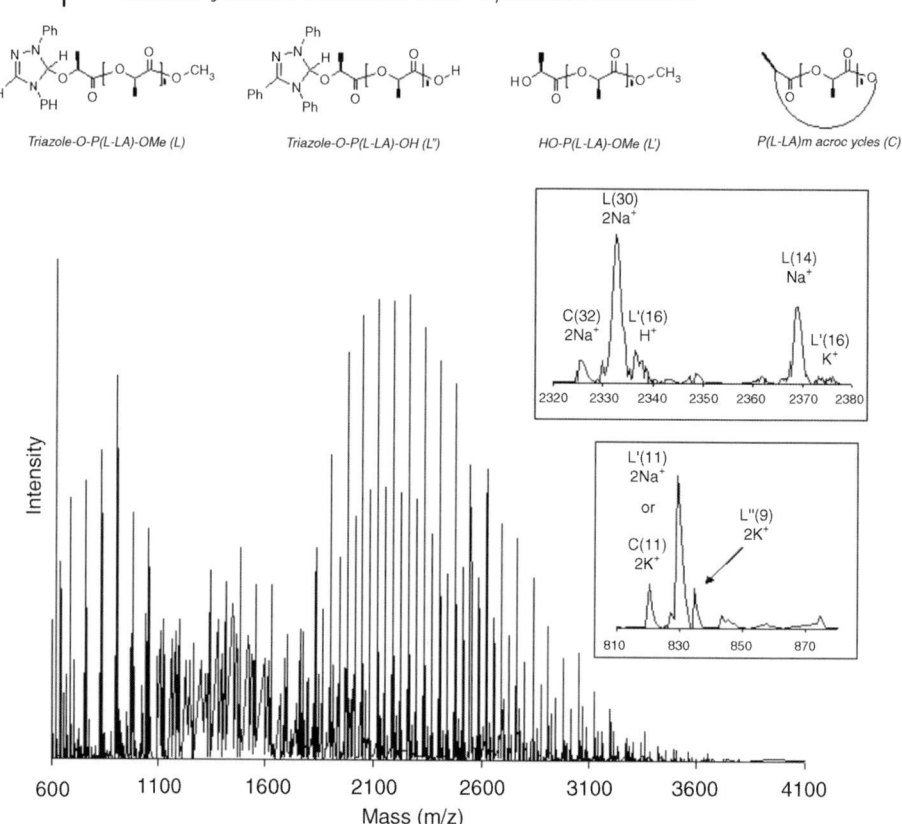

Figure 12.6 ESI mass spectrum of poly(L-lactide) no terminated by CS_2. DP_n of the poly(L-lactide) chains and macrocycles are given in parenthesis. Reproduced with permission from Ref. [37].

hydroxyl-terminated polymer–carbene adduct was confirmed by structural studies with the aid of ESI-MS of low-molecular-weight products obtained via ROP of L-lactide. The ESI-mass spectrum revealed the presence of two main series of periodically repeating signals. These signals represent the mono and doubly charged linear macromolecular triazole-O-P(L-LA)$_n$-OMe chains (L) (Figure 12.6). Moreover, the signals of lower intensities corresponding to the linear triazole-O-P(L-LA)$_n$-OH (L″) and HO-P(L-LA)$_n$-OMe (L′) molecular chains as well as P(L-LA) cyclic oligomers (C) were observed (Figure 12.6).

The studies performed by Coulembier *at al.* indicated that under optimum conditions the ROP of lactide in the presence of alcohol adducts of *N*-heterocyclic carbene occur upon complete consumption of the monomer with a concomitant reduction of undesirable transesterification reactions. However, when lactide is polymerized with an excess of carbene **1** to the alcohol, the polymerization is less controlled.

Moreover, these authors demonstrated that the application of the catalyst/initiator system for the ROP of β-butyrolactone enables the synthesis of PHB polymers with predictable molecular weight (up to 200 repeat units) and end groups. However, polymerization of this monomer targeting higher molecular weights is generally accompanied by long reaction times and the formation of macromolecules terminated by crotonate end groups.

Endo and coworkers studied the mechanism of anionic copolymerization of bis γ-lactone and epoxide (glicydyl phenyl ether) with potassium *tert*-butoxide based on MALDI-TOF-MS structural studies of the co-oligomers obtained [38]. The MALDI-TOF-mass spectra of these co-oligomers revealed the presence of three series of signals that can be assigned to the linear alternating co-oligomer macromolecules terminated by carboxylate chain ends or alkoxide chain ends and cyclic ones. The formation of three type of co-oligomers suggested that the copolymerization process studied was accompanied by two side reactions, namely intermolecular transesterification (causes the reduction of the molecular weight and transformation of alkoxide active chain ends into a carboxylate end group) and backbiting (formation of cyclic oligomers). Thus, from the information obtained from MALDI-TOF-mass spectra, the mechanism of the copolymerization was established.

The ROP of β-lactones in the presence of coordination–insertion initiators can proceed according to an O-alkyl and O-acyl bond cleavage mechanism. Although these initiators were abundantly employed for the polymerization of β-lactones, only limited mechanistic information could be found. In contrast, numerous studies of the polymerization mechanisms of larger lactone rings such as ε-caprolactone and lactide with various coordination–insertion initiators have been published. Two classes of initiators, that is, multivalent metal alkoxides and carboxylates were mostly used where ROP of larger lactones occurs selectively via O-acyl bond scission.

Although many studies of the polymerization mechanisms of larger lactones with a number of covalent metal alkoxides and carboxylates making use of ^{1}H and ^{13}C NMR spectroscopy have been published since 1960s, the first publications concerning mechanistic studies with the aid of MS techniques appeared only in the late 1990s.

Kricheldorf and coworkers applied classical MS, fast atom bombardment MS as well as MALDI-TOF MS in order to obtain direct experimental evidence for confirmation of the structure of the tin containing polylactone macrocycles formed via ring expansion polymerization of L-lactide, ε-caprolactone, β-D,L-butyrolactone, and 1,4-dioxane-2-one initiated by cyclic 2,2-dibutyl-2-stanna-1,3-dioxepane [39]. The authors in their previous studies established the formation of such macrocyclic polylactones by NMR techniques. The macromolecular peaks of the original polyesters were never detectable in chemical ionization mass spectra due to the complete thermal degradation via transesterification when the resulting macrocyclic oligolactones were subjected to MS. In the case of fast atom bombardment (FAB-MS) and (MALDI-TOF-MS) measurements, the alcohols or phenols used as matrices may cause hydrolysis of the Sn–O bonds of the macrocycles. Thus, the MALDI-TOF mass spectra showed the macromolecular peaks of linear OH-terminated telechelic polylactones. However, when the macrocyclic polylactones are stabilized by the formation of the more stable Sn–S bonds, MALDI-TOF mass spectra confirmed the macrocyclic structure of the polylactones studied.

Recently, the structure of multi- and triblock copolymers synthesized via ROP copolymerization of 1,5-dioxepan-2-one (DXO) with ε-caprolactone initiated by cyclic tin-alkoxide have been investigated by Albertson and coworkers [40]. MALDI-TOF mass spectra of different DXO/CL copolymers showed that regardless of the copolymer type or composition the main series of peaks in the mass spectra correspond to sodium-cationized linear chains terminated by hydroxyl and carboxylic end-groups. Moreover, OH-terminated telechelic macromolecules containing an $-O-CH_2-CH_2-O$ ether bridge in the middle, due to the insertion of monomer into the Sn−O bond and acyl-oxygen cleavage, as well as cyclic oligomers (formed as a result of intramolecular transesterification reactions) were identified. The high abundance of macromolecules terminated by hydroxyl and carboxylic acid end groups as well as cyclic oligomers indicate that side reactions take place during the synthesis as well as during the storage of the poly(ether-ester) samples studied.

Duda and coworkers applied MALDI-TOF MS for structural studies of poly(L-lactide) prepared via ROP of L-lactide initiated by tin(II) butoxide in THF at 80 °C [41]. The MALDI-TOF spectrum of the poly(L-lactide) indicated almost exclusively peaks corresponding to the $C_4H_9O(C(O)CH(CH_3)O)_nH$ polyester macromolecules doped with Na^+. Although tin(II) alkoxide active species belong to the most reactive ones in the case of lactide polymerization, they provide only a low extent of the intermolecular transesterification. The authors concluded that polymerization of LA initiated with $Sn(OBu)_2$ proceeds as with other covalent metal alkoxides.

Kricheldorf and coworkers presented the first results concerning the application a single-site bismuth initiator for the ROP of lactones [42]. The structural studies of the poly(ε-caprolactone) obtained with the use of diphenyl bismuth etoxide indicated that regardless of the polymerization temperature the MALDI-TOF mass spectra display the same pattern illustrated in Figure 12.7. The presence of one main series of signals corresponding to the linear polylactone chains La (having ethyl ester and CH_2OH end groups) was observed, which is in perfect agreement with the coordination–insertion mechanism proposed by these authors. Furthermore, a small amount of cyclic oligoesters C was found. The oligomers (regardless if cyclic or not) are not detectable as a separate maximum in the SEC elution curves and represent a weight fraction of <1%. The formation of cyclic oligoesters by backbiting was observed for the metal-alkoxide initiated polymerizations of lactones (and lactide) at temperatures of 100 °C. Yet, surprisingly cyclic oligolactones were found in this study even at a reaction temperature as low as 20 °C. However, only traces of cycles were detectable at this temperature.

Kricheldorf and coworkers also employed diphenyl bismuth etoxide as an initiator for the ROP of trimethylene carbonate [43]. The dependence of the molecular weight on the monomer/initiator ratio and MALDI-TOF-MS structural studies of the resulting poly(trimethylene carbonate) suggested that this polymerization obeys a coordination–insertion mechanisms. Due to the backbiting reaction cyclic oligomers were also formed and detected by MS.

Among the numerous catalytically active tin compounds, tin(II) 2-ethylhexanoate, (denoted as $SnOct_2$) plays a predominant role, as it is the most widely used initiator in the polymerization of cyclic esters for research purposes as well as for the technical

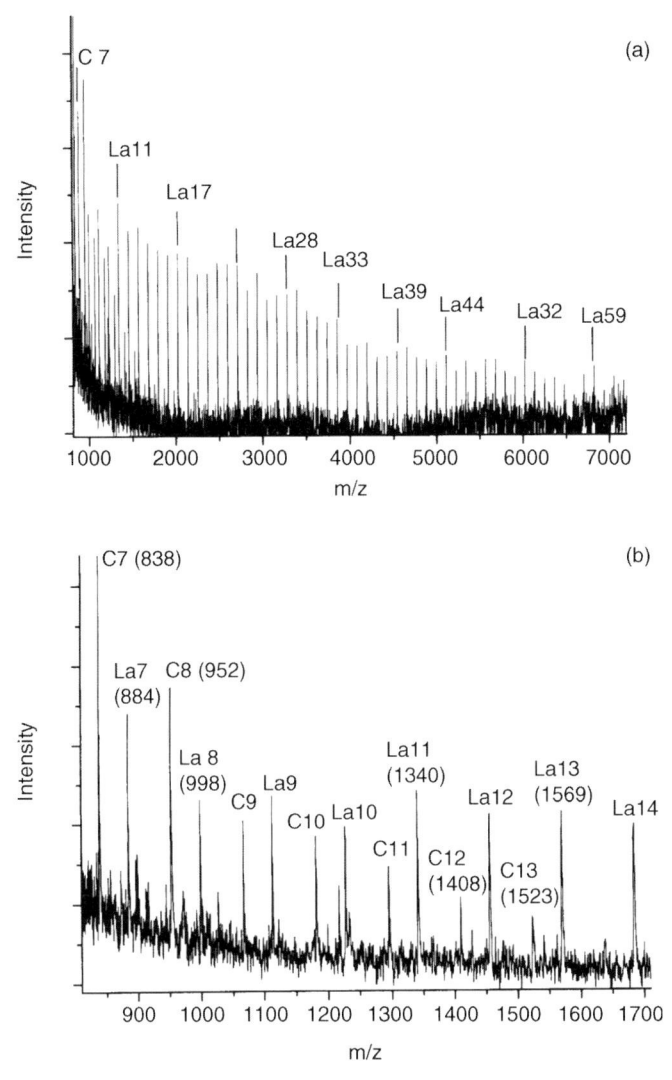

Figure 12.7 MALDI-TOF mass spectrum (a) and its extension (b) of poly(ε-caprolactone) initiated with Ph$_2$BiOEt at 60 °C in THF. Reproduced with permission from Ref. [42].

production. However, the reaction mechanism of Sn(Oct)$_2$ had not been fully elucidated for a long time. Therefore, the ROP of the lactones with Sn(Oct)$_2$ initiator has been systematically studied and several polymerization mechanisms were proposed. The proposed mechanisms can be divided into two categories. The first one involves the direct catalytic action of Sn(Oct)$_2$ [44, 45]. Sn(Oct)$_2$ has been proposed to activate monomer, forming a donor–acceptor complex which further participates directly in propagation. According to this mechanism, Sn(Oct)$_2$ is liberated in every propagation step. Thus, Sn(II) atoms are not covalently bound to the polymer chain at any stage of polymerization. The second group of authors has proposed an "active chain-end" mechanism. These authors assumed that Sn(Oct)$_2$ must be converted into tin(II) alkoxide in order to be able to initiate the polymerization. According to this mechanism, Sn(Oct)$_2$ reacts with compounds (purposely added or already present in the reacting mixture) containing hydroxide groups and gives a real initiator, that is, tin(II) alkoxide or hydroxide [46–51]. Then, the polymerization proceeds via the insertion of the monomer between the metal (tin (II)) alkoxide bond.

The studies of Sn(Oct)$_2$ as initiating system performed by Vert and coworkers stressed the complexity of this system and difficulties in understanding the detailed mechanism, because it is not possible to check the chemical nature of the end groups for high-molecular-weight polymers [52]. The experimental data which show that tin containing end groups could be detected in polyester macromolecules were provided by Duda and coworkers [53]. These authors used MALDI-TOF MS for structural studies of poly(ε-caprolactone) prepared via ROP of ε-caprolactone initiated with Sn (Oct)$_2$ in the presence of butyl alcohol (BuOH) or water as a coinitiator (conducted in tetrahydrofuran as a solvent at 80 °C).

It was assumed that for successful detection the Sn(II)-containing species connected with the polymer chains by MALDI-TOF-MS, at least two prerequisites should be fulfilled. A high starting concentration of Sn(Oct)$_2$ is needed to have macromolecules with M_n sufficiently low to detect Sn in isotopic profiles of individual macromolecules. In addition if there is too much water in the system the Sn(II)−O bonds will hydrolyze before the MALDI experiment is conducted.

The MALDI-TOF mass spectra of ε-caprolactone/Sn(Oct)/(butyl alcohol or water) reacting mixture recorded under living conditions enabled the direct observation of species with a tin atom covalently bonded with the polyester chain. The formation of the following populations of macromolecules was revealed: Bu[O(O)C(CH2)5]$_n$OSnOct (**A**), Bu[O(O)C(CH2)5]$_n$Oct (**B**), Bu[O(O)C-(CH2)5]$_n$OH,(**C**) H[O(O)C(CH2) 5]$_n$Oct, (**D**) H[O(O)C(CH2)5]$_n$OH, (**E**) macrocyclics [O(O)C(CH2)5]$_n$, (**F**) and macrocyclics with incorporated tin(II) alkoxide moieties [O(O)C(CH2)5]$_n$OSn (Figure 12.8).

The identification of the tin-containing macromolecules was not only based on the agreement between the observed m/z and the calculated molar mass values but also on the particular isotopic distribution provided by the tin atom (Figure 12.9).

The presence in the mass spectra signals representing a population of macromolecular chains containing Sn atoms in the chains (either linear and/or cyclic ones) constitutes a strong argument for the polymerization proceeding with Sn-alkoxides as active species.

In the further studies, the authors demonstrated that Sn(Oct)$_2$ itself also does not play an active role in the ROP of L,L-dilactide and that the L,L-dilactide/Sn(Oct)$_2$ system is mechanistically similar to the CL/Sn(Oct)$_2$ system [54]. The tin containing end groups were also observed in poly(L,L-lactide) macromolecules by MALDI-TOF MS. The authors conclude that L,L-dilactide/Sn(Oct)$_2$ polymerization proceeds by simple monomer insertion into the ...−Sn−OR bond, reversibly formed in the reaction ...−SnOct + ROH...−Sn−OR + OctH, where ROH is either the low-molar-mass coinitiator (an alcohol, hydroxy acid, or H$_2$O) or a macromolecule fitted with a

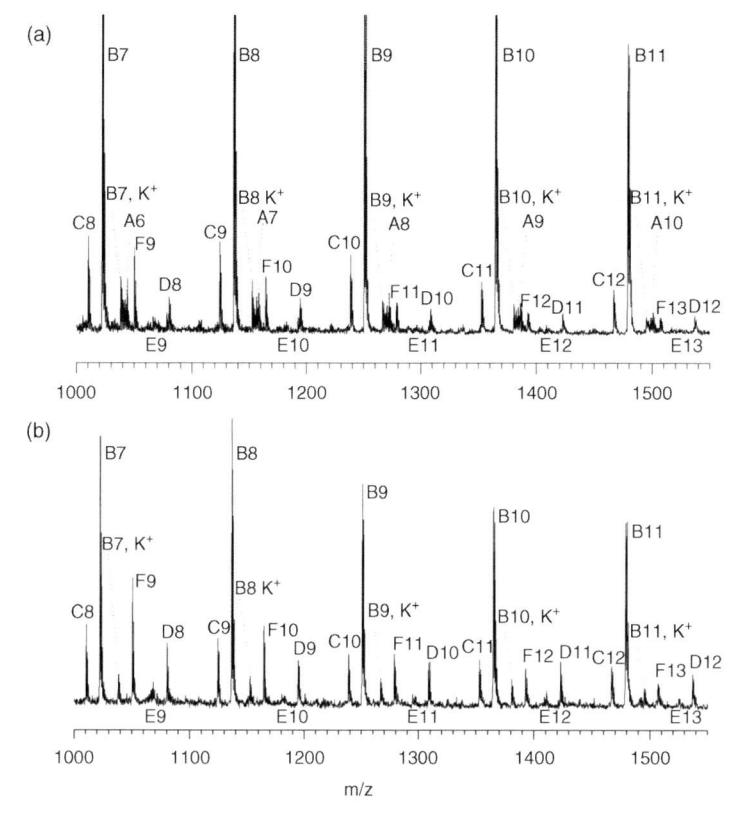

12.3 Ring-Opening Polymerization Mechanisms of Cyclic Esters and Carbonates

Figure 12.8 Comparison of the 1000–1550 m/z fragments in the MALDI-TOF mass spectra of the ε-caprolactone/tin octoate/butyl alcohol reacting mixture recorded under living conditions (a) and after the HCl aq treatment (b). Reproduced with permission from Ref. [53].

hydroxy end group. These interconversions take place throughout the entire polymerization process. Formation of the real initiator from $Sn(Oct)_2$ and a hydroxy group-containing compound (ROH) was also envisaged by kinetic arguments.

More recently, it was demonstrated that the mechanism of polymerization in the CL or LA by $Sn(Oct)_2$/primary amine (RNH_2) system does not differ appreciably from that coinitiated with alcohol (ROH) [55]. The MALDI-TOF mass spectra of the poly(ε-caprolactone) and poly(L-lactide) obtained with a $Sn(Oct)_2$/primary amine (RNH_2) system show almost exclusively the presence of signals corresponding to $C_4H_9NH-[C(O)(CH_2)_5O]_n-H$ and $C_4H_9NH-[C(O)CH(CH_3)O]_n-H$ chains (Figure 12.10).

Kricheldorf and coworkers investigated $SnOct_2$ in combination with 1,4-butanediol, 1,1,1-tris(hydroxymethyl)propane or pentaerythritol as initiator systems for the synthesis of telechelic polyesters or star-shaped homopolyesters of ε-caprolactone or L,L-lactide. Homo and copolymerizations of trimethylene carbonate (TMC) were also studied, yet they were plagued by intensive side reactions such as the formation of cyclic oligomers [56].

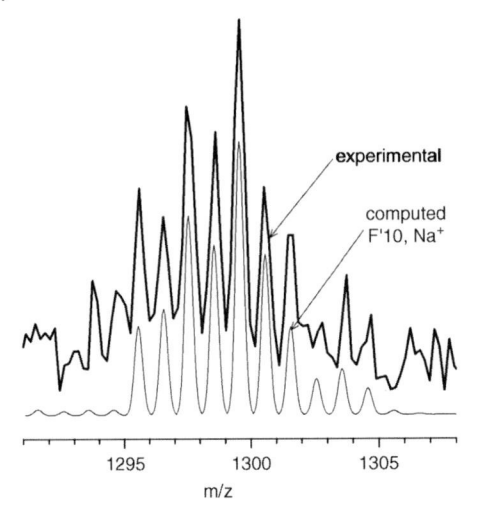

Figure 12.9 Comparison of the 1291–1308 m/z fragment of the MALDI-TOF mass spectrum of the ε-caprolactone/tin octoate/H$_2$O reacting mixture (bold line) with the isotopic distribution computed for the cyclic species: SnO-(CL)$_{10}$, Na$^+$ (F'10, Na$^+$) (thin line). Reproduced with permission from Ref. [53].

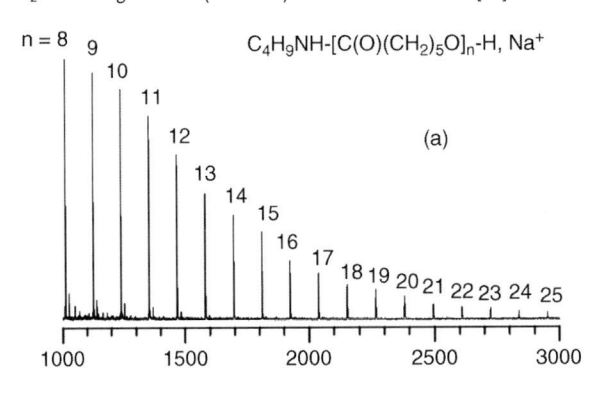

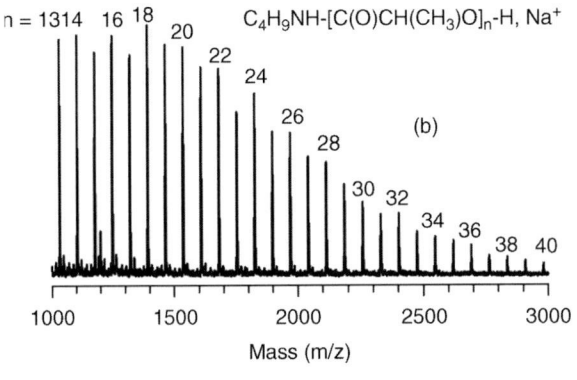

Figure 12.10 MALDI-TOF mass spectra (linear mode) of the ε-caprolactone/(Sn(Oct)2)]/(BuNH$_2$) (a) and (L,L-lactide)/Sn(Oct)$_2$/BuNH$_2$ (b) reacting mixtures. Reproduced with permission from Ref. [55].

Rokicki and coworkers investigated the influence of a catalyst and reaction conditions on the synthesis of oligocarbonate diols by the transesterification of propylene or ethylene carbonates with aliphatic diols [57]. Based on the information obtained from the MALDI-TOF-MS analysis of oligomeric products, these authors indicated that during the transesterification of cyclic propylene or ethylene carbonates with aliphatic diols (in the presence of coordination catalyst for example, tin carboxylate) only relatively small amounts of oxyethylene fragments were inserted into the oligocarbonate diols. Such information was valuable for these studies and the discussion of this reaction mechanism.

The reactions of the respective comonomer pairs that is, ethylene carbonate–propylene carbonate, ethylene carbonate-ε-caprolactone, and propylene carbonate-ε-caprolactone initiated by the *p-tert*-butylphenol/$KHCO_3$ system were investigated by Zsuga and coworkers [58]. The liquid chromatographic/electrospray ionization mass spectrometric characterization of the co-oligomers formed indicated the predominant presence of oligomers without carbonate linkages. Oligomers carrying carbonate linkages were also identified, however, their fraction was very small. Three major co-oligomer series were found in each case, which were identified as co-oligomers with *tert*-butylphenol and hydroxyl end groups as well as cyclic co-oligomers. Based on the LC-ESI MS results the authors proposed and discussed the mechanism of the formation of cyclic co-oligomers by backbiting reactions as well as formation of linear co-oligomers terminated by *tert*-butylphenol and hydroxyl end groups due to the chain degradation of co-oligomers containing carbonate linkages.

Rokicki and coworkers investigated the mechanistic aspects of the coordination polymerization of six-membered cyclic carbonates (5,5-dimethyl-1,3-dioxan-2-one, DTC) initiated by tin(II) alkoxide-based catalyst [59]. The MALDI-TOF mass spectrometric analysis of the polymerization products revealed that predominantly macrocyclic oligocarbonates are formed in the presence of a stannane catalyst. The lack of the mass spectrometric signals corresponding to cyclic oligocarbonates containing a residue diol originated from the catalyst used indicated that only one kind of Sn−O bond in the dimeric catalyst is active in the coordination–insertion of a cyclic carbonate monomer. Moreover, these authors observed that the addition of small amounts of diol or water into the system leads to the formation a linear product with hydroxyl end group in addition to the cyclic macromolecules.

12.4
Ring-Opening Metathesis Polymerization

ROMP is a chain growth polymerization process in which cyclic olefins are converted to polymeric macromolecules (Scheme 12.4). This metal-catalyzed redistribution of carbon–carbon double bonds belongs to generally known olefin metathesis, which stems from the Greek word meaning "changing places". ROMP of monomers containing strained, unsaturated rings was one of the earliest commercial applications

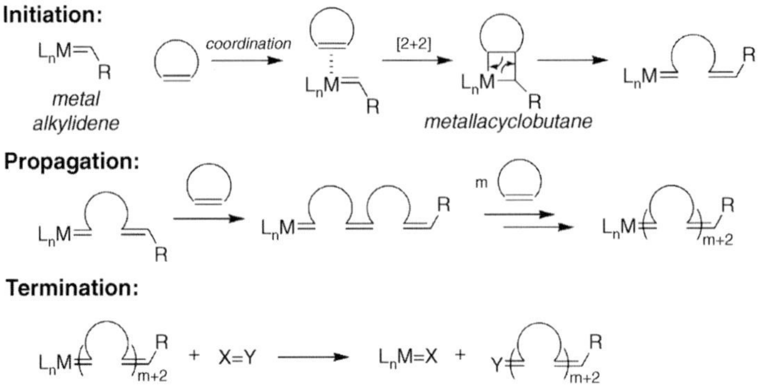

Scheme 12.4 General mechanism of a typical ring-opening metathesis polymerization. Reproduced from Ref. [60].

of olefin metathesis. A review of the fundamental aspects of living ROMP together with its historical development from a catalyst-design perspective has been published by Grubbs and coworkers [60]. The driving force for ROMP is the ring-strain release which determines the irreversible nature of ROMP. A general mechanism for ROMP obeys initiation which begins with coordination of a transition metal alkylidene complex to a cyclic olefin and following [2 + 2]-cycloaddition to a four-membered metallacyclobutane intermediate which starts the polymer chain growth.

Various organometallic ROMP catalysts have been studied including the so-called ill-defined catalysts and more defined titanium, talantum, tungsten, molybdenum, and ruthenium organometallic systems. Despite of the considerable progress in organometallics, there are major gaps in understanding of mechanistic details, due to the complexity of transition-metal catalysis in conjunction with difficulties to perform proper kinetic studies. Recently, transition-metal catalysts have been studied in the gas phase via ESI-MS, where either selected components of the catalytic systems or the complete reaction mixtures were evaluated and, in several cases, a direct online monitoring of reactions occurring in solution by coupling to ESI-MS has been realized [61]. For example Hofmann and coworkers applied this new screening methodology, which combines *in situ* synthesis of complexes with an assay by ESI-MS, to investigate highly active, cationic ruthenium–carbene catalysts in ring-opening metathesis polymerization. It was establish that the reactivity trends determined in the gas phase parallel solution-phase reactivity. Moreover, the overall rate in a solution is also determined by a favorable dimer/monomer preequilibrium providing the active catalyst by facile dissociation of dicationic, dinuclear catalyst precursors. The study demonstrated that the rapid assay of a variety of structural effects on metathesis rate, combined with mechanistic analysis enables optimization of the catalyst structure for given applications [62]. Electrospray ionization MS and subsequent MS/MS analyses were used by Metzger and coworkers to study two first-generation ruthenium catalysts. It has been shown that reactions of first-generation ruthenium olefin metathesis catalysts in a solution can be studied by ESI-MS. The authors detected and characterized 14-electron ruthenium intermediates directly from solution and demonstrated the catalytic activity of these species directly by gas phase reaction with ethylene [63]. Mass spectrometric methods have been applied by Chen and coworkers for the experimental determination of the phosphine binding energy in first- and second-generation ruthenium metathesis catalysts. The gas-phase values estimated by deconvolution of the energy-resolved collision-induced dissociation cross-sections of electrosprayed ruthenium carbene complexes were consistent with the quantum chemical calculations as well as solution-phase results [64]. MALDI methods have been used for monitoring of ROMP of various monomers using several catalytic systems. Binder *et al.* demonstrated that in comparison with the classical kinetic analysis (which enables the qualitative monitoring of the chain-growth reaction of the ROMP process) MALDI mass spectrometric method allows the monitoring of the reaction directly at the point of the crossover reaction, thus enabling a better evaluation of the polymerization [65]. The study was based on the crossover reactions of four structurally different monomers in a ROMP-based block copolymerization and were conducted via (a) kinetic analysis and (b) MALDI MS. While the kinetic analysis revealed large differences between the use of Grubbs first- and third-generation catalysts (with Grubbs third-generation catalyst being the significantly faster and more efficient catalyst during the crossover reaction), MALDI analysis revealed large differences in the quality and efficiency of the crossover process. Recently, MALDI-TOF-MS has been applied successfully for evaluation of grafting-through ROMP using ruthenium N-heterocyclic carbene catalysts. The synthesis of bottle-brush polymers has been performed by Grubbs and coworkers. Bivalent-brush polymers were prepared by grafting-through ROMP of drug-loaded poly(ethylene glycol) macromonomers. Using this synthetic approach anticancer drugs, i.e., doxorubicin and camptothecin were attached to a norbornene-alkyne-PEG macromonomers via a photocleavable linker. The MALDI spectra confirmed the expected mass increase after the coupling reaction [66].

As mentioned above, a very significant amount of new ROMP catalysts have been prepared but only a few structures provide excellent results. It is therefore reasonable to agree with Robert Grubbs that, "given the increasing rate at which new catalysts are now appearing, we look forward to further surprises and control mechanisms" [67]. There may be a hope that MS techniques will help to establish the correct ones.

12.5
Mechanisms of Step-Growth Polymerization

The mechanism of step-growth polymerization refers to a process where bi-functional or multifunctional monomers react to form oligomers and eventually long-chain polymers. However, a high extent of reaction is required to achieve high-molecular-weight polymer. Pioneering works of Montaudos on statistical modeling of the mass spectral intensities of copolymers provided information on the distribution of monomers along the copolymer chain and has been applied to determine the composition and microstructure of copolymers (both addition and condensation types) [68]. Since then the distribution of functionality and the determination of the extent of cyclization was predominantly studied using various MS techniques. The application of MS is of particular importance for the understanding of polycondensation reactions of esters and amides. In 2005, a review on MS characterization of polymers derived via step-growth polymerization including polycondensation and polyadditon has been published [69]. In this review pathways for dendritic and hyperbranched polymers, supramolecular polymers and nano condensates have been also discussed.

Recently, MALDI-TOF/TOF-MS/MS was employed by Montaudo and coworkers to analyze poly(butylene adipate) oligomers and to investigate their fragmentation pathways. Oligomers terminated by carboxyl and hydroxyl groups, methyl adipate and hydroxyl groups, dihydroxyl groups, and dicarboxyl groups were studied. It was found that different end groups do not influence the fragmentation of sodiated polyester oligomers and similar series of product ions were observed in all MALDI-TOF/TOF-MS/MS spectra. According to the structures of the most abundant product ions identified in this study, three fragmentation pathways have been proposed to occur most frequently: β-hydrogen transfer rearrangement, leading to the selective cleavage of the $-O-CH_2-$ bonds, $-CH_2-CH_2-$ (beta–beta) bond cleavage in the adipate moiety as well as ester bond scission [70].

MS studies were conducted to identify products obtained by heating blends of condensation polymers (such as polyesters–polycarbonate, polyester–polyamide, and polyester–polyester) where exchange reactions occurred. It was shown that MALDI TOF MS (and MS in general) can be used to monitor the yield of the reactive blending reactions by measuring the amount of unreacted homopolymer [71].

Ballistreri and coworkers employed MALDI-TOF MS to determine the end groups present in the samples of copolymers of (R)-3-hydroxybutyric acid (HB) and ε-caprolactone (CL) synthesized by transesterification of the corresponding homopolymers in solution in the presence of 4-toluenesulfonic acid. Due to the fact that the copolyesters studied possess a rather high dispersity, the authors fractionated the samples and subsequently recorded the spectra of a few selected fractions. Based on the results obtained from MALDI-TOF mass spectra of the SEC fractions of the studied copolyester, those authors concluded that samples rich in HB units contained mostly hydroxyl- and carboxyl-terminated species, and the samples rich in CL units contained mostly tosyl- and carboxyl-terminated species. These results strengthened the evidence gathered by NMR analysis, that is, that the tosyl groups are linked only to terminal CL units [72].

Similar copolyesters synthesized by transesterification of the corresponding homopolymers, i.e., atactic poly[(R,S)-3-hydroxybutyrate], a-PHB, and poly(L-lactide) (PLLA) or poly(ε-caprolactone) (PCL), respectively, were investigated by Adamus et al. [73]. The structure of the individual copolyester macromolecules, including end-group chemical structures, was established initially using MALDI-TOF-MS and then SEC/MALDI-TOF-MS.

The MALDI-TOF mass spectra of the investigated copolymers show the occurrence of linear hydroxyl- and carboxyl-terminated chains. In the case of poly[(R,S)-3HB-co-CL] copolyesters, the presence of small amounts of cyclic copolyester

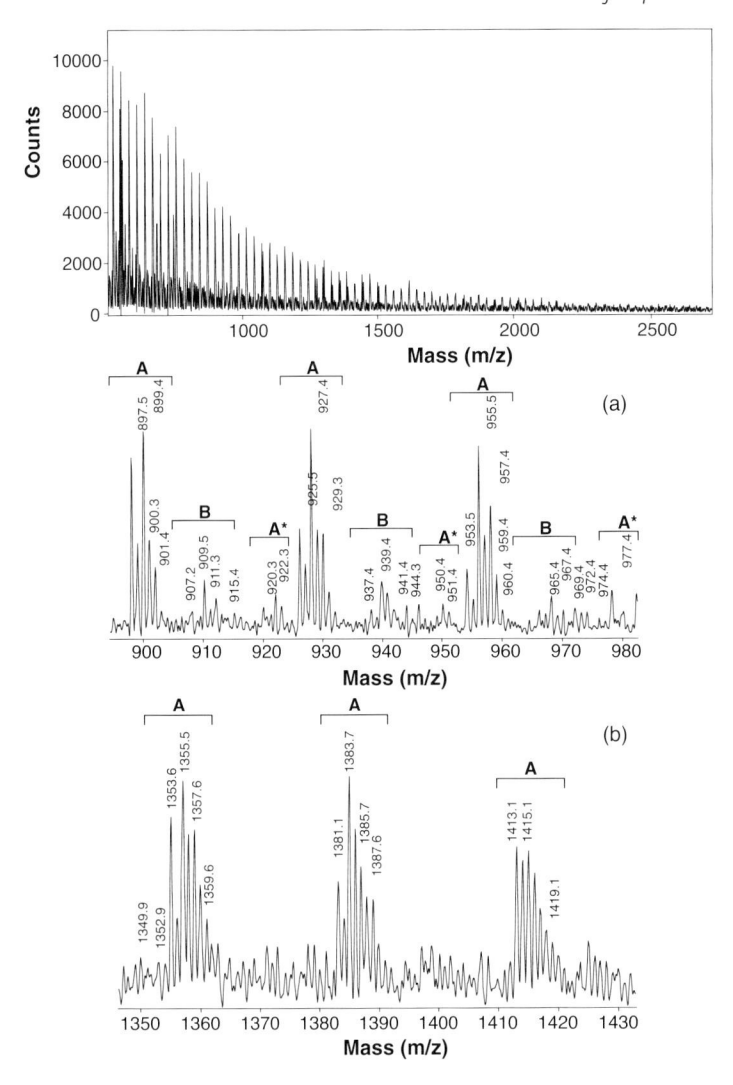

Figure 12.11 MALDI-TOF mass spectrum of unfractionated poly[(R,S)-3-hydroxybutyrate-co-(ε-caprolactone)] copolyester and spectral expansions in the mass ranges: (a) m/z 895–980 and (b) m/z 1345–1430. Series A corresponding to linear hydroxyl- and carboxyl-terminated copolyester macromolecules; series B corresponding to cyclic copolyester oligomers and/or linear copolyester chains terminated with tosyl and carboxyl end groups. Reproduced with permission from Ref. [65].

oligomers and/or linear copolyester chains terminated with tosyl and carboxyl end groups were also observed (Figure 12.11). However, in the latter case, the MALDI-TOF-MS technique was found to be of limited value for the differentiation of the cyclic byproducts and/or linear species with the tosyl moiety derived from the catalyst, due to the overlapping of the respective ions.

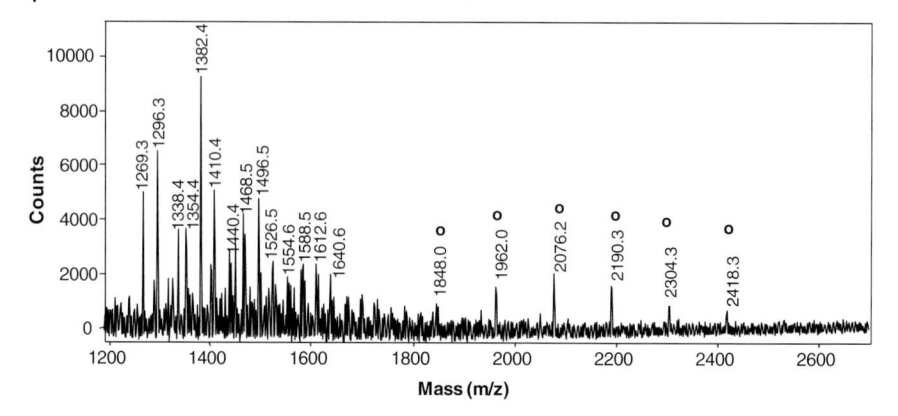

Figure 12.12 MALDI-TOF mass spectrum in reflector mode of the narrow SEC fraction of poly[(R, S)-3-hydroxybutyrate-co-(ε-caprolactone)]. Series denoted o represent cyclic oligomers. Reproduced with permission from Ref. [65].

The presence of linear and cyclic oligomers in poly[(R,S)-3HB-co-CL] copolyester samples was confirmed by MALDI-TOF-MS analysis of separated narrow fractions. Due to the lower hydrodynamic volume of cyclic species than of linear ones with the same molecular backbone, cyclic macromolecules with higher masses are eluted at the same retention volume as lower mass linear oligomers (Figure 12.12).

The sequence distribution was determined using the signal intensities of the ions, corresponding to individual copolyester chains, in the MALDI-TOF mass spectra. Furthermore, sequence analysis gave information about the degree of transesterification. The authors concluded that in this type of transesterification reaction the microstructure of the starting homopolymers as well as the experimental conditions of the reaction strongly affect the structure of the resulting copolyesters.

The development of synthetic strategies toward well-defined polyester oligomers is sometimes needed for understanding the performance of these materials in a variety of applications and allows deeper insight into their physical and biomaterial properties. Recently, a series of individual polyester oligomers up to the 64-mer was prepared via an exponential growth strategy [74]. The molecular weights and associated purity for all of the oligomers were verified by both electrospray and MALDI TOF MS. In all cases, molecular ions of the desired oligomeric species were observed with little or no contamination. High-resolution mass spectral analysis was also found to correspond with the anticipated molecular formulas.

Adamus et al. reported a detailed molecular and structural characterization of the copolyester oligomers obtained via facile transesterification of (R,S)-β-butyrolactone with two isomeric hydroxy acids (6-hydroxyhexanoic or (R,S)-2-hydroxyhexanoic acids) conducted in bulk without catalyst at 70 °C with the aid of ESI-MS/MS spectrometry [75]. The mass spectra of the studied oligocopolyesters have enabled the identification of their molecular structures. The linear copolyester oligomers terminated by hydroxyl and carboxylate end groups were identified in the ESI mass spectra. Moreover, the mass spectra provided information about the composition and sequence distribution. The random arrangements of co-monomer structural units along the oligocopolyester chains were confirmed by ESI-MS/MS experiments of the individual oligocopolyesters ions with different composition and an investigation of their fragmentation pathways. In further studies, Adamus characterized with the aid of ESI-MS/MS technique poly[(R,S)-3-hydroxybutyrate-co-L-lactide] oligocopolyesters prepared by equimolar reaction of (R,S)-β-butyrolactone with L-lactic acid and in a mirror experiment by reaction of (R,S)-3-hydroxybutyric acid with L-lactide, respectively, conducted in bulk without catalyst at 70 °C [76]. In both cases linear copolyester oligomers terminated by hydroxyl and carboxylate end groups were identified in the mass spectra. Moreover, the mass spectra of the copolyester obtained via the reaction of (R,S)-3-hydroxybutyric acid with L-lactide contained signals corresponding to the copolyester macromolecules with both even and odd numbers of LA units. The last ones were formed due to the intermolecular transesterification that occurs during the synthesis. However, it is interesting to note that no intramolecular transesterification to cyclic by-products was observed in either reaction. The structure of the end groups established by ESI-MS/MS analysis indicated that both acydolysis and alcoholysis reactions took place during the formation of the copolyester oligomers. However, in the equimolar reaction of (R,S)-3-hydroxybutyric acid with L-lactide the alcoholysis is the main reaction. The mass spectra provided information regarding the sequence distribution and indicated that despite the synthetic pathway applied, random poly[(R,S)-3HB-co-LA] copolyesters were predominantly formed. Small differences in the microstructure of the copolyester samples of P[(R,S)-3HB-co-LA] (depending on the synthetic pathway) have been detected based on the ESI-MS/MS experiment performed for the individual copolyester ions selected from both samples and the investigation of their fragmentation pathways.

Kricheldorf and coworkers investigated the role of cyclization in the synthesis of polyamide-6 by high-temperature polycondensation of ε-caprolactame with ε-aminocaproic acid [77]. The MALDI-TOF-MS analysis applied toward the characterization of products obtained after different reaction times indicated that only two types of polyamide products are formed, that is, cyclic and linear (terminated by amino and carboxylic end group) macromolecular chains. Moreover, the authors observed that the molar ratio of cyclic versus linear polyamide increased together with the increase of molecular weight, and the mass spectra of high-molar-mass samples exclusively displayed only the signals of the cycles. Regardless of the feed ratio of ε-caprolactame/ε-aminocaproic acid, after sufficiently long reaction time, all reaction products achieve the same molecular weight which indicated that the chain growth was limited by cyclization at the given reaction temperature (250 °C). The authors concluded that these results are in good agreement with the proposed theory of thermodynamically controlled polycondensations (TCPs). According to this theory, the content of cycles in polyamide-6 should varies from 0% at low conversions (<80%) up to 100% at 100% conversion.

MALDI TOF MS has been employed for studying the transurethanisation reaction of α,ω-bis(hydroxycthyloxycarbonylamine) alkanes with longer chain diols and the structure of the resulting product was confirmed [78].

Applying partial acid hydrolysis to polyester–polyurethanes, a series of hydroxy-terminated oligomers were obtained. This methodology was found to be selective and hydrolyses ester bonds while leaving the urethane groups containing polyurethane hard segments completely intact. Such a tool enables the characterization of polyurethanes using size exclusion chromatography and matrix-assisted laser desorption ionization MS (SEC/MALDI) [79].

Due to their three-dimensional topologies, easily tailored functionality, and unique solution, bulk, and self-assembly properties, dendritic polymers are subject of special scientific interest. Most applications of the physical and photophysical properties of dendrimers have been covered in a recent review [80]. MS has been widely used for the characterization of such structures to accurately calculate the repeat unit and end-group masses. For the preparation of dendrimers with polymeric repeat units characterization of well-defined azide-functionalized macromolecules is frequently needed. Recently, a variety of different monoazide functionalized polymers was examined, including polycaprolactone (PCL), poly(ethylene glycol), as well as polystyrene, and evidence of metastable ions was provided. Metastable ions are formed in MALDI-TOF MS when a parent ion is generated during laser desorption fragments at some point during the flight path in the field free region between the ion source and the detector. In this study the post-source metastable ions were identified by a non-uniform, non-integer mass off-set relative to the parent ion, and were confirmed by their disappearance using linear mode detection [81].

12.6
Concluding Remarks

ROP prepared polymers are extremely useful materials particularly in the area of biomedical and pharmaceutical applications. They can also be employed as carries in various delivery systems for drugs, bioactive cosmetic compounds, and pesticides. Thus, there is a continuous need for molecular level characterization of such materials with well-defined architecture including end groups. Moreover, opportunities of controlled synthesis of polymers with desired topology requires detailed verification of the precursor structures as well as the synthesized for example, hyperbranched macromolecules. For all such needs MS offers new opportunities due to continuous development of MS instrumentation. MS is nowadays a routine technique for the characterization of polymers and the expansion in its application for polymer analysis has been indicated in the scientific literature since the late 1980s after the revolutionary development of the MALDI and ESI ionization techniques [82]. This trend was also observed in the publications on the ROP of cyclic esters and ether as well as in research devoted to ROMP, polycondensation, and polyaddition processes. Among these studies only a limited number of published reports was focused on the elucidation of novel polymerization mechanisms or verification of already proposed mechanistic pathways. There are still open mechanistic or microstructural questions that are to be addressed especially for substituent's effects as well as the nature and structure of active intermediates. This particularly concerns the possibility of employing simple organic molecules as ROP catalysts or promoters, providing alternatives to organometalic catalysts as the removal of the metal contaminant is important for biomaterial applications. Resolving the outstanding questions besetting the ROP mechanism may benefit from the application of relatively new mass analyzer techniques such as Orbitraps and FT-ICRs, which feature a strongly improved sensitivity and mass resolving power. Multistage MS techniques contribute most to the field of structural characterization of ROP, ROMP and step-growth prepared (co)polymers. However, there is still a need to elucidate and understand the fragmentation mechanisms taking place during the MS^n experiments.

Successful examples of MS application in evaluation of some novel polymerization mechanisms as well as verification of already proposed ones in ROP of selected oxacyclic monomers and in ring-opening metathesis polymerizations as well as step-growth polymerization systems were discusses in the current chapter. Special emphasis has been given to carbene-organocatalyzed ROP of cyclic ethers and esters, anionic ROP of four membered ring lactones, coordination–insertion ROP of lactones and lactide as well as cyclic carbonates. The MS studies of ROMP and polycondensation reactions to polyesters and polyamides were also addressed.

References

1 Wesdemiotis, C., Solak, N., Polce, M.J., Dabney, D.E., Chaicharoen, K., and Katzenmeyer, B.C. (2011) Fragmentation pathways of polymer ions. *Mass Spectrom. Rev.* **30** (4), 523–559.

2 Nasioudis, A., Memboeuf, A., Heeren, R.M.A., Smith, D.F., Vkey, K., Drahos, L., and van den Brink, O.F. (2010) Discrimination of polymers by using their characteristic collision energy in tandem mass spectrometry. *Anal. Chem.*, **82** (22), 9350–9356.

3 Koleske, J.V. (1996) *Poly(ethylene oxide) (overview)* in Polymeric Materials Encyclopedia, vol. 8 (ed. J.C. Salamone), CRC Press, Boca Raton, FL, pp. 6036–6037.

4 Jackson, A.T., Scrivens, J.H., Williams, J.P., Baker, E.S., Gidden, J., and Bowers, M.T. (2004) Microstructural and conformational studies of polyether copolymers. *Int. J. Mass Spectrom.*, **238** (3), 287–297.

5 Raynaud, J., Absalon, C., Gnanou, Y., and Taton, D. (2009) N-heterocyclic carbene-induced zwitterionic ring-opening polymerization of ethylene oxide and direct synthesis of α,ω-difunctionalized poly(ethylene oxide)s and poly(ethylene oxide)-*b*-poly(ε-caprolactone) block copolymers. *J. Am. Chem. Soc.*, **131** (9), 3201–3209.

6 Kubisa, P. and Penczek, S. (1999) Cationic activated monomer polymerization of heterocyclic monomers. *Prog. Polym. Sci.*, **24** (10), 1409–1437.

7 Yahiaoui, A., Hachemaoui, A., and Belbachir, M. (2009) Cationic polymerization of ethylene oxide with maghnite-H as a clay catalyst in the presence of ethylene glycol. *J. Appl. Polym. Sci.*, **113** (1), 535–540.

8 Tang, L., Wasserman, E.P., Neithamer, D.R., Krystosek, R.D., Cheng, Y., Price, P.C., He, Y., and Emge, T.J. (2008) Highly active catalysts for the ring-opening

polymerization of ethylene oxide and propylene oxide based on products of alkyl aluminum compounds with bulky tetraphenol ligands. *Macromolecules*, **41** (20), 7306–7315.

9 Tokar, R., Kubisa, P., Penczek, S., and Dworak, A. (1994) Cationic polymerization of glycidol: coexistence of the activated monomer and active chain end mechanism. *Macromolecules*, **27** (2), 320–322.

10 Dworak, A., Walach, W., and Trzebicka, B. (1995) Cationic polymerization of glycidol. Polymer structure and polymerization mechanism. *Macromol. Chem. Physic.*, **196** (6), 1963–1970.

11 Sunder, A., Hanselmann, R., Frey, H., and Mulhaupt, R. (1999) Controlled synthesis of hyperbranched polyglycerols by ring-opening multibranching polymerization. *Macromolecules*, **32** (13), 4240–4246.

12 Wolf, F.K. and Frey, H. (2009) Inimer-promoted synthesis of branched and hyperbranched polylactide copolymers. *Macromolecules*, **42** (24), 9443–9456.

13 Hofmann, A.M., Wurm, F., Huhn, E., Nawroth, T., Langguth, P., and Frey, H. (2010) Hyperbranched polyglycerol-based lipids via oxyanionic polymerization: toward multifunctional stealth liposomes. *Biomacromolecules*, **11** (3), 568–574.

14 Rowley, J.M., Lobkovsky, E.B., and Coates, G.W. (2007) Catalytic double carbonylation of epoxides to succinic anhydrides: catalyst discovery, reaction scope, and mechanism. *J. Am. Chem. Soc.*, **129** (16), 4948–4960.

15 Schmidt, J.A.R., Lobkovsky, E.B., and Coates, G.W. (2005) Chromium(III) octaethylporphyrinato tetracarbonylcobaltate: a highly active, selective, and versatile catalyst for epoxide carbonylation. *J. Am. Chem. Soc.*, **127** (32), 11426–11435.

16 Church, T.L., Getzler, Y.D.Y.L., Byrne, C.M., and Coates, G.W. (2007) Carbonylation of heterocycles by homogeneous catalysts. *Chem. Commun* (7), 657–674.

17 Carpentier, J.-F. (2010) Discrete metal catalysts for stereoselective ring-opening polymerization of chiral racemic β-lactones. *Macromol. Rapid Comm.*, **31** (19), 1696–1705.

18 Coulembier, O., Degée, P., Hendrick, J.L., and Dubois, P. (2006) From controlled ring-opening polymerization to biodegradable aliphatic polyester: Especially poly(β-malic acid) derivatives. *Prog. Polym. Sci.*, **31** (8), 723–747.

19 Jedliński, Z. (1996) *Polyesters (synthesis from lactones) in Polymeric Materials Encyclopedia*, vol. 8 (ed. J.C. Salamone), CRC Press, Boca Raton, FL, pp. 5897–5902.

20 Jedliński, Z., Adamus, G., Kowalczuk, M., Schubert, R., Szewczuk, Z., and Stefanowicz, P. (1998) Electrospray tandem mass spectrometry of poly (3-hydroxybutanoic acid) end groups analysis and fragmentation mechanism. *Rapid Commun. Mass Spectrom.*, **12** (7), 357–360.

21 Adamus, G. and Kowalczuk, M. (2000) Electrospray multistep ion trap mass spectrometry for the structural characterisation of poly[(R,S)-3-hydroxybutanoic acid] containing a β-lactam end group. *Rapid Commun. Mass Spectrom.*, **14** (4), 195–202.

22 Juzwa, M. and Jedliński, Z. (2006) Novel synthesis of poly(3-hydroxybutyrate). *Macromolecules*, **39** (13), 4627–4630.

23 Zawidlak-Węgrzyńska, B., Kawalec, M., Bosek, I., Luczyk-Juzwa, M., Adamus, G., Rusin, A., Filipczak, P., Głowala-Kosińska, M., Wolańska, K., Krawczyk, Z., and Kurcok, P. (2010) Synthesis and antiproliferative properties of ibuprofen–oligo(3-hydroxybutyrate) conjugates. *Eur. J. Med. Chem.*, **45** (5), 1833–1842.

24 Buruiana, E.C., Kowalczuk, M., Adamus, G., and Jedliński, Z. (2008) Designing of dipeptide-based oligoconjugates as potential carrier for drug delivery. Pyroglutamil-S-glutamic acid bisoligo-3-hydroxybutyrates. *J. Polym. Sci. Pol. Chem.*, **46** (12), 4103–4111.

25 Yamashita, Y., Tsuda, T., Iida, H., Uchikawa, A., and Kuriama, Y. (1968) Anionic copolymerization of β-lactones in correlation with the mode of fission. *Die Macromoleculare Chemie*, **113**, 139–146.

26 Hofman, A., Słomkowski, S., and Penczek, S. (1984) Structure of active centers and mechanism of the anionic polymerization of lactones. *Die Makromolekulare Chemie*, **185** (1), 91–101.

27 Kricheldorf, H.R. and Scharnagl, N. (1989) Polylactones. 17. Anionic polymerization of β-D,L-butyrolactone. *J. Macromol. Sci. Chem.*, **A26**, 951–968.

28 Kurcok, P., Kowalczuk, M., Hennek, K., and Jedliński, Z. (1992) Anionic polymerization of β-lactones initiated with alkali-metal alkoxides: Reinvestigation of the polymerization mechanism. *Macromolecules*, **25** (7), 2017–2020.

29 Jedliński, Z., Kowalczuk, M., Kurcok, P., Adamus, G., Matuszowicz, A., Sikorska, W., Gross, R.A., Xu, J., and Lenz, R.W. (1996) Stereochemical control in the anionic polymerization of β-butyrolactone initiated with alkali-metal alkoxides. *Macromolecules*, **29** (11), 3773–3777.

30 Kurcok, P., Kowalczuk, M., and Jedliński, Z. (1994) Response to "On the ambident reactivity of β-lactones in their reactions with alcoholates initiating polymerization". *Macromolecules*, **27** (17), 4833–4835.

31 Kurcok, P., Kowalczuk, M., and Jedliński, Z. (1993) Reactions of β-lactones with potassium alkoxides and their complexes with 18-crown-6 in aprotic solvents. *J. Org. Chem.*, **58** (16), 4219–4220.

32 Arkin, A.H., Hazer, B., Adamus, G., Kowalczuk, M., Jedliński, Z., and Lenz, R.W. (2001) Synthesis of poly(2-methyl-3-hydroxyoctanoate) via anionic polymerization of α-methyl-β-pentyl-β-propiolactone. *Biomacromolecules*, **2** (3), 623–627.

33 Adamus, G. and Kowalczuk, M. (2008) Anionic ring-opening polymerization of β-alkoxymethyl substituted β-lactones. *Biomacromolecules*, **9** (2), 696–703.

34 Kawalec, M., Adamus, G., Kurcok, P., Kowalczuk, M., Foltran, I., Focarete, M.L., and Scandola, M. (2007) Carboxylate-induced degradation of poly(3-hydroxybutyrate)s. *Biomacromolecules*, **8** (4), 1053–1058.

35 Adamus, G. (2009) Molecular level structure of (*R,S*)-3-hydroxybutyrate/ (*R,S*)-3-hydroxy-4-ethoxybutyrate copolyesters with dissimilar architecture. *Macromolecules*, **42** (13), 4547–4557.

36 Connor, E.F., Nyce, G.W., Myers, M., Möck, A., and Hedrick, J.L. (2002) First example of N-heterocyclic carbenes as catalysts for living polymerization: organocatalytic ring-opening polymerization of cyclic esters. *J. Am. Chem. Soc.*, **124** (6), 914–915.

37 Coulembier, O., Lohmeijer, B.G.G., Dove, A.P., Pratt, R.C., Mespouille, L., Culkin, D.A., Benight, S.J., Dubois, P., Waymouth, R.M., and Hendrick, J.L. (2006) Alcohol adducts of N-heterocyclic carbenes: latent catalysts for the thermally-controlled living polymerization of cyclic esters. *Macromolecules*, **39** (17), 5617–5628.

38 Zhang, C., Ochiai, B., and Endo, T. (2005) Matrix-assisted laser desorption/ionization time-of-flight mass spectrometry study on copolymers obtained by the alternating copolymerization of bis(-lactone) and epoxide with potassium *tert*-butoxide. *J. Polym. Sci. Pol. Chem.*, **43** (12), 2643–2649.

39 Kricheldorf, H.R. and Eggerstedt, S. (1999) Macrocycles, 6. MALDI-TOF mass spectrometry of tin-initiated macrocyclic polylactones in comparison to classical mass-spectroscopic methods. *Macromol. Chem. Physic.*, **200** (6), 1284–1291.

40 Adamus, G., Hakkarainen, M., Höglund, A., Kowalczuk, M., and Albertsson, A.-C. (2009) MALDI-TOF-MS reveals the molecular level structures of different hydrophilic–hydrophobic polyether-esters. *Biomacromolecules*, **10** (6), 1540–1546.

41 Kowalski, A., Libiszowski, J., Duda, A., and Penczek, S. (2000) Polymerization of L,L-dilactide initiated by tin(II) butoxide. *Macromolecules*, **33** (6), 1964–1971.

42 Kricheldorf, H.R., Behnken, G., Schwarz, G., and Kopl, J. (2008) High molar mass poly(ε-caprolactone) by means of diphenyl bismuth ethoxide, a highly reactive single site initiator. *Macromolecules*, **41** (12), 4102–4107.

43 Kricheldorf, H.R., Behnken, G., Schwarz, G., Simon, P., and Brinkmann,

M. (2009) High molar mass poly (trimethylene carbonate) by Ph$_2$BiOEt and Ph$_2$BiBr-initiated ring-opening polymerizations. *J. Macromole. Sci., Part A: Pure Appl. Chem.*, **46** (4), 353–359.

44 Jun Du, Y., Lemstra, P., Nijenhuis, A.J., Van Aert, H.A.M., and Bastiaansen, C. (1995) ABA type copolymers of lactide with poly(ethylene glycol). Kinetic, mechanistic, and model studies. *Macromolecules*, **28** (7), 2124–2132.

45 Kricheldorf, H.R., Kreiser-Saunders, I., and Boettcher, C. (1995) Polylactones: 31. Sn(II)octoate-initiated polymerization of L-lactide: a mechanistic study. *Polymer*, **36** (6), 1253–1259.

46 Leenslag, J.M. and Pennings, A.J. (1987) Synthesis of high-molecular-weight poly (L-lactide) initiated with tin 2-ethylhexanoate. *Macromol. Chem. Phys.*, **188**, 1809–1814.

47 Storey, R.F. and Taylor, A.E. (1998) Effect of stannous octoate on the composition, molecular weight, and molecular weight distribution of ethylene glycol-initiated poly(ε-caprolactone). *J. Macromole. Sci., Part A: Pure Appl. Chem.*, **35** (5), 723–750.

48 Penczek, S., Duda, A., Kowalski, A., Libiszowski, J., Majerska, K., and Biela, T. (2000) On the mechanism of polymerization of cyclicesters induced by tin(II) octoate. *Macromol. Symp.*, **157**, 61–70.

49 Kowalski, A., Duda, A., and Penczek, S. (1998) Kinetics and mechanism of cyclic esters polymerization initiated with tin(II) octoate, 1. Polymerization of ε-caprolactone. *Macromol. Rapid Comm.*, **19** (11), 567–572.

50 Majerska, K., Duda, A., and Penczek, S. (2000) Kinetics and mechanism of cyclic esters polymerisation initiated with tin(II) octoate, 4. Influence of proton trapping agents on the kinetics of ε-caprolactone and L,L-dilactide polymerization. *Macromol. Rapid Comm.*, **21**, 1327–1332.

51 Kricheldorf, H.R., Kreiser-Saunders, I., and Stricker, A. (2000) Polylactones 48. SnOct$_2$-initiated polymerizations of lactide: a mechanistic study. *Macromolecules*, **33** (3), 702–709.

52 Schwach, G., Coudane, J., Engel, R., and Vert, M. (1997) More about the polymerization of lactides in the presence of stannous octoate. *J. Macromole. Sci., Part A: Pure Appl. Chem.*, **35** (16), 3431–3440.

53 Kowalski, A., Duda, A., and Penczek, S. (2000) Mechanism of cyclic ester polymerization initiated with tin(II) octoate. 2. Macromolecules fitted with tin (II) alkoxide species observed directly in MALDI-TOF spectra. *Macromolecules*, **33** (3), 689–695.

54 Kowalski, A., Duda, A., and Penczek, S. (2000) Kinetics and mechanism of cyclic esters polymerization initiated with tin(II) octoate. 3. Polymerization of L,L-dilactide. *Macromolecules*, **33** (20), 7359–7370.

55 Kowalski, A., Libiszowski, J., Biela, T., Cypryk, M., Duda, A., and Penczek, S. (2005) Kinetics and mechanism of cyclic esters polymerization initiated with tin(II) octoate. Polymerization of ε-caprolactone and L,L-lactide co-initiated with primary amines. *Macromolecules*, **38** (20), 8170–8176.

56 Kricheldorf, H.R., Ahrenstorf, K., and Rost, S. (2004) Polylactones. *Macromol. Chem. Physic.*, **205** (12), 1602–1610.

57 Rokick, G. and Kowalczyk, T. (2000) Synthesis of oligocarbonate diols and their characterization by MALDI-TOF spectrometry. *Polymer*, **41** (26), 9013–9031.

58 Kéki, S., Török, J., Déak, G., and Zsuga, M. (2005) Mechanism of ring-opening and elimination cooligomerization of cyclic carbonates and ε-caprolactone: Formation of cyclic cooligomers. *Eur. Polym. J.*, **41** (7), 1478–1483.

59 Rokicki, G., Piotrowska, A., and Pawłowski, P. (2003) Macrocyclic vs. linear polymer formation in the coordination–insertion polymerization of cyclic carbonates. *Polym. J.*, **35** (2), 133–140.

60 Bielawski, C.W. and Grubbs, R.H. (2007) Living ring-opening metathesis polymerization. *Progress Polym.. Sci..*, **32**, 1–29.

61 Agrawal, D. and Schroeder, D. (2010) Insight into solution chemistry from gas-phase experiments. *Organometallics*, **30**, 32–35.

62 Volland, M.A.O., Adlhart, C., Kiener, C.A., Chen, P., and Hofmann, P. (2001) Catalyst

screening by electrospray ionization tandem mass spectrometry: Hofmann carbenes for olefin metathesis. *Chem. A. Eur. J.*, **7** (21), 4621–4632.

63 Wang, H. and Metzger, J.O. (2008) ESI-MS study on first-generation ruthenium olefin metathesis catalysts in solution: direct detection of the catalytically active 14-electron ruthenium intermediate. *Organometallics*, **27**, 2761–2766.

64 Torker, S., Merki, D., and Chen, P. (2008) Gas-phase thermochemistry of ruthenium carbene metathesis catalysts. *J. Am. Chem. Soc.*, **130**, 4808–4814.

65 Binder, W.H., Pulamagatta, B., Kir, O., Kurzhals, S., Barqawi, H., and Tanner, S. (2009) Monitoring block-copolymer crossover-chemistry in ROMP: catalyst evaluation via mass-spectrometry (MALDI). *Macromolecules*, **42**, 9457–9466.

66 Johnson, J.A., Lu, Y.Y., Burts, A.O., Xia, Y., Durrell, A.C., Tirrell, D.A., and Grubbs, R.H. (2010) Drug-loaded, bivalent-bottle-brush polymers by graft-through ROMP. *Macromolecules*, **43**, 10326–10335.

67 Grubbs, R.H. and Vougioukalakis, G.C. (2010) Ruthenium-based heterocyclic carbene-coordinated olefin metathesis catalysts. *Chem. Rev.*, **110**, 1746–1787.

68 Montaudo, M.S. and Montaudo, G. (1992) Further studies on the composition and microstructure of copolymers by statistical modeling of their mass spectra. *Macromolecules*, **25** (17), 4264–4280.

69 Klee, J.E. (2005) Review: mass spectrometry of step-growth polymers. *Eur. J. Mass Spectrom.*, **11** (6), 591–610.

70 Rizzarelli, P., Puglisi, C., and Montaudo, G. (2006) Matrix-assisted laser desorption/ionization time-of-flight/time-of-flight tandem mass spectra of poly(butylene adipate). *Rapid Commun. Mass Spectrom.*, **20** (11), 1683–1694.

71 Montaudo, M.S. (2007) Analysis of melt copolymers. *Eur. J. Mass Spectrom.*, **13** (1), 61–67.

72 Impallomeni, G., Giuffrida, M., Barbuzzi, T., Musumarra, G., and Ballistreri, A. (2002) Acid catalyzed transesterification as a route to poly(3-hydroxybutyrate-*co*-ε-caprolactone) copolymers from their homopolymers. *Biomacromolecules*, **3** (4), 835–840.

73 Adamus, G., Rizzarelli, P., Montaudo, M.S., Kowalczuk, M., and Montaudo, G. (2006) Matrix-assisted laser desorption/ionization time of flight mass spectrometry with size-exclusion chromatographic fractionation for structural characterization of synthetic aliphatic copolyesters. *Rapid Commun. Mass Spectrom.*, **20** (5), 804–814.

74 Takizawa, K., Tang, Ch., and Hawker, C.J. (2008) Molecularly defined caprolactone oligomers and polymers: synthesis and characterization. *J. Am. Chem. Soc.*, **130** (5), 1718–1726.

75 Adamus, G., Montaudo, M.S., Montaudo, G., and Kowalczuk, M. (2004) Molecular architecture of poly [(*R*,*S*)-3-hydroxybutyrate-*co*-6-hydroxyhexanoate] and poly [(*R*,*S*)-3-hydroxybutyrate-co(*R*,*S*)-2-hydroxyhexanoate] oligomers investigated by electrospray ionisation ion-trap multistage mass spectrometry. *Rapid Commun. Mass Spectrom.*, **18** (13), 1436–1446.

76 Adamus, G. (2007) Structural analysis of poly[(*R*,*S*)-3-hydroxybutyrate-*co*-ʟ-lactide] copolyesters by electrospray ionisation ion trap mass spectrometry. *Rapid Commun. Mass Spectrom.*, **21** (15), 2477–2490.

77 Kricheldorf, H.R., Al Masri, M., and Schwarz, G. (2003) Cyclic polyamide-6 by thermal polycondensation of ε-caprolactam and ε-aminocaproic acid. *Macromolecules*, **36** (23), 8648–8651.

78 Rokicki, G. and Piotrowska, A. (2002) A new route to polyurethanes from ethylene carbonate, diamines and diols. *Polymer*, **43** (10), 2927–2935.

79 Murgasova, R., Brantley, E.L., Hercules, D.M., and Nefzger, H. (2002) Characterization of polyester–polyurethane soft and hard blocks by a combination of MALDI, SEC, and chemical degradation. *Macromolecules*, **35** (22), 8338–8345.

80 Astruc, D., Boisselier, E., and Ornelas, C. (2010) Dendrimers designed for

functions: From physical, photophysical, and supramolecular properties to applications in sensing, catalysis, molecular electronics, photonics, and nanomedicine. *Chem. Rev.*, **110** (4), 1857–1959.

81 Li, Y., Hoskins, J.N., Sreerama, S.G., and Grayson, S.M. (2010) MALDI-TOF mass spectral characterization of polymers containing an azide group: Evidence of metastable ions. *Macromolecules*, **43** (14), 6225–6228.

82 Hart-Smith, G. and Barner-Kowollik, C. (2010) Contemporary mass spectrometry and the analysis of synthetic polymers: trends, techniques and untapped potential. *Maromol. Chem. Phys.*, **211**, 1507–1529.

13
Polymer Degradation

Paola Rizzarelli, Sabrina Carroccio, and Concetto Puglisi

13.1
Introduction

Polymer degradation is one of the most important areas of polymer chemistry, being a major factor restraining application of these outstanding and versatile materials. The idiom "polymer degradation" takes account of different processes, induced by one or more environmental factors such as heat, light, microorganisms, or chemicals that deteriorate polymers producing alterations in their properties. The degradation is the result of irreversible changes that are usually undesirable or, in some cases, required, as in biodegradation or recycling, or else induced to support structure determination.

Degradation plays an important role in every life phase of a polymer, that is, during its synthesis, processing, use, and even after it has accomplished its planned purpose. As a result, stabilization is required to extend the life time of most polymers. Compounding with stabilizers is the preferred and well-established method for improving stability. However, monitoring and controlling degradation requires the understanding of many different phenomena, including the diverse chemical mechanisms underlying structural changes in macromolecules, the complex reaction pathways of additives, the interactions of fillers, as well as impurities, and the complicated relationship between molecular-level changes and macroscopic properties.

Various schemes to classify polymer degradation exist. Because of its complexity, with regard to both the causes and the response of the polymer, classification is usually performed on the basis of the dominating features. We will refer to the classification based on the main factors responsible for degradation: thermal, thermo-oxidative, photo, photo-oxidative, hydrolytical, chemical, biological degradation, and so on.

Modern mass spectrometry (MS) methods for their high sensitivity, selectivity, and speediness offer the opportunity to explore the finest structural details in polymer degradation. In fact, whatever the cause, the deterioration mainly yields degradation products frequently bearing characteristic end groups, which can be revealed and differentiated by MS, being indicative of specific degradation pathways [1–12].

13.2
Thermal and Thermo-Oxidative Degradation

The study of thermal the degradation of polymers is an active field of investigation. In the last decade, about 200 of papers (articles and reviews)[1)] concerning thermal degradation studies of synthetic polymers by MS have appeared in the literature. Understanding the thermal degradation of polymers is of paramount importance for developing a reasonable technology in polymer processing and higher-temperature applications. Most processing equipment restricts access of oxygen to the molten polymer and pyrolysis may be the predominating degradation reaction. However, the contribution of thermal oxidation cannot be ignored. For this reason, studies on thermal [10–48] and thermo-oxidative [5, 11, 13, 21, 30, 36, 41, 43, 48–57] degradation processes by MS will be discussed together.

Investigations of thermal decomposition are routinely performed by thermogravimetry (TGA) and differential scanning calorimetry (DSC). They have proved useful for defining suitable processing conditions as well as drawing up valuable service guidelines. Nevertheless, thermal analysis methods do not provide information about the degradation mechanisms involved. The "hyphenated" thermoanalytical techniques, TG-MS and TG-FTIR, have been demonstrated to be more powerful tools in providing information on decomposition products [2, 5–7, 10, 26, 33].

Gas chromatography/mass spectrometry (GC/MS) is an active method for identification of additives and contaminants as well as for investigating degradation products [2, 3, 5, 7, 10, 11]. Recently, in combination with solid-phase microextraction (SPME), it has been used successfully to study the degradation products of several polymers [5, 11, 51].

Due to the high heating speed, accurate temperature reproducibility, and a wide temperature range, pyrolysis-GC/MS (py-GC/MS) is regularly used for polymer characterization [47] and it is the most frequently method employed in thermal degradation processes [2, 3, 6, 7, 10, 13, 18, 19, 31, 37, 38, 41, 48].

Direct-pyrolysis mass spectrometry (DPMS) has been and is still used in several studies as a method to acquire structural information as well as to investigate degradation processes [1–3, 6, 7, 10, 12, 15, 21, 39, 40].

Even though all these techniques have been long employed, the instrumental progresses have been more and more focused and well-matched in studying degradation of polymers [11, 33, 34, 41].

However, modern soft ionization MS techniques emphasize the detection of primary thermal decomposition products, carrying detailed information on the

---

1) Scopus search, with restrictions to review and journal contributions in English, published in the last decade (2000–2010), using the keywords: "mass spectrometry" and "thermal degradation" and "polymer."

relationships between polymer end-chain structures and degradation mechanisms [4, 8, 9, 14, 16–18, 20–25, 27–30, 32, 34–36, 42–46, 49–57]. Since soft ionization techniques have been employed, polymer degradation has gained a dedicated section in relevant reviews [6–9]. For these reasons and in view of the very significant number of papers on thermal degradation by conventional MS methods, most of them usually related to the characterization of low-molecular-mass degradation products, this chapter will be focused on recent findings obtained mainly by soft ionization MS techniques.

A great difference exists between the thermal decomposition mechanisms of condensation and addition polymers. Thermal degradation pathways of condensation polymers are frequently dominated by the polarity and reactivity of the functional groups within their structure and their thermal decomposition reactions will be ionic and selective. Most addition polymers undergo thermal degradation through radical and unselective mechanisms. Macroradicals may decay through several simultaneous routes, involving bond cleavages, for example, β-scission, recombination, small stable molecules from side-groups elimination, or H-abstraction [1].

To simulate the reactions that take place during the processing of poly(ethylene terephthalate) (PET) under N_2 atmosphere, Samperi et al. [23] performed several isothermal degradation experiments. PET samples processed for 1 h at 320, 340, and 370 °C were analyzed by matrix-assisted laser desorption ionization-time of flight (MALDI-TOF) MS and relevant changes were observed in the spectra (Figure 13.1). The study revealed the formation of cyclic oligomers (symbol A, Figure 13.1) that become unstable as the heating temperature rises. Instead, a significant growth in the abundance of ions assigned to PET oligomers bearing two carboxyl chain ends was detected (symbol C, Figure 13.1). Moreover, MALDI spectra showed two new peak series attributed to sodiated ions bearing terephthalic anhydride units (symbols M, N, and P, Figure 13.1). On the other hand, the presence of vinyl ester groups was not detected. A recombination reaction between vinyl ester and carboxyl ended oligomers, leading to the formation of anhydride-containing species, was proposed as an alternative pathway to the vinyl ester hydrolysis reaction [23].

Similar degradation experiments were carried out on poly(butylene terephthalate) (PBT) [24]. The structural characterization of the reaction products was performed by MALDI and nuclear magnetic resonance (NMR). Experimental data showed that linear PBT chains as well as cyclic oligomers, formed at temperatures below 290 °C by ring-chain equilibration mechanism, underwent thermal decomposition at higher temperature by a β H-transfer mechanism. The formation of unsaturated oligomers was detected by MALDI and also by NMR techniques, whereas, in contrast with the thermally degraded PET sample and with the results of Manabe et al. [13], terephthalic anhydride-containing oligomers were not detected by MALDI [24].

Ciolacu et al. [53] investigated the oxidative degradation and the subsequent discoloration of PET by MALDI-TOF-MS combined with thin-layer chromatography (TLC), laser desorption/ionization on silicon-mass spectrometry (DIOS-MS), and time-of-flight secondary ion mass spectrometry (TOF-SIMS). MALDI spectra of the degraded polymer showed the formation of cyclic oligomers via two mechanisms. The generation of linear oligomers with carboxyl end groups was ascertained, and it was proposed that the process occurred via chain scission at the ether linkage present

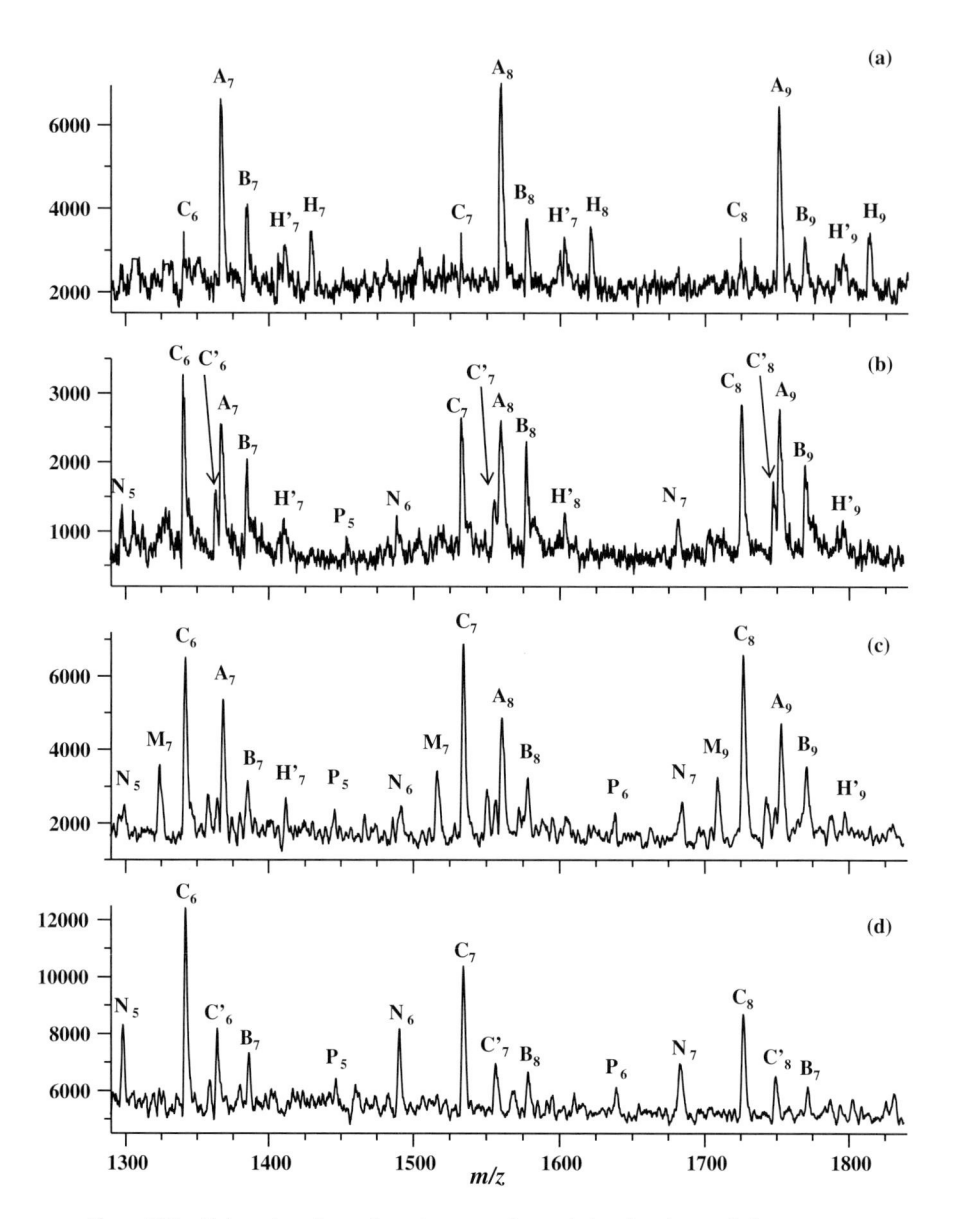

Figure 13.1 Enlarged sections of matrix-assisted laser desorption ionization-time of flight (MALDI-TOF) mass spectra of poly(ethylene terephthalate) (PET) samples heated for 60 min at (a) 30 °C, (b) 320 °C, (c) 340 °C, and (d) 370 °C. Cyclic oligomers (A) become unstable as the heating temperature rises. Instead, the abundance of oligomers bearing two carboxyl chain ends (C) increases. New peak series, attributed to sodiated ions bearing terephthalic anhydride units (M, N, P), were detected. Reproduced from Ref. [23] with kind permission of Elsevier.

in PET. Furthermore, a MALDI analysis, performed directly on the TLC plate, showed the presence of terephthalic acid, diethyl terephthalate, and of a small amount of mono- and dihydroxy diethyl terephthalate units, which were indicated as responsible for yellowing [53]. In addition, Ciolacu et al. tested a nanostructured organometallic macromer trisilanolisobutyl-POSS (T-POSS) to prevent discoloration of PET and to attain stabilization during melt processing. Thermal studies showed that the T-POSS additive improves the thermo-oxidative stability, resulting in better color stability of the material. The X-ray photoelectron spectroscopy (XPS) and MALDI results confirmed that the stabilization is achieved by covalent interaction of the nanostructured additive to PET [54].

Romão et al. studied the thermo-mechanical and thermo-oxidative degradation mechanisms of bottle-grade PET (btg-PET) by using ^{1}H NMR and MALDI-TOF-MS. In thermo-oxidative degradation, the concentration of low-molar-mass compounds increased with time and the main products were cyclic and linear di-acid oligomers. The diethylene glycol (DEG), one of the most important comonomer in btg-PET, was evidenced to be the precursor to color changes in btg-PET [36, 44].

Degradative reactions occurring in bisphenol A-polycarbonate (BPA-PC) have received continued attention over the past two decades. It is reported [5, 58] that thermo-oxidative degradation phenomena are relevant during the polymer service life, whereas thermolysis and hydrolysis are the main degradative processes occurring during injection molding operations.

Oba et al. proved by py-GC/MS the presence of branching and cross-linking structures on thermally oxidized PC samples. They postulated that the latter were formed by coupling of methylene and/or phenoxy radicals [48]. Further studies by MALDI-TOF-MS, size exclusion chromatography (SEC), SEC/MALDI-TOF, and NMR analysis [49, 50] showed that the cross-linking occurs through the coupling of bisphenol units to yield biphenyl structures. MALDI and SEC/MALDI spectra of the thermal oxidized PC samples (Figure 13.2) showed the presence of polymer chains containing a variety of terminal groups: acetophenone (symbols L and S, Figure 13.2), phenyl substituted acetone (symbols E, I, and Q, Figure 13.2), phenols (symbol O, Figure 13.2), benzyl-alcohol (symbol G, Figure 13.2), and biphenyl (symbols N, T, and V, Figure 13.2). The presence of biphenyl units justified the occurrence of cross-linking processes, which are responsible for the formation of the insoluble gel fraction.

The identification of thermal oxidation products of PC established the mechanisms for their formation: (i) hydrolysis of carbonate groups of PC; (ii) oxidation of the isopropenyl groups of PC; and (iii) oxidative coupling of phenols end groups to form biphenyl groups (Scheme 13.1).

Montaudo et al. [21] monitored also the thermal decomposition process of PC by MALDI-TOF and fast atom bombardment-MS (FAB-MS). MALDI spectra of the pyrolysis residues at 300 °C showed only a progressive reduction of the abundance of cyclic oligomers. At 350 °C, the occurrence of an extensive hydrolysis reaction producing phenol end groups was observed. Oligomers bearing phenyl and isopropenyl end groups were also detected at 400 °C together with condensed aromatic compounds such as xanthones that became the most abundant species at 450 °C [14, 21].

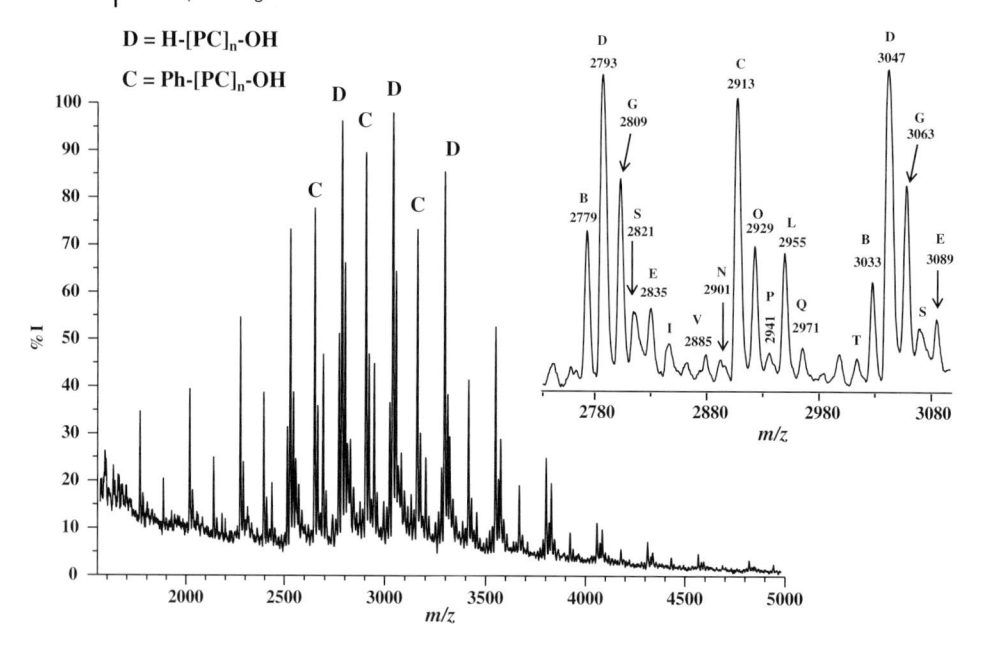

Figure 13.2 Matrix-assisted laser desorption ionization-time of flight (MALDI-TOF) spectrum of a size exclusion chromatography (SEC) fraction, collected at 28 mL, of a poly (bisphenol-A-carbonate) sample oxidized at 300 °C. SEC/MALDI spectrum showed the presence of polymer chains containing as terminal groups: acetophenone (L, S), phenyl substituted acetone (E, I, Q), phenols (O), benzyl-alcohol (G), and biphenyl (N, T, V). The presence of biphenyl units justified the occurrence of cross-linking processes, which are responsible for the formation of the insoluble gel fraction. Reproduced from Ref. [50] with kind permission of Elsevier.

Gies et al. exploited the evaporation-grinding method (E-G method) [59] in combination with MALDI-TOF-MS to check the chemical modifications involved in the synthesis of high-MW poly(p-phenylene sulfide)s (PPS). Results showed that the production of high-MW PPS involved linear chain extension, oxidative branching, and arylthio metathesis reactions, with the possibility of thermally induced free-radical chemical mechanisms [45]. Gies and Hercules also examined the thermal curing of poly(o-hydroxyamide) (PAOH) and a poly(benzoxazoleamide) precursor by combining the evaporation-grinding method for MALDI analysis and evolved gas analysis/GC/MS. Changes observed in the MALDI spectra were suggested to result from cyclodehydration and structure modifications due to decarboxylation and branching [28].

Carroccio et al. investigated the thermal [15] and thermo-oxidative [52] degradation mechanisms of commercial Ultem® polyetherimide (PEI), respectively, by DPMS [15] and MALDI-TOF [52]. The structure of the pyrolysis compounds detected in the DPMS analysis suggested that they were mainly formed by the scission of the isopropylidene bridge of bisphenol A; the oxygen-phthalimide bond; and the phenyl-phthalimide bond, which are apparently the weakest bonds of PEI. Extensive

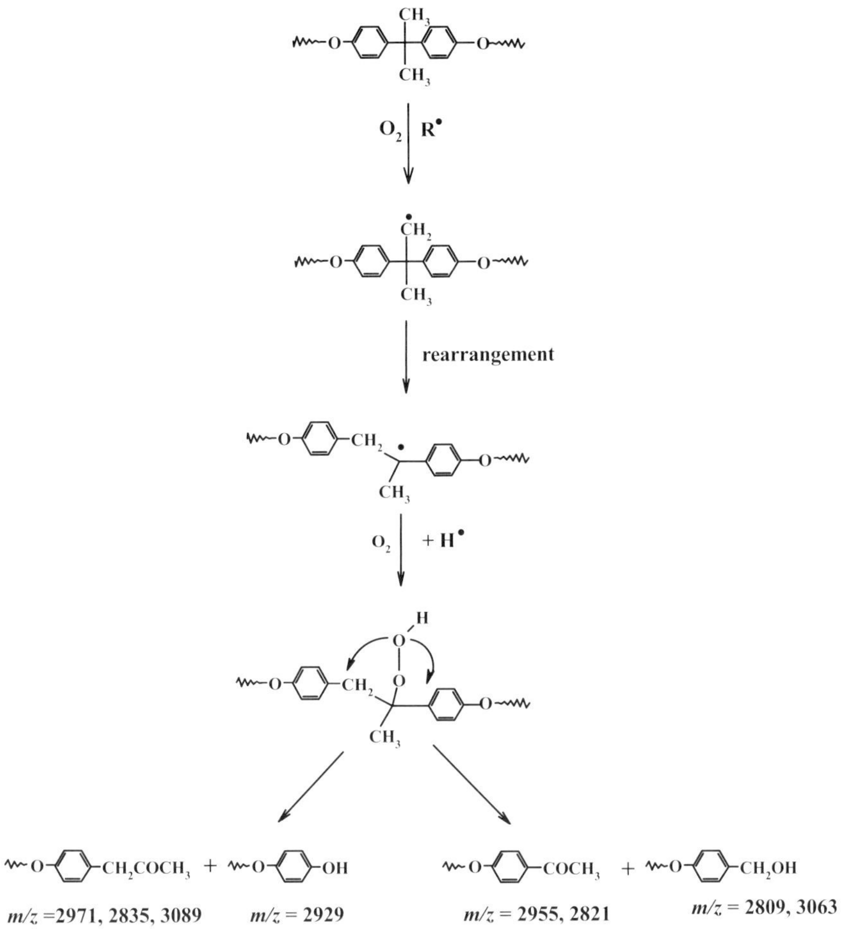

Scheme 13.1 Thermal oxidative degradation processes of poly(bisphenol-A-carbonate): (i) hydrolysis of carbonate groups of PC; (ii) oxidation of the isopropenyl groups of PC; and (iii) oxidative coupling of phenols end groups to form biphenyl groups. Adapted from Ref. [50] with kind permission of Elsevier.

H-transfer reactions and subsequent condensation reactions have been accounted for the high amount of char residue [15]. Highly valuable structural information was extracted from MALDI spectra of the thermally oxidized PEI soluble fractions. Identification of end groups established that the degradation mechanisms involved three cleavage reactions of the biphenyl ether units, oxidative degradation of the isopropylidene bridge of BPA units, and thermal cleavage of phenyl-phthalimide units [52].

Aliphatic polyamides (PAs) are widely used materials, especially for fiber production and engineering resins. When the polymer is subjected to thermal, light, or mechanical energy particularly in the presence of moisture or oxygen, a gradual increase of yellowing, and eventual embrittlements are clearly observable. Puglisi *et al.* [22] investigated the thermal decomposition processes occurring in nylon 66 (Ny66) by MALDI-TOF-MS. The formation of a gel fraction was observed after a short heating time. MALDI spectra of the soluble fraction evidenced that secondary amino groups and cyclopentanone chain ends originated from the degradation process. The gel fraction was partially hydrolyzed to break down the network structure. The MALDI spectra of the soluble material revealed the presence of *N*, *N*-substituted amide as side chains and of azomethine structures. These species were suggested to be responsible for the gel formation on heating Ny66 samples. Similar experiments on nylon 6 (Ny6) samples showed that only secondary amino groups were formed, leading to branching but not to cross-linking [22]. Gel formation was also evidenced in the melt mixing reactions of oxazoline-cyclophosphazene units (CP2OXA) and Ny6 samples, terminated with one specific reactive chain end (–COOH or –NH_2). MALDI spectra on the soluble part of partially acid hydrolyzed gel revealed that oxazoline rings can react with amide groups along the Ny6 chains and with secondary amino groups, formed by a condensation reaction involving the elimination of NH_3 from two amino chain ends [35].

Carroccio *et al.* determined the oxidation products of Ny6 and Ny66 films aged in air by using MALDI-TOF-MS. Structural identification of the reaction products in MALDI spectra, collected at different temperatures and exposure time, supported an α-H abstraction as the main thermo-oxidation process, with subsequent formation of hydroperoxides. The latter subsequently decompose, yielding oligomers containing aldehydes, carboxylic, amides, methyl, and *N*-formamides as terminal groups [55].

The chemical structures of pyrolysis products from polyether-based polyurethanes (PUs) have been investigated by using diverse MS techniques [20, 37] providing more detailed data and enabling to confirm prior findings on thermal degradation processes. Lattimer and Williams elucidated in detail the degradation mechanism of PUs by MALDI, direct probe chemical ionization MS (CI-MS), and attenuated total reflectance FTIR (ATR-FTIR). The formation of the initial degradation products was explained by three decomposition pathways. Diphenylmethane diisocyanate, an "amine-isocyanate," 1,4-butanediol, and hydroxyl-terminated polyurethane and polyether oligomers as pyrolyzates derived from dissociation to isocyanate and alcohol. Dissociation to amine, olefin, and CO_2 produced an amine-isocyanate, methylene dianiline, and butenyl-terminated polyurethane and polyether oligomers. Cyclic urethane oligomers were originated from an "intramolecular exchange" reaction [20].

Lattimer [16] also analyzed the low-temperature pyrolysis products from a poly (ethylene glycol) (PEG) sample. After pyrolysis, the residue was analyzed by MALDI-MS and by CI-MS. MS analysis showed that degradation of PEG started at 150 °C in inert atmosphere. Eight series of oligomeric pyrolyzates were identified with the support of tandem mass spectrometry (CI-MS/MS) and by deuteration of hydroxyl end groups in the pyrolyzate. The pyrolysis experiments indicated that the decomposition scheme is free radical in nature. Identification by MS of the end groups indicated that the most favorable initial degradation step was homolytic cleavage of C−O bonds, followed by H-abstraction to produce saturated (ethyl and hydroxyl) end groups. At higher temperatures, the increasing abundance of "methyl ether" terminated oligomers pointed out that C—C bond cleavage became more prevalent [16].

Gallet et al. investigated the thermal oxidative degradation of poloxamer 407, a poly(ethylene oxide–propylene oxide–ethylene oxide) triblock copolymer. The combination of MALDI-TOF-MS, SEC/NMR, SEC/MALDI-TOF-MS, and SPME/GC-MS allowed establishing a new mechanism of oxidation. The detection of 1,2-propanediol,1-acetate,2-formate as the first volatile degradation product to appear let the authors suggest that the degradation started via a six-ring intramolecular decomposition reaction of the poly(propylene oxide) block [51].

Using MALDI-TOF-MS, Kawasaki et al. [29] and Watanabe et al. [46] have studied the ultrasonic degradation respectively of PEG and of poly(ethylene oxide-block-propylene oxide) copolymers, PEG-b-PPG. Five degradation pathways involving free radical reactions were suggested. By applying chromatographic separation, it was evidenced that the initial ultrasonic degradation occurred mainly at the boundary of two PEG and PPG chains, followed by the degradation at the PEG or PPG block [46].

Thermal degradation of poly(L-lactide) (PLA) [6, 7, 10, 11] and poly(hydroxyalkanoic acid)s, particularly poly[(R)-3-hydroxybutyrate] (PHB) and its copolymers, has been widely studied also in the last decade with the support of MS [10, 11, 32, 38]. PHB has a narrow thermal processing window, the knowledge of thermal properties is important to improve the thermal processing conditions of this biopolymer. Kawalec et al. proposed a new mechanism of thermal degradation of PHBs, induced by carboxylate groups by using electro-spray ionization mass spectrometry (ESI-MS). The degradation via intermolecular R-deprotonation by carboxylate has been suggested to be the main PHB decomposition pathway at moderate temperatures [32]. Anhydride production was proposed by Ariffin et al. as an additional pyrolysis mechanism of PHB. To confirm the mechanisms out of the random scission, minor pyrolyzates from PHB were characterized by NMR, FTIR, and FAB-MS. As a result, crotonic anhydride and its oligomers were detected as minor products from condensation reactions between carboxyl groups [38].

Rizzarelli and Carroccio [43] described a map of the thermo-oxidative degradation processes occurring in synthetic (PBSu) and commercial (Bionolle 1001®) poly(butylene succinate). MALDI combined with the use of extremely thin polyester films provided a virtual magnifying glass on the thermal-oxidation products of PBSu. Degradation experiments were carried out on extremely thin films to maximize the percentage of degradation products with respect to the nonoxidized chains in the bulk. In fact, the MALDI spectra presented several new well-resolved peaks (Figure 13.3), which provided information on the end groups of the oxidation products.

An α-H abstraction process was suggested as the primary step during the oxidation of both samples (Scheme 13.2). The initial step in this process consists of a H-abstraction from the methylene group adjacent to the ester linkage, leading to the formation of a hydroperoxide intermediate. Remarkably, the hydroperoxide intermediate decomposes by radical rearrangement reactions via a hydroxyl ester or from a radical which may follow two different pathways [43].

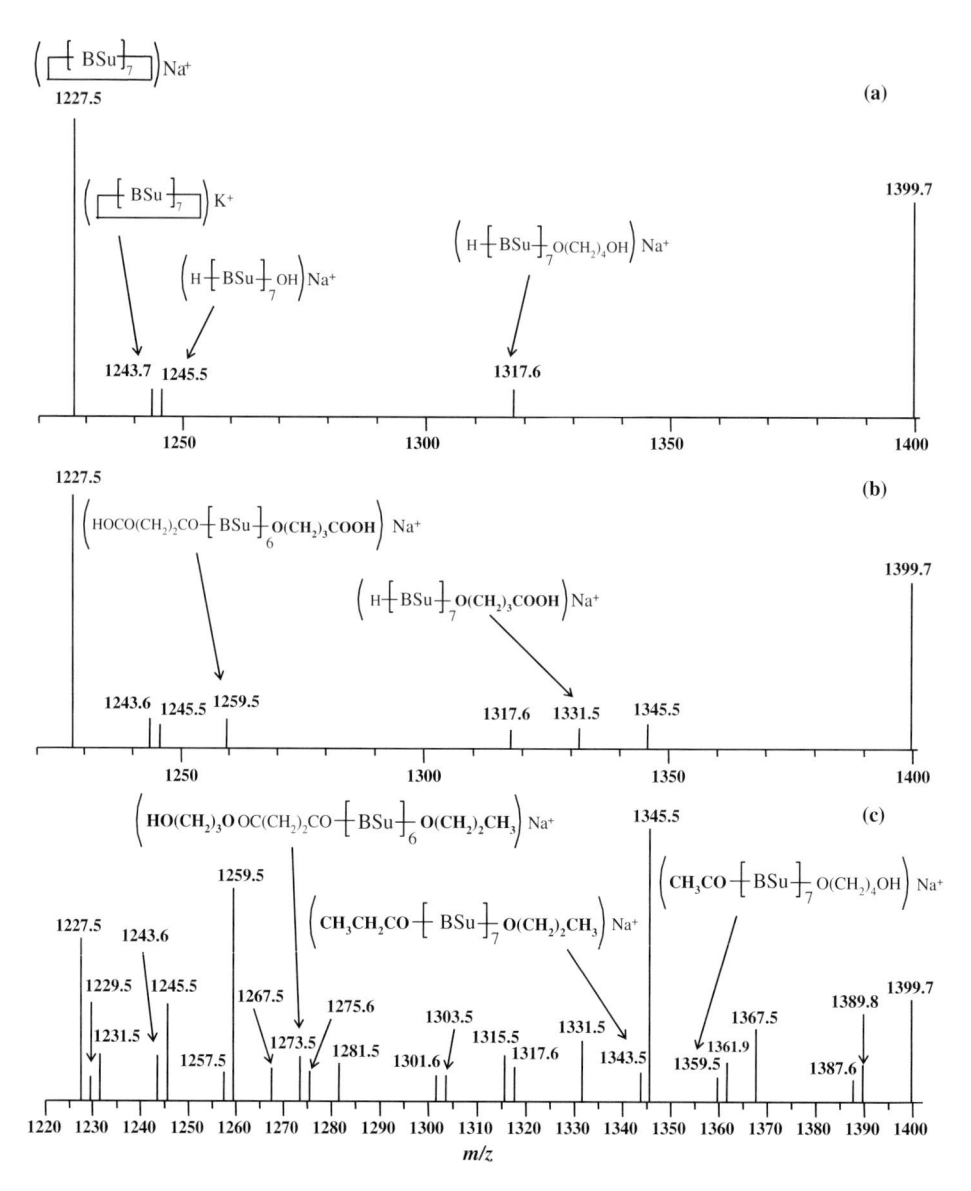

Figure 13.3 Matrix-assisted laser desorption ionization (MALDI) mass spectra after the deisotoping procedure, in the m/z range 1220–1400, of the (a) original poly(butylene succinate) (PBSu) sample and thermo-oxidized at 170 °C for (b) 1 h and (c) 2 h. Degradation experiments were carried out on extremely thin films to maximize the percentage of degradation products with respect to the nonoxidized chains in the bulk. To distinguish and separate between the contributions of isotopic peaks M + 1 and M + 2 and peaks due to isobaric structures, a deisotoping program (Data Explorer™ software) was used. Reproduced from Ref. [43] with kind permission of Elsevier.

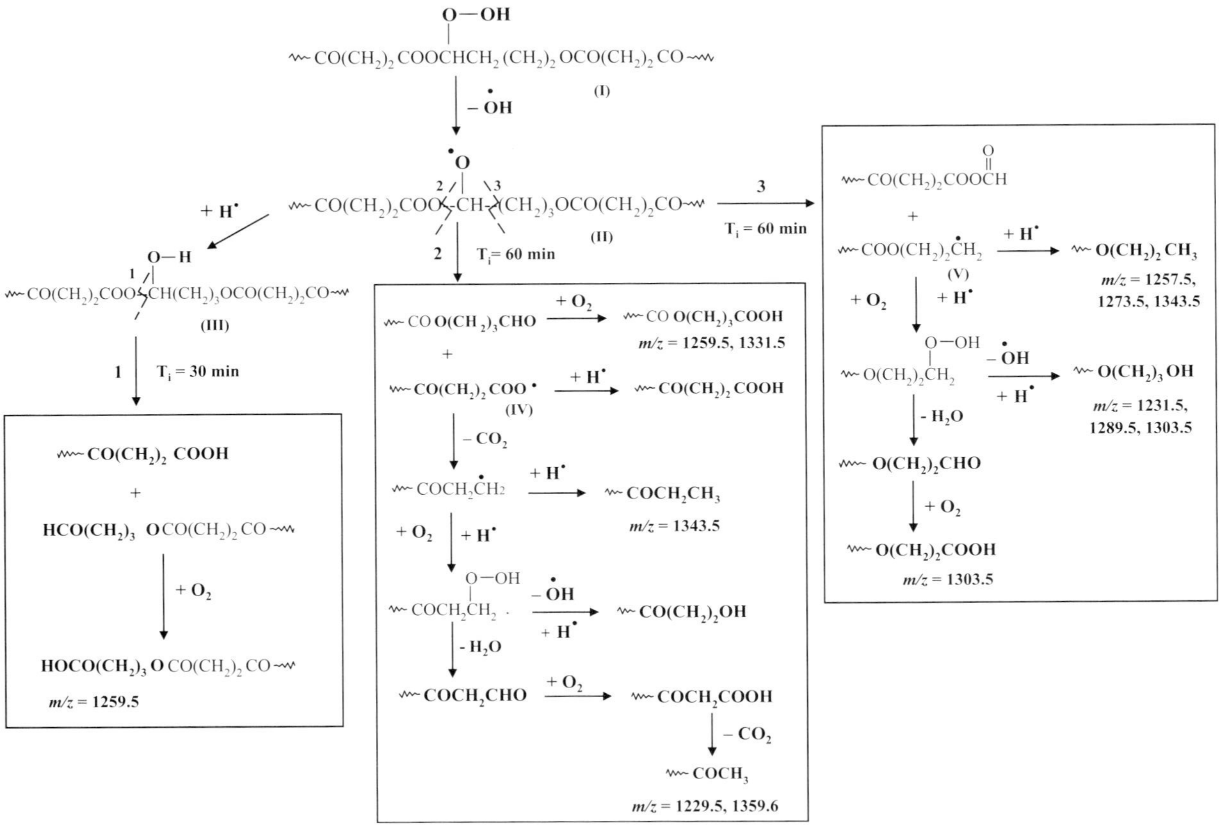

Scheme 13.2 Overall thermal-oxidation processes in poly(butylene succinate) (PBSu) (Ti = induction time). A α-H abstraction from the methylene group adjacent to the ester linkage, leading to the formation of a hydroperoxide intermediate, is the primary step. The hydroperoxide intermediate decomposes by radical rearrangement reactions via a hydroxyl ester or from a radical which may follow two different pathways. Reproduced from Ref. [43] with kind permission of Elsevier.

Moreover, thermal degradation experiments were also performed under N_2 at 240–260 °C. The new species identified in the MALDI spectra of both samples support a decomposition pathway taking place through a β-H-transfer bond scission, followed by the generation of succinic anhydride from succinic acid end molecules via a cyclisation decomposition mechanism [43]. This result was found in agreement with a previous study concerning the thermal degradation mechanism of poly (propylene succinate) (PPSu), investigated using py-GC/MS [31].

Poly(methyl methacrylate) (PMMA) is a polymer whose degradation has been widely investigated for its industrial value. Two radicalic mechanisms are generally accepted for the initiation of the degradation: main chain random scission and the homolytic scission of the methoxycarbonyl side group. Regardless of the mechanism of the initiation step, the degradation product is almost exclusively a methyl methacrylate (MMA) monomer (>99%) [58]. Latest findings highlight for the first time that PMMA degradation does not exclusively advance via radical intermediates. In fact, Bennet et al. identified by ESI-MS the thermo-oxidative degradation products of PMMA model compounds. Data showed that degradation of vinyl terminated PMMA proceeded via the incorporation of oxygen leading to the formation of ethylene oxide-type end groups. These epoxide end groups are subsequently eliminated via the formation of formaldehyde and 2-oxo-propionic acid methyl ester, resulting in the effective removal of a MMA monomer unit from the polymer chain [30].

MALDI-TOF-MS, NMR and py-GC/MS were used to study the effects of end groups on thermal degradation of poly(butyl cyanoacrylate). The NMR and MALDI results indicated that with cyanoacrylate polymers the initiator is present as the end group and that it can influence the degradation process [18].

In view of the wide use of poly(vinyl chloride) (PVC) and poly(styrene) (PS), considerable efforts have been made and several techniques have been applied to detect their thermal decomposition products. Adam et al. [27] employed single-photon ionization TOF-MS (SPI-TOF-MS) and resonance-enhanced multiphoton ionization TOF-MS (REMPI TOF-MS), and Roland et al. [19] used py-GC/MS. Recently, Li et al. [42] investigated the thermal decomposition of PVC and PS by synchrotron VUV photo-ionization MS at low pressure. Data confirmed two stages of the PVC thermal decomposition process: the low-temperature step during which HCl and benzene are produced, and the high-temperature stage to form various aromatic hydrocarbons. It has been reinforced that PVC pyrolysis involves significant cross-linking and aromatization reactions [42]. MALDI-TOF-MS has shown that the thermal degradation reaction in hexa-adducts $(PS)_6 C_{60}$ results from the breaking of the C—C bonds in α and β positions to fullerene C_{60}, the latter being about three times less favored than the direct fullerene arm bond [25].

Lattimer studied the low-temperature pyrolysis products from the residue of polytetrahydrofuran (PTHF) by MALDI and CI-MS. Assignment of 11 series of oligomeric pyrolyzates structures was aided by CI-MS/MS and by deuteration of −OH end groups in the pyrolyzate. A free radical mechanism was proposed to explain the main degradation products, bearing "ethyl ether," "propyl ether," "butyl ether," or "aldehyde" and one hydroxyl end group, which is retained from the low-molecular-weight starting polymer [17].

13.3
Photolysis and Photooxidation

The techniques for investigating polymer photo-degradation are numerous and they are often supported by accelerated testing methods. In the past, UV, IR and wet chemistry methods have mainly been used to identify the products formed and to follow the mechanisms of photodegradation [58, 60]. Traditional spectroscopic techniques continue to be the most employed in photodegradation studies. However, UV and IR supply structural information only on functional groups or segments of molecules, not on the entire molecule. Additionally, molecules formed in oxidation processes are often very reactive, do not accumulate, and are present only in minor amounts among the reaction products requiring a high sensitivity to be monitored. Applications of soft ionization technique, especially ESI and MALDI-MS for the study of polymer photo-oxidation are quite recent [56, 57, 61–68].

Photo-aging of PAs is of great concern since many security devices are manufactured using these high performance materials. Carroccio et al. [61, 62] determined by MALDI-MS the structure of the molecules produced during photo-aging of Ny6 and Ny66. Noteworthy results were obtained by the identification of almost 40 compounds in MALDI spectra that allowed drawing a detailed map of the photodecomposition mechanisms. Besides the H-abstraction and subsequent hydroperoxide formation, two other major processes appear to be operating in Ny6: chain-cleavage reactions Norrish type I and II (Scheme 13.3). These photo-oxidation processes had earlier been revealed exclusively in polyolefin samples.

Remarkably, the kinetic tracings of the oxidation products (Figure 13.4), representative of the three oxidation mechanisms found to be operating, showed that both Norrish I and II photoproducts appear after an induction period, whereas the oligomer generated by H-peroxide decomposition is formed immediately [61].

The Norrish I photo-cleavage was evidenced by MALDI-MS to be active in PBSu as well, together with the α-H abstraction and consequent hydroperoxide formation and the oxidation of hydroxyl end groups [63].

Carroccio et al. employed MALDI-MS to investigate and compare photo and thermal-oxidation processes in commercial PEI (Ultem® 1000). The structural changes were attributed to the cleavage of C—N bonds and two photo-cleavage pathways were suggested: the phthalimide unit cleavage to yield phthalic anhydride and the cleavage of the phthalimide N—CH_3 bond. On the contrary, the photoscission of the isopropylidene bridge justified the formation of degradation products similar to those observed in the thermal oxidation. In fact, the cleavage of the isopropylidene bridge is a nonspecific oxidation process, occurring each time the molecule is excited by any kind of energy (thermal or photo) [56].

In a similar way, Carroccio et al. [57] compared thermal and photooxidation processes occurring in PBT. The major oxidation products suggested that an α-H abstraction mechanism supplied to both the photo and thermal oxidation of PBT, leading to the formation of similar series of oligomers discriminated by MALDI-MS. In fact, the extraction of methyl hydrogen, yielding a methylene radical which reacts with oxygen to form a hydroperoxide intermediate activated both the photo and

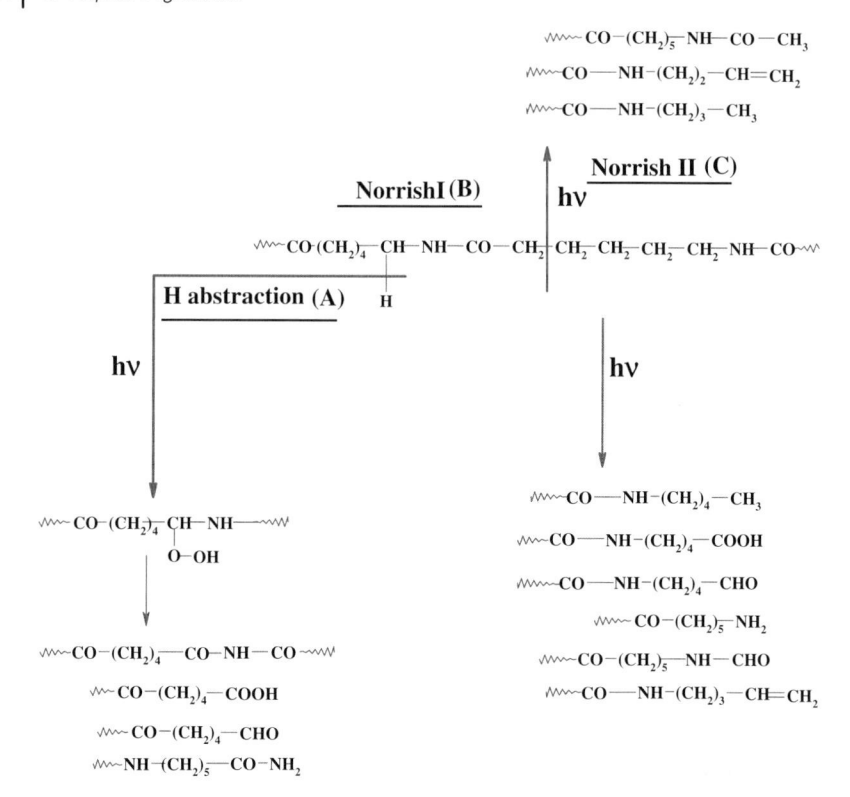

Scheme 13.3 Overall photo-oxidation processes in nylon 6. Besides the H-abstraction and subsequent hydroperoxide formation, chain-cleavage reactions Norrish type I and II appeared to be operating in nylon 6. These photo-oxidation processes had earlier been revealed exclusively in polyolefin samples. Reproduced from Ref. [61] with kind permission of the ACS.

thermal processes. The decomposition of this hydroperoxide, when occurring thermally, led to several end groups. Conversely, photooxidation was more selective, and the oxidation products bearing phenyl and phenol end groups, derived from loss of CO_2, were not observed. Despite the data reported in the literature [13, 60], the results gave evidence that the α-H abstraction plays an important role in the photo-oxidative degradation process and became the prominent mechanism at higher exposure times [57].

Nevertheless, MALDI-TOF-MS failed in the detection of small amounts of species originated from photo-Fries rearrangement in a PC sample, revealed by fluorescence spectroscopy [64].

Malanowski et al. studied the mechanism of photolysis and photo-oxidation of noncrosslinked poly(neopentyl isophthalate) (PNI) coatings. Structural changes induced by short ($\lambda > 254$ nm) [65] and long ($\lambda > 300$ nm) [66] UV irradiation were investigated using ATR-FTIR, SEC, and MALDI-MS. The mechanism proposed involves Norrish photo-cleavage (type I) of the ester group. It was suggested that the

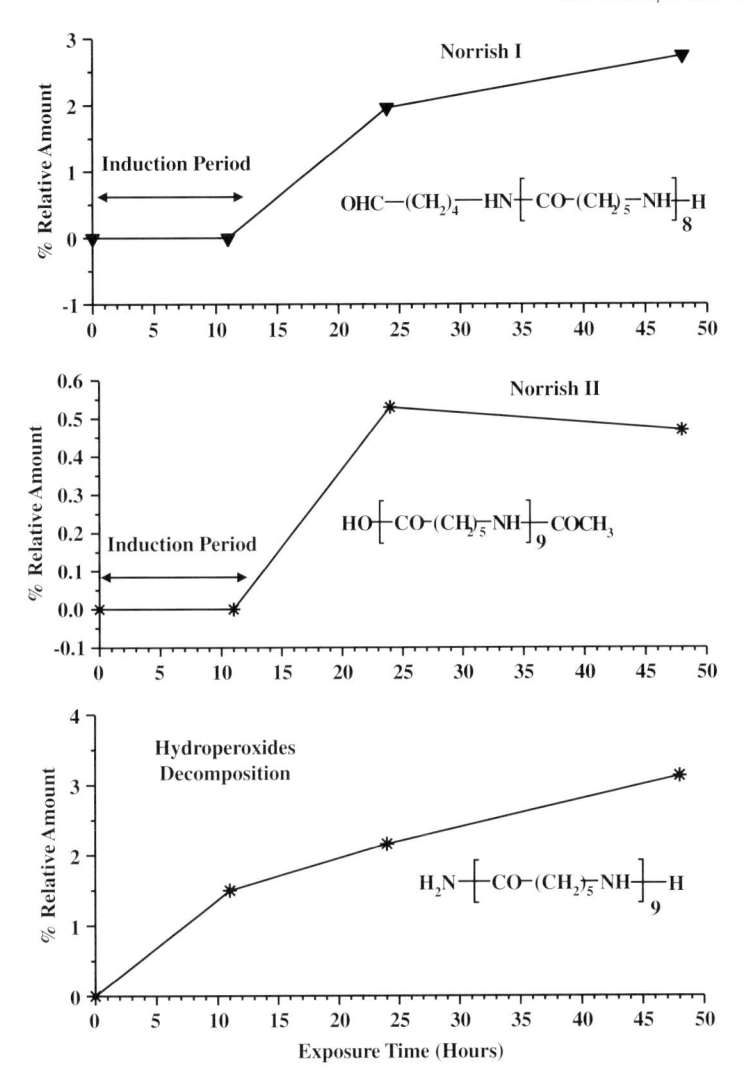

Figure 13.4 Relative amount versus exposure time of species at m/z (a) 1029, (b) 1101, and (c) 1058, as obtained from the matrix-assisted laser desorption ionization (MALDI) spectra of photo-oxidized 10 μm nylon 6 samples. The kinetic tracings of the oxidation products, representative of the three oxidation mechanisms found to be operating in nylon 6, show that both Norrish I and II photoproducts appear after an induction period, whereas the oligomer generated by H-peroxide decomposition is formed immediately. Reproduced from Ref. [61] with kind permission of the ACS.

six different radicals formed (photolysis) directly abstract hydrogen, producing mainly isobutyl and phthalic acid end groups, or react with oxygen, leading to the formation of primarily acid and –OH end groups (photooxidation). Moreover, an H-abstraction from the polymer backbone, followed by oxidation reactions played an important role in photodegradation too. The alkoxy radical formed as a consequence of this process can rearrange to an anhydride via a cage reaction [65]. The three analytical techniques proved that the Norrish I mechanism was operative in the photo-degradation of PNI both in laboratory and outdoor conditions [66].

Recently, Bennet *et al.* by using ESI-MS obtained significant results on the degradation of PMMA [67], and poly(butyl acrylate) (PBA), as well as poly(2-hydroxyethyl methacrylate) (PHEMA) model compounds [68], exposed to 95 °C and high UV radiation, separately as well as in combination. Vinyl-terminated polymers were compared to their saturated analogues; the terminal vinyl bond was found to be a source of instability and consistent changes are evident in ESI spectra of vinyl-ended PMMA exposed to UV radiation (Figure 13.5).

However, all PMMA degraded via the same mechanism under all conditions tested (Scheme 13.4).

UV and thermal radiation showed a synergistic effect on degradation of PMMA, with the vinyl-terminated sample also exhibiting cross-linking [67]. The cyclic degradation mechanism proposed for PMMA was found to be relevant in PBA and PHEMA as well. In addition, PBA and PHEMA were susceptible to other degradation and cross-linking reactions, while cross-linking was especially rapid in PHEMA exposed to UV radiation [68].

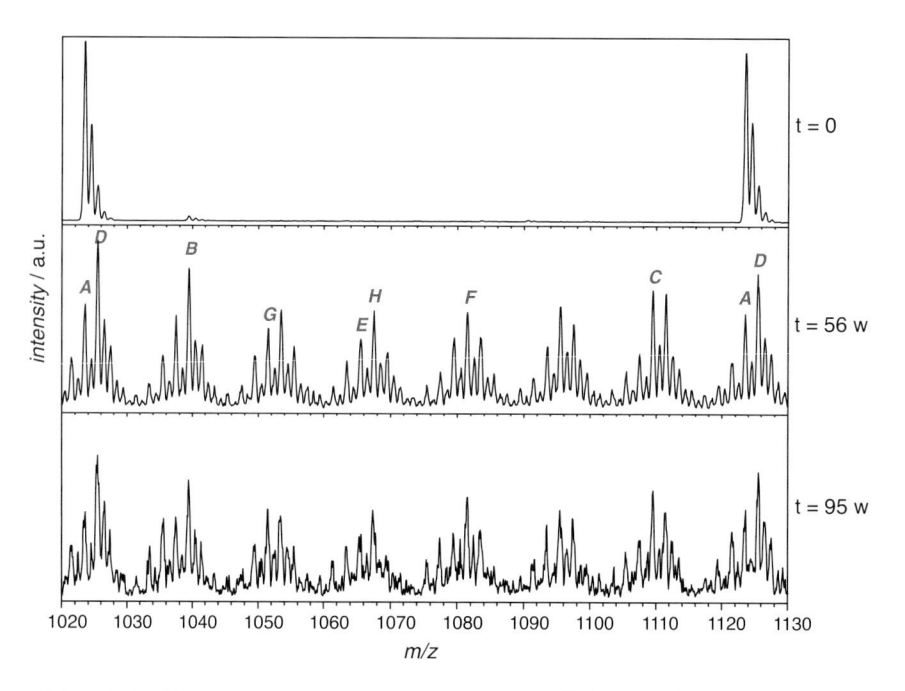

Figure 13.5 Electrospray ionization mass spectrometry (ESI-MS) spectra (with quadrupole mass analyzer) showing vinyl-terminated poly(methyl methacrylate) (PMMA) prior to, and after 56 and 95 weeks exposure (respectively) to UV radiation at 95 °C. Spectra were recorded using a Thermo Finnigan LCQ Deca quadrupole ion trap mass spectrometer. These spectra show advanced degradation at 56 weeks, and some crosslinking at 95 °C causing the lower quality spectrum. Reproduced from Ref. [67] with kind permission of Wiley-VCH.

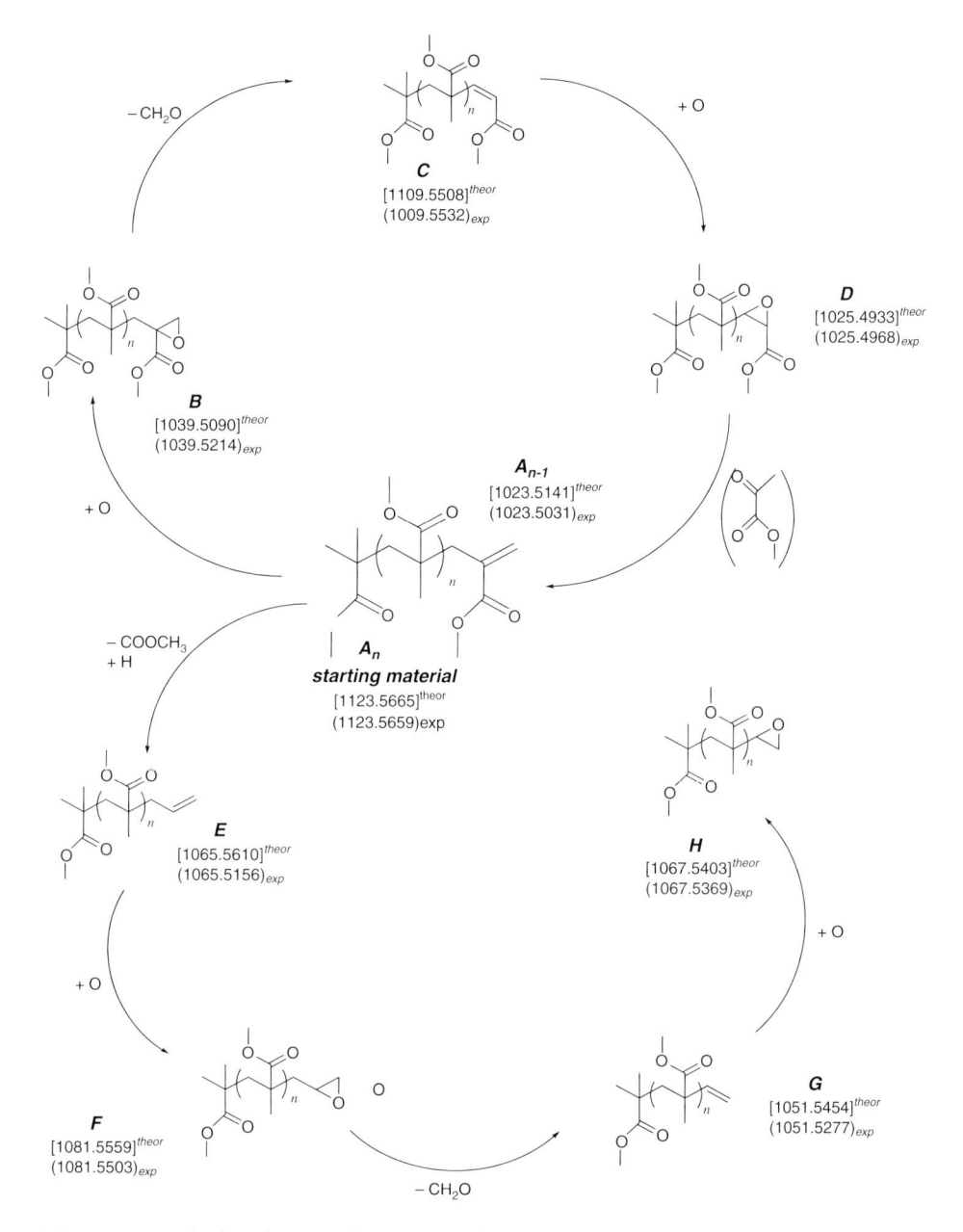

Scheme 13.4 Cyclic degradation mechanism for vinyl-terminated poly(methyl methacrylate) (PMMA) at 95 °C involving epoxidation of the terminal vinyl bond and subsequent elimination of formaldehyde and 2-oxopropionic acid methyl ester. The scheme also gives the theoretical and experimental (Q-ToF) isotopic masses for the monoisotopic peaks of each product. Reproduced from Ref. [67] with kind permission of Wiley-VCH.

13.4
Biodegradation

In the last decade, the attention and worldwide consumption of biodegradable polymers has increased even though competition with commodity plastics, which are cheaper and familiar to customers, hampers their commercialization. In recent years, several analytical methods were used to investigate the potential biodegradation of polymeric compounds in different environments (e.g., soil, compost, or aqueous media) [69]. Studies on biodegradation of polymers by MS are a new field in comparison with other degradation processes [11, 69–80]. However, several MS techniques have been recently tested to develop new methodologies in order to evaluate the biodegradation of polymeric materials [69–71, 80].

Aliphatic polyesters constitute the main class of environmentally friendly polymers [81]. It is known that polymers with ester bonds may be hydrolytically degraded under natural conditions through the cleavage of ester bonds. Different studies on hydrolytic degradation of poly(α-hydroxy acid)s, especially poly(L-lactic acid) (PLA) and their copolymers, by diverse MS methods have been published [10, 11, 79]. Höglund et al. determined by ESI-MS the water-soluble degradation products of PLA and PLA grafted with acrylic acid (PLA-AA) after different hydrolytic degradation periods. Surface modification significantly influenced the rate of degradation and resulting product patterns [79].

Currently, polyglycolide (PGA) and its copolymers with LA, poly(glycolide-*co*-caprolactone) with ε-caprolactone (CL), and poly(glycolide-*co*-trimethylene carbonate) with trimethylene carbonate are widely used as materials for the manufacturing of absorbable sutures. Kasperczyk et al. used ESI-MS together with NMR to follow the hydrolytic degradation of various PGA/PCL copolymers. ESI-MS resulted in revealing the detailed chemical structures of various sequences formed during degradation [78].

ESI-MS, ESI-MS/MS [74, 75], and GC/MS [11] were applied in hydrolytic degradation studies of biodegradable homo and copolyesters of 1,5-dioxepan-2-one (DXO), CL and of crosslinked polyester-ether networks. In particular, ESI-MS revealed the influence of hydrophilicity on the water-soluble degradation products of homo and copolyesters of DXO and CL. In fact, the product pattern of DXO-CL-DXO triblock copolymer mainly consisted of DXO-based oligomers, more hydrophilic [74].

High-performance liquid chromatography (HPLC) connected to ESI-MS (HPLC/ESI-MS) has been shown to be an effective tool for the analysis of the enzymatic degradation products of biodegradable polymers [72, 73, 77]. Rizzarelli et al. [72] separated and identified the water-soluble monomers and cooligomers from the enzymatic hydrolysis of poly(butylene succinate-*co*-butylene sebacate), P(BSu-*co*-BSe)s, and poly(butylene succinate-*co*-butylene adipate), P(BSu-*co*-BAd)s, films by online HPLC/ESI-MS. Optimization of the HPLC analysis allowed the separation of isobaric cooligomers, differing only for the comonomer sequence. Figure 13.6a shows the single ion current (SIC) trace of the B$_2$SSe oligomers, whereas parts b and c show the tandem mass spectra of the (M-H$^+$) parent ions at m/z 447 taken at the two maxima of the SIC trace. The ions at m/z 263 and 347 are diagnostic and are due to BSuB and BSeB species that can be exclusively produced from the fragmentation of BSuBSe and BSeBSu, respectively. Therefore, the predominant sequence was found to be BSuBSe, generated from the enzymatic hydrolysis of the sebacic ester bond. Similar conclusions were inferred from ESI-MS/MS of all of the sequences detected, and it was concluded that the sebacic ester bond was preferentially degraded with respect to the succinic ester bond [72].

In a similar fashion, Pulkkinen et al. performed the online characterization of the enzymatic degradation products of 2,2(-bis(2-oxazoline)-linked poly-ε-caprolactone (PCL-O) using HPLC/ESI-MS^n. Specific structures of the PCL-O oligomers evidenced that pancreatic enzymes cleaved mainly ester bonds and thus were largely unable to break down the amide bonds in the polymer chain [77].

13.5
Other Degradation Processes

In recent years, other degradation processes including chemical [82–84], radiation [85], ultrasonic [29, 46], and electron irradiation [86] induced degradation, have been investigated by MS techniques, especially ESI and MALDI-MS.

Osaka et al. used ESI-MS and MS/MS, and MALDI-TOF post source decay (PSD) fragmentation analysis for the characterization of solvolysis products of linear and cyclic PLAs [82]. ESI-QTOF-MS was applied to unambiguously confirm cyclic degradation products by the treatment of poly[(3,3,3-trifluoropropyl) methylsiloxane] (PTFPMS) with various solvents via backbiting reactions. Complete degradation of PTFPMS takes place in acetone at elevated temperature and longer incubation times. Degradation is distinctly retarded in ethyl acetate. Surprisingly, depolymerization occurs also in its non-solvent methanol [84]. Gruendling et al. presented a detailed MS analysis of the degradation products originated during storage of dithioester-functional PMMA and PS carrying cumyldithiobenzoate end groups in THF as well as in an inert solvent, dichloromethane. Based on the ESI-MS evidence, they postulated an unexpected radical degradation mechanism for the PMMA macro-RAFT agent stored in THF. In inert solvents, degradation was significantly less extended and mainly produced vinyl terminated polymer due to cleavage of the dithioester as well as some sulfine from reaction with dissolved singlet oxygen. These findings strongly suggest a substitution of cyclic ethers as solvents for RAFT polymers in their synthesis and analysis [83].

Polyurethane coatings exhibiting sustained physicomechanical properties were exposed to a mixed radiation field supplied by a nuclear research reactor. They were suggested as an additional barrier in the design nuclear waste disposal containers. FTIR, DSC, dynamic mechanical analysis (DMA), wide angle X-ray diffraction (WAXS), and MALDI were used to characterize the changes that occur because of radiation and to relate these changes to polymer structure and composition [85].

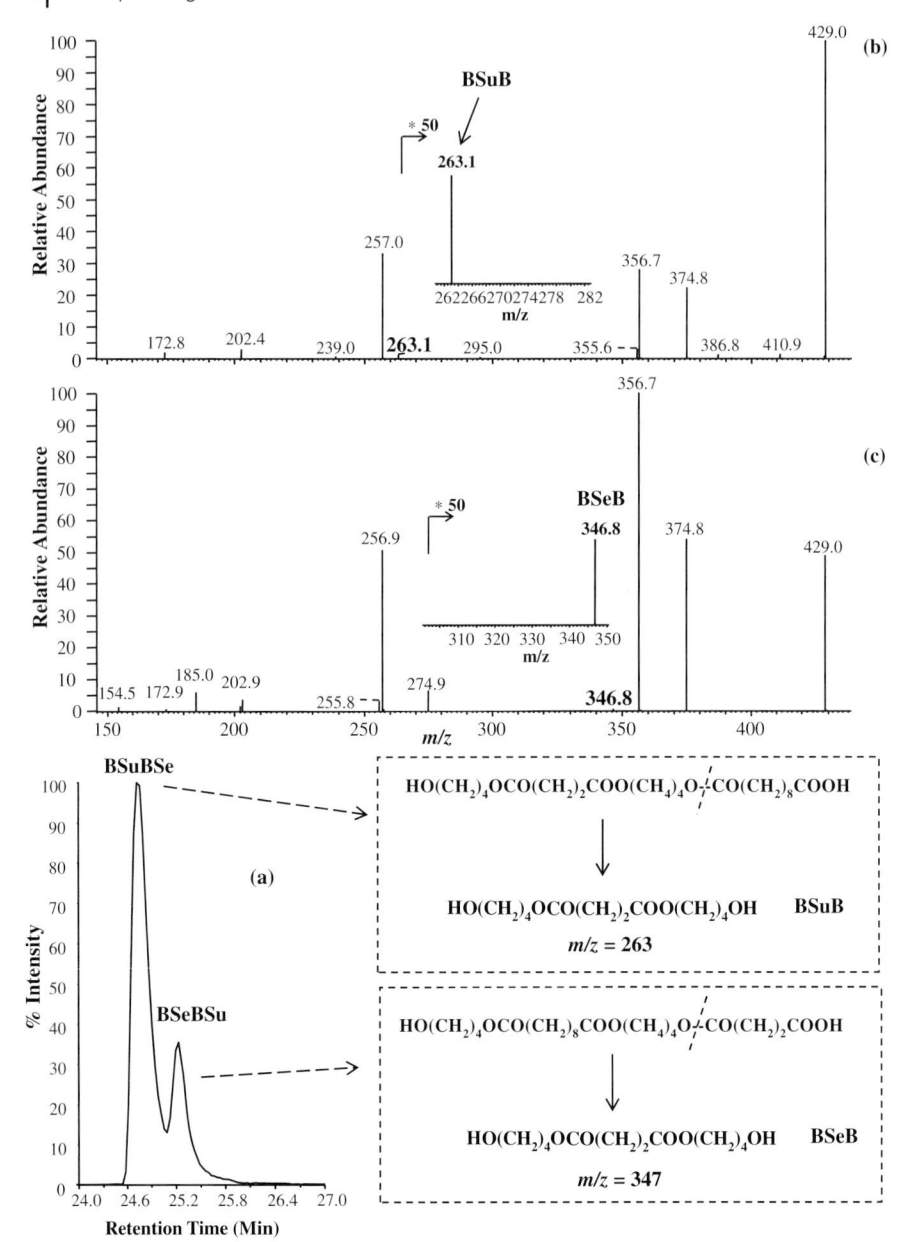

Figure 13.6 Single ion current (SIC) trace of B₂SSe oligomers extracted from the total ion current (TIC) of the water-soluble products obtained after 20 h of enzymatic degradation of poly(butylene succinate-co-butylene sebacate) (P(BSu-co-BSe)) 50/50 by lipase from *R. arrhizus* (a) and electrospray ionization tandem mass spectra (ESI-MS/MS) (+) of the two possible sequences BSuBSe and BSeBSu (b, c). The ions at m/z 263 and 347 can be exclusively produced from the fragmentation of BSuBSe and BSeBSu, respectively. Therefore, the predominant sequence was found to be BSuBSe, generated from the enzymatic hydrolysis of the sebacic ester bond preferentially degraded with respect to the succinic ester bond. Adapted from Ref. [72] with kind permission of the ACS.

13.6
Conclusions

Mass spectrometry continues to be extensively employed for the detection of products originating from degradation processes. Table 13.1 gives a picture of the selected polymer degradation studies carried out by MS in the last decade.

Thermal deterioration and changes induced in polymeric structures by heat in inert atmospheres have been investigated by MS more extensively than that of any other source of degradation. Even though soft ionization MS is acquiring more and more importance as a technique to study polymer degradation, a surprising number of papers concerning thermal degradation are based on traditional MS methods, including TG-MS, GC/MS, or py-GC/MS, as well as DPMS, and required the support of other spectroscopic techniques. Undoubtly, the thermal degradation pathways of a polymer chain can often be deduced from the pyrolysis products formed, and in such instance pyrolysis MS methods yield unique information, polyaddition polymers being more suited than polycondensates. The structural characterization of pyrolysis products by DPMS is particularly advantageous since the pyrolysis is accomplished "online" under high vacuum and nearby the ion source. The pyrolysis products are rapidly volatilized and immediately ionized, thus preventing the occurrence of secondary reactions and any further rearrangement [1, 12]. On the other hand, TG-MS, GC/MS or py-GC/MS, and DPMS fail to provide a direct observation of the chemical changes occurring during degradation on a molecular level within the polymer backbone, necessary for understanding the thermal degradation mechanisms. The applications of soft ionization techniques, especially MALDI and ESI-MS, opened new vistas onto thermo and photooxidative degradation mechanisms of diverse polymers [30, 53, 57, 61–63, 67, 68] and highlighted the great potential of the approach that should be more widely applied in thermal degradation investigations. Studies on polymer oxidation by MALDI and ESI-MS are quite recent and involve the collection of mass spectra at different irradiation times and/or temperatures to observe the structural changes induced by heat or light under an oxidizing atmosphere. The polymer sample can be directly analyzed, and the recorded mass spectrum arises from a mixture of nonoxidized and oxidized chains. The results obtained for the systems so far investigated are surprisingly highly informative, as compared with previous studies based on conventional techniques. In fact, the mass spectra yield precise information on the size, structure, and end groups of molecules originating from oxidation processes, allowing discrimination among possible oxidation mechanisms and providing remarkable information on their induction period. Obviously, MALDI as well as ESI-MS suffer significant restraints. In fact, only the soluble part of the degradation products generated in the processes can be directly analyzed. If some cross-linking occurs, the cross-linked portion becomes insoluble and cannot be analyzed. Furthermore, it is paramount for successful MALDI and ESI-MS analysis that the polymer backbone is readily ionizable and, whenever ESI-MS is utilized, the molecular weight of the polymer should be relatively small so that the observation of singly charged molecules is encouraged.

Table 13.1 Overview of mass spectrometry techniques employed in polymer degradation studies.

MS technique	Polymers	Additional techniques used	Degradation studied	Selected reference
Py-GC/MS	/	/	/	[2, 3, 6, 7]
	PBT	MALDI, NMR	Thermal, thermo-ox, photo-ox	[13]
	P(BCA)	TGA	Thermal	[18]
	PS	DSC, SEC, TGA	Thermal	[19]
	PPSu	TGA	Thermal	[31]
	Aliphatic PU	TGA, FAB, NMR, FTIR	Thermal	[37]
	PHB	GC/MS	Thermal	[38]
	PEO-co-PPO	/	Thermal, thermo-ox, photo-ox	[41]
	BPA-PC	/	Thermal	[48]
	P(BSu-co-BAd)		Biodegradation	[70]
DPMS	/	TGA	/	[1–3, 6, 7, 12]
	PEI	MALDI, FAB, SEC	Thermal	[15]
	BPA-PC	/	Thermal, thermo-ox	[21]
	Polyaniline and Polypyrrole	TGA, FTIR, TEM	Thermal	[39]
	PS-b-poly(2vinylpyridine)	/	Thermal	[40]
MALDI-MS	/	SEC	/	[4, 8, 9]
	PC	Cl-MS, Cl-MS/MS	Thermal	[14]
	PEO	Cl-MS, Cl-MS/MS	Thermal	[16]
	PTHF	Py-GC/MS, NMR	Thermal	[17]
	P(BCA)	Cl-MS,ATR-FTIR	Thermal	[18]
	Aromatic PU	MALDI, FAB, SEC	Thermal	[20]
	BPA-PC		Thermal, thermo-ox	[21]

Polymer	Techniques	Degradation type	Ref.
Ny66	Viscometry	Thermal	[22]
PET, PBT	NMR, viscometry	Thermal	[23, 24]
PS-C$_{60}$	SEC	Thermal	[25]
Aromatic Polybenzoxazole	EGA-GC/MS, TGA	Thermal	[28]
PEO, PMMA	/	Ultrasonic degradation	[29]
Ny6 and oxazoline cyclophosphazene	NMR, FT-IR	Thermal	[35]
PET	NMR	Thermo-mech., thermo-ox	[36]
PBSu	NMR, SEC	Thermal, thermo-ox	[43]
PET	Viscometry	Thermo-mech.	[44]
Poly(p-phenylene sulfide)	/	Thermal	[45]
PEO-co-PPO	LC-APCI-MS	Ultrasonic degradation	[46]
BPA-PC	SEC, NMR, SEC/MALDI	Thermo-ox	[49, 50]
PEO-b-PPO-b-PEO	SEC/NMR, SEC/MALDI, SPME/GC-MS	Thermo-ox	[51]
PEI	TLC, DIOS-MS, TOF-SIMS	Thermo-ox	[52]
PET	XPS, TGA, DSC	Thermo-ox	[53]
PET	Viscometry	Thermo-ox	[54]
Ny6, Ny6,6	/	Thermo-ox	[55]
PEI	Viscometry	Thermo-ox, photo-ox	[56]
PBT	Viscometry	Thermo-ox, photo-ox	[57]
Ny6, Ny6,6	SEC, UV	Photo-ox	[61, 62]
PBSu	FTIR, UV, chemiluminescence, fluorescence	Photo-ox	[63]
BPA-PC		Photo-ox	[64]
Poly(neopentyl-isophthalate)	UV, ATR-FTIR, DSC, SEC	Photo-ox	[65, 66]
Peptide based polymers	SEC, HPLC, NMR, TGA	Biodegradation	[76]
PU	FTIR, DMA, DSC, WAXS	Radiation-induced degradation	[85]

(Continued)

Table 13.1 (Continued)

MS technique	Polymers	Additional techniques used	Degradation studied	Selected reference
ESI-MS	/	/	/	[2, 3, 6–9, 11]
	PMMA	/	Thermal, thermo-ox	[30]
	PHB	FTIR, NMR, SEC, TGA	Thermal	[32]
	PMMA, PBA, PHEMA	/	Photo-ox	[67, 68]
	PBSu, PBAd, PBSe and copolymers	HPLC, NMR	Biodegradation	[72]
	PEOT-b-PBT	HPLC	Biodegradation	[73]
	DXO-CL-DXO	DSC, SEC	Biodegradation	[74]
	DXO-co-CL	DSC, FTIR	Biodegradation	[75]
	PCL	HPLC, NMR	Biodegradation	[77]
	PGA/PCL	NMR	Biodegradation	[78]
	PLA	AFM, SEC, FTIR, DSC	Biodegradation	[79]
	PLA	MALDI	Chemical degradation	[82]
	PMMA, PS	SEC, SEC/ESI	Chemical degradation	[83]
	PTFPMS	GC, NMR, XPS, fluorescence	Chemical degradation	[84]

References

1. Montaudo, G. and Lattimer, R.P. (eds) (2002) *Mass Spectrometry of Polymers*, CRC Press, Boca Raton, FL.
2. Peacock, P.M. and McEwen, N.C. (2004) Mass spectrometry of synthetic polymers. *Anal. Chem.*, **76** (12), 3417–3428.
3. Peacock, P.M. and McEwen, N.C. (2006) Mass spectrometry of synthetic polymers. *Anal. Chem.*, **78** (12), 3957–3964.
4. Montaudo, G., Montaudo, M.S., and Samperi, F. (2006) Characterization of synthetic polymers by MALDI-MS. *Prog. Polym. Sci.*, **31**, 277–357.
5. Gijsman, P. (2008) Review on the thermo-oxidative degradation of polymers during processing and in service. *e-Polymers*, **65**, 1–34.
6. Weidner, S.M. and Trimpin, S. (2008) Mass spectrometry of synthetic polymers. *Anal. Chem.*, **80** (12), 4349–4361.
7. Weidner, S.M. and Trimpin, S. (2010) Mass spectrometry of synthetic polymers. *Anal. Chem.*, **82** (12), 4811–4829.
8. Hart-Smith, G. and Barner-Kowollik, C. (2010) Contemporary mass spectrometry and the analysis of synthetic polymers: trends, techniques and untapped potential. *Macromol. Chem. Phys.*, **211**, 1507–1529.
9. Gruendling, T., Weidner, S., Falkenhagen, J., and Barner-Kowollik, C. (2010) Mass spectrometry in polymer chemistry: a state-of-the-art up-date. *Polym. Chem.*, **1**, 599–617.
10. Pielichowski, K. and Njuguna, J. (eds) (2005) *Thermal Degradation of Polymeric Materials*, Rapra Technology Limited, Shawbury, Shrewsbury, Shropshire SY4 4NR, UK.
11. Albertsson, A.N. and Hakkarainnen, M. (eds) (2008) Chromatography for sustainable polymeric materials, renewable, degradable and recyclable. *Adv. Polym. Sci.*, **211**, 1–191.
12. Hacaloğlu, J. (2007) Pyrolysis mass spectrometry, in *Encyclopaedia of Mass Spectrometry*, vol. **6** (eds M.L. Gross and R.M. Caprioli), Elsevier, Amsterdam, pp. 925–938.
13. Manabe, N. and Yokota, Y. (2000) The method for analyzing anhydride formed in poly(butylenes terephthalate) (PBT) during thermal and photo-degradation processes and applications for evaluation of the extent of degradation. *Polym. Degrad. Stab.*, **69**, 183–190.
14. Puglisi, C., Samperi, F., Carroccio, S., and Montaudo, G. (2000) MALDI-TOF investigation of polymer degradation. Pyrolysis of poly(bisphenol A carbonate). *ACS, Polym. Preprints, Div. Polym. Chem.*, **41** (1), 680–681.
15. Carroccio, S., Puglisi, C., and Montaudo, G. (2000) Thermal degradation mechanisms of polyetherimide investigated by direct pyrolysis mass spectrometry. *ACS Polym. Preprints, Div. Polym. Chem.*, **41** (1), 684–768.
16. Lattimer, R.P. (2000) Mass spectral analysis of low-temperature pyrolysis products from poly(ethylene glycol). *J. Anal. Appl. Pyrol.*, **56**, 61–78.
17. Lattimer, R.P. (2001) Mass spectral analysis of low-temperature pyrolysis products from poly(tetrahydrofuran). *J. Anal. Appl. Pyrol.*, **57**, 57–76.
18. Hickey, A., Leahy, J.J., and Birkinshaw, C. (2001) End-group identity and its effect on the thermal degradation of poly(butyl cyanoacrylate). *Macromol. Rapid. Comm.*, **22** (14), 1158–1162.
19. Roland, A.I. and Schmidt-Naake, G. (2001) Thermal degradation of polystyrene produced by nitroxide-controlled radical polymerization. *J. Anal. Appl. Pyrol.*, **58**, 143–154.
20. Lattimer, R.P. and Williams, R.C. (2002) Low-temperature pyrolysis products from a polyether-based urethane. *J. Anal. Appl. Pyrol.*, **63**, 85–104.
21. Montaudo, G., Carroccio, S., and Puglisi, C. (2002) Thermal and themoxidative degradation processes in poly(bisphenol a carbonate). *J. Anal. Appl. Pyrol.*, **64**, 229–247.
22. Puglisi, C., Samperi, F., Di Giorgi, S., and Montaudo, G. (2002) MALDI-TOF characterisation of thermally generated

gel from nylon 66. *Polym. Degrad. Stab.*, **78**, 369–378.

23 Samperi, F., Puglisi, C., Alicata, R., and Montaudo, G. (2004) Thermal degradation of poly(ethylene terephthalate) at the processing temperature. *Polym. Degrad. Stab.*, **83**, 3–10.

24 Samperi, F., Puglisi, C., Alicata, R., and Montaudo, G. (2004) Thermal degradation of poly(butylene terephthalate) at the processing temperature. *Polym. Degrad. Stab.*, **83**, 11–17.

25 Audouin, F., Nuffer, R., and Mathis, C. (2004) Thermal Stability of the Fullerene-Chain Link in 6-Arm PS Stars with a C60 Core. *J. Polym. Sci. Part A Polym. Chem.*, **42**, 4820–4829.

26 White, D.R. and White, R.L. (2005) Infrared and mass spectrometry studies of the thermal decomposition of a nitro-aromatic polymer. *J. Appl. Polym. Sci.*, **95** (2), 351–357.

27 Adam, T., Streibel, T., Mitschke, S., Mühlberger, F., Baker, R.R., and Zimmermann, R. (2005) Application of time-of-flight mass spectrometry with laser-based photoionization methods for analytical pyrolysis of PVC and tobacco. *J. Anal. Appl. Pyrol.*, **74**, 454–464.

28 Gies, A.P. and Hercules, D.M. (2006) MALDI-TOF MS study of aromatic polybenzoxazole fibers. *Macromolecules*, **39**, 2488–2500.

29 Kawasaki, H., Takeda, Y., and Arakawa, R. (2007) Mass spectrometric analysis for high molecular weight synthetic polymers using ultrasonic degradation and the mechanism of degradation. *Anal. Chem.*, **79**, 4182–4187.

30 Bennet, F., Lovestead, T.M., Barker, P.J., Davis, T.P., Stenzel, M.H., and Barner-Kowollik, C. (2007) Degradation of poly(methyl methacrylate) model compounds at constant elevated temperature studied via high resolution electrospray ionization mass spectrometry (ESI-MS). *Macromol. Rapid. Commun.*, **28**, 1593–1600.

31 Bikiaris, D.N., Chrissafis, K., Paraskevopoulos, K.M., Triantafyllidis, K.S., and Antonakou, E.V. (2007) Investigation of thermal degradation mechanism of an aliphatic polyester using pyrolysis-gas chromatography-mass spectrometry and a kinetic study of the effect of the amount of polymerisation catalyst. *Polym. Degrad. Stab.*, **92**, 525–536.

32 Kawalec, M., Adamus, G., Kurcok, P., Kowalczuk, M., Foltran, I., Focarete, M.L., and Scandola, M. (2007) Carboxylate-induced degradation of Poly(3-hydroxybutyrate)s. *Biomacromolecules*, **8**, 1053–1058.

33 Saraji-Bozorgzad, M., Geissler, R., Streibel, T., Mu1hlberger, F., Sklorz, M., Kaisersberger, E., Denner, T., and Zimmermann, R. (2008) Thermogravimetry coupled to single photon ionization quadrupole mass spectrometry: a tool to investigate the chemical signature of thermal decomposition of polymeric materials. *Anal. Chem.*, **80**, 3393–3403.

34 Whitson, S.E., Erdodi, G., Kennedy, J.P., Lattimer, R.P., and Wesdemiotis, C. (2008) Direct probe-atmospheric pressure chemical ionization mass spectrometry of cross-linked copolymers and copolymer blends. *Anal. Chem.*, **80**, 7778–7785.

35 Samperi, F., Bazzano, S., Battiato, S., Scaffaro, R., Botta, L., Mistretta, M.C., Bertani, R., and Milani, R. (2009) Reactions occurring during the melt mixing of nylon 6 and oxazoline-cyclophosphazene units. *Macromolecules*, **42**, 5579–5592.

36 Romão, W., Franco, M.F., Corilo, Y.E., Eberlin, M.N., Spinacé, M.A.S., and De Paoli, M.A. (2009) Poly(ethylene terephthalate) thermo-mechanical and thermo-oxidative degradation mechanisms. *Polym. Degrad. Stab.*, **94**, 1849–1859.

37 Zhang, Y., Xia, Z., Huang, H., and Chen, H. (2009) Thermal degradation of polyurethane based on IPDI. *J. Anal. Appl. Pyrol.*, **84** (1), 89–94.

38 Ariffin, H., Nishida, H., Shirai, Y., and Hassan, M.A. (2009) Anhydride production as an additional mechanism of poly(3-hydroxybutyrate) pyrolysis. *J. Appl. Polym. Sci.*, **111** (1), 323–328.

39 Hacaloğlu, J., Tezal, F., and Kucukyavuz, Z. (2009) The characterization of polyaniline and polypyrrole composites by pyrolysis mass spectrometry. *J. Appl. Polym. Sci.*, **113** (5), 3130–3136.

40 Elmaci, A., Hacaloğlu, J., Kayran, C., Sakellariou, G., and Hadjichristidis, N. (2009) Thermal decomposition of polystyrene-b-poly(2-vinylpyridine) coordinated to co nanoparticles. *Polym. Degrad. Stab.*, **94**, 2023–2027.

41 Watanabe, C., Tsuge, S., and Ohtani, H. (2009) Development of new pyrolysis–GC/MS system incorporated with on-line micro-ultraviolet irradiation for rapid evaluation of photo, thermal, and oxidative degradation of polymers. *Polym. Degrad. Stab.*, **94**, 1467–1472.

42 Li, J., Cai, J., Yuan, T., Guo, H., and Qi, F. (2009) A thermal decomposition study of polymers by tunable synchrotron vacuum ultraviolet photoionization mass spectrometry. *Rapid. Commun. Mass Spectrom.*, **23** (9), 1269–1274.

43 Rizzarelli, P. and Carroccio, S. (2009) Thermo-oxidative processes in biodegradable poly(butylene succinate). *Polym. Degrad. Stab.*, **94**, 1825–1838.

44 Romão, W., Franco, M.F., Iglesias, A.H., Sanvido, G.B., Maretto, D.A., Gozzo, F.C., Poppi, R.J., Eberlin, M.N., Spinacé, M.A.S., and De Paoli, M.A. (2010) Fingerprinting of bottle-grade poly(ethylene terephthalate) via matrix-assisted laser desorption/ionization mass spectrometry. *Polym. Degrad. Stab.*, **95**, 666–671.

45 Gies, A.P., Geibel, J.F., and Hercules, D.M. (2010) MALDI-TOF MS Study of Poly(p-phenylene sulfide). *Macromolecules*, **43**, 943–951.

46 Watanabe, T., Okabayashi, M., Kurokawa, D., Nishimoto, Y., Ozawa, T., Kawasakia, H., and Arakawa, R. (2010) Determination of primary bond scissions by mass spectrometric analysis of ultrasonic degradation products of poly(ethylene oxide-block-propylene oxide) copolymers. *J. Mass Spectrom.*, **45** (7), 799–805.

47 Rial-Otero, R., Galesio, M., Capelo, J.L., and Simal-Gándara, J. (2009) A review of synthetic polymer characterization by pyrolysis–GC–MS. *J. Chromatogr.*, **70**, 339–348.

48 Oba, K., Ishida, Y., Ito, Y., Ohtani, H., and Tsuge, S. (2000) Characterization of branching and/or cross-linking structures in polycarbonate by reactive pyrolysis-gas chromatography in the presence of organic alkali. *Macromolecules*, **33**, 8173–8183.

49 Carroccio, S., Puglisi, C., and Montaudo, G. (2002) Mechanisms of thermal oxidation of poly(bisphenol A carbonate). *Macromolecules*, **35**, 4297–4305.

50 Montaudo, G., Carroccio, S., and Puglisi, C. (2002) Thermal oxidation of poly(bisphenol A carbonate) investigated by SEC/MALDI. *Polym. Degrad. Stab.*, **77**, 137–146.

51 Gallet, G., Carroccio, S., Rizzarelli, P., and Karlsson, S. (2002) Thermal degradation of poly(ethylene oxide–propylene oxide–ethylene oxide) triblock copolymer: comparative study by SEC/NMR, SEC/MALDI-TOF-MS and SPME/GC-MS. *Polymer*, **43**, 1081–1094.

52 Carroccio, S., Puglisi, C., and Montaudo, G. (2005) New vistas in polymer degradation. Thermal oxidation processes in poly(ether imide). *Macromolecules*, **38**, 6849–6862.

53 Ciolacu, F.C.L., Choudhury, N.R., Dutta, N., and Voelcker, N.H. (2006) MALDI-TOF MS and DIOS-MS investigation of the degradation and the discoloration of poly(ethylene terephthalate). *Macromolecules*, **39**, 7872–7881.

54 Ciolacu, F.C.L., Choudhury, N.R., Dutta, N., and Kosior, E. (2007) Molecular level stabilization of poly(ethylene terephthalate) with nanostructured open cage trisilanolisobutyl POSS. *Macromolecules*, **40**, 265–272.

55 Carroccio, S., Scaltro, G., Puglisi, C., Ferreri, T., and Montaudo, G. (2007) MALDI TOF investigation on Ny6 and Ny66 thermo-oxidation products. *Eur. J. Mass Spec.*, **13**, 397–408.

56 Carroccio, S., Puglisi, C., and Montaudo, G. (2005) Comparison of

Photooxidation and thermal oxidation processes in poly(ether imide). *Macromolecules*, **38**, 6863–6870.

57 Carroccio, S., Rizzarelli, P., Scaltro, G., and Puglisi, C. (2008) Comparative investigation of photo and thermal oxidation processes in poly(butylen terephthalate). *Polymer*, **49**, 3371–3381.

58 Clough, R.L., Billingham, N.C., and Gillen, K.T. (eds) (1996) Polymer Durability. Degradation, Stabilization and Lifetime Prediction, *Advances in Chemical Series*, vol. 249, ACS, Washington (DC).

59 Gies, A.P., Nonidez, W.K., Anthamatten, M., Cook, R.C., and Mays, J.W. (2002) *Rapid. Commun. Mass Spectrom.*, **16**, 1903–1910.

60 Rivaton, A. and Gardette, J.L. (1998) Photo-oxidation of aromatic polymers. *Die Angew. Makromol. Chem.*, **261–262**, 173–188.

61 Carroccio, S., Puglisi, C., and Montaudo, G. (2003) New vistas in the photo-oxidation of nylon 6. *Macromolecules*, **36**, 7499–7507.

62 Carroccio, S., Puglisi, C., and Montaudo, G. (2004) MALDI investigation of the photooxidation of nylon-66. *Macromolecules*, **37**, 6037–6049.

63 Carroccio, S., Rizzarelli, P., Puglisi, C., and Montaudo, G. (2004) MALDI investigation of photooxidation in aliphatic polyesters: poly(butylene succinate). *Macromolecules*, **37**, 6576–6586.

64 Diepens, M. and Gijsman, P. (2007) Photodegradation of bisphenol A polycarbonate. *Polym. Degrad. Stab.*, **92**, 397–406.

65 Malanowski, P., Huijser, S., Scaltro, F., Van Benthem, A.T.M., Van der Ven, L.G.J., Laven, J., and De With, G. (2009) Molecular mechanism of photolysis and photooxidation of poly(neopentyl isophthalate). *Polymer*, **50**, 1358–1368.

66 Malanowski, P., Huijser, S., Van Benthem, A.T.M., Van der Ven, L.G.J., Laven, J., and De With, G. (2009) Photodegradation of poly(neopentyl isophthalate). Part I. Laboratory and outdoor conditions. *Polym. Degrad. Stab.*, **94**, 2086–2094.

67 Bennet, F., Hart-Smith, G., Gruendling, T., Davis, T.P., Barker, P.J., and Barner-Kowollik, C. (2010) Degradation of poly(methyl methacrylate) model compounds under extreme environmental conditions. *Macromol. Chem. Phys.*, **211**, 1083–1097.

68 Bennet, F., Barker, P.J., Davis, T.P., Soeriyadi, A.H., and Barner-Kowollik, C. (2010) Degradation of poly(butyl acrylate) and poly(2-hydroxyethyl methacrylate) model compounds under extreme environmental conditions. *Macromol. Chem. Phys.*, **211**, 2034–2052.

69 Eubeler, J.P., Bernhard, M., Zok, S., and Knepper, T.P. (2009) Test methodologies and procedures. *Trends Anal. Chem.*, **28** (9), 1057–1072.

70 Sato, H., Furuhashi, M., Yang, D., Ohtani, H., Tsuge, S., Okada, M., Tsunoda, K., and Aoi, K. (2001) A novel evaluation method for biodegradability of poly(butylene succinate-co-butylene adipate) by pyrolysis-gas chromatography. *Polym. Degrad. Stab.*, **73**, 327–334.

71 Lee, J.W. and Gardella, J.A. Jr. (2003) Simultaneous time-of-flight secondary ion MS quantitative analysis of drug surface concentration and polymer degradation kinetics in biodegradable poly(L-lactic acid) blends. *Anal. Chem.*, **75**, 2950–2958.

72 Rizzarelli, P., Impallomeni, G., and Montaudo, G. (2004) Evidence for selective hydrolysis of aliphatic copolyesters induced by lipase catalysis. *Biomacromolecules*, **5**, 433–444.

73 Hayen, H., Deschamps, A.A., Grijpma, D.W., Feijen, J., and Karst, U. (2004) Liquid chromatographic–mass spectrometric studies on the in vitro degradation of a poly(ether ester) block copolymer. *J. Chrom. A*, **1029** (12), 29–36.

74 Hakkarainen, M., Adamus, G., Höglund, A., Kowalczuk, M., and Albertsson, A.C. (2008) ESI-MS reveals the influence of hydrophilicity and architecture on the water-soluble degradation product patterns of biodegradable homo and copolyesters of 1,5-dioxepan-2-one and ε-caprolactone. *Macromolecules*, **41**, 3547–3554.

75 Höglund, A., Hakkarainen, M., Kowalczuk, M., Adamus, G., and

Albertsson, A.C. (2008) Fingerprinting the degradation product patterns of different polyester–ether networks by electrospray ionization mass spectrometry. *J. Polym. Sci.: Part A: Polym. Chem.*, **46**, 4617–4629.

76 Van Dijk, M., Nollet, M.L., Weijers, P., Dechesne, A.C., Van Nostrum, C.F., Hennink, W.E., Rijkers, D., and Liskamp, R.M.J. (2008) Synthesis and characterization of biodegradable peptide-based polymers prepared by microwave-assisted click chemistry. *Biomacromolecules*, **9**, 2834–2843.

77 Pulkkinen, M., Palmgren, J.J., Auriola, S., Malin, M., Seppala, J., and Kristiina, J. (2008) High-performance liquid chromatography/electrospray ionization tandem mass spectrometry for characterization of enzymatic degradation of 2,20-bis(2-oxazoline)-linked poly-ε-caprolactone. *Rapid. Commun. Mass Spectrom.*, **22**, 121–129.

78 Kasperczyk, J., Li, S., Jaworska, J., Dobrzyński, P., and Vert, M. (2008) Degradation of copolymers obtained by ring-opening polymerization of glycolide and 3-caprolactone: a high-resolution NMR and ESI-MS study. *Polym. Degrad. Stab.*, **93**, 990–999.

79 Höglund, A., Hakkarainen, M., Edlund, U., and Albertsson, A.C. (2010) Surface modification changes the degradation process and degradation product pattern of polylactide. *Langmuir*, **26** (1), 378–383.

80 Kunioka, M., Ninomiya, F., and Funabashi, M. (2009) Biodegradation of poly(butylene succinate) powder in a controlled compost at 58 °C evaluated by naturally-occurring carbon 14 amounts in evolved CO_2 based on the ISO 14855-2 method. *Int. J. Mol. Sci.*, **10**, 4267–4283.

81 Gross, R.A. and Kalra, B. (2002) Biodegradable polymers for the environment. *Science*, **297**, 803–807.

82 Osaka, I., Watanabe, M., Masashi, M., Masahiro, T., Murakami, M., and Arakawa, R. (2006) Characterization of linear and cyclic polylactic acids and their solvolysis products by electrospray ionization mass spectrometry. *J. Mass Spectrom.*, **41**, 1369–1377.

83 Gruendling, T., Pickford, R., Guilhaus, M., and Barner-Kowollik, C. (2008) Degradation of RAFT polymers in a cyclic ether studied via high resolution ESI-MS: Implications for synthesis, storage, and end-group modification. *J. Polym. Sci. Part A Polym. Chem.*, **46**, 7447–7461.

84 Kahlig, H., Zollner, P., and Mayer-Helm, B.X. (2009) Characterization of degradation products of poly[(3,3,3-trifluoropropyl) methylsiloxane] by nuclear magnetic resonance spectroscopy, mass spectrometry and gas chromatography. *Polym. Degrad. Stab.*, **94**, 1254–1260.

85 Mortley, A., Bonin, H.W., and Bui, V.T. (2008) Radiation effects on polymers for coatings on copper canisters used for the containment of radioactive materials. *J. Nucl. Mater.*, **376**, 192–200.

86 Ahn, H., Oblas, D.W., and Whitten, J.E. (2004) Electron irradiation of Poly(3-hexylthiophene) films. *Macromolecules*, **37**, 3381–3387.

14
Outlook

Christopher Barner-Kowollik, Jana Falkenhagen, Till Gruendling, and Steffen Weidner

The synthesis of macromolecules of an increasing architectural diversity and complexity is the lifeblood of modern macromolecular chemistry, allowing for specialized applications in fields ranging from industrial coatings to medical drug delivery systems. The high complexity of modern polymers with regard to molar mass, chemical functionality, composition, and topology requires a versatile technique for their comprehensive characterization. Today, mass spectrometry, in combination with the soft ionization techniques ESI and MALDI, has been established as an extremely valuable tool for the analysis of polymeric materials.

Mass spectrometry is unique in that it provides the capability to characterize individual molecular species in a polymeric material. Combined with the very high sensitivity and mass resolution of modern mass analyzers, the great chemical information content has made mass spectrometry a universal tool in functional polymer characterization.

In the current publication, a number of expert authors have presented an introduction to the state- of-the-art of modern polymer mass spectrometry. This technique, in addition to being an extremely useful tool for the everyday characterization of functional synthetic polymers, comprises a vibrant research field which is constantly reinventing itself. Recent technological developments, such as the introduction of solid-state lasers, novel high-resolution mass analyzers such as the Orbitrap, novel ionization techniques such as APPI and detectors of increased dynamic range and increased sensitivity at high molecular weights will further improve the applicability of mass spectrometry for the polymer analysis field. The developments benefit from the progress in experimental equipment on the one hand and from the ever increasing demand for high quality and detailed structural information which is needed to tackle the growing complexity of modern polymers, on the other hand.

The coupling to liquid-phase separation techniques allows the chemical heterogeneity and molar mass distribution of polymers to be more readily interpreted and therefore facilitates the sample identification process. The combination of MS with additional gas-phase separation techniques, such as ion mobility spectrometry (IMS), is a field of great potential. The many possible applications of IMS, providing what

Mass Spectrometry in Polymer Chemistry, First Edition.
Edited by Christopher Barner-Kowollik, Till Gruendling, Jana Falkenhagen, and Steffen Weidner.
© 2012 Wiley-VCH Verlag GmbH & Co. KGaA. Published 2012 by Wiley-VCH Verlag GmbH & Co. KGaA.

can be regarded as chromatography in the gas phase at the high speed and resolutions inherent to mass spectrometry, are still to be exploited.

Especially in the field of polymer synthesis and when monitoring polymer degradation processes, mass spectrometry provides a powerful analytical tool for polymer chemists. Living/controlled radical polymerization techniques have revolutionized polymer chemistry in the last decade. A great deal of our current mechanistic knowledge about these polymerization techniques is due to studies employing soft ionization mass spectrometry. The determination of the mechanistic details of nonradical polymerization methods (such as coordination polymerization, polycondensation, and polyaddition) has also strongly benefited from the development and application of suitable MS methods. Moreover, these sophisticated MS tools are providing new impulses for the investigation of copolymers and their functional building blocks, precursors, or macroinitiators. Similar to the field of proteomics – where tandem mass spectrometry has been widely used for sequence analysis of peptides and proteins – soft ionization tandem mass spectrometry is gaining in importance for the structural identification of copolymers.

In order to significantly improve the usability of MS techniques, especially in polymer characterization, software tools for automated data processing are being developed. Further research into this area, however, is highly desirable, with the ultimate goal being a fully automated (co)polymer identification and quantitative analysis pipeline providing information on the functionality, composition, and molecular weight of polymers in a reproducible and robust manner.

In a very short time-span, mass spectrometry (especially MALDI- and ESI-MS) has been shown to complement the existing spectroscopic methods, which has led to an increasing interest in its applications in both research and industry. We hope that the current book – in addition to being a valuable handbook for the interested polymer scientist – may provide important impulses for analytical chemists and students alike for the application and further development of this valuable technique in the macromolecular sciences.

Karlsruhe and Berlin, February 2011
Christopher Barner-Kowollik
Jana Falkenhagen
Till Gruendling
Steffen Weidner

Index

a

A–B linkages in a random copolymer of A-co-B (ABABAB)   154
ABm-type monomers, polymerization   407
abundance sensitivity   5, 6, 10–12
acetone   127
acoustic nebulization (AN)   187
acrylate polymerization
– macromonomer formation mechanism   361
– β-scission in   360
acrylonitrile/butadiene   390
N-acryloylmorpholine (NAM)
– polymerization   390
– RAFT-mediated polymerization   389
activated monomer (AM) mechanism   406
active chain end (ACE) mechanism   406, 420
aerosol interface, principle of   215
Ag cationization
– method of   156–162
alkoxyamine BlocBuilder®   377
alkylcarbonyloxy radicals
– decarboxylation   322
– disadvantage   322
alkyl peroxyacetates   327, 328
alkyl peroxypivalates   323–327
allyl butyl ether (ABE)   287
alpha-cyanocinnamic acid (CCA)   189
tert-amyl peroxyacetate (TAPA)   327
tert-amyl peroxypivalate (TAPP)   324
AP-MALDI applications   217
AP-MALDI/ion trap interfaces   216
AP-MALDI source
– schematic view of   218
architectural elucidation   33
atmospheric pressure (AP)   85
– sources   214

atmospheric pressure chemical ionization (APCI)   45, 49, 69, 283
– advantages   49
– limitations   49
atmospheric pressure glow discharge ionization (APGDI)   194
atmospheric pressure photoionization (APPI)   49
atmospheric solids analysis probe (ASAP)   69, 93
atomic force microscopy (AFM)   128
atomic ion beams   176
atom transfer radical polymerization (ATRP)   226, 284, 373, 378
– anionic polymerization to   382
– catalysts   381
– efficiency   381
– generated polymers
–– structure   378
– initiators
–– catalytic activity   381
ATRP-prepared PBAs
– MALDI-TOF-MS spectra   379
attenuated total reflectance FTIR (ATR-FTIR)   444
Au-cationized PS oligomeric distribution   166
automated data processing   468
2,2′-azobisisobutyronitrile (AIBN)   349, 381

b

Bayesian probability theory   268
α-benzyl, ω-hydroxy polyethylene oxide
– MALDI-TOF mass spectrum and SEC trace   408
benzyl radicals   326
bipolyesters poly(3-hydroxybutyrate)-co-poly (3-hydroxyvalerate)   306
bis γ-lactone

– anionic copolymerization mechanism   417
2,2(-bis(2-oxazoline)-linked poly-
   ε-caprolactone (PCL-O)   455
bisphenol A-polycarbonate (BPA-PC)   441
bis-3,5,5-trimethylhexanoyl peroxide
   (BTMHP)   350
bombarded region
– gel point   180
bottle-grade PET (btg-PET)
– thermo-mechanical/thermo-oxidative
   degradation mechanisms   441
bovine serum albumin (BSA)   182
bromine-terminated poly(t-BA)   385
B_2SSe oligomers
– single ion current (SIC) trace   456
n-butyl acrylate (BA)
– block copolymers   375
– RP   358
tert-butyl acrylate (t-BA)
– ATRP   379
butyl alcohol (BuOH)   420
tert-butyl mercaptan   351
– chain transfer mechanism   351
n-butyl methacrylate (BMA)   341, 343
– polymerization   395
N-tert-butyl-N-(1-diethylphosphono-2,2-
   dimethylpropyl nitroxide)   376
tert-butyl peroxyacetate (TBPA)   328
– initiation by   335
tert-butyl peroxypivalate (TBPP)   326
p-tert-butylphenol/$KHCO_3$ system   423
(R,S)-β-butyrolactone
– transesterification   428
β-butyrolactone
– anionic ring-opening polymerization   411
– ROP   410

c

capillary electrophoresis (CE)   214
ε-caprolactone (CL)   426, 454
ε-caprolactone/Sn(Oct)/(butyl alcohol
   or water)
– MALDI-TOF mass spectra   420, 421
carbon-centered radicals   322
α-carboxyl-terminated oligomers   354
catalytic chain transfer (CCT)   373
– effective monomers   395
– mechanism   395
– polymerization   393, 395, 397
catalytic chain-transfer agent (CCTA)   356
catalytic-chain-transfer polymerization (CCTP)
   mechanism   320
– catalytic cycle   394
catalytic cycle   393

– in catalytic chain transfer (CCT)
   polymerizations   394
cationization methods   164
C_{60}-capped polystyrenes   385
chain growth polymerization process   423
chain-length-dependent propagation (CLDP)
   phenomenon   339, 340
chain-length distribution (CLD)   324, 347
– determination of   336
chain-transfer reactions   321
chain transfer to polymer (CTP)   335
– facilitator   359
– tertiary radical from   360
characteristic collision voltage (CCV)   405
CHCA matrix   128
chemical ionization   36, 37
– limitation   37
cobalt cyclopentadienyl dicarbonyl   52
collisionally activated dissociation (CAD)   59
collision energy   28
collision-induced decay experiments   387
collision-induced dissociation (CID)   26, 59
– experiments   381
– fragmentation   85
condensation polymers
– degradation pathways   439
controlled/living radical polymerization
– simplified mechanisms   374
co-oligomers
– MALDI-TOF-mass spectra   417
copolymer characterization   281
– biological/(bio)medical application
   304–307
– MS spectrum   343
– reviews   282, 283
– scope   282
– separation prior MS
– – ion mobility spectrometry-mass
   spectrometry (IMS-MS)   299–301
– – LC-MS   297–299
– – quantitative MS   303
– – tandem MS (MS/MS)   301–303
– soft ionization techniques   283
– – APCI   294–297
– – ESI-MS   292–294
– – MALDI, application of   283–292
– software development   307–309
copolymers
– fingerprints   345
– formation processes   382
– mass spectra of   248
– MS analysis   343
copolymer structures   282
copper(I) thiocyanate (CuSCN)   381

cryptands β-butyrolactone   410
CuBr/2,2′-bipyridine (bpy) as a catalyst   378
Cu-catalyzed azide/alkyne reactions   385
4-cyanopentanoic acid-4-dithiobenzoate (CPADB)   294
cyclic carbonates
– ring-opening polymerization mechanisms   408–423
cyclic esters
– ring-opening polymerization mechanisms   408–423
– ring-openning polymerization   415
– ROP   415
cyclic ethers/esters
– ring-opening polymerization mechanisms   406–408
– ring-opening polymerization (ROP) mechanisms   405
cyclic oligolactones   418
cyclic oligomers   418
– formation   439
cyclic polymers   377
cyclic polystyrene (PS) oligomer   75

d

dead polymer molecules   353
decomposing initiators   331
degradation mechanisms, determination of   33
degradation process
– role   437
degree of polymerization   320, 394
design of experiment (DoE)   240
desorption atmospheric pressure photoionization (DAPPI)   93
desorption chemical ionization (DCI)   45
desorption electrospray ionization (DESI)   48, 69, 192, 193
– mass spectrometry   149
– source   48
desorption ionization on silicon (DIOS)   52, 385
desorption sonic spray ionization   93
diacyl peroxides   328–331
α,β-dialkyl-substituted-β-lactones
– ROP   413
α,α-dialkylsubstitutet-β-lactones
– ROP   410
di-benzoyl peroxide (DBP)   329
di-iso-butyryl peroxide (DIBP)   329
dicaprylcapryl adipate (DCA)   173
N,N-diethylacrylamide (DEAM)   377
– SEC/ESI-MS spectrum   377
diethylene glycol (DEG)   441

di-2-ethylhexyl peroxydicarbonate (EH-PDC)   331
di-ethyl peroxydicarbonate (E-PDC)   331
differential scanning calorimetry (DSC)   287, 438
2,5-dihydroxyacetophenone (2,5-DHAP)   96
2,5-dihydroxybenzoic acid   94
dihydroxybenzoic acid (DHB)   121, 122, 128
– matrix/polymer system   190
– optical images of   187
dihydroxy telechelic PMMA   395
diisooctylsebacate (DOS)   173
2,2-dimethoxy-2-phenylacetophenone (DMPA)   332
– photodissociation   332, 333
dimethyl itaconate (DMI)   332
dimethylphenyl silane (DMPS)   164
di-2-naphthoyl peroxide   330
di-n-dodecanoyl peroxide (DDDP)   328
di-n-tetradecyl peroxydicarbonate (TD-PDC)   331
D ion traps   17–19
– basic components   19
3D ion trap system   17
– abundance sensitivities   18
– components   19
– linear dynamic ranges   18
– mass selective instability   18
– performance characteristics   18
1,5-dioxepan-2-one (DXO)
– ROP copolymerization   418
direct analysis in real time (DART)   93, 194
direct introduction probe chemical ionization (DAPCI)   93
direct-pyrolysis mass spectrometry (DPMS)   438
disproportionation
– rate coefficient   348
– termination by   348
dithioester
– UV-induced radical β-cleavage   392
double focusing sector mass analyzer   14
DP-APCI measurements   297
drug-loaded poly(ethylene glycol) macromonomers
– ROMP   425
dynamic mechanical analysis (DMA)   455

e

ε-caprolactone (CL)   296
electrohydrodynamic ionization (EHI)   45
electron capture dissociation (ECD)   61
– applications   61
– feature of   61, 62

electronic noise 241
electron ionization (EI) 34, 35
– limitations 35
electron transfer dissociation (ETD) 61
– fragmentation 85
electrospray-assisted laser desorption/ionization (ELDI) 93
electrospray deposition (ESD) 219
– principle of 186
electrospray ionization (ESI) 21, 33, 46, 47, 57, 87, 186, 281
– advantages 46, 47
– limitations 47, 48
– mass spectra 210, 386
– mass spectrum 384
– sensitivity 329
– techniques 320
electrospray ionization mass spectrometry (ESI-MS) 238, 331, 340, 390, 406, 424, 445, 452
– chromatographic elution profiles of 268
– chromatographic setup 266
– instrument 332
– MS analysis 429
– MS experiments 406
– technique 410
emulsion polymerization (EP) 364
– mechanism 365
end-group analysis 33
energizing collisions with gaseous targets 59
enhanced spin capturing polymerization (ESCP) 373
entropic
– compensation of 212
epoxide (glicydyl phenyl ether)
– anionic copolymerization mechanism 417
ESI-MS^n
– fragmentation experiments
– – application 413
– techniques 410
ESI-Q/ToF tandem mass spectrometer 71
ethylene oxide
– zwitterionic ring-opening polymerization mechanism 407
ethylene oxide (EO)
– anionic polymerization 406
– polymerization 407
evaporation-grinding method (E-G method) 442
extensible markup language (XML) standard 238
– mzData 238
– mzML 238

– mzXML 238
extractive electrospray ionization (EESI) 94

f

fast atom bombardment (FAB) 42, 43
– limitations 43
fast atom bombardment-MS (FAB-MS) 417, 441
field desorption mass spectrometry (FD-MS) 38–40
– limitations 40
fingerprint approach 347
Flory-Fox equation 266
Fourier filtering 243
Fourier transformation (FT) 58
Fourier transform infrared (FT-IR) 288
Fourier transform-ion cyclotron resonance (FT-ICR) 287
– mass analyzers 223, 391
– mass spectrometers 61
Fourier transform ion cyclotron resonance mass analyzers 22–24
– cyclotron frequencies 22
– operating principles 23
– perfomance characteristics 22
fragmentation
– by ETD and CID 103, 104
fragmentation reactions 378
fragment ions in triple-stage MS experiments 58
free radical polymerization (FRP) 287

g

gas chromatography/mass spectrometry (GC/MS) 438
gas cluster ion beams (GCIB) 177
gas-phase separation of linear, of poly (ε-caprolactone) 91
– ESI-IMS-MS 2D plot of PCL 91, 92
– folding transitions 91
gas-phase separation techniques 467
Gaussian peak shape 267
gel permeation chromatography (GPC) 162
Gibbs–Helmholtz equation 212
glutamate film 176
glycidol 407
glycidyl methacrylate (GMA) 343
glycodentritic copolymers
– MALDI-TOF MS spectra of 304
glycolidyl, copolymer
– chemical structure, changes 295
glycosulated PEG-dendritic copolymers
– MALDI-TOF MS spectra of 305
gradient chromatography 226

gravimetric mixtures 265
G-SIMS process 174

h

$^1H/^{13}C$ NMR spectroscopy 417
heat for ionization 97
N-heterocyclic carbene catalysts (NHC) 406
N-heterocyclic carbenes
– alcohol adducts 416
– application 413
High-energy CAD 59, 60
higher molecular weight (MW) analyses 97
high-molecular-weight polymers 420
high-performance liquid chromatography (HPLC) 211, 454
– optimization 454
high-resolution mass analyzers
– Orbitrap 467
– FT-ICR 61, 223, 287, 391
high-resolution mass spectral analysis 428
^{1}H-NMR spectroscopic data 382
homopolymers 155
hydroperoxy-capped macromolecules formation 391
hydrophilic–lipophilic balance (HLB) value 142
3-hydroxybutyrate oligomers formation 411
(R)-3-hydroxybutyric acid (HB) 426
(R,S)-3-hydroxybutyric acid
– equimolar reaction 429
(R,S)-3-hydroxybutyric acid reaction 429
N-(2-hydroxypropyl) methacrylamide (HPMA)
– polymerization 354
– semitelechelic poly(HPMA) synthesis 354
hyphenated coupling methods 228
hyphenated techniques 209
– coupling principles
–– LAC/LC-CC with MALDI-/ESI-MS 224–228
–– off-line coupling devices 218–220
–– online coupling devices 214–218
–– of SEC with MALDI-/ESI-MS 220–224
–– transfer devices 214
– polymer separation techniques 210–214

i

Ill-conditioned problem 268
imaging mass spectrometry 149, 150
include desorption electrospray ionization (DESI) 93
induction-based fluidics (IBFs) 220
inductively coupled plasma mass spectrometry (ICP-MS) 185
infrared spectroscopy 33

inlet ionization methods 91
instruments for chemical analysis and imaging 94
ion abundance 91, 94
ion cyclotron resonance (ICR) trap 58
ionic coordination catalysts 406
ion intensity distributions 189
ion–ion reactions 61
– fragment ions arising from 61
ionization bias effects 10
ionization efficiency 162
ion mobility mass spectrometry (IMS) 185, 281
– applications 467
– mass spectrometry (MS) 85
–– instrument, for polymers analysis 88
ion traps (ITs) 58
IR-MALDI 398
IR multiphoton photodissociation (IRMPD) 61
– applications 61
isobaric components of nonionic surfactant 71, 72
isolation and fragmentation processes 62
isopropanol 127
N-isopropylacrylamide (NIPAM)
– 2,2,6,6-tetraethylpiperidin-4-on-N-oxyl-mediated polymerization 377
IUPAC-recommended method 336

k

kinetic energy 59

l

lactide
– ring-openning polymerization
β-lactones
– chemistry 409
– ring-openning polymerization 417
laser ablation (LA) 185
laser ablation electrospray ionization (LAESI) 93
laser desorption (LD) 43, 44
– limitations 44
laser desorption/ionization (LDI) 293
laser desorption/ionization on silicon-mass spectrometry (DIOS-MS) 439
laserspray ionization (LSI) 85, 96
laserspray ionization/inlet (LSII) 85, 95, 96
– practiced on high-performance instruments 94
laserspray ionization/vacuum (LSIV) 85
LC–MALDI coupling principles
– scheme of 215

LC-MS chromatograms 227
light scattering 33
linear dynamic range 5, 9, 10
linear ion traps 19, 20
– performance characteristics 20
liquid adsorption chromatography (LAC) 209
liquid adsorption chromatography at critical conditions (LACCCs) 3
liquid adsorption chromatography under critical conditions (LACCCs) 376
liquid chromatographic/electrospray ionization mass spectrometric characterization 423
liquid chromatography at critical conditions (LC-CCs) 209
liquid chromatography, 2-D
– schematic setup for 213
liquid chromatography mass spectrometry (LC-MS) 237, 284
liquid exclusion adsorption chromatography (LEAC) 228
liquid injection field desorption/ionization mass spectrometry (LIFDI-MS) 386
liquid metal ion gun (LMIG) 168
liquid-phase separation techniques 467
liquid secondary-ion mass spectrometry (L-SIMS) 42, 43, 375
– limitations 43
living/controlled radical polymerization
– CCT, protocols based on 393–397
– degenerative chain transfer, protocols based on 388–393
– novel protocols and minor protocols 397
– persistent radical effect, protocols based on 373–388
– reaction mechanisms and polymer structure elucidation 373–398
low-energy CAD 60
low-temperature plasma ionization (LTPI) 194
LSII-ETD, to identify myelin basic protein fragment 94
LSII-IMS-MS
– 2D plot 98
– of model polymer blend 97
LSII mass spectrum 95
– of PEG-6690 94, 95
LSII-MS analysis of PEG-970 using a dithranol and NaCl matrix 88
LSIV-IMS-MS imaging in reflection geometry (RG) 107–109
LSIV in reflection geometry at intermediate pressure (IP) 100–102
– LSIV-IMS-MS 2D plot of 102

– LSIV-MS spectrum 102
LTQ-Velos mass spectrometer 96
– LSII-MS analysis of polymers on 96

m

macrocyclic polylactones 417
macromolecular design via interchange of xanthates (MADIX) 386, 387
macromolecules
– synthesis of 467
macromonomer production machine 361
magnetic sector mass analyzer system 13
MALDI methods 379, 425
– 2,5-DHB matrix 191
– evaporation-grinding method for 442
– experiment 387
– imaging, desorption/ionization, process of 185
– imaging, mass spectrometry 186
– imaging, of peptides and proteins 185
– interface, scheme of 216
– mass segregation 189
– for small molecule analysis 99
– spectra 449
MALDI-MS
– CLDs 338
– quantification of polymers 52
– spectra 343
– spectrum 359
– techniques 374
MALDI-Q/ToF tandem mass spectrometer 69
MALDI sample preparation, for polymers 119
– absorption of laser light 121, 122
– basic solvent-based sample preparation recipe 127
– choice of matrix 125
– choice of solvent 125–127
– chromatography as sample preparation 138–140
– – rule of thumb for analysis of polymers 138, 139
– deposition methods 127–130
– – Venturi effect 130
– effective ionization 123–125
– efficient desorption 122, 123
– important aspects of sample preparation 143
– intimate contact 121
– matrix-to-analyte ratio 134–136
– – matrix-to-analyte plots for DDAVP and bovine insulin 135, 136
– predicting MALDI sample preparation 142, 143

– problems in MALDI sample
preparation   140–142
– roles of matrix   120, 121
–– fulfill different functions to generate
successful result   120, 121
– salt-to-analyte ratio   136–138
–– Peak area plotted *versus* salt-toanalyte ratio
for   137
– solvent-free sample preparation   130–132
– vortex method   132–134
–– basic recipe for   132, 133
–– mass spectra of different PEG
standards   133
–– SEM image of a vortex prepared
sample   134
MALDI techniques   374
MALDI-TOF
– analysis   395
– mass spectra   418
–– data, time-series segmentation   248
– mass spectrometer   297
– mass spectroscopy (MS)   288, 378,
391, 407
–– analysis   292, 305, 391, 429
–– application   382
–– CO/styrene copolymerization   288
–– methods   406
–– microstructure of   288
–– MS spectra, of precursor ions   302
–– NMR spectroscopy   290
–– spectrometry analysis   406
–– technique   427
–– TEMPO-capped polystyrenes to   375
– mass spectrum   419, 428
–– m/z fragment   422
– mass spectrum signals
–– isotope distributions   380
– measurement   385
– spectrum   244
MALDI-ToF/ToF MS2 mass spectra
– of $[M + Ag]^+$ ions from   74
Mark-Houwink parameters   266, 319
Mark–Houwink parameters   319
mass accuracy   5, 8, 9
mass analyzer performance   5
mass analyzer techniques   6
– ability   320
– FT-ICRs   431
– Orbitraps   431
mass analyzer technologies   5
MassChrom2D   251
mass range   5, 9
mass resolving power   5, 6, 7
– calculation   6

– 3D ion trap-derived mass spectrum of the
polymer   7
– full width at half maximum height
(FWHM)   7
– IUPAC recommendations   6
– peak width definition   6
mass spectra
– fingerprint region of   154
mass spectral peaks, definition of   241
mass spectrometer   271
– transmission   162
mass spectrometric analysis   376
mass spectrometric experiment
– historical concept   1–3
mass spectrometry (MS)   5, 57, 237, 319, 457,
467
– advanced fragmenting techniques   251
– for AP-MALDI, limited m/z range   94
– application of   33, 386, 426, 467
– applications to synthetic polymers   57
– basic components   33, 34
– containing trapping analyzers   58
– electron multiplier detection   337
– MALDI-TOF   284
– mass accuracy   319
– methods   437
– molecular weight distribution (MWD) of
polymer   57
– procedure for quantitation   261, 262
– purpose of   33
– role   339
– techniques, overview   458–460
– use   321
mass-to-charge ratio (m/z)   57, 58
matrix absorption   95
matrix assisted inlet ionization
(MAII)   85, 100, 101
– MAII-MS spectrum   100
matrix-assisted laser desorption electrospray
ionization (MALDESI)   93
matrix-assisted laser desorption ionization
(MALDI)   12, 20, 33, 49–51, 57, 87,
210, 281
– application of   283
– imaging   184
–– mass spectrometry, history of
184, 185
–– of polymers   188–192
–– sample preparation   185–188
– ion formation, conceptualization   51
– limitations   51, 52
matrix-assisted laser desorption ionization
mass spectrometry (MALDI-MS)   119, 237,
446

matrix-assisted laser desorption/ionization-quadrupol-time-of-flight tandem MS (MALDI-Q-TOF MS/MS)   301
matrix-assisted laser desorption/ionization-time of flight (MALDI-TOF)   374
– mass spectrometry (MS)   320, 439
matrix crystal sizes   87
matrix-enhanced SIMS (MA-SIMS)   167
matrix-free DIOS-MS   385
matrix-free ionization method   385
MaxEnt regularization   269
maximum entropy (MaxEnt) regularization   268
Mayo's mechanism   334
meta-SIMS
– molecular weight distributions   166
methanol   127
methoxy PEO (mPEO)   226
methyl acrylate (MA)   320, 379
– polymerization   394
methylmethacrylate(MMA)   320
– BMA system   343
methyl methacrylate(MMA)
– monomer   448
methylmethacrylate(MMA)
– oligomers   324
– – ESI-MS spectrum   324, 326–328, 330
– – water-soluble   364
– polymerization   347, 352
2-methyl-4′-(methylthio)-2-morpholinopropiophenone (MMMP)   332, 349
methyl poly(ethylene oxide) (mPEG)   296
– mass spectra   296
2-methyl-2-propanethiol (tBuS-H)   351
α-methylstyrene (AMS)   355
mid-chain radicals formation   390
molar mass distribution (MMD)   319, 320
– measurement, schematic illustration of   254
– types   337
molecular dynamics (MD) simulations   175
molecular mass determination   33
molecular mass distribution (MMD)   210, 237
– accuracy of a polymer   253
– model embodied
– – implications of   263
– SEC yield, classical methods   253
– two-dimensional quantity   254
molecular weight distribution (MWD)
– band-broadening effects   267
molecular weight distribution (MWD) determination   51
MS-based method   350

MS^2 in space   58
MS/MS^2 experiments with ion mobility spectrometry (IMS)   71
MS/MS of polymers   26
– mass analyzers and mass analyzer combinations used for   29
– scan types   27
MS^n of polymers   26
– mass analyzers and mass analyzer combinations used for   29
MS^2 (CAD) spectrum
– $[M + 2H]^+$ ions ions generated by ESI from   78
MS^2 studies, structural information from   75
– binding energies, assessment of   77, 78
– copolymer sequences   76, 77
– end-group analysis   75
– intrinsic stabilities, assessment of   77, 78
– isomer/isobar differentiation   75
– polymer architectures   75
multiply charged ions   97
multistage tandem mass spectrometry (MS^n)   405
multivariate curve resolution (MCR)   171
mzXML file   239

n

nanospray ESI source, analytical advantages   48
National Physics Laboratory (NPL)   181
– organic multilayer reference material, depth profiling in   181
Nd/YAG lasers   95
negative chemical ionization (nCI) source   61
negative ion ToF-SIMS images,
– polymer/polymer interface   171
neopentyl diol (NPG)   226
neutral polymers   123
new ionization method, capable of producing highly charged ESI-like ions   94
NHC-catalyzed polymerizations   406
^{63}Ni-based source   45
Ni grid   165
NIST MALDI-based method   273
NIST polystyrene molecular mass
– MMD of   274
nitroxide-mediated polymerization
– intermolecular CTP occurrence   358
nitroxide-mediated polymerization (NMP)   226, 358, 373
– PMAA–PMMA copolymer   294
noisy polymer mass spectrometry data   244
– autocorrelation   244
Norrish type II fragmentation   392

nuclear magnetic resonance (NMR) 33, 365, 439
– experiments 363
– spectroscopy 281
nylon-6
– monomer 153
– photo-oxidation processes 450
nylon 66 (Ny66) 444

o

1-octanethiol (OctS-H) 362
OH-terminated PDMS
– calibration curve of 225
– chromatogram of 224
oligocarbonate diols synthesis 423
online spray method
– principles of 217
orbitrap mass analyzer 24, 25
– perfomance characteristics 25
organometallic ROMP catalysts 424
oscillating capillary nebulizer (OCN) 129, 187, 219
oxazoline-cyclophosphazene units
– melt mixing reactions 444
oxiranes, polymerization 406
oxygen-centered radicals 329

p

[PBG]$_8$Na$^+$
– ion intensity distributions 190
penicillin G 410
(N,N,N',N',N-pentamethyldiethylenetriamine (PMDETA) 379
peptide biopolymers 26
performance, measures of 5
peroxyacetates 327
peroxydicarbonates 331
peroxyesters (POEs) 322, 323
– general structure 322
peroxypivalates 323
PGMA-PBMA trimer 300
– distributions of 300
2-phenylallyl alcohol (PhAA)
– use 397
photodissociation methods 60, 61
photoinduced conjugation reaction 393
photoinitiation 334
photopolymerization
– ESI-MS study 349
PLA grafted with acrylic acid (PLA-AA) 454
plasma desorption (PD) 44, 45
– limitations 45
plasma desorption ionization mass spectrometry (PDI-MS) 149

plasma desorption ionization techniques 194
plasma-polymerized polyethyleneglycol (pPEG)
– PCA, analysis of 172
– PCA for analysis of 172
PLP
– MS method 342
– MS studies 341
– size-exclusion chromatography (SEC) 336
pluronic (P104)
– 3D distribution of 182
– 3-D volumetric representations of 183
PMAA-PMMA, MALDI-TOF MS spectra of 293
PMMA
– ESI-MS spectra 396
– mass-weighted molecular mass distributions 272
– model, thermo-oxidative degradation products 448
– SEC retention time 269
– synthesized, commercial standards of 272
PMMA-*b*-PBAcopolymer 383
PMMA(CPDB)
– polymerization of 272
PMMA 10 900 Da
– 3D plot for 223
Poisson distributions 347
poly (ε-caprolactone)
– MALDI-TOF mass spectra 421
poly (3-hydroxybutanoic acid)
– ESI-mass spectrum 412
poly (butyl methacrylate)
– MS analysis of 335
poly (MMA)
– MALDI-MS spectrum 337
poly(acrylic acid) (PAA) 383
polyalkylcyanoacrylate (PBCA) nanoparticles
– formation of 222
poly(alkyl methacrylates) 151
polyamides (PAs) 225, 443
poly(BA)
– ESI-MS spectrum 362, 363
– MALDI-MS spectrum 359
poly(bisphenol-A-carbonate)
– thermal oxidative degradation processes 443
poly (ethylene oxide)-b-poly(styrene) copolymers 376
poly(butyl acrylate) (PBA) model 452
poly(butyleneglycol) (PBG) 189
– single ion intensity distribution of 191
poly(butylene succinate) (PBSu)
– thermal-oxidation processes 447

poly(butylene terephthalate) (PBT)   439
– thermal oxidation   449
PolyCalc
– software tool   245
polycaprolactone   430
poly(ε-caprolactone) (PCL)   426
polycarbonates (PCs)
– molecular weights of   154
poly(dimethylsiloxanen) (PDMS)   222
polydisperse components
– gravimetric mixture of   259
polydispersity indices (PDI)   289
polyester copolymer
– MALDI-TOF MS, LC-CC, 2D-plot   299
polyester oligomers   428
polyesters
– ESI-mass spectra   413
polyether-based polyurethanes (PUs)   444
poly(ethyl acrylate) (PEA)   379
polyethylene (PE14)
– side-view snapshots   175
poly(ethylene glycol) (PEG)   88, 164, 282
– positive ion DESI mass spectrum of   193
poly(ethylene glycol)-b-poly(acrylic acid)-b-poly(n-butyl acrylate)   290
poly(ethylene oxide) (PEO)   286, 406
– MALDI-TOF mass spectrum   409
poly(ethylene oxide)-b-poly(propylene oxide)
– block copolymers of   299
polyethylene (PE) surface   176
poly(ethylene terephthalate) (PET)   439
– samples   140
poly(ethylhexyl acrylate)   383
poly(3-ethyl-3-hydroxy-methyloxetane)-derived macroinitiators
– MALDI-TOF-MS   383
poly(ethyl methacrylate) (PEMA)   154
poly(glycolic acid) (PGA)   167, 454
– mass spectra of   174
poly[(R,S)- 3HB-co-CL] copolyesters   427
poly(β-hydroxyalkanoate)s (PHAs)   409
poly(o-hydroxyamide) (PAOH)   442
poly(3-hydroxybutyrate)
– ESI-MS analysis   411
– ESI-MS structural studies   410
– macromolecules   410
poly(3-hydroxybutyrate)-co-poly(3-hydroxyhexanoate)   306–307
poly(hydroxy-ethyl methacrylate) (PHEMA)   151, 452
poly(3-hydroxy-4-etoxybutyrate)
– ESI-mass spectrum   414, 415
poly(hydroxylethylmethacrylate) (PHEMA)

– positive secondary ion fingerprint spectrum of   152
poly(isobutylene) (PIB)
– binder   173
– containing plastic, PCA, analysis of   173
– PCA, image analysis of   173
polyisoprene (PIs)   177
poly(N-isopropylacrylamide)
– MALDI-TOF mass spectrum of   388
poly(lactic acid) (PLA)   167
– ESI mass spectrum   416
– linear and cyclic structures of   225
– matrix, quantitative depth profiling in   179
– thermal degradation   445
– thin film, representative depth profile of   177
poly(lactide)-block-poly(2-hydroxyethyl methacrylate) polymers   382
poly(L-lactide) (PLLA)   426
poly(L-lactide) multiarm star polymers
– molecular characterization   408
polymerator software
– screenshot of   252
polymer chains   180
polymer characterization   33
polymer crystallinities   153
polymer degradation   437–460
– biodegradation   453–455
– degradation processes   455, 456
– photolysis and photooxidation   449–453
– thermal and thermo-oxidative degradation   438–448
polymer depth profiling   180, 182
– limitations for   177
polymer distributions   265
polymer elution curves   212
polymer formation mechanism   320
polymerization process   249
polymerization techniques   405–431, 468
– cyclic esters and carbonates, ring-opening polymerization mechanisms   408–423
– cyclic ethers, ring-opening polymerization mechanisms   406–408
– radical structures in   349
– ring-opening metathesis polymerization   423–425
– step-growth polymerization mechanisms   425–430
polymer mass spectrometry
– automated data processing and quantification   237
– automated spectral analysis and data reduction   241–248
– chromatographic setup   266

– copolymer analysis   248–251
– file and data formats   237–239
– indistinguishable/overlapping mixture, schematic illustration of   258
– ionization conditions, optimization of   239–241
– MALDI-MS/ESI-MS
–– absolute MMD, determination of   262–266
–– monodisperse components, mixtures of   256, 257
–– oligomer, correction factor for   260, 261
–– polydisperse components, mixtures of   257–259
–– quantitation, procedure for   261, 262
–– quantitative   253–256
– MMD of homopolymers   266–270
–– comparison of two methods   273–274
–– components, mixtures of   270–273
– molar abundance of   274–276
– MS/MS, data interpretation   251, 252
– nonoverlapping mixture, schematic illustration of   259
– principal data processing, flow diagram of   271
– SEC, convolution process
–– graphical representation of   267
– SEC/ESI-MS
–– quantitative MMD measurement   266
polymer molecular weights   153
– determination of   152
polymers
– analysis   10
–– application for   430
– chemical structures   86
– chromatography of   211
– depth profiling of   176
– LC investigations of   211
– quantitative depth profiling   178
– RI detection   340
– schematic representation of   211
– sputter depth profiling   180
– thermal degradation   438
polymer science   26
polymers in electrolyte fuel cells (PEFCs)   87
polymer synthesis   468
poly(methacrylic acid)-poly(methyl methacrylate) (PMAA-PMMA)
– molar mass determination of random copolymers   292
poly(methyl methacrylate) (PMMA)   222, 239, 376, 448
poly(MMA)
– ESI-MS spectrum   348

– MALDI-MS spectrum   353
– samples   338
poly(NAM)-*block*-polystyrene copolymers   389
poly(neopentyl isophthalate) (PNI)   450
poly(paraphenylene sulfide)   52
poly(para-phenylene terephthalamide) (PPD-T)   52
poly (ethylene glycol) (PEG)   444
poly (ethylene terephthalate) (PET) samples
– matrix assisted laser desorption ionization-time of flight (MALDI-TOF) mass spectra   440
poly(p-phenylene sulfide)s (PPS)   442
poly (propylene succinate) (PPSu)
– thermal degradation mechanism   448
poly(propionyl-ethyl methacrylate) (PPEMA)   151
poly(propylene oxide) (PPO)   307
poly (tetrafluoroethylene) (PTFE)   165
poly[(R)-3-hydroxybutyrate] (PHB)   445
– polymers synthesis   416
poly[(R,S)-3- hydroxybutyrate-co-L-lactide] oligocopolyesters
– ESI-MS/MS technique   429
polystyrene (PS)   213, 286
– ion intensity distribution of   190
– MALDI-TOF mass spectrum of   189, 247
– positive ion mass spectra of   195
– ToF-SIMS spectrum of   165, 168
– use of   448
– variable separation techniques, coupling of   210
polystyrene-block-polyisoprene system   343
– copolymer fingerprints   346
– MALDI-ToF-MS spectra   344
polystyrene-block-poly(styrene-co-acrylonitrile) copolymers   376
poly(styrene-coisobutylene)
– positive secondary ion mass spectral images   170
poly(styrene-co-isobutylene) triblock copolymer matrix   169
polystyrene molecular weight
– positive secondary ion spectra of   163
polystyrene/poly(methylmethacrylate (PS/PMMA) polymer blend film
– nano-SIMS imaging of C andOcomponents   169
polytetrahydrofuran (PTHF)   448
poly[(3,3,3-trifluoropropyl) methylsiloxane] (PTFPMS)   455
poly(trimethylene carbonate)   418

polyurethane-co-dimethylsiloxane (PU-PDMS) 167
polyvinylchloride (PVC) 195
– pyrolysis 448
– use of 448
poly(vinylidene difluoride) (PVdF) topcoat 170
post-source decay (PSD) 62, 387
– fragmentation analysis 455
P104 polymer
– amphiphilic nature of 183
PPO–PEO–PPO triblock copolymer
– stepwise data treatment of 308
principal components analysis (PCA) 159
propagation rate coefficients 319, 341
propylene oxide (PO) 406
– polymerization 407
PS_{20}-b-PEO_{70}
– MALDI-TOF MS spectrum of 286
P(t-BA) transformation
– electrospray ionization mass spectra 392
pulsed laser-initiated radical copolymerization
– copolymer fingerprint plots 250
pulsed-laser polymerization (PLP) method 319
pyrolysis (Py)-GC/MS 225
pyrolysis-GC/MS (py-GC/MS) 438
pyrolysis mass spectrometry (Py-MS) 37
– limitations 37, 38

q

QqQ instruments 28
quadrupole ion trap (QIT)
– detector 390
– mass analyzer 381
quadrupole ion trap (QIT) mass spectrometers 63–68
– ESI-QIT MS^3 mass spectrum of b_{23} fragment 66
– ion–ion reactions 67
– ion motion inside the trap 63
– isolated precursor ions acceleration 65
– mass-selective axial instability mode 64
– Mathieu equation 63
– precursor ion chosen 64
– QIT MS^2 scan process for precursor ion 65, 67
– QIT with external ESI and CI sources 68
– shortcoming of CAD experiments in QITs 67
– stability region 64
quadrupole ion traps 17
quadrupole mass filter system 15–17
– basic components 15
– Mathieu stability diagram 16
– performance characteristics 16
quadrupole/time-of-flight (Q/ToF) mass spectrometers 69–72
– MS^2 (CAD) mass spectrum of 70
quadrupole-TOF (Q-TOF) mass spectrometer 28
quantification at trace levels 33
quantum-chemical calculations 330

r

racemic α-methyl-β-pentyl-β-propiolactone
– anionic ROP 412
radical polymerization (RP) process 319, 320, 347, 357, 373
radical polymerization reaction mechanisms
– basic principles and general considerations 320, 321
– chain transfer 351–364
– – transfer to small molecules 351–356
– elucidation 319–365
– emulsion polymerization (EP) 364, 365
– initiation 321–335
– initiator efficiency 335
– propagation 335–347
– – chain-length dependence 340–347
– – rate coefficients 335–340
– radical generation 321–335
– – photoinduced initiator decomposition 331–334
– – thermally induced initiator decomposition 321–331
– termination 347–351
random coupling hypothesis 347
rate coefficient 376
reactive MALDI MS 52
refractive-index (RI) detection
– use 319
reversible addition-fragmentation chain transfer (RAFT) 287, 334, 397
– advantages 395
– agent-derived end group containing chains 11
– generated polymers, propensity 387
– MADIX-generated polymers 389
– mechanism 390
– polymerizations 11, 223, 361, 373, 386, 387
– – analysis of 389
– – MALDI-TOF mass spectra 389
– – R-group approach 391
– – Z-group approach 391
– – star polymers formation (See R-group approach-RAFT polymerizations)
– radiolysis 335
– via ESI-MS 390

RI detector 267
ring-chain equilibration mechanism 439
ring-opening metathesis polymerization (ROMP) 405, 423–425, 424
– based block copolymerization 425
– chain-growth reaction 425
ring-opening polymerization (ROP) mechanisms 431
– of cyclic ethers/esters 405
rotational–vibrational degrees of freedom of the ion 59
ruthenium olefin metathesis catalysts 425

s

Savitsky-Golay smoothing 243
scanning electron microscopy (SEM) 128
scanning microprobe matrix-assisted laser desorption/ionization imaging mass spectrometry (SMALDI-MS) 188
Schulz–Flory theory 350
secondary ion mass spectrometry (SIMS) imaging 40, 41, 128, 150
– cluster ion beams, polymer depth profiling 174–182
– – optimized beam conditions, role 180–182
– correlation of 153
– data analysis methods 171–174
– limitations 41, 42
– PC molecular weight, quantitative analysis of 155
– polymer blends/multicomponent systems 168–171
– polymers, reference citations list 156–161
– polymers, static 150
– – fingerprint region 151–162
– – high-mass region 152–168
– polymer systems, 3-D analysis 182–184
– specific polymers characterization 156–161
sector mass analyzers 12–15
– performance characteristics 14
SG1-mediated polymerization 377
signal autocorrelation 243
silver tri-fluoroacetic acid (AgTFA) 195
single ion current (SIC) trace 454
single-pulse (SP) PLP experiments 332
singly charged copolymer
– molar mass 345
size exclusion chromatography (SEC) 3, 33, 90, 209, 238, 281, 319, 336, 390, 441
– ESI-MS method 273
– hyphenation 398
– internal calibration 340

– MALDITOF-MS
– – schematic view of 220
– workhorse technique 220
size exclusion chromatography and matrix-assisted laser desorption ionization MS (SEC/MALDI) 430
size exclusion chromatography electrospray ionization mass spectrometry ((SEC/)ESI-MS) analysis 374
size exclusion chromatography (SEC) fraction
– matrix-assisted laser desorption ionization-time of flight (MALDI-TOF) spectrum 442
$Sn(Oct)_2$ initiator 419
soft-ionization mass spectrometry
– analysis of 382
– use 374
soft-ionization mass spectrometry (MS) techniques 373
soft ionization technique
– applications 449
– MS techniques 385, 438
soft ionization techniques 283
– APCI 294–297
– applications 457
– ESI and MALDI 467
– ESI-MS 292–294
– histogram showing 283
– MALDI, application of 283–292
solid-phase microextraction (SPME) 438
step-growth polymerization mechanisms 425–430
Stochastic numerical optimization 240
styrene-MMA statistical copolymer 345
β-substituted-β-lactones 409
surface analyses by imaging MS 104–109
surface analysis technique. See desorption electrospray ionization (DESI)
surface-assisted laser desorption/ionization (SALDI) 52
surface-induced dissociation (SID) 60
surface mass spectrometry methods
– desorption electrospray ionization (DESI) 192–194
– electrospray droplet impact for 194, 195
– plasma desorption ionization techniques 194
switching nitroxide-capped polystyrene 378
symmetric diacyl peroxides
– scheme for decomposition 329
synthetic polymer
– analyses of 63
synthetic polymers
– MALDI imaging for 188

t

tandem mass spectrometry 444, 468
tandem ToF mass spectrometers 72
Taylor cone 46
Taylor expansion
– zeroth derivative of 260
termination reaction 348–351
tertiary radicals (TRs) 361
– formation 363
tetrahydrofuran (THF) 127, 391
1,1,3,3-tetramethylbutyl peroxyacetate (TMBPA) 328
1,1,3,3- tetramethylbutyl peroxypivalate (TMBPP) 325
2,2,6,6-tetramethylpiperidine-N-oxyl (TEMPO)
– capped polystyrene 51, 375
– mediated living radical polymerization 375
1,1,2,2-tetramethylpropyl peroxyacetate (TMPPA) 328
1,1,2,2-tetramethylpropyl peroxypivalate (TMPPP) 325
thermal degradation experiments 448
thermal energy for sublimation/evaporation 97
thermodynamically controlled polycondensations (TCPs) theory 429
thermo fisher scientific LTQ-ETD mass spectrometers 91
– from single laser shots 91
thermogravimetry (TGA) 438
thermo mass spectrometers 97
thermospray ionization (TSP) 45
thermospray (TSP) source 45
thin-layer chromatography (TLC) 439
thioketone-mediated polymerization (TKMP) process 397
timed ion selection (TIS) 72
time-of- flight (TOF) 338
– mass analyzer 11, 22
– – major advance in 21
– – orthogonal acceleration, basic components of an 21
– – perfomance characteristics 22
– mass analyzers 20–22
time-of-flight mass spectrometers (TOFMS) 119
time-of-flight secondary ion mass spectrometers (ToF-SIMSs) 151, 439
– secondary electron microscopy (SEM) imaging 168
time-series segmentation
– schematic representation of 246
tin-containing macromolecules
– identification 420
ToF/ToF instruments 72–75
p-toluenesulfonyl initiator fragment 382
total solvent-free analysis (TSA) 85
traditional chain-transfer agent (TCTA) 351, 355
transfer rate coefficient 354
transmission geometry (TG) ion source 96
trapping rate coefficients 342
traveling wave (T-wave) 71
trifluoroacetate (TFA) salts 124
trimethylene carbonate (TMC)
– homo/copolymerizations 421
2,2,5-trimethyl-4-(isopropyl)-3-azahexane-3-oxyl (BIPNO) 375
2,2,5-trimethyl-4-phenyl-3-azahexane-3-oxyl (TIPNO) 375
trimethylsilyl nucleophiles (NuE) 406
1,1,1-tris(hydroxymethyl) propane (TMP) 407
trisilanolisobutyl-POSS (T-POSS) 441
two-dimensional (2D) gas-phase separation 89, 90
– IMS dimension drift time $vs.$ MS m/z yields 89
– sigmoidal transition 90
– – drift times 90
two-dimensional liquid chromatography
– schematic setup for 213

u

Ultem® polyetherimide (PEI) 442
ultraffast LSII-MS imaging in transmission geometry (TG) 106, 107
ultra performance liquid chromatography (UPLC) 227
– use of 298
ultraviolet (UV) lasers 94, 121
ultraviolet/visible spectroscopy 33
unimolecular rearrangement mechanisms 59
universal calibration. See Mark-Houwink parameters
UV-induced fragmentation 387

v

vapor-deposited glutamate
– using Ar$^+$ 176
vapor pressure osmometry (VPO) 33
Venturi effect 130
vinyl acetate (VAc) 383
ω-vinyl heterotelechelic oligomers 397
vinylic monomers
– polymerization 390
vinyl-terminated polymers 387

vinyl-terminated poly(methyl methacrylate) (PMMA)
– cyclic degradation mechanism   453

w
Waters SynaptHDMS™ mass spectrometer   71
wide angle X-ray diffraction (WAXS)   455

x
X-ray photoelectron spectroscopy (XPS)   154, 441

z
Z-group-approach-RAFT polymerization   391